Special Publication of the European Association of Petroleum Geoscientists No. 4

Alain Mascle (Ed.)

Hydrocarbon and Petroleum Geology of France

With 327 Figures, Some in Color, and 22 Tables

Springer-Verlag
Berlin Heidelberg New York London Paris
Tokyo Hong Kong Barcelona Budapest

Alain Mascle
Institut Français du Pétrole
B. P. 311, 1 & 4, Avenue de Bois-Préau
92506 Rueil-Malmaison Cedex, France

Library of Congress Cataloging-in-Publication Data
Hydrocarbon and petroleum geology of France/Alain Mascle (ed.). p. cm. – (Special publication of the European Association of Petroleum Geoscientists; no. 4)
ISBN-13: 978-3-642-78851-2 e-ISBN-13: 978-3-642-78849-9
DOI: 10.1007/ 978-3-642-78849-9
1. Petroleum-Geology-France. 2. Natural gas-Geology-France. 3. Coal-Geology-France. I. Mascle, A. II. Series: Special publication of the European Association of Petroleum Geoscientists; 4.
TN874.F8H93 1994 553.2'8'0944-dc20 946892

Softcover reprint of the hardcover 1st edition 1994

Typesetting: Appl. Wemding
SPIN: 10097067 32/3130-543210 – Printed on acid-free paper

Preface

This volume contains proceedings from a special session devoted to the hydrocarbons of France, which was held during the 4th Annual Conference of the European Association of Petroleum Geoscientists and Engineers at the Palais des Congrés, Paris, from 1 to 5 May 1992. The conference took place in parallel with the 54th Meeting of the European Association of Exploration Geophysicists. Altogether there were some 4500 attendees at both. This volume also contains additional papers selected to provide a more complete review of regional and thematic research currently being undertaken to improve the efficiency of hydrocarbon exploration and production in France.

This volume is composed of 32 chapters corresponding to five regional topics. Chapters 1 to 4 (Part 1) are general introductions and statistics on exploration-production of hydrocarbons and underground gas storage in France Chapters 5 to 12 (Part 2) concern the Paris Basin (where about 61 and 3.7 % of domestic oil and gas are currently being produced) and the neighboring offshore French Western Approaches. Chapters 13 to 17 (Part 3) are devoted to the Aquitaine Basin, including the Pyrenees foothills and the Parentis offshore basin, where about 38.7 and 93.5 % of oil and gas are currently being produced. Chapters 17 to 22 (Part 4) concern the Gulf of Lions and the onshore South-East Basin where three tiny and now abandoned oil fields were developed in the past. Chapters 23 to 28 (Part 5) are devoted to eastern and northern France where about 0.3 % of oil and 2.8 % of gas are produced, the latter as desorbed methane in abandoned coal mines. Chapters 29 to 32 (Part 6) deal with four French Overseas Departments and Territories where some exploratory work has been undertaken in the past.

This fourth special publication by EAPG follows three previous volumes:

Spencer A M (ed) (1991) Generation, accumulation and production of Europe's hydrocarbons. Special publication of the European Association of Petroleum Geoscientists No 1. Oxford University Press, 459 pp

Spencer A M (ed) (1992) Generation, accumulation and production of Europe's hydrocarbons II. Special publication of the European Association of Petroleum Geoscientists No 2. Springer, Berlin Heidelberg New York, 294 pp

Spencer AM (ed) (1993) Generation, accumulation and production of Europe's hydrocarbons III. Special publication of the European Association of Petroleum Geoscientists No 3. Springer, Berlin Heidelberg New York, 380 pp

Reference to works in this volume should be made in the following style:

Gaulier JM, Burrus J (1994) Modeling present and past thermal regimes in the Paris Basin; petroleum application. In: Mascle A (ed) (1994) Hydrocarbon and petroleum geology of France. Special publication of the European Association of Petroleum Geoscientists No 4. Springer, Berlin Heidelberg New York, 464 pp

The Association gratefully acknowledges the assistance provided to the Editorial Committee by the technical referees: J. Benkhelic, B. Biju Duval, E. Brosse, E. Deville, B. Durand, N. Ellouz, J. P. Herbin, J. M. Gaulier, R. Gillcrist, P. C. de Graciansky, W. Kuhnt, D. Larue, A. Le Marrec, H. P. Luterbacher, J. A. Miles, Y. Philippe, J. P. Roy, G. Sambet, P M. Shannon, J. C. Sibuet, M. Tardy, B. Van den Berghe, R. Vially, H. F. Wang, J. Wannesson, W. Wildi.

Alain Mascle

List of Contributors

M. P. Angoulvant
Ecole et Observatoire de Physique du Globe (CNRS URA 0323)
5, Rue René Descartes, 67084 Strasbourg Cedex, France

A. Arnaud-Vanneau
Institut Dolomieu, URA, 69, Rue M. Gignoux, 38031 Grenoble, France

C. Bacchiana
ESSO-REP, 213, Cours Victor Hugo, 33323 Begles Cedex, France

C. Beck
Laboratoire de Géodynamique, URA CNRS, 69, Université de Savoie
B. P. 1104, 73011 Chambéry, France

J S Bell
Geological Survey of Canada
3303, 33rd St. N. W., Calgary, Alberta T2L 2A7, Canada

G Bertrand
SCGH-Direction des Hydrocarbures – Ministère de l'Industrie
366, Avenue Napoléon Bonaparte, 92501 Rueil-Malmaison, France

G. Bessereau
Institut Français du Pétrole, B. P. 311
1 & 4, Avenue de Bois-Préau, 92506 Rueil-Malmaison Cedex, France

E. Blanc
Institut Dolomieu, Université J. Fourier
Rue Maurice Gignoux, 38021 Grenoble, France

E. Blondin
Gaz de France, Direction des Etudes et Techniques Nouvelles
Département "Réservoirs Souterrains"
361, Avenue du Président Wilson, B. P. 33, 93211 La Plaine-St.-Denis Cedex, France

C. Bois
Institut Français du Pétrole, B. P. 311
1 & 4, Avenue de Bois-Préau, 92506 Rueil-Malmaison Cedex, France

Ph. Bouysse
Bureau de Recherches Géologiques et Minières
B. P. 6009, 45060 Orléans Cedex 2, France
Present adress: C.C.G.M., Maison de la Géologie
77, Rue Claude Bernard, 75 005 Paris, France

J.-G. Bréhéret
Laboratoire de Géologie des Systèmes Sédimentaires, Université François Rabelais
Parc de Grandmont, 37200 Tours, France

J P Brun
Géosciences Rennes, Campus de Beaulieu, 35 042, Rennes Cedex, France

M F Brunet
URA 1759 CNRS Tectonique, Département Géotectonique
Université P & M. Curie, Boîte 129
4, Place Jussieu, 75252 Paris Cedex 05, France

A Bruneton
Beicip-Franlab, 232, Avenue Napoléon Bonaparte
92500 Rueil-Malmaison, France

J Burrus
Institut Français du Pétrole, B. P. 311
1 & 4, Avenue de Bois-Préau, 92506 Rueil-Malmaison Cedex, France

G. Caillet
Elf Petroleum Norge, B. P 168, 4001 Stavanger, Norway

E. Carrio-Schaffhauser
L.G.I.T. – IRIGM, B. P. 53X, 38041 Grenoble, France

B Colletta
Institut Français du Pétrole, B P. 311
1 & 4, Avenue de Bois-Préau, 92506 Rueil-Malmaison Cedex, France

L. Courel
Centre des Sciences de la Terre, Université de Bourgogne et U.R A. 157 du CNRS 60,
Boulevard Gabriel, 21000 Dijon, France

M. Cousin
Laboratoire de Géodynamique, URA CNRS 69, Université de Savoie
B. P. 1104, 73011 Chambéry, France

A Curial
Société DIASTRATA, 20 bis Rue Songieu, 69100 Villeurbanne, France

M Daignières
Département des Sciences de la Terre, Université de Montpellier II
Place E. Bataillon, 34095 Montpellier Cedex 05, France

J F. Deronzier
Beicip-Franlab, 232, Avenue Napoléon Bonaparte
92500 Rueil-Malmaison, France

E. Deville
Institut Français du Pétrole, B. P. 311
1 & 4, Avenue de Bois-Préau, 92506 Rueil-Malmaison Cedex, France

G Dromart
Centre Sciences Terre, Université de Lyon, 69622 Villeurbanne Cedex, France

J Espitalié
Institut Français du Pétrole, B. P. 311
1 & 4, Avenue de Bois-Préau, 92506 Rueil-Malmaison Cedex, France

B Fritsch
Ecole et Observatoire de Physique du Globe (CNRS URA 0323)
5, Rue René Descartes, 67084 Strasbourg Cedex, France

O. Gariel
Institut Français du Pétrole, B. P. 311
1 & 4, Avenue de Bois-Préau, 92506 Rueil-Malmaison Cedex, France

J. M Gaulier
Institut Français du Pétrole, B. P. 311
1 & 4, Avenue de Bois-Préau, 92506 Rueil-Malmaison Cedex, France

C. Gorini
Laboratoire de Géologie Structurale et Tectonophysique
38, Rue des 36 Ponts, 31400 Toulouse, France

S. Gouyet
Université de Pau et des Pays de l'Adour
Present address: Schlumberger-Geoquest, 4001, Stavanger, Norway

P. Guennoc
BRGM, B. P. 6009, 45060 Orléans Cedex 02, France

F. Guillocheau
Géosciences Rennes, Campus Beaulieu, 35042 Rennes Cedex, France

F Hanot
Compagnie Générale de Géophysique, B. P. 65, 91301 Massy Cedex, France

M. Hanzo
Laboratoire de Géologie des Ensembles Sédimentaires, Université de Nancy I
B. P. 239, 54506 Vandoeuvre Les Nancy Cedex, France

F. Heritier
Elf Aquitaine Production
Present adress: 33, Rue du Mont Valérien, 92210 Saint-Cloud, France

J C. Hervet
Ministère de l'Industrie, Direction des Hydrocarbures, DHYCA
99, Rue de Grenelle, 75700 Paris, France

P Houel
Beicip-Franlab, 232, Avenue Napoléon Bonaparte
92500 Rueil-Malmaison, France

Y. Lafoy
Service des Mines et de l'Energie de Nouvelle Calédonie
B. P. 465, Nouméa Cedex, New Caledonia

Ch. Lamiraux
SCGH-Direction des Hydrocarbures – Ministère de l'Industrie
366, Avenue Napoléon Bonaparte, 92501 Rueil-Malmaison, France

J C. Lecomte
Institut Français du Pétrole, B. P. 311
1 & 4, Avenue de Bois-Préau, 92506 Rueil-Malmaison Cedex, France

B Le Gall
Laboratoire de Géologie Structurale, URA du CNRS 1278
Université de Bretagne Occidentale
29200 Brest, France

A. Le Marrec
Elf Aquitaine Production, DEPF, 31360 Boussens, France

X B. Liu
Centre des Sciences de la Terre de l'Université de Bourgogne et U R.A 157 du CNRS
60, Boulevard Gabriel, 21000 Dijon, France

A Mascle
Institut Français du Pétrole, B. P. 311
1 & 4, Avenue de Bois-Préau, 92506 Rueil-Malmaison Cedex, France

A Mauffret
Laboratoire de Géodynamique, Tectonique et Environnement, URA 718, Case 119
Université P. et M. Curie
4, Place Jussieu, 75252 Paris Cedex 05, France

G Ménard
Laboratoire de Géodynamique, URA CNRS, 69, Université de Savoie
B. P. 1104, 73011 Chambéry, France

A. Meunier
Laboratoire de Pétrologie des Altérations Hydrothermales URA 721 CNRS
Université de Poitiers
40, Avenue du Recteur Pineau, 86022 Poitiers Cedex, France

J Mondin
Compagnie Générale de Géophysique, 1, Rue L. Migaux, 91341 Massy, France

P. Monier
Société Diastrata, 20 bis, Rue Songieu, 69100 Villeurbanne Cedex, France

R. Moretto
Société Diastrata, 20 bis, Rue Songieu, 69100 Villeurbanne Cedex, France

M. Munschy
Ecole et Observatoire de Physique du Globe (CNRS URA 0323)
5, Rue René Descartes, 67084 Strasbourg Cedex, France

J. Parpant
ESSO-REP, 213, Cours Victor Hugo, 33323 Begles Cedex, France

Y. Philippe
Institut Français du Pétrole, B. P. 311
1 & 4, Avenue de Bois-Préau, 92506 Rueil-Malmaison Cedex, France

C M Ponsot
Cenergys, Centre d'Affaíres la Boursidiére
B. P. 162, 92357 Le Plessis Robinson Cedex, France

A Richard
Elf Aquitaine, Tour Elf, 92078 Paris-La Défense, France

F. Roure
Institut Français du Pétrole, B. P. 311
1 & 4, Avenue de Bois-Préau, 92506 Rueil-Malmaison Cedex, France

M. Schaming
Ecole et Observatoire de Physique du Globe (CNRS URA 0323)
5, Rue René Descartes, 67084 Strasbourg Cedex, France

R. Schlich
Ecole et Observatoire de Physique du Globe (CNRS URA 0323)
5, Rue René Descartes, 67084 Strasbourg Cedex, France

M. Séguret
Département des Sciences de la Terre, Université de Montpellier II
Place E. Bataillon, 34095 Montpellier Cedex 05, France

B. Smart
ESSO-REP, 213, Cours Victor Hugo, 33323 Begles Cedex, France

M. Specht
Institut Français du Pétrole, B. P. 311
1 & 4, Avenue de Bois-Préau, 92506 Rueil-Malmaison Cedex, France
Present adress: Total, Tour Total, 24 Rue Michelet, 92069 Paris-La Défense Cedex 47, France

M Tardy
Laboratoire de Géodynamique, URA CNRS, 69, Université de Savoie
B. P. 1104, 73011 Chambéry, France

P. Unternehr
Elf Aquitaine, Pau, France
Present address: Elf Petroland BV, Mariahoeveplein
2509 AG, Den Hague, Netherland

F. Verdier
Gaz de France, DETN, B. P. 33, 93211 La Plaine-St.-Denis, France

R Vially
Institut Français du Pétrole, B. P. 311
1 & 4, Avenue de Bois-Préau, 92506 Rueil-Malmaison Cedex, France

H. Wang
Centre des Sciences de la Terre, Université de Bourgogne et U.R.A. 157 du CNRS 60, Boulevard Gabriel, 21000 Dijon, France

Introduction to Hydrocarbon Exploration and Underground Gas Storage in France

Exploration for and Production of Oil and Gas in France: A Review of the Habitat, Present Activity, and Expected Developments

A. Mascle, G. Bertrand, and Ch. Lamiraux

Abstract Exploration for and production of hydrocarbons in France are at present concentrated in the central part of the Paris Basin, and in the southwest and western part of the Aquitaine Basin. Production in the Paris Basin in 1991 was about 38100 bopd distributed through 50 fields, the largest of them being Chaunoy (14800 bopd), Villeperdue (9300 bopd), and Vert-le-Grand (3600 bopd). In the Aquitaine Basin, petroleum activity has focused on two distinct areas. To the southwest, along the Pyrenean foothills, exploration of deep and complex traps has resulted in the early discovery of major gas fields, including the Lacq and Meillon gas fields, still producing 300 and 108 MMcfpd. To the west of the Aquitaine Basin, on the Parentis area, 9700 bopd are produced from less than ten fields, including the Les Arbousiers field discovered at the end of 1991. Exploration is developing much more slowly in the other sectors of French territory. Mention should be made of: (1) some activity in the Jura thrust belt following the testing of oil in the Chaleyriat exploratory well in 1989; (2) marine seismic surveys in the Bay of Biscay across the offshore extension of the Parentis Basin; and (3) the recent interest in coalbed methane in several late Carboniferous basins in eastern and central France.

Introduction

Three major onshore sedimentary basins of Mesozoic age are present in France (Fig. 1; Autran and Dercourt 1980; Curnelle and Dubois 1986; Weaver and Van Damme 1988; Lucazeau and Vasseur 1989; Bois et al. 1990). They result from three main stages of subsidence, in late Paleozoic, Mesozoic, and Paleogene times. Two of them, the Paris and Aquitaine Basins, have been intensively explored for about 40 years and contain a significant amount of oil or gas, which still accounts for 4 and 10%, respectively, of domestic needs (Fig. 2). The major discoveries in the Aquitaine Basin were made as early as 1951 for gas and 1954 for oil at Lacq and Parentis, respectively (Fig. 3). Success came in two steps in the Paris Basin: the first significant discoveries were made in the 1950s at a few localities such as Coulomnes, Chailly, and Chateaurenard, for instance (Fig. 3), but the two main oil fields, Villeperdue and Chaunoy, were discovered in 1982 and 1983, respectively (but an oil show was reported from Villeperdue as early as 1958). Intensive exploration is now proceeding in these two basins, with significant discoveries still being made (Fig. 4). The third basin is the South-East Basin, where the first phase of exploration in the 1950s to 1970s was disappointing. The present level of exploration is very low. A few smaller sedimentary basins are present at the edge or are concealed below these three major basins, and one of them, the Rhine Graben of Tertiary age, was intensively explored and is still producing a marginal amount of oil (Fig. 3). Major sedimentary basins are also present in the offshore area (Boillot et al. 1984). Whereas two of them, the Gulf of Lions and the Western Approaches (or Iroise Sea) basins are almost totally marine, the other two (Bay of Biscay and Channel basins) may be considered as the offshore extension of the nearby onshore sedimentary area. The first results of exploration in both cases have been unsuccessful, but only a limited number of wells have been drilled.

Three major thrust belts are present in France. One of them, the Variscan thrust belt, is of Paleozoic age and corresponds to the economic basement of most Mesozoic and Tertiary basins. Relatively thick late and/or post-orogenic continental troughs of late Carboniferous and Permian age are present, however, locally and have to be taken into account for conventional or coalbed methane exploration (Chateauneuf and Farjanel 1989; Robert 1989). The Pyrenean and Alpine-Jura thrust belts are of late Cretaceous and Tertiary age (Roure et al. 1990; Daignères et al., this Vol.). Both of them have been and are still being explored, with major results in the Pyrenean foothills (Lacq and Meillon gas field), and only scattered oil and gas shows in the Alps and Jura. Other offshore sedimentary basins are present around the French overseas territories (Orstom 1981). They are generally unexplored with the exception of the French Guyana margin, where unsuccessful wells were drilled in the 1970s.

The following chapters will be devoted to an overview of the petroleum geology of all these basins, with special emphasis on their most significant results and on present activity. Detailed statistics on the oil and gas fields are presented in the Appendix, and a description of further 34 fields can be found in a recently published atlas (Comité des Techniciens 1991). Some important geologic and geophysical surveys have also been performed in France for gas storage in artificial cavities (in salt diapirs, for instance) or in aquiferous reservoirs from conventional anticlinal traps (Jeanette 1985; Blondin and Verdier 1992). A special chapter in this Volume will be devoted to this specific activity (Blondin, this Vol.) and thus will not be developed here.

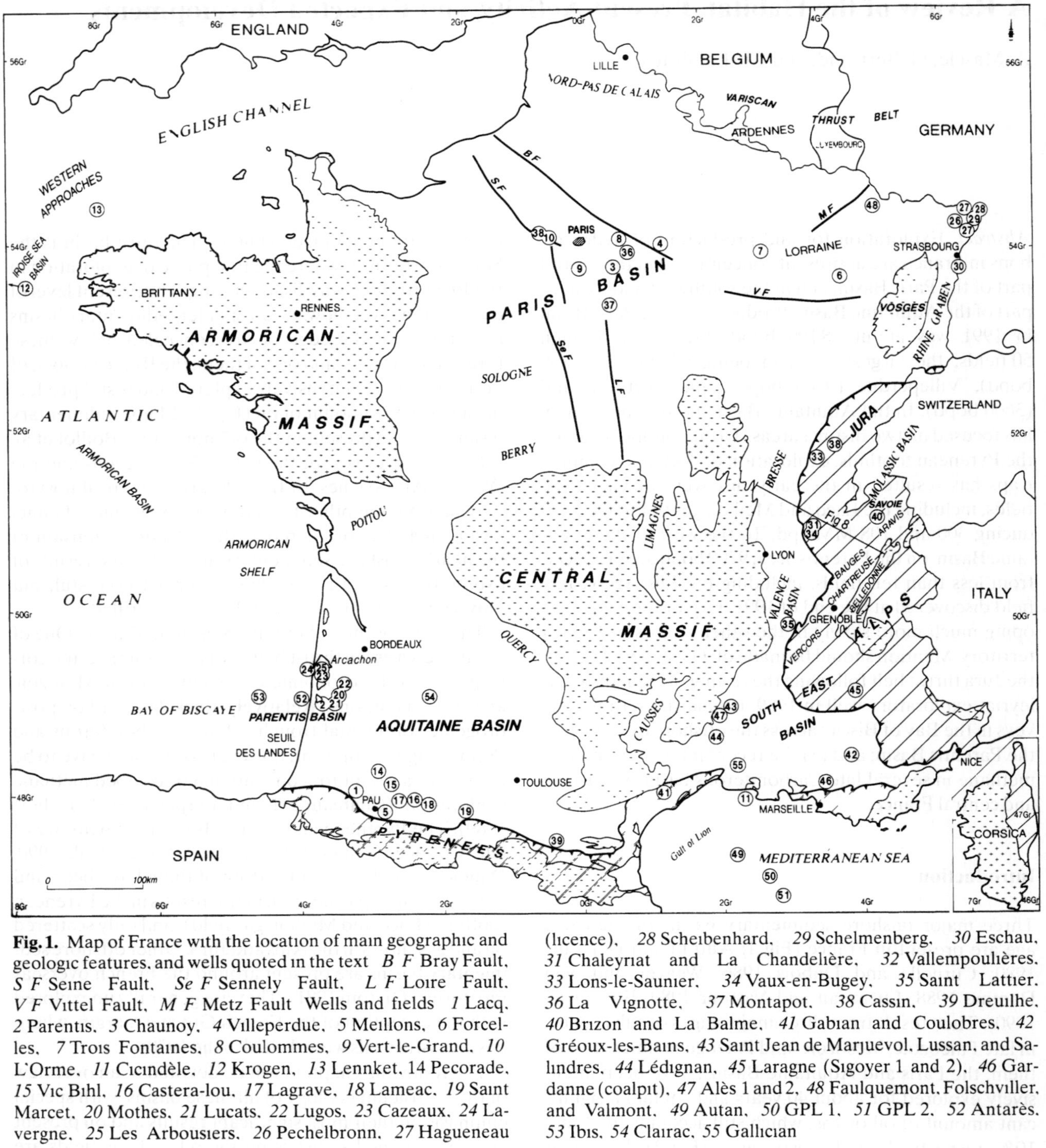

Fig. 1. Map of France with the location of main geographic and geologic features, and wells quoted in the text. *B F* Bray Fault, *S F* Seine Fault, *Se F* Sennely Fault, *L F* Loire Fault, *V F* Vittel Fault, *M F* Metz Fault. Wells and fields: *1* Lacq, *2* Parentis, *3* Chaunoy, *4* Villeperdue, *5* Meillons, *6* Forcelles, *7* Trois Fontaines, *8* Coulommes, *9* Vert-le-Grand, *10* L'Orme, *11* Cicindèle, *12* Krogen, *13* Lennket, 14 Pecorade, *15* Vic Bihl, *16* Castera-lou, *17* Lagrave, *18* Lameac, *19* Saint Marcet, *20* Mothes, *21* Lucats, *22* Lugos, *23* Cazeaux, *24* Lavergne, *25* Les Arbousiers, *26* Pechelbronn, *27* Hagueneau (licence), *28* Scheibenhard, *29* Schelmenberg, *30* Eschau, *31* Chaleyriat and La Chandelière, *32* Vallempoulières, *33* Lons-le-Saunier, *34* Vaux-en-Bugey, *35* Saint Lattier, *36* La Vignotte, *37* Montapot, *38* Cassin, *39* Dreuilhe, *40* Brizon and La Balme, *41* Gabian and Coulobres, *42* Gréoux-les-Bains, *43* Saint Jean de Marjuevol, Lussan, and Salindres, *44* Lédignan, *45* Laragne (Sigoyer 1 and 2), *46* Gardanne (coalpit), *47* Alès 1 and 2, *48* Faulquemont, Folschviller, and Valmont, *49* Autan, *50* GPL 1, *51* GPL 2, *52* Antarès, *53* Ibis, *54* Clairac, *55* Gallician

Paris Basin

The Paris Basin is a Mesozoic and Tertiary intracratonic basin superimposed on Carboniferous and Permian troughs, and on a Paleozoic basement which is generally interpreted as the northern branch of the Variscan thrust belt (Mégnien 1980; Brunet 1981; Pages 1987; Cazes and Torreilles 1988; Huai Fu Wei et al. 1990; Gely and Lorenz 1991; Perrodon and Zabek 1991, Piqué 1991). The more or less circular shape of the Mesozoic outcrops is related to the presence of successive depocenters often being located just east of the city of Paris and thinning towards the present edges (Pomerol 1989; Debrabant et al. 1992). This shape also results from the uplifts in Tertiary times of the Armorican, Central, Vosges, and Ardennes massifs on the western, southern, eastern, and northeastern borders of the basin, respectively. The subsidence of the basin was controlled by the reactivation of major late Variscan deep faults or arrays of faults, trending NW-SE in the west (Seine and Bray faults, for instance), N-S in the south (Sennely and Loire faults), and E-W to NE-SW in the east (Vittel and Metz). A tentative distribution of the stresses that have been applied to the basin from late Paleozoic to Present times is shown in Fig. 5. The most important stress, which best explains the present distribution and shape of structural traps, was probably the latest Cretaceous-Eocene (Pyrenean) shortening, in conjunction with one of the other events. The two main hydrocarbon reservoirs encountered so far in the Paris Basin are the late Triassic Chaunoy and Donnemarie sandstones, and the late Dogger (Callovian) oolitic limestones (Fig. 3). Smaller accumulations have been found locally in Neocomian and Rhaetian sandstones as well as, more recently, in Hettangian limestones (Morelot 1986; Leblanc et al. 1991; Sellier et al. 1991). The main source rocks are several intervals of Liassic black marls. Whereas the more prolific ones are undoubtedly early Toarcian "Schistes Cartons" with up to 10% of marine (Type II) organic matter, it is also believed that the slightly less organic-rich Lotharingian (Sinnemurian) and early Domerian (late Pliensbachian) marine marls played an important role in the quantity of oil expelled because of their deeper burial with respect to the overlying Toarcian shales (Espitalié et al. 1987; Poulet and Espitalié 1987; Bessereau et al. 1992). In the eastern Paris Basin, minor production of oil and gas is also provided by hydrocarbons generated from coal seams of late Carboniferous age in the Forcelles and Trois Fontaines fields, respectively. These coal measures belong to the Saar-Lorraine trough, the largest of the French late Carboniferous basins extending far to the northeast in Germany (Schafer 1986; Liu 1990; Pickel and Gotz 1991).

Whereas Coulommes was the first oil field to have been discovered as early as 1958 (Monchaux and Trémolieres 1985), the major developments occurred in the 1980s with the discovery of the three largest oil fields in the Paris Basin: Villeperdue (1982, Triton-Total; Duval 1992; Fitzeral and Mousset 1987; Pommier 1985), Chaunoy (1983, Essorep; see Bacchiana et al., this Vol.), and Vert-le-Grand (1985, Elf Aquitaine) which were still producing about 9400, 14800, and 3300 bopd, respectively, in the first half of 1992. The rate of success has been declining in recent years, however, with small oil fields still being discovered. Since December 1990, three discoveries (Elf Aquitaine operator) should be mentioned: L'Orme (110 bopd), La Vignotte (630 bopd), Montapot (135 bopd; all productions for April 1992). L'Orme 1 was probably the most promising discovery as it is the only successful well located west of the city of Paris in a new (Hettangian) reservoir, thus supporting the idea of long-distance migrations from the oil kitchen east of Paris. A second well (Cassin 1) drilled in the same area a few months later was dry, however.

Exploration in the Paris Basin is made difficult because of the very low vertical amplitude of structures. As velocity anomalies are common in the superficial late Cretaceous chalk and Tertiary levels (Hanot and Renoux 1991), very accurate calculation of primary static corrections has to be performed during seismic processing to avoid drilling pitfalls. This difficulty has recently been partly solved with the development of a shallow geology

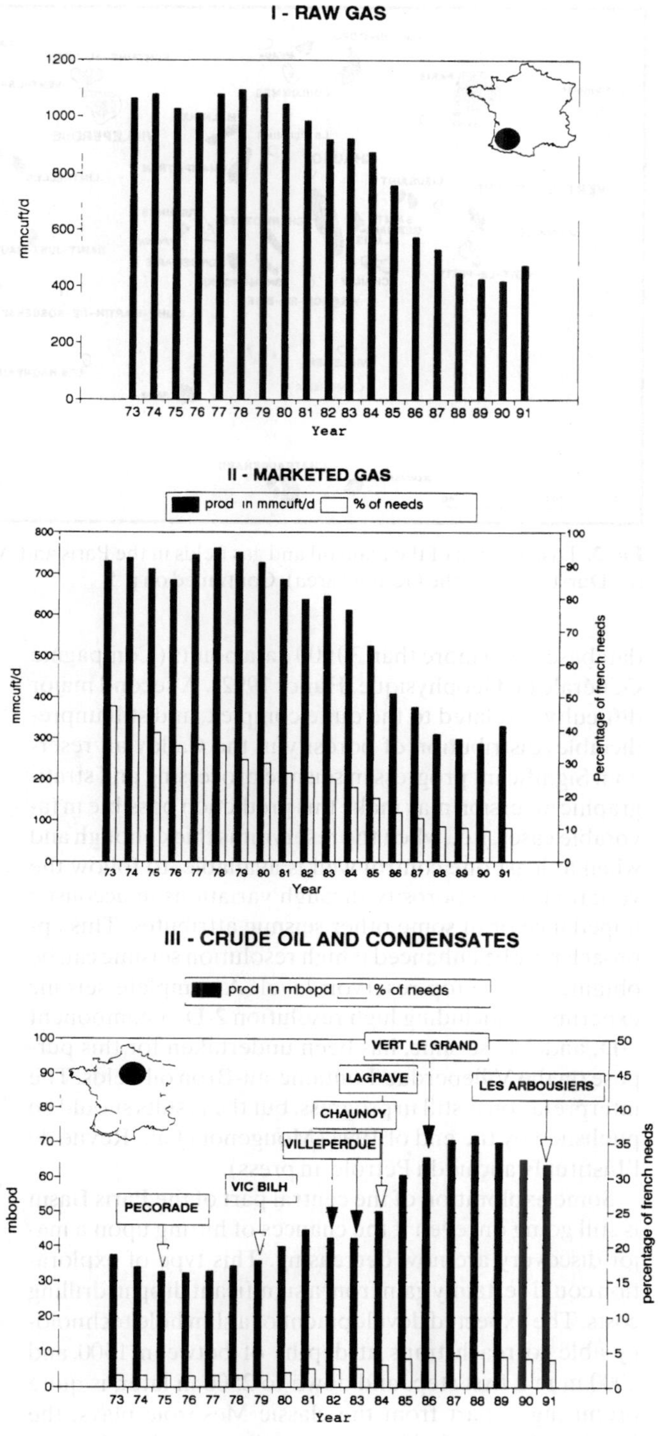

Fig. 2. Gas and oil production in France The Aquitaine Basin is the main gas producing area The *raw gas* is gas produced at the wellhead, including secondary components such as hydrogen sulfide and condensates. The *marketed gas* is refined gas from which the above-mentioned components have been removed The main oil fields appear on the graph the year they were put on-stream

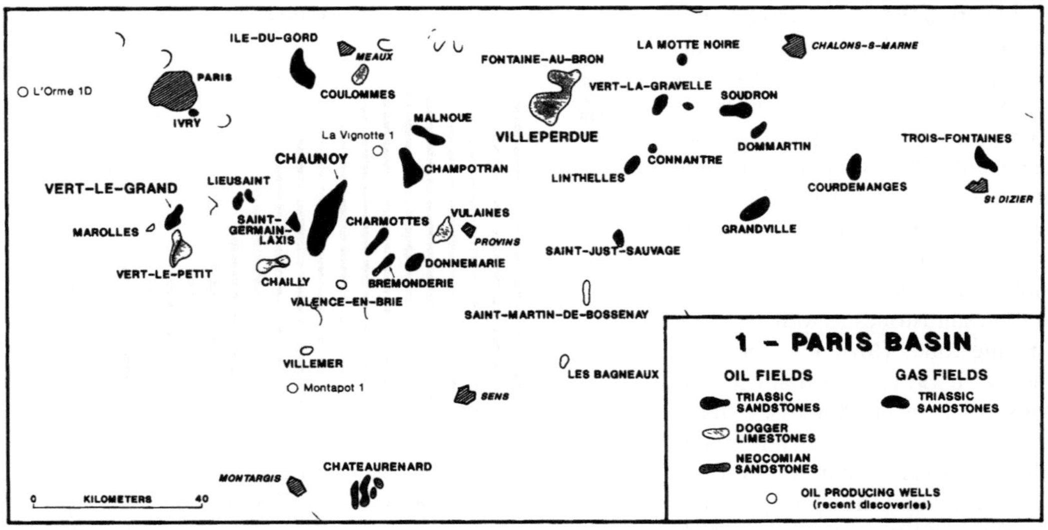

Fig. 3. Distribution of the main oil and gas fields in the Paris and Aquitaine Basins, and in the Rhine Graben (French territory only, see Durst 1991, for the German area) Continued on p 5

database using more than 30000 data points (Compagnie Générale de Géophysique, Hanot 1992). A second major difficulty is related to the quite complex, and still unpredictable, distribution of porosity in the Callovian reservoir. Significant progress in seismic processing and stratigraphic inversion may make this prediction possible in favorable cases, i. e., when the reservoir is thick enough and when at least one control well is available, to follow the variation of the porosity through variations in accoustic impedance or in some other seismic attributes. This approach may be enhanced if high resolution seismic can be obtained at the top reservoir level A complete seismic experiment, including high resolution 2-D, 3-component 2-D, and 3-D seismic, has been undertaken for this purpose on the Villeperdue/Fontaine-au-Bron oil fields. The interpretation is still in progress, but the results should be published by the end of 1994 (Mougenot et al., Revue de l'Institut Français du Pétrole, in press).

Some exploration of the central part of the Paris Basin is still going on, even if the chances of hitting upon a major discovery are now decreasing. This type of exploration could certainly gain from a significant drop in drilling costs. The expected development of a slim hole technology able to reach traps at depths of between 1500 and 2000 m in a first step, and down to 2500 m later, is quite promising. Apart from the classic Mesozoic plays, the Permian troughs below are virtually unexplored in the central and northern part of the Paris Basin. A deep stratigraphic well would be useful to document the potential distribution of reservoirs, seals, and source rocks. It should be mentioned, however, that the exploration of similar basins to the east (Lorraine) and southwest (Sologne-Berry) of the basin, below a thin Mesozoic cover, has been quite disappointing (Mascle 1990). Furthermore, there is for the moment no indication of any Paleozoic continental or lacustrine component in the oil produced from the main fields, with the exception of the tiny Forcelles oil field in the east (Lorraine, Fig. 1).

New plays could also be defined in the western part of the basin, following the success of the L'Orme discovery in Hettangian limestones. They imply a better understanding of the timing and pathways of migrating hydrocarbons and of the distribution of potential Triassic and Liassic reservoirs along the western edge of the basin.

To complete this overview, we should mention that original plays have been explored without success in the far south and north of the basin To the south, two isolated branches of the Tertiary Western Europe rift system are present in the Limagnes area. The Tertiary infilling of the troughs is up to 2500 m thick (Morange et al. 1971). Mature source rocks have been found in early Oligocene strata and may exist in isolated Stephanian or Autunian basins below Reservoirs and seals may locally be present, but only oil shows have been encountered so far. Stratigraphic traps, however, have never been tested. In the north of the Paris Basin, the Paleozoic basement is present below a thin late Cretaceous cover. Carbonate and coal measure plays have been tested at the deformation front of the Variscan thrust belt without any result, probably owing to the structural complexity and to an early (late Paleozoic) hydrocarbon generation (Cazes and Torreilles 1988; Le Gall 1992). In this area, a marginal attempt to produce the methane present in the old and now abandoned collieries is developing locally.

Aquitaine Basin

The Aquitaine Basin is a composite area that developed since the Triassic on a heterogeneous Paleozoic basement, which is generally interpreted as a segment of the southern branch of the Variscan orogeny (Curnelle et al 1982; Curnelle 1983; Mullan 1984; Paris et al. 1988; Bell et al. 1992; Espitalié and Drouet 1992). Late Carboniferous and Permian troughs up to a few thousand meters thick have developed locally as a result of post-Variscan stress

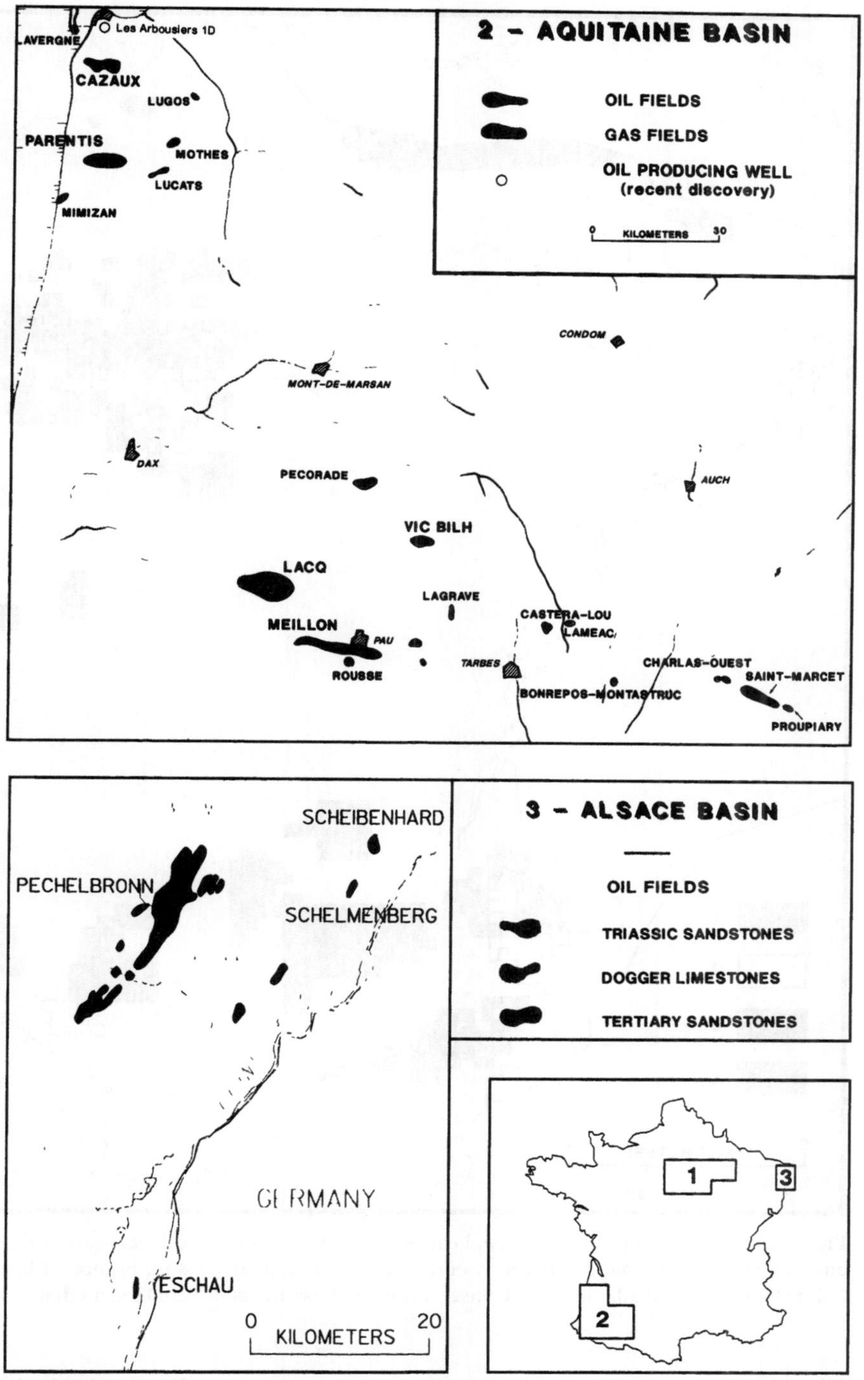

Fig. 3 *(continued)* Legend see p 6

releases. In Mesozoic times, a major hinge line (the "Celtaquitaine flexure"), running from Arcachon in the NW, to Toulouse in the SE, separates a stable and relatively less subsiding platform to the northeast from more rapidly subsiding basins and intervening highs to the southwest (Brunet 1991). This latter area furthermore should be subdivided into two distinct domains: in the south, the northern edge of the Pyrenees thrust belt and its moderately folded foreland that developed from late Cretaceous to late Eocene times (Baby 1988; Choukroune et al. 1989, Specht 1989); and the Parentis Basin to the west which has been individualized as a rapidly subsiding trough related to the opening of the Bay of Biscay in late Jurassic and early Cretaceous times (Mathieu 1986). Similar troughs were formed to the southeast that are now incorporated, or cancelled but partly preserved, below the Pyrenean thrust belt. Each of these areas has quite distinct petroleum potentials that will be discussed separately (Fig. 6).

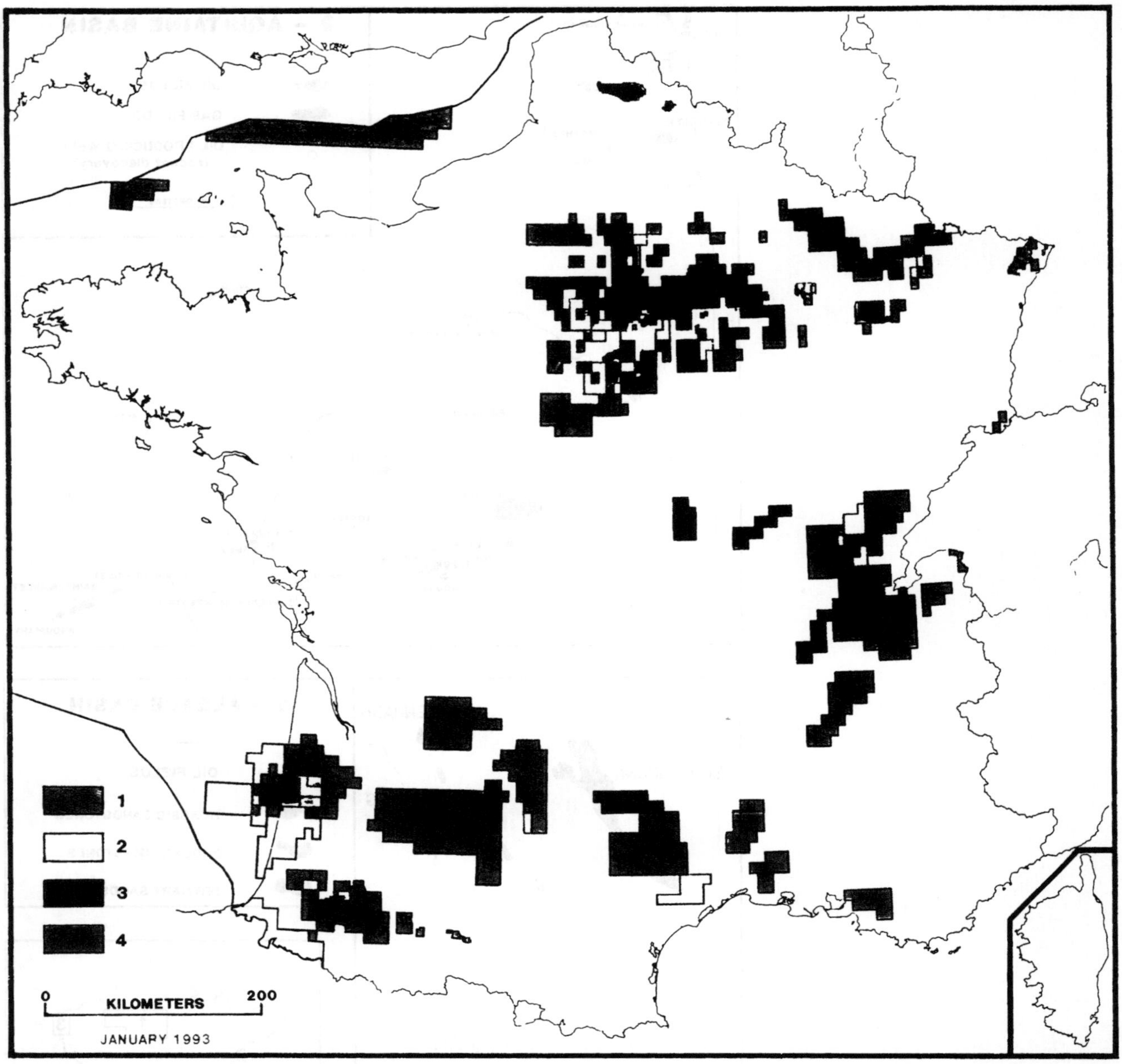

Fig. 4. Distribution of licenses and applications in France at the end of 1992 The complete half-yearly edited reference map at 1 1000000 scale is available at the Conservation des Gisements d'Hydrocarbure, 366 AV N Bonaparte, 92501, Rueil Malmaison, France *1* licences, *2* licence applications, *3* licences for coalbed methane, *4* licence applications for coalbed methane

The Northern Pyrenees and Their Foreland

The western half of the French northern Pyrenees and their immediate foreland is the most prolific gas-bearing province in France (Fig. 2). After the initial discovery of Saint Marcet in 1939, two major gas fields were subsequently discovered in deep and structurally complex traps. The Lacq field was discovered in 1951, has produced about 7.5 tcf since 1957, and is still producing 300 MMcfpd The Meillon field was discovered in 1966, has produced about 1.7 tcf since 1968, and is still producing 108 MMcfpd. The producing intervals are late Jurassic and Neocomian dolostones when the source rocks are Kimmeridgian marls and limestones (Espitalié and Drouet 1992). Initial porosities are low, but highly fractured zones greatly enhance the production (Haller and Hamon 1993). In the same area, six fields are also currently producing 19200 bopd (see Fig 3) from the same stratigraphic intervals (Pecorade, Vic Bihl, Castera-Lou fields), but also from younger Senonian levels (Lagrave and Lacq-Supérieur fields) or older Liassic ones such as at the small Lameac field. Intensive exploration is still going on in this area, with wide 3-D surveying in the complex foothills and deep drilling (6900 m in the last unsuccessful Bellevue exploratory well) below the frontal thrusts. Exploration up to now has been disappointing in the far foreland in the north, i e., towards the "Celtaquitaine flexure", possibly because of the disappearance of the above-mentioned classic reservoirs. Similarly, no success has ever been obtained from the eastern half of the Pyrenees, probably because of the absence of the Kimmeridgian source rocks and of the underlying Triassic-Jurassic section as well. In this area, significant gas shows at the Dreuilhe wells in very thick late Cretaceous clastic formations lying directly on the Paleozoic basement are believed to have originated from the basal interbedded black marls.

The Parentis Basin

The Parentis Basin developed as a rapidly subsiding E-W-trending trough as early as the Kimmeridgian and until Albian times. Almost 10000 m of sediments accumu-

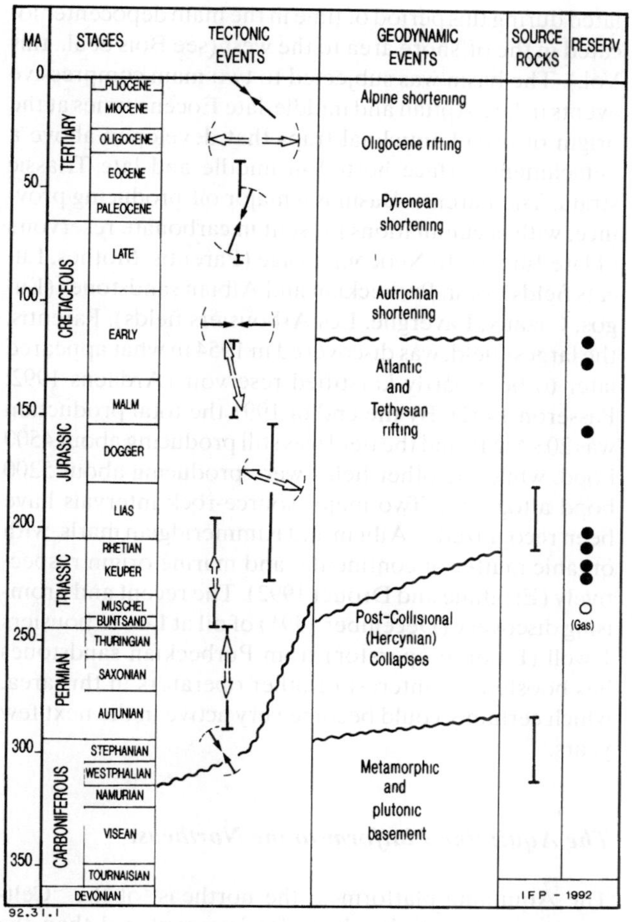

Fig. 5. Distribution of paleostress fields, main reservoirs and source rock intervals in the Paris Basin. A detailed account of the stratigraphy, lithology, source rock and reservoir distribution can be found in Perrodon and Zabek 1991

Fig. 6. Distribution of main source rocks and reservoir intervals in the Aquitaine Basin. A detailed account of this distribution in the Parentis Basin and along the western Pyrenean front and foreland can be found in Espitalié and Drouet 1992. Data from the southwest Aquitaine basin are from Paris et al. 1988, and the Clairac well area. Data from the eastern Pyrenean front and foreland are from the Dreuilhe wells area and data from the Quercy region are from field studies

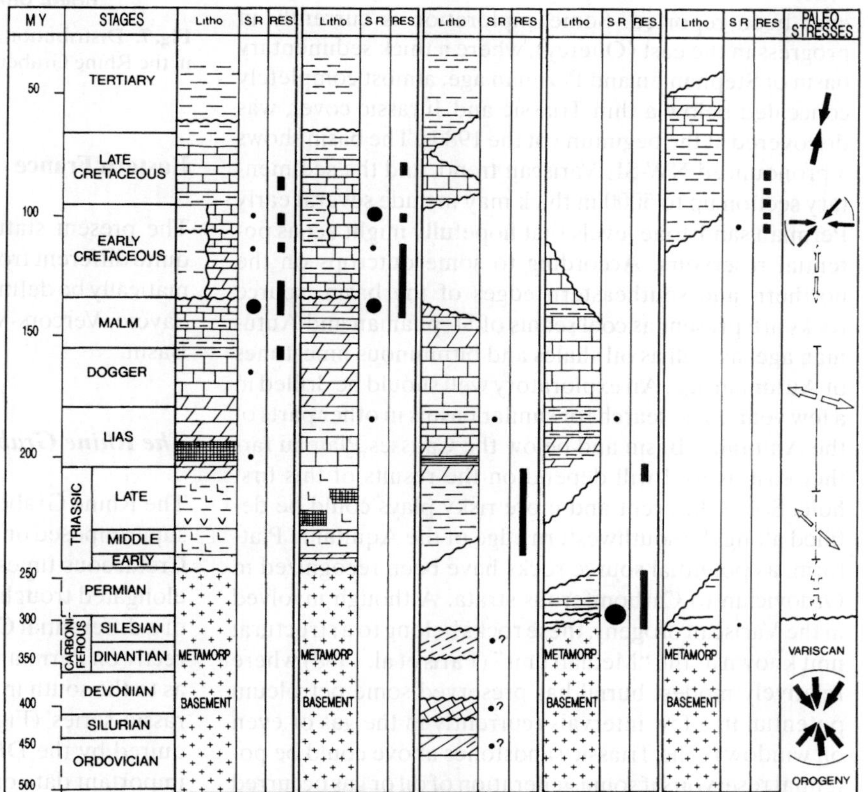

lated during this period of time in the main depocenter located in the offshore area to the west (see Bois et al., this Vol.). The basin was subjected to two main compressive events in late Aptian and middle-late Eocene times at the origin of broad anticlinal traps that developed above a detachment surface hosted in middle and late Triassic strata. The Parentis Basin is a major oil-producing province, with accumulations present in carbonate reservoirs of late Jurassic to Neocomian age (Parentis, Mothes, Lucats fields), or in Purbeckian and Albian sandstones (Lugos, Cazaux, Lavergne, Les Arbousiers fields). Parentis, the largest field, was discovered in 1954 in what appeared later to be a partly karstified reservoir (Ardaens 1992; Passeron 1992). By the end of 1991 the total production was 203 MMb and the field was still producing about 4500 bopd, while the other fields were producing about 5200 bopd altogether. Two major source-rock intervals have been recognized in Albian and Kimmeridgian marls, with organic matter of continental and marine origin respectively (Espitalié and Drouet 1992). The recent and promising discovery (December 1991) of oil at Les Arbousiers 1 well (Essorep operator) from Purbeckian sandstones has boosted the interest of other operators in this area, which certainly could become very active in the next few years.

The Aquitaine Platform to the Northeast

The Aquitaine platform to the northeast of the "Celtaquitaine flexure" has been far less explored than the other two areas, and no significant oil or gas show has ever been reported. Some exploration is currently in progress in the east (Quercy), where a thick sedimentary basin of Stephanian and Permian age, almost completely concealed below a thin Triassic and Jurassic cover, was discovered at the beginning of the 1980s. The basin shows a pronounced NW-SE Variscan trend, and the sedimentary section up to 5000 m thick may include several early Permian sandstone levels that hopefully might act as potential reservoirs. According to some outcrops on the northern and southeastern edges of the basin, source rocks are present as coal seams of Stephanian and Autunian age, as well as oil shales and bituminous limestones of Autunian age. An exploratory well should be drilled in a few years. The search for similar basins in other parts of the Aquitaine Basin and below the Causses Plateau farther east as well will depend on the results of this first hole. Some different and more risky plays could be defined along the southwestern edge of the Aquitaine Platform, as potential source rocks have been recognized in Ordovician to Carboniferous strata. Although involved in the Variscan orogeny, these rocks belong to a structural unit known as the "Median unit" (Paris et al. 1988) where relatively modest burial has preserved some petroleum potential in a few intervals (currently in the gas or even oil window). The Triassic sandstones above could be potential reservoirs if some generation of oil or gas occurred in Mesozoic times.

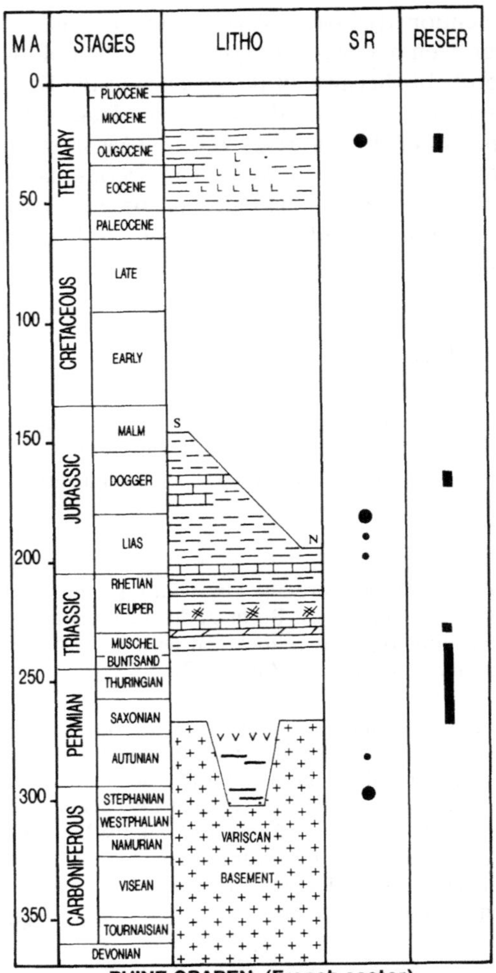

Fig. 7. Distribution of main source rocks and reservoir intervals in the Rhine Graben. For more details see Durst 1991

Eastern France

The present status of exploration in eastern France is quite different from north to south. Three zones can schematically be defined: the Rhine Graben, the Bresse-Jura, Savoie, Vercors-Valence Basin area, and the South-East Basin.

The Rhine Graben

The Rhine Graben is a north-south-trending Tertiary rift superimposed on a Mesozoic platform severely eroded in Cretaceous times, and on Permian-Stephanian NE-SW-elongated troughs (Villemin 1986; Durst 1991). Because of a differential Cretaceous erosion, Tertiary strata lie directly on Permian strata in Germany to the north, whereas to the south in France they unconformably overlie Jurassic series (Fig. 7). Recent deep seismic profiles acquired by the DECORP-ECORS group have provided important data on the regional and deep structure of the graben, emphasizing more particularly the importance of normal detachment faults on both sides of the graben (Brun et al. 1991; Wenzel and Brun 1991). The Rhine Graben is actually intensively fractured with major normal faults along the eastern and western edges of the basin, and synthetic or antithetic smaller normal faults in the center. The resulting structural traps are usually of limited extent. On French territory the closed areas are about a few km^2 and never reach 10 km^2. Moreover, some of these large faults may have acted as pathways for migrating or dismigrating hydrocarbons.

Although several potential reservoirs may be proposed in the Rhine Graben, the two main producing horizons encountered so far are: (1) The late Bajocian-early Bathonian "Great Oolith" sealed by late Bathonian marls or Tertiary claystones. The reservoir quality of these limestones however is hard to predict owing to the relative amount of primary and secondary porosities (fracturation or dolomitization) they contain. (2) The second and best reservoir is found in the Oligocene fluviatile and deltaic sandstones (upper Pechelbronn and Melettes formations) interbedded with and covered by clays of the same formations. Porosities and permeabilities are generally good to excellent (average 15–20% and 500 mD respectively), but their distribution is restricted to the northern part of the Rhine Graben. Exploration and production started as early as the 18th century when live oil and bitumen where mined in the Pechelbronn area (Sittler 1985). Conventional exploration started at the beginning of this century, with the first deep oil discovery being made at Pechelbronn in 1922. Unconventional mining extraction of oil went on, however, until 1963, when bitumen production stopped in 1950.

The present level of exploration and production is quite low. The last exploration licence has recently been abandonned (Haguenau, Triton), and present oil production does not exceed 250 bopd from the Scheibenhard, Schelmenberg, and Eschau oil fields (Elf Aquitaine operator, Fig. 3). The first two are producing in Tertiary limestones and were sourced by Oligocene (Sannoisian) shales, while the third is producing in Dogger limestones and was sourced by the Liassic black shales, as were most of the other, now abandoned, fields, such as the largest of them, the Pechelbronn field. The possibility that some activity could resume in the near future is slight, as the intensive exploration of the Mesozoic and Tertiary strata has been disapointing as a whole. Untested and high-risk plays do remain, however, in the late Carboniferous and Permian series, in areas where the Cretaceous erosion has preserved a cover of Triassic and Jurassic strata. Source rocks can actually be predicted in Stephanian or Autunian coal measures and oil shales, while reservoirs may be found in Permian sandstones if intraformational seals could have prevented direct communications with early Triassic sandstones. Local opportunities may also exist for some independent producers to reactivate now abandoned fields.

The Bresse-Jura-Savoie-Vercors-Valence Basin Area

This broad area is made up of four structural units that developed in Tertiary times on a Mesozoic platform, which in turn is superimposed on late Carboniferous-Permian troughs (Debrand-Passard et al. 1984; Guellec et al. 1990). They are, from west to the east (Fig. 1 and 8): the Bresse and Valence Basin rift grabens of late Eocene-early Miocene age, the Jura late Miocene-Pliocene thrust belt, the Eocene-Miocene Molasse basin and the early Tertiary Bauges-Aravis Subalpine massifs. These latter merge in the south with the Jura thrust belt to form the Chartreuse and Vercors Subalpine massifs (Welbon and Butler 1992). The Paleozoic basement reappears in the west as the autochthonous Massif Central, and in the east as the allochthonous Belledonne massif. The explored structural traps have, most of the time, been Tertiary horsts and tilted blocks in the Bresse and Valence Basins and the basal part of similar horsts preserved below the main decollement hosted in late Triassic evaporites in the Jura and Molasse Basin. Complex tectonic sheets have been explored in the Bauges massif.

The distribution of potential reservoirs and source rocks in these areas and in the South-East Basin is shown in Fig. 9. Most of the oil and gas shows encountered are in early and middle Triassic sandstones and dolomites. This is especially true for the recent (1989) and significant oil shows encountered in the southern Jura in the Chaleyriat and La Chandelière wells (Essorep operator), although the following two wells were dry. The tiny and now abandoned gas fields in the Jura (Valempoulières, Lons-le-Saunier north and south, Vaux-en-Bugey with total productions of 3.5, 0.3, 0.25, and 0.45 bcf, respectively) were also producing in similar reservoirs. Oil shows have been reported from Oligoncene sandstones in the Molase basin, but no commercial accumulation has ever been found. Minor gas and oil shows have also been encountered in Liassic and Dogger limestones or dolostones in the Bresse and Valence Basin, the more significant of them probably being obtained at the Saint Lattier 1 (1957), well east of the city of Valence. More promising gas shows have been found in the Bauges Subalpine massif in the Brizon (1987) and La Balme (1989) wells (Eurafrep operator), but the poor characteristics of the Tithonian and Neocomian carbonate reservoirs have prevented any subsequent production (Charollais and Jamet 1990). Natural oil and gas shows have also been reported from several localities in areas affected by the last Alpine Neogene deformations (Deville et al., this Vol.). Two stratigraphic intervals may contain the source rocks at the origin of these shows: coal seams and oil shales of Stephanian-Autunian age (Blanc et al. 1991), and the more widespread Toarcian marine black marls.

The Chaleyriat and La Chandelière discoveries boosted licencing activity in these areas in 1990, although the following two dry holes have slowed down exploration surveys. The most noticeable action at that time was the completion of a non exclusive seismic survey through the

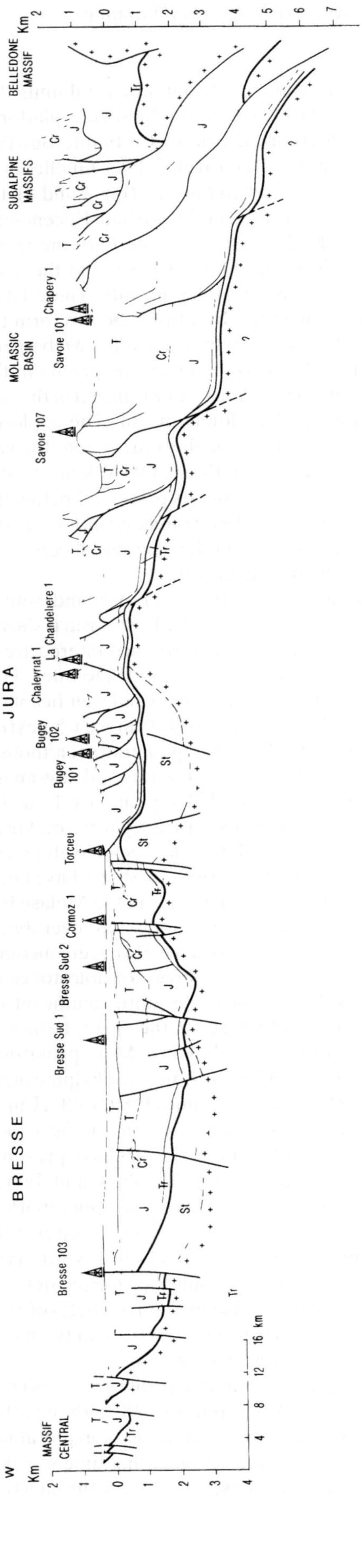

Fig. 8. Synthetic section through the Alpine thrust belt and foreland Location in Fig 1 *T* Tertiary, *Cr* Cretaceous, *J* Jurassic, *Tr* Trias, *St* Stephanian and/or Permian

Vercors and Chartreuse massifs (CGG-IFP commercial reports), which for the first time provided a clear understanding of the deepest part of these two Subalpine thrust belts. It is generally believed that the next step for a better evaluation of the petroleum potential of these areas will be a better assessment of the relative timing of hydrocarbon generation and migration with respect to the formation of structural traps (as well as stratigraphic traps in the Tertiary basins).

The South-East Basin

The South-East Basin is the thickest French sedimentary basin, since up to 10000 m of Mesozoic sediments may be found locally (Baudrimont et al 1977; Debrand-Passard et al. 1984, Dubois and Delfaud 1989). Several 100 m thick late Carboniferous and Permian basins are locally present below, or outcrop at the western or southeastern edge of the basin (Toutin-Morin 1987; Wang 1991; Saint Martin 1992; Toutin-Morin and Bonijoly 1992). Several major distensive events from Permian to early Cretaceous times led to the subsidence of the deep basins and of the more stable surrounding carbonate platforms; but the present structural complexity results mainly from the superimposition, since the late Cretaceous, of the two major Pyrenean and Alpine compressive events, with an intervening stretching event in Oligocene times (Roy and Trémolières 1992, Fig 9). Furthermore, salt diapirism (from the Keuper interval) was locally active throughout Tertiary times, but may have been initiated as early as the Bathonian (Dardeau and de Graciansky 1990; Dardeau et al. 1990). The Oligocene extension and subsequent Neogene subsidence were at the origin of rapidly subsiding troughs up to 5000 m thick (Arthaud et al. 1981; Valette 1990)

For the whole area possible plays include several intervals in the Triassic and Jurassic section with poor primary porosities enhanced by fracturation or karstification, and Oligocene sandstones or limestones in the corresponding troughs. Mature source rocks are known to occur in at least five stratigraphic intervals: coal measures and oil shales of Stephano-Autunian age, black shales of late Liassic age, silty black shales ("Terres Noires") of Bathonian-Oxfordian age, late early Cretaceous black marls, and Oligocene bituminous shales.

Although numerous oil and gas shows have been found, no commercial accumulation has been encountered so far The most significant shows have been obtained at different stratigraphic intervals. Oil was tested in late Permian-early Triassic sandstones at the Gabian and Coulobres wells (with a local production from a natural seep recorded as early as the 18th century, Barrabé and Schneegans 1935), gas was tested in Liassic dolo-

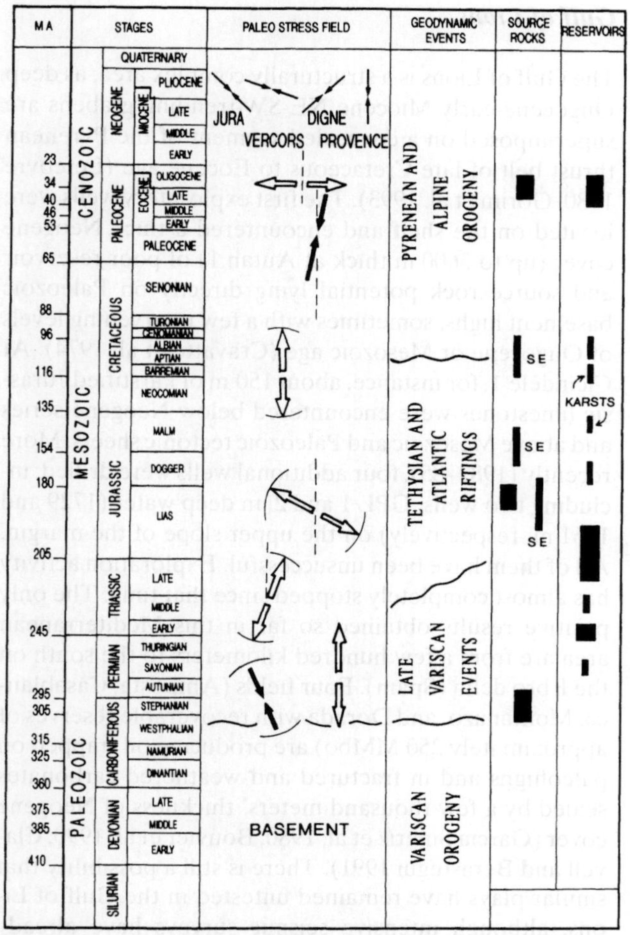

Fig. 9. Distribution of paleostress fields, potential reservoirs, and source rocks in eastern and southeastern France

stones at the Gréoux-les-Bains wells; oil was tested in Oligocene strata at Saint Jean de Maruéjols (where the Sannoisian source-rock is now being mined from shallow depths as bituminous limestone) and at Gallician. This lack of success in the first stage of exploration, which took place in the years 1950 to 1970, was probably due to the unexpected, very thick Mesozoic series being drilled (the available rigs were unable to reach the initial objectives), to the poor quality of the reservoirs encountered, but also to the limited resolution of complex structural traps by the seismic. Nonexclusive regional seismic profiles were shot by the Compagnie Générale de Géophysique throughout the South-East Basin in 1981–1984, including, in the Laragne area, the first 3-D ever shot in France. The generally good quality of these data will hopefully enable new operators to gain a better understanding of the structural framework and petroleum potential of the basin so that some exploration activity could start again (Roure et al. 1992).

Coalbed Methane

There has been a growing interest in the last few years in coalbed methane exploration, as elsewhere in the world, following the expanding successes in the United States. This still unconventional way to produce gas is linked to the possibility for coal to retain very large amounts of adsorbed methane. A content of up to 600 cf of gas per ton of coal has indeed been measured locally. The source-rock is thus also the reservoir, and no specific structural or stratigraphic trap is needed to preserve gas accumulations. The production of gas, however, is more complicated and uncertain than conventional production. It implies first decreasing the pressure in coal seams, usually after hydraulic fracturing, and pumping out the formation water. Gas is then able to become desorbed, and, in favorable cases, to migrate through the coal matrix, the natural cleat network and hydraulically induced fractures (Ayers and Keslo 1989; Kuuskraa and Brandenburg 1989; MacElhiney et al. 1989; Ayoub et al. 1991). Because numerous wells are necessary to develop a field, the currently producing coal seams in the United States are generally no deeper than 1200 m.

Although it is generally believed that the gas storage capacity of coal increases with increasing rank, this has not been totally confirmed by experiments and observations, as even bituminous coal may sometimes contain a fairly large amount of adsorbed methane. Numerous coal-bearing basins are present in France, some of them still being mined. A complete review of their distribution and potential for conventional coal extraction was made by the Bureau de Recherches Géologiques et Minières in the 1970s and early 1980s. Most of these basins are of Carboniferous-Autunian age. They are related to late stages of shearing and stress release on top of the Variscan chain. They show great variety in size, structural complexity, and coal content. The largest of them is the Saar-Lorraine basin (Donsimoni 1981; Ungerer et al. 1986; Liu 1990), where the total thickness of moderately deformed Westphalian coal in the Merlebach area, for instance, is in excess of 130 m. The second largest basin is the Nord-Pas de Calais basin of Carboniferous age (Becq-Giraudon 1983). This basin developed as a Variscan foredeep and was subsequently severely deformed when incorporated in the front of the thrust belt (Le Gall, this Vol.). Numerous basins of late Westphalian-Stephanian-Autunian age outcrop in the Massif Central and have been recognized below the Mesozoic cover on the edge (see for instance Robert 1985; Donsimoni 1990; Wang 1991; Courel et al., this Vol.). The rank of coal is highly variable from bituminous to anthracite, even in one single basin, probably owing to local volcanic and hydrothermal events superimposed on an overall high heat flow regime (Courel and Xing Bing Liu 1991). Mention should be made of the presence of little deformed low rank coals of Fuvelian (late Campanian) age in the Provence area, still being mined at depths in excess of 1000 m (Gardanne coalpit, Arcamone et al. 1980).

The present status of licenses for coalbed methane exploration is shown in Fig 4. At the time this chapter was written, very few wells had been drilled. Three of them (Faulquemont 1, Folschviller 1, and Valmont S-1) were spudded in 1988 by Windsor in the Lorraine basin They encountered the expected Westphalian coal seams at depths of between 500 and 1500 m, but production tests have been disappointing so far Three additional wells were drilled by Enron in 1993 in the same area. Also, two wells (Alès 1 and 2) were drilled by Kelt in the southeast of the Massif Central (Cevennes basin). They reached total depths of 560 and 811 m, respectively, through coal seams of Stephanian age concealed below Triassic and Liassic strata. Production tests have been said to be encouraging. There is no doubt that more significant results will be obtained in the near future, as a few more wells and tests are needed to confirm or not the economic potential of coalbed methane in France.

Offshore Area

Two types of continental margins may be defined around France. In the south, the Mediterranean margin is of Tertiary age and is related to the opening of the Provençal oceanic basin in Miocene times, following an initial phase of rifting of Oligocene to middle Aquitanian age (Burrus 1984). The continental shelf is usually narrow, with steep slopes down to the abyssal plain. In the Gulf of Lions, however, a broad continental shelf developed in response to the clastic input from the Rhone River and Delta in Pliocene and Pleistocene times. In the west of France the continental margin is of Cretaceous age and related to the accretion of oceanic floor in the Bay of Biscay from late Aptian to Campanian times, following the initiation of rifting as early as the late Jurassic. The continental shelf widens from south to north, and three segments can be distinguished. The southernmost faces the Aquitaine Basin and contains the offshore extension of the Parentis Basin (Bois et al., this Vol.). The central one is located along the coasts of Poitou and Brittany, and the sedimentary cover on the continental shelf is thin and of little interest for the petroleum industry. A relatively thick and unexplored sedimentary prism has, however, developed deeper on the continental slope and rise and is known as the Armorican Basin (Barbier et al. 1986). The northern segment faces the Western Approaches where thick Triassic and Jurassic NE-SW-trending basins developed prior to the opening of the Bay of Biscay (Montadert et al. 1979; Evans 1990, Bois et al. 1991). Farther east, the Channel Basins may be regarded as the offshore prolongation of the Paris and London Basins to the southeast and northwest respectively (Cope et al. 1992). The petroleum exploration has been restricted to three areas until now.

Gulf of Lions

The Gulf of Lions is a structurally complex area, as deep Oligocene-early Miocene NE-SW-trending grabens are superimposed on a concealed segment of the Pyrenean thrust belt of late Cretaceous to Eocene age (Lefebvre 1980; Gorini et al. 1993). The first exploratory wells were located on the shelf and encountered a thick Neogene cover (up to 3600 m thick at Autan 1) of poor reservoir and source rock potential lying directly on Paleozoic basement highs, sometimes with a few intervening levels of Oligocene or Mesozoic age (Cravatte et al. 1974). At Cicindèle 1, for instance, about 150 m of karstified Jurassic limestones were encountered below Neogene series and above Mesozoic and Paleozoic tectonic sheets. More recently (1982–85), four additional wells were drilled, including two wells, GPL 1 and 2, in deep water (1729 and 1261 m, respectively) on the upper slope of the margin. All of them have been unsuccessful. Exploration activity has almost completely stopped since that time. The only positive results obtained so far in this Mediterranean area are from a few hundred kilometers to the south on the Ebro delta (Spain). Four fields (Amposta, Casablanca, Montanazo, and Dorada with recoverable reserves of approximately 250 MMbo) are producing oil trapped on paleohighs and in fractured and weathered carbonates sealed by a few thousand meters' thickness of Neogene cover (Garcia Sineriz et al. 1980; Bouvier et al. 1990; Clavell and Berastegui 1991). There is still a possibility that similar plays have remained untested in the Gulf of Lions, although intensive seismic surveys have already been performed and some of them have already been drilled (Cicindèle well, for instance). Other plays could also be defined in the Oligocene series, where synrift clastic aprons can be expected and could be at the origin of stratigraphic traps.

Bay of Biscay

The exploration of the marine western extension of the onshore Parentis Basin started as early as 1966 and had slowed down considerably by the end of 1983 after 22 unsuccessful wells had been drilled. Two significant oil shows however have been encountered in late Barremian limestones (Ibis 2 well) and early to middle Jurassic limestones (Antares 101 well). Plays are probably more difficult to define than the ones already proven onshore. The distribution and quality of reservoirs is expected to change rapidly from the northern and southern edges of the basin to the much more quickly subsiding central depocenter, where the very thick late Jurassic to Albian sedimentary infilling probably also induced quite different timings of hydrocarbon generation and migration (Mathieu 1986). The structure of the basin is additionally made more complicated by a spectacular halokinesis originating from salt of Triassic age. A new phase of exploration is now developing, as the whole area is covered with exploration licenses or applications, and as a result

3-D seismic has recently been been shot. The sedimentary cover is much thinner to the south ("Seuil des Landes") and to the north (Armorican shelf), where more risky late Paleozoic plays still have to be defined (Pinet et al. 1987). It should be remembered that, at the southern edge of the Bay of Biscay and in Spanish waters, the small and structurally complex (as part of the Pyrenean thrust belt) Gaviota gas field is producing from a late Cretaceous carbonate reservoir with source rocks of possibly late Carboniferous age.

Western Approaches

The French side of the Western Approaches is better known as the Iroise Sea Basin. The first stage of exploration was completed in the mid 1980s without any significant result (if we except small oil and gas shows at Lennket and Krogen, in Purbeckian and Liassic to Dogger intervals, respectively), although 13 wells were drilled in French waters. However, at the same time, the near onshore Wytch Farm oil field was developed, and production was said to reach about 70000 bopd at the beginning of 1992 (Selley and Stoneley 1987). At the same time, the Kinsale Head and Ballycotton gas field in the southern Irish offshore area were also developed with reported reserves of 1.4 Tcf and 73 Bcf, respectively (Shannon 1991). It is thus believed that significant amounts of hydrocarbons could be present in French waters in still undisclosed structural or stratigraphic traps. The first structures initially drilled were actually broad inverted anticlines, whereas the Wytch Farm field has shown that the major oil accumulations are located in more subtly preserved Jurassic tilted blocks since the main phase of hydrocarbon generation predates the Oligocene inversion event (see Deronzier et al., this Vol.). Mature source rocks are present in the early Liassic interval, but show great lateral variation in both thickness and quality. Reservoirs are developed locally in early Triassic (main reservoir at Wytch Farm), Callovian, and Purbeckien sandstones. Good sandy reservoirs in Wealdian and Albian strata (producing horizon at Kinsale Head) are unfortunately not efficiently sealed. A new evaluation of the petroleum potential should be made in the near future, but will require high quality seismic profiles in order to delineate subtle traps, to map the distribution of organic-rich early Liassic intervals, and to evaluate the amount of erosion following the two inversion events of Aptian and Oligocene age. Two licenses are still valid in the central Channel area (Coparex and Hunt operators).

Overseas Departments and Territories

French overseas departments and territories are widespread throughout the world (Fig. 10). Their economic importance comes partly from their vast offshore areas, giving France a total of about 11 million km^2 of national oceanic domain (territorial waters plus exclusive economic zones). Some of them are of no interest for the purpose of oil exploration because of their volcanic origin and the excessive depths of the surrounding waters. A few others, however, should be considered as prospective areas in the near future. Two of them have actually been partly explored in the past:

The French Guyana Margin

The French Guyana margin is an Atlantic passive margin that has been developing in two main stages. In Jurassic and Cretaceous (pre-Albian) times, the northwestern part of the margin, together with the Surinam margin (Demerara Rise), is believed to represent the southern extremity of the Central Atlantic rift and Ocean. In middle Albian times, dextral shearing between Africa and South America led to the opening of the E-W-trending Equatorial Atlantic Ocean (Gouyet 1988). The offshore area of French Guyana is the only prospective area as the Paleozoic and older basement outcrop over most of the onshore area. Exploration on the Demerara Rise took place in 1960 and 1970 without any significant results, with most of the wells in Surinam waters. As a matter of fact, only two wells, Sinna Mary 1 and FG 1, were located in French waters. The first well (water depth 48 m) reached the basement at 2104 m below early Cretaceous sandstones and younger sediments, and the second one (water depth 820 m) bottomed out in basaltic lava flows of Barremian-early Aptian age. Expected plays were in Jurassic-early Cretaceous carbonates and sandstones, and the deeper ones have thus never been tested. The southeastern margin of French Guyana is still unexplored. A few seismic lines shot 15 years ago show, beneath the outer shelf and slope, a three-second-thick sedimentary wedge of presumed early Cretaceous to recent age. The presence or absence of older Mesozoic troughs beneath the 120-km-wide shelf cannot be inferred owing to the poor quality of the seismic. The area in the immediate vicinity of French Guyana have given no encouraging results, although it is not completely negative. In Brazilian waters and a few hundred kilometers from the French border, the Pirapema well on the Amapa platform tested gas of biogenic origin, but the field has not been developed since then (Campos 1983). In onshore Surinam, small heavy oil production is obtained from the Tambaredjo and Jossi Kreek fields (about 4500 bopd) from Paleocene sands of fluvio-deltaic origin (Jharap 1988). There is currently no petroleum activity in French Guyana, although some exploration is still in progress to the northwest in Surinam and Guyana waters.

New Caledonia

The geology of New Caledonia has been described in some detail by Paris (1981). The oldest terranes are of pre-Permian age. Subsequent Mesosoic strata were subjected to at least two metamorphic events of late Creta-

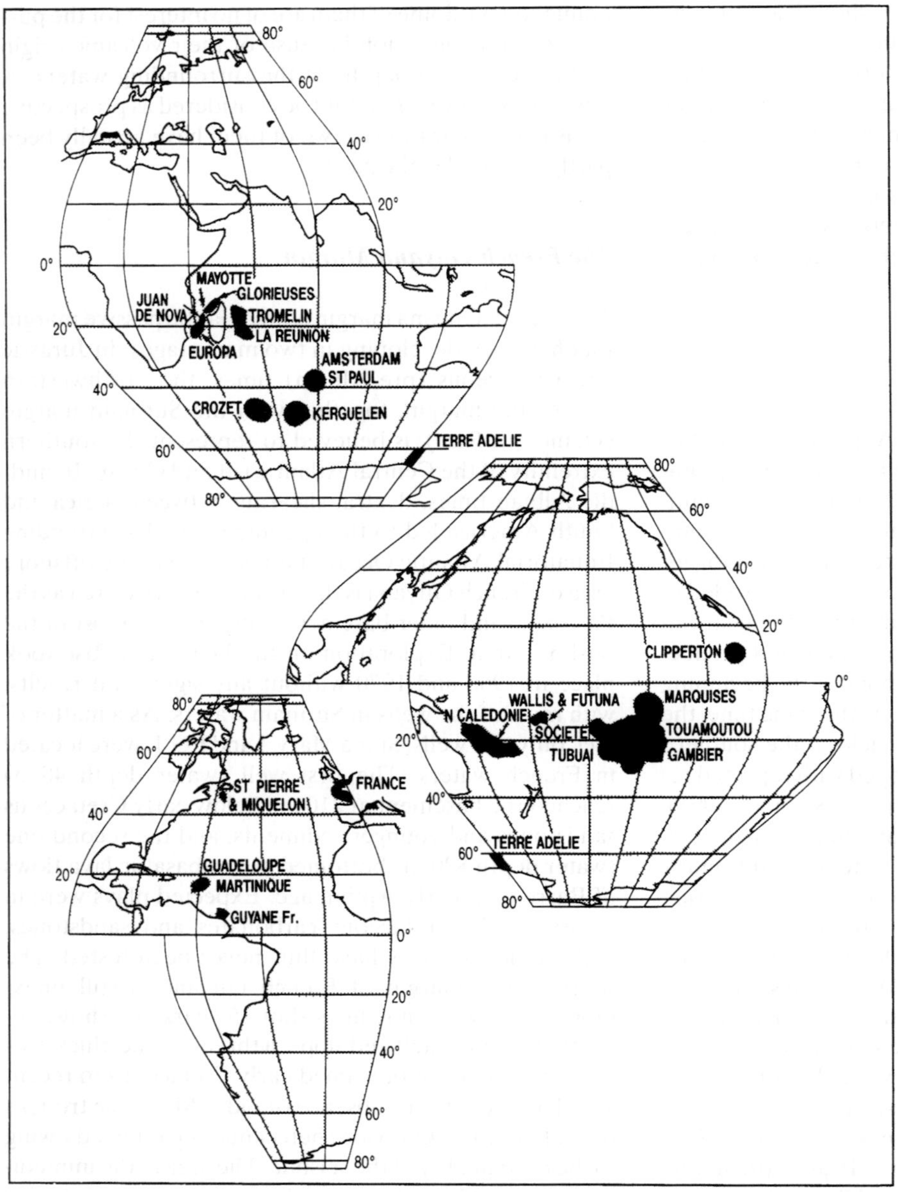

Fig. 10. Distribution of French overseas departments and territories (Ifremer)

ceous (pre-Senonian) and late Eocene age. A major event at that time was the obduction of ophiolites from the north (Loyalty Oceanic Basin) to the south over thick clastic sediments of Senonian to Eocene age with interbedded lava flows. These series are very locally and unconformably overlain by sediments a few hundred meters thick of late Oligocene to middle Miocene age. Onshore petroleum exploration has been limited to two shallow holes in the 1950s located along the western coast of the island (Gouaro anticline), where they encountered small oil shows in Eocene flysch (Tissot and Noesmoen 1958). Onshore licences have recently been awarded along the southern coast of the island. The offshore area around New Caledonia includes deep oceanic basins (Loyalty, New Caledonia Basins) and rises (Lord Howe, Loyalty Rises) which were recognized in the 1970s by regional seismic profiles (Ravenne et al. 1982; Lafoy, this Vol.) The shallow water area includes the shelf and lagoons of New Caledonia, but also the broad Chesterfield Bank to the west. Both of them are unexplored as very shallow, and locally poorly known water depths have prevented conventional seismic profiling from being shot.

Additional places of interest may be found in others seas. The presence of thick sedimentary basins has been recognized by regional seismic surveys shot in the 1970s and in the early 1980s. No well has been drilled, however They are succinctly described in the next paragraphs

Lesser Antilles

The Lesser Antilles island arc is an active Tertiary intra-oceanic volcanic arc resulting from the westward subduction of the Atlantic crust at a very slow rate (Bouysse and Westercamp 1990). A few of these islands are French departments, namely and from north to south: St Martin (pro parte), St Barthélemy, Guadeloupe, Désirade, Marie Galante, Les Saintes, and Martinique. These islands are generally built from volcanic rocks and intervening sediments of late Paleocene to Present age (Maury et al. 1990) and have only slight interest for petroleum exploration. Offshore basins up to three seconds thick have been discovered by seismic in water depths ranging from a few tens to a few thousand meters (Mascle and Letouzey 1990). The sediments contained in these basins are probably of Tertiary age, and the only indications we have about their nature come from the neighboring Saba Bank 1 and 2 wells, where reservoirs and source rocks (nonmature) have been reported in Paleogene strata (Despretz et al. 1985). These basins thus appear as a reasonably good prospective area, but only an exploratory well could now confirm or not their potential (Bouysse et al., this Vol.).

Saint Pierre and Miquelon

These small French islands along the southeast Atlantic coast of Canada are located in the vicinity of two major basins of Triassic to Recent age, the Sable and Jeanne D'Arc basins offshore from Nova Scotia and Newfoundland, respectively, where significant amounts of oil and gas have been found (MacLean and Wade 1992; Williamson 1992; Williamson and Smyth 1992). The French territories are restricted to the immediate vicinity of the islands on the shelf and to a narrow corridor 20 km wide extending south down to the Atlantic abyssal plain. Late Paleozoic basins underlie the waters surrounding the islands and the northern part of the corridor on the inner shelf, while thick Mesozoic series are present farther south below the outer shelf and continental slope.

Kerguelen

The Kerguelen islands in the Indian Ocean are of volcanic origin, but large submarine plateaus to the south contain up to 3500 m thick sedimentary basins of late early Cretaceous to Recent age (Schlich et al. 1992). Although a few source rock intervals have been encountered in Albian clayey siltstones (ODP site 750, Schlich et al. 1989), the great water depths (500–3000 m) and the hypothetical occurrence of potential reservoirs will probably delay any petroleum exploration activity for a long times. A similar statement can be made for all other French volcanic islands in the Indian and Pacific Oceans.

Terre Adélie, Antarctica

Although Terre Adélie cannot be considered as a French territory because the legal status of Antarctica has not been definitively defined, this part of the continent has, for a long time, been under some sort of French jurisdiction for scientific purposes. The continental margin facing Terre Adélie has been studied by conventional seismic surveys with a limited stratigraphic control from the nearby ODP well 269 (Wannesson 1990). The sedimentary wedge up to five seconds thick can be divided into Cenomanian to Recent post-rift units and an underlying Mesozoic (and possibly Paleozoic?) pre and syn-rift units. The absence of any outcrop along the coast, and of any well on the shelf or slope, prevents any further petroleum evaluation of this margin.

Conclusions

This introduction to the petroleum geology of France was prepared at a time when exploration activity was at a low stage, as shown by the present decline of drilling operations and seismic crews on the field. In 1992, only 9 exploratory and 18 development wells were drilled. The present international economic situation, i.e., the low oil prices and the very low exchange rate of the US dollar should probably be taken into account to explain this situation. The concentration of activity in more prolific areas has also induced the departure from France of some major oil companies, such as Shell, BP, and Fina, for instance. Mention should also be made of some disappointements from expensive exploration projects recently carried out in frontier areas such as the Jura thrust belt, or in the offshore domain a few more years ago. There are, however, still some encouraging results that have been obtained, such as the very promising Les Arbousiers 1 well discovery in the Parentis Basin, and a few smaller discoveries in the Paris Basin (L'Orme, La Vignotte, Montapot). Some exploration will probably still go on for a few years in these areas, such as possibly in the Pyrenean foothills, but at much greater depths.

Frontier areas now have some difficulty in attracting traditional and new players. The deep Permian basins explored so far have not produced any encouraging results, and any further exploration will depend on the results of the next hole to be drilled for such a play, probably on the Quercy Plateau. The Jura thrust belt has shown some promising initial results, however, which have not been confirmed by subsequent wells. The petroleum system, including the distribution of source rocks, the timing of migration, and characteristics of reservoirs, is perhaps not completely understood and would require additional regional syntheses. Regional seismic profiles such as those shot in 1991 through the Vercors and Chartreuse Subalpine massifs are certainly of great help in constraining the models of tectonic developments and of the hydrocarbon generation to migration, at both the present and pre-inversion stages of the thrust belt. Although progressing slowly, some exploration is still going on and we expect a few holes to be drilled in the near furture. A major challenge will now be to define new plays in presently neglected areas, such as the South-East Basin and the offshore domains around France and some of its overseas departments. Regional syntheses including the results of the first wells, new high quality regional seismic profiles, and tectonic and geochemical modeling, are probably some of the key elements for promoting new ideas and a revival of exploration Deep stratigraphic wells on onshore areas could certainly, in a few cases, provide the key data. Lastly, the coming years will also see the development of exploration for coalbed methane as about ten licenses have already been awarded and an increased number of wells is expected to be drilled soon

Acknowledgements The final version of this chapter has benefited from the comments of J. Wannesson (IFP) and J. P. Roy (ENSPM).

References

Arcamone J, Bideaud M et al (1980) Le gisement de charbon du bassin de l'Arc, Provence occidentale Mém BRGM 122 116

Ardaens R (1992) Le champ de Parentis Réservoir carbonaté complexe et paléokarst Pétrole Tech 372 34–42

Arthaud F, Ogier M, Seguret M (1981) Géologie et géophysique du Golfe du Lion et de sa bordure nord Bull BRGM 1, 3 175–193

Autran A, Dercourt J (1980) Géologie de la France Mém BRGM 107 355

Ayers W Jr, Keslo B (1989) Knowledge of methane potential for coalbed Oil Gas J Oct 23 64–67

Ayoub J, Colson L, Hinkel J, Johnston D, Levine J (1991) Learning to produce coalbed methane Oilfield Rev 3, 1 27–40

Baby P (1988) Chevauchements dans une zone à structure complexe la zone nord-pyrénéenne ariégeoise Thèse Univ Paul Sabatier de Toulouse, 149 pp

Barbier Fl, Duvergé J, Le Pichon X (1986) Structure profonde de la marge Nord-Gascogne Implication sur le méchanisme de rifting et de formation de la marge continentale Bull Centr Recherche Expl Elf-Aquitaine 10, 1 105–121

Barrabé ML, Schneegans MD (1935) Rapport d'ensemble sur le gisement pétrolifère de Gabian (Hérault) Ann Off Nat Combustibles Liquides 10, 4 595–665, 10, 5 819–897

Baudrimont AF, Dubois P (1977) Un bassin mésogéen du domaine péri-alpin le sud-est de la France Bull Cent Rech Explor Prod Elf-Aquitaine, 1, 1 261–308

Becq-Giraudon JF (1983) Synthèse structurale et paléogéographique du bassin houiller du Nord Mém BRGM 123 73

Bell JS, Caillet G, Le Marrec A (1992) The present-day stress regime of the southwestern part of the Aquitaine Basin, France, as indicated by oil well data J Struct Geol 14, 8/9 1019–1032

Bessereau G, Huc A, Carpentier B (1992) Distribution of the organic matter in the Liassic series of the Paris Basin an example of organic heterogeneity in a source rock interval In Spencer AM (ed) Generation, accumulation, and production of Europe's hydrocarbons II Springer, Berlin Heidelberg New York, pp 117–126

Blanc G, Doligez B, Lajat D, Mascle A (1991) Evaluation du potentiel pétrolier des formations paléozoiques de la Bresse et de sa bordure jurassienne, France Bull Soc Géol Fr 162, 2 409–422

Blondin E, Verdier F (1992) Interprétation sismique appliquée au stockage de gaz Mém Soc Géol Fr 161 77–88

Boillot G, Montadert L, Lemoine M, Biju-Duval B (1984) Les marges continentales actuelles et fossiles autour de la France Masson, Paris, 342 pp

Bois Ch, Gariel O, Pinet B (1990) Ecors deep seismic surveys across Paleozoic and Mesozoic basins in France and adjacent areas a review In Pinet B, Bois Ch (eds) The potential of deep sismic profiling for hydrocarbon exploration TECHNIP, Paris, pp 381–401

Bois Ch Gariel O Sibuet JCl (1991) Etude de la croûte terrestre par sismique profonde Mer celtique-Manche-Approches Occidentales Mém Soc Géol Fr 159 220

Bouvier JD, Gevers ECA, Wigley PL (1990) 3-D seismic interpretation and lateral prediction of the Amposta Marino field (Spanish Mediterranean Sea) Geologie Mijnbouw 69 105–120

Bouysse Ph, Westercamp D (1990) Subduction of Atlantic seismic ridges and late Cenozoic evolution of the Lesser Antilles island arc Tectonophysics 175 349–380

Brun JP, Wenzel F, ECORS-DEKORP Scientific team (1991) Crustal-scale structure of the southern Rhine Graben from Ecors-Dekorp seismic reflection data Geology 19 758–762

Brunet MF (1981) Etude quantitative de la subsidence du Bassin de Paris Thèse Univ Paris VI, 161 pp

Brunet MF (1991) Subsidence et géodynamique du bassin d'Aquitaine Relation avec l'ouverture de l'Atlantique Thèse de doctorat d'état, Univ Paris VI, 288 pp

Burrus J (1984) Contribution to a geodynamic synthesis of the Provençal Basin (North-Western Mediterranean) Mar Geol 55, 34 247–270

Campos CWM (1983) A Exploraçao de petroleo no Brasil situaço actual e perspectivas 2nd Congr Brasiliero de Petroleo, anais, Rio de Janeiro, pp 201–239

Cazes M, Torreilles G (1988) Etude de la croûte terrestre profonde, profil Nord de la France, structures hercyniennes TECHNIP, Paris, 260 pp

Charollais J, Jamet M (1990) Principaux résultats géologiques du forage Brizon 1 (BZN 1), Haute Savoie, France In Roure F, Heitzmann P, Polino R (eds) Deep structure of the Alps, Soc Géol Fr Mém 156 185–202

Chateauneuf JJ, Farjanel G (1989) Synthèse géologique des bassins permiens français Mém BRGM 128 288

Choukroune P, ECORS team (1989) The Ecors Pyrenean deep seismic profile reflection data and the overall structure of an orogenic belt Tectonics 8, 1 23–29

Clavell E, Berastegui X (1991) Petroleum geology of the Golfe of Valencia In Spencer AM (ed) Generation, accumulation and production of Europe's hydrocarbons Special Publ EAPG 1 Oxford Science Publication, Oxford, pp 355–368

Comité des Techniciens (1991) Monographies des principaux champs pétroliers de France Bull Cent Rech Expl Prod Elf-Aquitaine Mém 14 250

Cope JCW, Ingham JK, Rawson PF (1992) Atlas of paleogeography and lithofacies Geol Soc Lond Mem 13 153

Courel L, Xing Bing Liu (1991) Variation in the geothermal history in coal basins, relationship with basin dynamics Bull Soc Géol Fr 162, 2 363–370

Cravatte J, Dufaure P, Prim M, Rouaix S (1974) Les sondages du Golfe du Lion, stratigraphie et sédimentologie Not Mém CFP 2 209–274

Curnelle R (1983) Evolution structuro-sédimentaire du Trias et de l'infra-Lias d'Aquitaine Bull Cent Rech Expl Prod Elf-Aquitaine 7, 1 69–99

Curnelle R, Dubois P (1986) Evolution mésozoique des grands bassins sédimentaires français bassins de Paris, d'Aquitaine et du Sud-Est Bull Soc Géol Fr II, 4 529–546

Curnelle R, Dubois P, Seguin JC (1982) The Mesozoic-Tertiary evolution of the Aquitaine Basin Philos Trans R Soc Lond A 305 63–84

Dardeau G, de Graciansky PCh (1990) Halocinèse et rifting téthysien dans les Alpes Maritimes (France) Bull Cent Rech Expl Prod Elf-Aquitaine 14, 2 443–464

Dardeau G, Fortwengler D, de Graciansky PCh, Jacquin Th, Marchand D, Martinod J (1990) Halocinèse et jeu de blocs dans les Baronnies diapirs de Propiac, Montaulieu, Condorcet (département de la Drome, France) Bull Cent Rech Expl Prod Elf-Aquitaine 14, 111–159

Debrabant P, Chamley H, Deconinck JF, Récourt Ph, Trouiller A (1992) Clay sedimentology, mineralogy and chemistry of Mesozoic sediments drilled in the northern Paris Basin Sci Drill 3, 4 138–152

Debrand-Passard S, Courbouleix S, Lienhardt MJ (1984) Synthèse géologique de Sud-Est de la France Mém BRGM 125 615

Despretz J, Daly TE, Robinson E (1985) Saba Bank petroleum geology, NE Caribbean Oil Gas J nov 4 112–118

Donsimoni M (1981) Le bassin houiller lorrain, synthèse géologique Mém BRGM 117· 102

Donsimoni M (1990) Le gisement de charbon de Lucenay-les-Aix, essai de synthèse géologique d'après les sondages récents (1981–1986) Doc BRGM 179 195

Dubois P, Delfaud J (1989) Le bassin de Sud-Est In: Dynamique et méthode d'étude des bassins sédimentaires TECHNIP, Paris, pp 277–298

Durst H (1991) Aspects of exploration history and structural styles in the Rhine graben area In Spencer AM (ed) Generation, accumulation and production of Europe's hydrocarbons Spec Publ EAPG 1, Springer, Berlin Heidelberg New York, pp 247–261

Duval B (1992) Villeperdue Field Exploration of a subtle trap in the Paris Basin. In Halbouty MT (ed) Giant oil and gas fields of the decade 1978–1988 AAPG Mem 54 251–263

Espitalié J, Drouet S (1992) Petroleum generation and accumulation in the Aquitaine Basin In Spencer AM (ed) Generation, accumulation and production of Europe's hydrocarbons II Spec Publ EAPG 2, Springer, Berlin Heidelberg New York, pp 127–150

Espitalié J, Marquis F, Sage L (1987) Organic geochemistry of the Paris Basin In Brooks J, Glennie KW (eds) Petroleum geology of North West Europe Graham & Trotman, London, pp 71–86

Evans CDR (1990) The geology of the western English Channel and its western approaches British Geological Survey, United Kingdom Offshore Regional Report, London HMSO, 94 pp

Fitzeral M, Mousset E (1987) Triton France details its successes in the Paris Basin Oil Gas J May 23 69–72

Garcia Sineriz B, Querol R, Castillo F, Arribas JRF (1980) A new hydrocarbon province in the western Mediterranean 10th World Petroleum Congr Proc 2 191–197

Gély JP, Lorenz C (1991) Analyse séquentielle de l'Eocène et de l'Oligocène du Bassin Parisien (France). Rev Inst Fr Pétrole 46, 6 713–748

Gorini Ch, Mauffret A, Le Marrec A (1993) Contribution to the structural and sedimentary history of the Gulf of Lions (Western Mediterranean), from the Ecors profiles, industrial seismic profiles and well data Bull Soc Géol Fr, t 164, 3 353–364

Gouyet S (1988) Evolution tectono-sédimentaire des marges guyanaises et nord-brésiliennes au cours de l'ouverture de l'Atlantique Sud Thèse Univ de Pau et des Pays de l'Adour, 374 pp

Guellec S, Lajat D, Mascle A, Roure F, Tardy M (1990) Deep seismic profiling and petroleum potential in the Western Alps constraints with Ecors data, balanced cross-sections and hydrocarbon modelling In· Pinet B, Bois Ch (eds) The potential of deep seismic profiling for hydrocarbon exploration TECHNIP, Paris, pp 425–437

Haller D, Hamon G (1993) Meillon-Saint Faust gas field, Aquitaine basin a new insight of the structural scheme and understanding of water invasion In Parker JR (ed) Petroleum Geology of Northwest Europe, Proceeding of the 4th Conference. Geol Soc, vol 2, London, pp 1519–1526

Hanot F (1992) Seismic application of a Paris Basin shallow geology database First Break 10, 5 175–188

Hanot F, Renoux P (1991) Petrophysical variation in the Senonian chalk of the Paris Basin and their influence on static corrections First Break 9, 11 515–526

Huai Fu Wei, Ledoux E, De Marsily G (1990) Regional modelling of groundwater flow, salt and environmental tracer transport in deep aquifers in the Paris Basin J Hydrol 120 341–358

Jeannette A (1985) Les stockages souterrains Bull Soc Géol Fr 8, 1, 7 1143–1148

Jharap SE (1988) How Staatsolie conducts Suriname's oil operations Petrol Engineer Int June 30–36

Kuuskraa V, Brandenburg Ch (1989) Coalbed methane sparks a new energy industry Oil Gas J Oct 9 49–56

Leblanc M, Rouge J, Tallec E (1991) Reservoir development in a complex deposit environment Germigny-sous-Coulombs aquifer storage In Imarisio G, Frias M, Bemtgen JM (eds) The European oil and gas conference a multidisciplinary approach in exploration and production R and D Graham & Trotman, London, pp 70–87

Lefebvre D (1980) Evolution morphologique et structurale du Golfe du Lion Essai de traîtement statistique des données Thèse Univ Paris VI, 163 pp

Le Gall B (1992) The deep structure of the Ardennes Variscan belt from structural and ECORS seismic data J Struct Geol 14 531–546

Liu XB (1990) La qualité des charbons dans le bassin houiller de Lorraine Thèse de doctorat Univ Bourgogne, 258 pp

Lucazeau F, Vasseur G (1989) Heat flow density data from France and surrounding margins Tectonophysics 164, 2–4 251–258

MacElhiney J, Koenig R, Schraufnagel R (1989) Evaluation of coalbed methane reserves involves different techniques Oil Gas J Oct 30 63–72

MacLean BC, Wade JA (1992) Petroleum geology of the continental margin south of the islands of St Pierre and Miquelon, offshore Eastern Canada Bull Can Petrol Geol 40, 3 222–253

Mascle A (1990) Géologie pétrolière des bassins permiens français, comparaison avec les bassins permiens du nord de l'Europe In Potentiel économique des bassins permiens français Chronique de la recherche minière, 499, pp 69–86

Mascle A, Letouzey P (1990) Geological map of the Caribbean Scale 1 2 500 000 TECHNIP, Paris, 2 sheets

Mathieu C (1986) Histoire géologique du sous-bassin de Parentis Bull Cent Rech Expl Prod Elf-Aquitaine 10, 1 33–47

Maury R, Westbrook GK, Baker PE, Bouysse Ph, Westercamp D (1990) Geology of the Lesser Antilles In The geology of North America, vol H, the Caribbean region Geological Society of America, Boulder, CO, pp 141–166

Mégnien Cl (1980) Synthèse géologique de Bassin de Paris Mém BRGM 103, 3 vols

Monchaux A, Trémolières P (1985) Le gisement de Coulommes-Vauxcourtois historique et données nouvelles Pétrole Techn 313 19–27

Montadert L, Roberts DG, De Charpal O, Guennoc P (1979) Rifting and subsidence of the northern continental margin of the Bay of Biscaye Initial Reports of the DSDP 48 US Government Printing Office, Washington, pp 1035–1060

Morange M, Héritier F, Villemin J (1971) Contribution de l'exploration pétrolière à la connaissance structurale et sédimentaire de la Limagne dans le Massif Central. In: Symposium J Jung Géologie, géomorphologie et structure profonde du Massif Central français Plein Air Service, pp 295–308

Morelot G (1986) Bilan de l'exploration du Trias du Bassin de Paris Pétroles Techn 325. 10–15

Mullan HS (1984) Deep gas potential of Aquitaine Basin AAPG Bull 68 (12) 1857–1869
ORSTOM (1981) Outre-mer français et exploitation des océans Académie des sciences d'outre-mer ORSTOM, Paris, O608-2, 204 pp
Pages L (1987) Exploration of the Paris Basin In Brooks J, Glennie K (eds) Petroleum geology of North West Europe Graham & Trotman, London, pp 87–93
Paris JP (1981) Géologie de la Nouvelle Calédonie Mém BRGM 113 178
Paris F, Le Pochat G, Pelhate A (1988) Le socle paléozoique nord-aquitain caractéristiques principales et implications géodynamiques C R Acad Sci 306 II 597–603
Passeron JP (1992) Trente cinq ans d'exploration sismique sur le champ de Parentis (sud-ouest de la France) Mém Soc Géol Fr 161 97–102
Perrodon A, Zabek J (1991) Paris Basin In Leighton M, Kolata D, Oltz D, Eidel J (eds) Interior cratonic basins AAPG Mem 51 633–680
Pickel W, Gotz G (1991) Investigation on the petroleum generation potential of bituminous coals from the Saar region Org Geochem 17, 6 695–704
Pinet B, Montadert L, Ecors scientific party (1987) Deep seismic reflection and refraction profiling along the Aquitaine shelf (Bay of Biscaye) In Matthews (ed) Deep seismic reflection profiling of the continental lithosphere Geophys J R Astro Soc 89 305–312
Piqué A (1991) Les massifs anciens de France Sci Géol 44, 1–2 206, 3–4 390
Pomerol C (1989) L'évolution du Bassin Parisien In Dynamique et méthode d'étude des bassins sédimentaires TECHNIP, Paris, pp 165–178
Pommier G (1985) Le developpement d'un gisement complexe Montmirail-Villeperdue Ann Mines, mars–avril 33–42
Poulet M, Espitalié J (1987) Hydrocarbon migration in the Paris Basin In Doligez B (ed) Migration of hydrocarbons in sedimentary basins TECHNIP, Paris, pp 131–171
Ravenne Ch, Dunand JP, de Broin CE, Aubertin F (1982) Les bassins sédimentaires du Sud-Ouest Pacifique In Contribution à l'étude géodynamique du Sud-Ouest Pacifique, Trav Doc Orstom 147 461–477
Robert P (1985) Le bassin stéphanien de Messeix-Singles (Puy-de-Dome) Structuration, dynamique du remplissage, pétrographie Thèse Univ Bourgogne à Dijon, 221 pp
Robert P (1989) The thermal setting of Carboniferous basins in relation to the Variscan orogeny in Central and Western Europe Int J Coal Geol 13 171–206
Roure F, Brun JP, Colletta B, van den Driesche J (1992) Geometry and kinematics of extensional structures in the Alpine foreland basin of southeastern France J Struct Geol 14, 5 503–519
Roure F, Heitzmann P, Polino R (1990) Deep structure of the Alps Soc Géol Fr Mém 156 368
Roy JP, Trémolières P (1992) L'inversion du demi-graben oligocène d'Alès (France) données de terrain et modélisation C R Acad Sci 355, II 1777–1781
Saint Martin M (1992) Genèse et évolution structurale du bassin permien de Lodève Hérault-France) Cuad Geol Iberica 16 75–90
Schafer A (1986) Die Sedimente des Oberkarbons und Unterrotliegenden im Saar-Nahe-Becken Mainzer Geowiss Mitt 15 239–365
Schlich R, Wise SW et al (1989) Proc ODP Init Rep 120 College Station TX, Ocean Drilling Program, 650 pp
Schlich R, Angoulvant MP, Fritsch B, Munschy M, Schaming M (1992) Structure and tectonic history of the Kerguelen Plateau European Association of Petroleum Geoscientists and Engineers, Techn Progr Abstr Papers, pp 135–137
Selley RC, Stoneley R (1987) Petroleum habitat in South Dorset In Brooks J, Glennie KW (eds) Petroleum geology of North West Europe Graham & Trotman, London, pp 139–148
Sellier E, Bousquet Ph, Boischard R, Brun R (1991) Estimation of the effective porosity of the Dogger oolithic limestone in the Villeperdue oil field, Paris Basin, France In Spencer AM (ed) Generation, accumulation, and production of Europe's hydrocarbons Spec Publ European Association of Petroleum Geoscientists 1 Springer, Berlin Heidelberg New York, pp 241–246
Shannon P (1991) Tectonic framework and petroleum potential of the Celtic Sea, Ireland First Break 9, 3 107–122
Sittler C (1985) Les hydrocarbures d'Alsace dans le contexte historique et géodynamique du fossé rhénan Bull Cent Rech Expl Prod Elf-Aquitaine 9, 2 335–371
Specht M (1989) Tectonique de chevauchement le long du profil Ecors-Pyrennées un modèle d'évolution de prisme d'accrétion continental Thèse Univ Bretagne Occidentale de Brest, 353 pp
Tempier Cl (1987) Modèle nouveau de mise en place des structures provençales Bull Soc Géol Fr III, 3 533–540
Tissot B, Noesmoen A (1958) Les bassins de Nouméa et de Bourail (Nouvelle Calédonie) Rev Inst Fr Pétrole XIII, 5 739–760
Toutin-Morin N (1987) Les bassins permiens provençaux, témoins de l'orogenèse hercynienne et de l'ouverture de la Téthys dans le Sud-Est de la France Ann Soc Géol Nord CVI 183–187
Toutin-Morin N, Bonijoly D (1992) Structuration des bassins de Provence orientale à la fin de l'ère primaire Cuad Geol Iberica 16 107–121
Ungerer P, Espitalié J, Marquis F, Durand B (1986) Use of kinetic models of organic matter evolution for the reconstruction of paleotemperatures. Application to the case of the Gironville well (France) In Burrus J (ed) Thermal modeling in sedimentary basins TECHNIP, Paris, pp 531–546
Valette M (1990) Etude structurale du gisement salifère de Parrapon-Vauvert (Gard) et de son contexte géologique régional Thèse Univ des Sciences et Techniques du Languedoc, 220 pp
Villemin T (1986) Tectonique en extension, fracturation et subsidence le fossé rhénan et le bassin de Sarre-Nahe Thèse Univ Paris VI, 258 pp
Wang H (1991) Dynamique sédimentaire, structuration et houillification dans le bassin houiller stéphanien des Cévennes Thèse de doctorat Univ Bourgogne, 266 pp
Wannesson J (1990) Geology and petroleum potential of the Adelie coast margin, East Antarctica In Bill St John (ed) Antarctica as an exploration frontier AAPG Stud Geol 31 77–88
Weaver OD, Van Damme A (1988) France offers attractive EP opportunity Oil Gas J May 16 56–59, May 23 55–58; May 30 94–97
Welbon AI, Butler RWH (1992) Structural styles in thrust belts developed through rift basins a view from the western Alps In Larsen RM, Brekke H, Larsen BT, Talleraas E (eds) Structural and tectonic modelling and its application to petroleum geology NPF Spec Publ 1 Elsevier, Amsterdam, pp 469–479
Wenzel F, Brun JP (1991) A deep reflection seismic line across the Northern Rhine graben Earth Planet Sci Lett 104 140–150
Williamson MA (1992) The subsidence, compaction, thermal and maturation history of the Egret Member source-rock, Jeanne D'Arc Basin, offshore Newfoundland Bull Can Pet Geol 40, 2 136–140
Williamson MA, Smyth C (1992) Timing of gas and overpressure generation in the Sable Basin, offshore Nova Scotia implication for gas migration dynamics Bull Can Pet Geol 40, 2 151–169

Appendix 1. Main characteristics of crude oils produced in 1992

Producing field	Density (g/cm3) at 15°C	Gravity °API at 15°C	Barrels per metric ton	Viscosity (cPo) at 20°C	Sulfur content (weight %)	oil type
ELF AQUITAINE						
Brie (Trias)	0.858	33.5	7.33	2.70	0.15	paraffinic intermediate
Castéra-Lou	0.892	27.1	7.05	32.00	0.80	ud
Châteaurenard	0.890	27.5	7.07	68.29	0.19	paraffinic intermediate
Chuelles	0.890	27.5	7.07	65.53	0.24	paraffinic intermediate
Courdemanges	0.849	35.2	7.40			ud
Courtenay	0.892	27.1	7.05	101.30	0.18	paraffinic intermediate
Dommartin-Lettrée	0.850	35.0	7.40			ud
Eschau	0.864	32.3	7.28	24.50	0.11	paraffinic
Fontaine-au-Bron	0.848	35.4	7.42		very low	very paraffinic
Grandville	0.844	36.2	7.45	11.00	0.12	paraffinic intermediate
Grandville Est	0.844	36.2	7.45	11.00	0.12	ud
Itteville	0.864	32.3	7.28			ud
L'Orme	0.865	32.1	7.27			ud
La Vignotte	0.850	35.0	7.40			ud
Lacq supérieur	0.920	22.3	6.84	260.00	3.90	Mixed
Lagrave	0.813	42.5	7.74	3.20	2.60	paraffinic intermediate
Laméac	0.980	12.9	6.42		5.50	Asphaltic
Linthelles	0.851	34.8	7.39	10.00		ud
Montapot	0.868	31.5	7.25	28.40		ud
Pécorade	0.880	29.3	7.15	18.00	2.85	ud
St-Eloi	0.856	33.8	7.35	40.00 *	very low	very paraffinic
St-Firmin	0.890	27.5	7.07	92.07	0.21	paraffinic intermediate
Scheibenhard (Stampian)	0.868	31.5	7.25	20.00	0.21	paraffinic
Scheibenhard (Sannoisian)	0.822	40.6	7.65	7.70	0.14	paraffinic
Schelmenberg	0.828	39.4	7.60		0.11	paraffinic
Soudron (Dogger)	0.853	34.4	7.37	13.00 *	0.21	ud
Soudron (Rhetian)	0.853	34.4	7.37	22.00	0.14	ud
Valence-en-Brie	0.850	35.0	7.40	21.00	0.27	paraffinic intermediate
Vert-la-Gravelle	0.850	35.0	7.40	2.00		ud
Vert-le-Grand	0.851	34.8	7.39	20.00 *	0.15	very paraffinic
Vert-le-Petit	0.870	31.1	7.23	50.00	0.00	paraffinic
Vic Bilh	0.895	26.6	7.03			ud
Villemer (Dogger)	0.866	31.9	7.26	26.07	0.51	paraffinic intermediate
Villemer (Keuper)	0.855	34.0	7.36	2.70		ud
ESSOREP						
Brémonderie	0.870	31.1	7.23	13.00	0.17	paraffinic
Cabeil	0.840	37.0	7.49	7.30	0.39	paraffinic
Cazaux (Albian)	0.853	34.4	7.37	13.00	0.35	paraffinic
Cazaux (Purbeckian)	0.837	37.6	7.51	5.50	0.30	paraffinic
Champotran	0.870	31.1	7.23	14.50		ud
Champrose	0.853	34.4	7.37			ud
Charmottes (Dogger)	0.850	35.0	7.40	17.80	ud	ud
Charmottes (Keuper)	0.842	36.6	7.47	12.60	0.16	paraffinic
Chaunoy	0.839	37.2	7.50	13.20	0.16	paraffinic
Donnemarie	0.844	36.2	7.45	8.50	0.16	paraffinic
Lavergne	0.817	41.7	7.70	2.80	0.33	paraffinic
Les Arbousiers	0.853	34.4	7.37			ud
Lucats	0.833	38.4	7.55	5.80	0.32	paraffinic
Lugos	0.929	20.8	6.77	770.00	0.73	paraffinic
Malnoue (Dogger)	0.855	34.0	7.36	14.50	ud	ud
Malnoue (Keuper)	0.850	35.0	7.40	9.30		ud
Mothes	0.927	21.1	6.79	310.00	0.73	paraffinic
Parentis	0.858	33.4	7.33	12.40	0.42	paraffinic
Vulaines (Dogger)	0.869	31.3	7.24	ud	ud	ud
PETROREP						
Béchevret	0.863	32.5	7.29	ud	0.16	paraffinic
Coulommes	0.862	32.7	7.31	29.20	0.22	paraffinic
L'Ile du Gord	0.850	35.0	7.40	17.50	0.22	paraffinic
REPLOR						
Forcelles	0.800	45.4	7.86	6.00	0.19	paraffinic
SHELL FRANCAISE						
St-Martin-de-Bossenay	0.860	33.0	7.30	19.52	0.21	paraffinic
TOTAL EXPLORATION						
Hautefeuille	0.846	35.7	7.43	14.97	0.24	paraffinic
La Motte Noire	0.850	35.0	7.40	ud	ud	ud
Sancy-les-Provins	0.847	35.6	7.43			ud
Villeperdue/FAB South	0.853	34.4	7.37	14.97	0.24	paraffinic intermediate
TRITON FRANCE						
Bagneaux	0.875	30.2	7.19	45.70	0.21	paraffinic
Maincy	0.848	35.4	7.42	10.20	0.14	paraffinic
St-Germain	0.848	35.4	7.42	10.20	0.14	paraffinic
Sivry	0.864	32.3	7.28	29.70	0.18	paraffinic

ud, undefined
* From the measurements made at more than 20 °C.

Appendix 2. Physico-chemical characteristics of gas produced in 1992

Field	Density (air=1)	MOLECULE COMPOSITION (%)								
		C 1	C 2	C 3	C 4	C 5	C 6+	H2S	CO2	N2
COPAREX										
Trois-Fontaines	0 6590	75 08	2 23	0 24	0 07	0 02	0 03		0 19	22 14
ELF AQUITAINE										
Andoins	0 7640	72 58	2 03	0 46	0 34	0 18	0 11	13 36	10 51	0 43
Auzas	0 6250	89 86	5 09	1 66	0 70	0 24	0 10		0 43	1 92
Cassourat	0 7200	79 14	0 03	0 01			0 01	11 67	8 80	0 35
Lacq profond	0 8000	69 00	3 00	0 90	0 50	0 20	0 30	15 30	9 30	1 50
Le Lanot	0 7670	77 80	3 60	1 20	0 90	0 48	1 22	5 90	8 50	0 40
Mazères deep (1)	0 7310	78 80	3 50	0 90	0 80	0 30	0.10	5 90	9.20	0 50
Pécorade	0 8060	68 00	6 60	3 60	1 70	1 30		12 00	6 50	0 30
Pont d'As (1)	0 7560	76 76	2 98	0 79	0 97	0 56	1 06	7 04	9 46	0 38
Rousse	0 7100	76 52	4 57	2 04	1 62	0 89	8 26	0 77	4 58	0 75
St-Faust (1)	0 7560	76 30	3 46	1 11	0 99	0 54	1 20	6 26	9 62	0 52
St-Marcet	0 6350	89 30	5 00	1 85	0 87	0 37	0 29		0 37	1 95
Ucha	0 8900	78 79	4 97	2 21	1 83	1 29	4 68	0 22	5 26	0 75
Vic Bilh	0 7200	81 89	6 72	3 64	2 04	1 02	0 74	0 81	2 55	0 59
ESSOREP										
Chaunoy	0 8735	49 23	9 90	8 45	3 44	0 90			0 62	27 46
Parentis	0 8000	73 60	10 20	7 60	5 00	1 70	1 90			
Ledeuix	0 5650	97 50	1 60	0 21	0 14	0 10				0 45
Saucède	0 5210	94 93	0 19	0 01	0 01				2 62	2 24

(1). parts of Meillon field

Appendix 3. Main characteristics of producing reservoirs (1)

FIELD	YEAR OF DISCOVERY	YEAR OF STREAM	GEOLOGICAL STAGE	LITHOLOGY	TYPE OF TRAP
ANDOINS	10/1990	11/1991	Portlandian	dolomite	structural
AUZAS	06/1988	08/1988	Jurassic	limestone-dolomite	structural
BAGNEAUX	08/1988	08/1988	Dogger	calcaire	mixed
BECHEVRET	08/1988	11/1988	Keuper (Boissy)	sandstone	structural
BREMONDERIE	11/1984	04/1986	Lower Keuper	sandstone	structural
BRIE	03/1988	04/1988	Keuper	clayey sandstone	structural
CABEIL	01/1985	02/1985	Neocomian	limestone-dolomite	structural
CASSOURAT	02/1987	08/1989	Kimmeridgian	dolomite	structural
CASTERA-LOU	04/1976	07/1976	Upper jurassic - Portlandian	dolomite	structural
CAZAUX (1)	04/1959	07/1959	Albian	sandstone	stratigraphic
CAZAUX (2)	12/1961	01/1962	Purbeckian	sandstone	structural
CHAMPOTRAN	05/1985	08/1985	Upper Keuper	sandstone	mixed
CHAMPROSE	12/1990	02/1991	Keuper	sandstone	structural
CHARMOTTES	01/1985	03/1986	Dogger	limestone	mixed
CHARMOTTES	07/1984	08/1984	Lower Keuper	sandstone	mixed
CHAUNOY	04/1983	05/1983	Upper Keuper	sandstone	structural
COULOMMES	02/1958	04/1958	Dogger	limestone	structural
COURDEMANGES	08/1983	12/1983	Rhetian	sandstone	mixed
DOMMARTIN-LETTREE	10/1985	11/1985	Rhetian	sandstone	mixed
DONNEMARIE	07/1979	09/1979	Lower Keuper	sandstone	structural
EGRENAY	06/1987	10/1987	Upper Keuper	sandstone	mixed
ESCHAU	06/1956	12/1962	Dogger	limestone	structural
FONTAINE-AU-BRON	07/1988	10/1988	Dogger	limestone	mixed
FORCELLES	01/1978	06/1978	Lettenkohle	dolomite	structural
GRANDVILLE	01/1959	07/1965	Rhetian	sandstone	mixed
GRANDVILLE EST	08/1983	09/1983	Rhetian	sandstone	mixed
ITTEVILLE	12/1990	04/1991	Dogger	limestone	mixed
IVRY	08/1988	02/1989	Keuper	sandstone	mixed
HAUTEFEUILLE	02/1987	02/1987	Dogger	limestone	mixed
L'ORME	10/1990	01/1991	Hettangian	limestone	mixed
LA VIGNOTTE	04/1991	07/1991	Keuper	sandstone	structural
LACQ PROFOND	12/1951	05/1957	Barremian to Kimmeridgian	dolomite-limestone	structural
LACQ SUPERIEUR	12/1949	04/1950	Lower Senonian	limestone-dolomite	structural
LAGRAVE	09/1984	11/1984	Lower Senonian	limestone	mixed
LAMEAC	06/1970	02/1982	Lower Lias	limestone	mixed
LA MOTTE NOIRE	08/1987	09/1987	Rhetian	sandstone	structural
LAVERGNE	09/1982	09/1982	Purbeckian	sandstone	structural
LEDEUIX	11/1981	09/1982	Albian	episyenite	mixed
LE LANOT	07/1970	02/1981	Jurassic-Portlandian	dolomite	mixed
LES ARBOUSIERS	12/1991	12/1991	Purbeckian	sandstone	mixed
LIEUSAINT	09/1986	05/1987	Upper Keuper	sandstone	structural
L'ILE DU GORD	03/1986	06/1986	Upper Keuper	sandstone	mixed
LINTHELLES	09/1981	12/1981	Rhetian	sandstone	mixed
LUCATS	11/1956	12/1956	Neocomian-Portlandian	limestone-dolomite	structural
LUGOS	08/1956	08/1956	Aptian	sandstone	structural
LUGOS	11/1957	12/1957	Barremian	dolomite	structural
MAINCY	12/1987	03/1988	Upper Keuper	sandstone-dolomite	structural
MALNOUE	04/1985	03/1986	Dogger	limestone	structural
MALNOUE	05/1985	08/1985	Upper Keuper	sandstone	structural
MEILLON	03/1965	10/1968	Jurassic-Portlandian-Oxfordian	dolomite	structural
MONTAPOT	12/1991	04/1992	Upper Keuper	sandstone	structural
MOTHES	01/1955	05/1955	Néocomian	limestone-dolomite	structural
NEOCOMIEN (3)	04/1958	08/1958	Neocomian (Châteaurenard)	sandstone	mixed
NEOCOMIEN (4)	04/1958	08/1958	Neocomian (Chuelles)	sandstone	mixed
PARENTIS	03/1954	05/1954	Aptian-Neocomian	limestone-dolomite	structural
PECORADE	06/1974	06/1975	Barremian-Portlandian	limestone-dolomite	structural
PIERRE-MORAINS	07/1986	11/1987	Rhetian	sandstone	mixed
PROUPIARY	12/1944	05/1950	Cénomanien	calcareous breccia	structural
ROUSSE	03/1967	08/1972	Jurassic-Portlandian-Oxfordian	dolomite-limestone	structural
SAINT-ELOI	11/1985	08/1986	Dogger	limestone	mixed
SAINT-GERMAIN	11/1984	12/1984	Upper Keuper	sandstone-dolomite	structural
SAINT-MARCET	07/1939	11/1942	Cénomanian-middle and lower Jurassic	marno-limestone	structural
SAINT-MARTIN	03/1959	05/1959	Dogger	limestone	structural
SANCY-LES-PROVINS	02/1990	03/1990	Dogger	limestone	mixed
SAUCEDE	04/1979	08/1981	Upper Jurassic	dolomite-limestone	mixed
SCHEIBENHARD	05/1956	05/1963	Stampian	sandstone	mixed
SCHEIBENHARD	05/1956	05/1963	Sannoisian	sandstone	mixed
SCHELMENBERG	08/1982	01/1983	Stampian	sandstone	mixed
SIVRY	07/1984	10/1984	Upper Keuper	sandstone	mixed
SOUDRON	11/1976	04/1977	Dogger	limestone	mixed
SOUDRON	11/1976	01/1979	Rhetian	sandstone	mixed
TROIS-FONTAINES	08/1982	12/1984	middle Muschelkalk	sandstone	mixed
UCHA	06/1970	12/1971	Upper Jurassic-Oxfordian	dolomite	structural
VALENCE-EN-BRIE	11/1960	11/1960	Dogger	limestone	mixed
VERT-LA-GRAVELLE	05/1985	06/1985	Rhetian	sandstone	mixed
VERT-LE-GRAND	08/1986	12/1986	Keuper (Chaunoy)	sandstone	structural
VERT-LE-GRAND NORD	08/1987	10/1987	Keuper (Boissy)	sandstone	structural
VERT-LE-PETIT	01/1987	06/1987	Dogger	limestone	mixed
VIC BILH	06/1979	09/1979	Barremian-Upper Jurassic	limestone-dolomite	mixed
VILLEMER	11/1959	11/1959	Dogger	limestone	mixed
VILLEMER	12/1988	03/1989	Keuper (Boissy)	sandstone	structural
VILLEPERDUE	08/1958	05/1982	Dogger	limestone	mixed
VULAINES	12/1977	11/1986	Dogger	limestone	mixed

(1), altitude of "V" main level, (2), altitude of western top, (3), "sables de Chateaurenard" reservoir, (4), "Sables de Griselles" reservoir

Appendix 4. Main characteristics of producing reservoirs (2)

FLUID	SURFACE (km2)	TOP OF RESERVOIR (m/sl)	WATER TABLE (m/sl)	GROSS THICKNESS (m)	POROSITY (%)	SATURATION (%)	DRAINAGE
dry gas	1.80	-4717	ud	220	2	15	gas drive
wet gas	1.00	-1600	uk		2	-	water drive
undersaturated oil	ud	-1090	-1127	9	15	35	water drive
undersaturated oil	2.00	-1780	-1791	-	18	48	gas drive
undersaturated oil	8.00	-2500	-2540	55	6.5	34	gas drive
undersaturated oil	2.00	-2110	uk	63	15	30	gas drive
undersaturated oil	2.56	-1950	-2130	150	15	35	water injection
dry gas	10.00	-4532	uk	130	3	50	gas drive
undersaturated oil	3.00	-2611	-2734	225	9.3	50	water drive
undersaturated oil	18.00	-2380	-2600	500	18	20	water injection
undersaturated oil	8.00	-3040	-3170	80	20	20	water injection
undersaturated oil	20.00	-2385	-2446	65	12	28	water injection
undersaturated oil	3.20	-2202	ud	9	10.7	46	ud
undersaturated oil	20.00	-1670	-1695	ud	8.8	35	water drive
undersaturated oil	14.00	-2450	-2490	53	8.3	27	gas drive
undersaturated oil	60.00	-2150	-2205	65	12.5	32	water injection
undersaturated oil	14.50	-1678	-1732	200	9	35	water drive
undersaturated oil	4.50	-1510	ud	33	9.3	53	gas drive
undersaturated oil	7.00	-1970	ud	40	13	26	gas drive
undersaturated oil	10.00	-2555	-2595	28	10	28	water injection
undersaturated oil	7.00	-2004	-2012	ud	13	53	water drive
undersaturated oil	2.00	- 720	-775	50	17	30	gas drive
undersaturated oil	17.00	-1630	-1670	25	15	52	water drive
undersaturated oil	ud	+ 90	ud	7.5	20	40	ud
undersaturated oil	8.00	-1898	-1948	36	7.9	47	gas drive
undersaturated oil	4.00	-1895		36.5	10	56	gas drive
undersaturated oil	12.80	- 458	ud	20	15	58	ud
undersaturated oil	3.50	-1929	-1942	10	15	42	uk
undersaturated oil	5.00	-1640	-1660	25	14	50	water drive
undersaturated oil	4.80	-1439	ud	8.5	15.2	29	ud
undersaturated oil	4.50	-2334	ud	30	13.6	40	ud
wet gas	110.00	-3100	néant	450	3.5	15	gas drive
undersaturated oil	6.00	- 493	- 620	150	16.5	60	water drive
undersaturated oil	2.50	-1590	-1664	150	16	32	water drive
viscous oil	ud	-2531	ud	67	7.2	34	ud
undersaturated oil	ud	-2135	ud	43	13	30	gas drive
undersaturated oil	3.63	-3190	-3310	40	11	17	water injection
dry gas	ud	-2230	ud	10	11	23	gas drive
condensate gas	40.00	-4150	uk	200	4.5	20	gas drive
undersaturated oil	5.00	-2867	ud	49.5	16	20	ud
undersaturated oil	1.50	-2009	-2022	ud	13	50	gas drive
undersaturated oil	1.00	-2160	ud	ud	14	41	gas drive
undersaturated oil	5.00	-2350	ud	ud	9	37	gas drive
undersaturated oil	3.98	-2000	-2200	150	10	45	gas drive
undersaturated oil	4.50	-1440	-1630	80	15.5	20	water drive
undersaturated oil	2.40	-1490	-1630	40	fissures	-	water drive
undersaturated oil	ud	-2140	-2180	40	18	30	water + gas drive
undersaturated oil	7.30	-1635	-1861	40	8.5	51	water drive
undersaturated oil	9.30	-2540	-2605	85	12	35	water injection
wet gas	50.00	-3700	-4985	250	4	20	gas drive
undersaturated oil	11.80	-2033	ud	47	12	30	ud
undersaturated oil	3.00	-2350	-2450	50	13	20	water injection
undersaturated oil	24.80	- 395	- 406	0 à 6	25	35	water drive
undersaturated oil	24.80	- 426	- 466	0 à 4	27	25	water drive
undersaturated oil	43.00	-1985	-2400	346	11	30	water injection
saturated oil	12.00	-2320	-2740	340	5	40	water injection
undersaturated oil	3.30	-2060	ud	40	9.3	50	gas drive
condensate gas	3.00	-1495	-1550	50	4	30	gas drive
condensate gas	3.00	-4230	-4960	365	3	20	gas drive
undersaturated oil	5.50	-1430	-1445	ud	15	65	water drive
undersaturated oil	9.00	-2145	-2180	68	9.4	49	water injection
condensate gas	6.00	- 845	-1550	200	7	30	water drive
undersaturated oil	3.00	-1250	-1320	70	12	50	water drive
undersaturated oil	0.64	-1745	-1765	20	8	51	ud
dry gas	28.00	-4510	-	-	4.1	27	gas drive
saturated oil	1.00	- 390	- 447	25	28	30	gas drive
saturated oil	1.25	- 740	- 807	35	27	35	water drive
undersaturated oil	1.5	- 365	ud	8	22	47	gas drive
undersaturated oil	10.00	-2140	-2207	63	15	37	gas drive
undersaturated oil	5.00	-1223	-1236	-	12.5	50	water drive
undersaturated oil	10.00	-1972	-2032	40	9	50	gas drive
dry gas	18.00	-1425	-1510	17	30	36	gas drive
gaz à condensat	2.00	-4460	-4562	100	6.4	35	gas drive
undersaturated oil	1.50	-1630	-1655	25	8	40	water drive
undersaturated oil	7.00	-2280	ud	42	13	24	gas drive
undersaturated oil	8.60	-1800	ud	43	12.6	60	gas drive
undersaturated oil	12.00	-1800	-1843	-	12	58	water drive
undersaturated oil	6.20	-1450	-1476	-	16.5	50	water drive
saturated oil	17.00	-1900	-2430	500	10	40	gas drive
undersaturated oil	6.00	-1435	-1456	15	13	50	water injection
undersaturated oil	1.30	-2050	ud	5	15	19	gas drive
undersaturated oil	90.00	-1630	-1705	25	12	53	water injection
undersaturated oil	17.00	-1700	-1732	ud	8	40	water drive

ud, undefined, uk, unknown

Appendix 5. Evolution of oil production by region and field

(metric ton)

Field	Year of dis-covery	Year on stream	Cumulative production 12/31/1986	1987	1988	1989	1990	1991	1992	Cumulative production 12/31/1992
EAP										
Eschau	1956	1982	66 761	1 836	1 217	1 191	1 218	1 255	1 373	74 851
Scheibenhard (NDL)	1954	1983	59 165	10 360	9 498	8 686	7 684	6 832	5 932	108 157
Schelmenberg	1982	1983	21 841	4 200	4 403	3 723	3 299	3 257	3 098	43 821
Abandoned	-	-	373 661	-	-	-	-	-	-	373 661
SAEM										
Péchelbronn	1813	-	3 309 725	-	-	-	-	-	-	3 309 725
SHELL										
Staffelfelden	1952	1980	328	-	-	-	-	-	-	328
TOTALEX										
shut-in/abandoned	-	-	14 397	384	1 269	435	-	-	-	16 485
TOTAL ALSACE			3 845 878	16 780	16 387	14 035	12 201	11 344	10 403	3 927 028
EAP										
Lacq supérieur	1949	1950	3 686 740	48 275	43 318	51 710	49 654	46 997	31 153	3 957 847
Pécorade	1974	1975	1 138 861	126 137	125 272	97 084	78 346	90 922	77 380	1 734 002
Castéra-Lou	1976	1976	311 111	22 311	20 712	20 021	18 585	17 634	14 930	425 304
Vic Bilh	1979	1979	1 857 285	309 194	273 796	215 568	189 137	213 025	182 168	3 240 173
Laméac	1970	1982	6 647	909	178	302	200	210	142	8 588
Lagrave	1984	1984	329 626	196 397	228 450	283 768	267 470	220 045	209 024	1 734 780
shut-in/abandoned	-	-	755 242	-	-	27	-	-	-	755 269
ESSOREP										
Parentis	1954	1954	26 491 677	276 454	255 718	261 355	277 037	228 415	212 941	28 003 597
Mothes	1955	1955	1 124 123	8 247	7 931	7 451	6 603	5 808	4 472	1 164 635
Lugos	1956	1956	1 427 745	39 444	28 999	24 835	23 633	23 863	21 862	1 590 381
Lucats	1956	1956	97 315	1 800	1 359	1 555	1 162	989	822	105 002
Cazaux	1959	1959	8 915 984	214 677	228 282	218 095	228 953	202 900	189 966	10 198 857
Lavergne	1962	1962	1 560 239	18 454	16 288	16 018	14 696	17 145	16 317	1 659 157
Cabeil	1965	1965	232 764	1 831	1 811	1 584	1 669	2 089	1 724	243 472
Les Arbousiers	1991	1991	-	-	-	-	-	95	36 900	36 995
shut-in/abandoned	-	-	520 331	-	-	-	37	-	-	520 368
TOTAL AQUITAINE			48 455 690	1 264 130	1 232 114	1 199 373	1 157 182	1 070 137	999 801	55 378 427
BP										
Suatte	1988	1988	-	-	-	26	-	-	-	26
EAP										
Châteaurenard	1958	1958	651 715	18 302	21 434	19 127	18 218	16 561	13 288	758 645
Brie	1958	1959	441 283	-	-	-	3 115	3 666	2 854	450 918
Villemer	1959	1959	635 527	4 688	4 429	4 655	4 661	4 337	3 709	662 006
Grandville	1959	1959	209 394	5 257	4 666	3 685	4 207	4 675	5 795	237 679
Valence	1960	1960	99 230	1 035	1 032	900	869	803	750	104 619
Saint-Firmin	1960	1960	1 358 454	26 516	23 698	25 675	23 805	25 201	24 358	1 507 707
Chuelles	1961	1961	947 030	23 405	20 110	19 167	20 431	21 215	19 107	1 070 465
Courtenay	1964	1965	346 003	16 304	13 509	14 220	24 042	28 841	31 068	473 987
Soudron	1976	1977	239 007	22 047	21 452	18 839	18 337	16 955	16 461	353 098
Linthelles	1981	1981	2 756	563	517	435	427	415	414	5 527
Grandville Est	1983	1983	16 494	3 359	3 624	2 349	3 229	3 374	3 478	35 907
Courdemanges	1983	1983	16 889	6 120	5 274	4 625	4 326	4 470	4 067	45 771
Vert-la-Gravelle	1985	1985	26 453	10 169	7 901	7 373	5 395	5 044	3 843	66 178
Dommartin-Lettrée	1985	1985	13 026	16 338	21 224	28 543	35 905	37 620	26 780	179 436
Saint-Eloi	1985	1986	6 685	4 712	3 324	643	362	516	370	16 612
Fontaine-au-Bron	1986	1986	837	12 770	58 128	111 615	77 029	55 926	46 937	363 242
Vert-le-Grand	1986	1986	3 347	62 121	91 392	149 724	132 270	161 937	180 390	781 181
Vert-le-Grand Nord	1987	1987	-	1 513	10 591	9 553	13 281	14 764	13 678	63 380
Vert-le-Petit	1987	1987	-	6 025	10 521	8 438	17 355	21 486	17 366	81 191
L'Orme	1990	1991	-	-	-	-	-	7 489	4 377	11 866

Appendix 6. Evolution of oil production by region and field

(metric ton)

Field	Year of dis-covery	Year on stream	Cumulative production 12/31/1986	1987	1988	1989	1990	1991	1992	Cumulative production 12/31/1992
Itteville	1990	1991	-		-		-	14 322	49 078	63 400
Fay-les-Nemours	1991	1991	-			-	-	696	-	696
La Vignotte	1991	1991		-				9 365	21 777	31 142
Montapot	1991	1992	-		-		-	-	2 128	2 128
shut-in/Abandoned	-	-	1 366 186	4 858	15 008	8 479	7 626	3 640	-	1 405 797
ESSOREP										
Donnemarie	1979	1979	206 104	39 174	30 092	19 287	12 113	11 634	8 826	327 230
Chaunoy	1983	1983	1 386 879	768 470	795 689	733 007	700 484	720 140	746 610	5 851 279
Charmottes	1984	1984	19 309	7 310	8 061	7 729	8 961	6 648	2 920	60 938
Malnoue	1985	1985	15 180	20 802	19 564	16 197	13 467	9 339	19 570	114 119
Champotran	1985	1985	12 469	17 552	44 947	42 748	34 739	32 173	28 531	213 159
Brémonderie	1984	1986	3 490	3 526	2 829	2 385	2 067	1 852	1 553	17 702
Vulaines	1977	1986	1 861	7 519	8 054	8 115	8 626	12 552	12 034	58 761
Champrose	1990	1991	-	-	-	-	-	10 855	7 408	18 263
Abandoned	-	-	8 459	1 548	947	-	-	-	-	10 954
EURAFREP										
Abandoned	-	-	11 238	250	215	320	147	-	-	12 170
PETROREP										
Coulommes	1958	1958	1 786 453	19 539	18 215	17 197	17 841	15 216	15 321	1 889 782
Ile du Gord	1986	1986	3 465	5 798	9 752	26 611	31 511	40 257	49 967	167 361
Béchevret	1988	1988	-	-	475	4 101	5 407	7 477	20 465	37 925
Abandoned	-	-	-	524	435	280	-	-	-	1 239
REPLOR										
Forcelles	1978	1978	-	-	-		61	639	472	1 172
Abandoned	-	-	425	-	-		-		-	425
SHELL										
St-Martin	1959	1959	1 125 775	15 985	14 965	13 539	13 245	12 170	10 301	1 205 980
Abandoned	-	-	25 795	-	-	-	-	-	-	25 795
TOTALEX										
Villeperdue	1958	1982	819 757	714 308	704 711	601 654	510 100	451 019	374 390	4 175 939
FAB Sud	1986	1987	-	537	11 684	19 890	12 500	9 457	8 136	62 204
Hautefeuille	1987	1987	-	9 700	12 142	11 204	6 451	10 973	8 153	58 623
La Motte Noire	1987	1987	-	2 071	8 255	7 314	5 507	7 735	12 695	43 577
Sancy-les-Provins	1990	1990	-	-		-	5 647	4 380	872	10 899
shut-in/Abandoned	-	-	463	834	-	339	876	573	-	3 085
TRITON										
Sivry	1984	1984	23 357	8 707	8 665	8 918	8 951	6 435	5 538	70 571
Saint-Germain	1984	1984	140 010	54 201	67 745	46 912	29 328	24 627	20 597	383 420
Maincy	1987	1988	-	-	587	1 073	1 873	1 291	931	5 755
Bagneaux	1988	1988	-	-	733	4 068	8 985	9 414	8 175	31 375
Abandoned	-	-	64 598	8 870	-	-	-	-	-	73 468
TOTAL BASSIN DE PARIS			12 035 403	1 953 327	2 106 596	2 030 933	1 853 777	1 870 174	1 855 538	23 705 748
REPLOR										
Briod	1982	1982	49	-	-	-	-	-	-	49
TOTAL JURA			49	-	-	-	-	-	-	49
EAP										
Abandoned	-	-	30 554	-	-	-	-	-	-	30 554
TOTAL SUD - EST			30 554	-	-	-	-	-	-	30 554
TOTAL FRANCE			64 367 574	3 234 237	3 355 097	3 244 341	3 023 160	2 951 655	2 865 742	83 041 806

Appendix 7. Evolution of raw gas production by region, company and field

(thousands cubic meters)

Field	Year of discovery	Year on stream	Cumulative production 12/31/1986	1987	1988	1989	1990	1991	1992	Cumulative production 12/31/1992
EAP										
Schaffhouse	1958	1961	1 677	-	-	-	-	-	-	1 677
Schaeffersheim	1955	1977	40 350	16	-	-	-	-	-	40 366
TOTAL ALSACE			42 027	16	-	-	-	-	-	42 043
EAP										
Saint-Marcet	1939	1942	6 855 831	19 461	18 408	17 157	15 242	13 222	11 396	6 950 717
Proupiary	1949	1963	233 898	506	461	419	586	796	916	237 582
Lacq profond	1951	1957	198 611 969	3 764 191	3 191 107	2 459 581	2 617 405	3 095 936	2 824 170	216 564 359
Auzas	1968	1968	510 835	736	797	519	794	1 057	1 095	515 833
Meillon	1966	1968	42 225 804	1 159 840	974 880	1 353 101	1 046 577	1 100 201	1 228 314	49 088 717
Rousse	1967	1972	3 284 480	243 469	145 426	158 895	80 269	60 523	34 362	4 007 424
Ucha	1970	1971	1 391 906	86 319	62 215	77 803	52 337	22 578	44 678	1 737 836
Lacommande	1974	1976	28 649	5	-	-	-	-	-	28 654
Ger	1975	1977	104 454	2 339	5 087	831	-	-	-	112 711
Pécorade *	1975	1975	213 761	24 108	33 341	36 230	24 207	38 924	39 571	410 142
Vic Bilh *	1979	1979	174 049	41 226	45 271	48 026	47 416	60 493	71 414	487 895
Le Lanot	1970	1981	145 458	68 475	39 882	97 864	275 658	263 939	211 005	1 102 281
Charlas Ouest	1970	1981	43 030	1 324	230	-	-	-	-	44 584
Cassourat	1987	1989	0	-	-	26 026	13 282	66 891	45 928	152 127
Andoins	1990	1991	0	-	-	-	-	9 277	46 426	55 703
Abandoned			126 362	-	-	-	-	-	-	126 362
Total EAP			253 950 486	5 411 999	4 517 105	4 276 452	4 173 773	4 733 837	4 559 275	281 622 927
ESSOREP										
Parentis *	1954	1954	240 973	2 578	3 130	3 199	3 391	2 796	2 606	258 673
Saucède	1979	1981	45 903	-	2 695	941	761	625	440	51 365
Ledeuix	1981	1982	77 459	8 420	6 315	4 727	4 413	3 860	3 586	108 780
Total ESSOREP			364 335	10 998	12 140	8 867	8 565	7 281	6 632	418 818
TOTAL AQUITAINE			254 314 821	5 422 997	4 529 245	4 285 319	4 182 338	4 741 118	4 565 907	282 041 745
COPAREX										
Trois-Fontaines	1982	1982	0	-	-	-	11 843	154 516	180 610	346 969
ESSOREP										
Chaunoy *	1979	1979	27 350	15 554	16 105	14 900	14 168	14 577	15 112	117 766
EURAFREP										
Trois-Fontaines	1982	1982	101 871	47 463	99 004	107 257	126 019	0	0	481 614
TOTAL BASSIN PARISIEN			129 221	63 017	115 109	122 157	152 030	169 093	195 722	946 349
EAP										
Abandoned			98 258	-	-	-	-	-	-	98 258
TOTAL JURA			98 258	-	-	-	-	-	-	0
TOTAL FRANCE			254 584 327	5 486 030	4 644 354	4 407 476	4 334 368	4 910 211	4 761 629	283 030 137

* Gas associated with oil production

NB. only the gas associated with oil production and marketed is included

A History of Petroleum Exploration in France

F. Heritier

Abstract. Oil shows are not very common in France but in Alsace and in Languedoc they were exploited as early as Roman times. The "industrial" exploitation started in 1740 at Pechelbronn and continued until 1964.

Actively supported by the French Government, petroleum exploration in France by both state-owned and private foreign capital companies has been a success, as more than 100 Mt of oil and 325 Gm^3 of gas have been discovered since 1945.

Foreword

Oil shows are not very common France. Nevertheless, some oil seepages have been found in some sedimentary basins such as Alsace, Limagne, South East, and Aquitaine.

In most cases, these seepages are related to faults, salt diapirs, or even volcanic intrusions. Taking into account the distentional tectonics of the two basins of Alsace and Limagne, it is no wonder they are the richest in oil shows and seepages.

Exploration from the Origins to the Eve of the 20th Century (Fig. 1)

As shown by local names, some of these oil seepages were exploited as early as Roman times for medicinal oil, grease for wheel axes, and in some cases as oil for lamps. Some oil shows and tar sands outcrops were also exploited as a tar source for waterproofing or caulking.

Among these shows and seepages, two of them were particularly spectacular and well known because they were associated with mineral water springs. The first one, the spring of Pechelbronn (which means Pitch spring), was located in Alsace; the second, La Font de l'Oly (the oil spring) was located near the small village of Gabian in the southeastern part of France. A third oil seepage, which is associated with a mineral water spring and a volcanic intrusion, is the oil show of le Puy de la Poix (the pitch mount) in the Limagne basin. Old records mention that, around 1560, Catherine de Medicis and her son the future King Charles IX had visited the Puy de la Poix as one of the curiosities of the Kingdom.

Concerning the exploitation itself of these oil seepages, the oldest records that we have mention that since 1608 oil has been recovered from water at La Font de l'Oly near Gabian and that the average production was 36 quintals (around 2 metric tons) of oil per year. The oil flow varied with the water flow and the amount of rain!

In Alsace, near Merkwiller, in the "Pechel-Bronn", the country people have scooped up the heavy oil from the water spring since the 15th century. Exploitation of local tar sands was first attempted in 1627.

Time of the First "Industrial" Exploitations

In 1740, a permit for exploitation was granted by the Royal Authority in Alsace. This was the first exploitation permit ever granted in France by the national authorities, and this was the first step in industrial exploitation of tar sands by pits and galleries. The quarrying and transport of the sands were done with buckets and wheelbarrows. The tar was extracted by washing the tar sands with boiling water. In 1770, the galleries were deepened to about 100 m, following the sand layers.

In 1813, the first wells were drilled to guide the mine galleries. In one of those wells, some free oil that ran out of the sand was discovered at a depth of 77 m and demonstrated the presence of an oil field for the first time. The real discovery of the Pechelbronn oil field occurred in 1880 when two "deep" wells had an oil blowout.

During this first period, 45 km of galleries were dug out, 286 wells were drilled, and around 50000 t of product was extracted, 54% of which was oil. In some areas wells were drilled every 10 m. Exploitation continued this way until 1916, when 2860 wells had been drilled and around 720000 t of oil had been produced (Fig. 2).

Meanwhile, in the Gabian area, records mention that a permit demand for an industrial exploitation of the oil spring was rejected by the Mines Administration. Then in 1860, following the first oil rush in Pennsylvania, exploration for oil started again near Gabian, and numerous shallow wells were drilled around the old oil spring whose flow was increased by the boring of a gallery. In 1884, a well drilled 300 m updip from the oil spring found a good reservoir at a depth of 68 m and went up into a blowout of water and CO_2, which resulted in the definitive drying up of the oil spring. Then, with the exception of two other dry wells drilled in the immediate vicinity of the oil spring, no other exploration work was undertaken in the area until 1923.

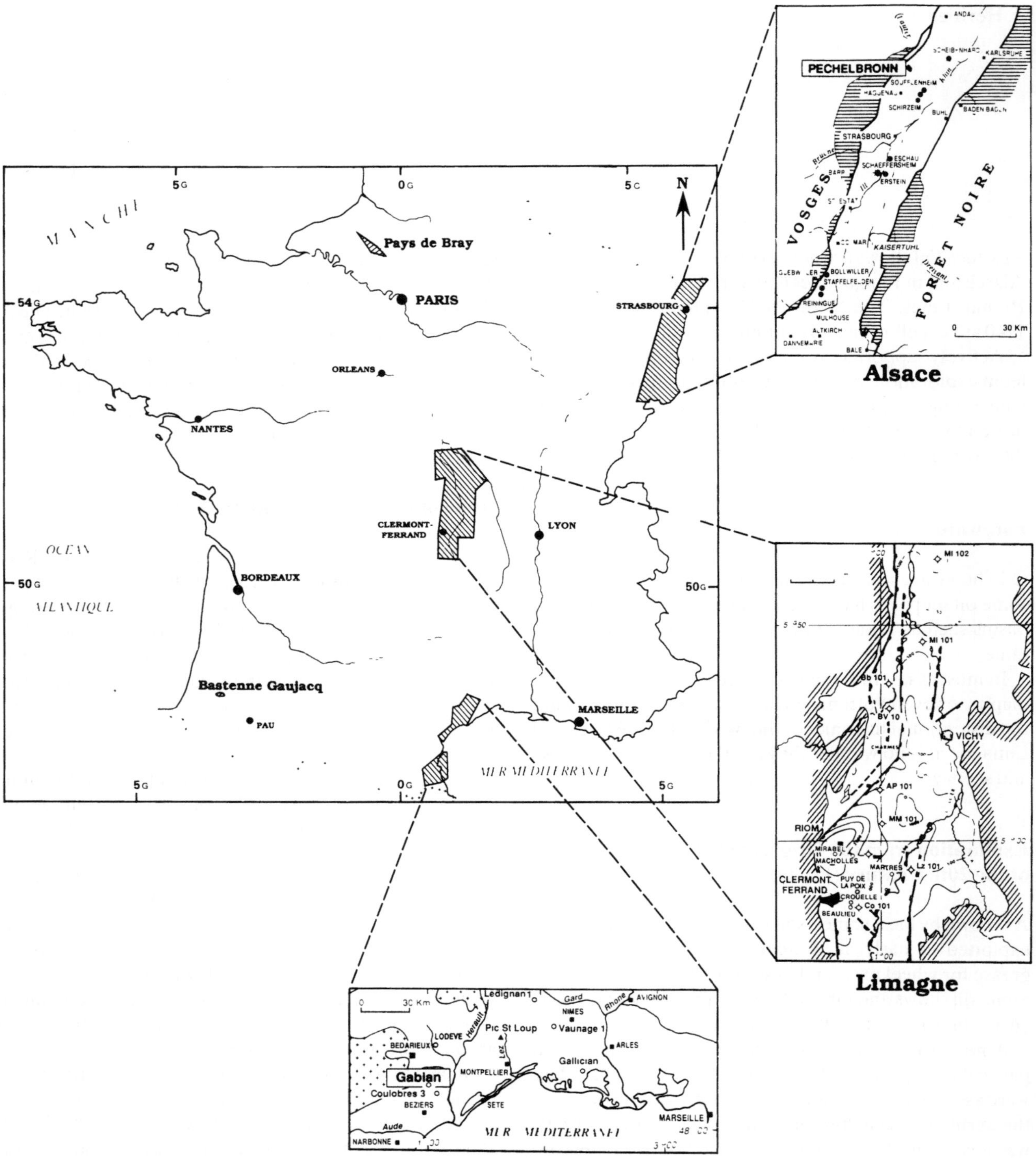

Fig. 1. Area of main oil shows in France

About the same time and until 1929, different permits for exploitation of tar were granted in Limagne by the Mines Administration.

In 1883, a succession of four to five shallow wells were drilled near the tar shows without any success. In 1891 a private company started a drilling campaign of five wells, the first one (Marolles 1) reaching a depth of 1164 m. All these wells found bituminous and gas shows but no production was possible (Fig. 3)

The history of the first oil exploration in South Aquitaine is not so well documented. Of course, the big outcrops of bitumen and tar sands were known probably for centuries near S[te] Suzanne, Salies, Dax, and Bastenne Gaujacq, but it was only in 1821 that oil seepages were described by a geologist at S[t] Boes near Orthez. Between 1896 and 1926, exploration companies were created by local leading citizens, and about 30 wells, some as deep as 750 m, were drilled in the vicinity of bituminous out-

Fig. 2. An old view of Pechelbronn oil field

Fig. 3. Old Raky rig used between 1918 and 1936

crops. These wells proved the existence of oil and/or gas in the Upper Cretaceous formations that are usually overlain by Triassic evaporites.

First Attempts at Modern Prospecting – 1920–1940

On the Pechelbronn field, after the first World War, five new pitwells and kilometers of galleries were bored to exploit oil still in place in the sand layers (Figs. 4, 5).

In 1936, the first rotary rigs were used, and Pechelbronn deep oil was found in the Dogger oolitic limestones and Triassic sandstones. By that time the Office National des Combustibles Liquides (ONCL) asked Dr. Barrabé and P. Vienot to make a complete geologic survey of all the Gabian area. From this survey a first exploration well was situated on a small faulted anticline located southwest of the old oil spring. The well was spudded in August 1924 and found, at a depth of about 100 m, an oil production (24 m^3/day) in sandstones and conglomerates of Muschelkalk age below a cover of Keuper gypsiferous marls. This was the discovery of the small Gabian field; over a period of 10 years, 57 wells were drilled on less than 0.5 km^2 and 23400 tons of oil were produced (Fig. 2).

At about the same time, in the Paris Basin geologists focused their attention on the Pays de Bray anticline north of Rouen, and in 1923 the ONCL spudded a first

Fig. 4. On Pechelbronn oil field the prayer before descending into the mine

Fig. 5. On Pechelbronn oil field. a view inside an oil production gallery

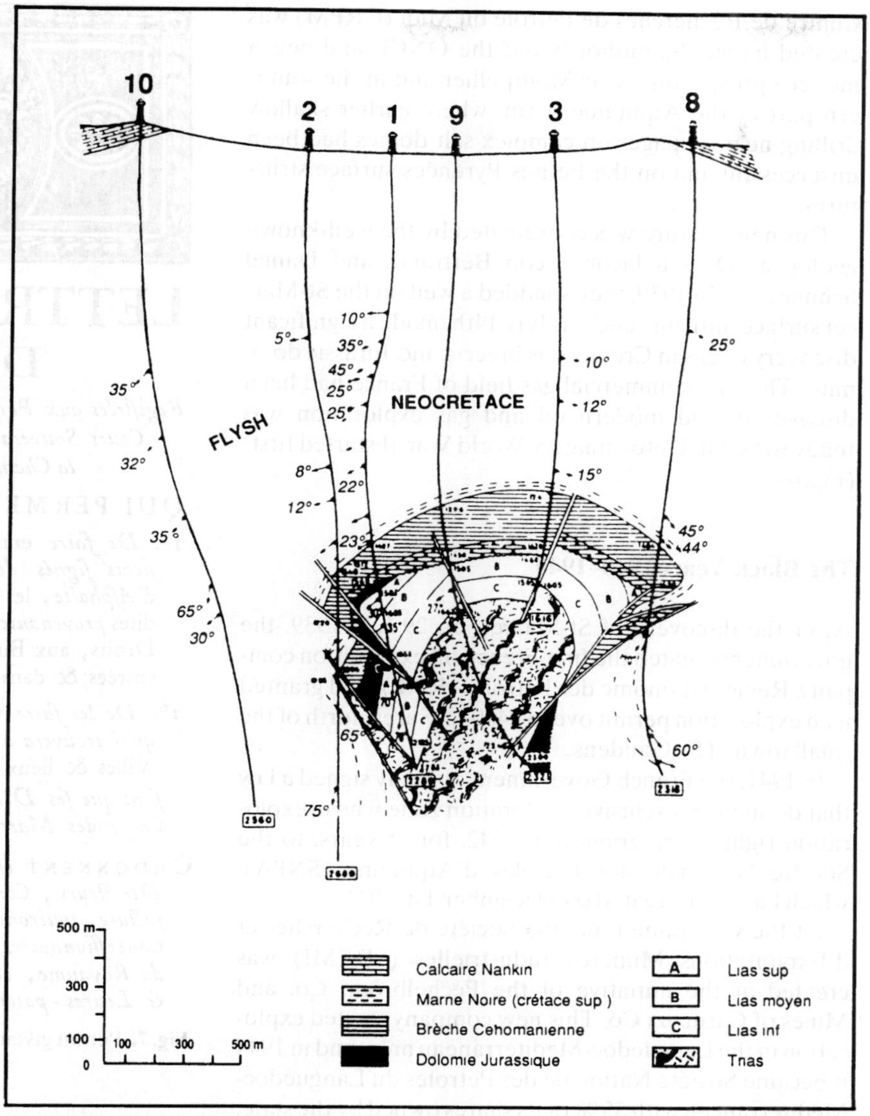

Fig. 6. Geological cross section of Saint-Marcet gas field (P Michel and D Schneegans)

exploratory well on the southern flank of the structure. This well drilled to a terminal depth of 1172 m found, in 1927, below the Mesozoic section, Carboniferous layers, and was a dry hole without any incentive shows (Fig. 1).

Farther south, in Limagne, just after the end of World War I, oil exploration resumed, and in 1919 the Geological Service and Mines Minister decided to spud new wells. The well of Martres d'Artières was drilled to a depth of 415 m and went off as a blowout of mineral water and CO_2. In 1920, another well at Puy de Crouelle was stopped at 856 m during the coring of impregnated sands. During the following 2 years, six shallows wells were drilled by local companies without any result. In 1923 ONCL decided to drill another well near the one at Puy de Crouelle. This well, Beaulieu, took 3 years, and finally reached the basement at 1164 m and had only some good bituminous shows in the marls and mineral water in the basal arkosic sands. Simultaneously, another deep well was spudded at Mirabel near the small town of Riom. Drilled down to 1320 m; this well encountered only bituminous shows in the upper marls and mineral water and CO_2 in the basal arkosic sands.

At the same time, some geophysical prospecting was done. This included a small electrical sounding survey and a torsion balance gravity survey, which were able to give a clear but pessimistic picture of the Limagne basin. Following these results, ONCL stopped its exploration in 1929.

On the western border of the Jura mountains, gas was found by accident in 1906 at 200 m depth in a well drilled for a salt search. The blowout flow was 11 000 m^3/day. After World War I, two small independent companies drilled 9 wells between 1920 and 1925. In 1920, the well SREP N. 2 found gas in dolomitic limestones of Triassic age. This locally useful gas well was completed and used by the small town of Amberieu up to 1941.

In 1937, under the pressure of the Front Populaire government, oil exploration was resumed in France. The

Centre de Recherches de Pétrole du Midi (CRPM) was created under the authority onf the ONCL and began modern prospecting near Montpellier and in the southern part of the Aquitaine basin, where earlier shallow drilling near seepages on complex salt domes had been unsuccessful and on the Petites Pyrénées surface structures.

This new venture was coordinated by the well-known geologists Dr. Ch. Jacob, Léon Bertrand, and Daniel Schneegans. In 1939, they spudded a well on the St-Marcet surface anticline and, on July 14th, made a significant discovery of gas in Cretaceous breccia and Jurassic dolomite. The first commercial gas field of France had been discovered, and modern oil and gas exploration was ready to begin. Unfortunately World War II started first! (Fig. 6)

The Black Years 1940–1945

After the discovery of St-Marcet, on 29 July 1939, the government created the first French oil exploration company, Régie Autonome des Pétroles (RAP), and granted it an exploration permit over a 2400 km^2 area north of the small town of St-Gaudens.

In 1941, the French Government at Vichy signed a law that defined an exclusive exploration zone whose exploration rights were granted in 1942, for 15 years, to the Société Nationale des Pétroles d'Aquitaine (SNPA), which had been created on December 1st, 1941.

At the very same time, the Société de Recherches et d'Exploitations Minières Industrielles (SREMI) was created at the initiative of the Pechelbronn Co. and Mines of Carmaux Co. This new company started exploration in the Languedoc-Mediterranean area, and in 1944 it became Société Nationale des Pétroles du Languedoc-Méditerranéen with 55 % of its shares owned by the state.

During all this period, RAP drilled some exploration wells on other geologically defined structures and some appraisal wells on the St-Marcet gas field, whose production started in 1942, to supply the city of Toulouse.

At the same time, a well was drilled on the Jura border and found gas in Triassic dolomitic limestones. The small production from this well was used by the nearby town of Revigny.

In 1944 SNPLM drilled nine new wells without any success in the vicinity of the old Gabian field.

After World War II–1945: Resumption of Petroleum Exploration

At this stage it would be interesting to pause and to relate a short history of mining legislation in France

At first only the King enjoyed the privilege of conceding a right for using mines for the benefit of the first occupier or of a preferential third party. Thus in 1772, King Louis XV licensed M. Lebel by royal letters for selling Pechelbronn oil and tar grease (Fig. 7).

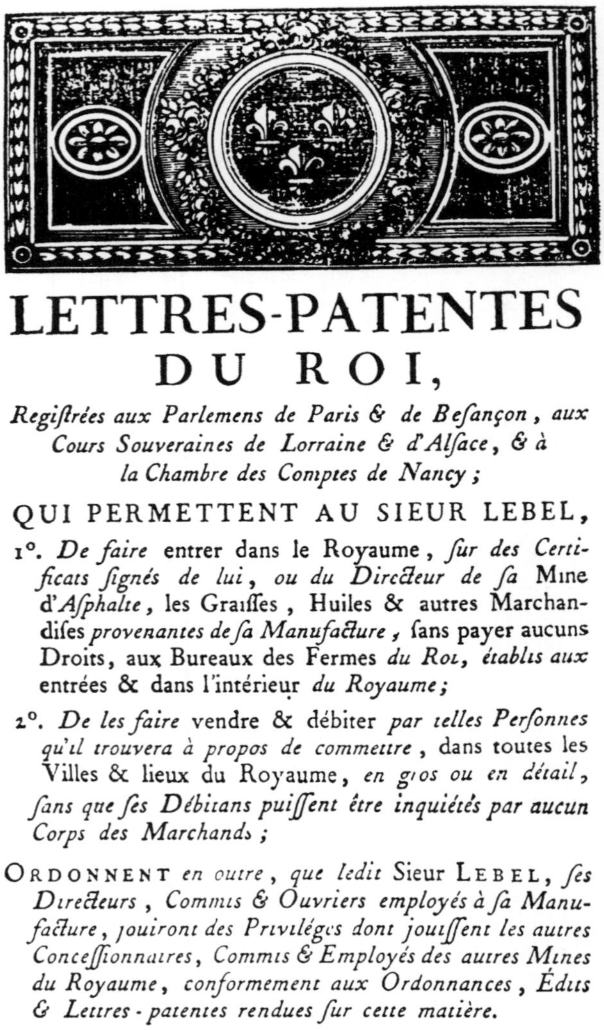
LETTRES-PATENTES
DU ROI,
Regiſtrées aux Parlemens de Paris & de Beſançon, aux Cours Souveraines de Lorraine & d'Alſace, & à la Chambre des Comptes de Nancy;
QUI PERMETTENT AU SIEUR LEBEL,
1°. *De faire* entrer dans le Royaume, *ſur des Certificats ſignés de lui, ou du Directeur de ſa* Mine d'*Aſphalte*, les Graiſſes, Huiles & autres Marchandiſes *provenantes de ſa Manufacture*, ſans payer aucuns Droits, aux Bureaux des Fermes *du Roi, établis aux* entrées & dans l'intérieur *du Royaume;*
2°. *De les faire* vendre & débiter *par telles Perſonnes qu'il trouvera à propos de commettre*, dans toutes les Villes & lieux du Royaume, *en gros ou en détail, ſans que ſes Débitans puiſſent être inquiétés par aucun Corps des Marchands;*
ORDONNENT *en outre, que ledit* Sieur LEBEL, *ſes Directeurs, Commis & Ouvriers employés à ſa Manufacture, jouiront des Priviléges dont jouiſſent les autres Conceſſionnaires, Commis & Employés des autres Mines du Royaume, conformement aux Ordonnances, Édits & Lettres-patentes rendues ſur cette matière.*

Fig. 7. Permit given by Royal Authority to Sir Lebel in 1772

In 1791, these royal rights ceased. The new revolutionary government, by a law of July 1791, placed the mines at the national disposal Up to 100 feet deep, the land-owner had free use of the mine product, below 100 feet deep, the government granted a permit to the ground landlord or to the discoverer.

In 1810, Napoleon enacted a new mining law which stated that the government could choose the license holder, and which established the General Mining Council, which gave advice on every administration ruling about mines

In 1871, a new law was enacted on oil imports. A duty tax of FF 12 per hectoliter was set for refined oil products The direct results of this law was the building of 17 oil refineries with a production capacity of 0.5 Mt per year.

After the First World War, in 1919, a new mining law was enacted, which suppressed the perpetual mine permits and linked oil and natural gas with other mine products. In 1922, this law and the old law of 1810 were improved to help oil and gas prospecting.

In 1925, the French Government realized the economic and strategic importance of domestic oil production or

Tabelle 1. Operating chart of national oil exploration industry in 1947

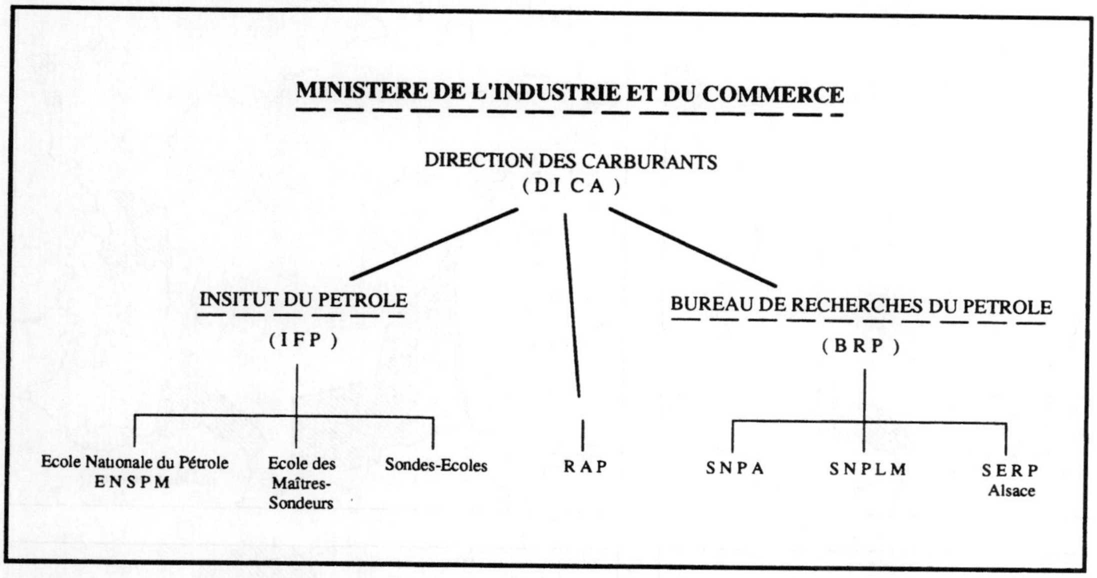

of a French-controlled oil production. So in March 1924, the Compagnie Française des Pétroles was created, which obtained 23.75 % share of the Turkish Petroleum Co. as the result of the San Remo agreement of April 1920. Then in 1925, the government created the Office National des Carburants Liquides (ONCL), whose aim was oil prospecting and refining, establishing a tanker fleet and a strategic oil reserve. It also created the Ecole du Pétrole, linked to Strasbourg University.

In 1939, the ONCL was replaced by the Direction des Carburants (DICA) and the 20th of July of the same year, after the discovery at St-Marcet, CRPM was replaced by the Régie Autonome des Pétroles (RAP).

In 1944, at the liberation of France and in 1945 at the end of the war, the government and most people were convinced of the major strategic importance of domestic oil and gas production for France. So, going on the same policy as its pre-war predecessors, from 1944 to 1947 the government created several public offices such as the Institut Français du Pétrole (IFP) (1944), which was a governmental establishment with responsibility for petroleum research and training of technical personnel, and the Bureau de Recherche du Pétrole (BRP) whose aim was to coordinate the exploration activities of the different oil companies (see Table 1). With the linking up of the Ecole Nationale Supérieure du Pétrole with IFP in 1947, everything was ready to again start up modern and efficient oil and gas exploration.

Postwar Exploration – 1945–1951 (Fig. 8)

Postwar exploration activity started with some geophysical surveys: electrical, telluric currents, gravimetric surveys, and the first modern seismic surveys. The number of deep wells drilled increased slightly with the use of 12 heavy rigs in 1945 and 19 at the end of 1946. Most of these rigs were old American rotary rigs, some German rigs left on the RAP permits in 1944, and at the end of the period five newly imported American rigs.

RAP continued the systematic exploration of the surface structures of the Petites Pyrénées around the St-Marcet field. All these exploration wells, except Proupiary 1 and 2, were dry with only some gas shows.

SNPA started drilling into the big structural axes of Garlin, Audignon, and Antin revealed by the first geophysical surveys, and began the exploration of the flanks of the Bastennes Gaujacq salt dome. A lot of light oil shows were found but no production was obtained.

In the Languedoc Mediterranean area, SNPLM carried on exploration by drilling into the big structures of La Gardiole, Quissac, Castelnau, and La Vaunage, where very strong shows of combustible gas and CO_2 were encountered. On the northern border of the basin, in 1950, the well of Lussan had a small (3000 m^3/day) production of combustible gas (Fig. 8 a).

In Alsace Pechelbronn SA drilled 50 exploitation wells, 26 of which were oil producers and produced 49000 t of oil (29000 t by wells and 20000 t by galleries).

It should be noted that in 1947 an exploration permit was requested for the northern part of the Aquitaine basin around Bordeaux by an American consortium led by Esso.

In 1948 the St-Marcet field had produced 688 Mm^3 of gas and around 2000 m^3 of oil.

At the same time, the results of the gravimetric surveys made in the southern part of the SNPA permit revealed a trend of positive anomalies between the towns of Pau and Orthez with the same orientation as the Petites Pyrénées. After a seismic survey which confirmed the existence of a big anticline, SNPA in 1949 spudded the first well on the Lacq structures. This well struck oil at a depth of 630 m in

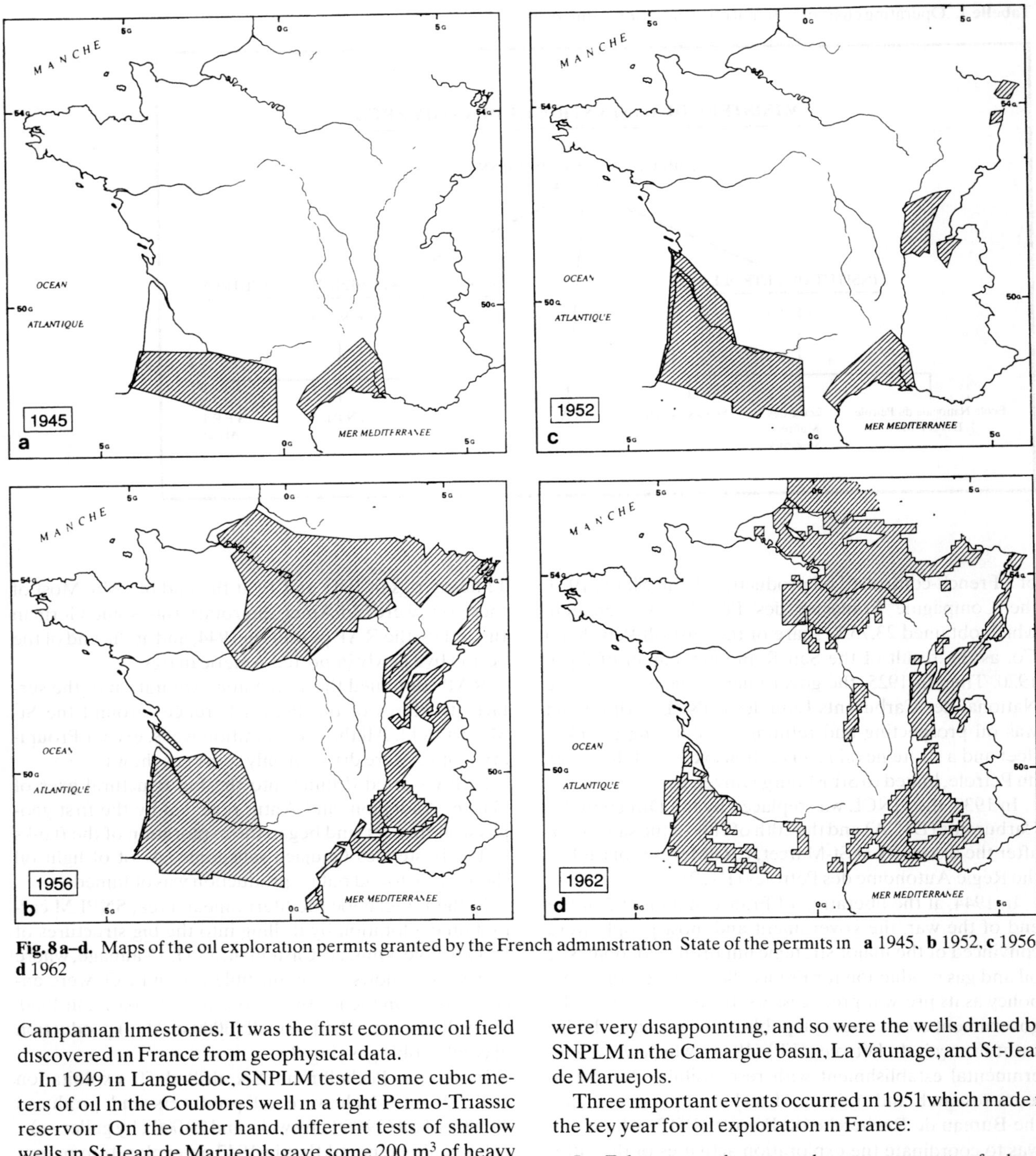

Fig. 8 a–d. Maps of the oil exploration permits granted by the French administration State of the permits in **a** 1945, **b** 1952, **c** 1956, **d** 1962

Campanian limestones. It was the first economic oil field discovered in France from geophysical data.

In 1949 in Languedoc, SNPLM tested some cubic meters of oil in the Coulobres well in a tight Permo-Triassic reservoir On the other hand, different tests of shallow wells in St-Jean de Maruejols gave some 200 m^3 of heavy oil, and exploration by geophysics and drilling started in the Camargue basin.

In 1950, RAP continued on its exploration of the Petites Pyrénées structures and, on the Muret high, started the exploration of the Mollasse basin and obtained, for the first time, new permits outside Aquitaine in Bresse, Jura, and Savoie (Fig 8b)

SNPA developed the Lacq oil field, and by June 1951, 20 wells had been drilled, 11 of which were flowing, and oil production reached 230000 t, and the total field reserves were estimated to 4.7 Mt of oil Meanwhile, the other deep wells on Bastenne Gaujacq, Pau, and Garlin were very disappointing, and so were the wells drilled by SNPLM in the Camargue basin, La Vaunage, and St-Jean de Maruejols.

Three important events occurred in 1951 which made it the key year for oil exploration in France:

- In February, a large exploration permit of about 1800 km^2 in the northern part of the Aquitaine basin and around the city of Bordeaux was granted to Standard Française des Pétroles, and affiliate of Standard Oil of New Jersey, with participation by the Atlantic and Gulf companies, which was later called Esso Rep. It was the first awarding of an exploration permit to a foreign company by the French authorities.
- BRP asked IFP to carry out the geological and petroleum reconnaissance of the Paris Basin, while CGG was making the first general gravimetric survey of the Basin.

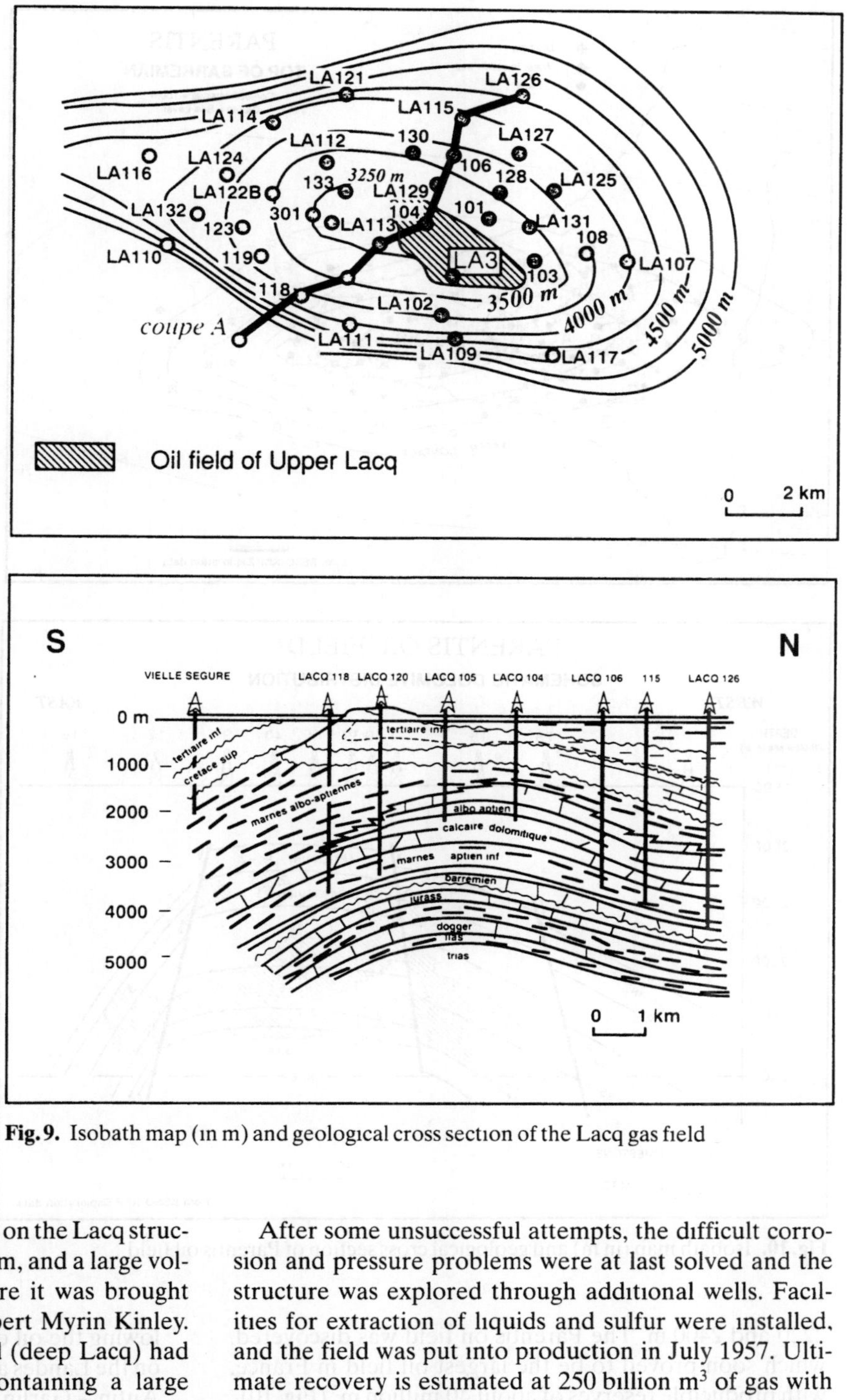

Fig. 9. Isobath map (in m) and geological cross section of the Lacq gas field

- In Décember 1951, the first deep test on the Lacq structures blew out from a depth of 3530 m, and a large volume of very toxic gas flowed before it was brought under control by the American expert Myrin Kinley. The large gas field of Lacq profond (deep Lacq) had thus been discovered. The gas, containing a large amount of condensate (5.4% of C_{2+} and 16% of H_2S, 9.8% of CO_2), was under high pressure (670 kg/cm^2 at a depth of 3530 m) and was highly corrosive for steel casing and drillpipes. That caused a major problem for SNPA because this type of corrosion had never been encountered or solved before (Fig. 9).

Nevertheless, in 1952 two appraisal wells (LA 101 and L1 102) were drilled and struck a major gas accumulation in Lower Cretaceous and Jurassic carbonate rocks. The reservoir thickness is 450 m and the productive column exceeds 1200 m.

After some unsuccessful attempts, the difficult corrosion and pressure problems were at last solved and the structure was explored through additional wells. Facilities for extraction of liquids and sulfur were installed, and the field was put into production in July 1957. Ultimate recovery is estimated at 250 billion m^3 of gas with 50 million metric tons of sulfur and around 15 million metric tons of condensate and LPG.

After Lacq, Parentis Took the Relay

In the northern part of the Aquitaine basin, Esso Rep started exploration by an intensive program of geophysical surveys, gravity, then seismic. The first wildcat was a failure, but a discovery well on a seismic anticline at Parentis in 1954 flowed 3500 barrels of 32° API gravity oil from Lower Cretaceous and Jurassic dolomites between

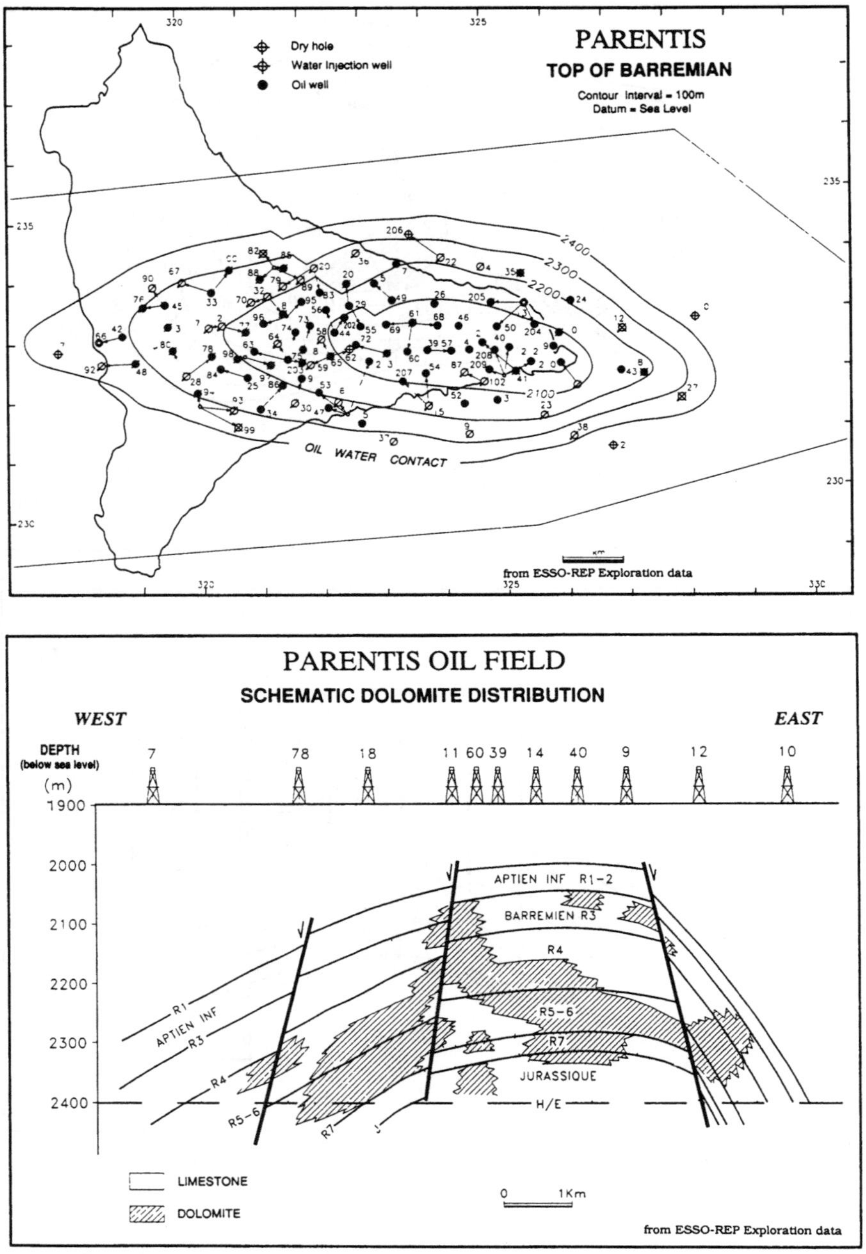

Fig. 10. Isobath map (in m) and geological cross section of Parentis oil field

2220 and 2400 m. The Parentis oil field was discovered, which soon proved to be the largest oil field in France, with producible reserves of about 40 million m^3 (Fig. 10).

Subsequent exploration on the SNPA and RAP concessions in the Aquitaine basin were, for 12 years, very disappointing:

- RAP on its first permit and on the permit of Société des Pétroles de la Garonne (SPG) (an association of SNPA and RAP) drilled about 40 dry wells with some oil shows which demonstrated that in most of the area north of St-Marcet the Jurassic limestone reservoirs were flushed by fresh water.
- On its permit SNPA drilled 7 wells on the Pyrenean foothills on the trend of the Lacq structure. Then, following the oil discovery of Parentis, it drilled 20 wells on the Landes area and on the north structural trend of Antin – Garlin – Bastenne Gaujacq. Except for some oil shows, all these wells were dry, but their results showed the existence of two separate Lower Cretaceous basins, basin of Parentis on the north and basin of Arzacq on the south. Most of the area on the structural high in between was flushed by fresh water.

Oil Discoveries in the Paris Basin

Meanwhile, RAP, then SNPA asked for large research permits in the Paris basin, and in 1953 RAP drilled two stratigraphic test wells in the eastern then central part of the permit. This last well, Courgivaux 1, reached the basement at 3810 m and so proved the deepness of the basin and the thickness of the sedimentary column and, for the first time, the presence of oil shows in Dogger limestones and Rhaetic sandstones. Then in 1954, RAP drilled a well in the Pays de Bray anticline and found a small production of heavy oil, which confirmed the oil production potential of the basin.

A surge of applications for exploration permits followed the Parentis discovery and the obvious oil shows in the Paris Basin. So the Compagnie d'Exploration Pétrolière (CEP which was the former SNPLM) and Fropex, an affiliated company of Caltex, obtained permits in the Paris Basin in 1955 when many companies, among them SNPA, CEP, CFP, Shell, and Mobil, had applied for permits in the Paris Basin and Aquitaine Basin. So most of the reasonably prospective areas in France were under permit by 1956 (Fig. 8c).

From 1955 to 1957, RAP was engaged in very active exploration of the Paris Basin but found only oil shows in Dogger limestones then in Rhaetic sandstones near Courgivaux and Mailly le Camp. At the same time, SNPA started the exploration of the eastern Carboniferous basin, but the two wells of Morhange and Royaumeix gave very disappointing results, while some oil was discovered in the Muschelkalk formation in the Forcelles 1 well In the western and southern parts of the basin, CEP also made a very impressive exploration effort without any success except for some small oil production associated with salt water in tests at Chateau Landon and Puiselet.

So in 1957, in the different companies, the mood of the exploration managers was very pessimistic. Indeed, from the first well results it appeared that the difficulties of exploration of the Paris Basin were much greater than expected. The very low relief of the structures associated with large seismic velocity variations, and in some places with very great and difficult-to-compute surface corrections, made these structures barely within the resolving power of seismic methods.

Finally, in 1958, three commercial discoveries were made by Petrorep at Coulommes in Dogger limestones, then by Fropex in Neoconian sandstones on the Chateaurenard anticline and at the end of the year in Rhaetic sandstones by RAP at Granville just South of Mailly le Camp. These first successes were rapidly followed in 1959 by the discovery of the field of Chailly en Bière by CEP and its extension Chartrettes by RAP, then Saint-Martin de Bossenay by Copesep and Villemer by RAP. All these fields were producing from upper Dogger limestones reservoirs.

All of the fields discovered during these 2 years represent about 30 Mt of oil accumulation and 10 Mt of producible reserves.

In 1962 and 1963 other small fields like Valence en Brie, Montmirail, Pont s/Seine, Avon la Pèze, Gizy des Nobles, and Auffervile were discovered in the Dogger. Yet the exploration of the wedgeout of Neoconian sandstones was carried on with some success by CEP, which had taken Amoco's assets in Fropex. On the other hand, RAP rested gas in Triassic reservoirs at Trois Fontaines 101 (Fig. 8d).

However, none of these discoveries was very prolific, and deeper Triassic and Paleozoic tests drilled by SNPA in the Paris Basin were failures, as were the Paleozoic tests in the northern part of France. So, with less than 3% of wildcat wells proving economically successful, exploration in the Paris Basin began a rapid decline in 1964.

Yet after the merging of RAP and CEP, which became the Elf group, exploration continued, and a small oil production was found in Triassic sandstones at Chailly and Villemer.

Exploration Elsewhere in France

Since 1950, exploration in France outside the Aquitaine and Paris basins has been active. From north to south oil and gas prospecting has been done:

In Alsace. After the merging of Pechelbronn S.A. and BRP into Prepa in 1953, the exploration of the Oligocene basin of Alsace continued on the same time in the north, center, and south

Up to 1960, more than 100 wells were drilled. Some small oil fields such as Scheibenhard, Schirein, and Souflenheim were discovered in Oligocene sands in the north. In the southern and central part of the basin, discoveries such as Staffelfelden and Eschau were made in the Dogger great oolite (Fig. 1).

Stopped for 10 years, exploration was resumed in the 1970s with poor results, and finally all prospection work ceased in 1986 after a last unsuccessful Triassic test.

In Bresse. In 1950, RAP obtained a permit in the Bresse Tertiary basin which also covered the western part of the Jura where some small gas accumulations had been found previously. In 1952, the exploration of the southern part of Bresse started. The two first wells went through the Oligocene and found salt and some reservoirs. The second well also found some gas shows in the Dogger limestones. So, three wells were drilled on the Chatillon s/Chalaronne high with the Mesozoic as the target. They were all very disappointing, as the Jurassic limestones were flushed by fresh water, and the Triassic dolomites, sometimes with oil shows, were tight. Then six dry wells were drilled in this same part of the basin and in the new Dombes permit.

The northern part of the Bresse permit and the Côte d'Or permit granted to Prepa were actively explored from 1956 to 1960. Seven deep wells were drilled, which found only some water-saturated reservoirs without any oil shows. After the first permit had been relinquished by RAP, in 1960 SNPA resumed exploration on the Bresse area and on the Jura border. Within 10 years, seven wells were drilled with very disappointing results, and in 1970 RAP and SNPA gave up exploration in this part of France.

In 1978, Shell – BP, then Total and SNEA (P), again started exploration in their Bourg en Bresse and Meximieux permits and drilled three deep wells without any success

In the Jura Since 1950, RAP, Safrep, Prepa, and SNPA have worked actively in all the area between the Franche Comté and the Swiss border and have drilled numerous shallow wells, and some deep wells, most of them unsuccessful. Yet in 1957, in Muschelkalk dolomites RAP found a small shallow gas field at Lons, then, in 1961, in Lettenkhole Prepa found another small gas field at Valempoulières. Finally, in 1964, RAP proved a southern extension of its Lons field.

Final attempted exploration was made in the 1980 by an association operated by BP, without any success.

In Savoie. From 1952 to 1958, RAP and its associates drilled some wells to test the Tertiary basin where some nice oil shows were known on some outcrops.

All the reservoirs encountered in the Tertiary section had been flushed by fresh water and, below the Mesozoic layers, showed only some light oil shows in poor reservoirs. Furthermore, the tectonics was very complex and no production was obtained.

In the Valence Area. The Tertiary basin of Valence has been explored since 1955 by an association operated by BP. The initial Oligocene objectives were found to have been flushed by fresh water, and as in Savoie, the Mesozoic carbonates showed only some oil stains. With the aim of finding reservoirs in the mesozoic carbonates, 8 deep wells were drilled without any success.

In 1961, SNPA resumed exploration in this area and at Montmiral 1 found a large CO_2 gas field in the Dogger limestones.

Farther south, near Montelimar, in 1959 and 1960, SNPA drilled two deep wells, Marsanne and Savasse, which found no reservoir and proved that the "Terres Noires" shale facies extended farther to the west than expected.

Camargue In this small Tertiary basin, in 1951 SNPLM found a small accumulation of heavy oil in poor Oligocene reservoirs at Galician Nearly 40 wells were drilled in this basin, trying to reach the Oligocene reservoirs; but the structural aspect of the basin was very difficult to figure out by seismic, and most of the Jurassic paleohighs had been flushed by fresh water. The exploration of this area continued up to 1960 without any noticeable results.

In the South East Basin Since 1955, six companies have worked in this basin and drilled many wells without any success Esso Rep found CO_2 in the Triassic reservoirs of Vacquières 1 and reached compact Dogger limestones over 5000 m deep in its Bouquet 1 well.

East of the Rhone, exploration also continued with no success. Either the reservoirs were absent or tight in the basin or, on the southern flank of the basin, the Dogger and Upper Jurassic limestone were good reservoirs but had been flushed by fresh water.

East of the major Durance fault the basin became shallower, and reservoirs were present in Jurassic limestones and in the Triassic but they were found to have been invaded by fresh water or CO_2 as at Jouques 1

Following these disappointing results, exploration has ceased almost completely since 1977.

In the Center of France in Limagne (Fig. 1). This Tertiary basin had been explored for a long time before 1932 In 1953, BRP resumed exploration by making detailed gravimetric and electric surveys. Then in 1959, in Limagne permit was granted to RAP, which shot a large seismic survey and drilled ten wells. Most of the deep wells drilled in the deepest part of the basin between Riom and Clermont-Ferrand showed Sannoisian reservoirs invaded by fresh water, CO_2 saturated (Vichy mineral water!), and some heavy oil show in the shales. RAP gave up its exploration in Limagne in 1962.

In 1981, Eurafrep resumed oil prospection in the deepest part of the Limagne basin and drilled two wells, which revealed some heavy oil shows and CO_2 The Cebazat 1 well tested at 4000 m^3/h of CO_2! All exploration ceased in Limagne in 1983.

New Discoveries in Aquitaine (Fig. 11)

In the Parentis Basin, Esso Rep made some noticeable discoveries of oil fields such as Mothes in 1955, Lugos in 1956, and Cazaux in 1959 and 1961 in the deeper Purbeckian field.

In the southern part of the Aquitaine basin, SNPA and RAP carried out exploration and, thanks to seismic improvements, SNPA could show the real structure below the large Pau anticline, and in 1965 drilled a well which discovered the Meillon gas field in Middle Jurassic dolomites. This field was successfully extended by the discoveries of Mazères, St-Faust then Pont-d'As.

Following this success, exploration in the Aquitaine basin started again in two main directions:

- In the Pyrenean foothills three gas fields were discovered in the vicinity of Lacq and Meillon: Rousse in 1967, Ucha in 1970, and Lacommande in 1974 One small gas field was discovered in 1971 in the vicinity of St-Marcet, and farther west the Lannemezan well No. 2 found a very small accumulation of gas in the Jurassic dolomites at a depth exceeding 7000 m, below the Pyrenean Front overthrust
To be noted also, the discovery in 1972 of the little oilfield of Montastruc-Bonrepos on the western flank of the large Gensac anticline.
- On the northern flank of the basin the two very small oil fields of Lameac and Osmets were discovered in 1970 and 1975. Farther west the Pecorade oilfield, whose reserves reached 2.8 Mt, was discovered in 1974.

Other than these small discoveries, however, most of the wells drilled during this period in the Aquitaine basin were failures. Nevertheless, the different OPEC crises and the sharp rise in oil prices after 1973 prompted all the oil companies to increase their exploration budged in France.

In the Aquitaine basin, Elf Aquitaine, associated with Essorep, discovered the Vic Bilh oilfield in 1978 on the northern flank of the heavily explored Garlin structure.

The intensive exploration that followed this 4-Mt oil discovery achieved its purpose in 1984 with the discovery of the Lagrave oil field by Elf Aquitaine associated with BP.

All the very deep wells drilled in the Pyrenean foothills to explore structural highs below the main front overthrust were expensives failures. Yet Essorep discovered two small gas fields, Ledeuix and Saucede, in the allochtonous formations above the overthrust.

Large Oil Field Discoveries in the Paris Basin

In the Paris Basin, the main permits of RAP and SNPA expired in 1969, and new permits were granted in 1975 and 1978 to Esso Rep, Elf, and Shellrex. In the Sezanne permit in January 1977, Essorep associated with Elf discovered the oil field of Soudron, which produced from Dogger and Rhaetic reservoirs. After this discovery many wells were drilled in this eastern part of the basin, but apart from three subeconomic oil accumulations in Rhaetic sandstones and a small gasfield in Triassic reservoirs at Trois Fontaines 102, all these wells were very disappointing.

Westward, in the central part of the basin, Esso Rep perseveringly explored its Brie permit for Triassic targets and finally get oil at Donnemarie, then at Chaunoy, where, in 1983, it found the largest oil field in France.

Meanwhile, in 1981 Triton and Total reopened the old RAP well, Montmirail 2, and, by a step-by-step appraisal work discovered the large Villeperdue oil field in the upper Dogger reservoirs. With about 8 Mt of reserves this oil field was one of the largest in France.

Incited by these two large successes, a lot of companies, often small independents, applied for permits of the Paris Basin. So, 10 companies were operating there in 1980, 20 in 1983 and about 50 at the end of 1987. Of course, the largest permits were held by Esso Rep, Total, BP, Triton, and Elf (Fig. 11 b).

Some small discoveries were made in Rhaetic sandstones and in upper Dogger reservoirs but the most interesting results were obtained on the Triassic reservoirs west of the Chaunoy field and south of Paris.

Also to be noted, the discovery in 1988 of a small oil accumulation at Ivry inside the Paris urban city limits, which has made after a very spectacular vibroseismic survey done at night in the city and suburban streets. It even took place along the Champs Elysées.

Since that time, and due to the latest OPEC crisis and to the fact that oil prices had fallen drastically, explora-

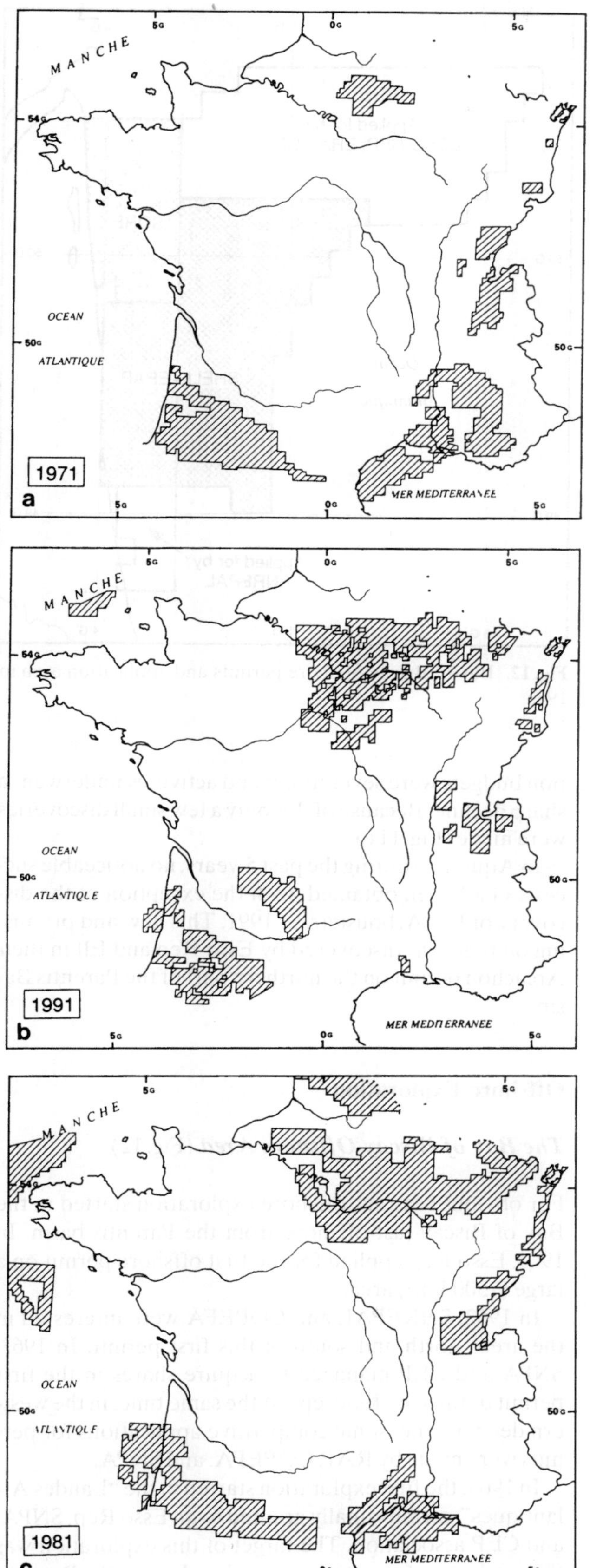

Fig. 11 a–c. Maps of the oil exploration permits granted by the French administration State of the permits in **a** 1971, **b** 1981, **c** 1991

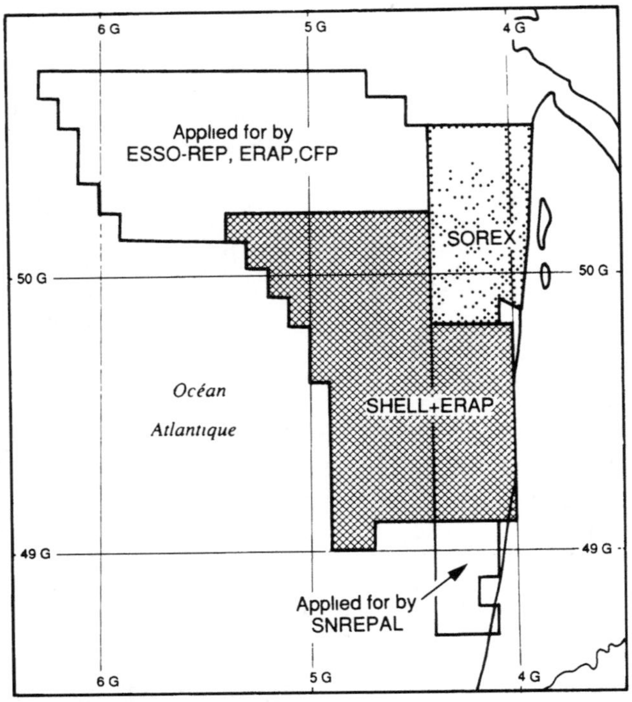

Fig. 12. Bay of Biscay offshore permits and application area in 1966

tion budgets were severely cut and activities underwent a sharp decline. Because of this only a few small discoveries were made (Fig. 11c).

In Aquitaine during the past 5 years, no noticeable successes had been obtained, with the exception of the discovery of Les Arbousiers, in 1991. This new and promising oil field was discovered by Esso Rep and Elf in their Arcachon permit on the northern part of the Parentis Basin

Offshore Exploration

The Bay of Biscay Offshore Area (Fig. 12)

For obvious reasons, offshore exploration started in the Bay of Biscay just offshore from the Parentis basin. In 1961, Esso Rep applied for the first offshore permit on a large (1500 km^2) area.

In 1962, SNREPAL and COPEFA were interested in the areas north and south of this first permit. In 1963, SNPA and CEP managed to acquire shares in the first permit defined by Essorep. At the same time, in the western deeper zone, some competitive applications for permits were made by RAP, COPEFA, and SNPA.

In 1964, the first exploration started in the "Landes Atlantiques" permits finally granted to an Esso Rep, SNPA, and CEP association. The target of this exploration was the productive formations proved onshore in the Parentis basin, i e., Albian and Purbeckian sandstones, Barremian carbonates, and upper Jurassic dolomites. A large seismic survey was shot throughout the permit. Some large structures were found and in 1966 the first well, Antares 1, was spudded 10 km in front of the Parentis oil field. The well went through low-permeability oil-impregnated Barremian limestones, which were thoroughly tested These tests proved that the amount of producible reserves was not sufficient to be worth being exploited. In fact, this Antares 1 well was the only one that found an oil accumulation in the Bay of Biscay Indeed, Essorep drilled ten other wells from 1961 to 1977 without making any discovery. Some time later, Shell in association with Elf drilled four dry wells from 1969 to 1975. Then, from 1977 to 1983, the Elf Aquitaine group alone, or associated with Esso Rep, drilled six other wells in the western extension of the Parentis basin and on the southern Tertiary basin. All these wells were dry holes.

This failure of offshore exploration of the western part of the Parentis basin was difficult to explain, and even now this unsuccessful exploration seems mainly to be related to the geodynamic history of the basin.

The Western Approaches and Mer d'Iroise Offshore Area (Fig. 13)

In this offshore area, the first reconnaissance work (an airmag survey and some experimental seismic surveys) had been done by BRP in 1964 and 1965.

Real exploration started in 1971 with some preliminary geophysical surveys done by a consortium of the main companies involved in oil exploration in France. This consortium was led by SNPA as operator

In May 1975, the Iroise, Armor, and Celtic Sea permits were granted to the consortium, and as soon as possible, three wells were drilled, whose results were very disapponting A source rock was present in the Lower Jurassic section, but no reservoirs were encountered After some improvements in seismic surveys and a new airmag survey, drilling was resumed in 1978, and through 1981 seven wells were drilled, without any success.

After the first relinquishment round, in 1980, a general geological and geophysical study was carried out This study showed that the area of interest was the western part of the permits. So the two eastern permits of Iroise and Armor were completely relinquished. The Celtic Sea and the newly granted Petite Sole permits were actively explored by a reduced association BP, Esso Rep, and Elf Aquitaine as operator.

Two hundred km from the coast of Brittany at more than 200 m of water depth, two wells were drilled on tilted blocks clearly defined by seismic. These wells encountered some good reservoirs in the lower Cretaceous and Callovian but they did not find any production.

These last disappointing results ended the Sea of Iroise exploration. An attempt to explore the French part of the English Channel by Elf Aquitaine and BP, which drilled one well, Nautile 1, was also unsuccessful

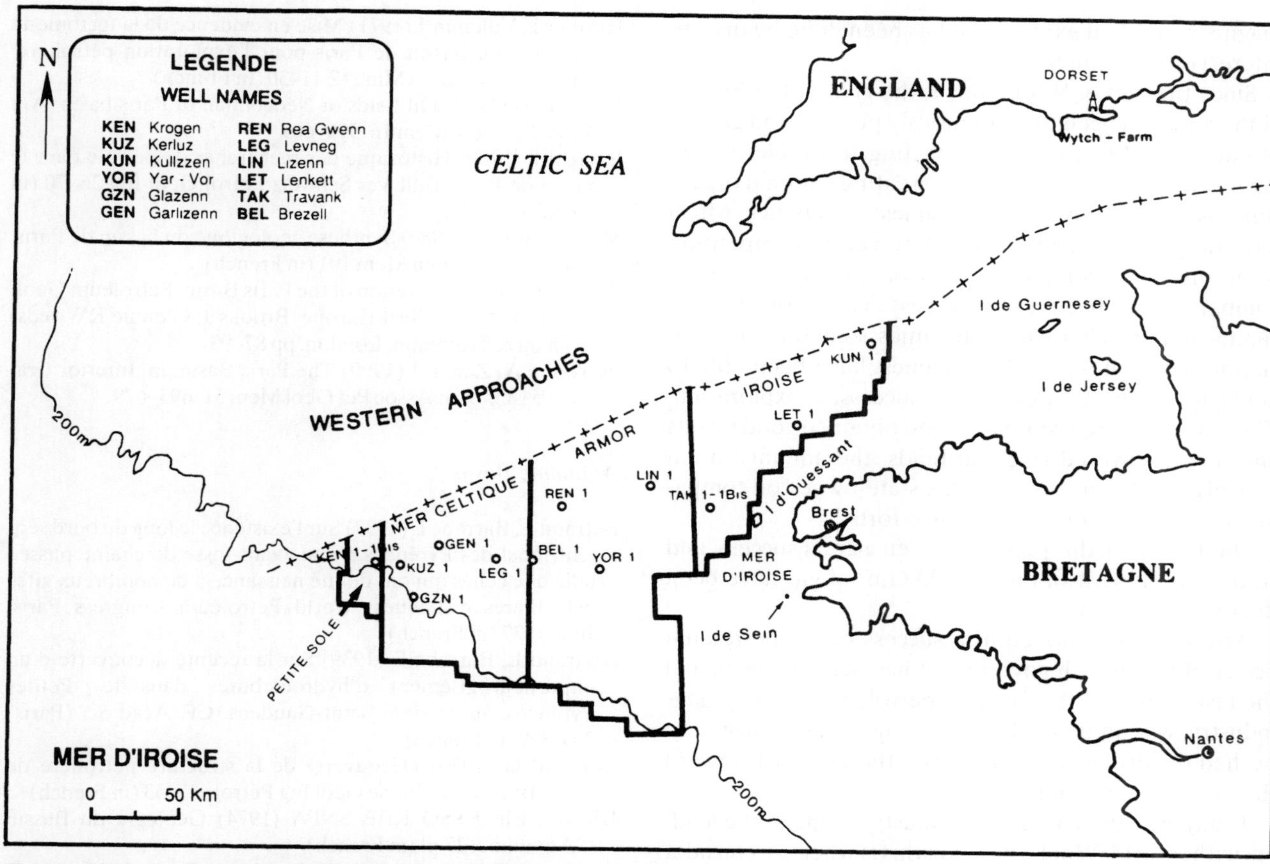

Fig. 13. Western Approaches and mer d'Iroise offshore permits in 1975

Mediterranean Offshore Area–Gulf of Lions (Fig. 14)

Oil and gas exploration on this offshore area was initiated in 1966 by Total, which drilled five wells (Istre, Autan, Mistral, Sirocco, Tramontane) and encountered a thick clayish section of Pliocene age lying directly on top of the metamorphic basement.

Some time later, in 1978 and 1984, two other wells, Cicindele and Grau du Roi, were drilled by Elf Aquitaine. The aim of these wells was to find an Amposta-like structure with a Mesozoic karstic reservoir below the Lower Miocene limestines. This attempt was a complete failure.

Meanwhile, Chevron had applied for permits in the relinquished Total permits. In 1979 and 1980, this company drilled two wells, Rascasse and Autan 2, which were as unsuccessful as the former ones.

In 1982, Total, Elf Aquitaine, and Esso Rep obtained two deep-water permits south of the Rhone delta. Two wells were drilled by the American drilling boat Discoverer Seven Seas. The first one, Golfe du Lion Profond 1, was drilled under 1714 m of water depth and reached 3607 m; the second, Golfe du Lion Profond 2, was drilled under 1246 m of water depth and reached a TD of 5291 m. As no encouraging results were encountered in these two wells, the association ceased its exploration and the permits were relinquished. The two last wells (Calmar and Agde Maritime), drilled in 1983 and 1985 by Union Texas and Elf Aquitaine, were dry holes.

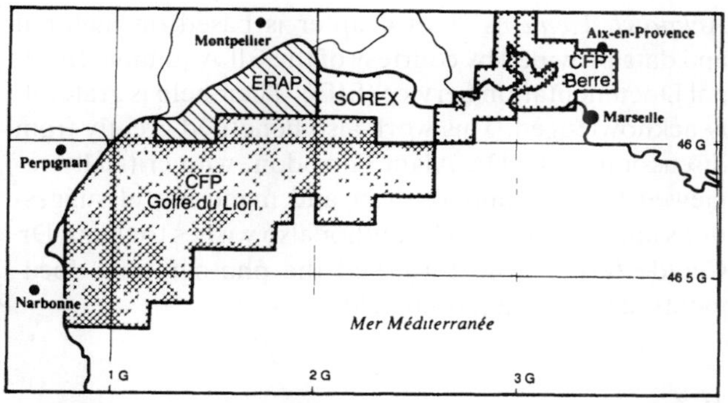

Fig. 14. Mediterranean Gulf of Lions offshore permits in 1986

On the whole, the exploration results in the Mediterranean offshore area were very disappointing, as were the results in the other French offshore basins, where 50 dry holes have been drilled.

Conclusions

Since French sedimentary basin are not very prolific and are not very rich in obvious oil shows, they did not appear worth being a target for oil prospecting for many years by most of the private oil companies. With the exception of some small-scale and short attempts in Aquitaine and Limagne, no real oil exploration has been done by private interest oil companies.

Since the end of World War I, the goal of the French authorities has been to continuously push oil and gas exploration of the country by enacting favorable mining laws, favorable tax rules, and by creating national organizations and state-owned companies This policy, which has always been flexible enough to accept competition from foreign oil companies and from private French oil companies, has been a real success. Thus, both foreign capital and French private companies, sometimes in association with state-owned companies, have been able to apply for permits to carry out successful exploration Then, when foreign companies had obtained good results and had discovered large oil fields, the administration played a subtle game to incite the state-owned oil companies to increase their exploration efforts

On the whole, this policy has been a great success, and since 1945, 100 Mt of oil and 325 Gm^3 of gas have been discovered in France.

Also to be mentioned as a success of the petroleum policy of the French government has been the birth and the fast growth of the offshore petroleum industry. This industry was initiated by public agencies, which also pushed the drilling companies and the shipyards toward this new and expanding market.

Today, the French offshore industry is one of the leaders in the world. What a paradox this is when we consider that there is no production at all in France's offshore area!

Acknowledgements This chapter is based on material and data provided by courtesy of the Elf Aquitaine Internal Documentation Service (SID). Their help is gratefully acknowledged. This work also benefited greatly from discussions with Dr. Alain Perrodon, who critically reviewed the preliminary draft and made useful suggestions and comments. The author also wishes to thank Dr Claude Sittler, who furnished the photographic documents of the Pechelbronn field.

References

Further Readings

B. R. P.	Annual reports	From 1946 to 1960
DICA	Annual reports – Tome II	From 1961 to 1967
Comité Professionnel du Pétrole	Annual reports	From 1968 to 1973
Service de Conservation des Gisements d'Hydrocarbures – Annual reports		From 1974 to 1991

Paris Basin

Espitalié J, Marquis F, Sage R (1987) Organic geochemistry of the Paris basin Petroleum geology of NW Europe Brooks J Glennie KW (eds) Graham & Trotmann, London, pp 71–86

Heritier F, Villemin J (1971) Mise en évidence de la tectonique profonde du bassin de Paris pour l'exploration pétrolière Bull Bur Rech Géol Min 1 2 11–30 (in French)

Heuillon B (1972) Oil fields of Neocomian of Paris basin Am Assoc Pet Geol Mem 16 599–609

Lavigne J (1959) Historique des recherches de pétrole dans le bassin de Paris Bull Ver Schweiz Petrol Geol Ing 26 70 (in French)

Megnien C (ed) (1980) Synthèse géologique du bassin de Paris Bur Rech Géol min Mem 101 (in French)

Pages L (1987) Exploration of the Paris Basin Petroleum Geology of the North West Europe Brooks J, Glennie KW (eds) Graham & Trotmann, London, pp 87–93

Perrodon A, Zabek J (1990) The Paris Basin, in Interior Cratonic basins Am Assoc Pet Geol Mem 51 693–679

Aquitaine Basin

Betrand L, Barrabe L (1937) Sur l'existence le long du bord septentrional des Pyrénées d'une avant-fosse de chaîne plissée telle que celles qui ont donné naissance à de nombreux gîtes pétrolifères classiques World Petroleum Congress Paris, June 1937 (in French)

Bertrand L, Barrabe L (1939) Sur la récente découverte d'un important gisement d'hydrocarbures dans les Petites Pyrénées, au nord de Saint-Gaudens CR Acad Sci (Paris) 209 399 (in French)

Bonnard G (1950) Découverte de la structure pétrolière de Lacq Bull Assoc Suisse Géol Ing Pétrole 17 53 (in French)

BRGM, Elf, ESSO REP, SNPA (1974) Géologie du Bassin d'Aquitaine, 27 pl (in French)

de Chevilly F, Colo G, Henry G, Nicolai A, Quoix J, Winnock E (1969) Le champ de gaz de Meillon-Rousse et la structure profonde de l'anticlinal de Pau – France Sud-Ouest Rev Inst Fr Pétrole 24, 5 564–578 (in French)

Curnelle R, Dubois P, Seguin JC (1982) The Mesozoic-Tertiary evolution of the Aquitaine basin Philos Trans R Soc Lond A 305 63–84

Michel P et al (1959) Quinze ans d'exploitation de gaz naturel à St-Marcet In Atti del convegno di Milano, 30 set–5 ott 1957 Su i Giacimenti Gassiferi dell'Europa Occidentale – Roma – Accademia nazionale dei Lincei (in French)

Schneegans D (1948) Gas-bearing structures of southern France Bull Am Assoc Pet Geol 32 2

Winnock E, Pontarlier Y (1968) Lacq gas field, France In Geology of giant petroleum fields Am Assoc Pet Geol Mem 14 370–387

South East Basin

Barrabe L, Schneegans D (1935) Rapport d'ensemble sur le gisement pétrolifère de Gabian Annales de l'Office National des combustibles liquides N° 4 – juillet–août 1935 et N° 5 – sept –octobre 1935 (in French)

Baudrimont AF, Dubois P (1977) Un bassin mésogéen du domaine péri-alpin le Sud-Est de la France Bull Cent Rech Expl Prod Elf Aquitaine 1, 1 261–308 (in French)

Debrand-Passard S, Courbouleix S, Lienhardt MJ (ed) (1984) Synthèse géologique du Sud-Est de la France Bur Rech Géol Min Mém 125, 126 (in French)

Dubois P, Delfaud J (1989) Le bassin du Sud-Est In Dynamique et méthodes d'étude des bassins sédimentaires Technip, Paris, pp 277–297 (in French)

Flandrin J, Weber C (1966) Geophysical data on the deep-seated structures of the Diois and the Baronnies region Bull Soc Géol Fr 8, 3 87–392

Mainguy M (1962) Vingt ans d'exploration pétrolière dans le bas Rhône et le Languedoc Bull Assoc Fr Tech Pétrol 155 681–692 (in French)
Valery P, Berthon JL, Montadert L (1984) The Mediterranea and petroleum exploration – The deep Rhone delta Petrol Technol 309 13–18

Limagne Basin

Morange A, Héritier F, Villemin J (1971) Contribution de l'exploration pétrolière à la connaissance structurale et sédimentaire de la Limagne dans le Massif Central In Jung J (ed) Symposium Géologie, géomorphologie et structure profonde du Massif Central français Plein Air Service, Clermont-Ferrand, pp 295–308 (in French)
Jung J (1938) Historique des recherches de pétrole en Limagne Rev Sci Nat Auvergne IV 129–135 (in French)
Jung J (1946) Géologie de l'Auvergne et de ses confins bourbonnais et limousins Mem Carte Géol de France SGF Paris 1–372 (in French)

Alsace Basin

Sittler C (1972) Le pétrole dans le département du Haut-Rhin Bilan d'un siècle et demi de recherches et d'exploitation Bull Sci Géol 25, 2–3 151–161, Strasbourg (in French)
Sittler C (1974) Sous-sol et ressources pétrolières en Basse Alsace Saison d'Alsace 52, 31–62, Strasbourg (in French)
Sittler C (1985) Les hydrocarbures d'Alsace dans le contexte historique et géodynamique du fossé rhénan Bull Cent Rech Expl Prod Elf Aquitaine 9, 2 335–371 (in French)

Additional Readings

Comité des Techniciens (1992) Monographies des principaux champs pétroliers de France Bull Cent Rech Expl Prod Elf Aquitaine Mem 14 (in French)

(Up-to-date information and data on the main oil and gas fields in France compiled by a team of senior geologists from all the major oil companies working in France)

Regulations of Hydrocarbon Exploration and Production in France

J. C. Hervet

Extended Abstract. In France, soil and subsoil belong to the state. This is a very important rule, enacted by the Napoleon Code. Thus, the government using discretionary power, but according to the Mining Code, and its application texts, delegates companies the right to explore and produce oil and gas. The purpose of the present section is to give an overview of the categories of permits granted, and to outline the main steps of procedure to be followed. First, a distinction should be drawn between exploration and development.

Exploration

The exploration permit gives its holder an exclusive right of temporary occupation of land. Any area which is not covered yet by such a permit may be subject to application.

How to Make an Application for an Exploration Permit

First, it should be noted that there is no requirement for the applicant to establish a French company. This obligation is required only if a concession is granted. Thus, companies may be foreign companies (EEC or non-EEC). The application contains as much information as possible about the technical and financial capacities of the applicant (articles of association, annual report, etc.) Various documents should also be included with the application:

- a map showing the requested area,
- a geological memorandum,
- a financial commitment referring to a work program
- a "notice d'impact", showing that the work program respects environmental regulations. This document has to be adapted to the specific area applied for, and should not be a standard form.

This application is then submitted to a public enquiry, which is carried out during a month's period in the concerned area. During that period, the application (including the map and the "notice d'impact" is brought to the attention of the public by a publication in the Journal Officiel and in the local press. Thus, anyone may submit comments or even declare opposition to the project. Competitive applications are acceptable up to 30 days after the end of the public enquiry.

Meanwhile, the concerned DRIRE ("Direction régionale de l'industrie, de la recherche et de l'environment" – Regional Department of Industry, Research, and Environment, a decentralized subdivision of the Ministry of Industry) consults the local administrative services in order to establish a list of various constraints that may affect exploration work (urbanization projects, archeological or historical sites, military areas, etc.). After this phase of consultation, the local procedure ends with the engineers' report of the concerned DRIRE and the advice-note of the "Prefet" (representative of the Government in each Département).

At the same time, a procedure of arbitration between different applicants is carried out at a national level by the Hydrocarbons Department (DHYCA). This department, under the control of the General Department for Energy and Raw Materials, is responsible for the elaboration and bringing into operation of the governmental policies for the supply of hydrocarbons in France and the development of the oil industry. Then, the DHYCA asks each company (sometimes, more than ten companies are involved in the same area) to make a presentation of their projects (work program, geological approach, financial aspects, etc.). These presentation enable the DHYCA to appreciate the seriousness of the applicants.

At this stage, it is possible that, due to the great number of applicants, the DHYCA informs them that the concerned area will be divided into several blocks and that permits will be granted to coholders. Of course, the DHYCA takes into account the desires of the applicants to be associated with other companies. Thus, if granting of permits is a discretionary power, this does not mean arbitrary.

Besides, selected companies are asked to sign a letter formalizing their agreement about the project of attribution (list of coholders, area, duration, financial commitment). Then the application is submitted to the General Council of Mines, "Conseil Général des Mines", a consultative authority, competent in matters such as mines, quarries, energy, mineral substances, etc.), and finally, to the Council of State ("Conseil d'Etat", the higher French administrative Court).

The exploration permit is granted by a "décret" (ordinance), signed by the Prime Minister and the Minister of Industry, and published in the Journal Officiel. This exploration permit is awarded for 5 years or less, provided that a financial commitment referring to a work program has been subscribed by companies. The duration of the whole procedure can be estimated at between 12 and 18 - months. A permit is granted to a holder, or, usually, several coholders. Coholders are the sole entities responsible towards the Administration for the fulfilment of the different obligations attached to the permit However, coholders may sign a joint-operating agreement with other companies under the terms of which those companies purchase a percentage interest in the permit without being coholders.

In a close future (may be as soon as 1994), some of these procedures should be significantly simplified. In particular, and in order to shorten the delivery delays, exploration permits will be granted by a single ministerial "arrêté" without public enquiry. At the same time, the technical and financial criterions for the appraising of exploration permit applications will be official defined.

These operations (called farms-in or farms-out) have to be subjected to the Administration, which disposes of a 2-month period (to oppose this project of agreement) should the occasion arise.

Renewal of an Exploration Permit

The first validity term may be renewed twice, for a maximum period of 5 years each time. However, the permit area is reduced to half for the first renewal and to a quarter of the remaining area for the second renewal. It is up to the companies to choose these remaining areas. In any event, the above-mentioned reductions cannot reduce the surface of a permit to less than 175 km^2.

The application for a renewal is submitted to the local administrative authorities, without any new public enquiry, and then to the General Council of Mines, before being awarded by decree.

The renewal of a permit is automatic for a period at least equal to 3 years or to the previous period if that period was less than 3 years, as long as the holder has met all his obligations and agreed, in his application for renewal, to financial undertakings at least equal to the financial commitments made during the previous period.

Extension of an Exploration Permit

The perimeter(s) of a permit may be extended to additional areas. The procedure is the same as for initial granting of the permit.

Change in Licensing "Mutation"

Costs of exploration works are high and, even where a single company is the holder of a permit, frequently several other companies will share the costs of operations (seismic surveys, drillings, etc.), in exchange for a percentage interest. The companies are not coholders, but only partners. In the case of discovery, only coholders have an exclusive right to obtain a development permit. Moreover, a partnership has to be submitted to the agreement of the Administration.

When a company has entered into a permit initially as a partner and decides later to become coholder, the procedure is called "mutation", and this operation is awarded by decree (no public enquiry, no local administrative consultation except for a report of the DRIRE), after consultation with the General Council of Mines.

Development

The Mining Code stipulates that hydrocarbon mines can only be developped when a concession or a "permis d'exploitation" has been granted.

According to the Mining Code, during the period of validity of an exploration permit, the holder has an exclusive right to obtain a hydrocarbon exploitation permit or a concession.

Hydrocarbon Exploitation Permit

This exploitation permit is awarded by decision of the Minister of Industry, after a public enquiry, a local administrative consultation, and a submission to the General Council of Mines.

Such an exploitation permit is granted for 5 years or less, and may be renewed twice for 5 years or less each time, without modification of area. However, the exploitation permit licensee must apply for a concession when a field cumulative production has reached 300000 tons of oil or 300 millions m^3 or gas.

This Hydrocarbon Exploitation Permit could disappear in a close future and only concession lincesees will be delivered.

Concession

A concession licensee must be a corporation, established under the law of a Member State of the EEC. Such a development permit is granted by decree, after a public enquiry, a local administrative consultation, and a submission to the General Council of Mines, then to the Council of State. A concession-granting decree encloses conditions of the contract that must be signed by the holder(s) on the terms and conditions specified ("cahier des charges"). The first validity term cannot exceed 50 years and may be renewed several times for 25 years or less each time.

Mining Royalties

Progressive Mining Royalties

Claimholders of liquid or gaseous hydrocarbon mines and of production licences must pay the Government a royalty proportional to the value of the products extracted. These royalty rates are as follows:

Oil: Annual production levels:

– < 50000 t	0
– 50000 to 100000 t	6%
– 100000 to 300000 t	9%
– 300000 t and more	12%

Gas: Annual production levels:

– < 300 million m³	–
– > 300 million m³	5%

Departmental and Communal Mining Taxes

These are proportional taxes that are determined by multiplying the amounts produced by a unit rate. This rate is revised each year by regulatory decision.

These two taxes will not have to be paid for offshore oil or gas produced from fields developped in 1994 and later on.

The rates of these taxes for 1993 are as follows: (in FF/t):

Communal taxes

– Crude oil	11.60
– Natural gas (for 1000 m³ extracted)	3.40
– Butane–Propane	22.90
– Gasoline from gasoline extraction	20.80

Departmental taxes

– Crude oil	15
– Natural gas (for 1000 m³ extracted)	4.3
– Butane–Propane	17.70
– Gasoline from gasoline extraction	15.60

All these mining taxes are deductible when determining the profit subjected to corporation taxes.

Depletion Allowance

The purpose of the depletion allowance is to encourage liquid or gaseous hydrocarbon mining companies to reinvest in exploration all or part of the profits they make from the development of discovered fields. Likewise, and so as to continue the development of marginal economic fields, the depletion allowance may also be reused for expenditures aiming to improve the recovery of oil and gas from fields. Oil companies are thus authorized to deduce from their net operating profits, to the limit of 50% of these profits, a depletion allowance equal to 23.5% of the amount from sales of marketable products extracted from the fields they exploit. This allowance must necessarily be reused within the space of 1 year.

Underground Gas Storage Facilities in France

E. Blondin

Abstract. Natural gas consumption varies greatly according to the day and the season. The distribution system alone, from producer to consumer, cannot provide all the flexibility required to handle such fluctuations, due to constraints arising from natural gas purchase agreements and the technical possibilities of processing, liquefaction, and transmission facilities.

As early as 1954, Gaz de France developed underground storage facilities for natural gas in salt cavities and aquifers. These facilities have proven to be the only effective way to store the large volumes of gas required. For nearly 40 years, these facilities have fully lived up to their potential with confirmed high performance. They have also demonstrated that they are environmentally safe and reliable (Table 1).

Two examples are discussed: the first one at Cere-la-Ronde is sited in an early Lias aquifer reservoir in the Paris basin; the second one at Etrez (Bresse basin) is a cavities storage in a salt layer of Oligocene age.

Introduction

Unlike all other energy sources, natural gas, distributed through a network of pipes, cannot be stored by consumers. Consequently, the distributor has the responsibility of adjusting supply to meet demand. Gaz de France continuously ensures a balance between resources and consumer needs. In addition, Gaz de France must be prepared for the ever-growing share of natural gas in France's energy balance. It now accounts for 12.3 %, and is likely to increase, in that the average for the Western European nations is 18 %.

Production and transmission equipment must, for economic reasons, be utilized at close to optimum capacity. Supply sources, however, tend to be located further and further away, making it more difficult to adapt deliveries to seasonal demand and increasing the cost of operations.

Purchase and transport agreements reflect the seller's desire to recover the large investments needed to bring a gas field into production. These agreements contain obligations that place considerable burdens on the buyer.

Table 1. Underground gas storage facilities operated by Gaz de France (status as of November 1, 1992)

Storage	Type	Year facility commissioned	Depth of reservoir top (m)	Total volume stored (million m³(n))	Effective energy TWh (billion kWh)
Beynes Supérieur (Yvelines)	Aquifer	1956	405	445	2
Saint-Illiers (Yvelines)	Aquifer	1965	470	1,294	6 8
Chémery (Loir-et-Cher)	Aquifer	1968	1,120	6,759	35 7
Tersanne (Drôme)	Salt cavity	1970	1,400	401	2 9
Cerville-Velaine (M -et-Moselle)	Aquifer	1970	470	1,450	7.2
Beynes Profond (Yvelines)	Aquifer	1975	740	776	4 3
Gournay-sur-Aronde (Oise)	Aquifer	1976	750	2,211	10 2
Etrez (Ain)	Salt cavity	1979	1,400	648	4 3
Saint-Clair-sur-Epte (Val d'Oise)	Aquifer	1979	742	688	3 3
Soings-en-Sologne (Loir-et-Cher)	Aquifer	1981	1,135	746	2 2
Germigny-sous-Coulombs (S -et-Marne)	Aquifer	1982	892	2,185	6 2

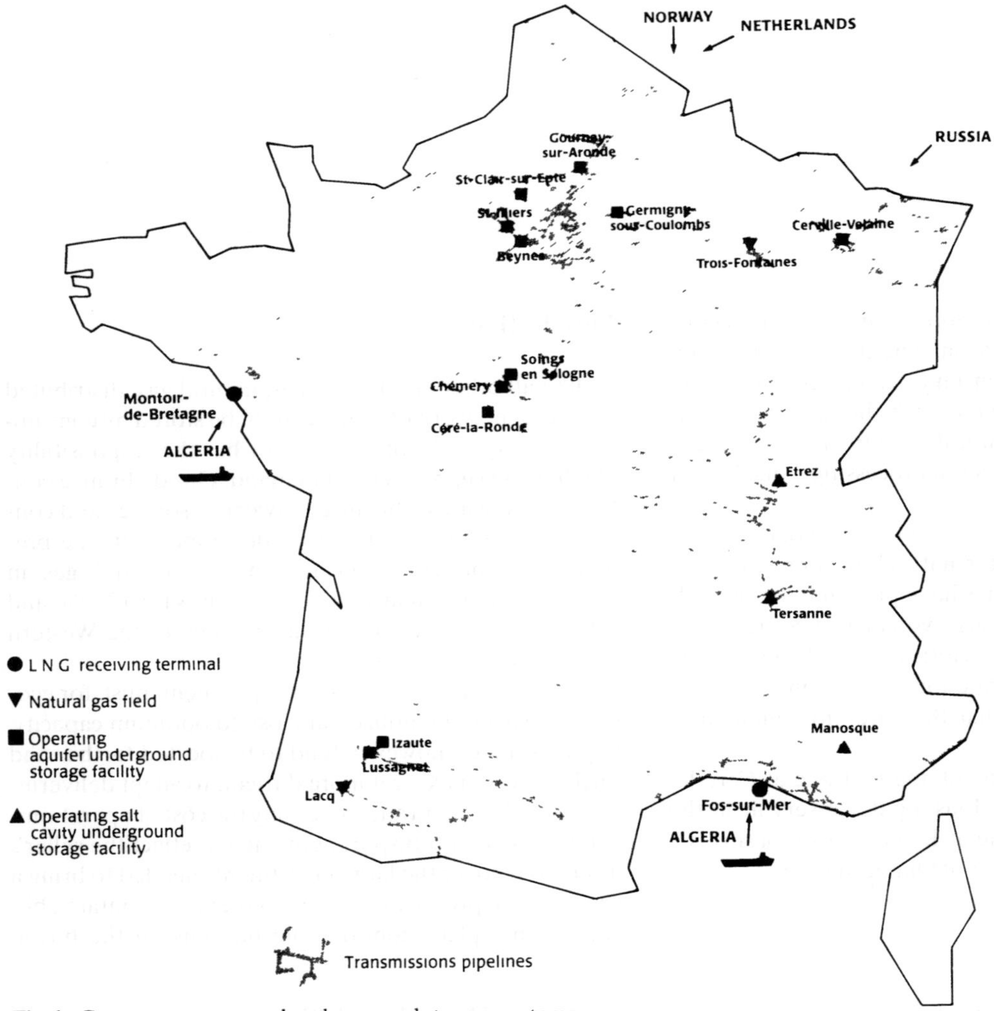

Fig. 1. Gas transmission and underground storage systems

Among these, the "take or pay" clause guarantees the seller a minimum payment even if the gas is not taken by the client. Gaz de France is therefore required to recover any gas at a later date that could not be taken at the time specified in the contract.

Gaz de France disposes of two major methods for matching supply to demand.

The first is interruption, whereby certain large industrial consumers, able to count on other energy sources, agree to interrupt their use of gas voluntarily. This procedure is used to compensate for a failure of one or more supply sources.

The second is underground storage (Figs. 1 and 2).

Statistical data highlight the correlations existing between the gas consumption recorded for a given period and the outdoor temperatures for the same period.

In France, heating accounts for most of the natural gas consumed every year and 40% of total consumption is a direct function of climatic conditions. In January, gas consumption is roughly seven times higher than in August. Such fluctuations in demand cannot be readily met in sofar as natural gas is supplied at a constant rate all year long.

Underground storage helps balance out the fluctuating demand for gas around the year. In addition, storage facilities satisfy unforeseen demands in unusually cold winters. During such winters, which may only occur twice in a century, Gaz de France would need to withdraw sufficient gas from storage to cover a 15% increase over an average year.

Moreover, underground storage has other uses. While gas withdrawal operates primarily in winter, storage also serves as a backup supply to satisfy daily peaks in demand. Daily adjustments, partially covered by network "breathing", in which pressure changes serve to buffer flow-rate variations, are also facilitated by having a standby storage system ready to come on line if necessary.

Underground storage also helps provide a safety net for supply sources. The combination of large consumer interruption and underground storage limits the consequences of a supply failure or technical problems within the system.

On February 7, 1991, the coldest day of the winter 1990-1991, temperatures reached −7.7 °C in Paris. Gas consumption rose to 2.42 billion kWh, about 210 million

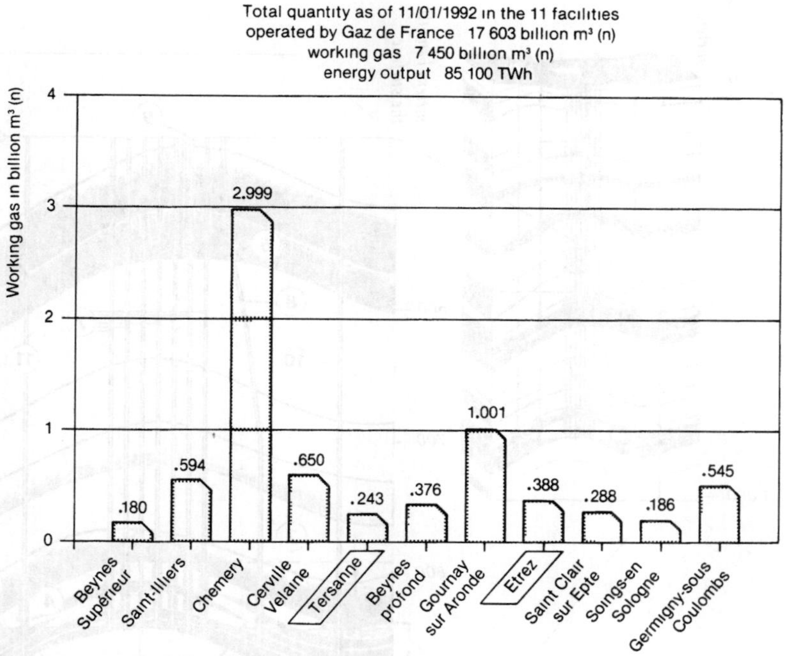

Fig. 2. Status of the underground storage facilities at the outset of the 1992/1993 winter season

m³ of natural gas. Underground storage facilities provided 57.5 % of the gas consumed (Fig. 3).

This means that the quantity of gas consumed that day was equivalent to 100000 megawatts, compared to the 70000 megawatts generated by Electricité de France on the same day. Indeed, the maximum available daily send-out – a function of pressure in the facility and the number of operating output wells – would have enabled Gaz de France to handle even colder weather.

Some 40 years of experience in developing and operating natural gas underground storage units has given Gaz de France considerable expertise in the many different techniques used in the field.

Much of this experience has been acquired through what we have learned from our 11 sites currently in operation. These sites are located in both aquifers and salt cavities, at depths ranging from 400 to 1500 m, making it possible to store gas at the appropriate pressure.

Near the Lacq natural gas field (Pyrénées-Atlantiques), Elf Aquitaine Production runs two aquifer storage facilities, one to serve its own needs (Lussagnet), and the other operated for Gaz de France (Izaute).

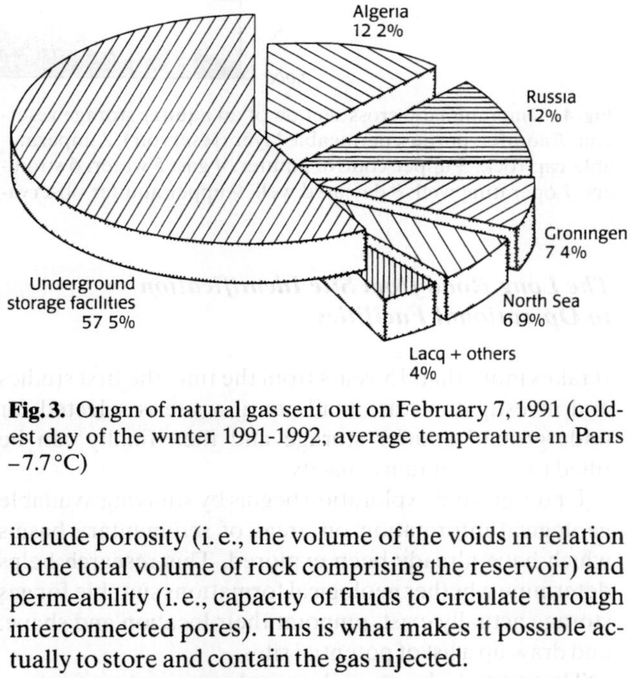

Fig. 3. Origin of natural gas sent out on February 7, 1991 (coldest day of the winter 1991-1992, average temperature in Paris −7.7 °C)

Aquifer Storage Sites

Geologically speaking, natural gas is stored in an aquifer in the same way as a natural gas field. The only difference is that the reservoir rock, which originally contained water (without liquid or gaseous hydrocarbons) is filled by injecting natural gas (methane). This gas partially takes the place of the water.

If an aquifer is to be used for underground storage, it must offer the same characteristics as a gas field. These include porosity (i. e., the volume of the voids in relation to the total volume of rock comprising the reservoir) and permeability (i. e., capacity of fluids to circulate through interconnected pores). This is what makes it possible actually to store and contain the gas injected.

In addition, such aquifers must be capped with an impermeable cap rock. The reservoir should also have an appropriate dome or closed anticline shape. This creates a trap that effectively contains the gas and prevents it from escaping either laterally or vertically.

This type of structure exists in sedimentary basins, in the Paris and Aquitaine basins in France.

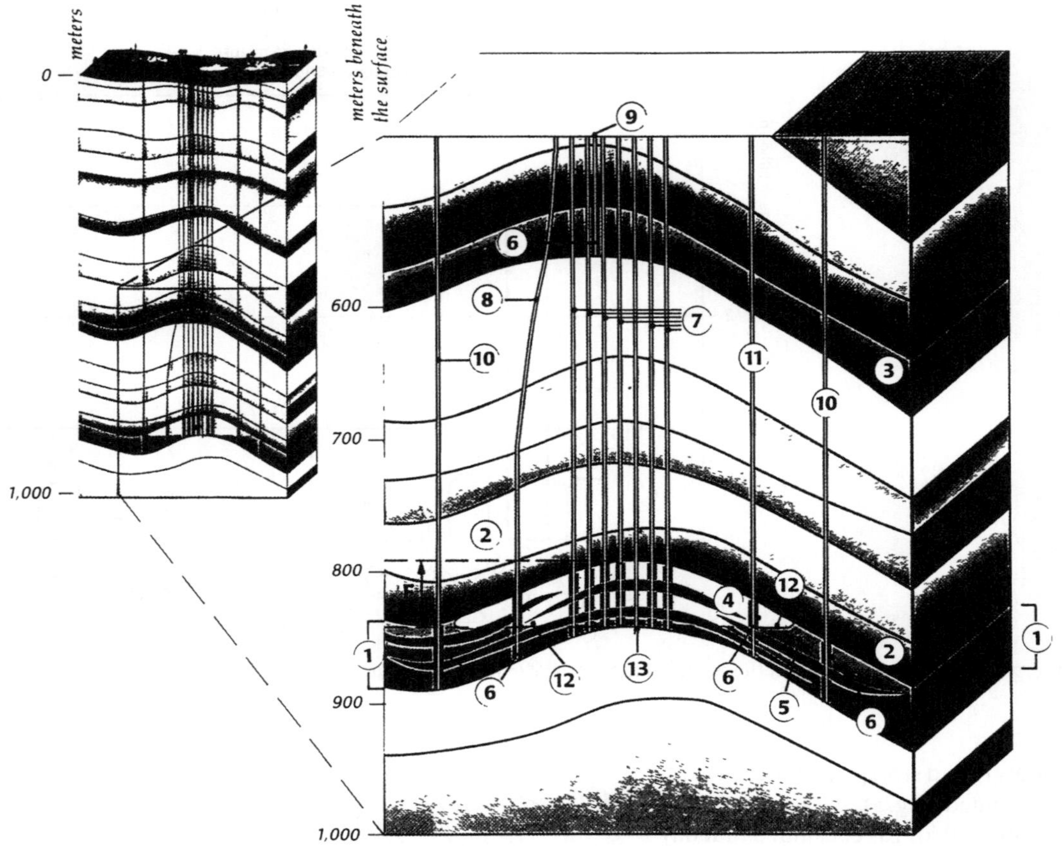

Fig. 4. Diagrammatic cross section of an aquifer storage reservoir *1* aquifer, porous, permeable layer (reservoir), *2* impermeable cap rock, *3* upper control aquifer, *4* gas, *5* water, *6* strainers, *7* operating well, *8* deviated well, *9* upper aquifer observation well, *10* peripheral observation well, *11* water level monitoring well, *12* water level gas/water interface, *13* neutron logging well, *F* closure

The Long Road from Site Identification to Operational Facilities

It takes more than 15 years from the time the first studies and investigations are made until a site is selected for underground aquifer storage and then finally can be filled to its maximum capacity.

Underground exploration begins by studying available geological information on areas of sedimentary basins which have already been explored. This research helps determine whether geological formations suitable for gas storage actually exist, pinpoint their location and shape, and draw up a list of potential sites.

The second phase involves making geophysical measurements on the sites which seem to have good potential. This provides more detailed information about the structure of the underground layers.

An underground storage facility requires two types of wells: operating wells are used both to inject and to withdraw gas, and are connected to a central station by buried pipelines; observation wells are used to perform the measurements required to monitor the operation of the storage site (Figs. 4 and 5).

The storage site is interconnected with the Gaz de France national transmission network via the central station, which includes the gas compression and processing installations. Gas withdrawn from storage contains water and, in some cases, traces of sulfur compounds, and consequently, it must be treated. At this stage, its water content and purity are adjusted to meet transmission and distribution requirements.

The compressors used to inject gas into the storage site or to move it through the transmission network are usually driven by gas engines or turbines.

Precise Operations Control

To fill an aquifer storage facility to its maximum capacity, the gas must be injected relatively slowly. In this way, the full potential of the reservoir rock is used.

This is why it is important to monitor the progress of the filling operation. Observation wells are used for this purpose, to take measurements that detect the movement of the gas and the water it displaces in the pores of the reservoir For this task, Gaz de France owns several mobile logging units to periodically and systematically check operation at the observation wells with its own teams

Information is collected daily throughout the well's operation, and this provides a clear picture of all the storage

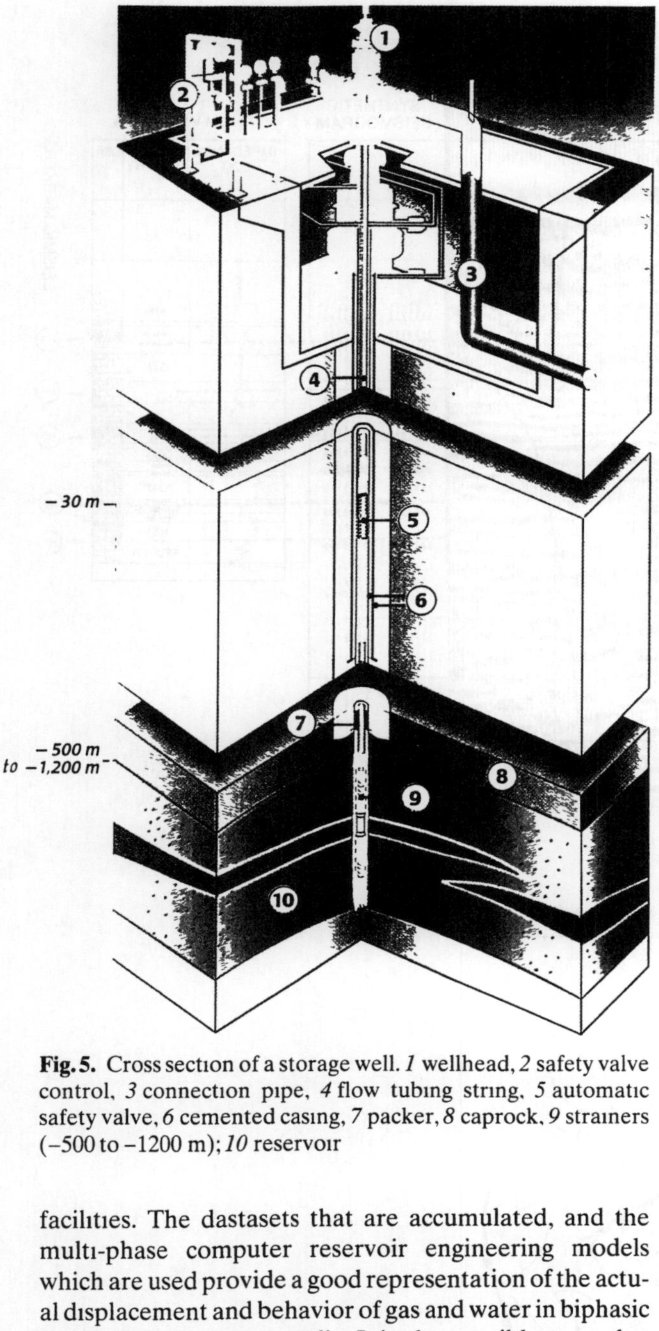

Fig. 5. Cross section of a storage well. *1* wellhead, *2* safety valve control, *3* connection pipe, *4* flow tubing string, *5* automatic safety valve, *6* cemented casing, *7* packer, *8* caprock, *9* strainers (-500 to -1200 m); *10* reservoir

facilities. The dastasets that are accumulated, and the multi-phase computer reservoir engineering models which are used provide a good representation of the actual displacement and behavior of gas and water in biphasic flows through porous media. It is also possible to predict reservoir behavior, especially pressure changes. By comparing the results of simulations of various operating strategies, we can adopt the optimal operating method for each facility.

Innovative Technology: Using Inert Gas as "Cushion Gas"

In an underground aquifer storage facility, a significant part of the total gas volume stored cannot be recovered during regular operation. This cushion gas is trapped and, unlike the working gas, does not contribute to making the reservoir breathe. It is nevertheless kept in the reservoir for technical reasons.

The cushion gas represents about half the total storage, making it the largest investment involved in developing a storage site. This explains the attraction of replacing at least part of the natural gas that forms this cushion with another, less expensive gas that would be chemically inert with regard to the facility installations and the geological environment. According to studies and experiments done by Gaz de France, at least 20% of the cushion gas can be replaced in this way.

Gaz de France uses a PSA (Pressure Swing Adsorption)-type inert gas generator and injector. Nitrogen is produced continuously, using two adsorbers in which the pores of the activated carbon are arranged so as to capture oxygen molecules, which are slightly larger than nitrogen molecules. Because the activated carbon becomes gradually saturated with oxygen, if must be periodically regenerated to maintain its full capacity. This is why two adsorbers are used, operating alternately.

The generator, which runs on electrical energy, is composed of two elements. One, an air-compressor set, supplies the adsorbers, in which the air's constituents are separated. These adsorbers alternate between the production and regeneration modes by means of a set of valves. Another set compresses the nitrogen generated so that it can be injected into the storage facility at appropriate pressure.

This unit is the largest of its kind in the world. With it, Gaz de France will benefit from favorable conditions in which to pursue its program of partially replacing cushion gas with inert gas.

Cere-la-Ronde Example

The structure of the Cere-la-Ronde aquifer in the southwest part of the Paris basin (Fig. 1) was first explored in 1964 by the oil industry and more recently evaluated by detailed seismic surveys and exploratory drilling (including core drilling) undertaken by Gaz de France. The interpretation of the seismic sections was constrained with the well data, as shown in Fig. 6. An isobath map (Fig. 7) drawn from all the data collected from the exploration works shows the extension of the structure at the level of the Early Lias shaly-sandstone and sand reservoir.

The leading features of this anticlinal structure, where the gas reservoir lies 900 m below ground level, and is about 30 m thick, and is overlain by a cap rock whose imperviousness has been tested, are favorable for use as a new storage facility. Its commissioning was carried out in Octobre 1993.

Particular attention was given to the structural closure to the western edge, and the effect of faults detected by the seismic survey and drillings.

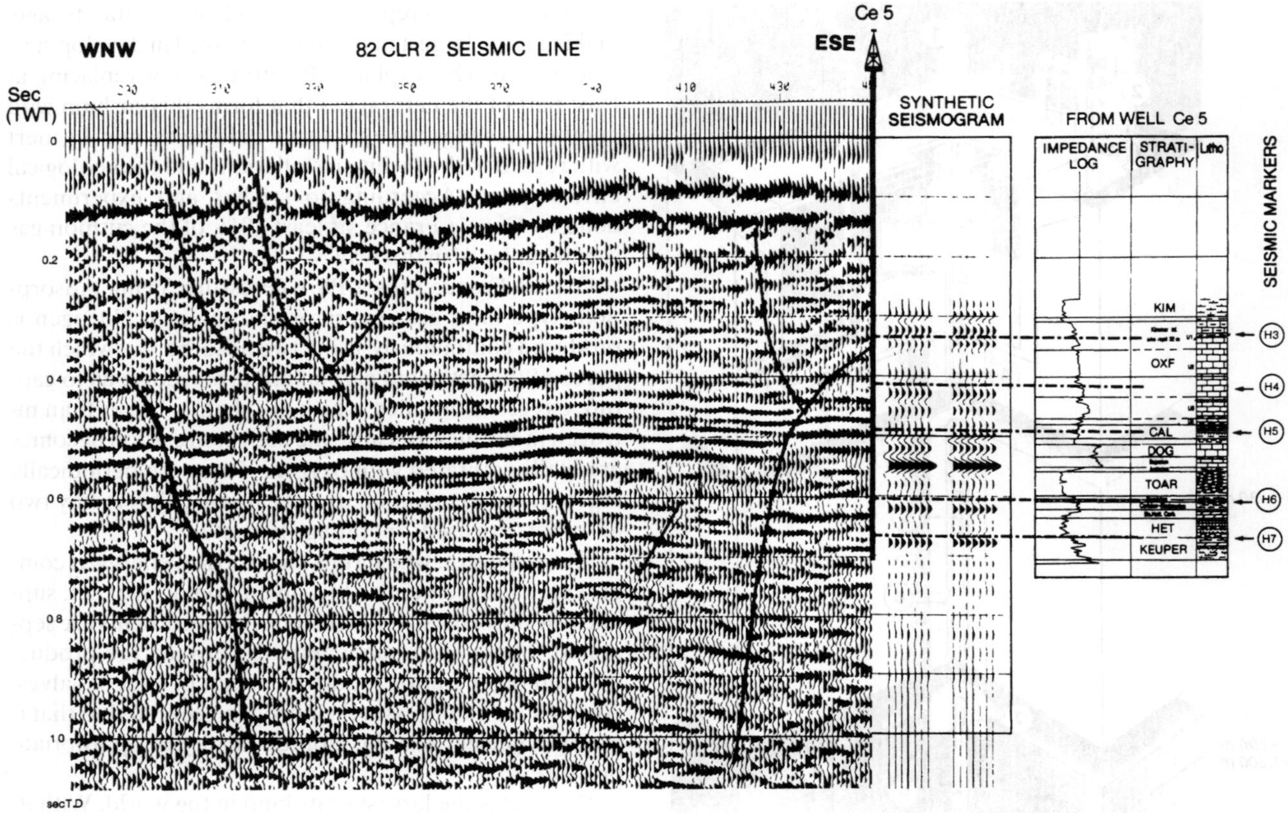

Fig. 6. Céré-la-Ronde correlation seismic well data

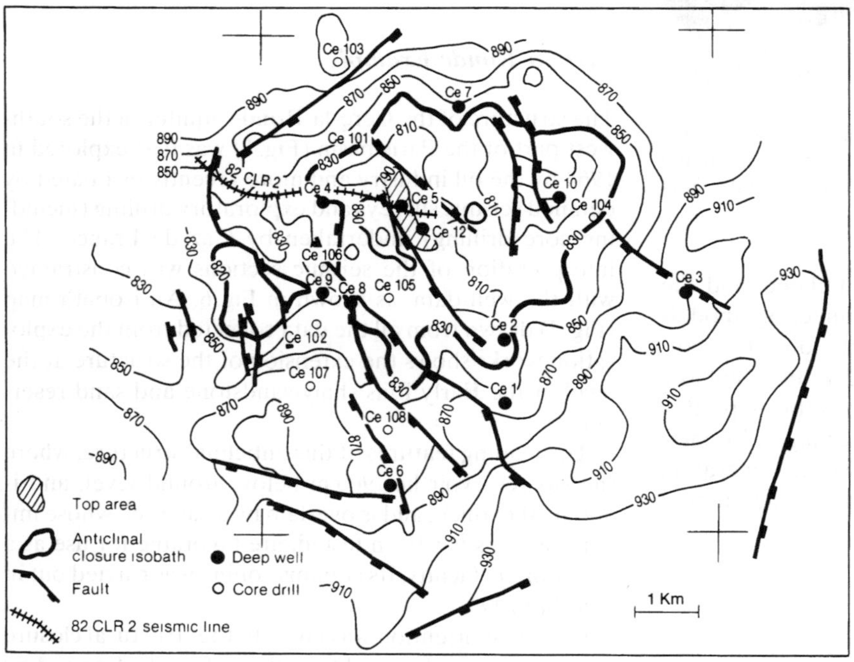

Fig. 7. Céré-la-Ronde contour map of the top of the reservoir, isobaths in m below sea level

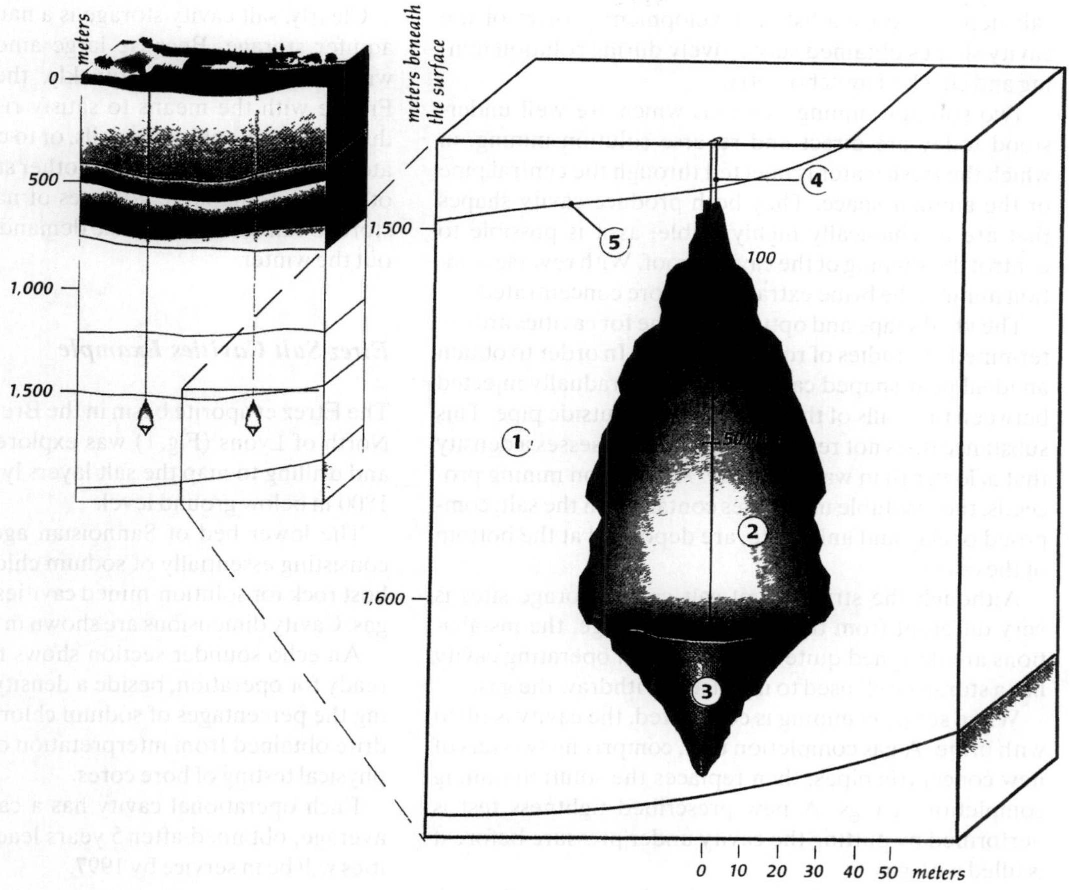

Fig. 8. Diagrammatic cross section of a salt cavity *1* rock salt layer, *2* open cavity, *3* insoluble residues, *4* cemented casing, *5* rock salt roof

Salt Cavity Storage

Unlike storage in aquifers, where the injected gas fills the pores of the rocks, salt cavity storage requires creating actual cavities, the unit volume of which ranges from 100 000 to 500 000 m^3. These are located more than 1000 m underground. Natural gas is stored there under pressure in its gaseous state (Fig. 8).

The cavities are created by solution mining. This process consists in dissolving the layers of sedimentary rock salt (sodium chloride) lying deep underground by injecting pressurized fresh water into the well, which is fitted with two concentric pipes. The salt then rises to the surface as brine (Fig. 9). The resulting brine is sold to an industrial user. This method can extract several hundred thousand tons of salt per year.

To dig 1 m^3 of cavity space, some 10 m^3 of fresh water must be injected. It can take several years to make the five to eight successive solution mining steps required to create a cavity with an available volume of 300 000 m^3. After each step, the concentric pipes are raised one level.

Simulation models have been developed, using the data collected on-site during exploratory drilling in the

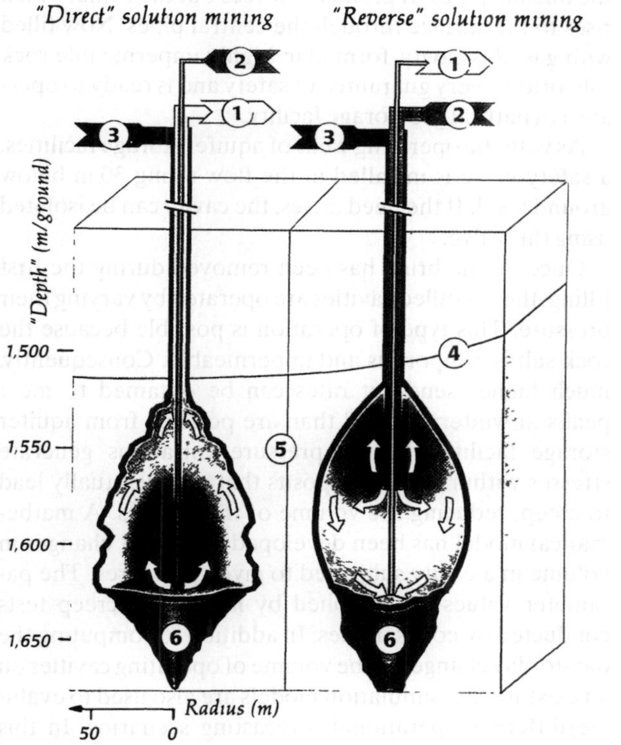

Fig. 9. Solution mining principle *1* brine extraction; *2* fresh water injection, *3* fuel oil injection, *4* salt caprock, *5* rock salt; *6* insoluble residue

salt deposit, to establish a development project of the cavity shapes obtained successively during solution mining and checked by echometry.

Two solution mining methods which are well understood today are direct and reverse solution mining, in which the fresh water is injected through the central pipe or the annular space. They both produce cavity shapes that are mechanically highly stable, as it is possible to control the shaping of the cavity's roof. With reverse solution mining, the brine extracted is more concentrated.

The ideal shape and optimal volume for cavities are determined by studies of rock mechanics. In order to obtain an ideal pear-shaped cavity, fuel oil is gradually injected between the walls of the cavity and the outside pipe. This substance does not react with salt and possesses a density that is lower than water density. As solution mining proceeds, the insoluble impurities contained in the salt, composed of clay and anhydrite, are deposited at the bottom of the cavity.

Although the structure of salt cavity storage sites is very different from that of aquifer storage, the installations are designed quite similarly. Each operating cavity has a storage well used to inject and withdraw the gas.

When solution mining is completed, the cavity is filled with brine. A gas completion unit, comprising two sets of new concentric pipes, then replaces the solution mining completion strings. A new prescribed tightness test is performed by putting the cavity under pressure before it is filled with gas.

The natural gas is then injected under pressure through the outside pipes. It gradually forces out the brine, which rises to the surface through the central pipes. Now filled with gas, this cavity, formed in highly impermeable rock salt, offers every guarantee of safety and is ready to operate as a natural gas storage facility.

As with the operating wells of aquifer storage facilities, a safety valve is installed in the flow string 30 m below ground level. If the need arises, the cavity can be isolated using this valve.

Once all the brine has been removed during the first filling, the gas-filled cavities are operated by varying their pressure. This type of operation is possible because the rock salt is nonporous and impermeable. Consequently, much higher send-out rates can be obtained to meet peaks in winter demand than are possible from aquifer storage facilities. These pressure variations generate stresses within the salt deposits that can eventually lead to creep, reducing the volume of the cavities. A mathematical model has been developed to predict changes in volume in a cavity subjected to given pressures. The parameter values are supplied by laboratory creep tests conducted on core samples. In addition to computing the day-to-day changes in the volume of operating cavities on an existing site, simulation models are also used to evaluate different operational forecasting scenarios. In this way, guidelines can be identified for preserving the capacity of the cavity, and an optimal usage strategy for the population of salt cavities can be established.

Clearly, salt cavity storage is a natural complement to aquifer storage. Because large amounts of gas can be withdrawn from them quickly, they provide Gaz de France with the means to satisfy rises in daily demand during exceptional cold spells, or to compensate immediately for the sudden loss of another supply source. On the other hand, the large volumes of natural gas in aquifer storage can satisfy the basic demand that exists throughout the winter.

Etrez Salt Cavities Example

The Etrez evaporite basin in the Bresse region of France, North of Lyons (Fig. 1) was explored by seismic survey and drilling to map the salt layers lying between 700 and 1800 m below ground level.

The lower bed of Sannoisian age (Early Oligocene) consisting essentially of sodium chloride was selected as host rock for solution-mined cavities to store pressurized gas. Cavity dimensions are shown in Fig. 10.

An echo sounder section shows the shape of a cavity ready for operation, beside a density log and scale showing the percentages of sodium chloride, shale, and anhydrite obtained from interpretation of the logs and petrophysical testing of bore cores.

Each operational cavity has a capacity of 300000 m^3 average, obtained after 5 years leaching. Nearly 20 cavities will be in service by 1997.

Safety and the Environment

The underground natural gas storage facilities operated by Gaz de France are noted for their reliability and discretion.

Underground storage of natural gas is particulary safe for a number of reasons. The first is the depth at which the facilities are located – between 400 and 1500 m underground. Secondly, underground storage units do not interfere with human life. The third reason is that only a very small percentage (1 %) of the total area of the storage facility is located above ground. Lastly, stored natural gas is harmless and lighter than air.

Gaz de France has nonetheless taken numerous precautions in order to ensure virtually total safety and security of its installations.

Surface equipment is in strict compliance with the general regulations governing such installations. The most important safety factor for Gaz de France, however, is to be able to control a possible gas blow-out under any circumstances.

Automatically closing safety valves, located 30 m underground, are placed within any well that may come into contact with gas.

Hundreds of underground storage facilities are in operation throughout the world. The fact that accidents are very rare proves, if proof is needed, that underground storage is reliable.

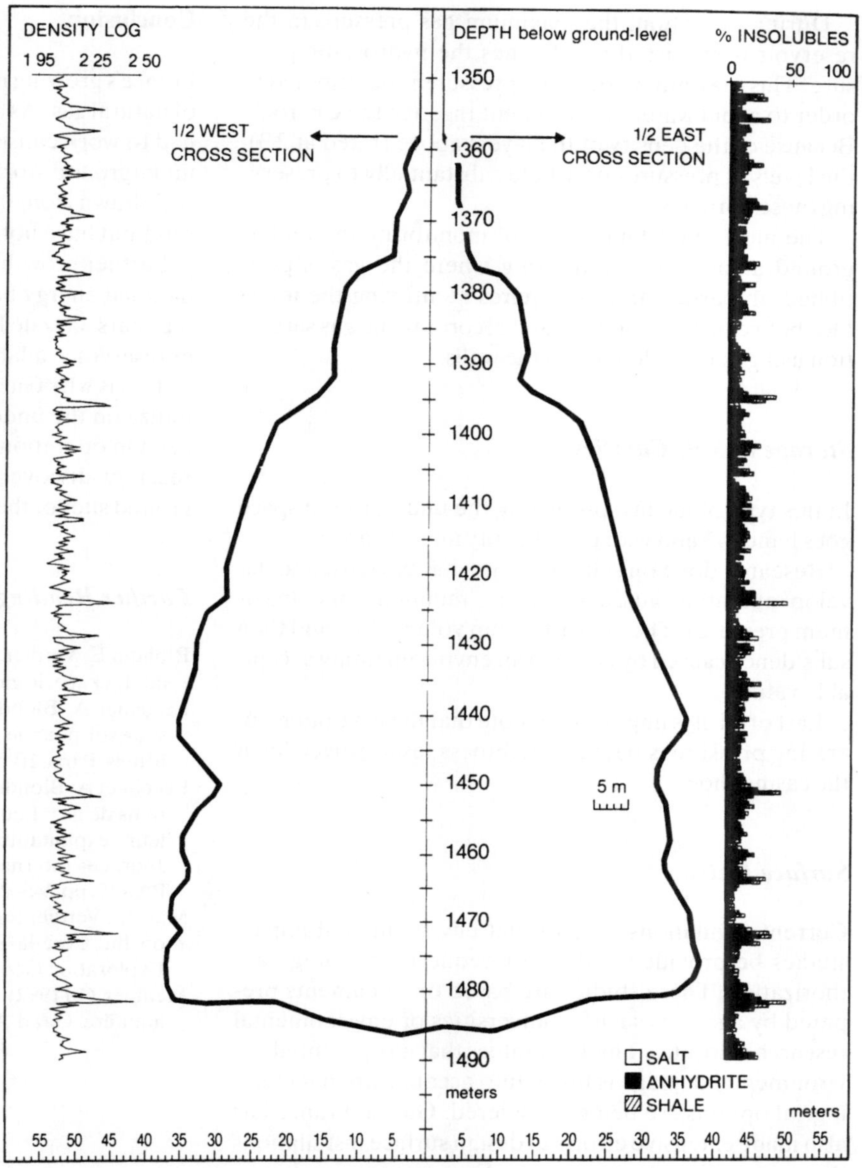

Fig. 10. Etrez salt cavity, vertical cross section

Aquifer Storage Facilities

One of the primary issues in aquifer storage is protecting the underground environment. This includes not only safeguarding the aquifers themselves, but also conserving and maintaining the stability of the geological layers covering them. It also requires monitoring the position of the gas-filled zone within a reservoir.

During both drilling and operation, special attention is paid to the aquifer intervals. It is important not to affect water quality in the uper horizons or to interfere with hydrostatic pressure at lower levels.

While the casings are being cemented, all appropriate methods are used to restore the impermeability of the geological layers. The types of muds, cements, and technologies used must be adapted to the properties of the rock encountered. When the facility is being equipped, specific efforts are made to avoid excessive high pressure, and the amounts of cement injected are carefully measured. The twofold goal is to avoid causing pollution, while ensuring that the casing is completely leakproof.

In the operating facility, the gas-filled area is also monitored by direct measurements taken in the different wells, and by modeling, which is used to track changes in the reservoir's hydraulic status. Aquifers lying above the cap rock are monitored systematically to check their pressure and water quality. Other tests are done to detect any gas dissolved in the water.

The water in the reservoir aquifer is generally not intended for human consumption because of its salinity. The inherent properties of the natural gas tend to limit physical and chemical interactions between this water and the gas. Furthermore, experience has shown that there is little movement in the water located near the gas-filled area once the storage facility is in regular use.

During operation, the maximum gas pressure in the reservoir is restricted to 1.5 times the hydrostatic pressure. This maximum operating pressure is determined in order to avoid water displacement through the cap rock. Because of the density of the layers above (rated at 2.3), the levels of pressure contribute substantially to preserving these horizons.

The most important aspect of monitoring the underground areas involves knowing where the gas is positioned. In particular, this requires identifying the interface between water and gas and recording the gas saturation using neutron logging in the wells.

Storage in Salt Cavities

In this type of facility, protecting the underground space goes hand in hand with good facility maintenance.

Research done on salt mechanics has been used to develop operating guidelines (e.g., minimum and maximum pressure). The goal is to keep volume loss and then subsidence caused by creep to an environmentaly acceptable value.

Last of all, it is important to note that the maximum operating pressure is based on tightness tests, especially of the casing shoe.

Surface Units

Current regulations require that environmental impact studies be provided with every request for storage authorization. These studies are based on documents prepared by agencies such as universities or environmental research institutes. This guarantees that any potential environmental damage is taken into account when technological options are being considered. Gaz de France has also made extensive efforts to design surface installations that are as discrete as possible. For instance, landscape architects, who may cooperate with local authorities, work to improve the design of well platforms.

The fact that compressor and treatment units are of small size facilitates efforts to improve landscape installations. These units are carefully soundproofed: they house all the infrastructure necessary to process and evacuate both liquid and gaseous effluents.

The pipes running from the wells to the station and from there to the transmission network, quickly become buried. Vegetation of all types can be allowed to grow back above the pipes.

Finally, experience has shown the importance of founding relationships based on trust with people living near these installations; they appreciate receiving complete and sincere information about the site.

Conclusion

France's growing need for energy has led to increased use of natural gas. As a result of this trend, Gaz de France has had to work consistently to develop its transmission and underground storage systems. The volume which can be withdrawn from underground facilities and the peak day send-out have both increased considerably.

Further growth in the share of natural gas in France's national energy balance may be expected over the coming years. Gaz de France will therefore need to have larger reserves available.

This is why Gaz de France is pursuing its efforts to capitalize on the underground storage facilities, which have been in operation for several years, develop facilities on recently discovered sites, and explore for the underground sites of the future.

Further Reading

Blondin E, Verdier F (1992) Interprétation sismique appliquée au stockage de gaz Mém Soc Géol Fr 161 77–88

Lecoanet A, Blondin E (1987) Les possibilités offertes par le sous-sol pour le stockage de grands volumes de gaz Ann Mines, Paris, 10 52–63

Lecoanet A, Blondin E, Gimenez J (1988) Les stockages souterrains de gaz Leur rôle, les modalités de leur recherche et de leur exploitation Jigastock 88, Versailles (France) III, Journées Internationales de Géothermie Appliquée AFME PARIS, pp 245–257

Mari JL, Verdier R (1991) VSP and acoustic logging case history on the Céré-la-Ronde structure European Association of Exploration Geophysicists, Florence (Abstr) pp 118–139

Meunier G (1992) Modèles pour le stockage de gaz en nappe aquifère Gaz d'Aujourd'hui, Paris 7, 8, 9 421–425

Modeling Present and Past Thermal Regimes in the Paris Basin: Petroleum Implications

J. M. Gaulier and J. Burrus

Abstract. The present-day heat flow density distribution along a E-W regional cross section in the Paris Basin is investigated by means of continuous temperature profiles analysis and thermal conductivity determinations. The shale-dominated Liassic strata have thermal gradients that are consistently higher (55–65 °C/km) than the carbonate-dominated Dogger and Lower Cretaceous strata (37–40 °C/km) or Upper Cretaceous chalk (31–34 °C/km). IFP's 2-D numerical model TEMISPACK is used to adjust the crustal heat flow and magnitude of regional hydrodynamics that best fit the observed thermal profiles. A good adjustment between observed and calculated temperatures is obtained, and indicates the following:

- The present-day crustal heat flow varies laterally between 70 and 90 mW m^{-2} along the section studied. The variations are probably explained by substratum heterogeneities.
- Vertical changes in the thermal gradients cannot entirely be explained by thermal conductivity contrasts between Liassic and Dogger/Cretaceous strata. An additional explanation is found by east-to-west regional circulations in active aquifers, mainly Dogger and Albian. The computed water flow seems compatible with estimated regional Darcy velocities found in the literature.

Extrapolation of the present thermal heat flow to the past reveals the following conclusions:

- Data on Liassic source rock maturity and published fluid inclusions consistently indicate that the maximum temperatures were warmer (up to 30–40 °C in Liassic Triassic strata to the east of the basin) than at present.
- Data on modeling kerogen maturity and fluid inclusions suggest that these higher temperatures were reached at the end of Cretaceous times. The cooling is attributed to several causes: Tertiary erosion to the east of the basin accounts for 10–15 °C, surface temperature decrease between Cretaceous and Tertiary for 15 °C, and convective flow for 7–10 °C. The cooling could be more progressive if the uplift were more recent than considered here.

From the petroleum exploration point of view, it is interesting to notice that our data and our modeling results do not indicate significant variation of basement heat flow with time since the Upper Jurassic.

Introduction

The purpose of this chapter is to study the present and past thermal regimes of the Paris Basin. The thermal history of source rocks is indeed of paramount importance for effective kitchen delineation and timing of hydrocarbon generation and expulsion. This study is based on thermal profiles acquired between 1988 and 1989 by BRGM, EAP, and IFP in nonproductive petroleum and geothermal wells (Gable et al. in prep.). It focuses along a 300-km east-west regional cross section located between Versailles and Vittel (Fig. 1). IFP's 2-D regional model TEMISPACK is used both to explain the observations and determine the present thermal heat flow density variations along the section, and also to reconstruct the past thermal regime. The modeled temperatures are discussed considering the maturity observed in the measurements of the Liassic source rocks and fluid inclusions.

One of the objectives of this chapter is to establish that, in contradiction with views commonly found in the literature, available paleothermal indicators in the Paris Basin are consistent with a time-invariant basement heat flow.

Short Geological Setting

The intracratonic Paris Basin (Fig. 1) developed mostly during Mesozoic times above a fractured substratum. Major faults that are believed to have guided the Mesozoic subsidence include:

1. the curved Bray-Vittel and the Sancerre-Rouen fault systems inherited from Hercynian history,
2. the north-south faults which dominate in the south of the basin,
3. the NE-SW faults that characterize the Sarre-Lorraine basin.

The origin of the tectonic subsidence in the Paris Basin is not yet well understood. Brunet and Le Pichon (1982) stressed the concept of a Permian-Triassic rifting followed by post-rift sagging until Barremian-Aptian times. Others argue for purely flexural processes (e. g., Mégnien and Pomerol 1980). Subsidence ceased in the Tertiary, which is represented by a thin layer of Eocene-Oligocene sediments. During late Pyrenean and Alpine orogenies, the eastern part of the section was uplifted and eroded. The complex inherited fault system has been continuously reactivated throughout the Mesozoic and the Tertiary.

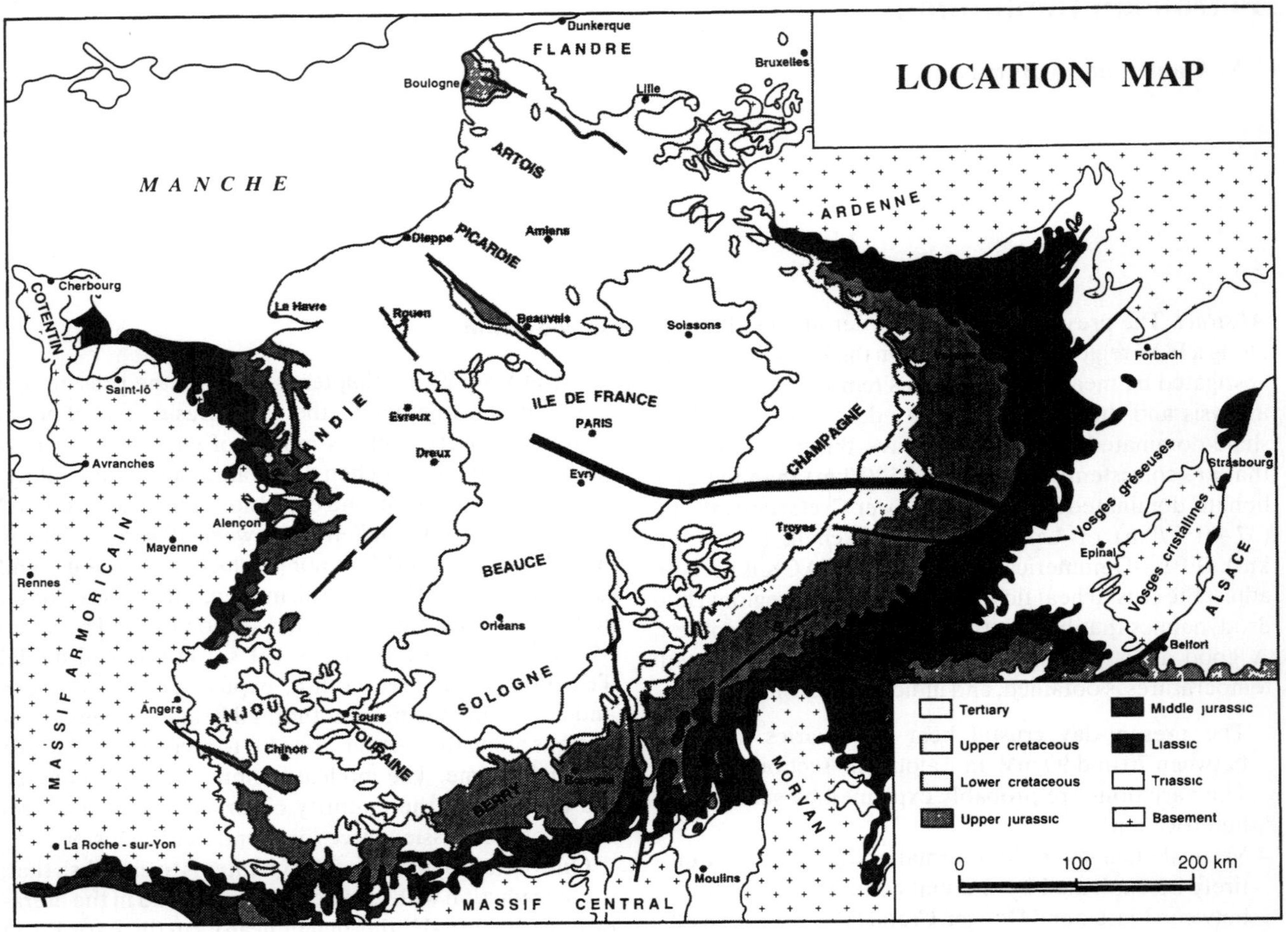

Fig. 1. Location map of the regional section investigated (Simplified geological map after Mégnien 1980)

Methodology and Principal Inputs to the Modeling Procedure

Thermal Profile Acquisition

Conventional thermal data in sedimentary basins consist in bottom hole temperatures (BHT), which are severely affected by drilling, or in rarely available drill stem tests (DST). Continuous thermal profiles in thermally equilibrated wells present the unique advantage of providing a reliable measurement of formation gradients (e.g., Coleno 1986), whereas BHTs or DSTs provide only local information.

In general, a continuous thermal profile in a well can be divided in linear segments which correlate with homogeneous lithologies.

In the Paris Basin, shale-dominated strata, located in the deepest part, and having the lowest thermal conductivities, can be expected to show the highest thermal gradients, whereas carbonate-dominated formations can be expected to show the lowest ones. The thermal regime is said to be conductive if the heat flow is constant throughout the sedimentary column.

heat flow=thermal gradient × thermal conductivity

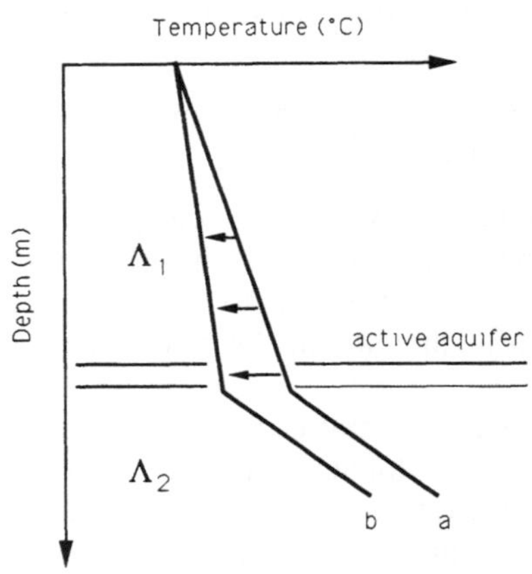

Fig. 2. Theoretical effect of convective cooling on steady-state thermal gradients Below the aquifer, steady-state thermal profiles are shifted, but thermal gradients are not changed *a* conductive thermal gradient, *b* convective thermal gradient, $\Lambda_1 > \Lambda_2$ thermal conductivities

If active hydrodynamism contributes to modify the conductive regime, the interpretation of thermal gradients is more complex (e.g., Smith and Chapman 1983; Vasseur et al. 1993). Temperature gradients are modified within and above the active aquifer (Fig. 2). Below the aquifer,

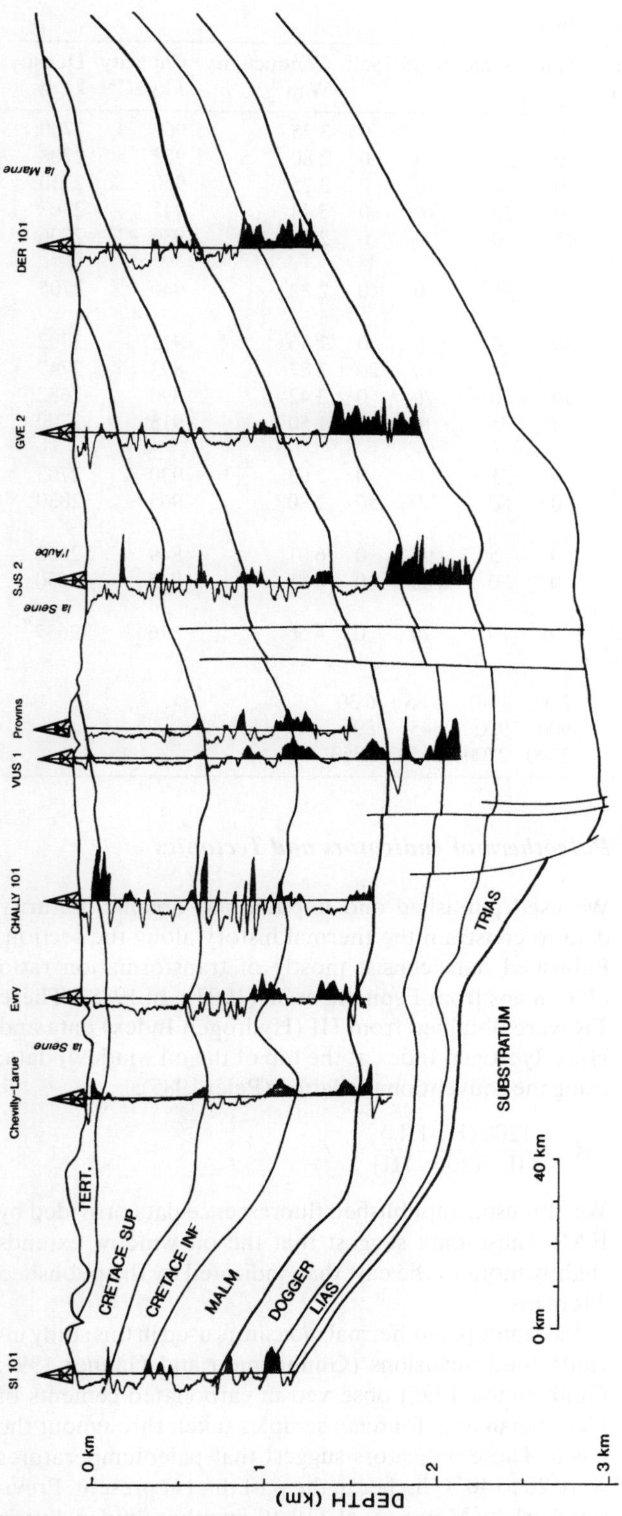

Fig. 3. Thermal gradient profiles measured in nine wells along the modeled section

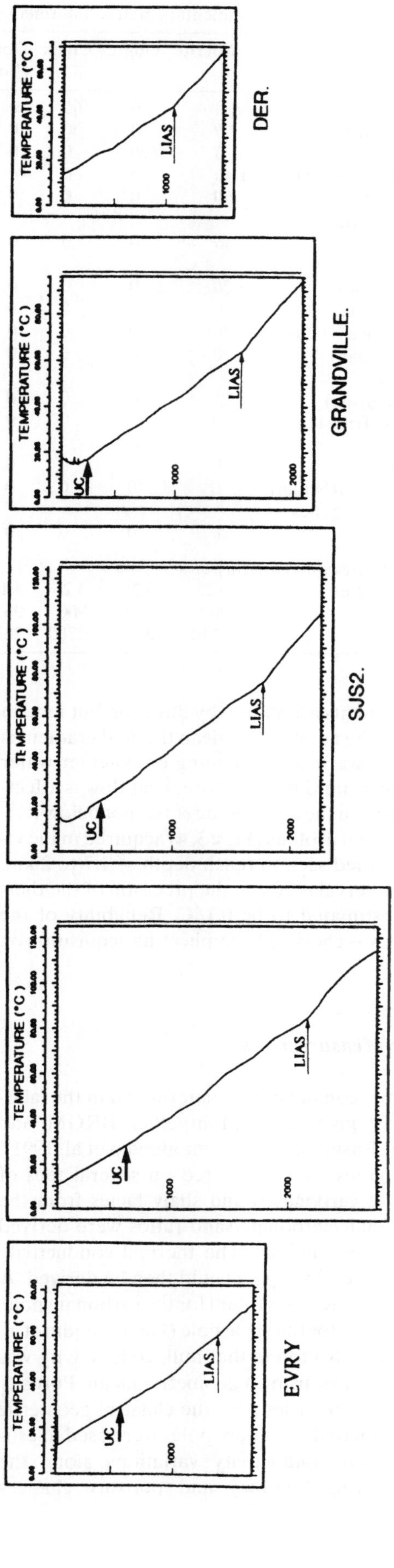

Fig. 4. Temperature-depth profiles measured in five wells along the section. These profiles have been used to calibrate the present-day crustal heat flow values along the section *UC* base of Upper Cretaceous; *Lias* top of the Liassic

Table 1. Thermal parameters of sedimentary matrix (intermediate scenario)

Lithologies	Micrite	Oolite	Chalk	Dolomite (Vol %)	Marl	Shale	Sand	Salt	Conductivity W m^{-1} °C^{-1}	Capacity J/kg °C^{-1}	Density kg m^{-3}
Chalk (Upper Cretaceous)	0	0	*100*	0	0	0	0	0	3.25	960	2700
Shaly limestone (Malm)	*80*	0	0	0	0	*20*	0	0	2.80	952	2708
Oolithic limestone (Dogger)	0	*100*	0	0	0	0	0	0	3.25	960	2710
Dolomitic limestone (Portlandian)	*40*	0	0	*40*	0	*10*	*10*	0	3.44	945	2767
Marly limestone (Malm)	*60*	0	0	0	*40*	0	0	0	2.80	936	2706
Interbedded shales and limestones (Malm)	*50*	0	0	0	0	*50*	0	0	2.32	940	2705
Calcareous marls (Triassic and Jurrassic)	*20*	0	0	0	*80*	0	0	0	2.46	912	2702
Anhydritic marls (Triassic)	0	0	0	0	*70*	0	*10*	*20*	2.87	893	2747
Sandy-calcareous marls (Triassic)	*20*	0	0	0	*40*	0	*40*	0	3.42	890	2682
Marly shales (Toarcian)	0	0	0	0	*25*	*75*	0	0	1.50	915	2700
Calcareous shales (Hettagian-Sinemurian)	*25*	0	0	0	0	*75*	0	0	2.03	930	2703
Evaporitic shales (Triassic)	0	0	0	0	0	*50*	0	*50*	2.80	905	2830
Shaly sandstones (Lower Cretaceous)	0	0	0	0	0	*5*	95	0	6.01	849	2653
Sandstones (Albian – Rhetian)	0	0	0	0	0	0	100	0	6.85	845	2650
Shaly calcareous sandstones (Tertiary)	*20*	0	0	0	0	*10*	70	0	4.56	876	2667
Characteristics of pure mineralogical poles											
Thermal conductivities	3.25	3.25	3.25	4.10	2.32	1.40	6.85	6.30			
Heat capacity	960	960	960	960	900	920	845	890			
Density	2710	2710	2700	2870	2700	2700	2650	2960			

the thermal gradients are shifted by diffusion but remain parallel. Above the aquifer, the mean thermal gradient is either cooled or warmed, depending on whether water flows down-dip or up-dip. Therefore, heat flow is affected above and within the active aquifer but not below it.

The nine thermal profiles (Figs. 3, 4) acquired in the vicinity of the studied section reach depth of 1.7 to 2 km. The sampling interval is 20 cm, the precision of the thermal probe is estimated to be 0.1 °C. Reliability of the measurements was checked by duplicating acquisition on some sites.

Conductivity Measurements

We used thermal conductivities, determined in the laboratories of the University of Montpellier, BRGM, and IFP, from Paris Basin samples (Demongodin et al. 1991). Bulk conductivities were measured on several tens of samples in both carbonated and shaly facies from the studied area. Shale/carbonate/sand ratios were derived from mineralogical analysis. The thermal conductivity for each pure mineralogy pole could then be derived. A value of 3.25 W m^{-1} K^{-1} was found for the carbonate pole, 1.2 to 1.4 W m^{-1} K^{-1} for the shale pole (Demongodin et al. 1991). This study also found that bulk conductivity was best approximated by using a geometric mean. Porosity was accounted for by using also the classical geometric mean. The conductivities of pure poles were used to estimate the thermal conductivity variations along the section, considering lithology and porosity changes (Table 1).

Paleothermal Indicators and Tectonics

We used published and unpublished organic maturity data to constrain the thermal history along the section. Published data consist mostly of transformation ratio (TR) maps from Espitalié et al. (1987 a, b; 1988). These TR were obtained from HI (Hydrogen Index) data and HI_0 (Hydrogen Index at the top of the oil window) data, using the conventional relation (Pelet 1985):

$$TR = \frac{1200\,(HI - HI_0)}{HI_0\,(1200 - HI)}$$

We also used unpublished fluorescence data provided by EAP. These data suggest that the oil window extends slightly more to the east than indicated by the published TR maps.

Inorganic paleothermal indicators used in this study include fluid inclusions (Guilhaumou and Gaulier 1991; Guilhaumou 1993) observed in carbonated cements of Hettangian and Toarcian samples taken throughout the basin. These indicators suggest that paleotemperatures were 20 to 40 °C higher in the past than at present. Previous work by Matray et al. (1989) on other fluid inclusion data had already come to similar conclusions. Mossman (1987, 1991) used K/Ar dating of authigenic illites to document three regional diagenetic events dated respectively at 190, 150, and 80-55 Ma. Maximum temperatures found by Guilhaumou ought correspond to the last mineralization event. Other workers (e.g., Guillocheau 1991) have stressed the idea that several episodes of accelerated subsidence occurred during the Mesozoic. The episodic character of the subsidence history, together with the evidence of past higher temperatures during mineralization events, have lead to the idea that the Paris Basin could have been affected in the Mesozoic and the Tertiary by discrete thermotectonic events. If this were the case, major consequences for both the extension and timing of petroleum generation could be expected.

Table 2. Permeabilities of the main aquifers (for 10% porosity)

	K_h (mD)	K_v (mD)
Lusitanian limestones	2	0 002
Dogger oolithic limestones	125	25
Albian shaly sands	125	0 0025

Modeling Procedures

The numerical model used in this study is IFP's 2-D finite differences code TEMISPACK (Ungerer et al. 1990; Burrus et al. 1991). This program reconstructs:

1. the geometric evolution along the section taking into account sedimentation, erosion, and compaction, starting in this study in Triassic times,
2. the thermal transfer due to conduction and convection, and the maturity of the organic matter, assumed to obey a first-order Arrhenius scheme (Tissot and Espitalié 1975),
3. the development of pore pressures, compactional flow, and regional discharge in aquifers.

Other possibilities of TEMISPACK are not addressed in this chapter (petroleum generation and migration).

The modeling procedures involved the following steps.

Mesh Design. A regional time-stratigraphic and litho-stratigraphic model was established using public wells available at the Direction des Hydrocarbures (Fig. 5). This detailed model was simplified into a processable mesh consisting of 20 strata ranging in age from Keuper to Quaternary, and 25 vertical columns (Figure 6c). Sixteen litho-facies were considered from variable proportions in carbonate/shale/sand/salt and organic content (Table 1). Using porosity-depth function attributed to each bulk lithology, taken from the literature, the section was backstriped and past geometries restored (Fig. 6).

The eroded thickness to the east of the basin (Champagne, Lorraine) was taken from Goy (1979), who studied Sonic Log anomalies observed in the Schistes Cartons. According to this interpretation, the maximum amount of Tertiary erosion is estimated around 300 m along the studied section. In particular, the Upper Cretaceous chalk is assumed to have been isopaque in Champagne west of the GV101 well and to have progressively thinned out in the vicinity of the LLR well (Fig. 7).

Present-Day Thermal Modeling. Thermal conductivities were attributed to each lithology as indicated above. TEMISPACK was used to adjust the basement heat flow by trial and error against the observed temperature profiles, assuming pure conduction first. We then investigated the contribution of convective cooling related to regional discharge in active aquifers: Dogger and Lusitanian carbonates, Albian sandstones. The permeabilities of these aquifers were taken from often imprecise or uncomplete transmissivity maps (Maget 1983) following Table 2. The recharge zone is constituted by the uplifted basin east of Champagne (Fig. 1). The topographic difference between the eastern and western extremities of the section studied is around 350 m. We compared the computed Darcy velocities with reported values found in the literature. Although Aubertin et al. (1987), interpreting tracers experiments in the Dogger around Fontainebleau, found darcy velocities to be as high as 10 m a^{-1}, Wei et al. (1990) argue for much lower velocities in the range 0.1 to 1 m a^{-1}, which we considered to be more representative (G. Marsily, pers. comm.). Since reported ages of the Dogger waters in Fontainebleau area are 23±3 ka (cited in Wei 1990), the initial temperature of waters at the recharge zone was probably affected by the Wurm glaciation. We therefore introduced in our model the short-term surface temperature variations described by Vasseur and Lucazeau (1983) for the last 60 ka.

Past Thermal Modeling. We tested whether the assumption that the present-day basement heat flow has been invariant since Late Jurassic is consistent with available paleothermal indicators. Inputs needed are the following:

- Organic matter maturity was modeled with kinetic parameters derived from experimental Rock Eval pyrolysis following Tissot et al. (1974). We distinguished from the kinetic point of view the Type II Toarcian Schistes Cartons from the mixted Type II/Type III Hettangian-Sinemurian marls (Espitalié et al. 1987a, b).
- Long-term paleosurface temperatures since the Triassic were taken from Habicht (1979).
- While extrapolating to the past the present-day hydrology, we assumed that the present-day topographic profile has existed for 50 Ma (Early Eocene). Tertiary erosion was assumed to be restricted between 66-50 Ma. The possibility that the Tertiary tectonic events responsible for topographic uplift, subsequent hydrodynamism, and erosion could have lasted until Miocene will be discussed later.

Model Results and Discussion

Present-Day Thermal Regime

One of the most important modeling results is that several scenarios explain equally well the observed temperature profiles. Figures 8 and 9 compare observed (dotted line) and calculated (thick line) temperature profiles with the following assumptions. Figure 8 corresponds to a

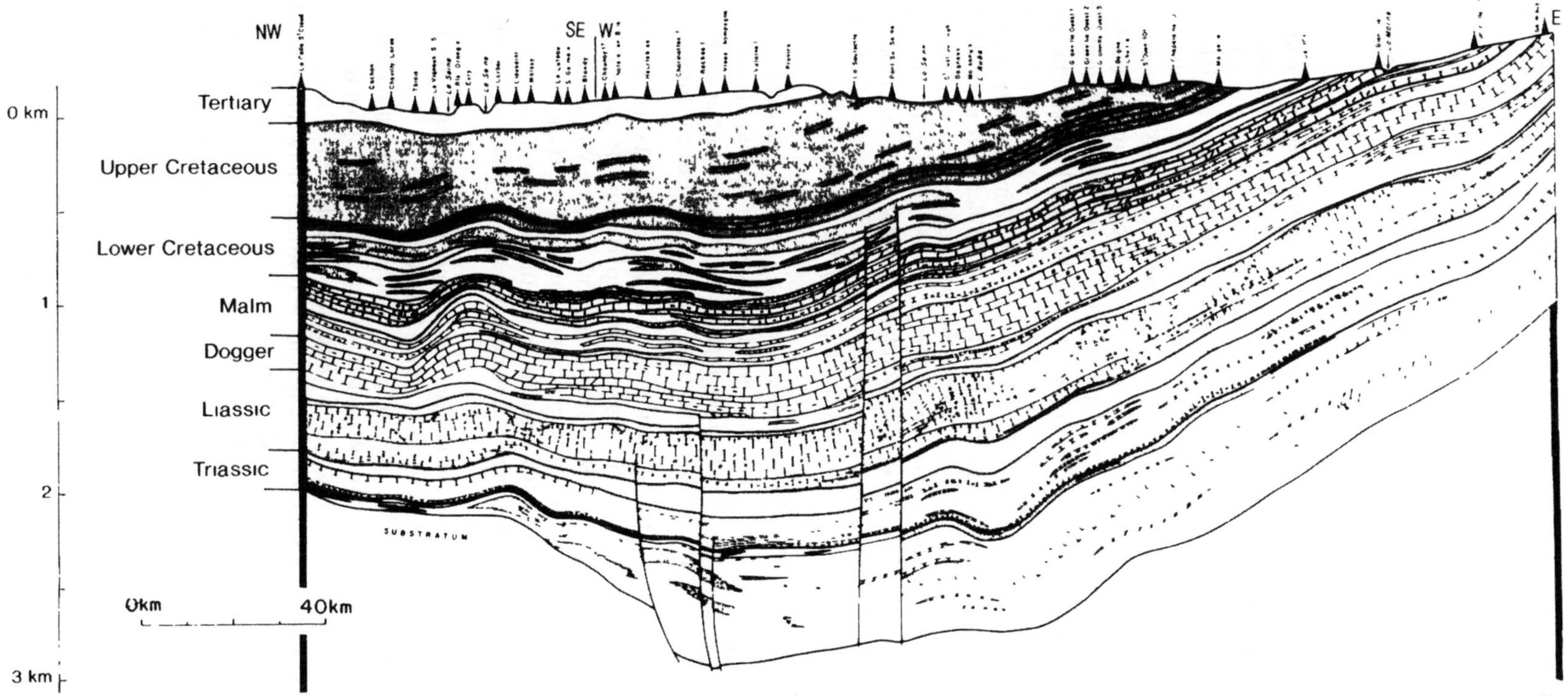

Fig. 5. Detailed geological cross section constructed from public wells and geological maps The modeled section extends 40 km more to the east This eastern part has no well The cross section was constructed from surface geology and subsurface maps

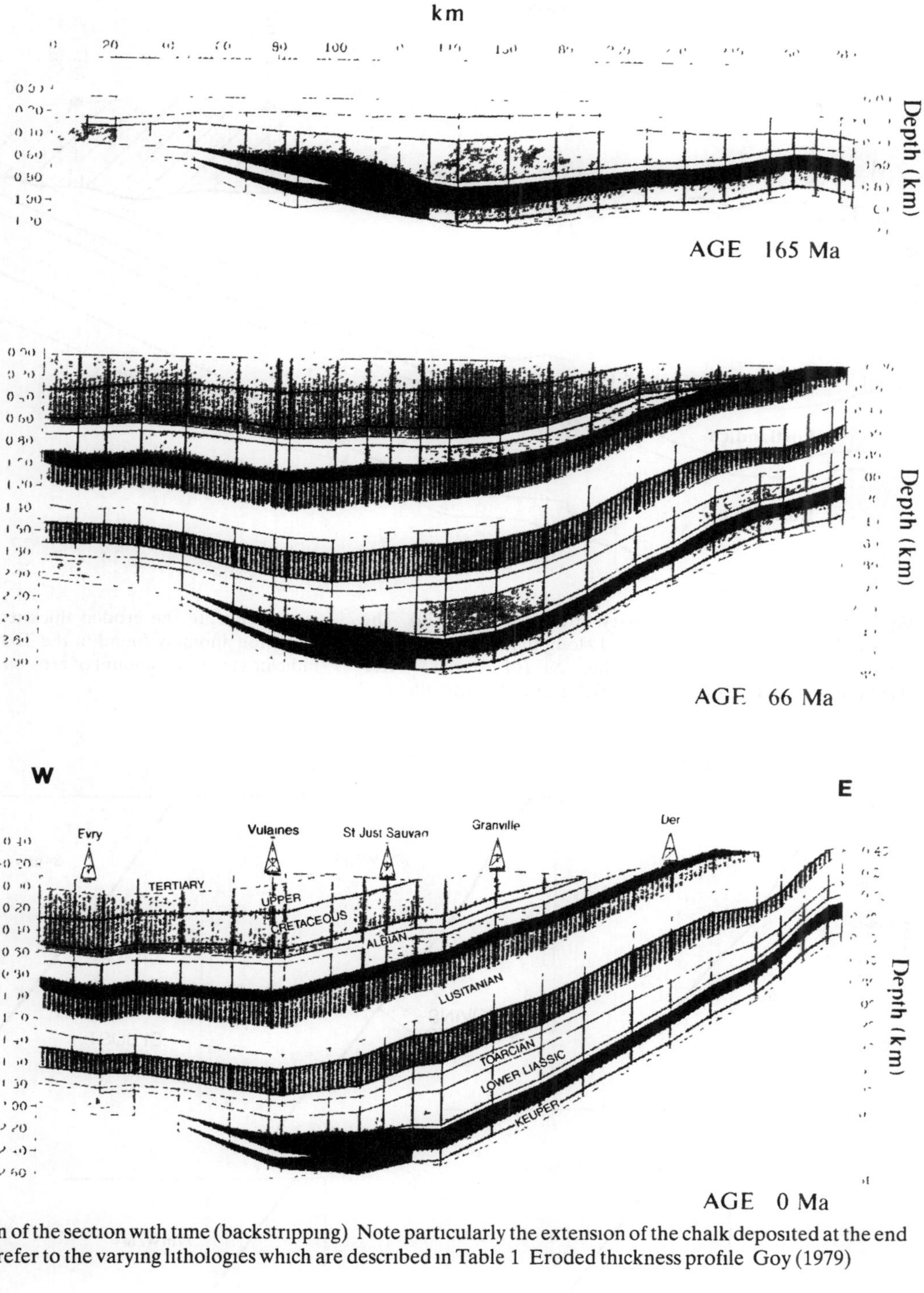

Fig. 6. Geometric evolution of the section with time (backstripping) Note particularly the extension of the chalk deposited at the end of cretaceous *Gray colors* refer to the varying lithologies which are described in Table 1 Eroded thickness profile Goy (1979)

pure conductive solution. The change in thermal gradient observed at the top of the Liassic is entirely due to high thermal conductivity contrasts between carbonate-rich post-Liassic strata and shale-dominated Liassic units. In contrast, Fig. 9 corresponds to a highly convective regime characterized by Darcy velocities higher than 5 m a^{-1} in the Dogger. The change of thermal gradient is only in part due to the thermal conductivity contrast, which is smaller than before. It is also related to cooling due to circulations in active aquifers (Albian and Dogger mostly).

It is impossible to discriminate the two solutions from the quality of the fit, which is equally good in both cases. Each solution is obtained with a given set of: (1) basement heat flow distribution along the section studied (Fig. 10), (2) magnitude of regional flow in the aquifers obtained with different permeabilities (Fig. 11), (3) value of the thermal conductivity of the shale pole. We discuss our preferred solution later.

The best-fitting basement heat flow profiles of Fig. 8 and 9 (see also Fig. 10) differ by around 10 mW m^{-2} east

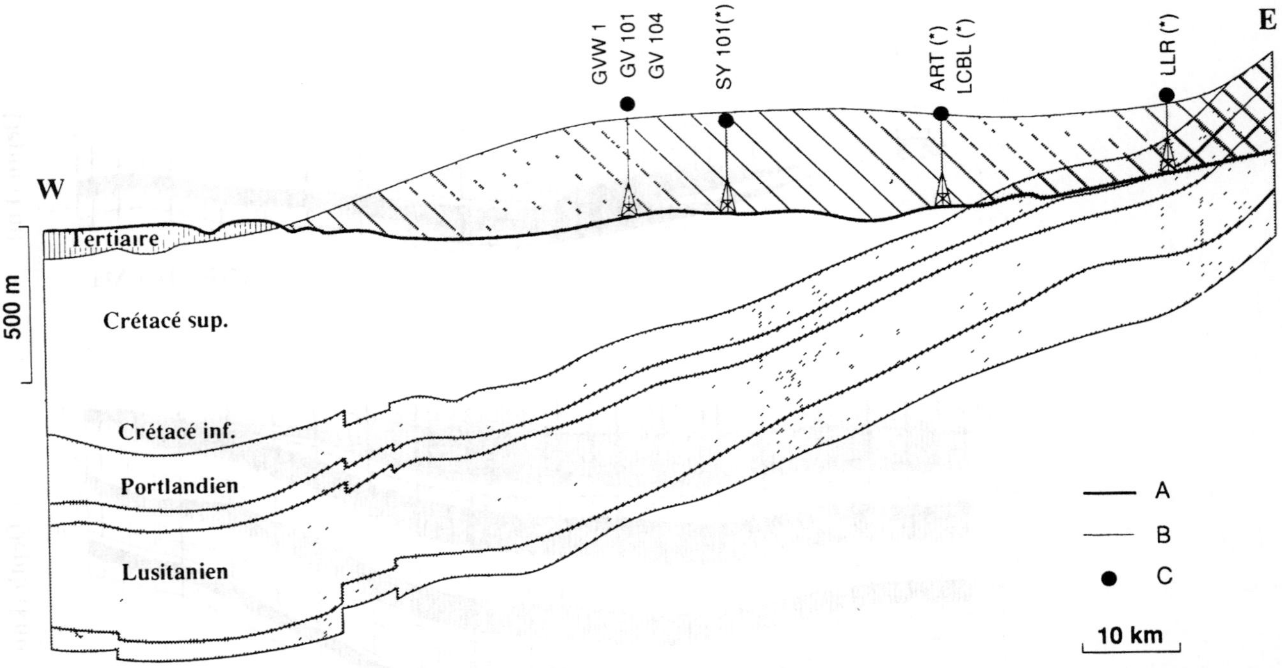

Fig. 7. Constraints on the Tertiary erosion (Goy 1979) The wells indicated by (*) are projected along the section *GV 101, GVW 1,* and *GV 104* Grandville, *SY 101* Songy, *ART* Arrentières, *LCBL* Colombey-les-Belles, *LLR* Lérouville *Full circles* indicate the eroded thickness estimated by Goy from Sonic log anomaly found in the Toarcian shales *Hatched area* indicates the total amount of erosion

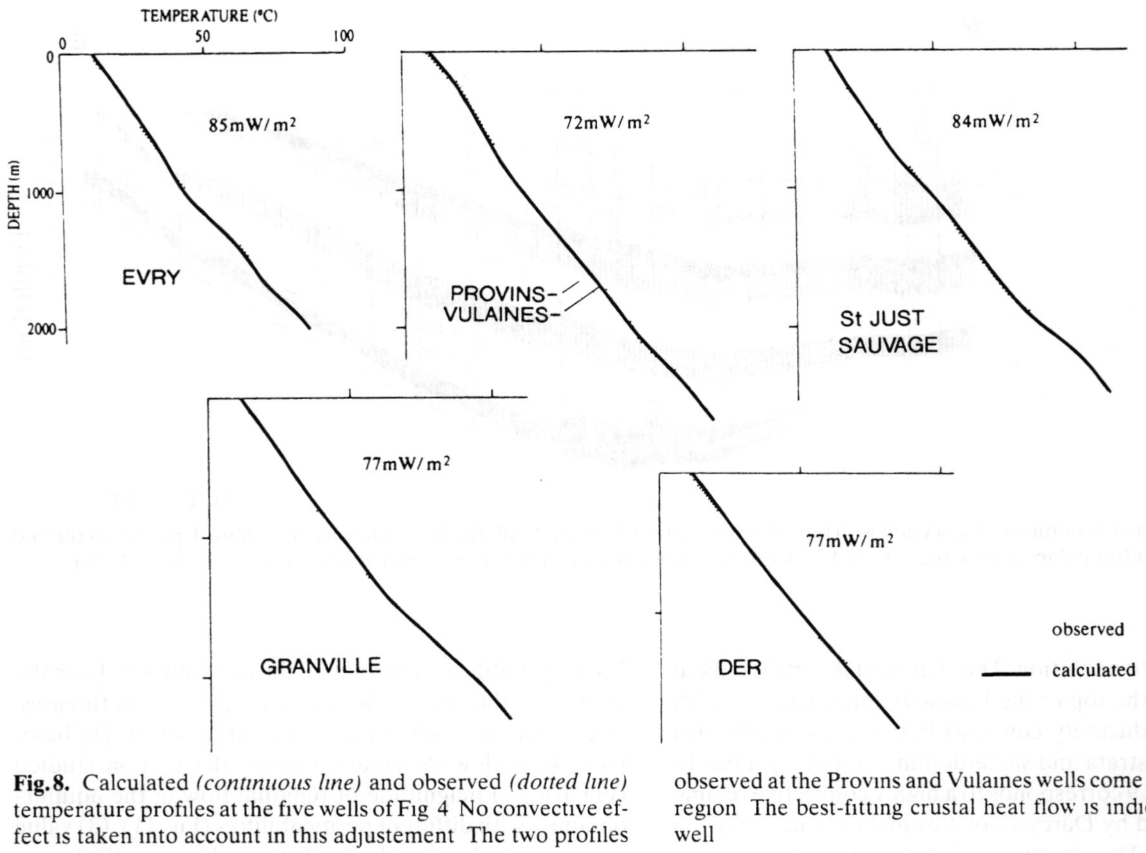

Fig. 8. Calculated *(continuous line)* and observed *(dotted line)* temperature profiles at the five wells of Fig 4 No convective effect is taken into account in this adjustement The two profiles observed at the Provins and Vulaines wells come from the same region The best-fitting crustal heat flow is indicated for each well

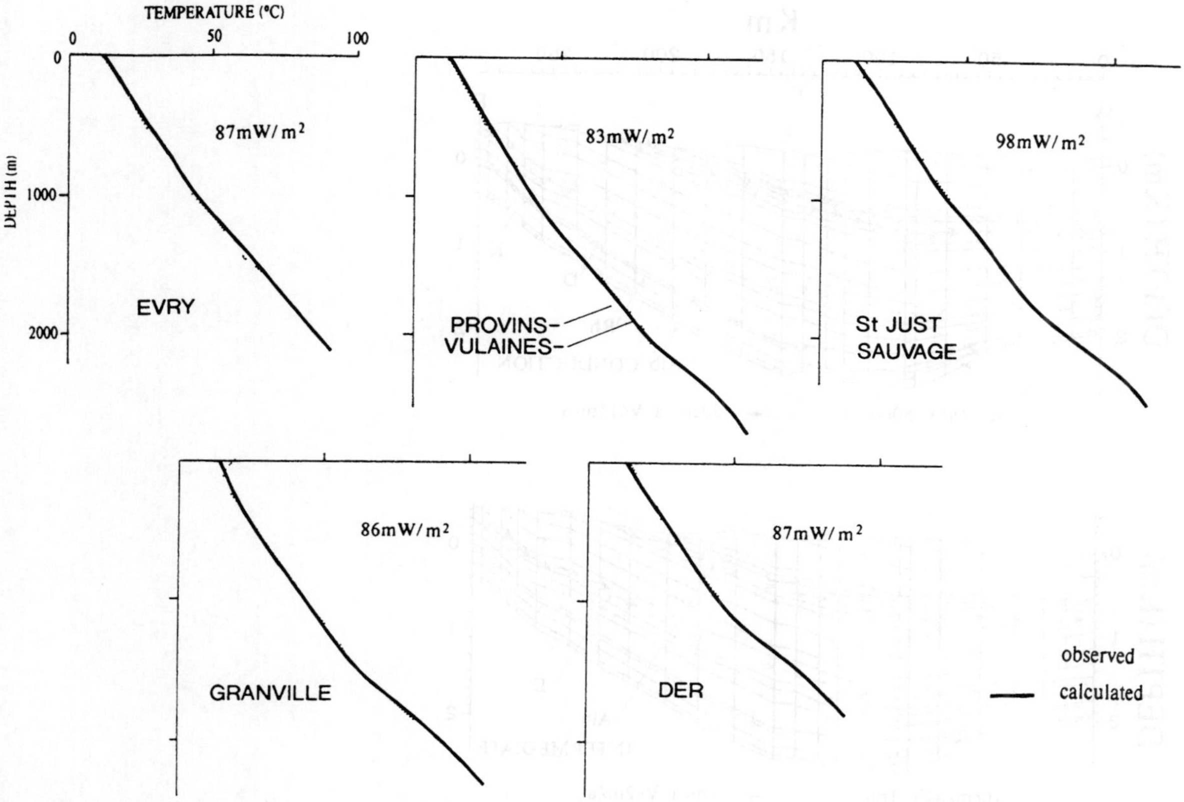

Fig. 9. Calculated *(continuous line)* and observed *(dotted line)* temperature profiles at the five wells of Fig 4 This adjustment corresponds to a highly convective scenario The adjustment is slightly better for the easternmost well Other comments see Fig 8

of km 75, but insignificantly to the west. This indicates that convective cooling is inefficient to the west of the basin. In the center of the basin (km 75 to 150), it is interesting to note that the best-fitting crustal heat flow varies significantly laterally. In the region of Vulaines (km 75–100), the heat flow is found to be around 15 mW m^{-2} lower than in the region of St-Just-Sauvage (km 100–140). This variability does not depend upon the magnitude of convective effects. It is most probably due to the heterogeneity of the crustal basement of the region study. The map of the pre-Triassic substratum (Mégnien 1980) indicates indeed that the high heat flow area coincides with gneiss-type basement, whereas the low heat flow area is underlained by mafic basement, expected to be less radiogenic.

We have compared the theoretical transmissivities obtained with the conductive and convective scenarios with values reported in the literature. Reported transmissivities in the Dogger (5×10^{-5} to 5×10^{-4} m^2 s^{-1}, Maget 1983) are slightly above modeled transmissivities in the conductive scenario (2×10^{-4}) and 10 to 100 times smaller than in the convective scenario. Reported values in the Albian are highly variable (5×10^{-4} to more than 10^{-2} m^2 s^{-1}); Albian modeled transmissivities for both scenarios are compatible with the observed range of variations. Accounting for Dogger transmissivity characteristics, we find the convective scenario to be unrealistic. On the other hand, considering the thermal conductivity constraints

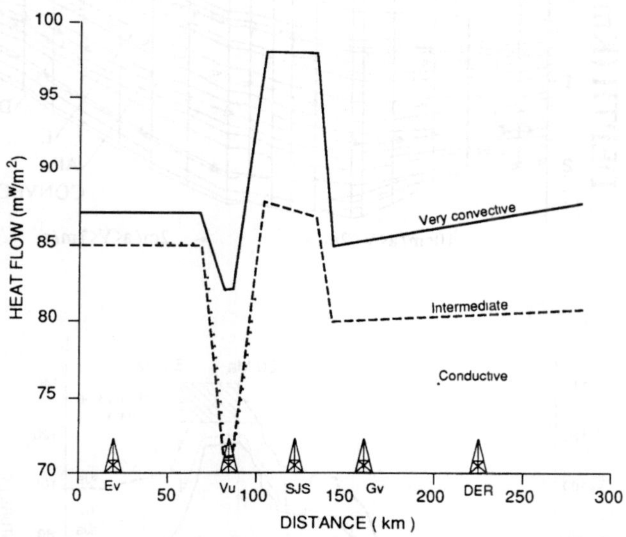

Fig. 10. Best-fitting crustal heat flow variations along the section studied for three thermal scenarios (purely conductive, intermediate, highly convective) The intermediate scenario provides the best compromise it satisfies the constraints on shale conductivity, on aquifer transmissivities, and on observed temperature profiles Note the 25–30-km wave-length crustal heat-flow variation in the center of the basin This is believed to be due to heterogeneous radiogenic production

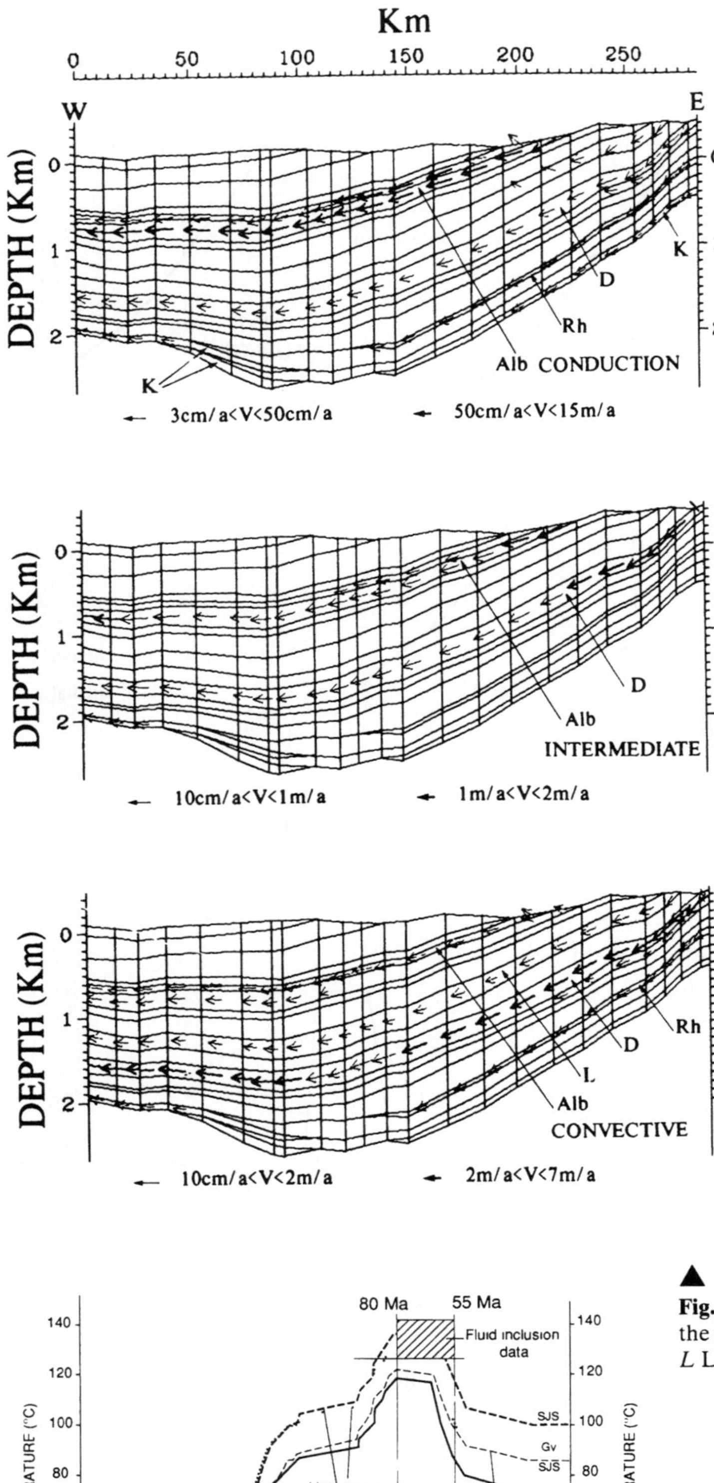

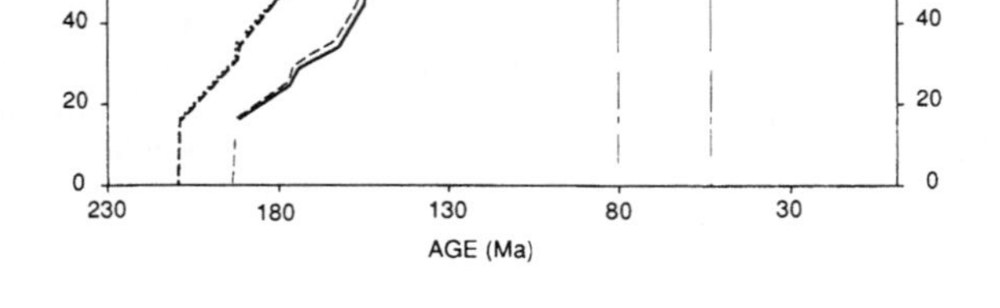

▲

Fig. 11. Computed Darcy velocities in the aquifers obtained for the three different solutions *K* Keuper, *Rh* Rhetian, *D* Dogger, *L* Lusitanian, *Alb* Albian

◀

Fig. 12. Temperature-time pathways computed in the vicinity of Saint-Just-Sauvage (*SJS* labels) and Granville (*GV* label) wells following the intermediate thermal scenario of Figs 10 and 11 *Shaded domain* represents the paleotemperatures indicated by fluid inclusions measurements in carbonate cements and K/Ar dating in diagenetic illites (see text) The modeled pathways appear to be consistent with these constraints

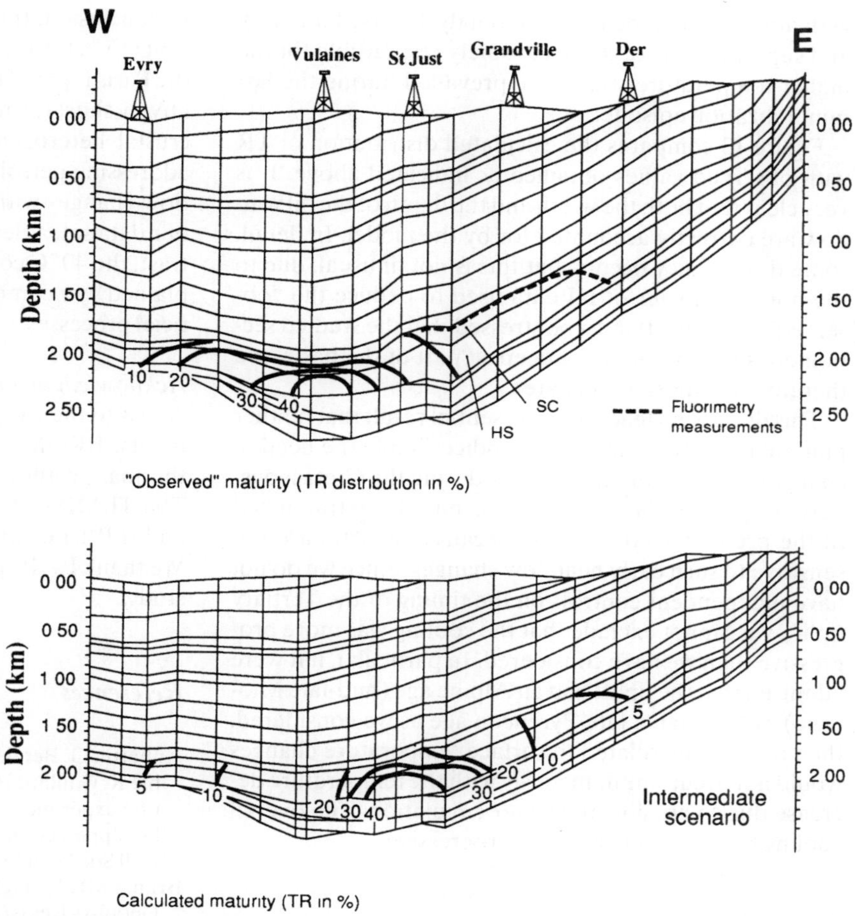

Fig. 13. Observed (*top* Espitalié et al 1988) and calculated *(bottom)* kerogen transformation ratio following the intermediate scenario The general agreement is good Discrepancies are believed to be due to the noisy maturity data (derived from dispersed IH/depth plots) The modeled eastward extension of the kitchen is consistent with unpublished fluorimetry data

tends to dismiss the pure conductive scenario. The fit of Fig. 8 implies indeed a thermal conductivity as high as 1.8 W m^{-1} K^{-1} for the shale pole, which contradicts the value of 1.1 to 1.2 calibrated in the laboratory. We have therefore defined an intermediate scenario which leads to equal fit of temperatures profiles, but better agrees with constraints on the thermal conductivities simultaneously with aquifer transmissivities. This "intermediate" scenario considers a low thermal conductivity for the shale pole (1.4 W m^{-1} K^{-1}), a value more consistent with the measured one. The model Dogger transmissivities are estimated about 10^{-3} m^2 s^{-1}. These transmissivities are only two times higher than the estimated maximum, which is an acceptable difference.

It is interesting to note that the Dogger Darcy velocity in the center of the basin is then 0.2 to 0.3 m a^{-1}, which is consistent with the results of Wei et al. (1990).

Past Thermal Regime

We modeled the past temperatures, taking into account sedimentation, erosion, and compaction along the section as well as the "intermediate" convective scenario assumed to apply to the post-Eocene period. The basement heat flow varied along the section according to Fig. 10 and was kept invariant with time. Figure 12 shows the temperature-time pathways calculated for the four cells representing Hettangian-Sinemurain and Schistes Cartons strata located in the vicinity of St-Just-Sauvage and Granville wells. Between time of deposition and end Cretaceous, modeled temperatures never decrease with time. The rapid temperature buildup between 100 and 80 Ma is due to the higher sedimentation rate of the chalk. Between 66 Ma and Present, modeled temperatures decrease by 35–40 °C. Most of the decrease takes place between 66 and 50 Ma, which is the period of erosion in our model. A sensitivity analysis, not shown here, indicates that the most important causes of cooling are by order of decreasing importance: the surface temperature decrease between Cretaceous and Tertiary (15 °C), the Tertiary erosion (10–15 °C), and convective cooling in the regional aquifers (7–10 °C).

Figure 12 shows that the modeled cooling is consistent with paleotemperatures derived from fluid inclusions as indicated above. This would suggest that the last mineralization event, described in the literature to have occurred between 80 and 55 Ma (see above), was not associated with any basement heat flow anomaly. Intrasedimentary and superficial processes seem solely responsible for the higher temperatures that have prevailed during the last mineralization episode.

Figure 13 compares the calculated distribution of TR with observed values obtained as indicated above. It is very clear that both the width and the depth of the oil window are correctly accounted for by the model. In detail, some discrepancy occurs, but this is not unusual, due to the noisy character of HI data used to deduce the "observed" maturity. It must be stressed that the studied section does not intersect the deepest part of the basin and therefore the most mature area.

This discussion leads to the essential result that neither mineral indicators nor organic indices imply the need of changing the basement heat flow during the Cretaceous-Tertiary period. This conclusion cannot be extrapolated to the pre-Cretaceous period because our data are not sensitive to such early heat flow changes. Since we do not have independent controls on the timing of the Tertiary uplift, we cannot exclude that the cooling was more progressive than we have considered. In particular, if it were admitted that the uplift had an Alpine age (mid-late Miocene) rather than a late Pyrenean age as we considered, then the cooling related to surface temperature changes would not occur simultaneously with the temperature decrease due to erosion and water circulation. The overall cooling history would be more progressive.

Conclusions

Modeling the present-day thermal profiles and the thermal history of the Paris Basin leads to the following conclusions.

- The present-day crustal heat flow varies laterally between 70 an 90 mW m^{-2} along the section studied. The variations are probably explained by substratum heterogeneities and different radiogenic production.
- Vertical changes in the thermal gradients cannot be entirely explained by thermal conductivity contrasts between Liassic and Dogger/Cretaceous strata. An additional explanation is found by east-to-west regional circulations in active aquifers, mainly Dogger and Albian. The computed water flow seems compatible with estimated regional Darcy velocities found in the literature.
- Data published on Liassic source rock maturity and fluid inclusions indicate consistently that the maximum temperatures were warmer (up to 30–40 °C in Liassic Triassic strata to the east of the basin) than in the present.
- Data on modeling kerogen maturity and fluid inclusions suggest that these higher temperatures were reached at the end of Cretaceous times. The cooling is attributed to several causes: Tertiary erosion to the east of the basin accounts for 10–15 °C, surface temperature decrease between Cretaceous and Tertiary for 15 °C, and convective flow for 7–10 °C.

In conclusion, this chapter stresses the idea that significant (15 %) crustal heat flow variations exist today across the Paris Basin The wave length of these variations is relatively short (around 35–50 km) and seems controlled by crustal heterogeneity Another important result when addressing petroleum generation history is that no heat flow changes with time appear necessary to account for our data set, at least for the post-Jurassic period. In contrast, the 40 °C cooling of Tertiary age seems entirely explained by a combination of intrasedimentary and superficial processes.

Acknowledgment We thank the companies that allowed access to their wells during the thermal profile measurements: BRGM, EAP, and TOTAL. Acquisition of the thermal profiles was funded by BRGM, EAP, and IFP. The TEMISPACK modeling study was funded by EAP and IFP. I. Le Nir helped design the stratigraphic model. We thank EAP and BRGM for permission to publish this study.

References

Aubertin G, Benderitter Y, Cordier Y, Doillon F, Fabris H, Gable R, Gaillard B, Ledoux E, de Marsill G (1987) Détermination expérimentale de la vitesse naturelle d'écoulement dans la nappe géothermique du Dogger de la région parisienne Cull Soc Géol Fr 111, 5 991–1000

Brunet MF, Le Pichon X (1982) Subsidence of the Paris Basin Geophys Res 87, B 10 8547–8560

Burrus J, Kuhfuss A, Doligez B, Ungerer P (1991) Are numerical models useful in reconstructing the migration of hydrocarbons? A discussion based on the northern Viking Graben In Petroleum migration England WA, Flett AJ (eds) Geol Soc Spec Publ 59 89–109 Oxford Blackwell Scientific Publications

Coleno B (1986) Diagraphies thermiques et distribution du champ de température dans le Bassin de Paris Thèse, Univ Bretagne Occidentale, Brest, Spécialité Géosciences, 220 pp

Demongodin L, Pinoteau B, Vasseur G, Gable R (1991) Thermal conductivity and well logs a case study in the Paris Basin Geophys J Int 105 675–691

Espitalié J, Marquis F, Sage L (1987 a) Organic geochemistry of the Paris Basin In Brooks J, Glennie K (eds) Petroleum geology of North-West Europe Graham & Trotman, London, pp 71–86

Espitalié J, Marquis F, Sage L, Barsony I (1987 b) Géochimie organique du Bassin du Paris Rev Inst Fr Pétrole 42, 3 271–302

Espitalié J, Maxwell JR, Chenet Y, Marquis F (1988) Aspects of hydrocarbon migration in the Paris Basin as deduced from organic geochemical survey In Advances in organic geochemistry 1987, Org Geochem, vol 13, 1–3 pp 467–481

Goy G (1979) Les "schistes carton" (Toarcien Inférieur) du Bassin de Paris, en affleurements et en sondages Thèse de 3ième Cycle Univ P M Curie Spécialité Géologie, Paris VI

Guilhaumou N (1993) Paleotemperatures inferred from fluid inclusions in diagenetic cements implications for the thermal history of the Paris Basin Eur J Mineral 5 1217–1226

Guilhaumou N, Gaulier JM (1991) Détermination de paléotempératures dans les roches mères du bassin de Paris C R Acad Sci Paris 313, II 773–780

Guillocheau F (1991) Mise en évidence de grands cycles transgression-régression d'origine tectonique dans les sédiments mésozoiques du Bassin de Paris. C R Acad Sci Paris 312 II 1587–1593

Habicht JKA (1979) Paleoclimate, paleomagnetism and continental drift AAPG Stud Geol 9

Maget P (1983) Potentiel géothermique basse température en France Rapp BRGM 83SGN375SPG

Matray JM, Meunier A, Thomas M, Fontes JC (1989) Les eaux de formations du Trias et du Dogger du Bassin Parisien histoire et effets diagénétiques sur les réservoirs BCREP, Elf, vol 13, 2 483–504

Mégnien C (1980) Synthèse géologique du Bassin de Paris. Stratigraphie et paléogéographie (vol I) Atlas (vol II) Lexique des noms et des formations (vol III) Mém BRGM 101 - 102 - 103. BRGM, Orléans, France

Mégnien C, Pomerol C (1980) Subsidence of the Paris Basin from the Lias to the Late Cretaceous In Dynamics of plate interior, Vol 1 Geological Society of America, Boulder, CO, pp 91–92

Mossmann JR (1987) Conditions physico-chimiques de réservoirs gréseux Approche pétrologique, minéralogique et isotopique Application aux grès rhétie du Bassin de Paris Doctoral thesis, l'Université de Strasbourg, France

Mossmann JR (1991) K-Ar dating of authigenic illite-smectite clay material application to complex mixtures of mixed-layer assemblages Clay Min 26 189–198

Pelet R (1985) Evolution quantitative des produits formés lors de l'évolution de la matière organique Rev Inst Fr Pétrole 40, 5 551–562

Smith L, Chapmann DS (1983) On the thermal effects of groundwater flow, 1 Regional scale systems. J Geophys Res 88 B 593–608

Tissot B, Espitalié J (1975) L'évolution thermique de la matière organique des sédiments. applications d'une simulation numérique Rev Inst Fr Pétrole 30, 5 743–777

Tissot B, Durand B, Espitalié J, Combaz A (1974) Influence of nature and diagenesis of organic matter in formation of petroleum Am Assoc Pet Geol Bull 58; 3 499–506

Ungerer P, Burrus J, Doligez B, Chenet PY, Bessis F (1990) Bassin evaluation by integrated two-dimensionnal modelling of heat transfer, fluid flow, hydrocarbon generation and migration Bull Am Assoc Pet Geol 74 309–335

Vasseur G, Lucazeau F (1983) Bounds on paleotemperatures and paleoclimatic corrections Zentralbl Geol Paleontol 1 17–24

Vasseur G, Demongodin L, Bonneville A (1993) Thermal modelling of fluid flow effects in thin-dipping aquifers Geophys J Int 112 276–289

Wei HF (1990) Modélisation tridimensionnelle du transfert d'eau, de chaleur et de masse dans l'aquifère géothermique du Dogger dans le Bassin de Paris Thèse de l'Ecole Nationale des Mines de Paris Spécialité Hydrologie et Hydrogéologie Quantitative, 300 pp

Wei HF, Ledoux E, Marsily G (1990) Regional modeling of groundwater and salt environmental tracers transport in deep aquifers in the Paris Basin J Hydrol 120 341–358

Paris Basin and Sea of Iroise

Sequence Stratigraphy of the Jurassic Series of the Paris–London Basin

C. M. Ponsot

Introduction

This chapter is the result of different studies done either for private companies or in close collaboration with universities, and represents thus a synthesis of works.

The sequence stratigraphic terminology used in this chapter was defined by the Exxon research staff (Vail et al. 1977; Van Wagoner et al. 1988, etc.); the interpretation procedure for subsurface data was largely described by Vail (1987) in the SEPM special publication no. 42 (1988), by Van Wagoner et al. (1990) etc; and by Vail et al. (1987) and Rioult et al. (1991), etc. for the outcrops procedure.

Most of the depositional sequences described are third-order sequences, i.e., whose duration is between 0.5 to 3 Ma, except for a few of them, which are fourth-order sequences, i.e., with a duration of 0.1 to 0.5 Ma. There is no difference between the two types, just a matter of time duration; they both have sequence boundaries and systems tracts. On the other hand, the difference between a fourth-order sequence and a parasequence is significant: a parasequence belongs to a systems tract and represents the deposition between two flooding surfaces; it is a "stratal building block of a sequence" (Van Wagoner et al. 1990). A second-order cycle corresponds to the regressive and transgressive trends of series of deposits,

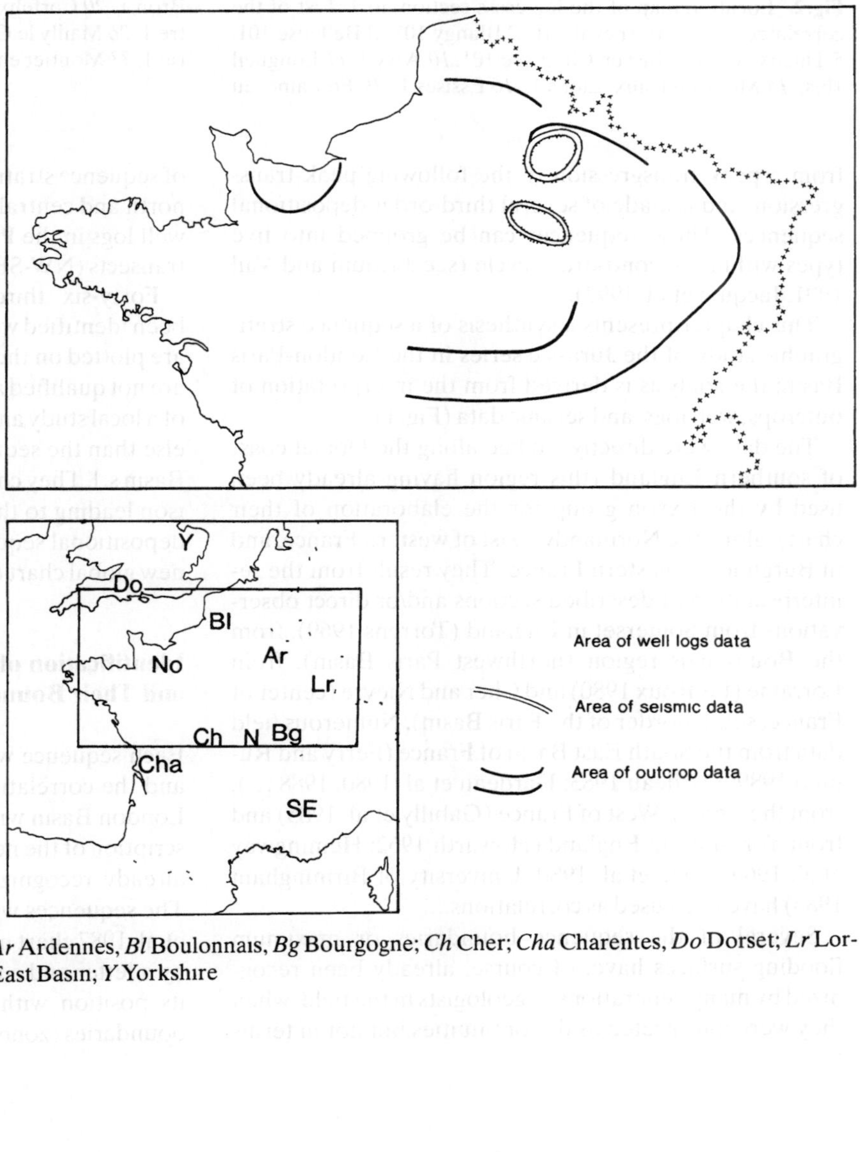

Fig. 1. Location map of the studied areas *Ar* Ardennes, *Bl* Boulonnais, *Bg* Bourgogne; *Ch* Cher; *Cha* Charentes, *Do* Dorset; *Lr* Lorraine; *N* Nièvre, *No* Normandy; *SE* South East Basin; *Y* Yorkshire

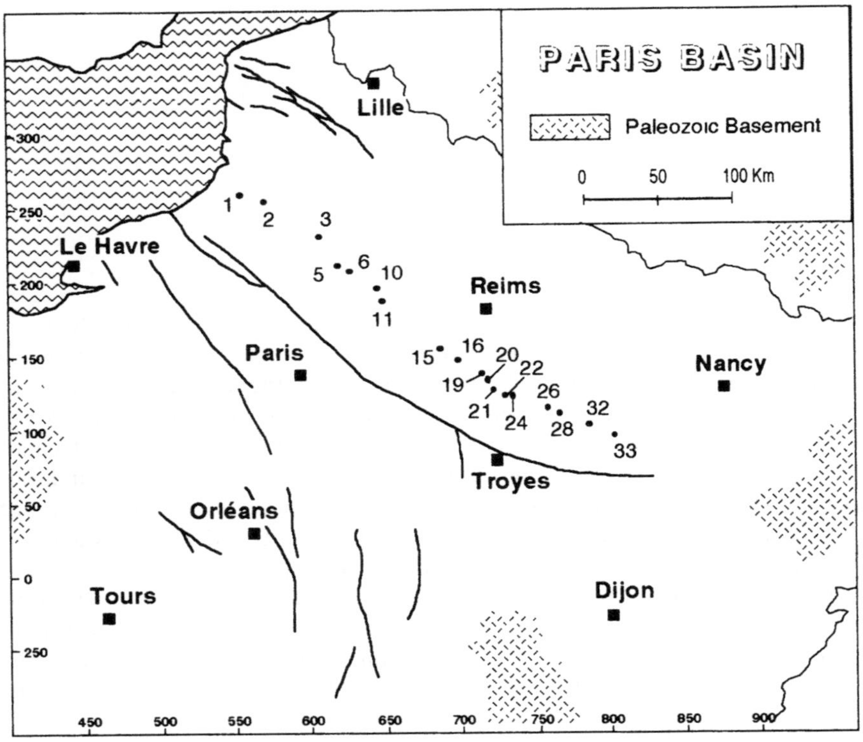

Fig. 2. Location map of the log cross section no 1 List of the correlated logs *1* Berneval 101, *2* Blangy 101, *3* Belleuse 101, *5* Thieux 101, *6* St Just en Chaussee 101, *10* Arsy 1, *11* Longueil 1bis, *15* Montreuil aux Lions 1, *16* Essises 1, *19* Fontaine au Bron 1, *20* Corfelix 1, *21* Chapton 1, *22* St Loup 1, *24* Connantre 1, *26* Mailly le Camp 103, *28* Mailly le Camp 102, *32* Margerie 1, *33* Montier en Der 1

from a peak transgression to the following peak transgression, and is made of several third-order depositional sequences. These sequences can be grouped into five types within a second-order cycle (see Jacquin and Vail 1991; Jacquin et al. 1992).

This chapter presents a synthesis of a sequence stratigraphic study of the Jurassic series in the London-Paris Basin; the analysis is derived from the interpretation of outcrops, well logs, and seismic data (Fig. 1).

The data were directly studied along the Dorset coast of southern England (this region having already been used by the Exxon group for the elaboration of their chart), along the Normandy coast of western France, and in Burgundy in eastern France. They result from the reinterpretation of described sections and/or direct observations from Somerset in England (Torrens 1969), from the Boulonnais region (northwest Paris Basin), from Lorraine (Le Roux 1980) and Cher and Nievre (center of France, south border of the Paris Basin). Numerous field data from the South East Basin of France (Ferry and Rubino 1989; Dardeau 1983; Dardeau et al 1980, 1988 ...), from the Center-West of France (Gabilly et al. 1985) and from Yorkshire in England (Howarth 1962; Hemingway et al. 1969; Cope et al 1980, University of Birmingham 1984) have been used as correlations.

Several of the sequence boundaries or maximum flooding surfaces have, of course, already been recognized by many generations of geologists in the field, when they were interpreted as discontinuities but not in terms of sequence stratigraphy. The seismic data come from the north and central parts of the Paris Basin; several tens of well logs in the Paris Basin have been studied along two transects (NW-SE and NE-SW), (Figs. 2–4, pp. 78, 80–83)

Forty-six third-order depositional sequences have been identified within the Jurassic series, whereas only 29 are plotted on the Haq et al. 1987 chart. These sequences are not qualified as global; this chapter presents the result of a local study and does not claim to represent something else than the sequence stratigraphy of the Paris-London Basin s.l. They could be (and are being) used for comparison leading to the recognition of the global versus local depositional sequences, for the purpose of establishing a new global chart (Hardenbol, 1992).

Identification of the Sequences, the Systems Tracts, and Their Bounding Surfaces

Each sequence will be described in at least one outcrop and the correlation with other places around the Paris-London Basin will be given. The focus will be on the description of the new identified sequences, but some of the already recognized sequences will also be reinforced The sequences will be referred to (1) following the Haq et al. 1987 chart nomenclature (supercycle and cycle), (2) by a letter and a number corresponding to the stage and its position within the stage, and (3) by their faunal boundaries (zone or subzone, when possible).

They are summarized in Table 1 (pp. 86–90). One well log has been taken as an example for the log responses (Chapton, central Paris Basin – Fig. 5, pp. 84, 85).

Lias

The first recognized sequence boundary is, in fact, of upper Rhetian age and is dated at 211 Ma. It fits the Haq B. V. et al. chart, and was reinforced by work in the South East Basin (Dumont 1992; Rohl et al. 1991).

Sequence 1: UAB 2-2.1/Ht. 1/Pre-Planorbis – Angulata zone

This sequence was already recognized in the Haq et al. chart (for the lower and middle Lias, the Dorset chart is summarized in Table 2, p. 91).

In Dorset (Lyme Regis, Pinhay bay), it corresponds to the White Lias Fm. and the lower half of the Blue Lias Fm.; the lower sequence boundary being the 211 and the upper sequence boundary coinciding at the top of the Brick ledge bed with an important gap (at least the lower half of the Angulata zone). This gap is known all round the Paris-London Basin and the South East Basin. The White Lias would represent a limy shelf margin complex, its upper erosional surface an important ravinement surface corresponding to the transgressive surface (within pre-Planorbis) overlain by a transgressive alternation of shales (usually bituminous) and marls or thin limy beds. The top of Lang's bed H 66 (Lang, 1924) represents the maximum flooding surface (within Portlocki subzone).

On some well logs of the central and northern part of the Paris Basin, this sequence could be apparently divided into two sequences, but this breakup was not documented by either cores or outcrops, so is not taken into account here.

Sequence 2: UAB 3-3.1/Ht-Si/Complanata subzone – top of Scipionianum subzone

In Dorset (Lyme Regis, Pinhay bay), this sequence comprises the upper Blue Lias Fm. and lower part of the Shales with Beef Fm., the upper sequence boundary being at the Table ledge bed level. The upper Blue Lias Fm. presents a progressive decrease in the amount of limy beds versus black shales in an alternating pattern, indicating a deepening of the depositional environment, and thus corresponding to a transgressive systems tract. The lower Shales with Beef Fm. is an organically very rich shale interval (TOC up to 50%) interpreted as a condensed section-early highstand systems tract, the maximum flooding surface being perhaps the Blue Lias/Shales with Beef limit (i. e., the grey ledge bed which has an erosional surface and an important faunal content – top of the Bucklandi zone). From dated well logs of the northern Paris Basin, the maximum flooding surface is here in the upper part of the Bucklandi zone. In the South East Basin of France, the maximum flooding surface seems to be identified on top of the Rotiforme subzone (P. E. Graciansky and G. Dardeau, pers. comm.).

Sequence 3: UAB 3-3.2/Si. 2/base of Sauzeanum subzone basal part of Birchi subzone

In Dorset (Lyme Regis), this sequence is entirely contained within the Shales with Beef Fm. Its upper sequence boundary is very tenuous. From the lower sequence boundary (Table ledge bed), the section shows an alternation of blue-gray marls and layers of beefs (i. e., fibrous calcite beds) with an upward increase in the amount of marl until the Little ledge bed, where the marls are replaced by thin laminated bituminous shales still alternated with beefs and nodules until the Tabular limestone bed (Brooki/Birchi subzone limit). Above this marker bed lies a thin interval of dark shales (1.25 m) which represent a condensed section, the Tabular limestone bed being its basal surface, ending the very progressive transgressive systems tract made of marl and beef alternation, with marls dominating, and finally bituminous shales.

This sequence was also very clear on the logs, especially the clayey peak of the maximum flooding surface, with a highstand systems tract a little more developed in the Paris Basin than on its periphery.

Sequences 2 and 3 of the lower Sinemurian stage are given as one in the Haq et al. 1987 chart; they are both characterized by condensed sections often rich in organic content.

Sequence 4: UAB 3-3.3/Lt. 1/within Birchi subzone – upper Obtusum zone

In Dorset (Cain's Folly) this sequence corresponds to the upper part of the Shales with beef Fm. and part of the Black ven marls Fm. The about 5 m of marls and nodules of the upper part of the Shales with beef ends by the Birchi nodular bed and represents a thin shelf margin complex and its upper transgressive surface. The overlying 30 m of shale and nodule alternation, pyrite- and organic-rich (TOC up to 6–8%), correspond to the transgressive systems tract until the Coinstone marker bed. This erosional Coinstone bed is characterized by a hardground with burrows, abundant pyrite, and an important gap (three subzones). This winowing surface is a major maximum flooding surface (top of Stellare subzone). It is widely recognized in western Europe, as in Yorkshire (Robin Hood Bay) at the calcareous shale/siliceous shale limit, in Burgundy (Dommergues et al. 1991), or on the logs of the Paris Basin (Ponsot and Vail 1991). It represents in the mean time the upper sequence boundary, the highstand being missing, certainly eroded and corresponding in time to some of the three missing subzones (see also Hesselbo and Jenkyns 1992 for details).

Sequence 5: UAB 3-3.4/Lt. 2/top of Obtusum zone – Raricostatum zone

In Dorset, the upper part of the Black ven Marls Fm. corresponds to a sequence bounded at the top by the Hummocky limestone marker bed. An important gap is identified at this level, the subzones Macdonnelli and Aplanatum being missing.

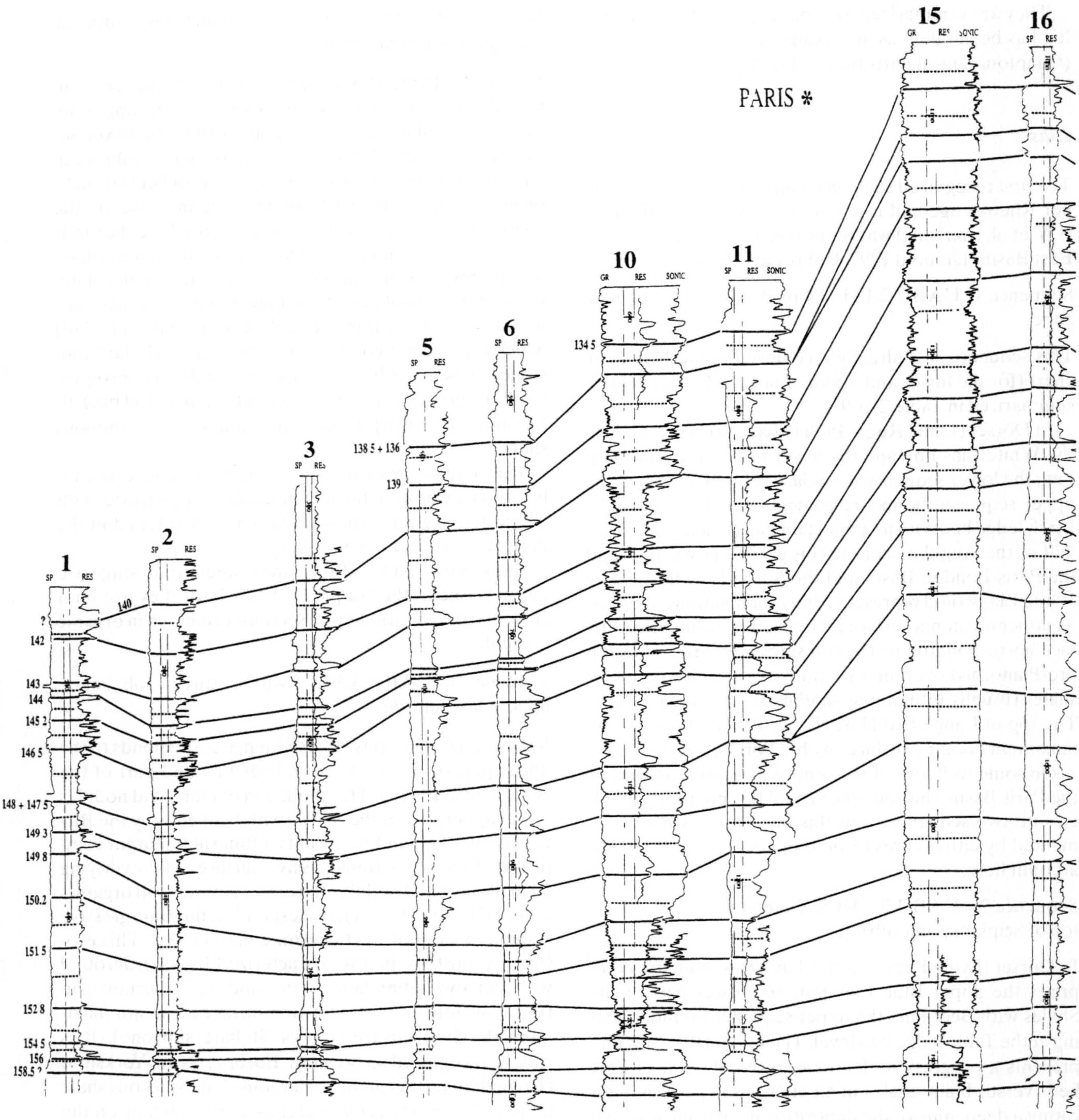

Fig. 3. Cross section no 1 Upper Jurrassic part *Left* NW part of Paris Basin, *right* SE part of Paris Basin (for more details, see Fig 4) (Ponsot 1991)

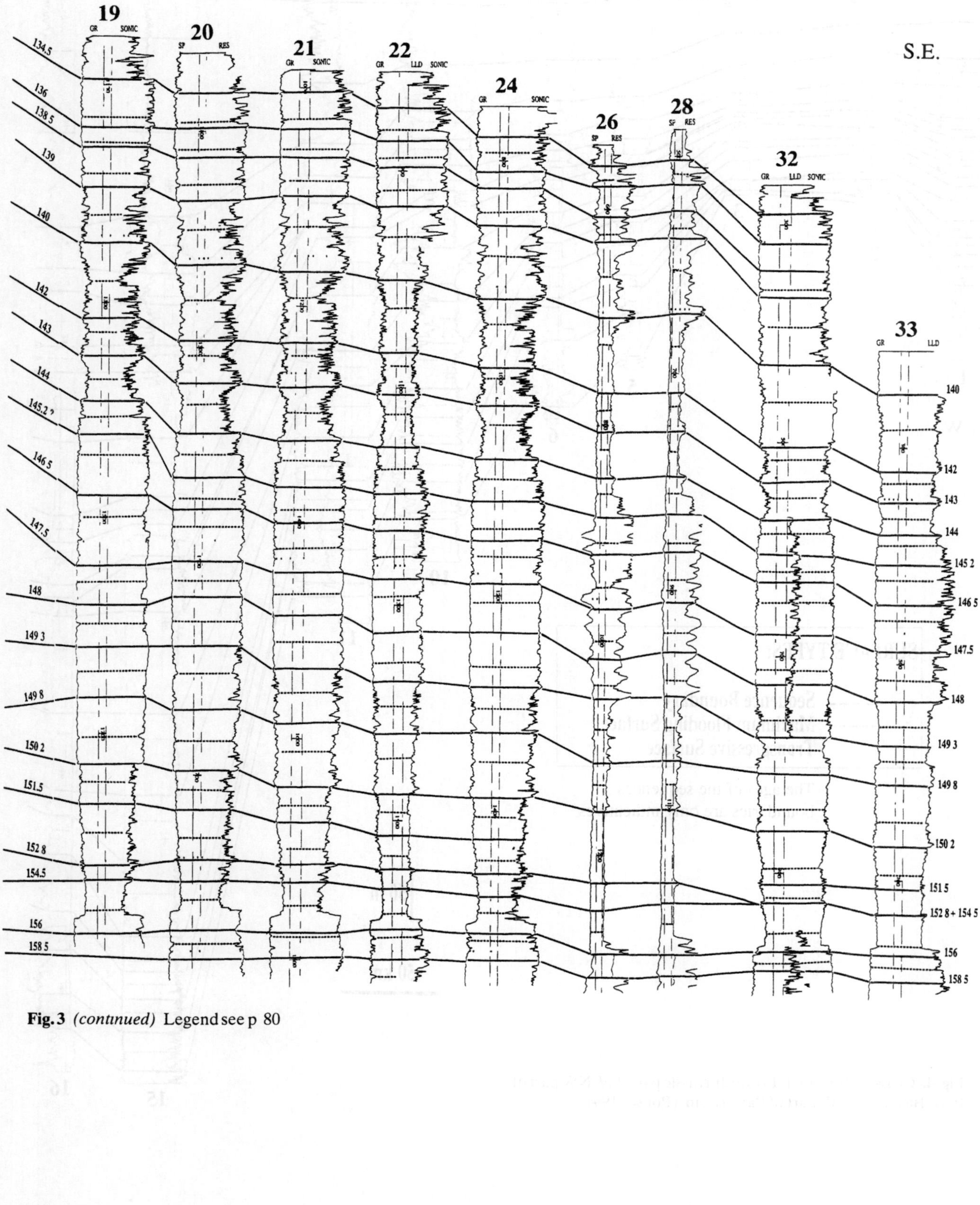

Fig. 3 *(continued)* Legend see p 80

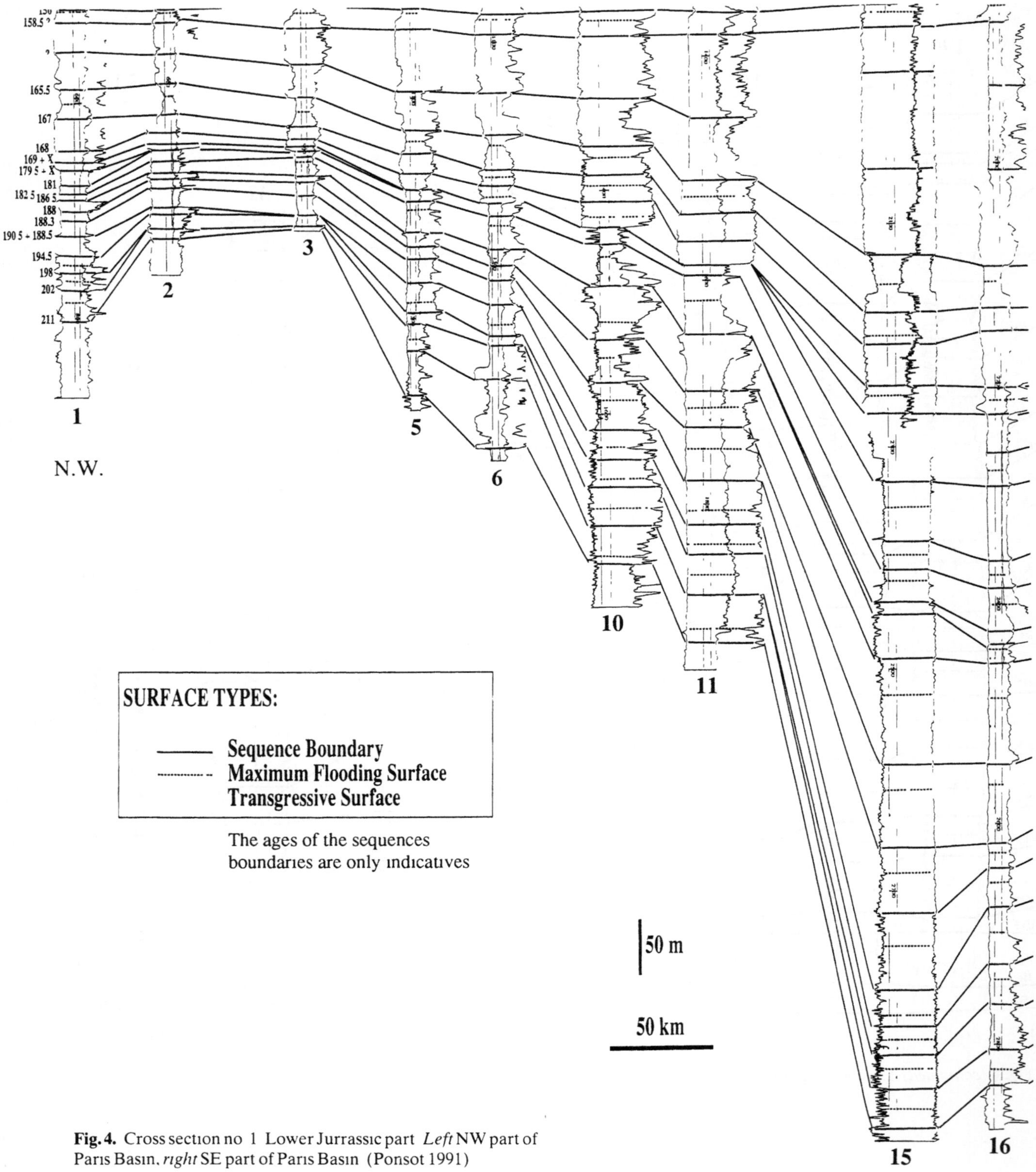

Fig. 4. Cross section no 1 Lower Jurrassic part *Left* NW part of Paris Basin, *right* SE part of Paris Basin (Ponsot 1991)

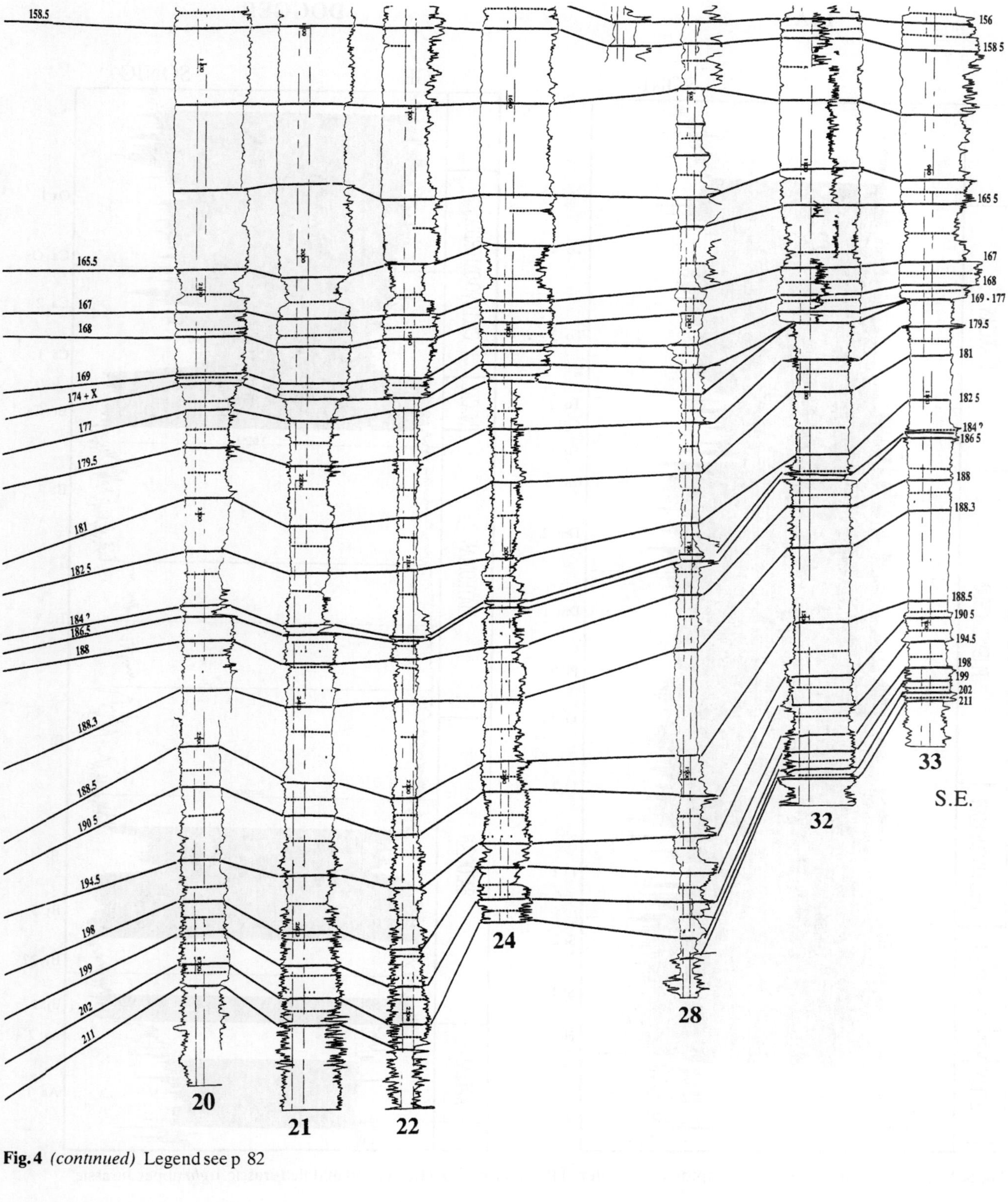

Fig. 4 *(continued)* Legend see p 82

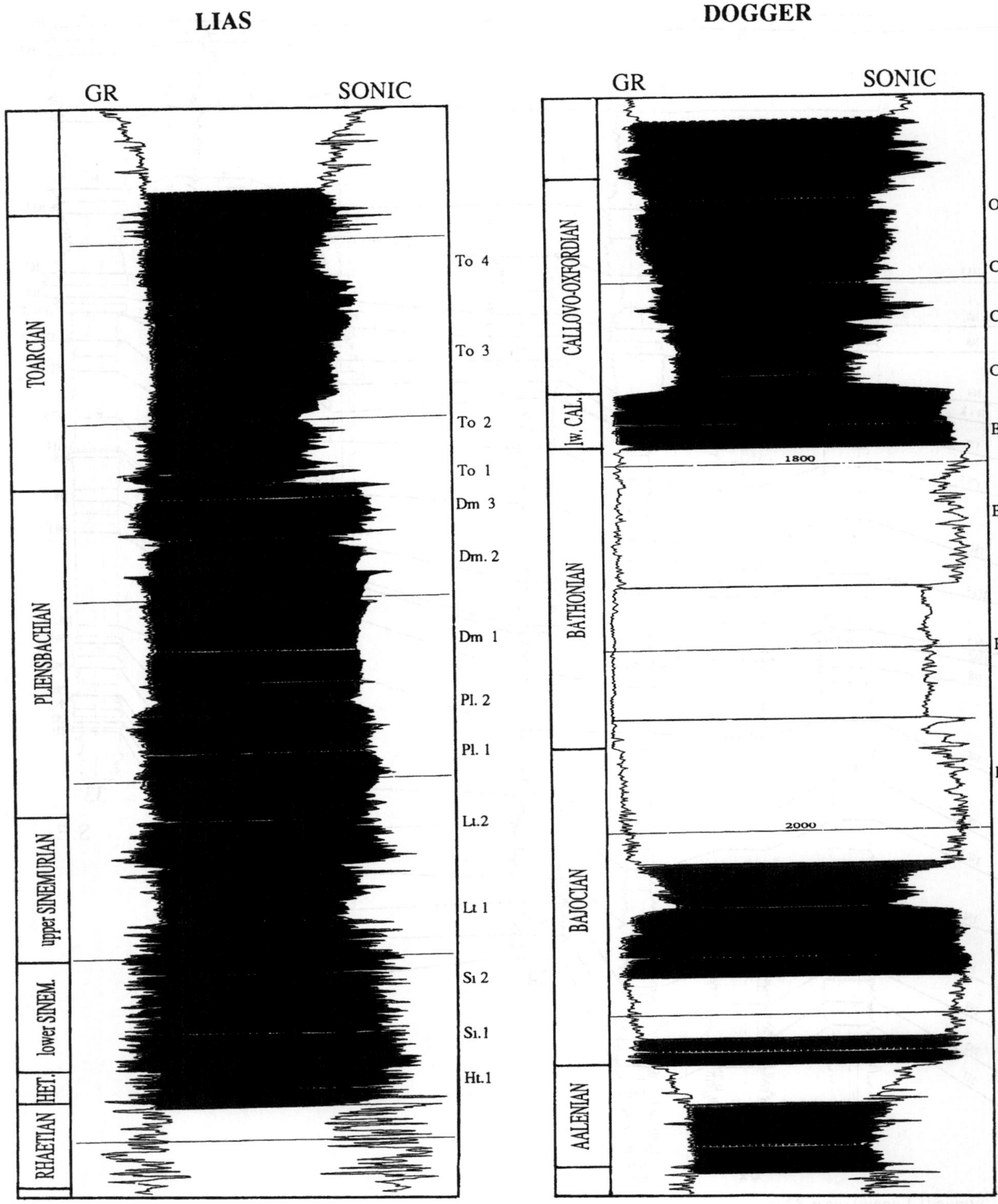

Fig. 5. Sequence stratigraphy of the Chapton well (center of Paris Basin) *Left* Lower and middle Jurassic, *right* upper Jurassic

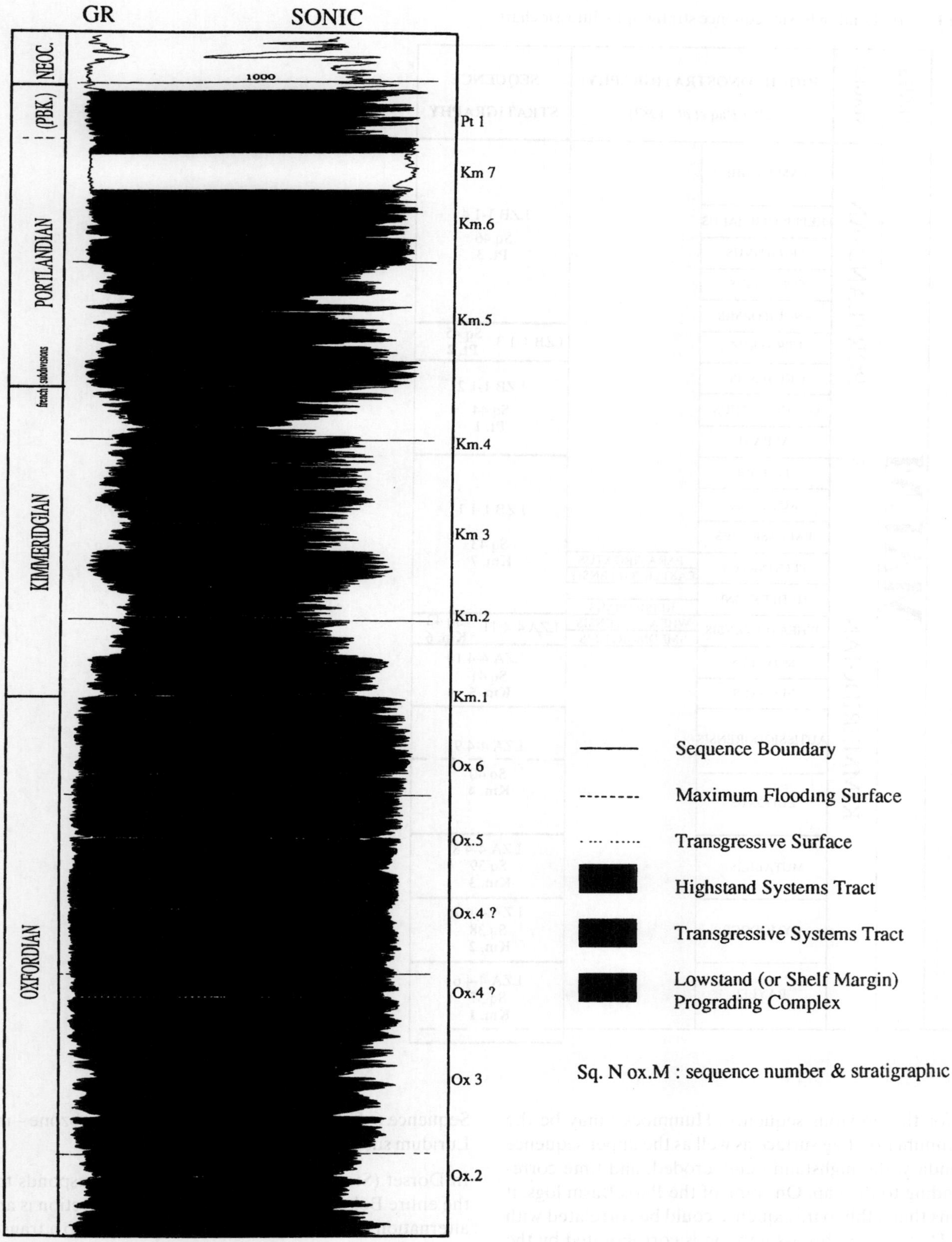

Fig.5 *(continued)* Legend see p 84

Table 1. Paris-London Basin sequence stratigraphy Jurassic chart

TIME MY	SERIES	STAGES	BIOCHRONOSTRATIGRAPHY (after Haq *et al* 1987)		SEQUENCE STRATIGRAPHY
131	MALM	PORTLANDIAN	LAMPLUGHI		LZB 1-1 4 Sq 46 **Pt. 3**
			PREPLICOMPHALUS		
			PRIMITIVUS		
			OPPRESSUS		
			ANGUIFORMIS		
			KERBERUS		LZB 1-1 3 Sq 45 **Pt. 2**
			OKUSENSIS		LZB 1-1 2 Sq 44 **Pt. 1**
135			GLAUCOLITHUS		
			ALBANI		
		136			
		KIMMERIDGIAN	FITTONI		LZB 1-1 1 Sq 43 **Km. 7**
			ROTUNDA		
			PALLASIOIDES		
			PECTINATUS	PARAVIRGATUS	
				EASTLECOTTENSIS	
			HUDLESTONI	REISIFORMIS	
			WHEATLEYENSIS	WHEATLEYENSIS	LZA 4-4 11 Sq 42 **Km. 6**
				SMEDMORENSIS	
			SCITULUS		LZA 4-4 10 Sq 41 **Km. 5**
			ELEGANS		
140			AUTISSIODORENSIS		LZA 4-4 9 - - - Sq 40 **Km. 4**
			EUDOXUS		
			MUTABILIS		LZA 4-4 8 Sq 39 **Km. 3**
			CYMODOCE		LZA 4-4 7 Sq 38 **Km. 2**
			BAYLEI		LZA 4-4 6 Sq 37 **Km. 1**
145					

As for the previous sequence, Hummocky may be the maximum flooding surface as well as the upper sequence boundary, the highstand being eroded, and time corresponding to the gap. On some of the Paris Basin logs, it seems that a thin parasequence could be correlated with this Dorset gap: this assumption is corroborated by the faunal distribution on the northwestern Europe scale (Dommergues and Meister 1991). In this case, we may here have the expressed highstand.

Sequence 4 and 5 of Lotharingian age are given as one in the Haq B. U. et al. 1987 chart.

Sequence 6: UAB 3-3.5/Pl. 1/base of Taylori subzone – in Luridum subzone

In Dorset (Stonebarrow), this sequence corresponds to the entire Belemnite Marls Fm. Most of the section is an alternation of dark and pale marls equivalent to a transgressive systems tract. The upper part [Lang's bed 120 (1928), pyritic marls] may be a condensed section including an early highstand, the Belemnite stone marker bed being the upper sequence boundary (dated as Luridum subzone). Several of the Paris Basin logs record a lowstand prograding complex (up to 10 m) at the base of this sequence (Trouiller et al. 1992).

Table 1 *(continued)*

TIME MY	SERIES	STAGES	BIOCHRONOSTRATIGRAPHY (after Haq *et al.* 1987)		SEQUENCE STRATIGRAPHY
145	lw. MALM	OXFORDIAN	ROSENKRANTZI	BAUHINI	LZA 4-4 5 Sq 36 Ox. 6
				MARSTONENSE	
			REGULARE		
			SERRATUM	SERRATUM	LZA 4-4 4 Sq 35 Ox. 5
				KOLDEWEYENSE	
			GLOSENSE	GLOSENSE	
				ILOVAISKII	or
			TENUISERRATUM	BLAKEI	LZA 4-4 3
				TENUISERRATUM	Sq 34 Ox. 4
			DENSIPLICATUM	MALTONENSE	
				VERTEBRALE	LZA 4-4 2 Sq 33 Ox. 3
150			CORDATUM	CORDATUM	LZA 4-4 1 Sq 32 Ox. 2
				COSTICARDIA	
				BUKOWSKII	LZA 3-3 5 Sq 31 Ox. 1
			MARIAE	PRAECORDATUM	
				SCARBURGENSE	LZA 3-3 4 Sq 30 Ca.-Ox.
152	up. DOGGER	CALLOVIAN	LAMBERTI	LAMBERTI	
			ATHLETA	HENRICI	
				SPINOSUM	
				PRONIAE	
				PHAEINUM	LZA 3-3 3
			CORONATUM	GROSSOUVREI	Sq 29 Ca. 2
				OBDUCTUM	
			JASON	JASON	
				MEDEA	
155				ENODATUM	LZA 3-3 2
			CALLOVIENSE	CALLOVIENSE	Sq 28 Ca. 1
				KOENIGI	
			MACROCEPHALUS	KAMPTUS	
				MACROCEPHALUS	LZA 3-3 1
			DISCUS	DISCUS	Sq 27 Bt.-Ca.
				HOLLANDI	

This sequence was already recognized in the Haq B. U. et al. 1987 chart.

Sequence 7: UAB 3-3.6/Pl. 2/Luridum subzone – within lower Stokesi subzone

In Dorset, this sequence corresponds to the entire Green Ammonite Beds Fm. The upper sequence boundary is a major, very sharp, erosional surface just beneath the lowest Three Tiers Fm.

The lower half of the section presents the typical transgressive pattern with clay and nodule or limestone bed alternation until the Red Band bed [Lang's bed 126 (1936)], which is interpreted as the maximum flooding surface. Then follows a clayey highstand systems tract with an increasing loamy and silty trend.

From the English biostratigraphic data, the upper sequence boundary would not be at the Carixian/Domerian limit but already within the base of the Domerian stage. This sequence can be well developed within the Paris Basin subsurface, with notably a clear highstand made of decreasing (in size) – prograding parasequences.

This sequence was already recognized in the Haq B. U. et al. 1987 chart.

Sequence 8: UAB 4-4.1/Dm. 1/Lower Stokesi subzone – upper Stockesi subzone?

The early Domerian appears to be a period of thick sedimentation in relation with high subsidence rate. In Dorset the Stokesi subzone is recognized on more than 100 m of thickness.

Table 1 *(continued)*

TIME MY	SERIES	STAGES	BIOCHRONOSTRATIGRAPHY (after Haq *et al* 1987)		SEQUENCE STRATIGRAPHY
	DOGGER	BATHONIAN	ASPIDOIDES		LZA 2-2 8 Sq 26 **Bt.5**
			HODSONI		
160			MORRISI		LZA 2-2 7 Sq 25 **Bt.4**
			SUBCONTRACTUS		LZA 2-2 6 Sq 24 **Bt.3**
			PROGRACILIS		LZA 2-2 5 Sq 23 **Bt 2**
			TENUIPLICATUS		
			ZIGZAG	YEOVILENSIS	LZA 2-2 4 Sq 22 **Bt 1**
				MACRESCENS	
				CONVERGENS	
165		165	PARKINSONI	BOMFORDI	
		BAJOCIAN		TRUELLI	LZA 2-2 3
			GARANTIANA	ACRIS	Sq 21 **Bj 4**
				TETRAGONA	
				SUBGARANTI	
				DICHOTOMA	
			SUBFURCATUM	BACULATA	LZA 2-2 2 Sq 20 **Bj.3**
				POLYGYRALIS	
				BANKSI	
			HUMPHRIESIANUM	BLAGDENI	or / and ? LZA 2-2 1 Sq 19 **Bj 2**
				HUMPHRIESIANUM	
				ROMANI	
			SAUZEI		LZA 1-1 3
170			LAEVIUSCULA	LAEVIUSCULA	Sq 18
				OVALIS	**Bj 1**
			DISCITES		
		171	CONCAVUM	FORMOSUM	LZA 1-1 2 Sq 17 **Aa.3**
		AALENIAN		CONCAVUM	
			MURCHISONAE	GIGANTEA	
				BRADFORDENSIS	↓ LZA 1-1 1 Sq 16 **Aa.2**
175				MURCHISONAE	
				HAUGI	
			OPALINUM	SCISSUM	UAB 4-4 8
				OPALINUM	Sq 15 **Aa.1**

This sequence 8 in Dorset (Golden Cap/Seatown) corresponds to the Three Tiers and the Eype Clay Fm. The Three Tiers is interpreted as a shelf margin systems tract. The first about 40 m of the Eype Clay Fm. are made of micaceous clays and marls alternated with a few silty beds, and represent the transgressive systems tract. The overlying Eype Nodule Bed is the maximum flooding surface; it is either a gray clay bed with small nodules of calcite or locally a calcareous sandstone, but in both cases characterized by an important faunal distribution (ammonites, bivalves, gastropods, corals, etc). The following about 20 m of blue marls are the highstand systems tract until the Shell Bed, which marks the upper sequence boundary

Sequence 9· UAB 4-4.2/Dm.2/upper Stokesi subzone? – Subnodosus subzone?

In Dorset (Seatown) this sequence corresponds to the extreme top of the Eype Clay Fm., the Downcliff Sands Fm., and the Margaritatus stone and clay Fm The small interval comprising the Shell bed and the Starfish bed is interpreted as a small shelf margin complex. The about 30 m of silts and sands of the Down Cliff Sands represent the transgressive systems tract; the massive Margaritatus stone with abundant ammonites and ironshots is the maximum flooding surface overlain by a few meters of shaly crowded clays considered as a condensed early highstand systems tract.

The precise datation (subzone level) of these two sequences is not very sure, but they are both of Margaritatus zone age.

From the Paris Basin subsurface data, the datation is not better detailed, and the two Margaritatus sequences are not recognized on all logs; they both seem to be identifiable only in what would at that time have been the deeper parts of the basin (particularly the actual central-northern Paris Basin).

Sequence 10: UAB 4-4.3/Dm.3/Subnodosus subzone? – top? of Hawskerense subzone

In Dorset (Seatown/Eypemouth) the sequence corresponds to the Thorncombe Sands, the Thorncombiensis Bed, and part of the Marlstone bed of the Junction beds Fm. The Thorncombe sands and doggers represent a shelf margin complex; the transgressive surface is the thin crowded marly limestone at the base of the Thorncombiensis grey clays which are interpreted as the small transgressive systems tract. The overlying gap of nondeposition (part of the Apyrenum subzone) corresponds to the maximum flooding surface and the Marlstone Rock bed to the condensed small highstand.

In Yorkshire (near Straithes) the maximum flooding surface would correspond to the uncomformity below the Main Seam ironstone which cuts out successively older beds (Powell J. H. 1984), and the upper sequence boundary may correspond to the Howarth bed 56 (Howarth M. K. 1955), which presents a very irregular top surface below the overlying shales.

This important sequence boundary also appears very clearly on the Paris Basin Logs (cf. Fig. 4).

The maximum flooding surface can be very well marked and underlined by a condensed period on some of the subsurface data (verified from cores)

The Domerian stage is now subdivided in three sequences, whereas two global ones are indicated in the Haq B. U. et al. 1987 chart.

Table 1 *(continued)*

TIME MY	SERIES	STAGES	BIOCHRONOSTRATIGRAPHY (after Haq *et al* 1987)		SEQUENCE STRATIGRAPHY
	LIAS	TOARCIAN	LEVESQUEI	AALENSIS	
				MOOREI	UAB 4-4 7
180				LEVESQUEI	
				DISPANSUM	Sq 14 To. 4
			THOUARSENSE	FALLACIOSUM	
				STRIATULUM	UAB 4-4 6 Sq 13 To. 3
			VARIABILIS	VITROSA	
				ILLUSTRIS	
				VARIABILIS	
			BIFRONS	CRASSUM	UAB 4-4 5
				FIBULATUM	
				COMMUNE	Sq 12 To. 2
			FALCIFERUM	FALCIFERUM	
				EXARATUM	UAB 4-4 4 Sq 11 To. 1
			TENUICOSTATUM	SEMICELATUM	
185				TENUICOSTATUM	
				CLEVELANDICUM	
				PALTUM	
186		PLIENSBACHIAN	SPINATUM	HAWSKERENSE	UAB 4-4 3
				APYRENUM	
			MARGARITATUS	GIBBOSUS	Sq 10 Dm.3 ↑?
				SUBNODOSUS	UAB 4-4 2 Sq 9 Dm. 2
				STOKESI	UAB 4-4 1 Sq 8 Dm. 1
			DAVOEI	FIGULINUM	UAB 3-3 6
				CAPRICORNUS	Sq 7 Pl. 2
190				MACULATUM	
			IBEX	LURIDUM	UAB 3-3 5 Sq 6 Pl. 1
				VALDANI	
				MASSEANUM	
			JAMESONI	JAMESONI	
				BREVISPINA	
				POLYMORPHUS	
				TAYLORI	

Sequence 11: UAB 4-4.4/To. 1/Lower Tenuicostatum zone – top of Exaratum subzone

This new sequence is very particular. It is not recognized everywhere in the Paris-London Basin and surrounding areas. However, on several logs coming from the central part of the Paris Basin, a thin but clear sequence appears at the basal part of the so-called schistes carton. It thins and disappears toward the W, the NW, but seems to persist ab specific points toward the SW and NE (cf. Fig. 4).

It has been very clearly identified on logs of the central-northern part of the Basin, with even a small sandy lowstand (a few meters), and organic-rich transgressive and early highstand system tracts. From the given data, the maximum TOC values are found within the transgressive systems tract (Trouiller A. et al. 1992). We found this sequence also on logs of the Couy borehole in the southern Paris Basin (Lorenz C. et al. 1991) and on outcrops in Vendée (outcrop section from Anse Saint Nicolas, Jard-Gabilly J. 1976).

This sequence does not exist in Dorset, where the Tenuicostatum zone and certainly part of the Exaratum subzone seem to be missing (following the very precise work of Howarth M. K. 1980, it may not be absent but

Table 1 *(continued)*

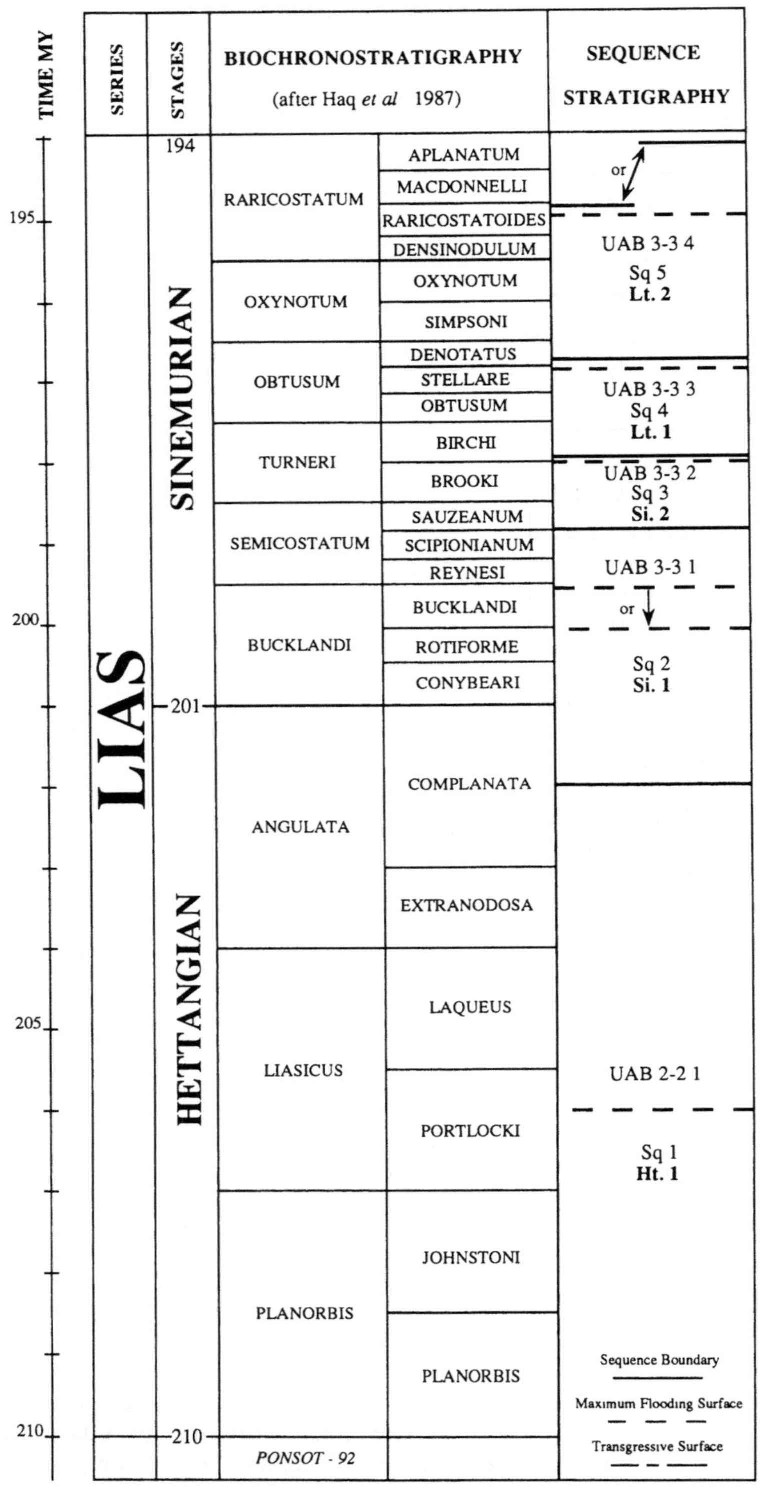

very condensed, the Marlstone bed being divided in three, the lower layer R a coarse conglomeratic and oolitic limestone belonging to the Apyrenum subzone – maximum flooding surface of sequence 10, the middle layer Px a hard grey and pink limestone of the Hawskerense subzone – highstand of sequence 10, and the third layer, layer P, being highly condensed, very rich in ammonites from the Hawskerense subzone to the Semicelatum subzone). However in looking at the Yorkshire coast (Blea Wyke, Saltwick Nab, Port Mulgrave), it seems possible to identify this sequence and give some datation: following Powell's nomenclature (Powell J. M. 1984), it corresponds to the Grey Shale member and part of the Jet Rock member of the Whitby mudstone Fm. The Grey

Table 2. Lower and middle Lias chart for the Dorset region

SERIES	STAGES	BIOCHRONOSTRATIGRAPHY (after Haq *et al* 1987)		SEQUENCE STRATIGRAPHY	SEQ. N°	MARKER BED	FORMATION
LIAS		FALCIFERUM	EXARATUM				Junction Beds
		TENUICOSTATUM	SEMICELATUM				
			TENUICOSTATUM				
			CLEVELANDICUM				
			PALTUM				
	PLIENSBACHIAN	186 — SPINATUM	HAWSKERENSE	Dm.3		Marlstone bed	
			APYRENUM	— —			
		MARGARITATUS	GIBBOSUS		10		Thornc. Bed
			SUBNODOSUS	Dm.2			Thornc. Sands
					9	Marg. stone Shell bed	Marg. Clays
				Dm.1			Down Cliff Sands
			STOKESI	Pl.2	8	Eype Nodule bed	Th Tiers/Eype Cl.
		DAVOEI	FIGULINUM	— —		Red Band	GREEN AMMONITE BEDS
			CAPRICORNUS		7		
			MACULATUM	Pl.1		Belemnite Stone	
		IBEX	LURIDUM				BELEMNITE MARLS
			VALDANI				
			MASSEANUM				
		JAMESONI	JAMESONI		6		
			BREVISPINA				
			POLYMORPHUS				
			TAYLORI				
	SINEMURIAN	194 — RARICOSTATUM	APLANATUM	Lt.2 or			
			MACDONNELLI			Hummocky	
			RARICOSTATOIDES				
			DENSINODULUM				
		OXYNOTUM	OXYNOTUM		5		BLACK VEN MARLS
			SIMPSONI				
		OBTUSUM	DENOTATUS	Lt.1		Cornstone	
			STELLARE				
			OBTUSUM		4		
		TURNERI	BIRCHI	Si.2		Birchi Nodular Tabular Limestone	SHALES WITH BEEF
			BROOKI		3		
		SEMICOSTATUM	SAUZEANUM	Si.1		Table Ledge	
			SCIPIONIANUM				
			REYNESI			Grey Ledge	
		BUCKLANDI	BUCKLANDI	or ↓			BLUE LIAS
			ROTIFORME		2		
			CONYBEARI				
	HETTANGIAN	201 — ANGULATA	COMPLANATA	Ht			
			EXTRANODOSA			Brick Ledge	
		LIASICUS	LAQUEUS				
			PORTLOCKI	— —		Lang's bed H66	
		PLANORBIS	JOHNSTONI		1		
			PLANORBIS				
	RHAE.	210 —		— —			WHITE LIAS
				211			

C M PONSOT 92

Shale would represent the transgressive systems tract with its alternating pattern of increasing shales versus nodules beds, the shell bed and the underlying few decimeters of pyritized nodules on top of the Grey Shale would be the maximum flooding surface. These grey Shales are dated as Tenuicostatum zone. The overlying lower part of the Jet Rock member is interpreted as the highstand systems tract until the Millstones which are the expression of the upper sequence boundary. This lower Jet Rock consists of bituminous silty mudstones with a series of nodule beds (Cannon ball, Whale stone, Curling stone), and progresses upward into argillaceous limestones. The highstand is dated as Exaratum subzone and the sequence boundary Millestones is at the Exaratum – Falciferum subzone limit. It seems that the maximum of anoxia, the most bituminous paper shales of this lower Toarcian period in Yorkshire, would correspond to this small highstand.

Sequence 12: UAB 4-4.5/To.2/base of Falciferum subzone – near top of Bifrons zone

The lower sequence boundary of this sequence is often characterized by an important hiatus, the previous sequence 11 being frequently missing.

This sequence 12 then corresponds to the schistes carton and equivalents.

In the central part of the Paris Basin, a small shelf margin complex is sometimes identified on the logs, but it disappears on the borders of the actual Paris Basin. A clear transgressive systems tract is well observed on the logs with an alternating and backstepping pattern in relation to a relatively important TOC content.

In Dorset, this sequence is not recognizable within the Junction beds; but on the Yorkhire coast (Blea Wyke), it comprises the upper part of the Jet Rock member and the Alum Shale member, the sequence boundary being at the top of the Cement shale unit and associated with the gap of the Crassum subzone. It is difficult to identify precisely the maximum flooding surface on these outcrops, the Alum Shale member being a condensed section. This maximum flooding surface is observed in the Paris Basin logs, but is not precisely dated except as within the Bifrons zone.

This sequence exists also in the South East Basin of France but is sometimes very thin and very condensed (a few meters), it is dated as Falciferum/Sublevisoni subzones (de Graciansky et al. 1993); both sequences 11 and 12 can also be identified there.

Sequence 13: UAB 4-4.6/To.3/base of Variabilis zone – top of Striatulum subzone

This sequence is rather well expressed on the Paris Basin logs within the Toarcian marls, with a clear progressive enrichment in clays towards the maximum flooding surface and an opposite trend during the highstand. Its upper sequence boundary corresponds to the top of the Junction Beds in Dorset and to the top of the Peak Mudstone member in Yorshire.

This sequence was already recognized in the Haq B U et al. 1987 chart.

Sequence 14: UAB 4-4.7/To.4/base of Fallaciosum subzone – top of Moorei subzone

This upper Toarcian sequence also appears clearly on some of the Paris Basin logs; but it can be missing, a major hiatus existing from early upper Toarcian to the top of Aalenian because of nondeposition and/or erosion.

This sequence is well expressed also in Yorkshire (Ravenscar), where the Fox Cliff siltstone member represents a shelf margin complex followed by a transgressive systems tract and the two sandstone members of the overlying Blea Wyke Sandstone Fm. two packages of the highstand systems tract (Knox 1984). The upper sequence boundary corresponds there to the erosinal surface and gap of the Dogger Fm. In Dorset, this upper sequence boundary is tenuous within the lower Birdport Sands, but has been identified and dated in south Somerset (Ham Hill limestone, 7 km W of Yeovil) (Torrens H. S. 1969).

This sequence was already recognized in the Haq B. U. et al. 1987 chart.

Dogger

Sequence 15: UAB 4-4.8/Aa. 1/Aalensis subzone – top of Scissum subzone

This first sequence of Dogger is usually observed in the deepest parts of the Basin, i. e., on the central Paris Basin and in the Norman "marly ditch" logs and on outcrops in Dorset, but is not identified on the borders, except sometimes as reworked or very condensed material (for example, the oolite ferrugineuse aalenienne of the Normandy coast outcrops – Rioult M. et al 1991).

It has been observed on outcrops in Dorset (Burton Cliff), and corresponds to a large part of the Birdport Sands, the Rosty or Foxy Bed, and the Scissum Beds, these sands indicating a relatively shallow environment, with an important sedimentation rate, at least for the Birdport sands. The Foxy Bed is a highly faunal condensed clayey bed and represents the maximum flooding surface, the transgressive systems tract being perhaps very thin, even comprised within the Foxy bed. In this case, the Birdport Sands would be a shelf margin complex. The upper sequence boundary is clearly marked by the erosional surface of the overlying Yellow Conglomerate.

This sequence was already recognized in the Haq B. U. et al. 1987 chart.

Sequence 16: LZA 1-1.1/Aa.2/base of Haugi subzone – top of Bradfordensis subzone?

This sequence is not often identified because of the important and widespread hiatus of late Lias/early Dogger in the studied area; but in Burgundy (Solutre), this sequence is recognized as the lower member of the Calcaire à entroques Fm. (unit 1 a, Morestin 1986), which may, in

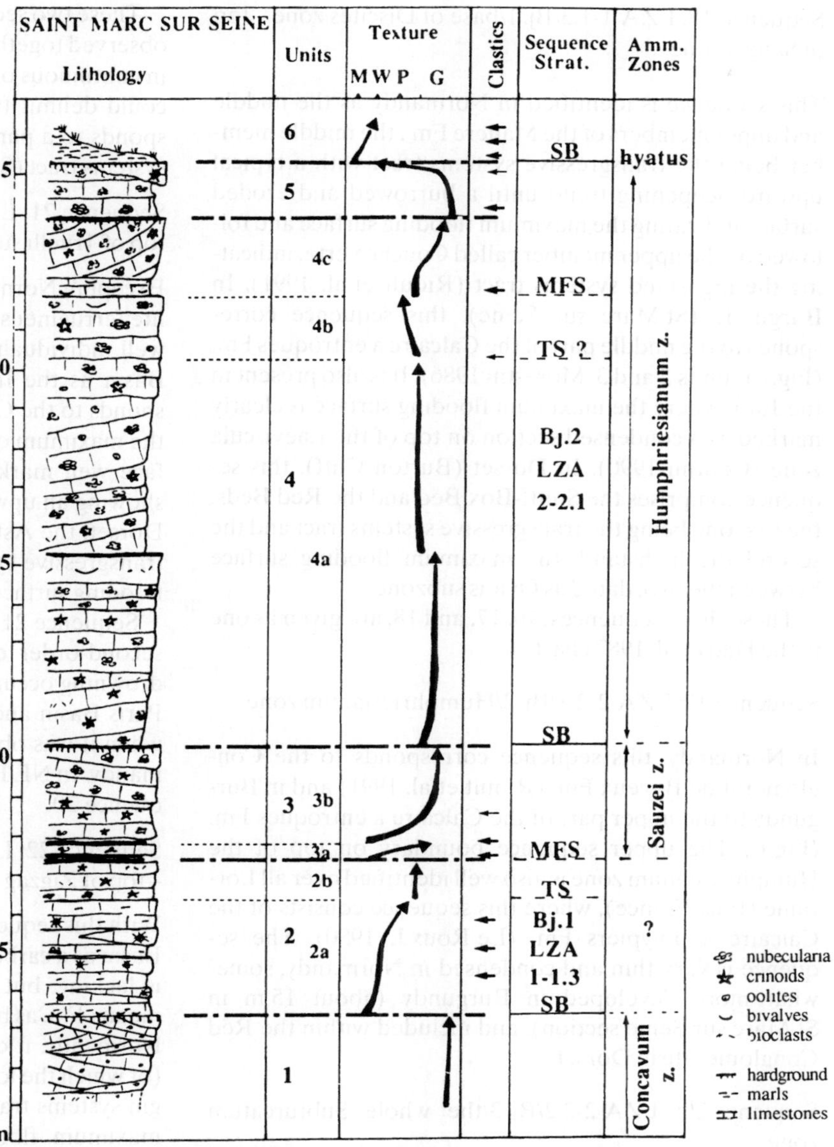

Fig. 6. Sequence stratigraphic interpretation of some Burgundian Bajocian Series (Modified after Morestin 1986)

fact, represent only the highstand systems tract. In Normandy, this sequence may correspond to the lower member of the Maliere Fm., but it is too thin and too condensed to be clearly identified as a depositional sequence. The upper sequence boundary has been precisely dated also in the Jura area as within the Murchisonae zone, before the Gigantea subzone and is recognized over the whole eastern France (Contini 1990). On the Dorset coast (Burton Cliff) the sequence is contained within the Yellow Conglomerate and thus is not identifiable; but its upper part may exist a few miles northward (Horn Park, Seawington) as a thin, hard, nonoolitic, sometimes crinoidal limestone overlain with a sharp contact by oolitic limestones of Concavum zone age.

Sequence 17: LZA 1-1.2/Aa. 3/Gigantea subzone – top of Concavum zone

This sequence is also rarely identified for the same reasons as the previous one. It will be totally missing in Normandy as a gap of this ammonite zone occurs within the Maliere Fm. In south Burgundy (Solutre), it is identified as unit 1 b of the Calcaire à entroques Fm. with transgressive and highstand systems tracts and is found also in north Burgundy (St Marc sur Seine, Forges section) as unit 1 of the same formation (Morestin 1986). In Dorset, it is also included within the Yellow Conglomerate.

On the Paris Basin logs, at least one of the sequences 16 and 17 can be recognized, but the given biostratigraphic data did not allow a datation other than early Dogger.

Sequence 18: LZA 1-1.3/Bj. 1/base of Discites zone – top of Sauzei zone

This sequence is identified in Normandy as the middle and upper members of the Malière Fm., the middle member being the transgressive systems tract with a typical upward deepening trend until a burrowed and eroded surface indicating the maximum flooding surface and followed by the upper member called Couche verte, indicating the highstand systems tract (Rioult et al. 1991). In Burgundy (St Marc sur Seine), this sequence corresponds to the middle part of the Calcaire à entroques Fm. (Fig. 6) (units 2 and 3, Morestin 1986). It is also present in the Jura, where the maximum flooding surface is clearly marked as a condensed section on top of the Laeviscula zone (Contini 1990). In Dorset (Burton Cliff), this sequence comprises the Snuff-Box Bed and the Red Beds, the first one being the transgressive systems tract and the second the highstand, the maximum flooding surface between the two, dated as Ovalis subzone.

These three sequences, 16, 17, and 18, are given as one in the Haq et al. 1987 chart.

Sequence 19: LZA 2-2.1/Bj 2/Humphriesianum zone

In Normandy, this sequence corresponds to the Conglomerat de Bayeux Fm. (Rioult et al. 1991) and in Burgundy to the upper part of the Calcaire à entroques Fm. (Fig. 6) The upper sequence boundary on top of the Humphriesianum zone is also well identified over all Lorraine (East France), where this sequence consists of the Calcaire à polypiers Fm. (Le Roux J. 1980). The sequence is very thin and condensed in Normandy, somewhat more developed in Burgundy (about 15 m in St Marc sur Seine section), and included within the Red Conglomerate in Dorset.

Sequence 20: LZA 2-2 2/Bj. 3/the whole Subfurcatum zone

This sequence, in contrast to the previous ones, seems not to have been deposited on the shallower areas, a hiatus of that ammonite zone time interval being usually recognized there. It seems to be identified mostly on the central Paris Basin logs, but not with absolute certainly, biostratigraphic data on the logs being not very precise. However, it could correspond in northeastern France (Lorraine and Marne) to the Marnes de Longwy Fm. and to its equivalent, the lower part of the Marno-calcaires à Ostrea acuminata Fm. The upper sequence boundary corresponds in Dorset to the top of the Red Conglomerate, in Burgundy to the base of the Ostrea acuminata Marls Fm. In Normandy, it is very difficult to identify within the condensed Oolite ferrufineuse de Bayeux Fm. This sequence has also been recognized on outcrops of the South East Basin (Ferry and Rubino 1989; de Graciansky et al. 1993) with even a lowstand systems tract in the deep part of the basin dated from the Blagdeni subzone, which seems not to have been recorded in the Paris Basin area.

These two sequences, 19 and 20, have not been directly observed together in the field (except in Lorraine but not in continuous outcrops): but on some logs, the responses could delimit two distinct sequences unless one corresponds to a parasequence set of the other transgressive systems tract (Sambet, pers comm.)

Sequence 21. LZA 2-2.3/Bj. 4/base of Garantiana zone – top of Truelli subzone

Except in Normandy, where it is very thin within the Oolite ferrugineuse de Bayeux Fm., this sequence is usually well individualized and easily recognizable in the Paris Basin as the first Dogger marl-rich interval. It corresponds to the Ostrea acuminata Marls Fm. in Burgundy, the maximum of marls being the maximum flooding surface well marked by the log responses, the highstand showing an upward-increasing trend of the limy beds. In Dorset, the Astarte bed would represent the condensed transgressive systems tract and the overlying maximum flooding surface, the Truelli Bed the highstand.

Sequence 21 marks the maximum drowning phase of a second-order cycle. It is interesting to note that this drowning occurred at that time principally in the central Paris Basin and on its eastern and southeastern border when it was obscured in Dorset and even more in Normandy. In NE France, it may have begun already with sequence 20.

Sequence 22: LZA 2-2 4/Bt. 1/base of Bomfordi subzone – top of Zigzag zone

With this sequence begins the sedimentation of the Paris Basin as a carbonate platform. In Burgundy, the datation is lacking, but we could consider the Calcaire de Premeaux Fm. as belonging to this depositional sequence. In Normandy, it comprises the Calcaire à spongiaires Fm. (of which the lower member is interpreted as a shelf margin systems tract) and the Couches de Passage Fm., the maximum flooding surface and the upper sequence boundary being perhaps fused on top of them. In Dorset, the equivalents are the Sponge beds and the Zigzag Bed (interpreted as a condensed section), the upper sequence boundary being on top of the so-called Scroff.

Most of the following Bathonian sequences have been dated from Normandy data (Rioult et al. 1991) and were not all identified in the rest of the area studied. The Burgundy and central Paris Basin Bathonian series are poorly dated and only major lithological units (oolitic and lagunal) are usually individualized; but the same number of sequences has also been identified on the logs of the NE Paris Basin. The correlation with the Lorraine outcrops has not yet been possible because the given datation is not as precise as in Normandy (at the subzone scale). It seems that Normandy and Lorraine were then in a deeper environment. The south Paris Basin (Cher, Nievre) is at present being restudied and reinterpreted (J.C. Menot, D. Marchand, J.P. Garcia, University of Burgundy), and already some sequence stratigraphic correlations can be outlined, this area being also in a deeper environment than the "Burgundy" Bathonian carbonate platform. The description of the Normandy sequences is given in great detail in Rioult et al. 1991 and thus will not be repeated here.

Sequence 23: LZA 2-2.5/Bt. 2/lower Tenuiplicatus zone – within Progracilis zone

The recognition of this sequence is limited to Normandy. It corresponds to the Marnes de Port en Bessin Fm. and its equivalent (Fig. 7). In Burgundy, the middle Bathonian has not yet been dated within the bathonian series. In Dorset, it is missing but the upper sequence boundary may have been recognized in Somerset (Torrens 1969) below the Milborne Beds on top of the Lio Ac. Bed dated from the upper Progracilis zone.

Sequence 24: LZA 2-2.6/Bt. 3/upper Progracilis zone – within Morrisi zone

Also missing in Dorset, this sequence in Normandy consists of the Calcaire de Saint Pierre du Mont Fm. (shelf margin and transgressive systems tracts) and the Caillasses de Longues Fm. (highstand systems tract). It is now recognized in Cher and Nievre and on the Burgundy platform: the so-called Oolite blanche (s. s.) can be interpreted as its equivalent (Garcia, pers. comm.). (Fig. 8)

Sequence 25: LZA 2-2.7/Bt. 4/within the Morrisi zone – within the Hodsoni zone

This is also a "Norman" depositional sequence comprising the Calcaire de Blainville Fm. and the Caillasse de Blainville Fm. It is clearly identified in the south Paris Basin (Nievre, Cher) (Roy 1983; Menot, pers. comm.). Its upper sequence boundary is, however, recognized in Dorset (Watton Cliff) as the base of the Wattonensis Beds

Sequence 26: LZA 2-2.8/Bt. 5/within the Hodsoni zone – lower Hollandi subzone

This upper bathonian sequence seems to mark the beginning of a new second-order transgressive phase that will be emphasized during Callovian time: in Normandy, it corresponds to the Calcaire de Ranville Fm., the Caillasse a cephalopodes, and the Caillasses de la Basse Ecarde Fm., in which two widespread marly intervals (Marnes à G. boueti and Digona beds) clearly indicate the transgressive advance. These marls exist also in Dorset within the relatively thick upper Fuller's Earth clay series (more than 45 m in Watton Cliff) which composes this sequence 26, the basal Wattonensis beds being perhaps a shelf margin systems tract. In the south Paris Basin, this sequence is present and well developed (up to 25 m at Chateauneuf Val de Bargis area, Nievre), the lower sequence boundary being there a well-correlated sharp erosional discontinuity (Menot, pers. comm.). In

▶

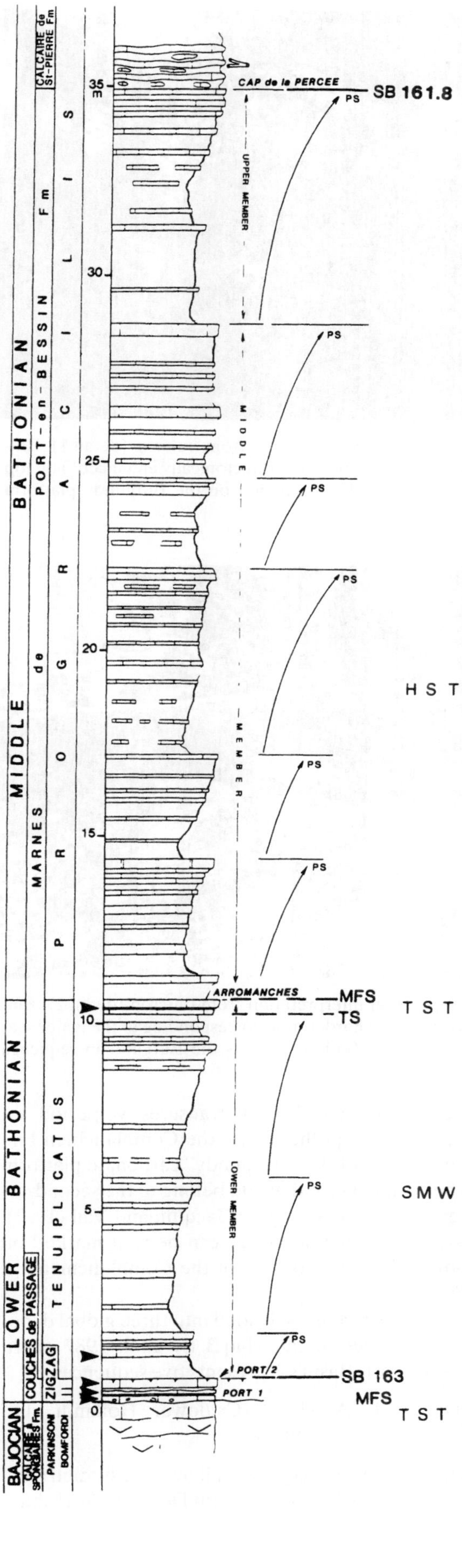

Fig. 7. Sequence stratigraphic interpretation of the Norman Marnes de Port en Bessin Fm (Modified after Rioult et al. 1991)

Fig. 8. Sequential interpretation from outcrop (detail 1) Emmersive angular discontinuity ("tectonically enhanced") on top of the Oolite Blanche Formation on the Burgundy platform (Plombières les Dijon, Highway) It corresponds to the second-order maximum of regression of late middle Bathonian *SB* sequence boundary

Fig. 9. Sequential interpretation from outcrop (detail 2) Emmersive burrowed hardground corresponding to the major sea-level drop of late Bathonian It is the Burgundian sequence boundary on top of the Comblanchien Fm prior to the overall early Callovian retrogradation, and late Callovian drowning *SB* sequence boundary

the case of an effectively early transgressive period, we could make the hypothesis that the Comblanchien Fm (or formations?) of the "Burgundy" carbonate platform would be of that age and would belong to the second-order aggrading sequences sensu Jacquin and Vail (1992). The upper sequence boundary can be well marked on outcrops in Burgundy on top of the Comblanchien s. s. Fm. (Fig. 9)

The Bathonian stage is divided into three global depositional sequences in the Haq B. U. et al. 1987 chart, whereas the Paris Basin s. l. presents five sequences.

Sequence 27: LZA 3-3.1/Bt.-Ca./lower Hollandi subzone – base of Koenigi subzone

This sequence is well recognized all over the studied area. In Dorset, and more generally from Dorset to Yorkshire, it corresponds to the Cornbrash Fm., the lower Cornbrash, now renamed Berry Member, being the transgressive systems tract and the upper Cornbrash or Fleet Member the highstand; the maximum flooding surface would be at the junction between the two members, at the level of the Astarte-Trigonia Bed and its correlative hiatus (Page 1989). The underlying clays of the Forest Marble Fm. could be considered as a lacustrine shelf margin systems tract. In Normandy, this sequence begins with a shelf margin systems tract consisting of the Calcaire de Langrune Fm. which ends by a widespread transgressive surface. The Argiles du Lion Fm composes the transgressive systems tract, its top is a sharp bioturbated discontinuity representing the maximum flooding surface, and the lower member of the Marnes d'Escoville Fm. the highstand.

Fig. 10. Detail of the Callovo-Oxfordian log response Vert le Grand well

In Burgundy and the central Paris Basin, this sequence 27 usually corresponds to the Pierre de Dijon-Corton System clearly limited by discontinuities D1 and D2 (Floquet et al. 1989); but it can begin by a shelf margin systems tract: the underlying uppermost Comblanchien Fm., which could be considered as a lateral equivalent of the Calcaire de Langrune. It is divided into a transgressive systems tract and a highstand by the maximum flooding surface corresponding to discontinuity d5 (Marnes à Digonella divionensis). The Marnes à Eudesia Fm., lying on top of discontinuity D1, is interpreted as an important flooding surface indicating the beginning of the transgressive backstepping. It is thus clearly identified on some well logs (Fig. 10).

The recent very precise work of Garcia (1992) and Jacquin et al. (1992) details these Bathonian-Callovian series; going down to a scale of fourth-order sequences, and thus define more sequences (for instance, the Marnes à Eudesia are also interpreted as a maximum flooding surface and a sequence boundary cuts the lower Pierre de Corton between the Marnes à Eudesia and the Marnes à Digonella divionensis).

Sequence 28: LZA 3-3.2/Ca. 1/base of Koenigi subzone – basal Jason subzone

This new sequence is now well confirmed and recognized in the studied area. Sequences 27 and 28 are of special importance because of petroleum potential within the Paris Basin. In Dorset, the data are not outcropping easily, but this sequence could correspond roughly to the Kellaways Fm. (clays and sands). In Normandy, it comprises the upper member of the Marnes d'Escoville Fm. (shelf margin systems tract), the Marnes d'Argences Fm. (transgressive systems tract) and the lower member of the

Marnes a Belem. latesu Fm (highstand systems tract). In Burgundy and the central Paris Basin, it is represented by the Pierre de Ladoix System and ends by discontinuity D3, the maximum flooding surface being at the so-called unit D (or Marnes à spongiaires) (Floquet et al 1989). In the subsurface data, a small oolitic shelf margin systems tract can be recognized on some logs, and the maximum flooding surface usually ends a clearly backstepping transgressive systems tract. On seimic, the flooding is well recognized and followed by a clear highstand. This sequence is also correlated within the South East Basin (Aiglun, Dourbes sections, Dardeau, Marchand).

The top of this sequence and the following sequences are not always recognized and dated on the studied area because of an important hiatus which exists in some places from Callovian to lower, and even middle Oxfordian times (cf. the ferruginous oolite of the proximal platform in the Dijon area).

Sequence 29: LZA 3-3.3/Ca 2/Jason subzone – upper Athleta zone

The lower sequence boundary is now well recognized and dated in some distal parts of the studied "basin" (Normandy, Cher, Nievre, southern Burgundy) and clearly expressed on numerous logs of the central Paris Basin. It is also identified in Charentes (SW France, Cariou et al. 1989). This sequence could correspond in England, as seen in Dorset, to the lower and middle Oxford Clay Fm., the lower member being the transgressive systems tract and the middle member the highstand with the maximum flooding surface at the junction between the two members, dated as the lower Athleta zone. In Normandy, a question remains about the highstand. The systems tracts are not really individualized within the sequence, which is there characterized by the upper member of the Marnes à Belem. latesu. Fm. and the Marnes et Calcaires de Crevecoeur Fm. (Rioult et al. 1991); or we might consider the overlying Couches du mauvais Pas Fm. as part of the highstand of this sequence also. This formation always outcrops badly, but it consists, like its southward equivalent the Assise des Carreaux Fm., of a series of shallowing-upward parasequences bounded on base and top by two sharp discontinuities. The maximum flooding surface could in that case be on top of the hiatus of Phaeinum subzone age which exists between the Crevecoeur and Couches du mauvais Pas Fm. An argument for this possibility is that in the Dijon and Chatillon areas (Burgundy-proximal part of the platform), the Proniae and Spinosum subzones are present within thin marl-argilaceous limestone beds, and it seems more logical to find highstand than lowstand sediments in such a proximal area. In Nievre (central France), this sequence would correspond to the Pierre de Nevers Fm. and equivalents, the upper sequence boundary dated as at least the top of Proniae subzone, the following Spinosum subzone being missing. In the northern part of the Paris Basin, the maximum flooding surface has been dated from cores of Proniae subzone age (by D. Marchand). In the South East Basin, this sequence has been identified in the Terres Noires with the maximum flooding surface at the base of the Athleta zone and the upper sequence boundary at the limit Athleta/Lamberti zone (by D Marchand and G. Dardeau). The same upper limit is also recognized in SW France (Peybernes 1990).

Sequence 30. LZA 3-3.4/Ca.-Ox /top of the Athleta zone – Preacordatum subzone

This sequence is not clearly identified within the upper Oxford Clay Fm. in Dorset. In Normandy, the previous Couches du mauvais Pas Fm. can also be consider as the shelf margin systems tract of this sequence with a transgressive surface as top, this interpretation being thus in relation with the hypothesis of a similar (in age) lowstand in Charentes (Cariou et al. 1989). The transgressive and highstand systems tracts are identified within the Marnes de Dives and Marnes de Villers Fm. In Nièvre, as in Burgundy, this sequence is not developed but is included within a thin upper Callovian/middle Oxfordian ferruginous oolite. However, in Cher, it would correspond to argilaceous limestone beds dated as the Lamberti zone and Scarburgense subzone (cf. Chateauneuf map notice)

Malm

Sequence 31: LZA 3-3.5/Ox. 1/Preacordatum subzone – Cordatum subzone

In Dorset, this sequence belongs to the upper part of the upper Oxford Clay and the upper sequence boundary is clearly at the junction Oxford Clay/Nothe Grit Fm. In Normandy, it corresponds to the condensed Oolite ferrugineuse de Villers Fm. (Rioult et al. 1991). In the northern Paris Basin, the Bukowskii subzone has been identified on core within a condensed ferruginous oolite level (by ammonites, D. Marchand). In Jura, the relatively deep marly sedimentation that began during the Lamberti zone reached during the Cordatum zone the borders of the trough that characterized the Jura area at that time (Contini C. 1989). This sequence marks the peak transgression of the late Bathonian to early Oxfordian transgressive phase of second order.

In Yorkshire, the regressive trend seems to begin at that time, i. e., a little ealier than in the Paris-London Basin with the deposition of the Lower Calcareous Grit Fm. (of Bukowski subzone age, D. Marchand). Sequence 31 in Hambleton area would include the Lower Calcareous Grit Fm and the Lower Leaf of the Hambleton Oolite Fm.

This Callovo-Oxfordian interval ("158.5-150.5") was divided into two depositional sequences in the Haq et al. 1987 chart, whereas five sequences have been identified in the studied area.

Sequence 32: LZA 4-4.1/Os. 2/Cordatum subzone – within the Vertebrale subzone

With this sequence begins a second-order regressive phase of middle and upper Oxfordian age. In Dorset, it corresponds to the Nothe Grit (shelf margin systems tract), the Preston Grit (transgressive systems tract) and the Nothe Clays (highstand systems tract). In Normandy, it comprises the Argiles à Lopha greg. Fm. (shelf margin systems tract) and the lower and middle member of the Calcaire d'Auberville Fm., respectively transgressive and highstand systems tracts. A clear change in sedimentation rhythm is recorded on the central Paris Basin logs and could be correlated with this sequence, but an uncertainty remains because of imprecise biostratigraphic data. On the seismic data, the lower sequence boundary is clearly recognized in some places as clearly erosive. The Burgundy proximal platform is still drowned and the sedimentation as ferruginous oolites. This sequence was already defined in the Haq et al. chart.

Sequence 33: LZA 4-4.2/Ox. 3/within the Vertebrale subzone – lower Maltonense subzone

In Dorset, this new sequence begins clearly at the base of the Bencliff Grit, which is interpreted as a shelf margin systems tract. The Upton member of the Osmington Oolite Fm. corresponds to the transgressive and highstand systems tracts, the pisolite bed being the maximum flooding surface. In Normandy, it comprises the upper member of the Calcaire d'Auberville Fm. (shelf margin systems tract) and the lower member of the Calcaire Oolitique de Trouville Fm. The sedimentation resumption is not yet recorded on the Burgundy proximal platform, but northward, in Ardennes and north Lorraine, some thin deposits of biodetrital limestones and marls testify to this resumption. Is the Middle Calcareous Grit of Yorkshire correlatable with this sequence?

Sequence 34: LZA 4-4.3a/Ox. 4/lower Maltonense subzone – upper Blakei subzone

In Dorset, this sequence corresponds to the Shortlake and Nodular Rubble members of the Osmington Oolite Fm., the basal middle white oolite being considered as a shelf margin systems tract, the rest of the shortlake member the transgressive systems tract with the maximum flooding surface on top of the upper white oolite and the Nodular Rubble member the highstand. In Normandy, the upper member of the Calcaire de trouville Fm. is interpreted as the shelf margin systems tract, the Coral Rag Fm. the transgressive, and the Calcaire de Blangy Fm. the highstand. In Lorraine, this sequence is essentially composed of constructed recifal formations. In Burgundy, it could correspond to the Marnes de Talant Fm. and the Calcaires à Pholadomya lineata Fm. in the Dijon area (Menot et al. 1982).

Sequence 4th order: LZA 4-4.3b/Ox. 4b/Ilovaiskii subzone

This sequence is not positively defined because it corresponds to a biostratigraphic gap very common within the studied area, as in Dorset and in Normandy. Nevertheless, from the logs of the central and eastern Paris Basin, it seems that a supplementary sequence could be identified as equivalent to this gap. The northeastern Paris Basin logs have been correlated with the Lorraine outcrops and could be assimilated to equivalents of the Calcaires à chaux grasse siderurgique Fm. and in Ardennes to the time-equivalent marly deposits (Marchand and Menot 1980). In SW France, the upper sequence boundary is also recognized in the middle of the Bifurcatus zone (French equivalent of the Glosense zone) within the Dalle à Perisphinctes Fm. (Peybernes B. 1990). The correlation with the lower part of the Upper Calcareous Grit of Yorkshire is a hypothesis.

Sequence 35: LZA 4-4.4/Ox. 5/base of Glosense subzone – Serratum subzone

After a gap of 0.5 Ma, the sedimentation resumption is terrigenous-rich as in Dorset with the deposit of the Trigonia clav. Beds, of which the basal Sandy Block is considered as a shelf margin systems tract, the three overlying members being the transgressive systems tract, and the maximum flooding surface on top at the junction with the Sandsfoot Clay which represent the highstand deposits. In Normandy, it corresponds to the Calcaire greseux de Hennequeville Fm., its lower and middle members are interpreted as shelf margin deposits, and its upper member as transgressive systems tract, and to part of the Marnes de Villerville Fm. characterizing the highstand. This basal terrigenous input increase is correlatable further east in the Jura area (Enay 1980) at the base of the Marnes de Besançon Fm. (Contini 1989); and also further north in Lorraine with the Calcaires argilo-sableux à trigonies. On some subsurface data, the highstand of this sequence corresponds to the installation of reefs.

Sequence 36: LZA 4-4.5/Ox. 6/Regulare zone – within the Bauhini subzone

In Dorset (Black Head, near Osmington), this sequence begins at the discontinuity between Sandsfoot Clay and Sandsfoot Grit members of the Sandsfoot Fm. and the upper sequence boundary is recognized within the Ringstead Waxy Clay member of the Passage Beds Fm., in its upper half, at the junction between clays and marly limestones. The maximum flooding surface has not been identified on the field but could be at the top of the Sandsfoot Grit which presents a faunal abundance peak. This sequence appears clearly on numerous logs of the Paris Basin transects as the transitional sequence between the middle-upper Oxfordian calcareous package and the kimmeridgian marly one. This Bauhini sequence boundary is also identified as a significant discontinuity on outcrops in the south Paris Basin (Berry and Loire region, top of the Calcaire de Tonnerre Fm. and base of the Calcaires de Montierchaume Fm., Debrand Passard 1982; Bernard 1988). In the South East Basin, it could be assimilated to the top of the second Oxfordian Reef.

The lower sequence boundary was already defined in the Haq et al. chart.

Sequence 37: LZA 4-4.6/Km 1/within the Bauhini subzone – top of Baylei zone

In Dorset, the maximum flooding surface of this sequence corresponds to the limit Oxfordian/Kimmeridgian on top of the Ringstead Coral Bed and the upper sequence boundary is the discontinuity between the calcareous mudstone of Baylei z age and the Wyke Siltstone of Cymodoce z. age of basal Kimmeridge Beds. (for the Kimmeridge Clay references, see Cox and Gallois 1981). This first kimmeridgian sequence appears very clearly on the Paris Basin logs. It corresponds in Normandy to the first lithological unit of Kimmeridgian age (Rioult in Megnien, 1980). In south Paris Basin (Berry for instance) it consists of the Calcaires de Montierchaume Fm. and Calcaires de Levroux (maybe just the lower member, a hiatus of upper Baylei zone being recorded in some place) (Bernard, 1988). In Burgundy, from north to south, the upper sequence boundary is the clear discontinuity on top of a series of formations of shallow carbonate platform (Menot et al., in Debrand-Passard 1984). An important tectonically enhanced boundary is recognized on the seismic data which could perhaps (no precise datation) correspond to the lower sequence boundary of this sequence and which limits a clear change in depositional environment (no more reefs after) and in paleotopography (subsidence "basinward"). In eastern France (Jura region) the upper sequence boundary would correspond to the discontinuity on top of the Calcaire de Besançon Fm. (Contini 1989). In southeastern France (Ardèche region), this sequence is also identified within the time interval there called Galar subzone and Platynota zone (Atrops 1982). In the Aquitaine Basin (southwest France), the maximum flooding surface would correspond to Hantzpergue's discontinuity D1 and the upper sequence boundary to the discontinuity D2 (Hantzpergue 1985) Sequences 36 and 37 were given as one in the Haq B. U et al. 1987 chart.

Sequence 38: LZA 4-4.7/Km. 2/whole Cymodoce zone

This sequence is not clearly individualized within the Kimmeridge Beds of Dorset (except for its lower sequence boundary), but it is recognizable on the Paris Basin logs. In the south Paris Basin, it comprises the Calcaires de Buzancais Fm. and part of the Marnes de St Doulchard Fm., the maximum flooding surface being at the glauconitic-fossil-rich bed located at the base of these marls and dated as lower Chatelaillonensis subzone. (Debrand-Passard 1982; Bernard 1988). In Normandy, the lower sequence boundary corresponds to the glauconitic conglomerate on top of the underlying Banc de Plomb, and the sequence is recognizable within the thin Marnes d'Harpagodes Fm. which is cut on top by an erosional discontinuity (Rioult in Megnien 1980). The upper sequence boundary can be recognized in Lorraine as the discontinuity and glauconitic conglomerate on top of the Calcaires rocailleux à pteroceres Fm., as well as in Burgundy with the glauconitic level on top of the Calcaire à Astartes Fm., which corresponds to this sequence, and also in southeastern France (Ardèche, Drome) on top of the massive limy beds overlying the "Vire à Crussoliceras" and dated as the Divisum zone (equivalent to the upper Cymodoce zone) (Atrops 1982). In the Aquitaine Basin, as well as in Quercy, then in Charentes, P. Hantzpergue (1985) recognizes several discontinuities within the Cymodoce zone, his discontinuity D5 at the base of the Roquedure Fm. could be a transgressive surface, and in that case the maximum flooding surface would be the limy-ammonite rich bed of Chatelaillonensis horizon age located on top of the marly lower tiers of this Roquedure Fm. and the upper sequence boundary would be discontinuity D7 on top of the more homogenous clayey-limy upper part of this formation (Hantzpergue and Lafaurie 1983).

Sequence 39: LZA 4-4.8/Km 3/whole Mutabilis zone (or up to lower Eudoxus zone?)

This sequence is also obscured in Dorset within the thick Kimmeridge beds and more generally over England, but again seems to appear on the Paris Basin logs. However, the lower sequence boundary may correspond to the minor erosional surface recognized by Cox and Gallois (1981) at the boundary between the two ammonites zones Cymodoce and Mutabilis, and characterized by a silty influx, the upper sequence boundary being perhaps at the base of the shelly siltstones of lower Eudoxus age. In Lorraine, this sequence seems to be only Mutabilis in age and corresponds to the Marnes inférieures and the Calcaires blancs inférieurs Fm. ended by a glauconitic discontinuity. In the Aquitaine Basin, this sequence could be recognized between Hantzpergue's discontinuities D7 and D10, the maximum flooding surface being within the Pont de Rhodes Fm. (Quercy), the discontinuity D9 (Hantzpergue and Lafaurie 1983). The question of age is multiple: (1) it comes from the fact that the Orthocera subzone on top of what is identified the discontinuity D10 can be considered either as the last subzone of the Mutabilis zone or as the first subzone of the Eudoxus zone (2) The upper sequence boundary in the north Burgundy region may be the top of the Calcaires d'Arc les Gray Fm and is dated as top of the Mutabilis zone. (3) In Ardeche, however, this boundary may be considered on top of the Calcaires de la Beaume Fm. dated as lower Eudoxus zone (Atrops, in Debrand-Passard 1984).

On the borders of the Paris Basin, the Kimmeridgian and younger deposits become more and more rare on outcrops so the references will become less and less numerous. The seimic data shows very clearly the late Jurassic/early Cretaceous differential erosion with upper Oxfordian to late Kimmeridgian strata progressively cut and toplapping.

Sequence 40: LZA 4-4.9/Km. 4/lower Eudoxus zone – top of Autissiodorensis zone

Again, this sequence is not clearly expressed within the Dorset Kimmerigde Clay; however, its upper sequence boundary is on top of the calcareous mudstone alternation at the level of Blake's bed 42 (Autissiodorensis/Elegans zone boundary of Cox and Gallois 1981). It is well individualized on numerous logs of the Paris Basin. In Burgundy and surrounding eastern areas, this sequence would be included within the upper half of the Marnes and Calcaires à Exogyra virgula Fm. In Lorraine, it corresponds to the Marnes moyennes, the Calcaires supérieurs, and the Marnes superieures, which end in bituminous levels rich in lignitic and vertebrate remains (Megnien 1980).

Sequence 41: LZA 4-4.10/Km.5/base of Elegans zone – basal part of Smedmorensis subzone

In the Kimmeridge Clay of Dorset, this sequence is included between Blake's bed 42 and the Gray Ledge bed, the Cattle Ledge bed being perhaps the maximum flooding surface. In France this sequence belongs to the so-called Gravesia zone. It consists in NE Burgundy of the Calcaires compacts à tubulures Fm. and in W Burgundy of the Calcaires du Barrois Fm., the marly-limy alternation (oyster-rich) of the lower limestones being probably the transgressive systems tract. In Jura, the Marnes des Abergements Fm. may be interpreted as the first deposits of this sequence in the relatively deep parts of the area (shelf margin system tract?), the overlying Couches du Chailley Fm. being the major deposits and presenting upward a clear emersion trend (highstand?) (description from Bernier 1984). In the NW of the Paris Basin (Boulonnais region) the basal sequence boundary is clearly expressed with the Gres de la Crèche Fm.

Sequence 42; LZA 4-4.11/Km.6/basal Smedmorensis subzone – basal part of Reisiformis subzone

In Dorset, this thin and short sequence corresponds to the deposits between the Gray Ledge Bed and the base of the uniform calcareous mudstone succession dated as lower Reisiformis subzone, the massive oil shale and concretional Blackstone bed being the maximum flooding surface on top of the underlying rythmic alternation of oil shales and mudstones of the transgressive systems tract. The correlation with the "Tithonique" of the South East Basin and its borders has not been made in this study. In the surrounding areas of the Paris Basin, the given datation of the different formations was not precise enough to allow a sequential subdivision, even if the outcrops present certain possible sequential cuttings out, notably in the French Boulonnais region.

Sequence 43: LZB 1-1.1/Km.7/basal Reisiformis subzone – top of Fittoni zone

From the given data, it has been not possible to differentiate any stratigraphical layout within the upper part of the Kimmeridge Clays of Dorset, but we may consider the uniform calcareous mudstones of middle Hudlestoni zone as part of a lowstand systems tract and the Rotunda Nodules as part of an early highstand. The upper sequence boundary is, however, clearly expressed as the base of the Massive Bed or its equivalent the Black Nore Sandstone. The French data also gave no further information about this sequence except also on its upper sequence boundary, represented in Boulonnais by a hard ground on top of the Calcaires et Argiles à Ostrea expansa Fm., in Berry, as in Lorraine, at the base of sandy limestones.

Sequence 44: LZB 1-1.2/Pt.1/base of Albani zone – top of Okusensis zone

This sequence is quite well expressed within the Portand Beds of Dorset: the Massive Bed and Black Nore Sandstone indicating the lower sequence boundary; the overlying marls represent the transgressive systems tract, the maximum flooding surface is recognized on top of the White Cementstone or equivalent Exogyra Beds, and the upper sequence boundary corresponds to the limit between the Lower Cherty Beds (and equivalent Portland Clay) and the Basal Shell Bed and equivalents. This sequence may be identified in Boulonnais by the thin Calcaires graveleux Fm. representing there only a condensed section.

Sequence 45: LZB 1-1.3/Pt.2/base of Kerberus zone – basal part of Anguiformis zone

This small sequence corresponds in Dorset to the Basal Shell Bed, whose top is the maximum flooding surface and thus represents a very thin transgressive systems tract, and to the Upper Cherty Beds, the upper sequence boundary being at the base of the overlying Freestone Member. In Boulonnais, this sequence consists of the Gres à Trigonia gibbosa Fm., the burrowed surface recognized in the middle of this formation possibly being the maximum flooding surface. This marks the end of the marine sediments in that area, the overlying deposits being lacustral limestones.

Sequence 46: LZB 1-1.4/Pt.3/basal Anguiformis zone – top of Lamplughi zone

This sequence could, in fact, possibly be several sequences, but as these deposits outcrop only in Dorset and not in the surrounding areas of the Paris Basin, we did not gather enough data to make a good assumption on the sequence stratigraphic cutting-out. However, we may have to consider the Freestone Beds as a sequence of its own, the top Shrimp Bed being in that case both the maximum flooding surface and the upper sequence boundary; the depositional environment of the overlying Purbeck Beds being nonmarine, it seems obvious that they record a different context and thus underline a sequence boundary. The last Jurassic sequence corresponds to the lower Purbeck Beds, so-called Lulworth Beds, whose upper sequence boundary coincides with the stage limit on the base of the Cinder Beds. The Dirt Beds, consisting of paleosoils and fossil forest, may be interpreted as a lowstand systems tract, the maximum flooding surface of this single sequence may be on top of the Broken Beds; but we have to keep in mind that this apparently thick highstand might represent several sequences.

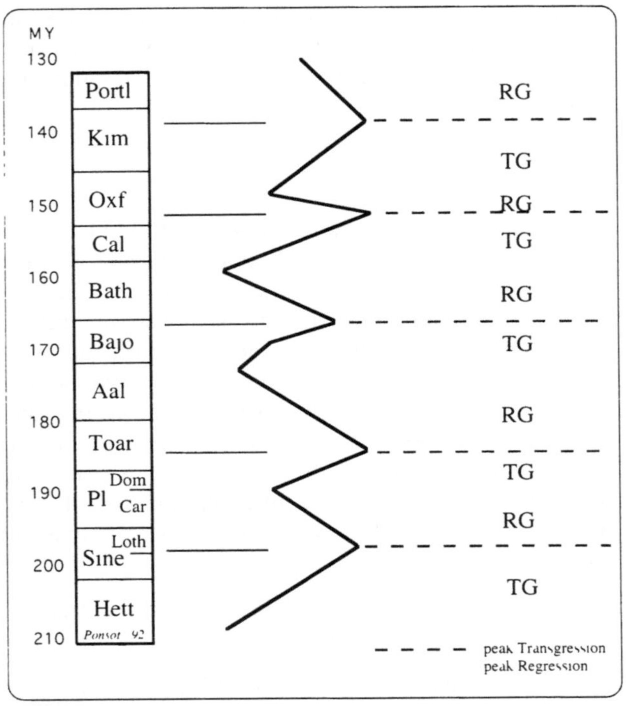

Fig. 11. Second-order evolution

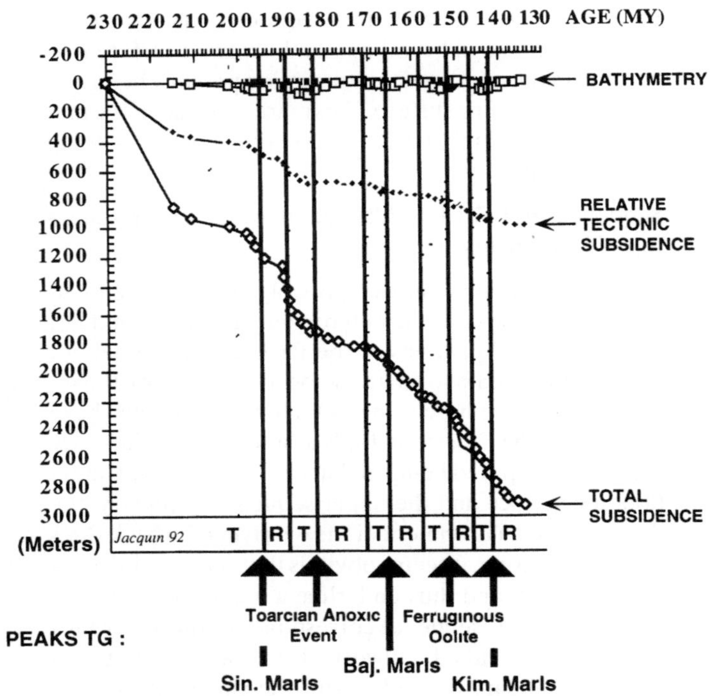

Fig. 12. Subsidence curve for the central Paris Basin (Chapton well, without sea-level fluctuations)

Conclusion

In conclusion, the studied Jurassic series are replaced within the second order evolution context (Fig. 11):

- The Jurassic series begin within the transgressive trend of a cycle whose regressive phase occurred during upper Triassic time The clayey and marly deposits of lower Lias age correspond to this transgressive phase, whose peak is of Lotharingian age, possibly of the Obtusum zone in the Paris-London Basin
- The regressive phase of the next cycle is recorded by the Carixian deposits; the regressive peak coincides with the Carixian/Domerian limit.
  The transgressive trend increases during Domerian time, but seems to be discontinuous, the upper Domerian being slightly regressive within this transgressive global trend; its peak is reached during lower Toarcian with the "paper shales" deposits and over.
- The following regressive phase begins within upper Toarcian time and is clearly indicated by the great Aalenian unconformity.
  The transgression is roughly of Bajocian age, with the installation of marls or ferruginous oolites, and its peak is reached within the Acris zone.
- The Bathonian platform corresponds to the regressive phase of the next cycle, which ends at the base of the Discus zone.
  The transgressive trend is clearly recorded by the Callovian drowning of the platform, and its peak often occurs as ferruginous oolites during lower Oxfordian time, most likely within the lower Cordatum zone.
- The following regression is short and ends within middle Oxfordian time.
  The progressive drowning of the upper Oxfordian platform and the installation of the Kimmeridgian mixed deposits correspond to the transgressive phase, the peak being of upper Kimmeridgian age.
- The Jurassic series ends in the Paris-London Basin by a regressive phase which is often partly eroded by the major early Cretaceous unconformity.

A subsidence curve has been constructed by a backstripping program (Marco Polo software) applied to log data from the central part of the Paris Basin (Fig. 12).

The total subsidence is the result of the combined effects of thermal cooling, crustal stretching, and basement faulting.

The total subsidence curve corresponds to the thickness of the decompacted sediments, the sediment load, and the paleobathymetry.

The relative subsidence curve is obtained by subtracting the sediment load from the total subsidence.

Corrections for sediment compaction with depth of burial are made depending on lithology.

In this case, the eustatic sea-level changes of the second order have not been taken into account, i.e., only the accommodation space is considered, and the risks of incertitude due to the evaluated water column are minimized.

The resulting curve confirms these regressive/transgressive cycles, and shows that the slope of the relative subsidence curve is more significant during transgressive periods.

Comparison with the Haq et al. 1987 chart

- Only one sequence is also observed for the Hettangian stage (even if on some logs a second one seems to be identifiable, it is not taken into account because of lack of datation and identification on outcrops).
- The lower Sinemurian stage is now divided into two sequences (versus one), both characterized by condensed sections often rich in organic content.
- The Lotharingian substage comprises two sequences (versus one), both ending usually by a significant time gap.
- There are still two depositional sequences of Carixian age.
- The Domerian is now subdivided in three sequences (versus two), the first two being within Margaritatus zone.
- Two sequences of lower Toarcian age are now recognized in the studied area (versus one); the late Lias also comprises two sequences.
- Early Aalenian is also characterized by one sequence, but an important gap is often recorded at the Lias/Dogger boundary.
- The middle-late Aalenian and early Bajocian period is now divided into three sequences (versus one).
- Possibly three sequences are observed during middle-late Bajocian time (versus one) (some doubt because the two first have not been observed in continuity).
- Five Bathonian sequences have been recognized on the borders of the basin (not all on the proximal patform), versus three in the Haq et al. chart.
- The middle-late Callovian and early Oxfordian transgression is now divided into four sequences (versus one).
- The middle-upper Oxfordian time is now characterized by five sequences (versus three and a half).
- The Kimmeridgian deposits consist of the studied area of seven sequences (versus four) with focus made on the lower half of the stage now divided into five (versus two), – The overlying time deposits (upper Kimmeridgian – Portlandian) often being missing or partially eroded, it has not been possible to differentiate more sequences than in the Haq et al. chart, even Haq sequences being not all identified (except Dorset).

Our objective has been to describe the apparent Jurassic depositional sequences of the Paris-London Basin and surrounding areas as part of a review of the Haq B. U. et al. 1987 chart. The purpose of this chapter was not to determine the causes of the relative sea-level changes for each described sequence, i.e., which of the major combined factors (eustasy, tectonism, sediment supply, climate) was or were predominant on the relative sea-level variation, or to address the magnitude of the falls and rises.

Acknowledgments: The author is particularly grateful to all the people with whom she worked: MM J. L. Dommergues; D. Marchand and J. C. Menot, University of Bourgogne; MM O. Dugue, G. Fily, and M. Rioult, University of Caen; Mr G. Dardeau, University of Nice; MM T. Jacquin and P. C. de Graciansky, Paris School of Mines; Mr R. Jan du Chene, Consultant in Bordeaux; and most of all Dr P. R. Vail, Rice University, Texas. She thanks the different companies from whom she received data or for whom she worked on this topic, such as Conoco-Houston, Triton-Paris, Andra-Fontenay aux Roses. Thanks are also due to her compagny, Cenergys, which allowed her to take time for this publication. G. Sambet is thanked for his helpful and constructive comments and suggestions, and Ms Nakou for her valuable help in drafting.

References

Atrops F (1982) La sous-famille des Ataxioceratinae dans le Kimméridgien inférieur du Sud Est de la France Doc Lab Geol Lyon 83

Bernard P (1988) La plateforme carbonatée niverno-icaunaise à l'Oxfordien supérieur et au Kimméridgien inférieur centre des Sciences de la Terre, Université Bourgogne, thèse de doctorat

Bernier P (1984) Les formations carbonatées du Kimmèridgien et du Portlandien dans le Jura méridional Doc Lab Geol Lyon 92

Cariou E, Hantzbergue P, Jan du Chêne R, Vail P R (1989) Excursion in the Jurassic of the Charentes and Poitou areas (France), Guide book

Contini D (1989) L'Oxfordien du Jura septentrional, définition des formations, évolution paléogéographique Ann Sci Univ Fr-Comte, Besançon, Geologie 4 (9)

Contini D (1990) Les crises sédimentaires à la limite Lias-Dogger. Cah Univ Catho Lyon Ser Sci 4

Cope J C W, Duff K L, et al (1980) A correlation of Jurassic rocks in the British Isles, part two Middle and upper Jurassic Geol Soc Lond Spec Rep 15

Cox B M, Gallois R W (1981) The stratigraphy of the Kimmeridge Clay of the Dorset type area and its correlation with some other Kimmeridgian sequences Inst Geol Sci Lond Rep 80/4

Dardeau G (1983) Le Jurassique des Alpes Maritimes (France) Thèse Doct Sci, Univ Nice

Dardeau G, Gauthier H, Marchand D, Montenat C (1980) Le Dogger et la base du Malm dans la partie Est de l'Arc de Castellane Geologie Mediterranéenne, T VII 3

Dardeau G, Atrops F, Fortwengler D, de Graciansky P C, Marchand D (1988) Jeu de blocs et tectonique distensive au Callovien et à l'Oxfordien dans le bassin du Sud-Est de la France Bull Soc Geol Fr (8) IV, 5

Debrand-Passard S (ed) (1982) Le Jurassique supérieur du Berry Bur Rech Geol Min Fr Mem 119

Debrand-Passard S (ed) (1984) Synthèse géologique du Sud-Est de la France. Bur Rech Geol Min Fr Mem 125

de Graciansky P C, Dardeau G, Dumont T, Jacquin T, Marchand D, Mouterde R, Vail P R (1993) Depositional sequence cycles, transgressive-regressive facies cycles, and extensional tectonics: example from the southern Subalpine Jurassic basin, France: Bull Soc Geol Fr 164, 5 709–718

Dommergues J L, Meister Ch (1991) Fréquences de présence des zonules et mise en évidence de fluctuations de l'environnement, exemple dans le Lias nord-ouest europeen C R Acad Sci Paris

Dommergues J L, Floquet M, Garcia J P, Marchand D, Thierry J, Jan du Chêne R, Vail P R (1991) A sequence stratigraphy excursion in the Jurassic of Burgundy (France) Jan du Chêne geological excursions-Guide Book

Dumont Th (1992) The late Triassic (Rhetian) sequences of Wombat Plateau (NW Australia, ODP Leg 122) Sea-level changes, tethyan rifting, and overprint of Indo-Australian breakup In von Rad U, Haq B U et al (eds) Proc ODP, Sci Results, 122, College Station

Enay R (1966) L'Oxfordien dans la moitié sud du Jura francais Nouv Arch Mus Hist Nat Lyon 8, 1, 2

Enay R (1980) Indices d'émersion et d'influences continentales dans l'Oxfordien supérieur – Kimméridgien inférieur en France Bull Soc Geol Fr (7) XXII, 4

Ferry S, Rubino J L (1989) Mesozoic eustacy record on western Tethyan margins, post meeting field trip in the Vocotian trough, Lyon, Nov 1989, Guide book

Floquet M, Laurin B, Laville P, Marchand D, Menot J Cl, Pascal A, Thierry J (1989) Les systèmes sédimentaires bourguignons d'âge Bathonien terminal-Callovien Bull Cent Rech Expl Prod Elf-Aquitaine 13, 1

Gabilly J (1976) Le Toarcien à Thouars et dans le centre-ouest de la France Les stratotypes francais, vol 3, C N R S, 217 pp

Gabilly J, Cariou E, Hantzpergue P (1985) Les grandes discontinuités stratigraphiques au Jurassique Bull Soc Geol Fr (8) I, 3

Garcia J P (1992) Les variations du niveau marin sur le Bassin de Paris au Bathonien-Callovien Thèse Univ Bourgogne, 295 pp

Hantzpergue P (1985) Les discontinuites sédimentaires majeures dans le Kimmeridgien français Geobios 18, 2

Hantzpergue P, Lafaurie G (1983) Le Kimméridgien quercynois un complément biostratigraphique du Jurassique supérieur d'Aquitaine Geobios 16, 5

Haq B U, Hardenbol J, Vail P R (1987) The chronology of fluctuating sea-level since the Triassic Science 235

Haq B U, Hardenbol J, Vail P R (1988) Mesozoic and cenozoic chronostratigraphy and cycles of sea-level change sea-level changes – an integrated Approach Soc Econ Paleontol Mineral Spec Publ 42

Hardenbol J (1992) The Mesozoic-Cenozoic global sequence stratigraphic framework Int Symp Sequence Stratigraphy of the European Basins, Dijon, France, Abstr vol

Hemingway J E, Wright J K, et al (1969) Excursion no 3, Guide for the north Yorkshire Int Field Symp British Jurassic, London

Hesselbo S P, Jenkyns H C (1992) Lower Jurassic sequence stratigraphy of the UK area Int Symp Sequence Stratigraphy of the European Basins, Dijon, France, Abstr vol

Howarth M K (1955) Domerian of the Yorkshire coast Proc Yorkshire Geol Soc 30, 2 (10)

Howarth M K (1962) The Jet Rock Series and the Alum Shale Series of the Yorkshire coast Proc Yorkshire Geol Soc 33, 4, 18

Howarth M K (1980) The Toarcian age of the upper part of the Marlstone Rock bed in England Palaeontology 23, 3

Jacquin Th, Vail P R (1991) The sequences and systems tracts that build transgressive/regressive facies cycles in carbonate settings Am Assoc Pet Geol Bull 75, 3, march 1991

Jacquin Th, Garcia J P, Ponsot C M, Thierry J, Vail P R (1992) Séquences de dépôt et cycles régressif/transgressifs en domaine marin carbonaté C R Acad Sci Paris 315, Ser II

Knox R W O'B (1984) Lithostratigraphy and depositional history of the late Toarcian sequence at Ravenscar, Yorkshire Proc Yorkshire Geol Soc 45

Lang W D (1924) The Blue Lias of the Devon and Dorset coasts P G A, 35

Lang W D (1928) The Belemnite Marls of Charmouth, a series in the Lias of the Dorset Coast Q J G S, 82

Lang W D (1936) The Green Ammonite Beds of the Dorset Lias Q J G S, 92

Le Roux J (1980) In Synthèse géologique du Bassin de Paris Bur Rech Geol Min Mem 101

Lorenz C, Lefavrais A, Lorenz J, Marchand D, Millon R (1991) Calage stratigraphique précis des diagraphies du Jurassique du Sud du Bassin parisien à partir de la serie carottée du sondage de Sancerre-Couy Bull Soc Geol Fr t 162, 5 947–952

Marchand D, Menot J Cl (1980) Oxfordien, Ardennes et Lorraine In Synthèse géologique du Bassin de Paris Bur Rech Geol Min Mem 101

Megnien Cl (ed) (1980) Synthèse géologique du Bassin de Paris Bur Rech Geol Min Mem 101

Menot J Cl, Drouet J J, Marchand D, Pascal A (1982) Structuration et sédimentation à l'est de Morvan au Callovien supérieur et à l'Oxfordien Bull Sci Bourg 35, 2

Morestin B (1986) Sedimentation et diagénèse dans le calcaire à Entroques de Bourgogne (France) Centre des Sciences de la Terre, Univ Bourgogne, thèse de troisième cycle

Page K N (1989) A stratigraphical revision for the English lower Callovian Proc Geol Assoc Lond 100, 3

Peybernes B (1990) Interprétation eustatique des principales discontinuités et lacunes stratigraphiques du Callovo-Oxfordien à ammonites du SW de la France C R Acad Sci Paris 310, II

Ponsot-Jacquin C (1992) Sequence stratigraphy in the Jurassic of the Paris Basin Int Symp Sequence Stratigraphy of the European Basins, Dijon, France, Abstr vol

Ponsot C M, Vail P R (1991) Sequence stratigraphy of the Jurassic new data from the Paris-London Basin Am Assoc Pet Geol Bull 75, 3, march 1991

Powell J H (1984) Lithostratigraphical nomenclature of the Lias group in the Yorkshire Basin Proc Yorkshire Geol Soc 45

Rioult M, Dugue O, Jan du Chene R, Ponsot C M, Fily G, Moron J M, Vail P R (1991) Outcrop sequence stratigraphy of the Anglo-Paris Basin middle to upper Jurassic (Normandy, Maine, Dorset) Bull Cent Rech Expl Prod Elf Aquitaine 15, 1

Rohl U, Dumont Th, von Rad U, Martini R, Zaninetti L (1991) Upper Triassic Tethyan carbonates off Northwest Australia (Wonbat Plateau, ODP leg 122) Facies 25, Erlangen

Roy B (1983) Géologie, géophysique, hydrogéologie des formations d'âge Jurassique moyen dans le nord-ouest de la Nièvre Centre des Sciences de la Terre, Univ Bourgogne, thèse de troisième cycle

Torrens H S (ed) (1969) International field symposium on the British Jurassic Excursion no 1 Guide for Dorset and south Somerset Geology Department, Keele University

Trouiller A, Jacquin-Ponsot C, Jan du Chêne R, Fauconnier D, Le Strat P, Andre-Jehan R (1992) Lias Sequence stratigraphy Northern Paris Basin – borehole A 901 Int Symp Sequence Stratigraphy of the European Basins, Dijon, France, Abstr vol

University of Birmingham, Dept of Geological Sciences (1984) Fieldtrip to Dorset and Yorkshire Guide book

University of Oxford (1989) Jurassic sequence stratigraphy, Dorset Coast, Wessex Basin Guide book

Vail P R (1987) Seismic stratigraphy interpretation using sequence stratigraphy In Bally A W (ed) Atlas of seismic stratigraphy Am Assoc Pet Geol 1 Stud Geol 27

Vail P R, Mitchum R M, Todd R G, Widmeri J W, Thompson S, Sangree J B, Bubb J N, Hatelid W D (1977) Seismic stratigraphy and global changes of sea level In Seismic stra-

tigraphy, application to hydrocarbon exploration Am Assoc Pet Geol Mem 26
Vail P R, Colin J P, Jan du Chene R, Kuchly J, Mediavilla F, Trifilief V (1987) La stratigraphie séquentielle et son application aux corrélations chronostratigraphiques dans le Jurassique du Bassin de Paris Bull Soc Geol Fr (8), 3, 7
Van Wagoner J C, Posamentier H W, Mitchum R M, Vail P R, Sarg P R, Loutit T S, Hardenbol J (1988) An overview of sequence stratigraphy and key definitions In Wilgus C K et al (eds) Sea level changes an integrated approach Soc Econ Paleontol Mineral Spec Publ 42
Van Wagoner J C, Mitchum R M, Campion K M, Rahmanian V D (1990) Siliciclastic sequence stratigraphy in well logs, cores and outcrops concepts for high-resolution correlation of time and facies Am Assoc Pet Geol Methods in Exploration Ser 7

Sequence Stratigraphy and Organic Matter Distribution of the Lias of the Paris Basin

G. Bessereau and F. Guillocheau

Abstract. Sequence stratigraphy is a powerful tool for the study of basin fill histories. Here, it was used for a better understanding of the relation between basin evolution and organic matter distribution in the Lias of the Paris Basin. The study consisted of a three-dimensional investigation and the main results are synthesized on a SW-NE cross section. It required the combination of two methods – the stacking pattern method and the Carbolog method – both applied to wells using wireline logs. The stacking pattern method, which necessitated the logs being calibrated on outcrops and on core material, and being biostratigraphically controlled, led to an accurate subdivision of the Liassic series into three superimposed sequences, traceable throughout the basin, inside the major transgressive-regressive Lias cycle. These three sequences are: the genetic units, the groups of genetic units, and four transgressive-regressive cycles with a duration of 5 to 10 Ma, and clearly of tectonic origin. The study pointed out the role played by the variations in subsidence regimes and by the reactivation of major basement faults for controlling the basin evolution at this "stage scale". The combination of these results to the characteristic pattern of organic matter distribution, as issued from Carbolog analysis, demonstrated that this distribution is in keeping with the general sequence stratigraphic framework, at a basin scale; especially noteworthy is the general good correlation which exists, for the lower order sequences, between organic-rich horizons and maximum flooding surfaces, when deposition occurs below the storm wave base. The study also brought out the role of local conditions (dilution effect due to high sedimentation rate, for instance) which act within the sequence stratigraphic framework but which may also impose their own dominant influence.

Introduction

Sequence stratigraphy is the best tool to reconstruct sediment geometries and to appreciate their origins. The fundamental units – the sequences – are interpreted to be formed in response to the interaction between the rates of eustasy, subsidence, sediment supply, and carbonate production (Posamentier et al. 1988; Van Wagoner et al. 1988). Therefore, as this approach provides both a genetic and a chronostratigraphic framework for the interpretation of the deposit succession in the sedimentary record, it can be considered as a compulsory tool for the study of basin fill histories. It allows the factors controlling the sequence architecture to be discriminated as well as the reservoir geometries to be predicted. This approach can be used at different scales. Here we use it for a better understanding of the relation between basin evolution and organic matter distribution. The study was performed on a basin scale – the Paris Basin – and focussed on a sedimentary series covering several tens of millions of years – the Lias – which is known as the major source rock for the oil pools in the basin (Espitalié et al. 1987).

The study consisted of a three-dimensional investigation and the main results were synthesized in a SW-NE transect, nearly 200 km long, comprising 16 wells, chosen to cover the major paleogeographic and tectonic features of this basin (Fig. 1).

The study required the combination of two methods, both applied to wells using wireline logs: (1) the Carbolog method to estimate the organic carbon content; (2) the stacking pattern method to identify and classify the different orders of superimposed sequences. Consequently, the two sets of data obtained had the same accuracy and thus enabled reliable comparisons to be made.

Geological Setting

The Paris Basin is a typical intracratonic basin, essentially of Mesozoic age. Originated during the Permo-Triassic distensional phase (Perrodon and Zabek 1991), it developed over three major basement blocks – Ardennes, Morvan Vosges, and Armorican massif – limited by major fault systems – Bray fault, Seine Sennely fault, St Martin de Bossenay, and other meridian faults. These major tectonic features, in response to the geodynamic events which affected the areas bordering the Paris Basin (Tethys and Atlantic), strongly influenced the Mesozoic structural and sedimentary history of the basin. During this period of time, six major transgressive-regressive cycles have been identified, which correspond to phases of acceleration of the subsidence (Guillocheau 1991). The Lias covers most of one of these cycles, namely the Carnian-Toarcian. Within this period, the basin was downwarped, readjustments took place between basement blocks, and the faults were strongly reactivated. An extensive transgression developed westwards, progressi-

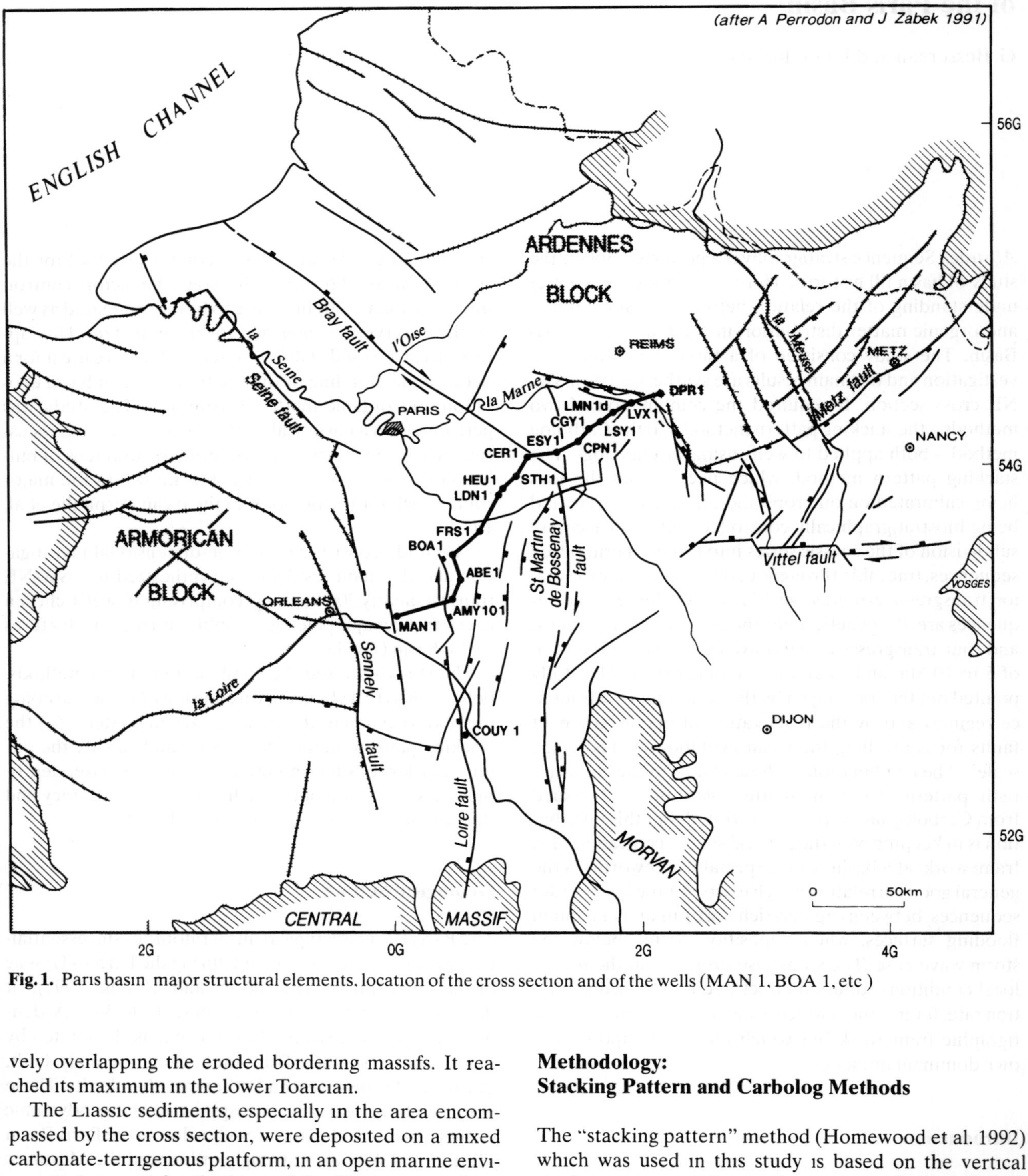

Fig. 1. Paris basin major structural elements, location of the cross section and of the wells (MAN 1, BOA 1, etc)

vely overlapping the eroded bordering massifs. It reached its maximum in the lower Toarcian.

The Liassic sediments, especially in the area encompassed by the cross section, were deposited on a mixed carbonate-terrigenous platform, in an open marine environment, except for the base of the Hettangian, where sediments were deposited in a restricted marine to continental environment. Within this open marine environment, a major distinction can be made between sediments deposited in lower offshore, that is to say under the storm wave base, and sediments deposited above it.

The serie, which is considered globally as a source rock, is characterized by strong vertical variations of the Total Organic Carbon (TOC) with several occurrences of organic-rich horizons beside the well-known Schistes Carton (Bessereau et al. 1992 and Fig. 7).

Methodology: Stacking Pattern and Carbolog Methods

The "stacking pattern" method (Homewood et al. 1992) which was used in this study is based on the vertical stacking pattern of the smallest stratigraphic units which can be identified in a sedimentary series and traced on a sub-basin scale. These so-called genetic units correspond to the smallest progradation-retrogradation cycles defined from sedimentary facies analysis. Such a method comprises three main phases: (1) identification of the genetic units, previously implying the determination of a sedimentological model: (2) stacking of these genetic units, which gives the global trends to a seaward stepping (progradation), a vertical stacking (aggradation) and a landward stepping (retrogradation) (Cross 1988: Mitchum

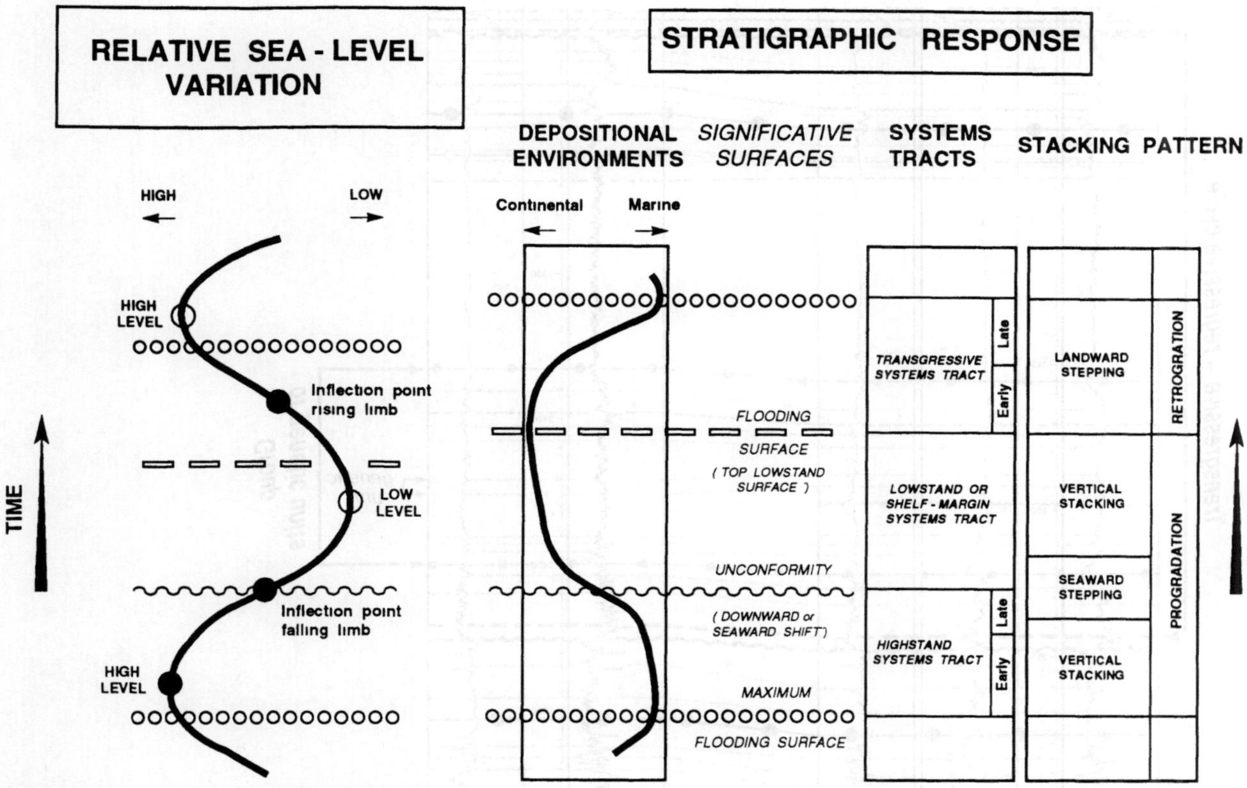

Fig. 2. Stratigraphic response of a cycle of relative sea-level variation· significative surfaces, relationship between systems tracts, and stacking pattern (Homewood et al 1992)

and Van Wagoner 1991; Homewood et al. 1992); (3) classification of the depositional sequences and their significative surfaces – maximum flooding surface (MFS), flooding surface (FS), and unconformity (Fig. 2) – which have to be validated by correlation of the different vertical profiles.

In this study, the sequence stratigraphy analysis was based on wireline logs, after calibration on outcrops, located east and south of the Paris Basin, and cores, including those from the fully cored well Couy 1 (Gely et al. 1991; Guillocheau et al. 1992) located in the southern part of the basin. The well log signatures were calibrated at the scale of the genetic units in terms of sedimentary environments (Guillocheau et al., in prep.).

The genetic units are here defined as the smallest units which could be identified on wireline logs and traced at the basin scale. Their average thickness ranges from 4 to 10 m, depending on the resolution of the wireline logs. They are limited by two maximum flooding surfaces. At this scale of stratigraphic organization, the MFS can be assumed to be isochronous because they correspond to turn-around periods between retrogradation and progradation. In these dominantly open marine Liassic sediments, these surfaces correspond to maxima of depth and are expressed by maxima of shale content. It has been checked, by comparing cores to log signatures, that maxima of shale content correspond in most cases to maxima of gamma ray. The stacking of these genetic units allowed the global trends to be defined, the different orders of stacked sequences to be classified and the significative surfaces to be validated by correlation of the different vertical profiles (Fig. 3).

The genetic units were tentatively dated using ammonite zonation from old wells synthesized by Serra (1971) and the dating of the stratigraphic well Couy 1 (Gely et al. 1991; Guillocheau et al. 1992).

The Carbolog method (Carpentier et al. 1989) estimates the in situ organic carbon content of the sediment from the combination of two conventional logs, the sonic and the resistivity, considering that organic matter is characterized by high sonic transit time and high resistivity (Fig. 4). This method has already been successfully applied to the Lias of the Paris Basin (Bessereau et al. 1992), resulting in an accurate view of the organic matter distribution.

The Sequence Stratigraphic Framework

The result of the stacking pattern approach is an accurate subdivision of the Liassic series in different orders of sequences, which can be easily traced throughout the basin. Three orders of superimposed sequences were defined within the major transgressive-regressive Liassic cycle: the genetic units, the groups of genetic units (i. e., the genetic unit sets of Van Wagoner et al. 1988) and four minor transgressive-regressive cycles (Figs. 5 and 6, pp. 108–110).

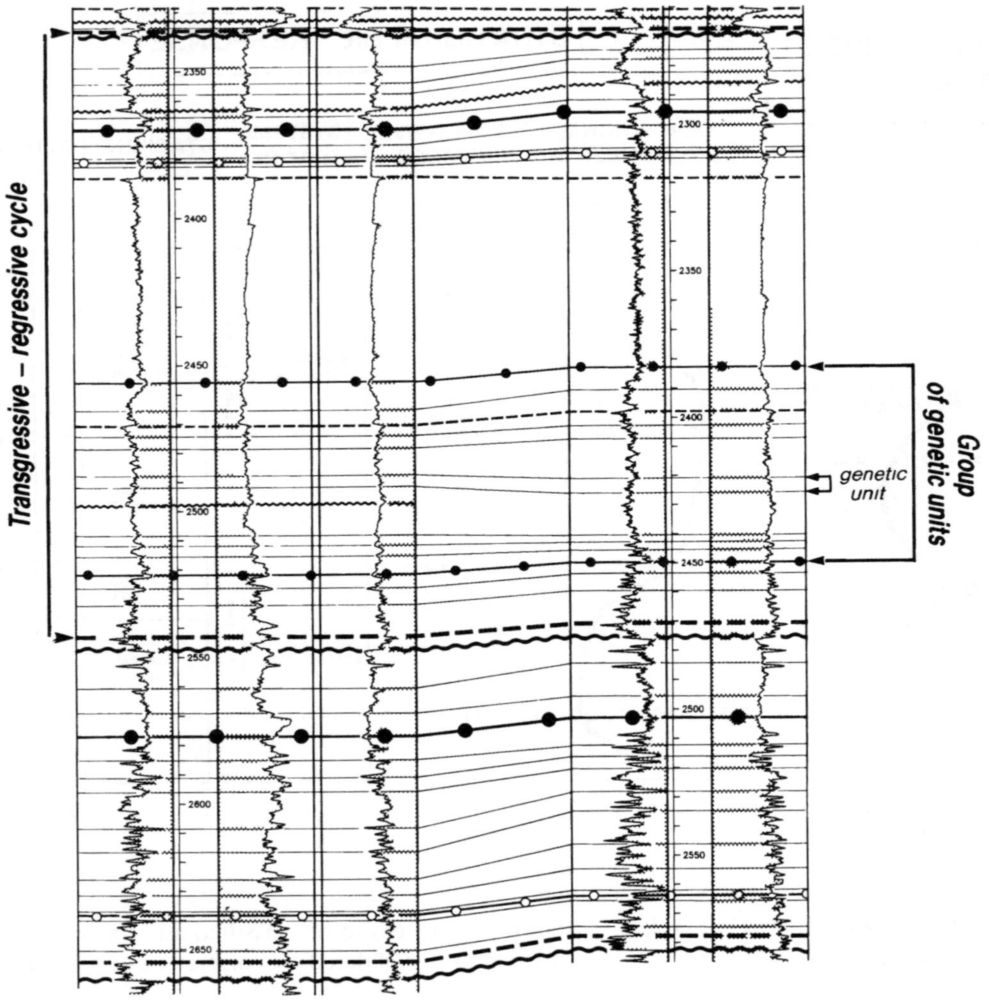

—— Maximum flooding surface of genetic units
—●— major } Maximum flooding surfaces
—•— minor
〰 Unconformity
- - - major } Flooding surfaces
- - - - minor

Fig. 3. Depositional sequence classification based on well logs

1. The genetic units have a duration ranging from 100000 to 400000 years. Where the series are the thickest, they can be subdivided into four or five smaller units. Thus, they do not always correspond to the same absolute duration; nor can a definite sequence order be ascribed to them. They are merely the smallest traceable units on a basin-wide scale. However, according to their duration and to the ratio of stacking pattern of higher frequency sequences, they can be compared to Milankovitch cycles. They are probably of climatic origin (glacio-eustatism and/or fluctuation of carbonate productivity).
2. The groups of genetic units are made up of a varying number of genetic units They have a duration ranging from 0.6 to 1 Ma. Consequently, they record subperiodic cyclic variations of the relative sea level. Moreover, the geometry of these sequences does not indicate any structural control. Both facts suggest that they could be of eustatic origin.
3. The four minor transgressive-regressive cycles result from the stacking of the previous sequences which are disposed in large cycles of increasing depth (transgressive phase), then decreasing depth (regressive phase) They are limited by a flooding surface. Their thickness varies (from about 20 or 30 m to 200 m) as well as their degree of symmetry (the transgressive phase can be much longer than the regressive one). As depicted in Fig. 6, where each cycle is taken independently with the flooding surface as the horizontal reference surface, the shift of the depocenters from one cycle to another clearly demonstrates their tectonic origin. Their duration ranges from 5 to 10 Ma – this is the "stage scale" They correspond to second-order cycles in Vail's nomenclature (Vail et al. 1991). Although their limits do not fit totally with the standard stage limits, they will be given, for simplification, the name of the stage with which they mostly coincide.

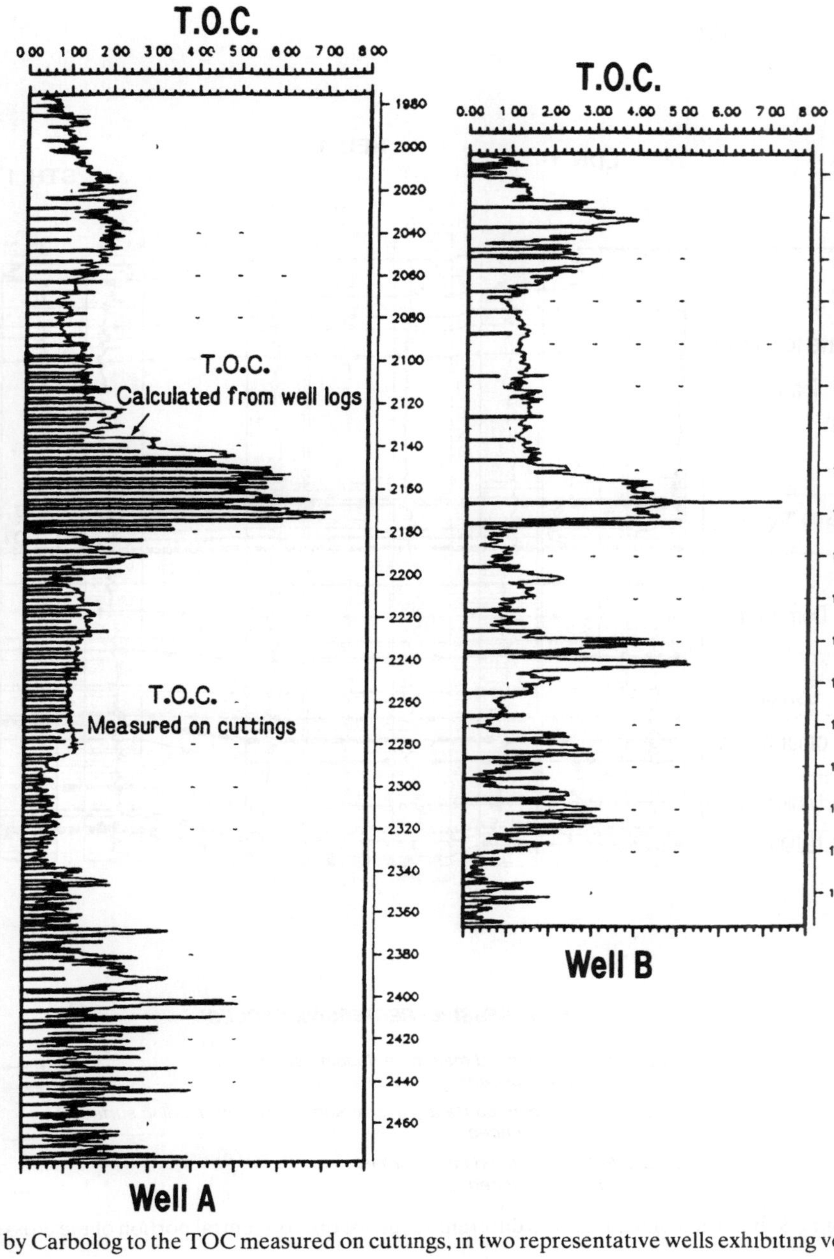

Fig. 4. Comparison of the TOC calculated by Carbolog to the TOC measured on cuttings, in two representative wells exhibiting very different sampling rates

The "Hettangian" cycle is rather thin and slightly dissymmetrical. In contrast, the "Sinemurian" cycle is thicker but, above all, it is very dissymmetrical: 3 Ma for the transgressive semicycle up to the maximum flooding surface, which is located within the Obtusum zone, 1.5 Ma for the regressive semicycle up to the lower part of the Raricostatum zone. Nevertheless, both cycles exhibit a common characteristic: their unconformity and flooding surfaces are practically merged and the interval between these surfaces is represented by only a few metres. Laterally, i.e., in the deepest part of the basin, this thin interval never seems to pass into thicker deposits which could be interpreted as lowstand wedge or shelf margin wedge. Nor is it possible to look for it in a deeper domain, the Tethys for instance: the involved bathymetric variations would necessarily induce strong emersions and stratigraphic gaps, and this has never been observed.

The "Pliensbachian" cycle is extremely dissymmetrical (its transgressive semicycle being globally three times longer than the regressive one). However, it is particularly characterised by a diachronous major MFS: southwest of the Bray fault, the major MFS (number 3 in Fig. 5) is located at the base of the Margaritatus zone (Stokesi subzone); northeast, this MFS is no more the major MFS of the Pliensbachian cycle, but only corresponds to a MFS of a higher sequence order (i.e., the groups of genetic units); the major MFS of the cycle is situated higher in the series (number 4) near the top of the Margaritatus zone. This shift of the major MFS is accompanied by a spectacular increase in the thickness of the sedimentary series

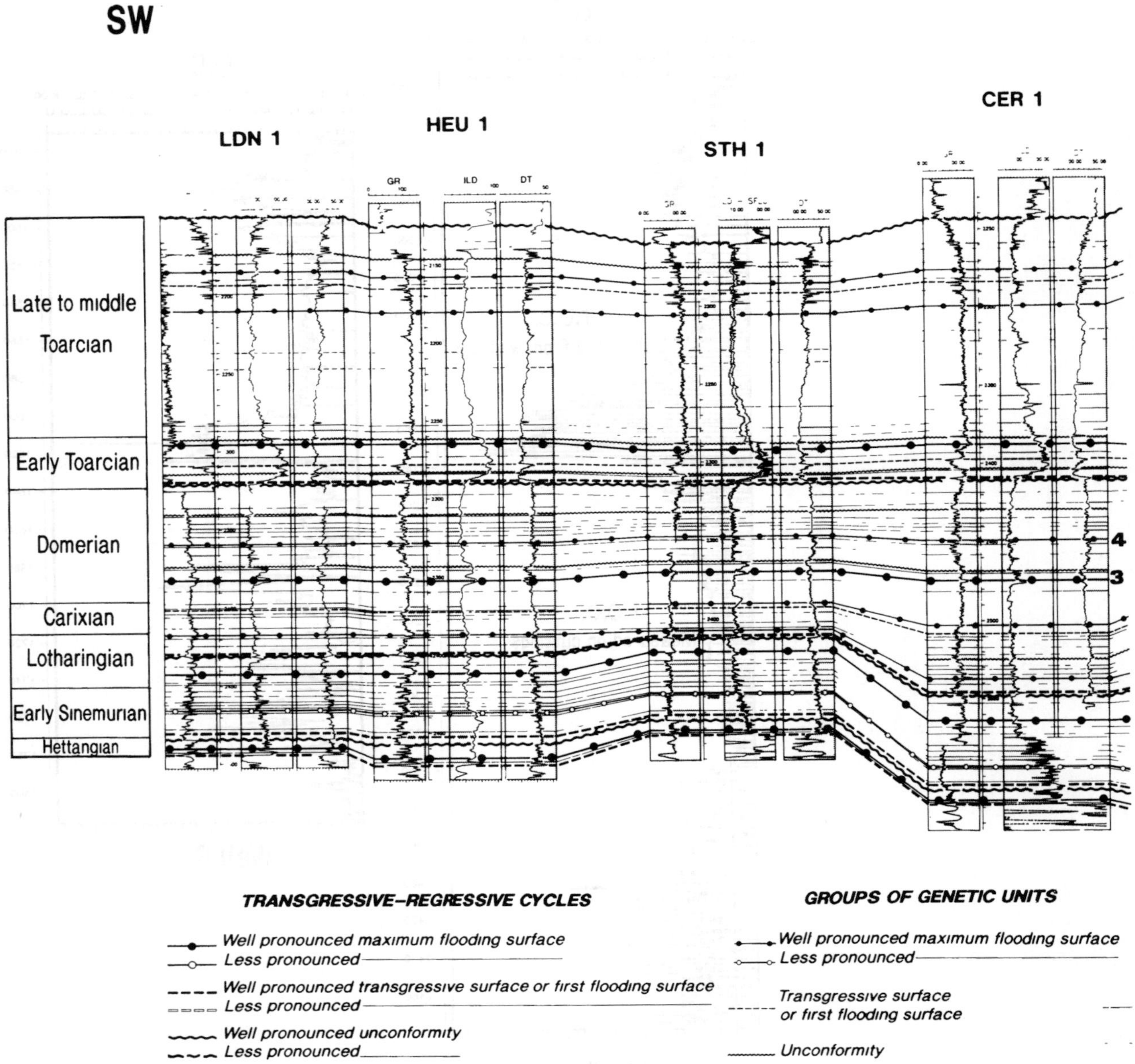

Fig. 5. Subdivision of the Lias into different sequence orders (central portion of the cross section)

between MFS 2 and MFS 3: this takes place between wells CER 1 and CPN 1 and characterizes the northeast part of the basin.

Moreover, the transition from the Pliensbachian cycle to the Toarcian cycle is rather different on either side of the Bray fault: in the SW part, the unconformity is immediately overlain by the flooding surface, while a lowstand systems tract develops to the northeast.

The "Toarcian" cycle depicts a completely different pattern: it is dominated by its regressive phase, the transgressive one being restricted to the early Toarcian. The top of this cycle is still badly dated. It corresponds to the base of the major transgressive-regressive cycle of Bajocian-Bathonian age (Guillocheau 1991) and is clearly a tectonic disconformity with erosion, especially on the Armorican side of the basin.

Thus, the variations in the subsidence regimes and the reactivation of major basement faults appear as the main parameters for controlling the basin evolution at this time order, i.e., "stage scale". Both elements are identified from the Hettangian to the Sinemurian cycle and to the Pliensbachian cycle. In the latter, the time shift of the major MFS can be explained by a lateral change in the subsidence rate, which is much higher in the NE block. This implies a decoupling in the tectonic regime of the blocks on either side of the faults. Furthermore, the lack of any lowstand systems tract development in the Hettangian, Sinemurian, and Pliensbachian (SW part) cycles can also be interpreted as due to variations in subsidence regimes: firstly, the disruption of the previous subsidence regime leads to a noticeable decrease in its rate and so to a shallowing (subsidence rate<sediment supply rate or

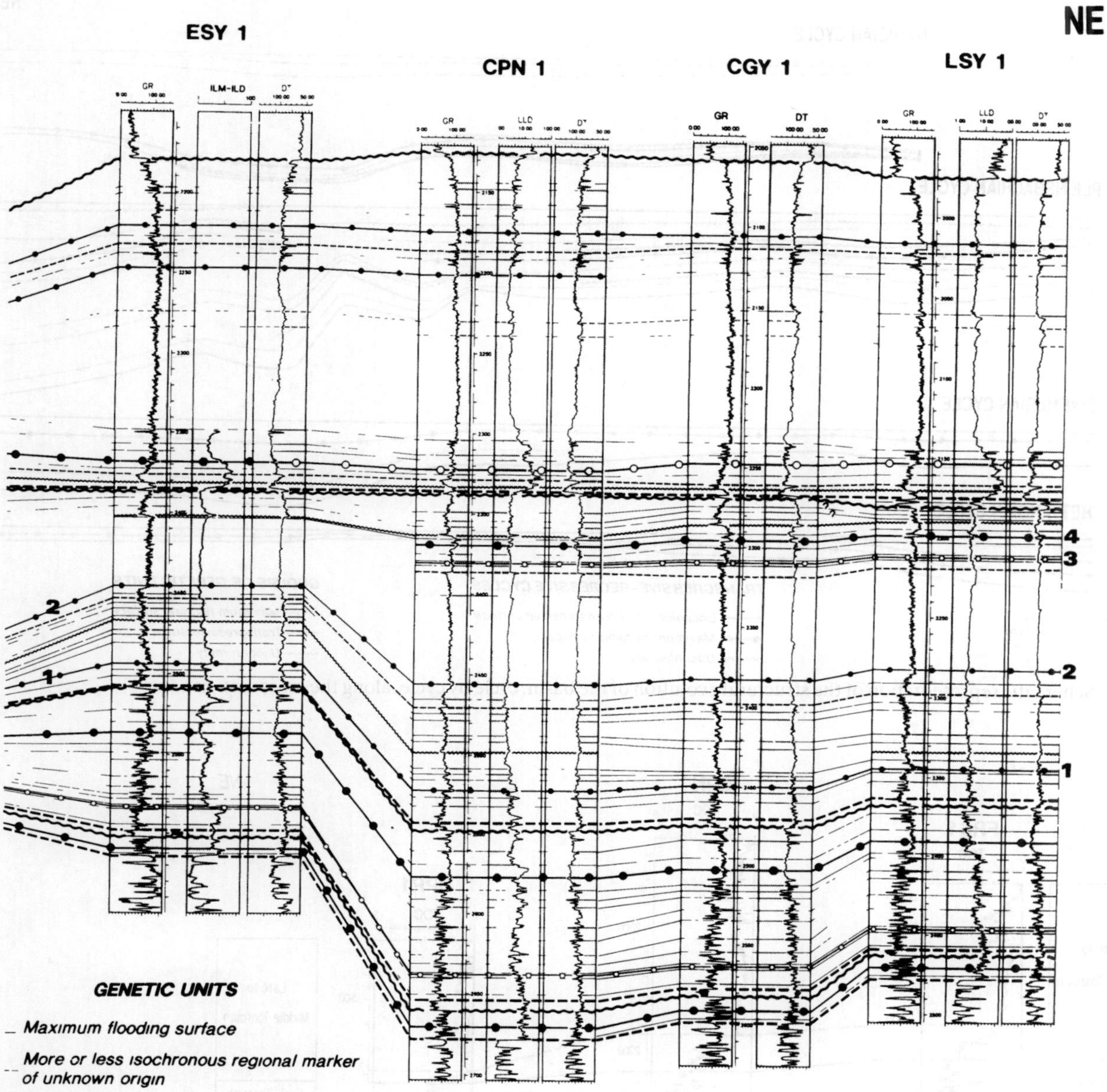

Fig. 5 *(continued)* Legend see p. 112

carbonate production rate); secondly, the onset of a new subsidence regime, with an increase of its rate concomitant with a new paleogeography, induces a deepening (subsidence rate>sediment supply or carbonate production rates). The development of a lowstand systems tract in the NE part of the basin in the Pliensbachian cycle shows up the decoupling of the blocks at that time. In contrast with the first three cycles, the Toarcian cycle is characterized by a homogeneous subsidence regime within the whole basin. The faults are no longer active, the basin is flexured, and the Schistes Carton onlap on both sides of the basin.

The overall transgressive-regressive cycle which covers the whole Lias is superimposed on the four T-R cycles. It is strongly asymmetrical: beginning in the Carnian, the maximum flooding occurs at the top of the Serpentinus zone, the regressive phase being restricted to the Middle-Late Toarcian. The distortion observed in the four cycles could have originated in this asymmetry, according to their location in this lower-order sequence. Also of tectonic origin, the Lias cycle results from an acceleration of the subsidence during the rifting of the Neo-Tethys.

The Organic Matter in the Sequence Stratigraphic Framework

As stated earlier (Bessereau et al. 1992), the Lias serie is characterized by strong variations in organic content. In this way, it exhibits very characteristic patterns of vertical and lateral organic matter distribution, as illustrated in Fig. 7, where the three representative continuous logs of TOC show:

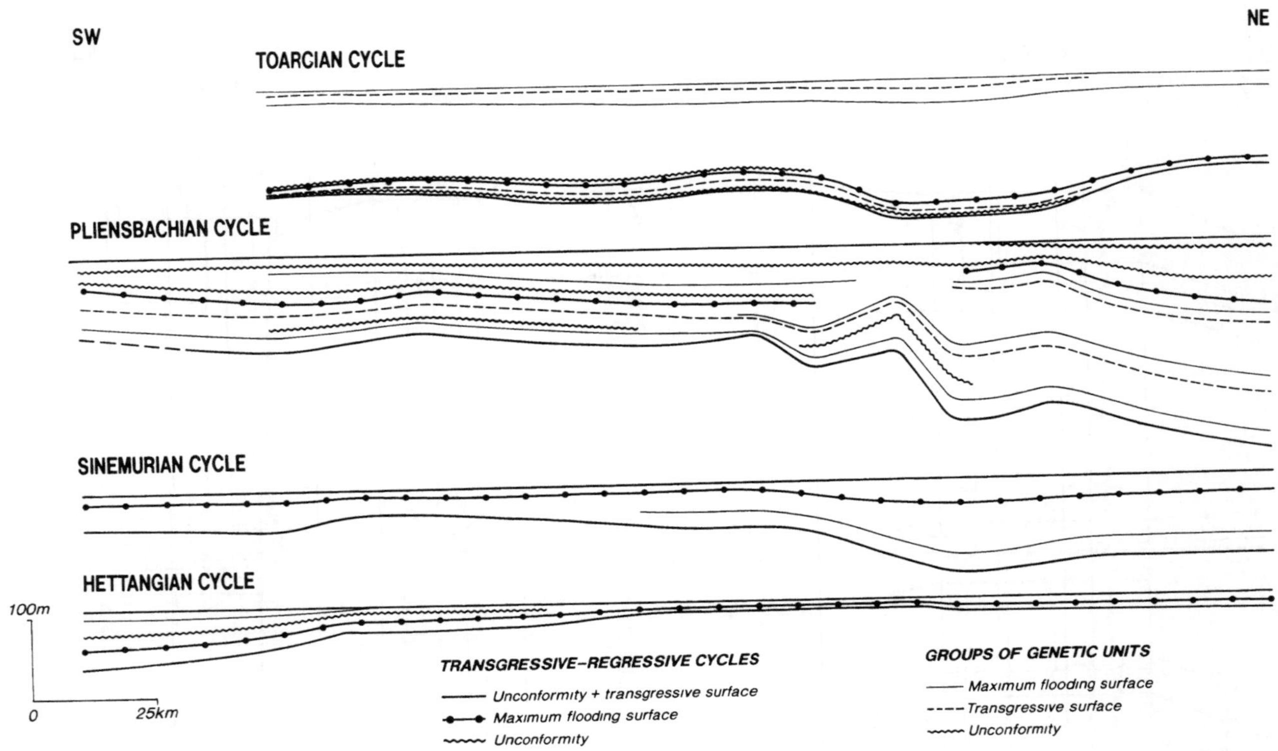

Fig. 6. Schematic representation of the structural evolution of the basin, cycle by cycle, along the cross section

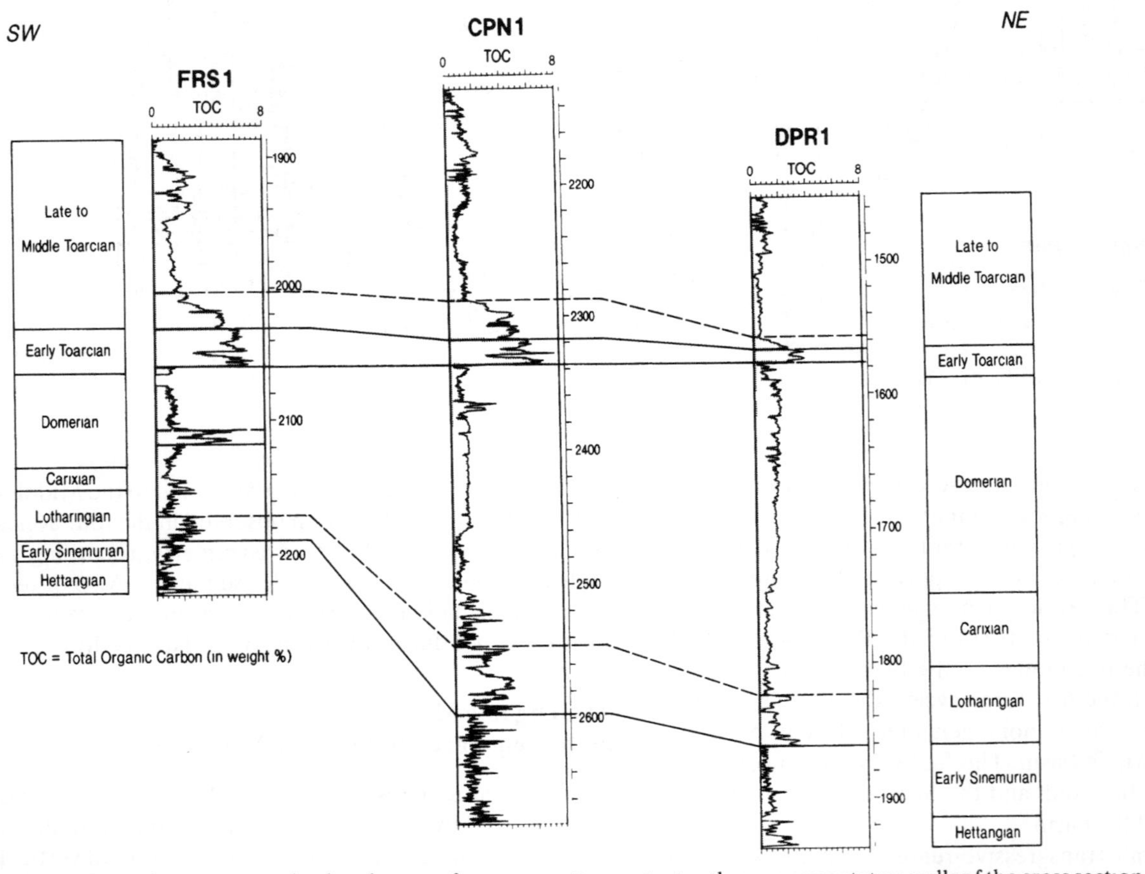

Fig. 7. Typical variations in the distribution of organic matter content in three representative wells of the cross section

- strong vertical TOC variations within the interval,
- the occurrence of several organic-rich horizons beside the well-known Schistes Carton; noteworthy are the Lotharingian interval and the doublet of high TOC layers near the base of the Domerian,
- the lateral continuity of these horizons over the basin, but with variations in thickness and in organic matter content. Especially noteworthy is the Domerian doublet, the signature of which is remarkably identical from AMY 101 to LDN 1 to CER 1 and which disappears to the NE, where the Pliensbachian is uniformly poor.

The superimposition of the main sequence boundaries (i.e., of the groups of genetic units and of the four minor transgressive-regressive cycles), previously obtained, on these TOC logs (Fig. 8) clearly shows:

1. A generally good correlation exists between organic-rich horizons and MFS. This is true for the Transgressive-Regressive cycles as illustrated by the Sinemurian and Toarcian cycles where the organic-richest interval systematically corresponds to this MFS. This is also confirmed in the Pliensbachian cycle but only in the SW part of the cross section. This also holds true for the higher-order sequences, i.e., the groups of genetic units as illustrated by the MFS of the Carixian.
2. In a wider sense, organic matter content tends to be high in the late stage of landward stepping (=the transgressive systems tract), and early stage of vertical stacking (=the highstand systems tract). This is particularly well expressed in the Sinemurian cycle: the very rapid variations of TOC which are inscribed in a global increasing TOC trend, culminate in an interval of maximum TOC coinciding with the MFS of the cycle.
3. The sequence hierarchy also seems to be reflected in the organic matter distribution (Fig. 9, p. 114). In other words, the amount of organic carbon stored in the organic-rich horizon at a MFS of a given order cycle could be determined by the position of this MFS in the lower-order sequence: the Schistes Carton occur at the MFS of the Toarcian cycle, which also corresponds to the maximum of the transgression in the Lias major transgressive-regressive cycle.

These features are well documented in the basin. Thus they provide a global framework for the distribution of organic matter and they can be used as guidelines for the appraisal of the petroleum potential of a basin. Does this necessarily mean that they can be used as rules which could be applied systematically? A very noticeable exception is given in the Pliensbachian cycle, in the NE part of the basin: no organic-rich interval is associated with its MFS (Fig. 10, p. 115). Less spectacular but also clear is, in the SW part of the basin, the lack of an organic-rich bed at the MFS of higher sequence order (i.e., the groups of genetic units) located in the upper part of the Margaritatus zone (MFS 4). In other respects, in the Toarcian cycle, the whole transgressive systems tract is very rich, always starting by a very high TOC bed (values up to 8%), which marks a major break with the underlying organic poor Domerian.

This problem may be partially answered by considering the factors which directly control the accumulation of organic matter – essentially its productivity and/or its preservation. The sequence stratigraphy appears to be especially appropriated for their identification.

In the general case, where the organic matter is located on either side of the MFS, especially in the Sinemurian cycle and in the Pliensbachian cycle – SW part of the basin – the MFS is recorded in sediments deposited in lower offshore, i.e., under the storm wave base, as mentioned above. In addition, the MFS is assumed from the sequence stratigraphy model to be characterized by slow sedimentation rates. The combination of these two conditions of depth and sedimentation rate appears as especially propitious to the preservation of the organic matter.

In contrast, the disappearance of the organic-rich interval associated to the major Pliensbachian MFS is very likely related to a dilution effect due, in this part of the basin, to a very high sedimentation rate of dominantly terrigenous sediments deposited nevertheless in lower offshore. The lack of the organic-rich bed associated with the MFS 4 in the SW part of the basin is most likely due to the combination of these two factors: higher rate of sedimentation and depositional environment in the upper offshore.

The deposition of bituminous shales of early Toarcian age has been the topic of great debate. Even when considering only the black shales of Northern Europe (Posidonienshеifer – Germany, Jet Rock Shale – Great Britain, Schistes Carton – France), numerous models have been proposed to explain their accumulation (see general discussion and references in Hollander et al. 1991, but also Morris 1979; Fleet et al. 1987; Myers 1987; Wignall 1991). This study does not claim to provide a new model for the deposition of the Schistes Carton, by using only the sequence stratigraphic framework. Nevertheless, this framework emphasizes some key facts: the Toarcian transgression, which is of major extension, takes place within a very short period of time (around 1 Ma) and in a completely different tectonic setting with respect to the previous times. The high rate of the transgression, which could be the origin of depositional conditions favorable to the preservation of the organic matter (see, for instance, Hollander et al. 1991), and quite homogeneous depositional settings created by the flexuration of the basin, could account for an organic-rich transgressive systems tract exhibiting analogous patterns at basin scale. In contrast, the occurrence and pattern of TOC above the major MFS cannot be directly inferred from the sequence stratigraphic framework. Here, the decrease in TOC might be hypothetically inferred from changes in paleo-oceanographic conditions (Hollander et al. 1991).

In addition, the present study shows (Fig. 10) that, in this basin and at the scale of the major transgressive-regressive cycles as well as the groups of genetic units, the organic-rich horizons associated with the MFS do not present the thinning pattern towards the basin which is usually proposed for these so-called condensed intervals:

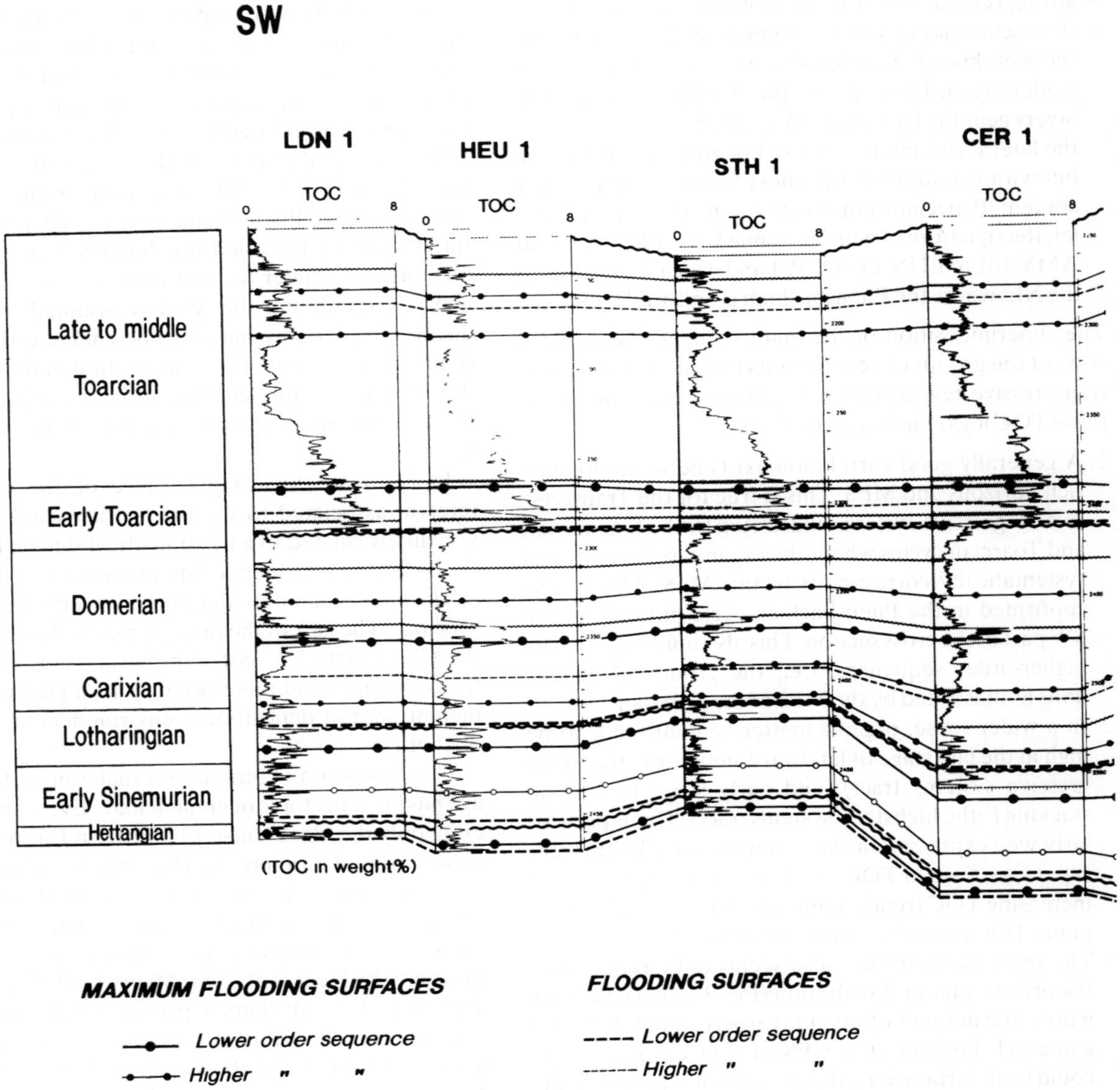

Fig. 8. Organic matter distribution in the sequence stratigraphic framework (central portion of the cross section)

in fact, the interval with the highest organic content is also the thickest. This organic enrichment towards the depocenters can be explained by the combination of the preservation and hydrodynamic processes: there will be a tendency for organic matter to accumulate in the most quiescent parts of the basin (Huc 1988).

All these results show that the organic matter can be interpreted in terms of sequence stratigraphy at a basin scale. Some general guidelines are provided for the distribution of the organic matter to be predicted, mainly – in this study – at lower sequence orders. However, they also bring out the influence of local conditions (dilution effect due to high sedimentation rate linked to the subsidence, for instance) which act within the sequence stratigraphic framework but which may also impose their own dominant influence. In addition, it also brings out the fact that keys or constraints can be given by the sequence stratigraphy for the identification of the different factors which control the accumulation of the organic matter.

Conclusions

The study undertaken on the Paris basin demonstrates the ability of sequence stratigraphy analysis to provide a better understanding of the basin evolution. It especially points out the role of tectonics in controlling this evolution at the "stage scale" and provides a precise stratigraphic framework to depict it. It also shows that this approach, through the combined use of the stacking pattern and Carbolog methods, provides a global framework for predicting the distribution of the organic matter at a ba-

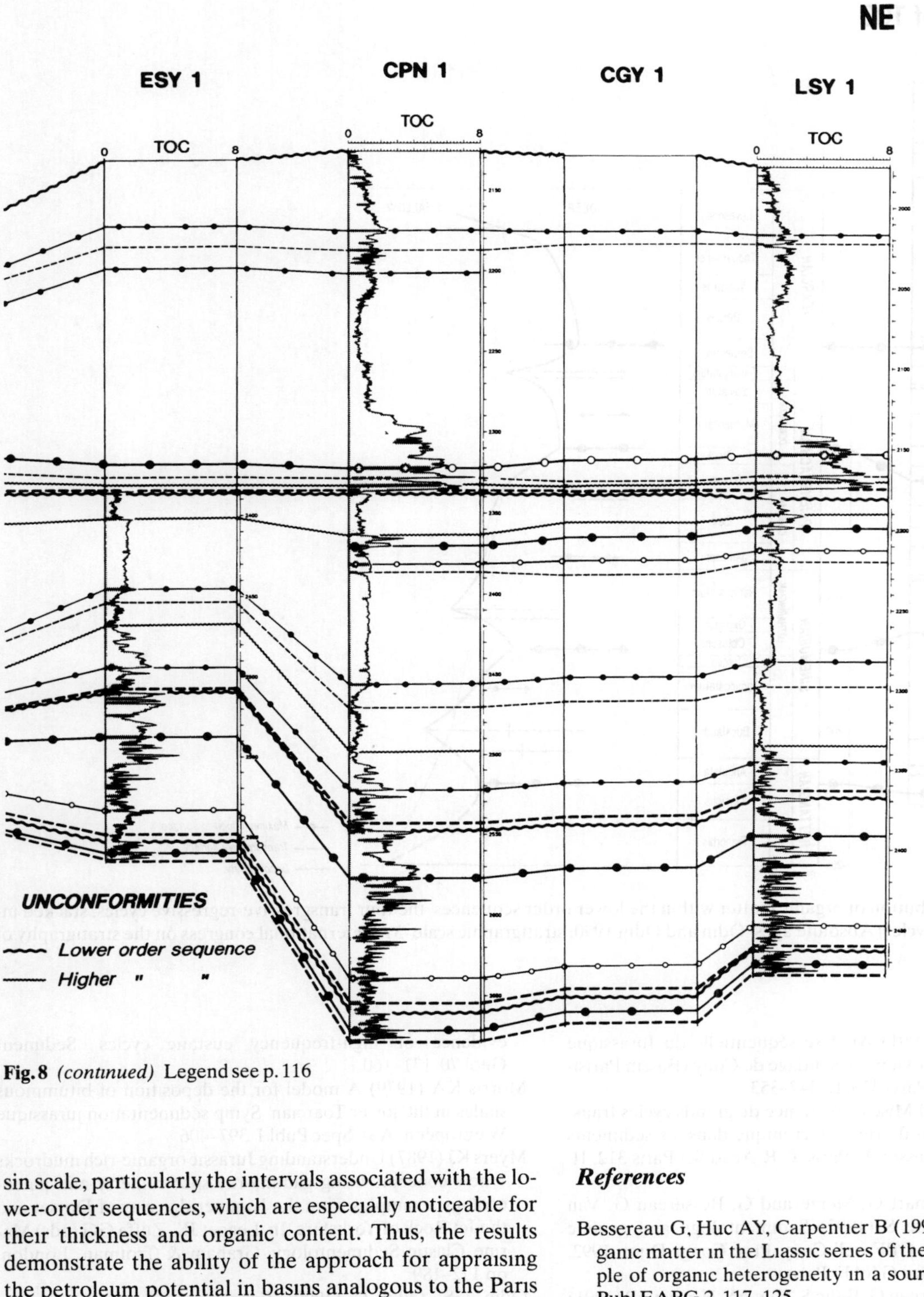

Fig. 8 *(continued)* Legend see p. 116

sin scale, particularly the intervals associated with the lower-order sequences, which are especially noticeable for their thickness and organic content. Thus, the results demonstrate the ability of the approach for appraising the petroleum potential in basins analogous to the Paris basin (i.e., intracratonic basins). This study also demonstrates the influence of local conditions which have to be carefully considered when applying this approach. Last of all, it is suggested that the sequence stratigraphy could provide a framework for further additional studies which aim at a better understanding of organic sedimentology.

References

Bessereau G, Huc AY, Carpentier B (1992) Distribution of organic matter in the Liassic series of the Paris basin an example of organic heterogeneity in a source rock interval. Spec Publ EAPG 2 117–125

Carpentier B, Huc AY, Bessereau G (1989) Diagraphies et roches-mères Estimation des teneurs en carbone organique par la méthode Carbolog Rev Inst Fr Pétrole 44, 6, 1 699–719

Cross TA (1988) Controls on coal distribution in transgressive-regressive cycles. SEPM Spec Publ 42· 371–380

Espitalié J, Marquis F, Sage L, Barsony I (1987) Géochimie organique du Bassin de Paris. Rev Inst Fr Pétrole 42, 3 271–302

Fleet AJ, Clayton CJ, Jenkins HC, Parkinson DN (1987) Liassic source rock deposition in western Europe In Brooks J, Glennie K (eds) Petroleum Geology of North West Europe Graham & Trotman, pp 59–70

Composite log of TOC

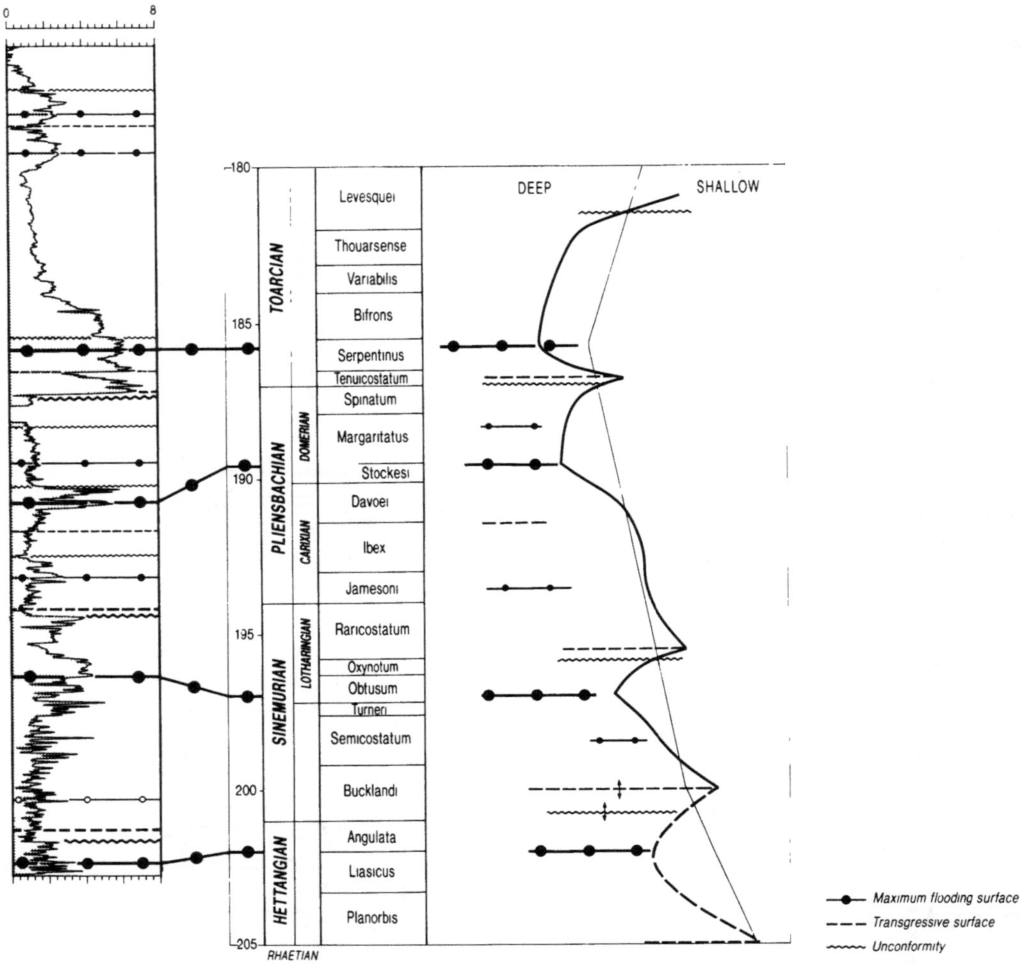

Fig. 9. Vertical distribution of organic matter within the lower order sequences the four transgressive-regressive cycles, stacked inside the whole Lias cycle (Absolute ages Odin and Odin 1990, stratigraphic scale 3rd international congress on the stratigraphy of the Jurassic)

Gely JP, Lorenz J (1991) Analyse séquentielle du Jurassique (Hettangien à Callovien) du sondage de Couy (Bassin Parisien) C R Acad Sci Paris 313, II 347–353

Guillocheau F (1991) Mise en évidence de grands cycles transgression-régression d'origine tectonique dans les sédiments mésozoiques du Bassin de Paris C R Acad Sci Paris 312, II 1587–1593

Guillocheau F, Dromart G, Merzeraud G, Bessereau G, Van Buchem F, Derasse S (1992) Sedimentology and genetic stratigraphy of the GPF well Couy Proc Congr Dijon 1992, Abstr (registered Soc Géol Fr Paris)

Hollander DJ, Bessereau G, Belin S, Huc AY, Houzay JP (1991) Organic matter in the early Toarcian shales, Paris basin, France a response to environmental changes Rev Inst Fr Pétrole 46, 5 543–562

Homewood P, Guillocheau F, Eschard R, Cross T (1992) High resolution correlations and genetic stratigraphy an integrated approach Bull Rech SNEA P 16, 2 357–381

Huc AY (1988) Sedimentology of organic matter In Frimmel FH, Christman RF (eds) Humic substances and their role in the environment John Wiley, Chichester, pp 215–243

Mitchum RM, Van Wagoner JC (1991) High-frequency sequences and their stacking patterns sequence-stratigraphic evidence of high-frequency eustatic cycles Sediment Geol 70 131–160

Morris KA (1979) A model for the deposition of bituminous shales in the lower Toarcian Symp sédimentation jurassique W européen ASF Spec Publ I 397–406

Myers KJ (1987) Understanding Jurassic organic-rich mudrocks – new concepts using gamma-ray spectrometry and paleoecology examples from the Kimmeridge clay of Dorset and the Jet Rock of Yorkshire In Legget JK, Zuffa GG (eds) Marine Clastic Sedimentology Graham & Trotman, London, pp 172–189

Odin GS, Odin C (1990) Echelle numérique des temps géologiques Géochronique 35, 1990 12–21

Perrodon A, Zabek J (1991) Paris Basin AAPG Mem 51, Interior Cratonic Basins, pp 633–679

Posamentier HW, Jervey MT, Vail PR (1988) Eustatic controls on clastic deposition I – Conceptual framework SEPM Spec Publ 42 109–124

Serra O (1971) Apports des sondages pétroliers à la connaissance du Lias du Bassin de Paris Mém BRGM 75 481–488

Vail PR, Audemart F, Bowman SA, Eisner PN, Perez-Cruz (1991) The stratigraphic signatures of tectonics, eustasy and sedimentology – an overview In Einsele G, Ricken W, Sei-

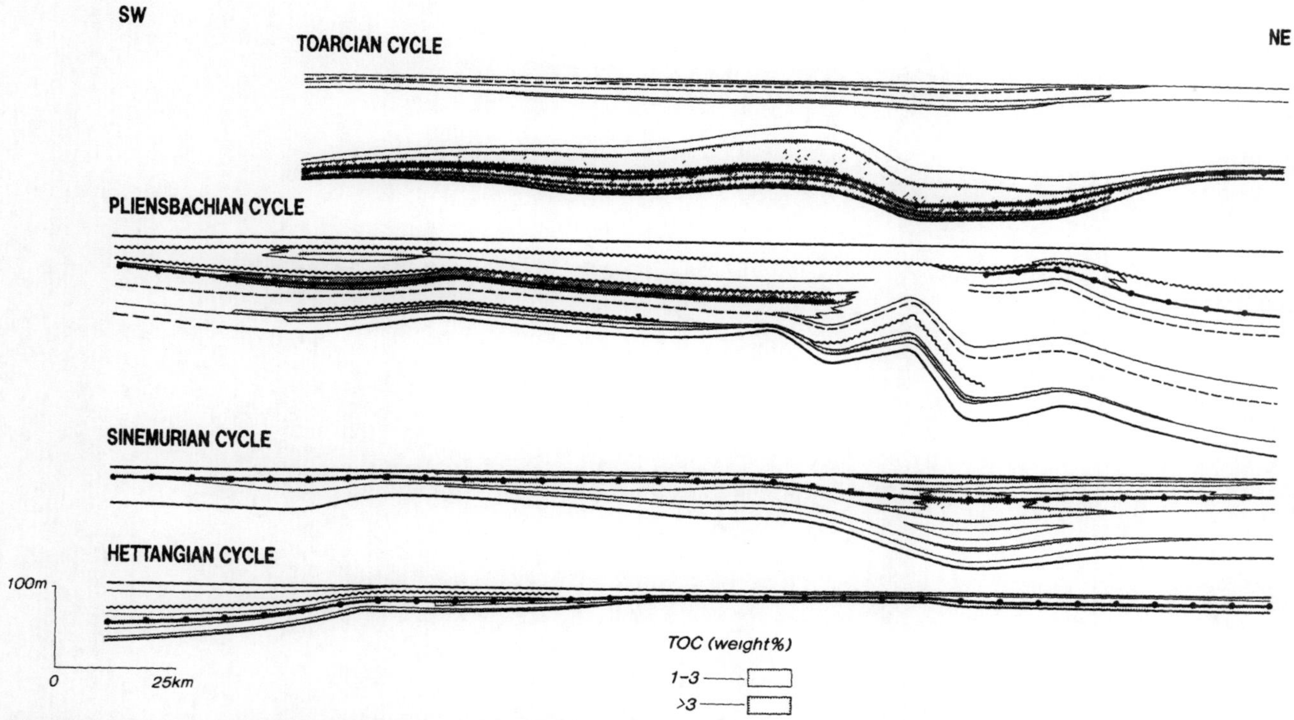

Fig. 10. Schematic representation of the organic matter distribution, cycle by cycle

lacher A (eds) Cycles and events in stratigraphy Springer, Berlin Heidelberg New York, pp 617–659

Van Wagoner JC, Posamentier HW, Mitchum RM, Vail PR, Loutit JF, Hardenbol J (1988) An overview of fundamentals of sequence stratigraphy and key-definitions SEPM Spec Publ 42: 39–45

Wignall PB (1991) Model for transgressive black shales? Geology 19 167–170

Relationship Between Organic Matter and Sedimentation in the Lias of Lorraine (France)

M. Hanzo and J. Espitalié

Abstract. The organic matter of the Lorraine Lias (eastern margin of the Paris Basin) has been analyzed in all its formations, from the north to the south of the area.

The results obtained serve to highlight the major geochemical variations at this study scale. Here, in time and in space, the contents and type of the organic matter result from paleogeographic features inherent in the situation of a zone of high subsidence with high sedimentation rate and in the proximity of the terrigenous sources. The highest values exist during transgressive pulsations, especially where the sedimentation rate is high and the environment confined. The units where the organic matter is richest and best preserved are found in northern Lorraine, outside the large coarse-grained terrigenous depositions.

Another objective of this analysis is to add to the geochemical analytical data and to provide further information on the source rocks of the Paris Basin.

Introduction

The aim of this chapter is to analyze the Lias of the eastern margin of the Paris Basin (Lorraine) by trying to interpret the results of organic geochemistry in relationship with the environmental variations during the Lias. The results obtained also serve to supplement the earlier studies (Espitalié et al. 1987 a, b) which showed that the Lias of the Paris Basin contained the main oil source rocks of the Trias and the Dogger, and that the center of the Basin (Brie/Champagne area), the most subsident zone, with very high sedimentation rate, generally contained the best organic matter contents and the hydrocarbon formation zone (Hanzo and Espitalié 1993).

The Lias outcrops along a westward concave aureole extending nearly 200 km from north to south (from Luxembourg to Langres) and 50 km from East to West (Fig. 1).

The axis of the zone of maximum subsidence of the Paris Basin trends approximately ESE/WNW throughout the Lias; it is located north of the area analyzed. Three areas are distinguished to the south of this axis in the following discussion, exhibiting varied paleogeographies and hence different genetic contexts for the deposition and preservation of organic matter, with shallow burial (Fig. 1). Northern Lorraine and central Lorraine are separated by the Metz Fault, a major tectonic feature, which controls the liassic sedimentation along a flexure zone associated with the high subsidence area of northern Lorraine (Fig. 2). On the contrary, another major fault, the Vittel Fault (between Nancy and Langres) has no effect on the sedimentation (Le Roux 1980). Accordingly, the central Lorraine/southern Lorraine areas are separated by a boundary located north of Fécocourt, and specified by some authors (Alpern and Huc 1976) as the limit between two geochemical provinces during early Toarcian times.

Study Method

The sections were selected in order to identify the variations in environmental conditions over time, throughout Lorraine. In addition to 12 outcrops (rare in Lorraine), 26 boreholes (mostly cored) supplied by EDF, Scetauroute, and BRGM, were analyzed (Fig. 1). Two of these boreholes, Cattenom (Hanzo and Guérin-Franiatte 1985) and Tranqueville, together cover nearly the whole of the Lias.

The sampling on each section was carried out taking account of the variety of lithofacies, fluctuating between argilites and limestones. Studies aimed at a better understanding of the paleoenvironment in which the organic matter was deposited.

Samples were analyzed by using following techniques:

- lithologic analysis, based on microscopic observations, shale analyses, and determinations of inorganic carbon;
- paleontological analysis of macroorganisms, benthic foraminifera (C. Ruget 1985, pers. comm., Ruget and Hanzo 1992), and microflora (data from M. C. Adloff, and Hanzo et al. 1990);
- organic content study of 600 samples, analyzed by Rock-Eval pyrolysis (Espitalié et al. 1985).

Bio-Lithostratigraphic Setting

In Lorraine, the sedimentological context of the Liassic series is that of a vast platform open to the Swabian and Alpine Basins, affected by a marine transgression from the base of the Jurassic (Maubeuge 1955; Allouc 1980; Groupe Français d'Etude du Jurassique 1980; Hanzo

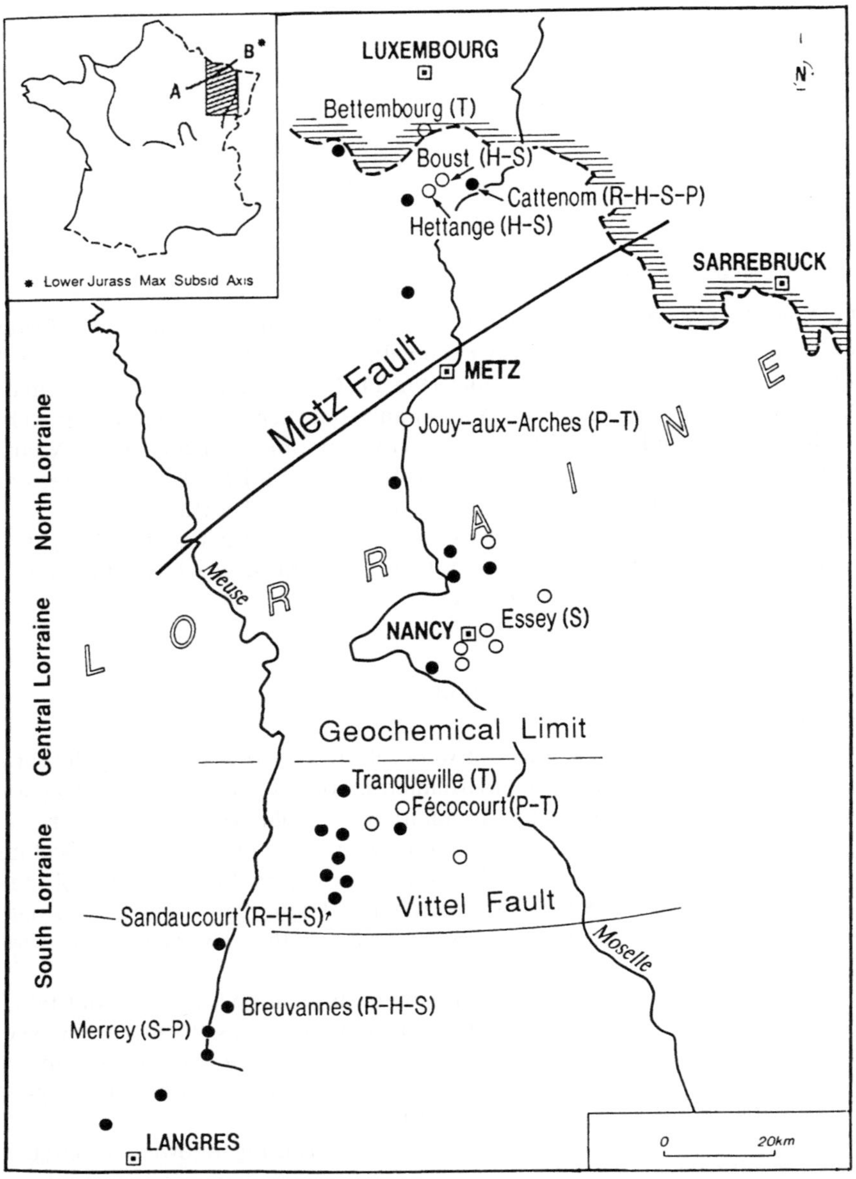

Fig. 1. Situation map of sites in northern, central and southern Lorraine ○ outcrops, ● boreholes, *R* Rhaetian, *H* Hettangian, *S* Sinemurian, *P* Pliensbachian, *T* Toarcian

1980). The bio-lithostratigraphic series is shown in Table 1, with 550 m of Lias to the North and 200 m in the center and south. After an incipient transgression during Lower Rhaetian, from the base of the Lias, the sea advanced from east to west, associated with the opening of the basin.

Classical Stratigraphy

In classical stratigraphy, the lithologic units obtained are grouped in klupfelian sequences which grade upwards into filling up facies: argilites are followed by shelly or sandy limestones. Following the Calcaire à gryphées (including the Grès d'Hettange to the North), they are:

- Argiles à Promicroceras and Calcaire ocreux;
- Marnes à Numismalis and Calcaire à Davoei (northern area);
- Argiles à Amalthées and Grès médioliasique;
- the series ranging from Schistes carton to Marnes à Voltzi (at the peak of the transgression) and Grès supraliasique;
- ferriferous layers attributed to the Minette complete the liassic series.

Sequence Stratigraphy

Four second-order transgressive-regressive cycles can be identified (Hanzo et al. 1992) in the Lorraine liassic sequence, from lithology, disconformities, and biostratigraphy:

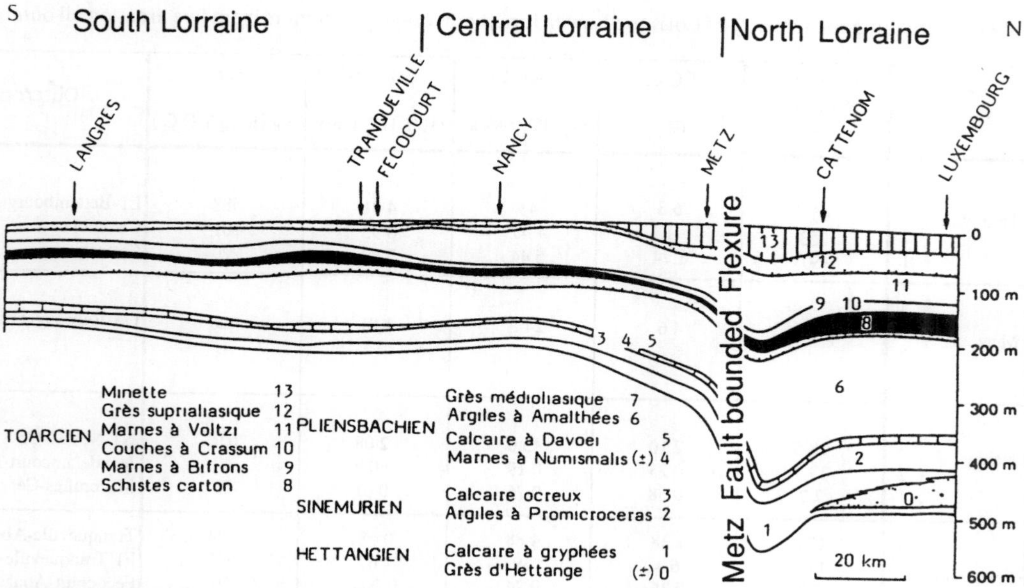

Fig. 2. Schematic section of Lorraine Lower Jurassic (After M Hanzo 1980)

Table 1. Bio- and lithostratigraphy of Lorraine Lower Jurassic

BIO-LITHOSTRATIGRAPHY					FORMATION THICKNESS		
	Ammonite Zones	Formations	Lithological Sequences	Genetical Sequences (2 order)	NORTH	MIDDLE	SOUTH
AALENIAN	*Aalensis*						
TOARCIAN	*Pseudoradiosa*	Minette			15 - 60 m	10 - 12 m	a few m
	Dispansum	Grès supraliasique			1 - 25 m	10 m	a few m
	Thouarsense	Marnes à Voltzi			20 - 60 m	65 m	20 - 60 m
	Variabilis	Couches à Crassum			1 - a few m	1 - a few m	1 - a few m
	Bifrons	Marnes à Bifrons			20 m	7 - 10 m	15 - 20 m
	Serpentinus *Tenuicostatum*	Schistes Carton			25 - 60 m	20 m	2 - 20 m
PLIENSBACHIAN – DOMERIAN	*Spinatum*	Grès médioliasique			40 m	8 - 15 m	30 m
	Margaritatus	Argiles à amalthées			100 - 180 m	30 m	30 m
CARIXIAN	*Davoei*	Calcaire à Davoei			0,5 - 3 m	1 - 2,5 m	3 - 5 m
	Ibex *Jamesoni*	± Marnes à Numismalis			a few - 13 m		3 - 6 m
SINEMURIAN SL – LOTHARINGIAN	*Raricostatum* *Oxynotum*	Calcaire ocreux			0,2 - 1 m	1,5 m	0,5 - 1,5 m
	Obtusum *Turneri*	Argiles à Promicroceras			50 m	30 m	15 - 30 m
SINEMURIAN ss	*Semicostatum* *Bucklandi*	Calcaire à gryphées ± Grès d'Hettange			20 - 70 m	10 - 20 m	4 - 30 m
HETTANGIAN	*Angulata* *Liasicus* *Planorbis*				0 - 30 m		

Table 2. Organic geochemistry of Lorraine Lower Jurassic average parameters from five areas on all outcrops and boreholes

Areas		Average Thickness (m)	T.O.C. (%)	Mean S2 (mg HC /g rock)	Total S2 (10^6t HC /km²)	H I (mg HC /g T O C)	Outcrops and Boreholes
Thionville	4						
	3	45 7	6 8	43	4 91	588	L1-Bettembourg-G1-G2
	2	253	1	1 47	0 93	101	Cattenom
	1	60	1 74	5 44	0 82	220	Cattenom
Metz	4						
	3	16 5	4 6	24 5	1 01	426	G3-Jouy aux Arches
	2						
	1						
Nancy	4						
	3	18 5	7 10	45 0	2 08	636	N1-N2-G4-G5
	2	13 1	0 25	0 19	0 006	40	Jeandelaincourt-Heillecourt-Essey-Ludres
	1	7 7	0 58	0 76	0 01	53	Mazerulles-Cerville
Fécocourt	4	47 8	1 38	4 88	0 58	174	Tranqueville-Aboncourt
	3	17 2	6 4	39	1 67	575	F1-Tranqueville-Fécocourt-G6-G7-Aouze
	2	7 3	0 36	0 24	0 004	59	Fécocourt-Aouze-Châtenois-Balléville-Mannecourt
	1	15 5	0 94	2 63	0 10	70	Poussay-Sandaucourt
Langres	4	15	1 01	1 63	0 06	138	Dampierre
	3	4 3	5 24	29 78	0 32	510	G8-Dampierre-G9
	2	8 9	0 47	1 71	0 04	79	Breuvannes-Merrey-Dampierre
	1	4 4	0 33	0 17	0 002	33	Breuvannes-Provenchères

4 = Middle-Upper Toarcian --> Marnes à Bifrons - Couches à Crassum - Marnes à Voltzi - Grès supraliasique
3 = Lower Toarcian --> Schistes carton
2 = Upper Sinemurian-Pliensbachian --> Argiles à Promicroceras - Calcaire ocreux - Marnes à Numismalis (±) - Calcaire à Davoei - Argiles à Amalthées - Grès médioliasique
1 = Hettangian - Lower Sinemurian --> Calcaire à gryphées (- Grès d'Hettange)

- Lower Calcaire à gryphées and Grès d'Hettange (only in northern Lorraine), with transgressive system tract and highstand system tract; they correspond to subzones from Planorbis to Angulata;
- upper Calcaire à gryphées, Argiles à Promicroceras, Calcaire ocreux, Marnes à Numismalis and Calcaire à Davoei, with a rhythmic succession of transgressive and highstand system tracts (with still some uncertainties); they correspond to subzones from Conybeari to Figulinum;
- Argiles à Amalthées and Grès médioliasique, with two episods transgressive-highstand system tracts; they correspond to subzones from Stokesi to Apyrenum;
- Schistes carton, Marnes à Bifrons, Couches à Crassum, Marnes à Voltzi, and Grès supraliasique, with a rhythmic succession of transgressive and highstand system tracts (with still some incertainties); they correspond to subzones from Hawskerense to Gruneri.

Analysis of the Organic Matter

Quantitative and qualitative analyzes of the organic matter have been carried out in the different sedimentation environments, characterized mainly by their lithology and their paleontological content. Immaturity is one characteristic of the organic matter found in the different units Only its content and its state of preservation vary in accordance with the sedimentation environment.

Characteristics of the Organic Matter in the Different Formation (Table 2)

Calcaire à Gryphées. This is represented by alternating limestones and marls or calcareous argillites, deposited in a medium energy reducing environment. Considering the northern Grès d'Hettange sandstone member which is part of the Calcaire à gryphées formation, we can note, from south to north, that the quantity of terrigenous material rises progressively and reflects the increasingly marked influence of the Ardennes Range. The presence of coarse bioclasts argues in favor of a weak bathymetry, also reported by other authors (Hallam 1960). The monospecific combinations of encrusted foraminifera indicate an environment in the course of opening in the Hettangian. Progressively, their variety and their abundance increase. The palynological content, in northern Lorraine, reveals a mixture of continental forms and marine forms, with the initial predominance of humic matter, followed progressively by amorphous matter (Hanzo et al. 1990).

The organic carbon contents in these facies show that, in the south, the few samples analyzed in the argilites of the transition zone (at the Rhaetian/Hettangian boundary) have uniformly low TOC contents (about 0.3 %). In the north, the high sedimentation rate zone, the values are higher and more variable (0.5 to 1.8 % from one section to the next). For the Calcaire à gryphées, the data are far more numerous and the ranges of values between north and south are approximately the same. On a given section, at least in a borehole, the variations are wide, ranging from 0.2 to 9 % TOC in the south (at Sandaucourt) and from 0.2 to 5.3 % in the north (at Cattenom). At Cattenom, the TOC contents show an overall increase with the carbonate contents at the top of the formation. They are lower in central Lorraine. In Grès d'Hettange, where coarse-grained terrigenous deposits are considerable, the thick sandstone facies of Hettange and Boust (in the subsident north zone) have low TOC values, lower than at Cattenom with only 2.5 m of sandstone.

Throughout the Calcaire à gryphées, the organic matter is marine, as is clearly demonstrated by the symmetrical shape of the S_2 pyrolysis peaks, and immature, but probably poorly preserved, judging from the low HI values (Espitalié et al. 1985; Crumière and Espitalié 1989). However, variations exist. In the north, at Cattenom, the organic matter definides a continental influence at the base of the formation; on the other hand, the organic matter is marine and well preserved at the top (with HI from 200 to 500) (Fig. 3 a). The samples analyzed in Grès d'Hettange, with frequent lignite levels, are very fine shale seams, the only manifestations of marine sedimentation. The organic matter there cannot be characterized because of the low TOC contents and HI. Elsewhere in Lorraine, with the exception of Sandaucourt, which demonstrates units with good organic matter, the HI values are low (less than 150 on average), characteristic either of the deposition of continental organic matter, or of the deposition of marine organic matter weathered to varying degrees. The analysis of the shape of the S_2 pyrolysis peaks (Espitalié et al. 1985) shows that they probably concern a marine organic matter weathered in the course of its deposition. This may be connected with the low sedimentation rate of these series, deposited in an environment that did not allow good preservation of the organic matter after its deposition.

Argiles à Promicroceras. With the relative rise of the sea level, the sedimentation took place in a calmer environment. The vertical facies distribution is, in fact, progressive, first marly, and then mainly shaly.

Macrofossils are rare, and the very irregularly distributed foraminifera are elongated and costulate individuals, which nevertheless attest to a sufficient oxygen in the aquatic environment. At Cattenom, they are very numerous and highly diversified, and the microflora prove to be rather poorly preserved there, with more and more humic matter, but also dinoflagellate cysts at the base of the series (Hanzo et al. 1990).

The organic carbon contents measured in these facies are low (0.2 to 0.9 % on average). They are generally uniform over a given section. At Cattenom, the transition from the Calcaire à gryphées to the Argiles à Promicroceras is progressive, not only lithologically, but also in terms of organic matter content (1 to 2.3 % in the base marls which also contain the highest organic matter). In this thick series composed of fine-grained terrigenous material, with very low organic matter contents, it is nevertheless in the northern subsidence zone that the highest TOC are located.

To the north, especially at Cattenom, the organic matter is generally of marine origin (symmetrical S_2 peaks) but poorly preserved, as indicated by the HI values (Fig. 3 a) which do not exceed 100: with an average of 13 to 60 (except in the neighborhood of the Calcaire à gryphées). As again demonstrated by the S_2 peaks, the organic matter of continental origin is visible in some samples. At the center (Essey) and south (Breauvannes and Merrey), the organic matter is even more poorly preserved (HI generally lower than 50), only allowing its terrigenous component to subsist.

Calcaire Ocreux, Marnes à Numismalis, Calcaire à Davoei. A regressive trend can be observed, reflected by a detrital carbonate episode, containing a marly insertion in northern Lorraine.

The facies are suddenly very rich in benthic macrofauna (sedimentation rate being low), whereas the evolution of the foraminifera is more progressive. In the Carixian, the microfauna is abundant and well preserved, displaying strong ribs which attest to major action of the waves. In these three formations, the palynological content reveals very few continental forms (spores and pollens) and marine forms (acritarches), which are poorly preserved throughout.

The organic matter contents of the limestone facies, which are more or less uniform throughout a given section, do not really appear to have changed in comparison with the above contents (0.06 to 0.8 % on average). The lowest values are found in central Lorraine. In the Marnes à Numismalis, they rise sharply (1.9 % on the average, with a maximum of 2.7 %).

Throughout the area investigated, the organic matter is very poorly preserved in the Calcaire Ocreux and the Calcaire à Davoei (HI generally below 60). In the Marnes à Numismalis (only in northern Lorraine, at Cattenom), the organic matter, of marine origin, is far better preserved (HI up to 335) (Fig. 3 a, b).

Argiles à Amalthées, Grès Médioliasique. The opening of the Basin then increased and the dominant energy is fairly low. This concerns shaly and silty facies, where the quantity and size of the silts are greater than previously (Allouc and Hilly 1979), increasing particularly from the Argiles à Amalthées to the Grès médioliasique.

Macrofossils and various traces of life are numerous in the Argiles à Amalthées, with new, abundant, rapidly evolving forms of foraminifera. At Cattenom, a few samples reveal several spores and pollens, confirming the continental influence aready marked by the extent of the silts, and a few rare acritarches. In the Grès médioliasique, the foraminifera tend to decrease, fol-

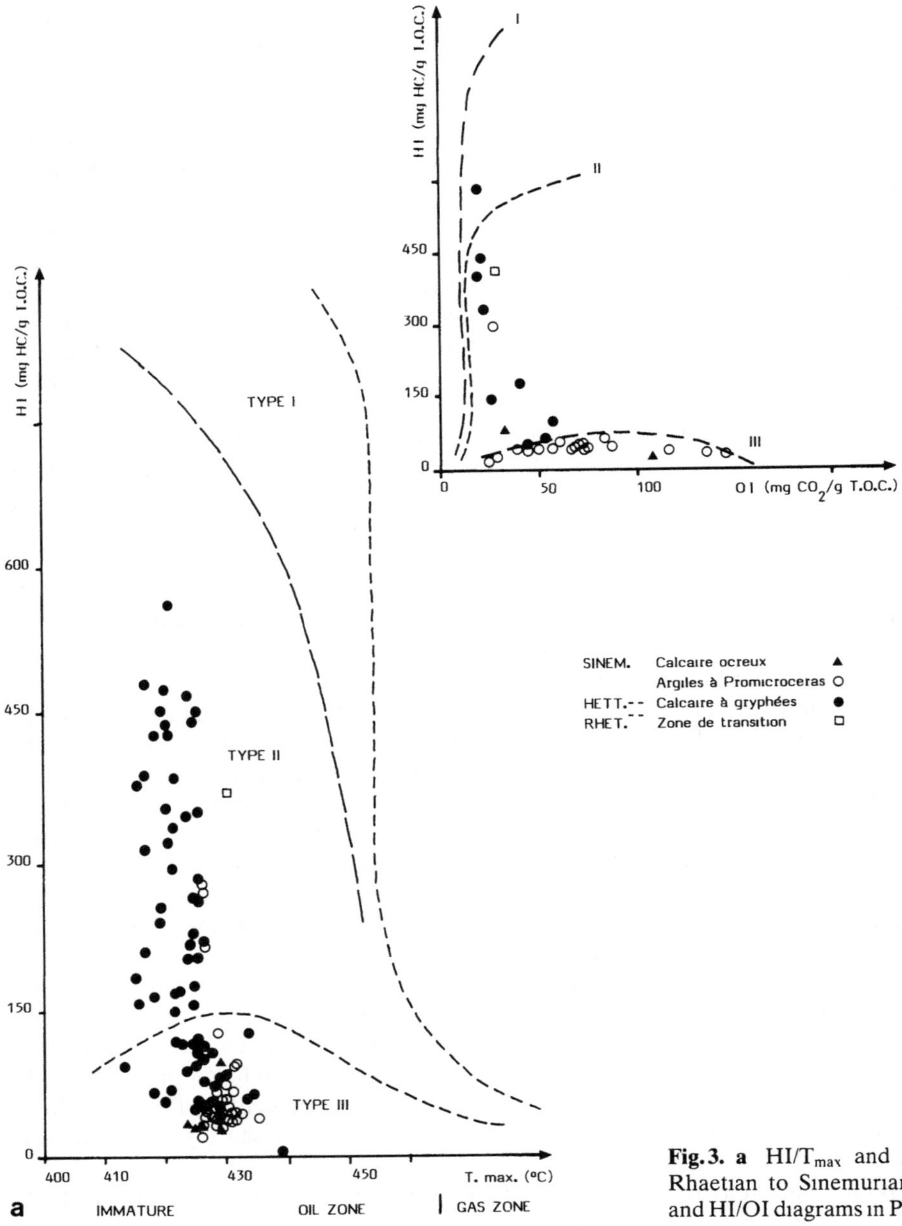

Fig. 3. a HI/T_{max} and HI/OI diagrams in formations from Rhaetian to Sinemurian, in Cattenom boreholes. **b** HI/T_{max} and HI/OI diagrams in Pliensbachian formations, in Cattenom

lowed by the extinction of the previous forms, except in northern Lorraine, where the series is thicker.

The preservation of the organic matter in these very silty facies is not good (0.3 to 0.8 % TOC on average). The lowest values are found in northern Lorraine. There is no variation in any given section, apart from a slight decrease in the Argiles à Amalthées, at the top of certain boreholes. Note the extremely low values on the Jeandelaincourt section.

In the Argiles à Amalthées in general, the organic matter exhibits a dominant marine character, except at the top of the Cattenom series, marked by the more pronounced terrigenous deposition of the next formation. The HI values do not exceed 130, revealing a weathered character (Fig. 3b). This preservation decreases further in the boreholes of southern Lorraine. In the Grès médioliasique where the values are less dispersed than in the Argiles à Amalthées, the hydrogen index values do not exceed 145. The continental influences are not clearly marked, probably because of dilution or oxidation of the transported terrigenous materials.

Schistes Carton and Base of Marnes à Bifrons. The sea transgressed and reached its maximum extension, depositing shales essentially with so-called bituminous facies.

The environment, with stratified waters, has become euxinic at the sea floor, and is marked by the absence of benthos. Macrofauna are absent and the Domerian foraminifera reveal many extinctions at the base of the Schistes cartons. This environment favors the preservation of organic matter. The observation of the microflora reveals very few pollens, spores, and acritarches at Tranqueville, and very few coccoliths at Bettembourg This probably results from a dilution by the terrigenous mate-

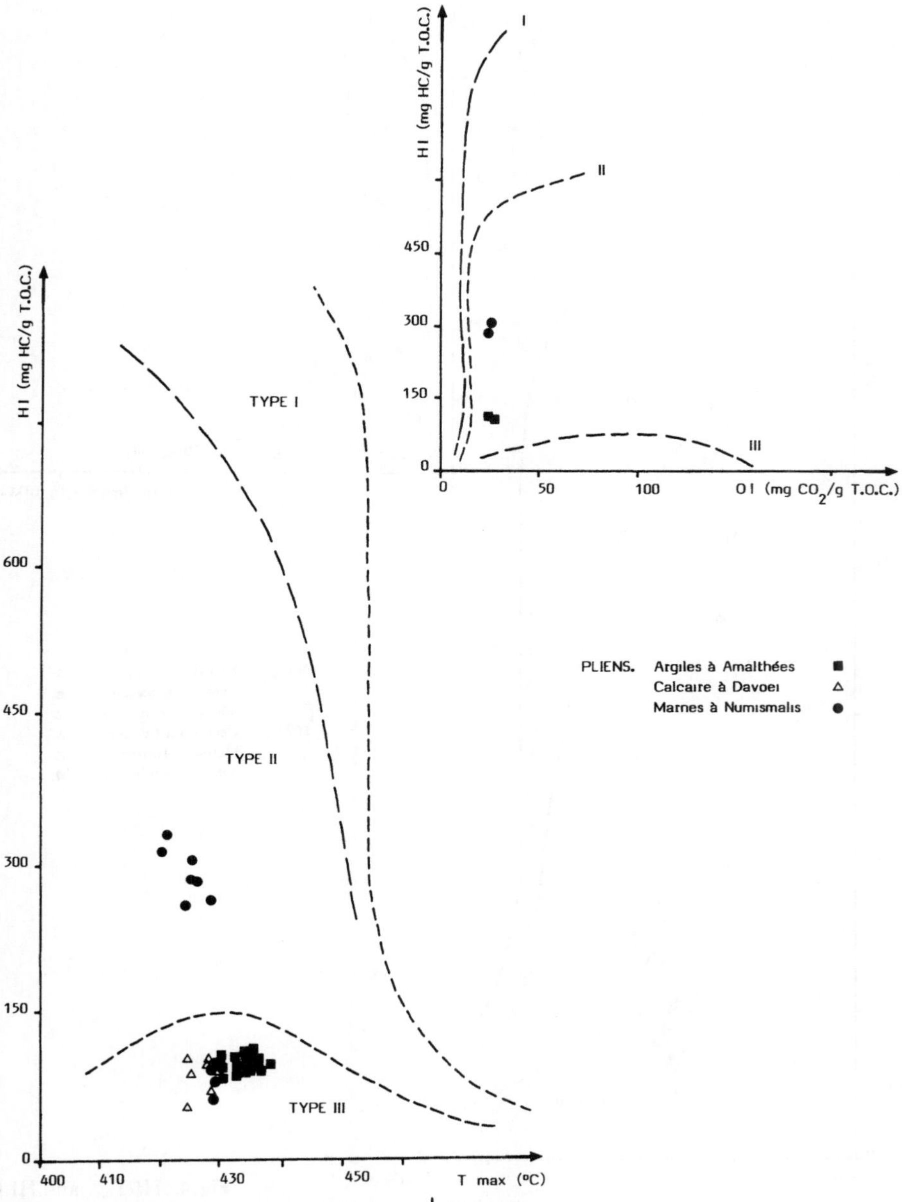

Fig. 3 *(continued)* Legend see p. 126 **b**

rial, because, elsewhere, the Schistes cartons are particularly rich in very well-preserved varieties of coccoliths, thanks to the confined character of the depositional environment (Goy 1979).

These facies are very rich in organic matter, sometimes up to 12.5 %, with averages of 2.4 to 8.5 %, with the maximum in the north, the highest sedimentation rate zone, and the minimum at Jouy aux Arches (north central Lorraine). On a given section, the range of values is sometimes rather wide, varying by a factor of 2 at Tranqueville, for example.

The organic matter is clearly marine and well preserved, with HI values up to 720 (Fig. 4) (271 to 720, with a random distribution). Nevertheless, some facies reveal the effect of weathering at the outcrop.

Marnes à Bifrons, Couches à Crassum, Marnes à Voltzi, and Grès Supraliasique. The environment progressively becomes more open and the deposits more oxygenated. They yield shales with carbonate laminites, with increasing energy in the environment, and enrichment in silts and carbonates in the Grès supraliasique.

The foraminifera demonstrate a deep renewal, up to the Grès supraliasique, where they disappear. At Tranqueville, the microflora, generally poorly preserved, reveal very few spores and pollens. Acritarches and dinoflagellate cysts exist from the Marnes à Voltzi onward. At their top, the Tasmanaceae indicate the generally regressive character of the sequence.

The organic carbon contents decrease, with an average of 1.2 to 0.8 %. The limestone facies are less rich than the shale facies. The Grès supraliasique facies are marked by decreasing organic matter content up to the Minette unit,

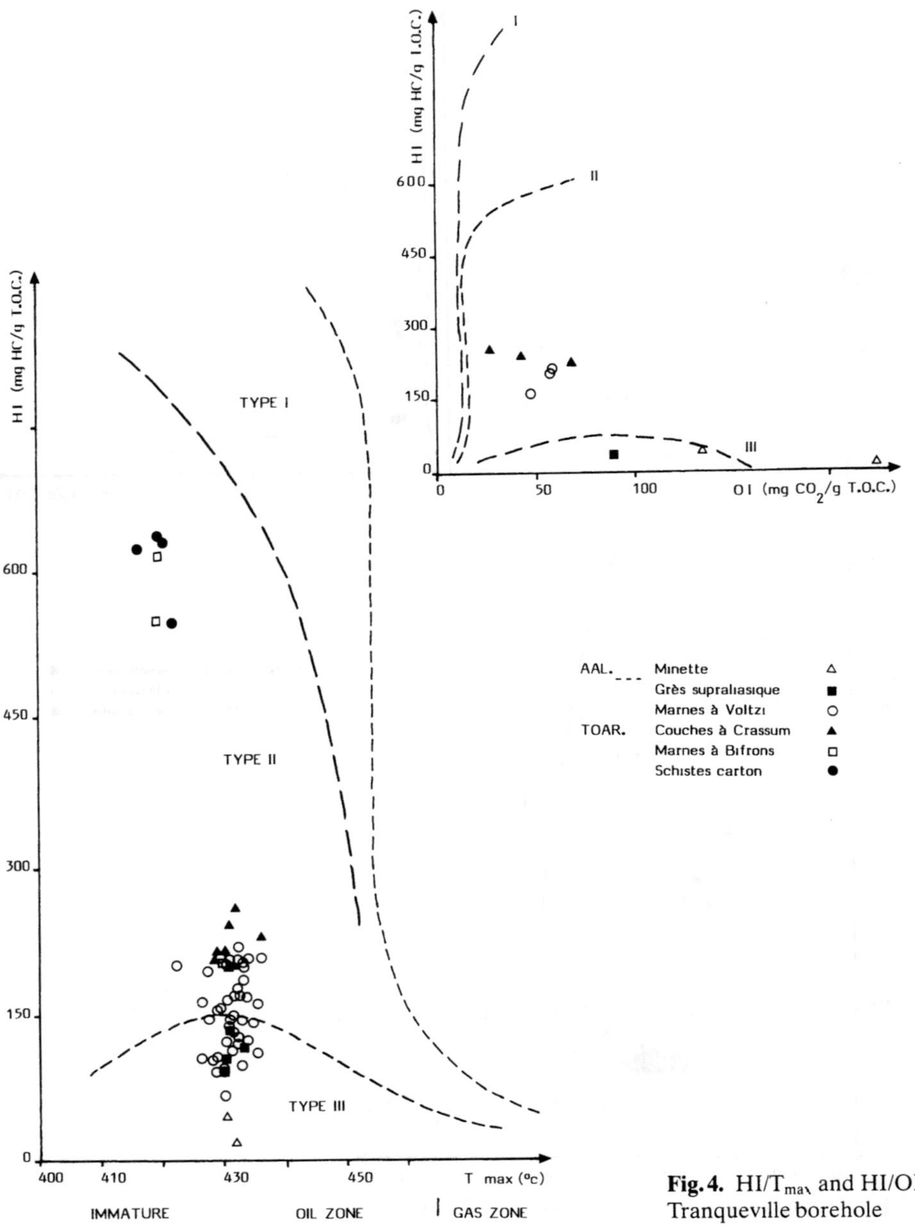

Fig. 4. HI/T_{max} and HI/OI diagrams in Toarcian formations, in Tranqueville borehole

0.6 and 0.4% average TOC, respectively. From the base of the Toarcian to the Aalenian, the decline in the organic carbon contents is progressive.

The hydrogen index drops progressively from the Marnes à Bifrons to the Marnes à Voltzi and from the Grès supraliasique to the Minette (100 to 44). On the whole, the organic matter is rather of marine origin, or possibly of continental origin masked by weathering. In fact, the continental influence is progressively demonstrated as one rises in the series, with an increasingly poorly preserved organic matter.

Figures 5, 6 and 7 (pp. 129–132) show the organic geochemistry results in the three regions of Lorraine. Table 3 (pp. 135, 136) gives the relevant data in all the formations: average values of the geochemical parameters.

Complement to the Study of the Source Rocks of the Paris Basin

A comparison of the source rocks of Lorraine and of the overall Paris Basin provides chronostratigraphic details concerning the source rock units, and helps to extrapolate, on the eastern margin of the Basin, the geochemical curves already available.

The chronostratigraphic subdivision is not always easy at the center of the Basin, considering the additional thickness of the sediments in comparison with the eastern margin of the Basin. Nevertheless, the Calcaire à gryphées facies, although with a more marly predominance at the center than to the east, date from the Hettangian/Lower Sinemurian (Poujol 1961). At the center of the basin, it is more difficult to fit the Marnes à Numismalis than in the east, where the facies is of Lower Carix-

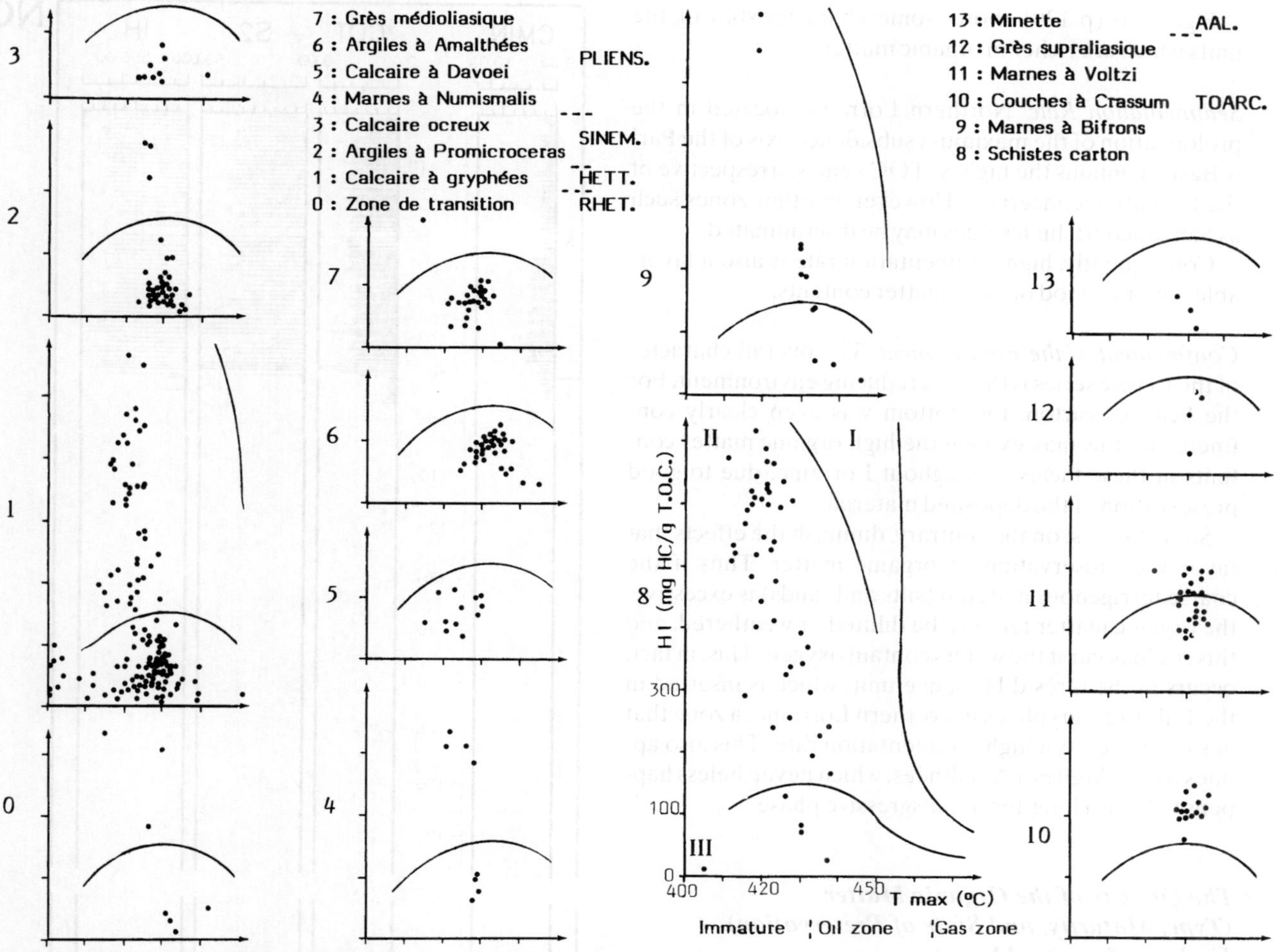

Fig. 5. Organic geochemistry of Lorraine Lower Jurassic classification of different formations, according to HI/T_{max} diagrams

ian age. The Schistes carton and the base of the Marnes à Bifrons, to the East, date from the Lower and pro parte Medium Toarcian. At the center of the basin, only the lithologic term of the Schistes carton is used for these two formations, and it is highly probable that the corresponding ammonite zones are the same in the east and at the center of the basin.

The TOC isocontent curves are plotted in Fig. 8 (p. 133), and supplement the existing maps (Espitalié et al. 1987 a). Whereas the source rocks are the same in the east and at the center of the Paris Basin, they are nevertheless even richer in organic matter at the center than in the east. This enrichment can be associated with a much greater subsidence rate in the Paris area than in Lorraine, and with predominantly more shaly facies at the center than in the east. At similar sedimentation rate, the organic matter is better preserved (or less diluted) at the center than in the east of the basin, because not affected by the Ardennes coarse terrigenous supply.

Relationship Between Organic Matter and Sedimentation

Some sedimentological and paleontological factors may or may not be favorable to the preservation of organic matter, in terms of quantity and quality (Fig. 9, p. 133).

The Levels Rich in Organic Matter Depend on Three Main Types of Geological Event

Variations in Sea Level In fact, the three Lias units richest in organic matter correspond to low to medium energy environments, i. e., to incipient lithologic sequences, irrespective of the lithologic predominance: limestones and marls of the first (Calcaire à gryphées) and marls or shales in the next two (Marnes à Numismalis and then Schistes carton).

During these transgressive and regressive cycles, the transgressive pulsations are especially favorable to large quantities of organic matter, principally when the transgression reaches its peak (Schistes carton). The exception is when the input of coarse-grained terrigenous diluted the organic matter (Argiles à Amalthées).

Fıgure 10 (p 130) shows some characterıstıcs of the unıts whıch are rıcher ın organıc matter

Sedımentatıon Rate Northern Lorraıne, located ın the prolongatıon of the maxımum subsıdence axıs of the Parıs Basın, contaıns the hıghest TOC values, ırrespectıve of the formatıon concerned. However, ın other zones such as Sandaucourt, hıgh values may be dıscrımınated.

Consequently, hıgh sedımentatıon rate ıs also a favorable factor to good organıc matter contents.

Confınement of the Envıronment The overall character of the Lıassıc serıes ıs that of a reducıng envıronment. For the Schıstes carton, the bottom was even clearly confıned, and thıs may explaın the hıgh organıc matter contents ın these facıes, throughout Lorraıne, due to good preservatıon of the deposıted materıal.

Some factors, on the contrary, dımınısh the effects that favor the preservatıon of organıc matter. Thus ıf the coarse terrıgenous materıal (sılts and sands) ıs excessıve, the organic matter tends to be dıluted or weathered, and thıs could occur ıf the waters contaın oxygen. Thıs, ın fact, occurs ın the Grès d'Hettange unıt, whıch ıs ınserted ın the Calcaıre à gryphées ın northern Lorraıne, a zone that has nevertheless a hıgh sedımentatıon rate Thıs also applıes to the Argıles à Amalthées, whıch nevertheless happens to be ın a long-term transgressıve phase

The Quality of the Organic Matter (Type, Maturity, and State of Preservation) Varies in Time and Space

It ıs always an ımmature matter, gıven the shallow burıal of the serıes (less than 1000 m)

It ıs predomınantly marıne ın the rıch unıts, whereas the terrigenous ınfluence ıs ıncreasıngly felt ın the poorer unıts, dependıng on the quantıtıes and types of deposıt supplıed by the emerged lands ın the north. Thıs occurs at every end of a sequence, ın the Calcaıre ocreux, the Calcaıre à Davoeı, and the Grès médıolıasıque/Mınette.

The state of preservatıon of the organıc matter ıs varıable. It ıs faırly good ın the rıch levels but ıncreasıngly poor as the contınental ınfluence ıncreases (the hydrogen ındex drops, not countıng matrıx effects). These observatıons are sometımes confırmed by the poor state of the mıcroflora. Thıs probably results from a weatherıng ınherent ın the transport by currents and/or the exıstence of an oxygenated sea floor, as attested by the benthıc fauna, at least ın the central and southern areas of less sedımentatıon rate. Thıs ıs the case of the regressıve phases.

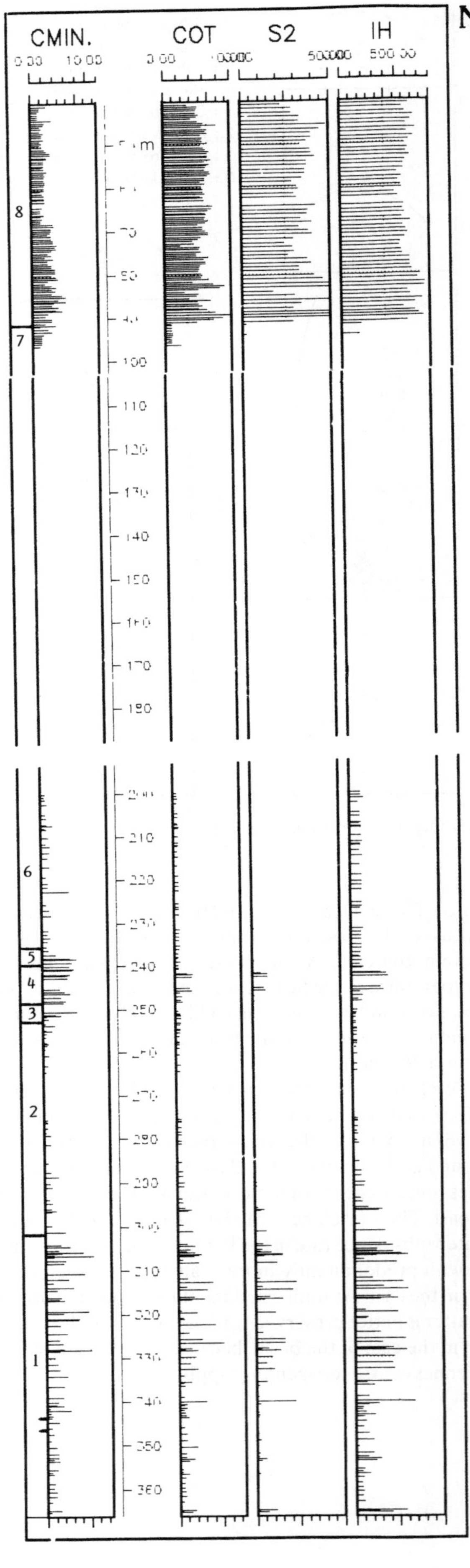

NORTH

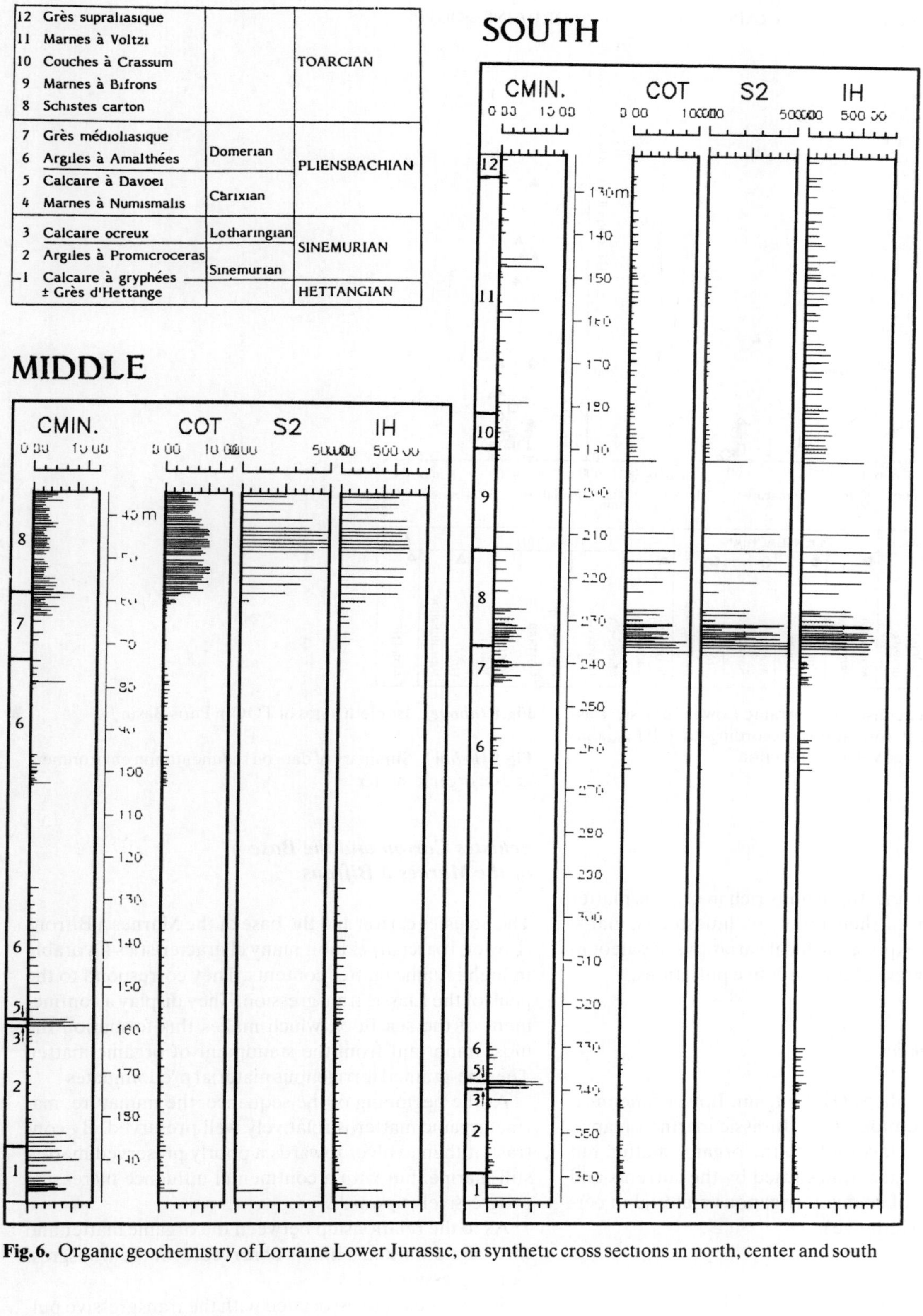

Fig. 6. Organic geochemistry of Lorraine Lower Jurassic, on synthetic cross sections in north, center and south

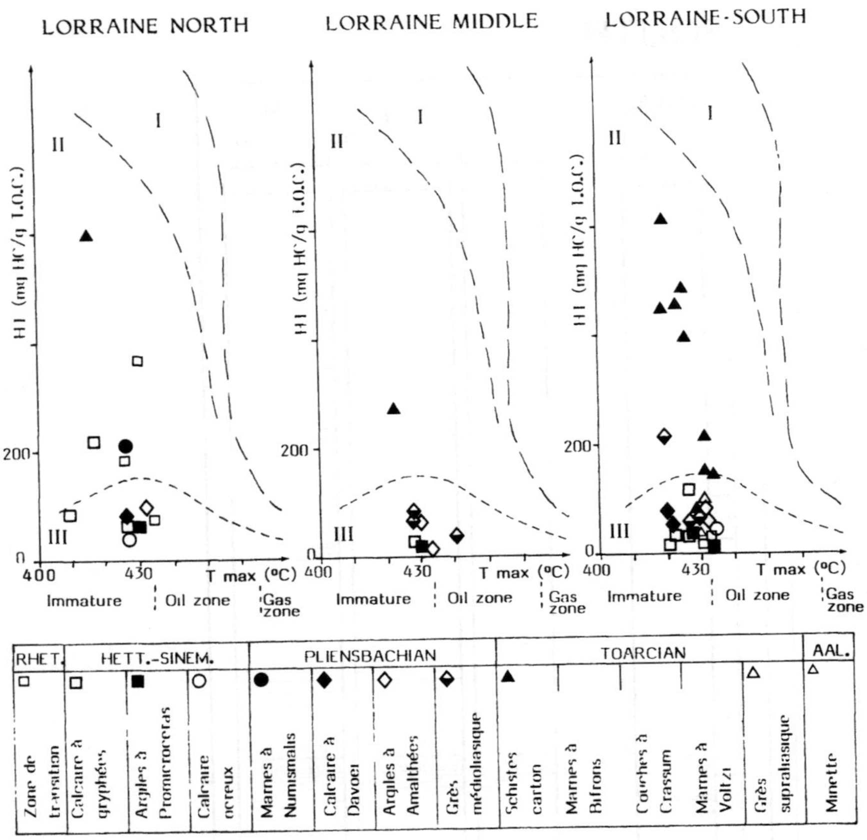

Fig. 7. Organic geochemistry of Lorraine Lower Jurassic classification of different formations, according to HI/T_{max} diagrams, based on average values per section

Fig. 8 ***(above).*** Isopleth maps of TOC in Paris Basin ▶

Fig. 9 ***(below).*** Summary of data on sedimentation environment and on organic matter

Conclusions

In the Lias of Lorraine, three units rich in organic matter exist, independent of their lithology: limestones, marls, or argilites, which express the local variations of paleoenvironment in a context of transgressive pulsations.

Calcaire à Gryphées

The Calcaire à gryphées (Hettangian, Lower Sinemurian) marks the inception of the Jurassic marine onlap, a transgressive phase favorable to the organic matter, but lessened by the oxygenation caused by the currents and by the abundance of coarse terrigenous material in certain units (Grès d'Hettange).

Marnes à Numismalis

The Marnes à Numismalis (Lower Pliensbachian) also mark the inception of a sequence, hence in a transgressive phase (slightly more than above), and the high sedimentation rate further favors the preservation of organic matter, especially in the absence of the coarse terrigenous influence connected with the Ardennes.

Schistes Carton and the Base of the Marnes à Bifrons

The Schistes carton and the base of the Marnes à Bifrons (Lower Toarcian) exhibit many characteristics favorable to high organic matter contents. They correspond to the peak of the Liassic transgression. They display a confinement of the sea floor, which makes this formation the most important from the standpoint of organic matter. The fine-grained terrigenous material predominates

At the beginning of the sequence, the immature, marine organic matter is relatively well preserved. By contrast, it then evolves towards a poorly preserved matter, still marine, but with a continental influence increasing progressively upwards.

As to the relationship between the organic matter and the sedimentation, a link is observed with the transgression/regression cycles

Source rocks are associated with the transgressive pulsations, chiefly at the maximum transgression (Schistes carton). That may be due to a higher planktonic biomass contents at the time, or the case may simply be one of better preservation after the deposition (confined environment).

On the contrary, associated with the regressions are the units poorest in organic matter, which are often badly

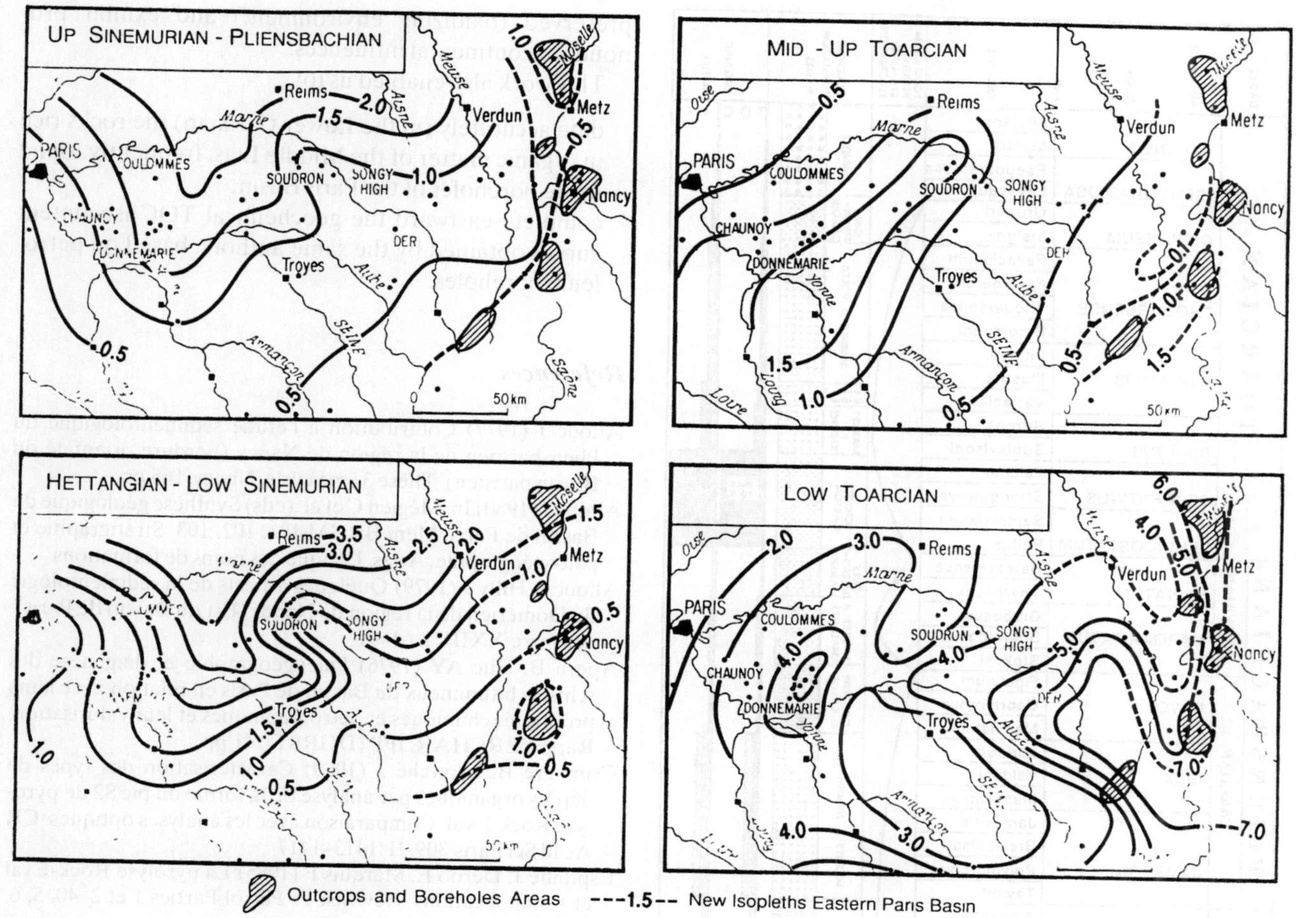
UP SINEMURIAN - PLIENSBACHIAN
MID - UP TOARCIAN
HETTANGIAN - LOW SINEMURIAN
LOW TOARCIAN
Reims
Verdun
Metz
Nancy
PARIS
COULOMMES
SOUDRON
SONGY HIGH
CHAUNOY
DONNEMARIE
DER
Troyes
Marne
Aisne
Meuse
Moselle
Aube
Seine
Yonne
Armançon
Saône
Oise
Loire
50 km
Outcrops and Boreholes Areas
---1.5--- New Isopleths Eastern Paris Basin

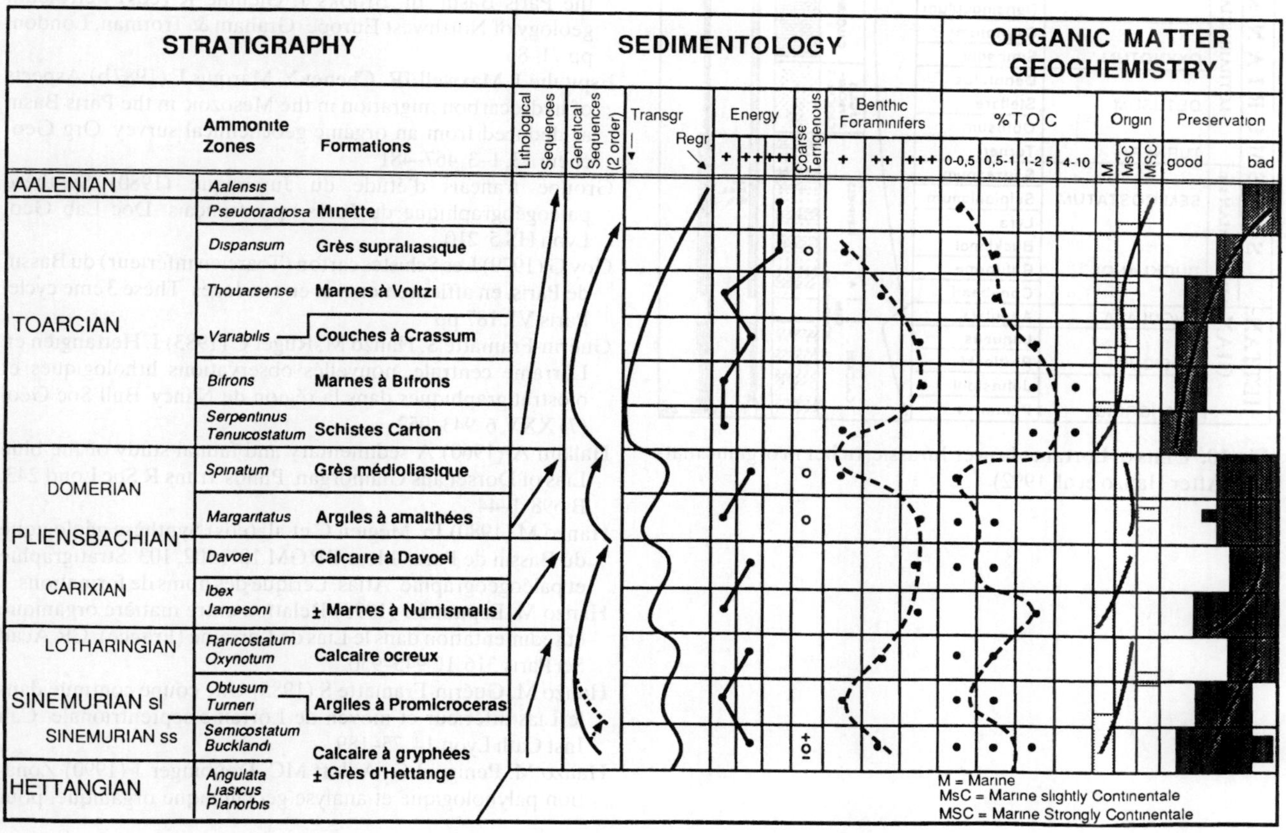
STRATIGRAPHY
SEDIMENTOLOGY
ORGANIC MATTER GEOCHEMISTRY
Ammonite Zones
Formations
Lithological Sequences
Genetical Sequences (2 order)
Transgr
Regr
Energy
Coarse Terrigenous
Benthic Foraminifers
%TOC
0-0,5
0,5-1
1-2 5
4-10
Origin
M
MsC
MSC
Preservation
good
bad
AALENIAN
Aalensis
Pseudoradiosa
Minette
Dispansum
Grès supraliasique
Thouarsense
Marnes à Voltzi
TOARCIAN
Variabilis
Couches à Crassum
Bifrons
Marnes à Bifrons
Serpentinus
Tenuicostatum
Schistes Carton
Spinatum
Grès médioliasique
DOMERIAN
Margaritatus
Argiles à amalthées
PLIENSBACHIAN
Davoei
Calcaire à Davoei
CARIXIAN
Ibex
Jamesoni
± Marnes à Numismalis
LOTHARINGIAN
Raricostatum
Oxynotum
Calcaire ocreux
SINEMURIAN sl
Obtusum
Turneri
Argiles à Promicroceras
SINEMURIAN ss
Semicostatum
Bucklandi
Calcaire à gryphées
± Grès d'Hettange
HETTANGIAN
Angulata
Liasicus
Planorbis
M = Marine
MsC = Marine slightly Continentale
MSC = Marine Strongly Continentale

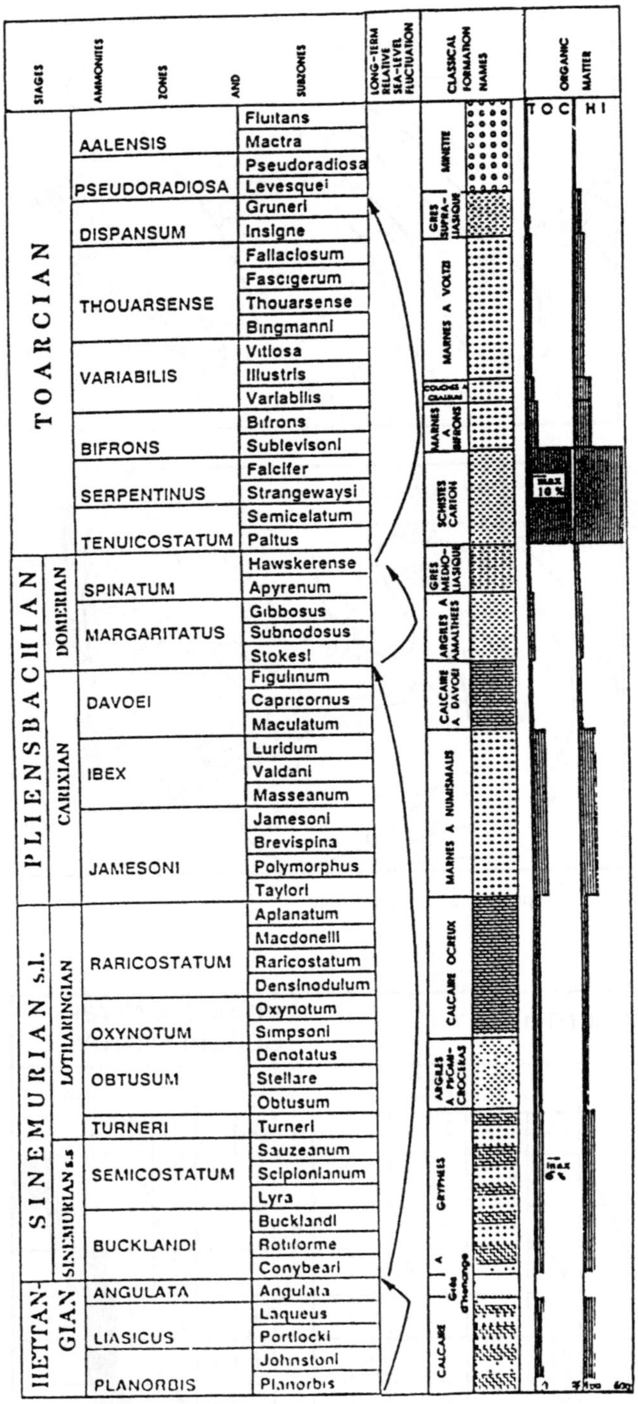

Fig. 10. Units of Lorraine Lower Jurassic, richer in organic matter (After Hanzo et al 1992)

preserved (oxidizing environment) and exhibit pronounced continental influences.

This work also enabled us to:

- date accurately (in the Lower Carixian) the rocks rich in organic matter of the Middle Lias, frequently found in the boreholes of the Paris Basin,
- complete eastward the geochemical TOC isocontent curves obtained by the same authors, based on petroleum boreholes.

References

Allouc J (1977) Contribution à l'étude sédimentologique du Pliensbachien de la région de Nancy (bordure orientale du Bassin parisien) Thèse 3ème cycle, Nancy, 165 pp

Allouc J (1980) In Mégien C et al (eds) Synthèse géologique du Bassin de Paris Mém BRGM 101, 102, 103 Stratigraphie et paléogéographie, Atlas, Lerique des noms de formations

Allouc J, Hilly J (1979) Quelques aspects de la sédimentologie du Domérien de la région de Nancy (Est du Bassin de Paris) Sci Terre XXIII, 1 61–91

Apern B, Huc AY (1976) Paléogéographie et diagenèse des schistes bitumineux du Bassin de Paris en relation avec leurs propriétés chimiques et pétrographiques et leur valorisation Rapp CERCHAR, IFP (DGRST), 41 pp

Crumière B, Espitalié J (1989) Caractérisation des types de dépôts organiques par analyse de la forme du pic S2 de pyrolyse Rock-Eval Comparaison avec les analyses optiques CR Acad Sci Paris 309, II 1413–1417

Espitalié J, Deroo E, Marquis F (1985) La pyrolyse Rock-Eval et ses applications Rev Inst Fr Pétrol Parties 1 et 2 40, 5, 6 563–784, Partie 3 41, 1 73–79

Espitalié J, Marquis F, Sage L (1987 a) Organic geochemistry of the Paris Basin In Brooks J, Glennie K (eds) Petroleum geology of Northwest Europe Graham & Trotman, London, pp 71–86

Espitalié J, Maxwell JR, Chenet Y, Marquis F (1987 b) Aspects of hydrocarbon migration in the Mesozoic in the Paris Basin as deduced from an organic geochemical survey Org Geochem 13, 1–3 467–481

Groupe français d'étude du Jurassique (1980) Synthèse paléogéographique du Jurassique français Doc Lab Géol Lyon HS 5 210

Goy G (1979) Les Schistes carton (Toarcien inférieur) du Bassin de Paris, en affleurements et en sondages Thèse 3ème cycle, Paris VI, 187 pp

Guérin-Franiatte S, Hanzo M, Ruget C (1983) L'Hettangien en Lorraine centrale nouvelles observations lithologiques et biostratigraphiques dans la région de Nancy Bull Soc Géol (7) XXV, 6 943–952

Hallam A (1960) A sedimentary and faunal study of the blue Lias of Dorset ans Glamorgan Philos Trans R Soc Lond 243, B 698 1–44

Hanzo M (1980) In Mégien C et al (eds) Synthèse géologique du Bassin de Paris Mém BRGM 101, 102, 103 Stratigraphie et paléogéographie, Atlas, Lerique des noms de formations

Hanzo M, Espitalié J (1993) Relation entre matière organique et sédimentation dans le Lias de Lorraine (France) CR Acad Sci Paris 316, II 945–950

Hanzo M, Guérin-Franiatte S (1985) Une coupe continue dans le Lias inférieur et moyen de Lorraine septentrionale Cah Inst Cath Lyon 14 75–189

Hanzo M, Peniguel G, Adloff MC, Doubinger J (1990) Zonation palynologique et analyse géochimique organique, pour

Table 3. Organic geochemistry of Lorraine Lower Jurassic average values in all formations (of outcrops and boreholes)

Region	Calcaire à gryphées	Argiles à Promicro.	Calcaire ocreux	Marnes à Numismalis	Calcaire à Davoei	Argiles à Amalthées	Grès médioliasi.	Schistes carton	Marnes à Bifrons	Couches à Crassum	Marnes à Voltzi	Grès supraliasique	
	HETT. / SIN. ss. (SINEMURIAN)	LOTH. (SINEMURIAN)	LOTH. (SINEMURIAN)	CAR. (PLIENSBACHIAN)	CAR. (PLIENSBACHIAN)	DOM. (PLIENSBACHIAN)	DOM. (PLIENSBACHIAN)	TOARCIAN	TOARCIAN	TOARCIAN	TOARCIAN	TOARCIAN	
Lorraine-north	1.24	0.85	0.83	1.92	0.6	0.7		6.97					T.O.C.
	0 12	0 03	0.03	0 11	0.05	0.04		1.25					S_1
	3.31	0.68	0.38	5	0.55	0.74		43 3					S_2
	417	429	427	426	427	433		416					T max
	155	61	43	217	87	98		588					HI
	28	68	70	27		32		31					OI
Lorraine-middle	0.12	0.35	0.06	/		0.32	0.35	6.17					T.O.C.
	0.02	0	0	/		0.03	0.01	0 14					S_1
	0.04	0.08	0.02	/		0 47	0 18	38.2					S_2
	428	430		/		431	429	423					T max
	30	23	23	/		19	74	566					HI
	361			/			924	35					OI
Lorraine-south	0.6	0.61	0.44	/	0.23	0 48	0.53	5.53	1 78	1.25	0.87	0.64	T.O.C.
	0.05	0	0.01	/	0.02	0.03	0.03	0.57	0.56	0.06	0.03	0.04	S_1
	0.99	0.21	0.2	/	0.16	0.39	1.09	32 1	6 73	2.72	1.33	0.73	S_2
	424	430	433	/	421	430	428	424	423	431	431	431	T max
	40	34	50	/	66	73	85	497	212	217	151	100	HI
	158	38		/		341	113	66		46	56		OI
LORRAINE	0.7	0.6	0.4	1.9	0.4	0.5	0.4	6.2	1 8	1.2	0.9	0.6	T.O.C.
	0.06	0.01	0,02	0.11	0.03	0.03	0.02	0.65	0.56	0.06	0.03	0.04	S_1
	1.45	0.3	0.2	5	0.4	0.5	0.6	37 9	6.7	2.7	1.3	0.7	S_2
	423	430	430	426	424	431	429	421	423	431	431	431	T max
	75	39	42	217	76	63	80	550	212	217	151	100	HI
	182	53	70	27		186	519	44		46	56		OI

T.O.C. : % ; S_1 et S_2 : mg HC/g rock ; Tmax : °C ; HI : mg HC/g T.O.C. ; IO : mg CO_2/g T.O.C.

préciser les milieux lors de la transgression liasique à Cattenom (Moselle, France) Cah Micropal 5, CNRS, pp 55–87

Hanzo M, Guérin-Franiatte S, Espitalié J, de Graciansky PCh (1992) Source rocks and liassic sequence stratigraphy in Lorraine (Eastern Paris Basin, France). Cong Mesozoic and Cenozoic Seq Stratigr of European Bas, Abstr Vol Dijon, pp 344–345

Haq BU, Hardenbol J, Vail PR (1987) Chronology of fluctuating sea levels since the Triassic Science 35: 1156–1167

Le Roux J (1980) La tectonique de l'auréole orientale du Bassin de Paris. Ses relations avec la sédimentation Bull Soc Géol Fr (7) XXII, 4 655–662

Maubeuge PL (1955) Observations géologiques dans l'Est du Bassin de Paris Nancy, 2 vols, 1082 pp, unpubl thesis

Poujol P (1961) In Colloque sur le Lias La série liasique du bassin de Paris Essai de corrélations entre les sondages de la R A P, Technip, Chambéry 4 577–604

Ruget C (1985) Les foraminifères (Nodosariides) du Lias de l'Europe occidentale Doc Lab Géol Lyon 94 273

Ruget C, Hanzo M (1992) Les foraminifères et la sédimentation du Lias inférieur et moyen des forages de Cattenom (Lorraine, France) première approche Cah Univ Cath Lyon Sér Sci 5 81–91

Table 3 *(continued)*

		HETT.	SINEMURIAN		PLIENSBACHIAN				TOARCIAN				
			SIN. ss.	LOTH.	CAR.		DOM.						
		Calcaire à gryphées	Argiles à Promicro.	Calcaire ocreux	Marnes à Numismalis	Calcaire à Davoei	Argiles à Amalthées	Grès médioliasi.	Schistes carton	Marnes à Bifrons	Couches à Crassum	Marnes à Voltzi	Grès supraliasique
Lorraine-north	T.O.C	1 41	0 85	0 83	1 92	0.6	0.7		6 47				
	S_1	0 13	0.03	0.03	0.11	0.05	0 04						
	S_2	3,83	0.68	0.38	5	0 55	0 74		31				
	T max	421	429	427	426	427	433						
	HI	172	61	43	217	87	98		582				
	OI	28	68	70	27		32						
Lorraine-middle	T.O.C.								7 01				
	S_1												
	S_2								43.8				
	T max												
	HI								625				
	OI												
Lorraine-south	T.O.C.	0.9	0.61	0.44		0.23	0.48	0.57	5.47	1 78	1.25	0.87	0 64
	S_1	0.05	0	0.01		0.02	0.03	0.04	0.5	0.56	0.06	0.03	0.04
	S_2	2.57	0.21	0.2		0 16	0 4	1.3	32 1	6.73	2.72	1 33	0.73
	T max	427	430	433		421	430	428	425	423	431	431	431
	HI	76	34	50		66	73	91	506	212	217	151	100
	OI	44	38				341	113	67		46	56	
LORRAINE	T.O.C.	1.16	0.7	0.6	1.9	0.4	0.6	0 6	6.3	1.8	1 2	0.9	0.6
	S_1	0.09	0.02	0.02	0.1	0.03	0.03	0.04	0.5	0.6	0 06	0.03	0 04
	S_2	3.2	0 4	0.3	5	0.4	0.6	1.3	39	6 7	2 7	1 3	0.7
	T max	424	430	430	426	424	431	428	425	423	431	431	431
	HI	124	47	46	217	76	85	91	571	212	217	151	100
	OI	36	53	70	27		186	113	67		46	56	

T.O.C. : % ; S_1 et S_2 mg HC/g rock ; Tmax : °C ; HI : mg HC/g T.O.C. ; IO mg CO_2/g T.O.C.

An Integrated Approach to Reservoir Characterization for a Gas Storage in the Paris Basin

J. C. Lecomte, P. Houel, and F. Verdier

Abstract. The main focus of this chapter is the geometric characterization of a reservoir in the South of the Paris Basin, which is exploited by Gaz de France as a gas storage. In this Gaz de France case study, we show that the combination of geologic, well logs, and seismic data can lead to a better interpretation of the reservoir.

Seismic inversion with a priori model is an efficient way to integrate different types of available data. We compare, on a 2-D seismic line, the traditional interpretation of true seismic amplitude with acoustic impedance section.

The goal of this interpretation is to determine the main lithostratigraphic boundaries. This kind of high resolution interpretation on a 50-ms-thick reservoir is not very usual because of the low vertical resolution of the seismic data at 1000 m deep. Seismic stratigraphic inversion increases the signal-to-noise ratio and allows the determination of the internal geometric shapes of the main lithostratigraphic boundaries using the concept of acoustic impedance pattern.

Introduction

Reservoir characterization is an important topic in petroleum companies and in research teams. Stochastic modeling techniques, based on geostatistical simulation, have been developed in order to take into account scare data like well logs and cores. Geologic, well logs, and cores data are traditionally the most commonly used data sets in reservoir characterization (Eschard et al. 1991).

It would be interesting to directly constrain the stochastic simulation by seismic data which provides spatial continuity information but a low resolution. If relations between seismic data and lithology or petrophysical properties exist, they are not obvious, so that the narrow band filtering of the seismic signal means that the correlations between reservoir and seismic data are often weak.

Another way to constrain the stochastic simulations for reservoir characterization with seismic data is to determine the lithostratigraphic unit boundaries of the target. With a combination of the high vertical resolution of the well logs and the spatial continuity of the seismic data, an effective approach to lithostratigraphic unit geometry can be made. This is an indirect way to introduce geometrical constraints into stochastic simulations.

In the Soings en Sologne case study, in a siliciclastic environment, using a seismic stratigraphic inversion software (Brac et al. 1988), we will show that it is possible to improve the two-way travel time shape of the main lithostratigraphic boundaries

Case Study: Soings en Sologne

The structure of Soings en Sologne is exploited by Gaz de France in order to store gas during summer and use it during winter when the consumers' demand is greater than the supply. Soings en Sologne is located in the south of the Paris Basin, in the Sologne depocenter (Fig. 1). The reservoir is located in upper Triassic continental red beds, at about 1100 m deep.

Between 1984 and 1988, more than 150 km of 2-D seismic lines were recorded by the Compagnie Générale de Géophysique at the request of Gaz de France. The seismic data were processed to preserve true amplitudes. The sample rate of the studied seismic line 88CY104 is 2 ms, and the theoretical coverage is 48. More than 30 wells have been drilled in this area with a complete set of well logs, and most of them are fully cores (Fig. 2).

A Brief Description of the Reservoir

In this conceptual cross section (Fig. 3), we can observe the different surfaces and units which have been distinguished by sedimentologists on well logs and cores.

From base to top, we observe the following surfaces and units (Houel et al. 1991; Mathieu et al. 1993):

- Surface S0 is an erosional sequence boundary above karstified dolomites.
- Units A and B between surfaces S0 and S1 correspond to proximal fluviatile deposits of a highstand systems tract.
- Surface S1, generated by a base level fall of the fluviatile system.
- Unit C between surfaces S1 and S2, which is the less proximal fluviatile series which has been deposited during a sea-level highstand.
- Surface S2 which is an another erosional surface associated with a base level fall of the fluviatile system.

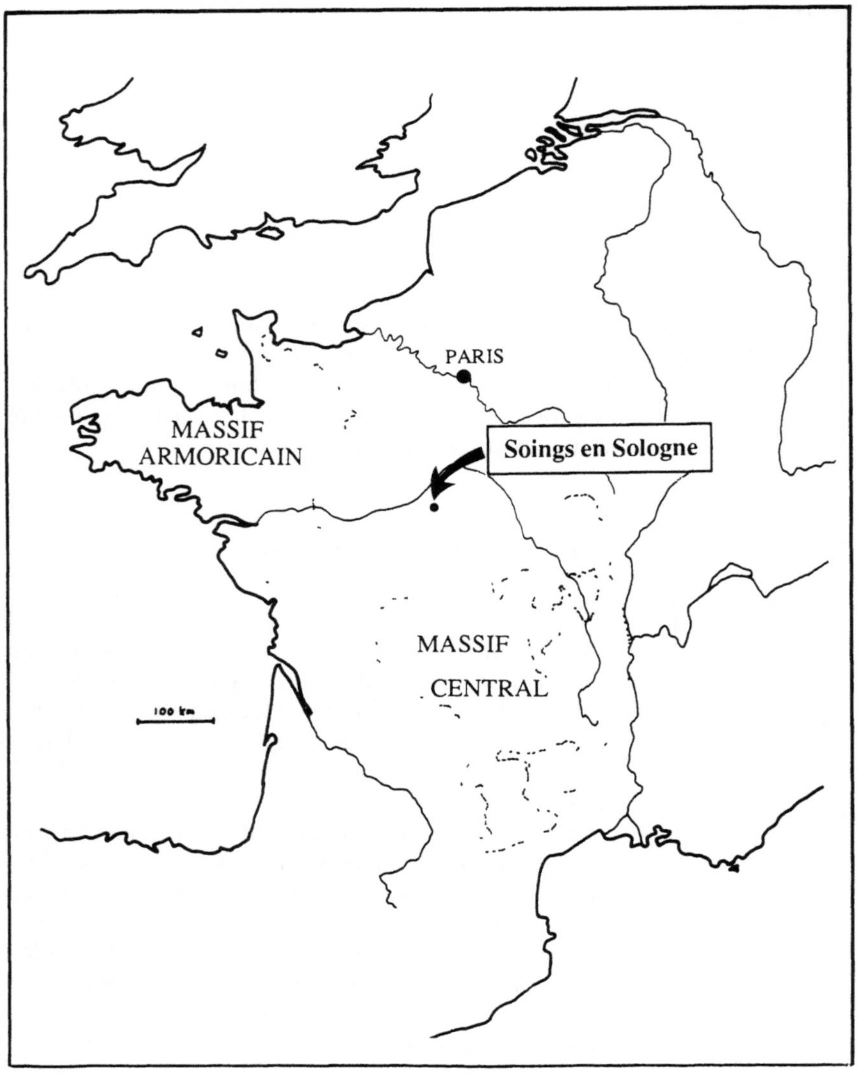

Fig. 1. Soings en Sologne geographic position

- Unit D which is not present all over the area and corresponds to the filling of an incised valley by fluvial sediments during a base level rise.
- Unit E which is a distal fluviatile prograding systems tract.
- Surface S3 marks the top of the reservoir formation and corresponds to the first Hettangian flooding surface. Above this surface, Hettangian series are made up of shallow marine carbonated deposits.
- Surface S4 is a discordant flooding surface emphasizing the start of the Sinemurian transgression.
- N1 and N2 correspond to paleohorizontal markers (key beds).
- N3 is the top of the Sinemurian which corresponds to the picked reflector in Fig. 4.

Interest of Seismic Stratigraphic Interpretation

We assume that seismic stratigraphic interpretation can significantly improve the detailed knowledge of a reservoir The seismic section 88CY104 is a good example of what we are able to do with such a data set (see Fig. 4).

The first step is a stratigraphic tie between time acoustic impedance log (well data) and seismic traces near the well (SG16).

The methodology used, described in many publications (Brac et al. 1988; Richard and Brac 1988), begins by extracting a zero phase wavelet from seismic data. Then, this zero phase wavelet is convoluted with the reflection coefficients calculated from the well. This first convolution is modified by testing the influence of different values of the wavelet phase to find the best fit between synthetic seismogram and a seismic trace near the well. This step is crucial because the accuracy of our seismic stratigraphic interpretation depends on this calibration.

The stratigraphic interpretation of this optimal fit can be detailed by making comparisons between interpreted

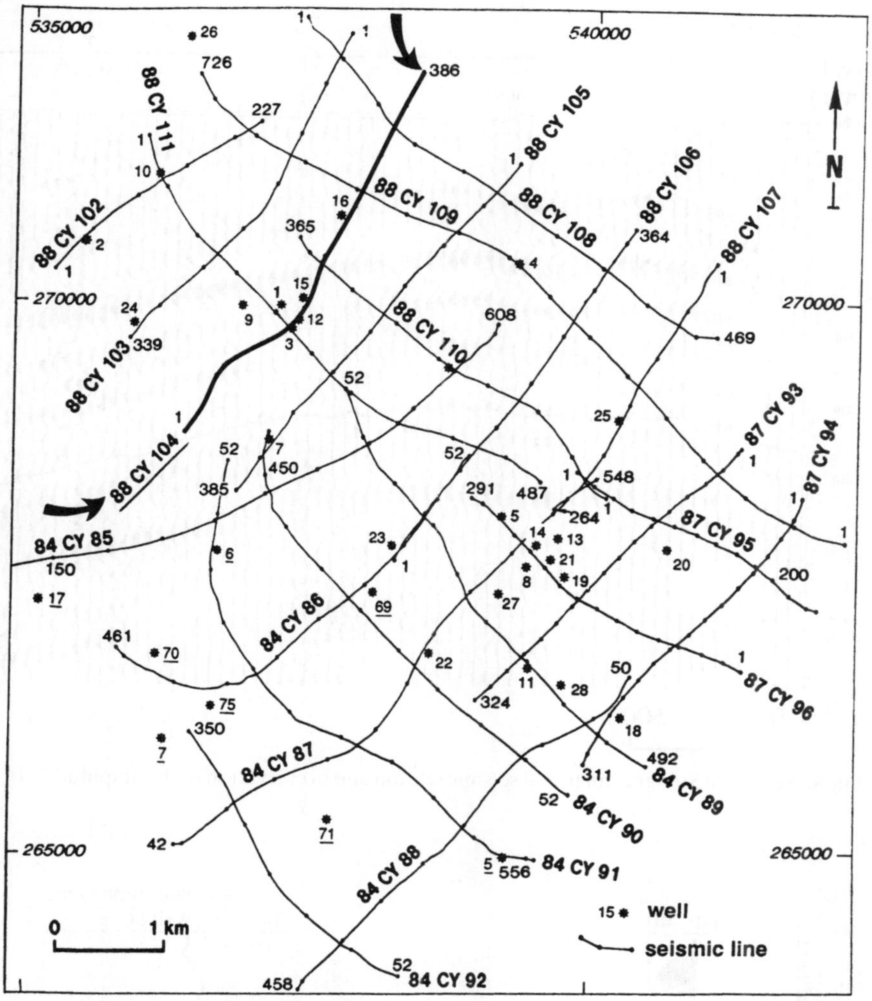

Fig. 2. Soings en Sologne position map

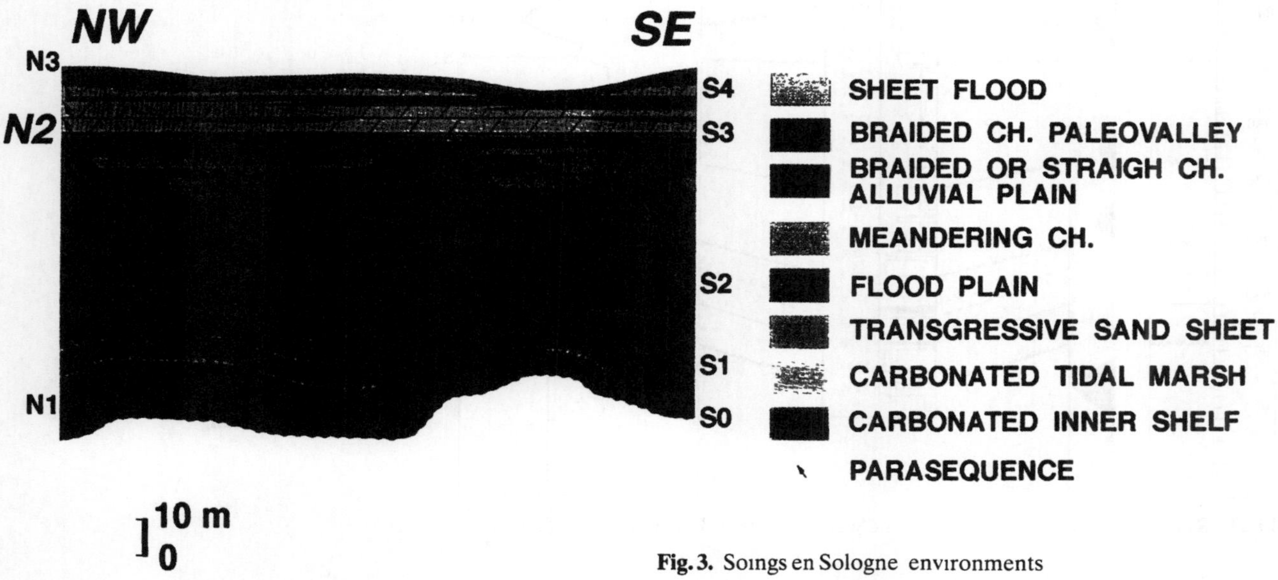

Fig. 3. Soings en Sologne environments

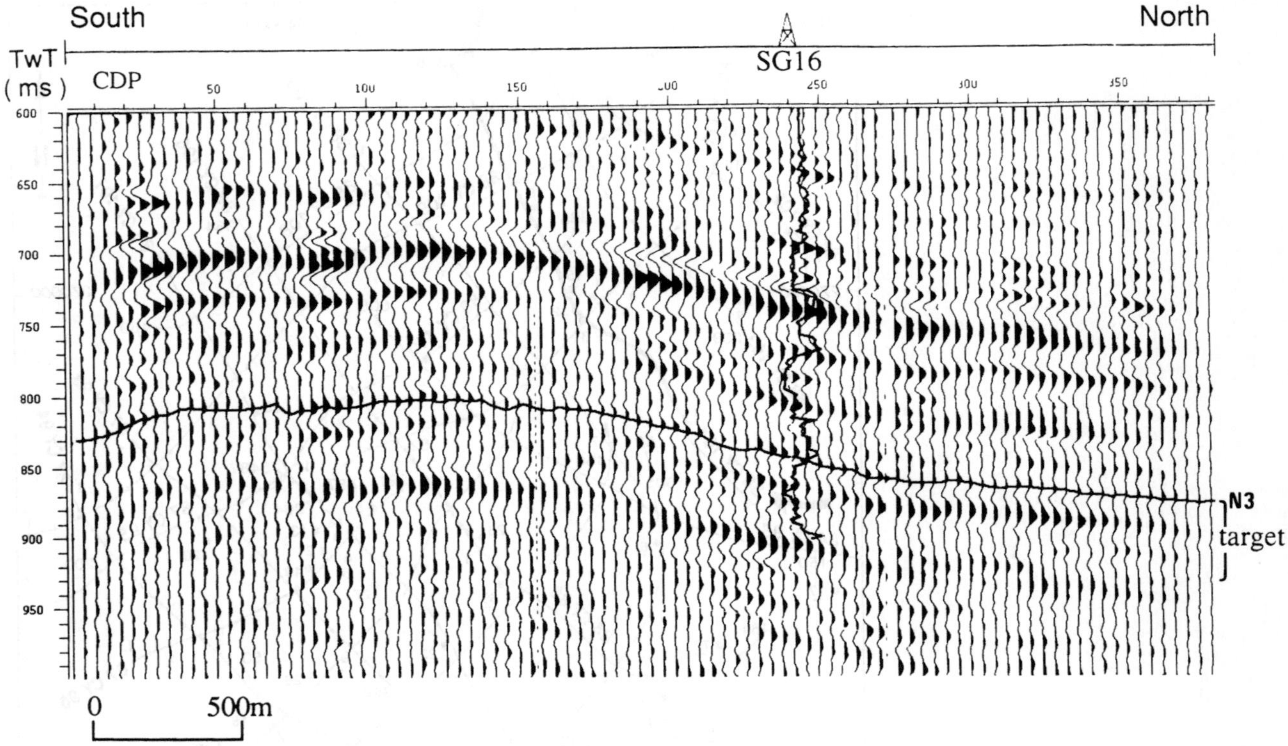

Fig. 4. Soings en Sologne migrated seismic section and preservation of the amplitudes (Courtesy Gaz de France)

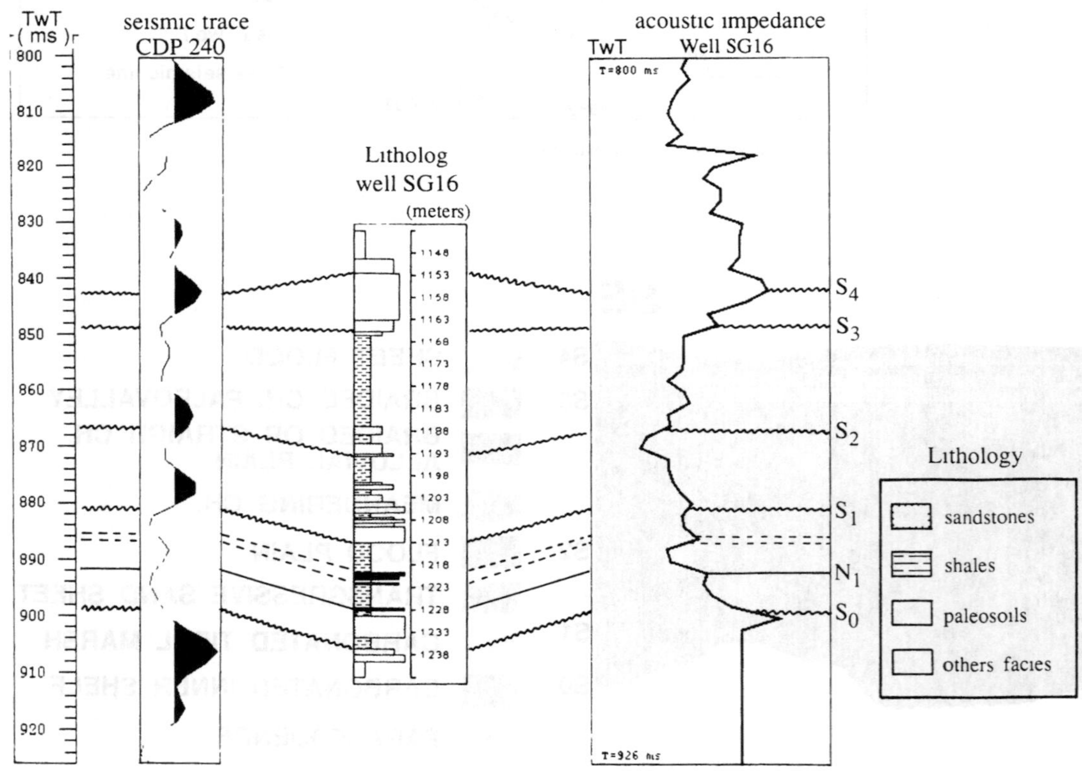

Fig. 5. Relations between seismic data and well data around the borehole SG16

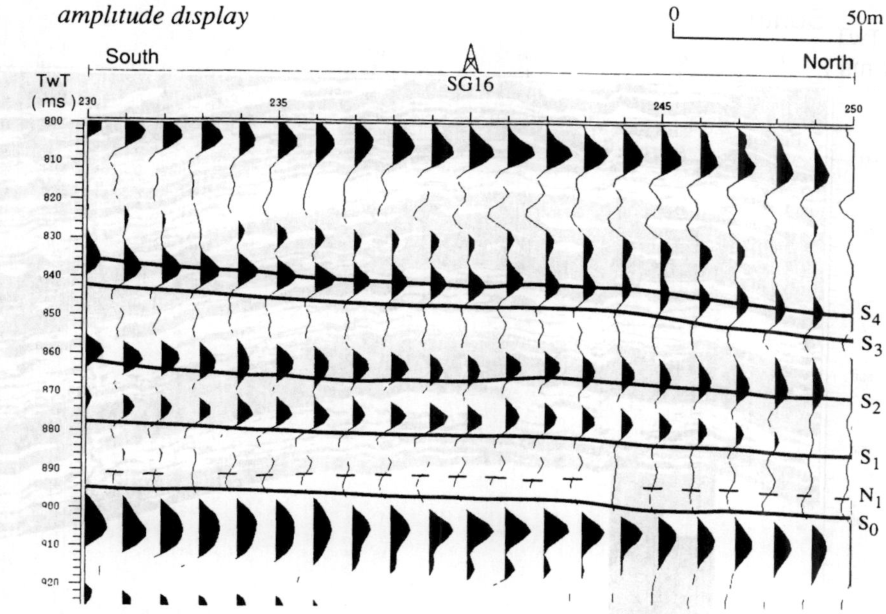

Fig. 6. Stratigraphic interpretation around well SG16, amplitude display

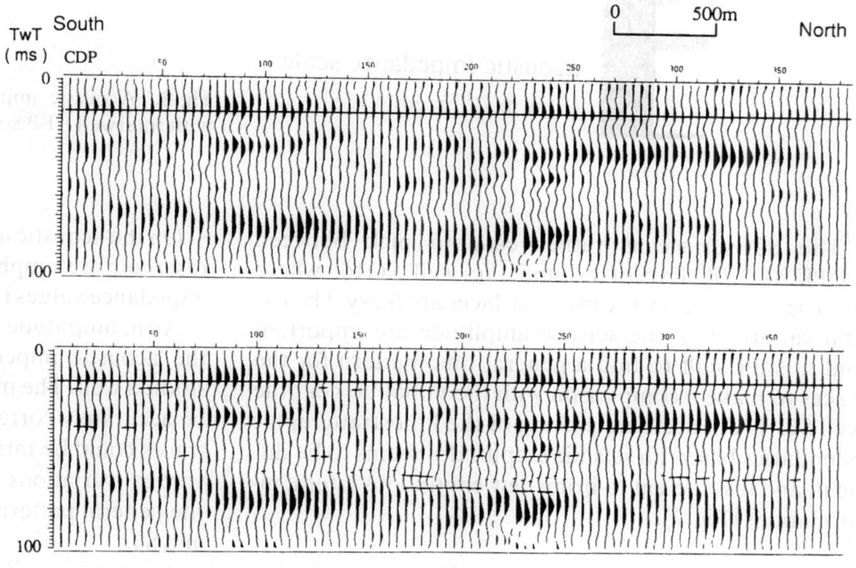

Fig. 7. Interpretation of the flattened seismic section

depth data (well logs or cores) and the time acoustic impedance log used in the previous step. We can propose some stratigraphic attributions to a seismic trace (CDP240) near the well SG16 (best fit with the synthetic seismogram). It becomes possible to identify in time on the seismic trace (CDP240), the lithostratigraphic surfaces which have been determined in depth (Fig. 5).

In Fig. 6, we just extrapolate on a few seismic traces around the well, the previous time stratigraphic interpretation on seismic trace (CDP240).

This local extrapolation is not difficult, and it can be seen that the main stratigraphic surfaces can be easily followed.

Seismic Stratigraphic Interpretation of Line 88CY104

In order to decrease the influence of the structuration, a flattening has been applied to the seismic section (Fig. 7). This method makes sense in sedimentary basins with weak structural deformations. Moreover, this method allows easier seismic stratigraphic correlations in thin reservoirs. Before going further, it must be noted that gas is absent in the reservoir layers near the well SG16 and present close to the structural high which can be observed on the seismic section. It appears that southward, where gas is present, the stratigraphic interpretation becomes more difficult.

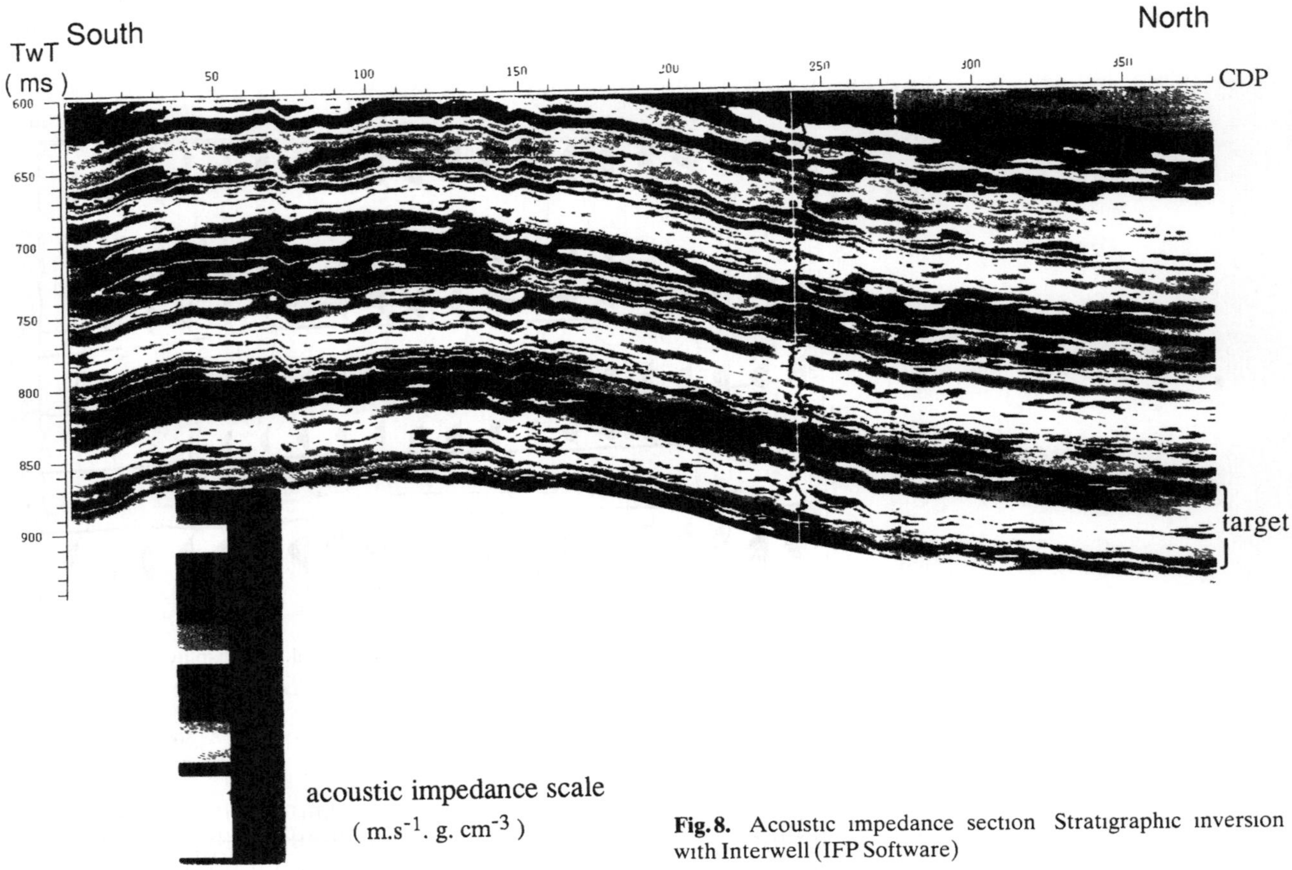

Fig. 8. Acoustic impedance section Stratigraphic inversion with Interwell (IFP Software)

The S2 surface is rather easy to pick between CDP[220–270]. Except around the well SG16, where there is a calibration, the other surfaces are fuzzy. The lateral variations of the seismic amplitude are important and it is difficult to find a general time pattern for the traces to make a comparison with the initial tie near the well SG16. An accurate picking of the lithostratigraphic boundaries based on seismic amplitudes seems to be difficult and insufficient to bring new continuous lateral informations (Fig. 7).

Seismic Stratigraphic Inversion

On the time acoustic impedance log SG16 (Fig. 5), we observe that the main stratigraphic surfaces correspond to some significant vertical changes in acoustic impedance values.

So, we can hope that it would be easier to pick these surfaces on an acoustic impedance section. The seismic stratigraphic inversion theory using well data and more particularly the software INTERWELL[1] (Dequirez et al. 1990, 1991) allows one to obtain such a display. In addition, even if there are some huge lateral or vertical variations of acoustic impedance values in the target, the general reservoir acoustic impedance pattern will be less modified than the vertical pattern of equivalent seismic trace.

On the acoustic impedance section, top and base of the reservoir are emphasized by strong vertical changes in impedance values (Fig. 8).

As in amplitude seismic display, it is easy to interpret the acoustic impedance display (wiggles) around the studied well. The main lithostratigraphic boundaries can be accurately correlated from trace to trace. Impedance traces could be interpreted like pseudologs (Fig. 9). The vertical variations of acoustic impedance slowly modify the general pattern of the pseudologs between levels S4 and S0.

Stratigraphic Interpretation of the Flattened Acoustic Impedance Section 88CY104

As in the interpretation of well logs, the pattern concept is crucial for the stratigraphic correlations of acoustic impedance wiggles. In Fig. 10, we observe that with this notion, the determination of the stratigraphic surfaces is easier than on the seismic amplitude section. In this case study, the stratigraphic boundaries are emphasized by strong vertical variations, and this fact makes the stratigraphic interpretation easier. From north to south, above the S2 surface, we observe a pattern variation which is ob-

[1] IFP SOFTWARE

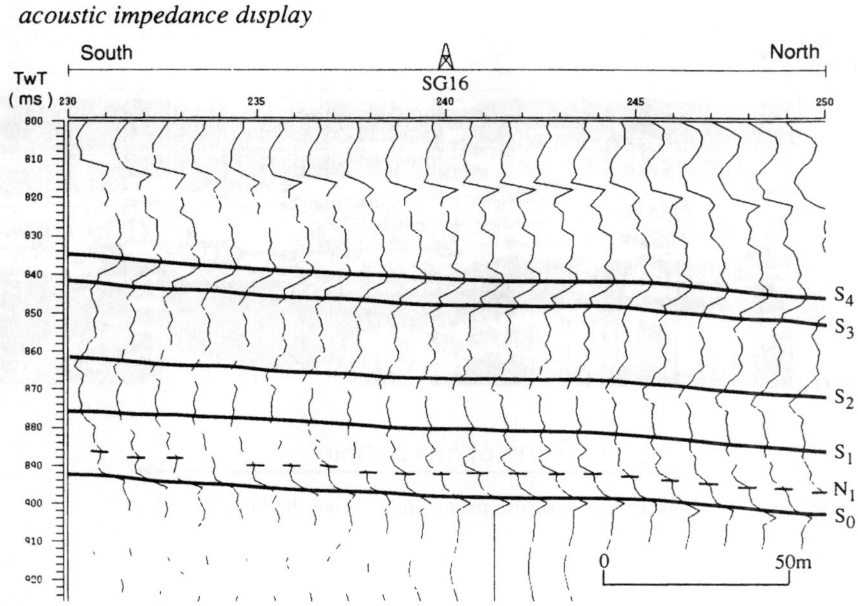

Fig. 9. Stratigraphic interpretation around well SG16, acoustic impedance display

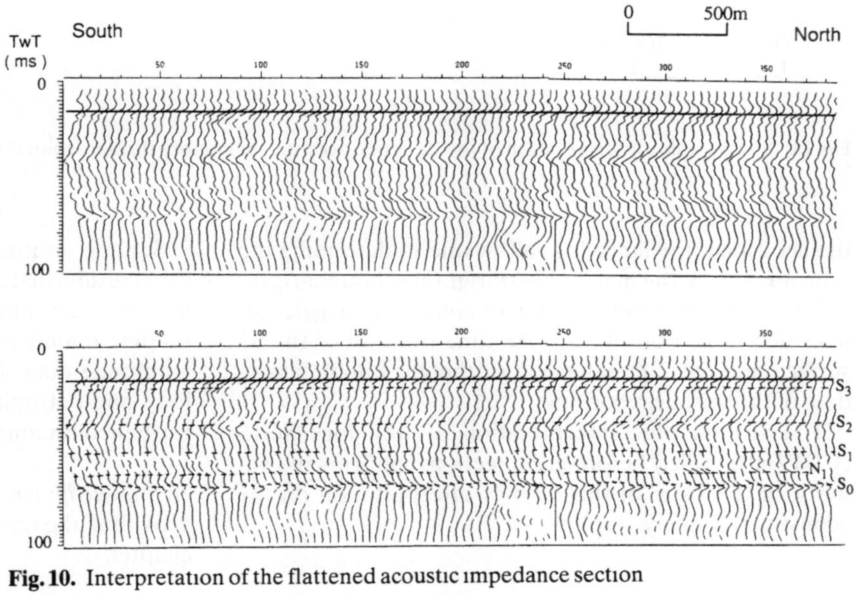

Fig. 10. Interpretation of the flattened acoustic impedance section

vious between CDP220 and CDP30. This general decrease of acoustic impedance value is due to the presence of gas in overlaying sandstones bodies. In fluviatile environment, the acoustic impedance contrast between channel sandstones and flood plain shales is weak (J. Sangree and B. Bally, pers. comm.), and this fact is confirmed for the studied reservoir. In the presence of gas, this sandstone/shale acoustic impedance contrast becomes null or negative, so that it becomes difficult to make a differentiation between these two lithologies. Here, we can understand that the concept of acoustic impedance pattern takes all its meaning because the stratigraphic picking (for instance of S2 surface) is not made by following the main acoustic impedance contrast. If this option had been chosen, we would have had a wrong interpretation or an undetermined time position for this S2 surface.

The last Fig. 11 is a color display of the flattened acoustic impedance section with a superposition of the acoustic impedance wiggles. This kind of display shows that it is dangerous to interpret only the color display. The huge vertical variations of impedance (emphasized by a very detailed color scale) do not necessarily correspond to the lithostratigraphic boundaries. In this example, this fact is highlighted between CDP30 and CDP220, where the presence of gas induces a modification of the impedances above the S2 surface. The time position of this surface is picked on the wiggle display with the notion of pattern. The vertical shape and not the amplitude variations of

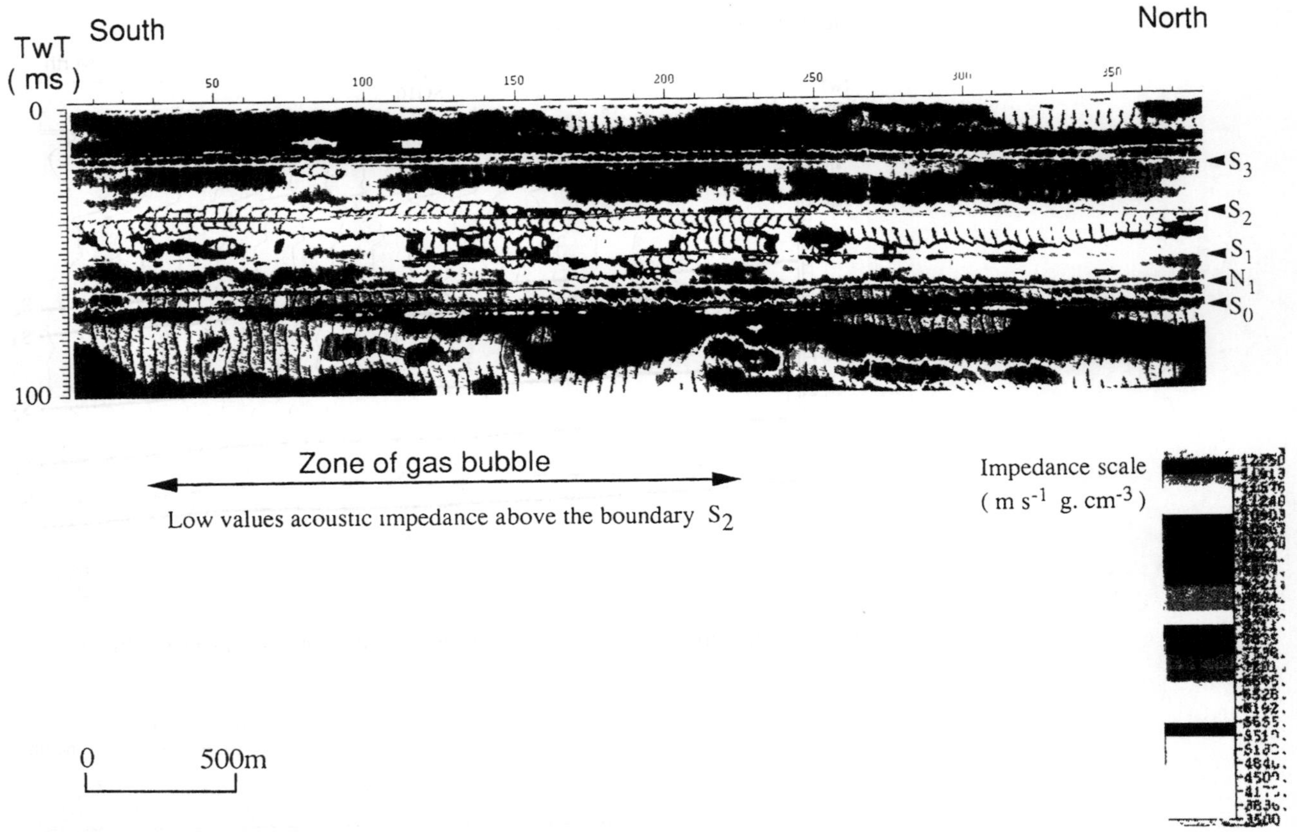

Fig. 11. Stratigraphic interpretation of the flattened acoustic impedance display color display

the acoustic impedance logs allow one to determine the time position of the main lithostratigraphic boundaries.

Thus, this composite display of colors and wiggles allows determination of both the time position of the S2 surface and the 2-D extension of the gas bubble at the date of the seismic recording.

The stratigraphic interpretation of well logs SG15 (CDP150), drilled before gas injection, confirms the validity, in this case, of this kind of interpretation on the acoustic impedance display.

Conclusions

This case study of the Soings en Sologne anticline shows that an integrated study, taking into account stratigraphic and seismic knowledge, leads to a better geometrical characterization of the reservoir Furthermore, these geometrical results would, after an appropriate time depth conversion, improve the stochastic simulations of the reservoir lithology and petrophysical properties inside the lithostratigraphic units. This kind of input can be considered as trivial, but it is obvious that the lithostratigraphic boundary shape given by this method will be better than a simple interpolation between the well data

With the development of 3-D seismic, this kind of integrated interpretation will become more and more important

Seismic stratigraphic inversion using well data provides results that can be interpreted in geometrical terms and so really improves the picking of the main stratigraphic boundaries, in comparison with initial seismic amplitude data. Relations between seismic data, lithology, and petrophysical properties will be developed in further communications

Acknowledgments We wish to thank Gaz de France for providing the data set and for permission to publish this chapter.

References

Brac J. Dequirez PY, Hervé F, Jacques F, Lailly P, Richard V, Tran Van Nhieu D (1988) Inversion with a priori information an approach to integrated stratigraphic interpretation 58th Annu SEG Meeting, Expanded Abstr, vol 2, Tulsa, pp 841–844

Dequirez PY, Fournier F, Richard V (1990) Application of inversion to integrated stratigraphic interpretation Revue de l'Institut Français du Pétrole, vol 45 (3) pp 397–416, Technip, Paris

Dequirez PY Fournier F, Richard V (1991) The advantage of using calibrated and integrated processing for a better reservoir description 13th World Petroleum Congress, Buenos Aires, 1991

Eschard R, Doligez B, Rahon D, Ravenne C, Leloch G, HERESIM group (1991) A new approach for reservoirs description and simulation using geostatistical methods In Proceedings Advances in reservoir technology characterization, modelling and management, Edinburgh, 21-22 February, 1991

Houel P, Lamiraux C, Mathieu Y, HERESIM group (1991) Etude sédimentologique du Trias de Soings en Sologne, Préparatoire à l'analyse géostatistique IFP Internal Rep ref 38591

Mathieu Y, Verdier F, Houel P, Delmas J, Beucher H (1993) Reservoir heterogeneity in fluviatile Keuper facies A subsurface and outcrop study In Eschard R, Doligez B (eds) Subsurface reservoir characterization from outcrop observations Technip, Paris

Richard V, Brac J (1988) Wavelet analysis using well log information In 58th Annu SEG Meeting, Expanded Abstr, vol 2, Tulsa, pp 946–949

Management of Chaunoy Oil Field Multilayered Reservoir

C. Bacchiana, J. Parpant, and B. Smart

Abstract. A 3 D full field numerical model is a powerful tool for the management of a multilayered reservoir

By providing adequate visualization of the waterflood performance at the end of the history-match, the model can be used to select actions and define the strategies to be evaluated in the predictive mode. Through this process, injection strategy refinements implemented at Chaunoy in 1991 raised the daily oil production by more than 500 m³/day (25 %) in a few months

Introduction

The Chaunoy field is one of the major oil fields in France and currently first in terms of production, with about one quarter of the French oil output.

Chaunoy was discovered in May 1983 and is the largest Triassic field in France. It is located in the central part of the Paris Basin near Melun, about 50 km southeast of Paris (Fig. 1).

The Liassic shales located above the Chaunoy formation are the source rock for both Triassic and Dogger fields (Fig. 2) currently producing from this basin.

The structural trap is a smooth northeast to southwest elongated anticline, 15 km long by 5 km wide, faulted to the west and with a low dip of about half a degree to the east. To date, 95 wells have been drilled and are represented by small dots on the structural map (Fig. 3). The closure is about 60 m and the original oil water contact (OWC) was found at 2205 m subsea, that is about 2300 m below the ground level.

The Chaunoy reservoir is a multilayered reservoir composed of alluvial fan deposits, braided streams, and lacustrine sandstones. A ten layer geological model based on the continuity of the most extensive shales on a field basis was adopted in 1987.

A west to east cross section (Fig. 4) shows this ten layer geological split as well as the great heterogeneity of the Chaunoy reservoir. In fact, it is almost impossible to correlate the sandstones which compose the porous part of the reservoir.

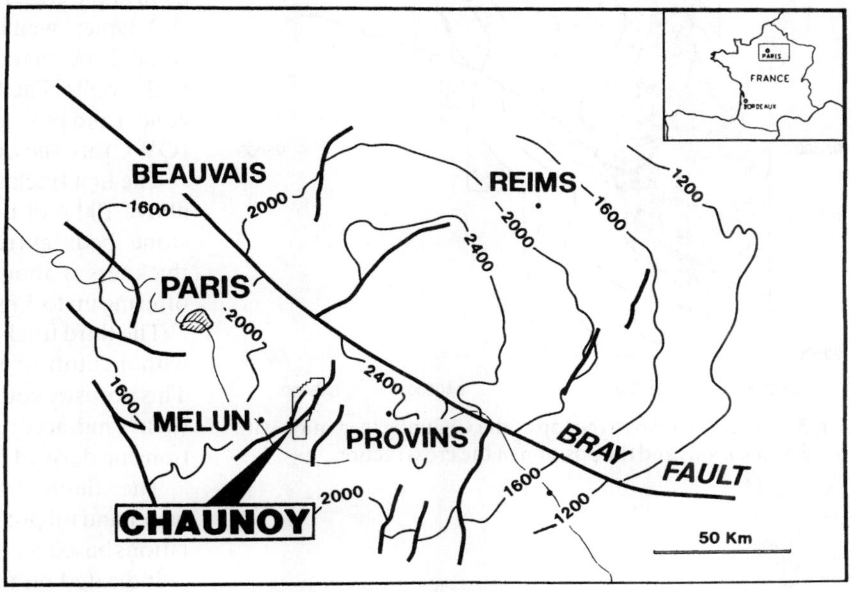

Fig. 1. Location map of Chaunoy Field in the central part of the Paris Basin Contours represent the depth to the top of the Triassic

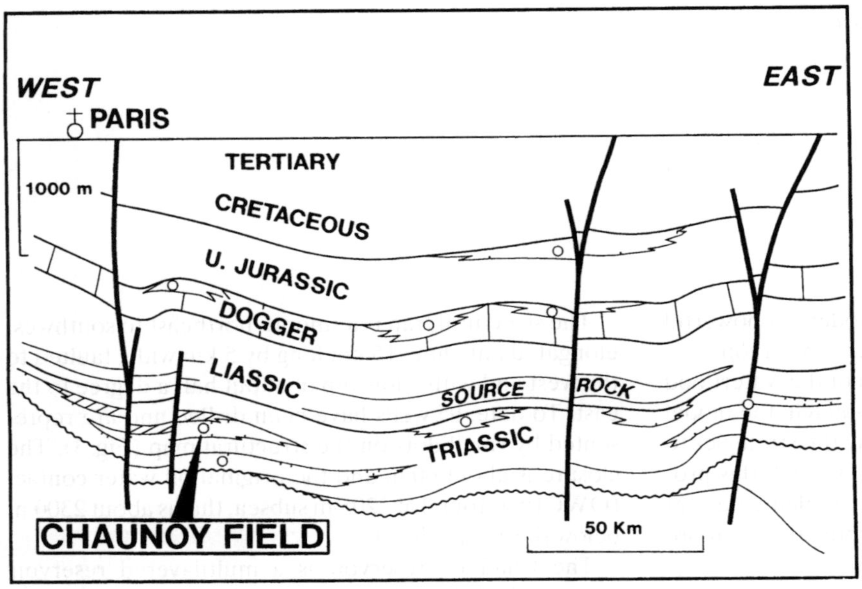

Fig. 2. Schematic cross section showing the different reservoirs and the position of the Chaunoy field

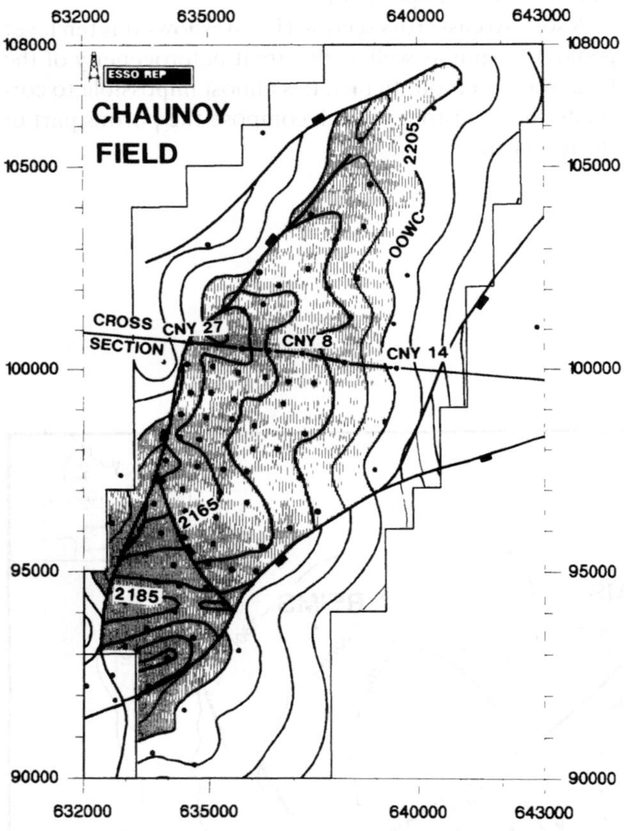

Fig. 3. Structural map to the top of the Chaunoy formation with the well locations and the position of the cross section

Table 1. Reservoir characteristics

Gross Thickness	66 M
Net to gross	38 %
Porosity avg	13 %
Permeability avg	150 MD
Net pay sands avg	11 M

Average Reservoir Characteristics

The formation is of nearly uniform thickness, averaging 66 m, with porous sands comprising 38 % of the interval. The net oil pay averages 11 m with a maximum of about 25 m at the crest near the west major fault (Table 1).

A typical well log is provided in Fig. 5 to better understand those characteristics and their vertical distribution in the wells. The layering (only for the seven oil-bearing zones) and position of the field original oil-water contact (OWC) are shown on the extreme left

The first track is the gamma-ray log with shales shaded above 100 API units The second track shows the sandstone beds currently perforated. The individual bed thickness is about 2 m but sometimes beds are stacked, offering up to 5 m of porous sands.

The third track is the porosity log scaled from 0 to 24 % with a cutoff of 8 % to determine the productive beds. This porosity cutoff corresponds to a permeability of less than 1 md according to the porosity permeability relationship derived from core data.

The "fluids" track indicates the relative proportions of water and oil present in the porous bed, with these calculations based on resistivity logs not displayed here

Indicated on the extreme right are the pressure measurements taken during the open hole logging program, which proved to be an important acquisition in the management of this multi-layered reservoir. For instance, for

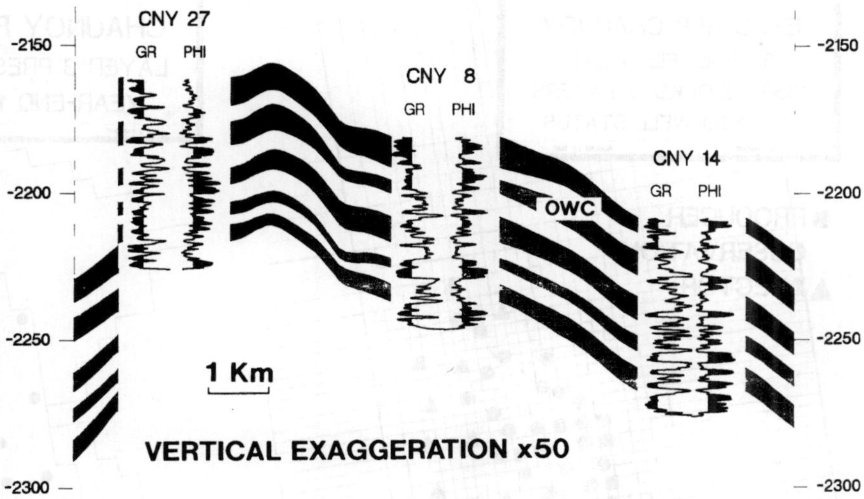

Fig. 4. East-west structural cross section showing the layering used in the simulation Well logs are gamma ray and effective porosity (shades below 100 API and above 8 %)

this well, the pressure measured in the upper layers was found to be only 50 % of the initial pressure represented by the heavy solid line and equal to 23 MPa (almost hydrostatic). This pressure depletion, which was observed in many wells drilled since 1986 despite an early peripheral injection, is of primary concern.

Multilayered Reservoir Management Issues

The reservoir management focus at Chaunoy is on three main issues: pressure monitoring, wellbore management, and flowmeter surveys.

Reservoir pressure surveillance is a key activity required to understand the behavior of a multilayered field and assess the waterflood efficiency. Fortunately, drilling was almost continuous from 1983 through 1991, thus providing a pressure database containing about 1000 records well distributed in space and time. Most of these pressure measurements are individual sand pressure measurements acquired with the Schlumberger RFT (Repeat Formation Tester) tool during open hole logging operations.

Wellbore management is also an important issue because most of the wells are perforated in several layers for economical reasons. When possible, layers having similar reservoir pressure are perforated to avoid crossflow problems. This was the case with most of the crestal wells, which were first perforated only in the lower, oil-bearing layers (Fig. 5). Injectors are also perforated in more than one layer in order to maximize wellbore utility. Therefore fluid distribution per layer must be measured or estimated. That is the purpose of flowmeter surveys which are performed annually on injectors to monitor the vertical distribution of injection fluids.

Production profiles, while initially captured on flowing wells, are no longer measured directly, as all wells are now pumped. The vertical conformance is determined

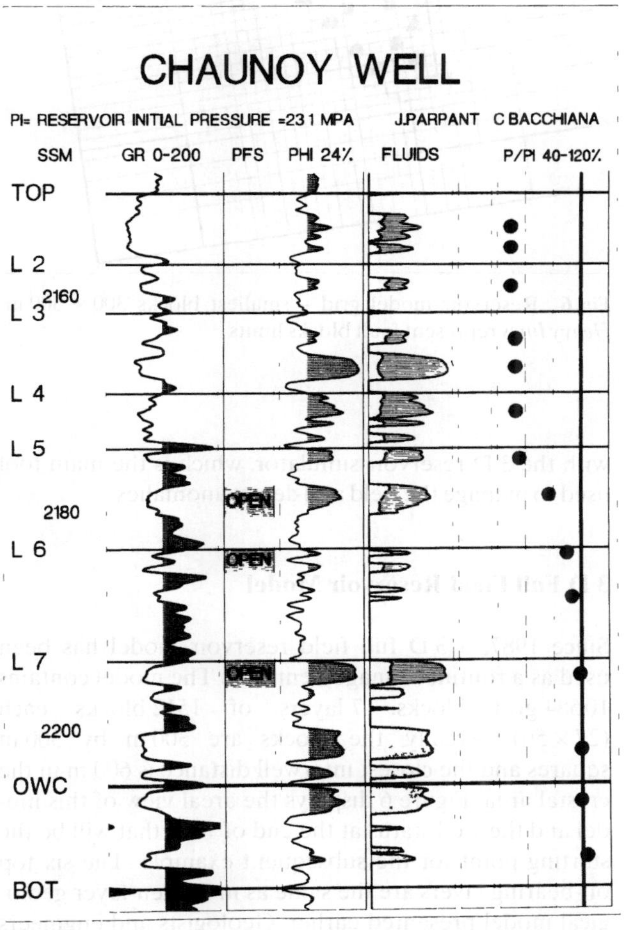

Fig. 5. Example of well logs *GR* Gamma ray (shaded above 100 API), *PFS* perforated zones, *PHI* effective porosity (scale 0–24 % – shaded above cut off 8 %), *Fluids* nature of the fluids in porous zones, *shade with horizontal bars* porosity times SW, *solid shade* porosity times So, *scale* 0–24 %, *P/PI* local pressure at the drilling date/field initial pressure, *L2-L3* top of geological layering, *OWC* original oil water contact

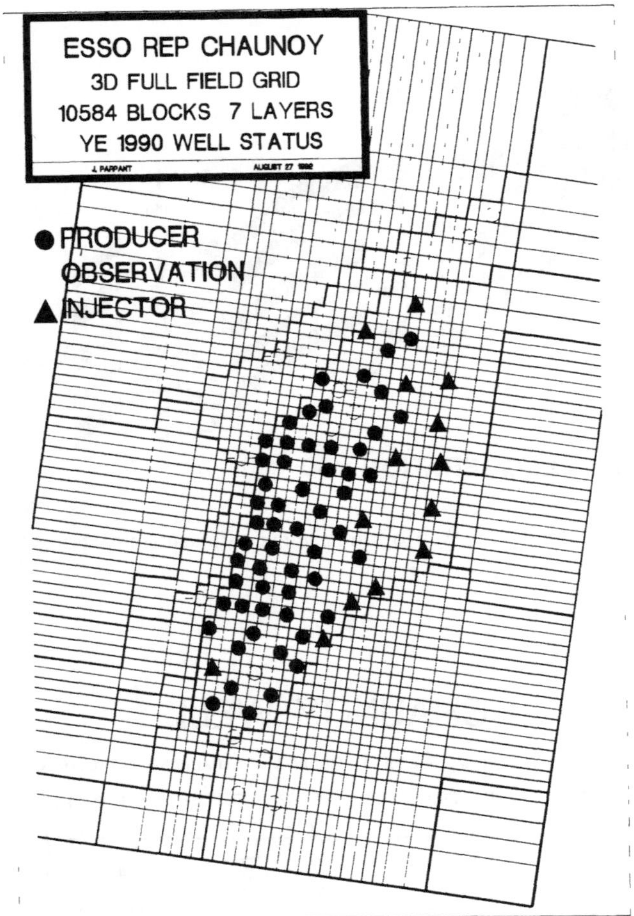

Fig. 6. Reservoir model grid – smallest blocks 300 × 300 m *Heavy lines* represent fault blocks limits

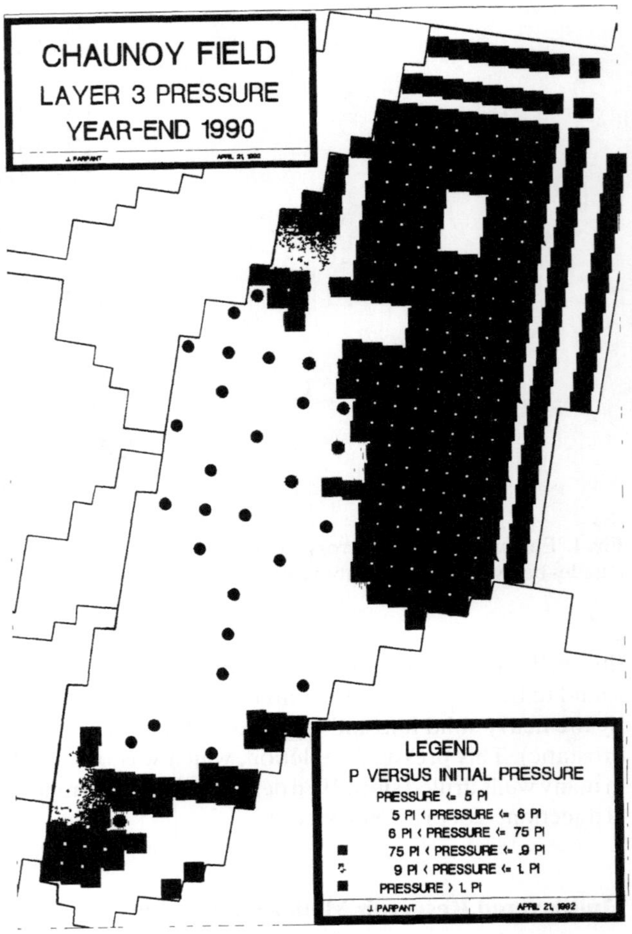

Fig. 7. Year end 1990 layer 3 pressure expressed in % of initial pressure The *black areas* represent a pressure increase from initial conditions. *black dots* represent wells completed in this layer

with the 3 D reservoir simulator. which is the main tool used to manage the field and detect anomalies

3 D Full Field Reservoir Model

Since 1987. a 3 D full field reservoir model has been used as a routine management tool. The model contains 10584 grid blocks. 7 layers of 1512 blocks each (27 × 56). Areally. the blocks are 300 m by 300 m squares and the closest interwell distance is 600 m in the crestal area. Figure 6 displays the areal view of this model and the well status at the end of 1990 that will be the starting point for the subsequent example The six top oil-bearing layers are the same as in the ten-layer geological model presented earlier. Geologists and engineers are jointly responsible for collecting. reviewing. and integrating data into the simulation model built with the in-house EXXON software MARS. This team effort has proved to be very effective. especially during the history-match process which is an interactive phase with many discussions about potential data modification to improve the match

In our case. data assumed to be more reliable: structure and oil in place. fluid characteristics. and relative permeabilities. were rarely modified. On the other hand. permeability barriers and vertical conformance of producers were regularly adjusted to fine-tune the match. Oil rates. cumulative production. and water cuts were matched at the field level. At the well level. the parameters selected to be matched were the pressure measurements. the water cut history of producers. and the distribution of injection per layer on injectors In addition to the previously mentioned individual sand RFT measurements. composite bottom-hole pressures calculated using routine annulus level measurements are also matched with the model.

Successful Management Using the Reservoir Model

Once a satisfactory history-match is achieved. predictions are made to compare different strategies and prepare long-term forecasts and reserves estimates Moreover. by providing adequate vizualization of the waterflood performance at the end of the history match. the

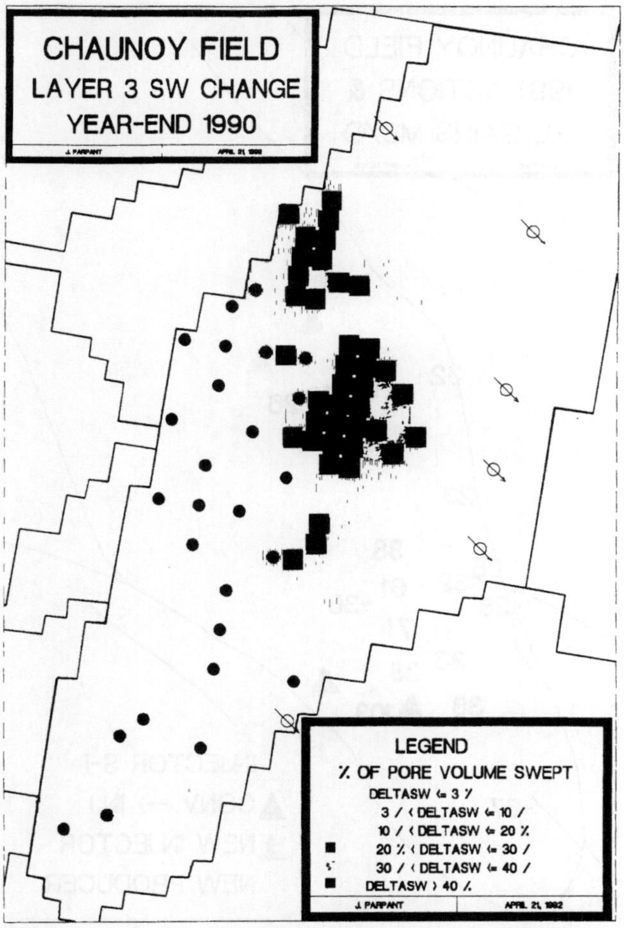

Fig. 8. Layer 3 water saturation variations between initial conditions and year's end 1990 *Black squares* represent initial oil areas swept by water, *white areas* represent areas without significant charges, *black dots* represent wells completed in this layer

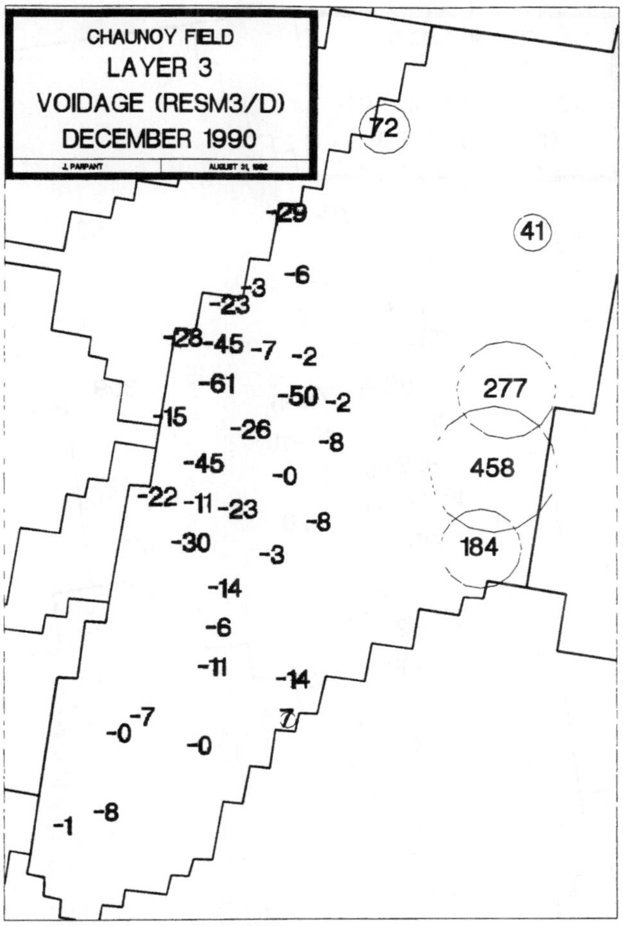

Fig. 9. December 90 voidage in layer 3 according to the 3 D reservoir model *Gray circles* are proportional to the daily withdrawals in December 90, *open circles* are proportional to the daily injections in December 90, rates in reservoir m^3/day are posted at the *center of the circle*

model has also been used to select actions and to define the strategies to be evaluated in the prediction mode. The example which follows will demonstrate the effectiveness of this approach.

All maps presented have been built using extracted data from the simulator. At the end of 1990 the pressure map for layer 3 (Fig. 7) shows clearly that a large area in the center part of the field is very depleted. It reveals the need to inject more and/or closer to the producers, especially in the center south area (producers from that layer are represented by circles on the map).

The water saturation changes map (Fig. 8) confirms the poor sweep efficiency in the center south area for this layer.

The instantaneous voidage map (Fig. 9) shows the location and volume of injections (positive figure) and withdrawals (negative figure) for December 1990. It is obvious that, despite a positive balance, layer 3 injection was not properly placed at that time. Remedial action, such as implementing new injectors, or performing workovers on existing injectors to improve layer 3 injection distribution was necessary.

In order to improve the pressure maintenance process, and thereby increase the efficiency of the waterflood, injection needed to be placed closer to production. By looking carefully at the three maps simultaneously, it was easier to identify most of the specific well actions required. This was done for all of the layers and compatible actions were defined, evaluated, and implemented.

Figure 10 shows the voidage situation in layer 3 one year later (December 1991), following the remedial actions. Fortunately, it looks much better than before with injection closer to the producers and significant withdrawals increase in the center south area.

Globally, soon after actions were implemented, the reservoir pressure increased in the depleted upper layers. Better behavior of wells with commingled zones, lift optimizations, workovers, and infill drilling (Fig. 11) converted this pressure gain into a production gain of about 500 m^3/day from 2000 m^3/day up to 2500 m^3/day: Fig. 12 shows the quarterly averaged oil production in m^3/day and the evolution of the water cut in percentage. Figure 13 summarizes the Chaunoy field history since its discovery.

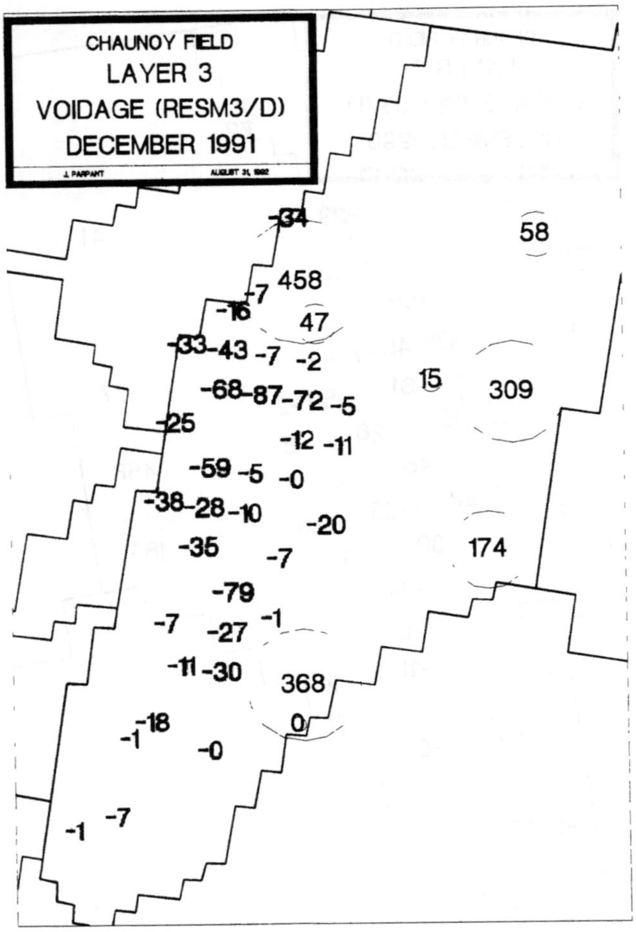

Fig. 10. December 1991 voidage in layer 3 – for legend see Fig 9

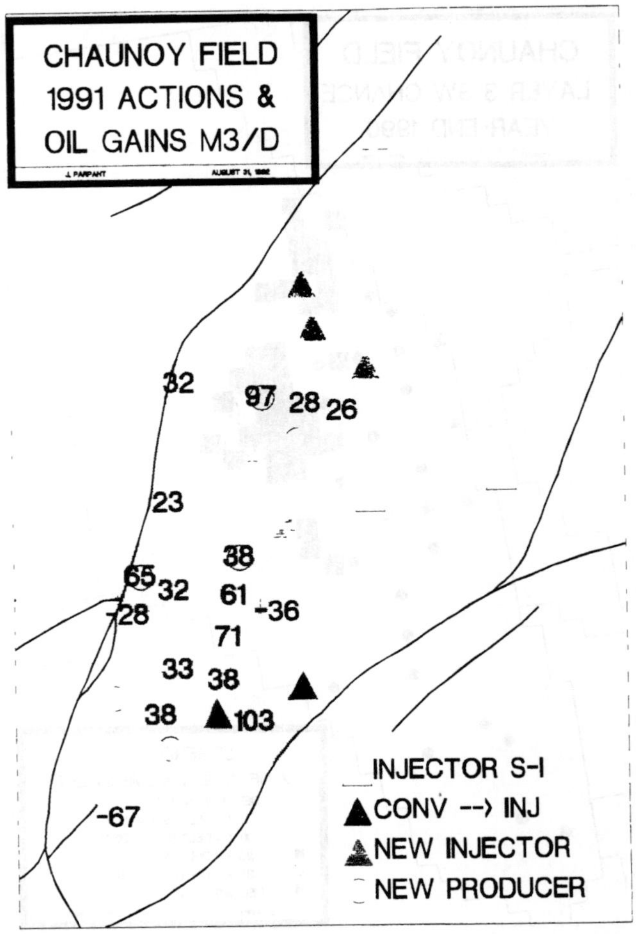

Fig. 11. 1991 Actions and oil gains *Open triangles* injector shut in, *black triangles* producer converted to injector, *gray triangles* newly drilled injector, *open circles* newly drilled producer

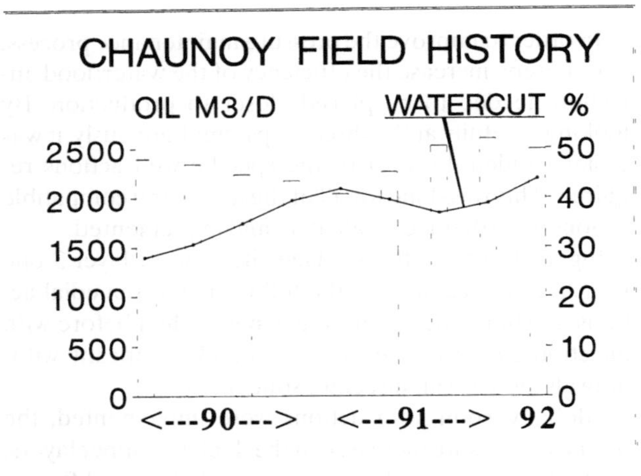

Fig. 12. Oil and water productions since beginning of 1990 *Bars* represent the oil production, connected points indicate the watercut in %

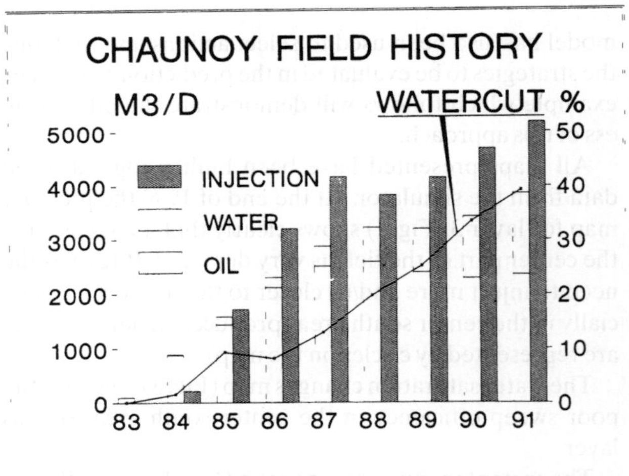

Fig. 13. Oil and water production and injection since 1983 *Left bar* represents total fluid production in m³/day with oil production at the bottom, *right bar* represents water injection in m³/day, *connected points* indicate the water cut in %

Conclusions and Plans

In conclusion, we can say that the current Chaunoy reservoir model integrates all data available, promotes geoscience teamwork, and makes reservoir management easier.

A new model based on a finer geological model will be built soon to better assess infill drilling. This work, done in collaboration with the IFP (Institut Français du Pétrole), will involve geostatistical modeling.

References

Comité des Techniciens (1992) Monographies des principaux champs pétroliers de France Chambre Syndicale de la Recherche et de la Production du pétrole et du gaz naturel – Mars 1992

Morelot G (1985) Bilan de l'exploration du Bassin de Paris – Pétrole et Techniques 327 – pp 10–15 – Mars 1985

Morelot G – Pagès L (1989) Objectif Trias à la recherche des gisements pétrolifères du Bassin de Paris Les roches au service de l'homme – géologie et préhistoire du Bassin Parisien – Association des géologues du Bassin de Paris Mém hors série 7, pp 198–200

Pagès L (1987) Exploration of the Paris Basin Petroleum geology of North West Europe Graham & Trotman, London, pp 87–93

Seismic Applications of a Paris Basin Shallow Geology Data Base

F. Hanot

The main purpose of the Paris Basin data base is to solve the very specific problem of the statics in this area. The difficulty consists in mastering the long-wave-trend component of the primary statics (Booker et al. 1976; Garotta 1986). The 3D geological method plays a leading part here in acquisition, by providing the best adjustment of the field parameters in relation to the shallow geological covering, as well as in processing and interpretation.

In order to provide reliable processing and hence work with an accuracy compatible with the real dimensions of the target structures, the method involves a number of shallow surface geological plays like Tertiary (Hanot 1992), (Fig. 1) Upper and Lower Cretaceous, Middle Jurassic, etc.), and is based on a series of structural and isovelocity maps generated from drill-holes and subsurface velocity survey data contained in a unique collection of 30000 control points spread over the entire Basin (Figs. 2, 3).

By using these comprehensive documents, any company operating in the area has access to information such as:

- extracts from structural maps of the various horizons (stratigraphic, lithological, phreatic) which are relevant to the company's concession;
- a previsional program of up-hole surveys, adequately designed to complement existing data;
- velocity cross sections up to the consolidated layer (chalk or other) in any requested direction, especially along the seismic line.

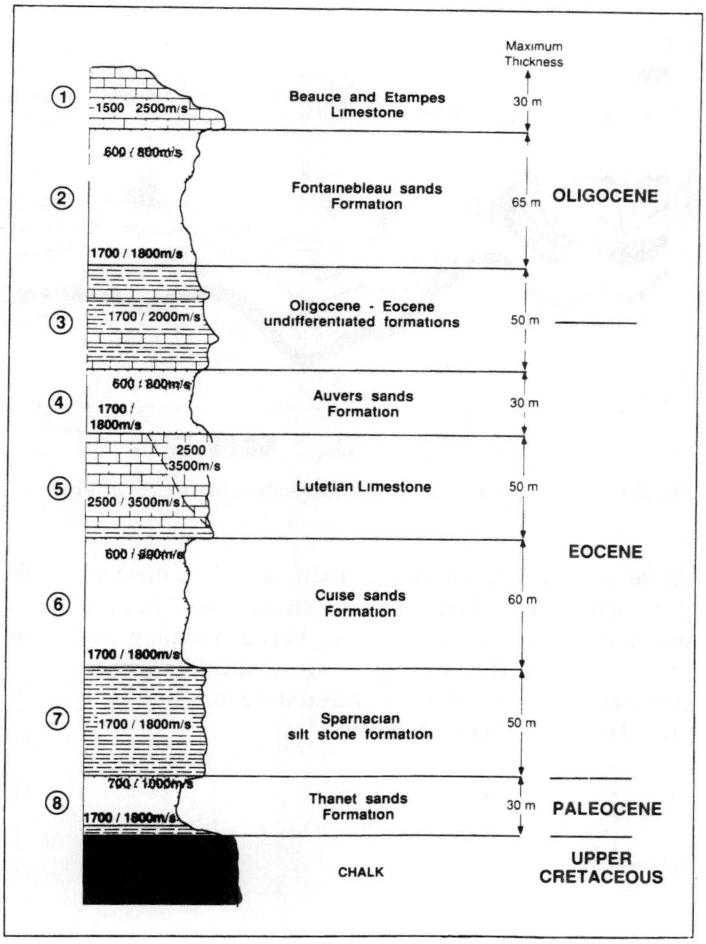

Fig. 1. Composite log of Paris Basin Tertiary formations which are crucial for statics

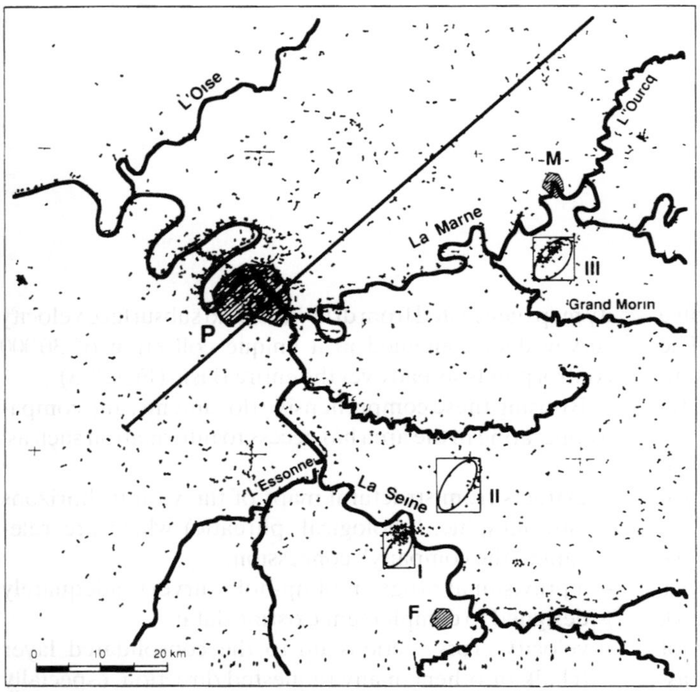

Fig. 2. Sample of the database control point distribution around Paris *P* Paris, *M* Meaux, *F* Fontainebleau, *I* Chailly field, *II* Chaunoy field, *III* Coulommes field, location of the geological cross section

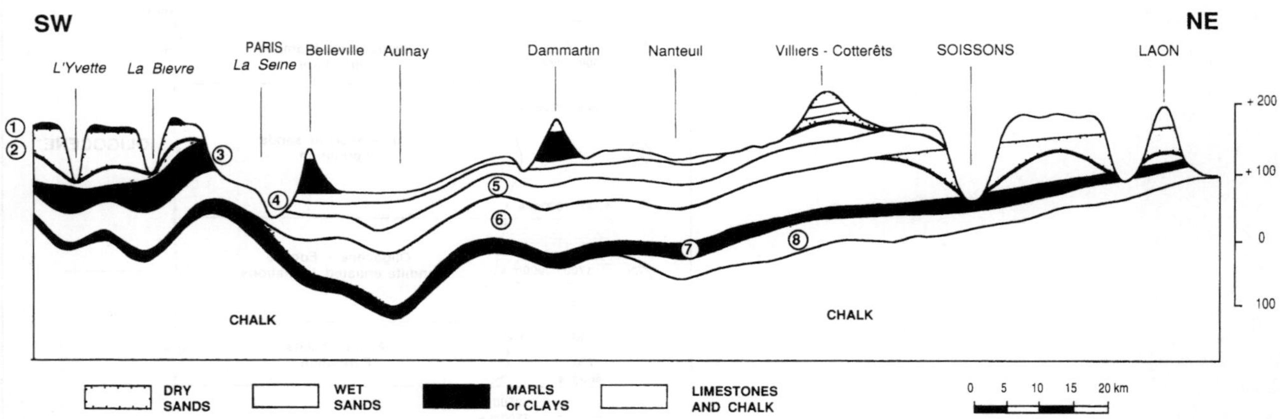

Fig. 3. Geological section of the Tertiary sediments of the Paris Basin

As far as interpretation is concerned, one of the main aspects of this comprehensive approach consists in the evaluation of the relationships existing between shallow and deep structures. This is a major point of interest when we consider that most of the Paris Basin structures were generated as recently as the Eocene.

References

Booker AH, Linville AF, Wason CB (1976) Long wavelength static estimation Geophysics 41 939–959

Garotta R (1986) Long wavelength static definition through partial common midpoint stacks CGG Tech Ser 559 86 09

Hanot F (1992) Seismic application of a Paris Basin shallow geology database First Break 10, 5, Mai 1992

Hanot F, Obert D (1992) Tectonique du Bassin Parisien par les déformations du Toit de la Craie Bull Inf Géol Bassin Parisien 29 4 1992

Hanot F, Renoux P (1991) Petrophysical variations in the Senonian chalk of the Paris Basin and their influence on static corrections First Break 9, 11, November 1991 515

Liassic Source Rocks from Three Wells from the Sea of Iroise (French Western Approaches) and Dorset (UK) Outcrops: Nature and Modeled Maturity History

J. F. Deronzier, A. Bruneton, J. Mondin, and A. Mascle

Abstract. The first stage of exploration in the Western Approaches and Channel Basins was completed in the middle of the 1980s without any significant results (apart from one oil and one gas condensate show) although 12 wells were drilled in French waters. However, the onshore Wytch Farm oil field was developed, and production was said to reach about 60000 bopd at the end of 1991. At the same time, the Kinsale Head and Ballycotton gas fields, offshore from southern Ireland, were developed with reported reserves of 1.4 Tcf and 73 Bcf respectively. Significant amounts of hydrocarbons may thus be present in as yet undiscovered structural or stratigraphic traps. Present studies will benefit from the results of the first wells, as well as from the much improved quality of present seismic data with respect to those available in 1970 It is now also possible to gain a better understanding of the quality, distribution, and maturation history of the potential source rocks, the most widespread and mature of which are the Liassic black shales. Sections of these shales have been sampled along the classic field exposures of the Dorset coast, and from cuttings in three wells. The Dorset section is immature and the average organic content is 3%. At Levneg 1 the Liassic section is about 1250 m thick, with the best potential in the Pliensbachian to Hettangian intervals. Present TOCs are low (about 1%), but these rocks are overmature so that their initial organic content can be estimated to have been in the range of at least 2%. The present depth of these levels and the present heat flow cannot explain this present maturation so that their tectonic history must also be taken into account. The IFP-CGG IS 88 seismic line clearly shows, at this locality, the inversion in Tertiary times of the basin where up to 4000 m of sediment hat accumulated in Mesozoic times. The relatively good Liassic source rocks at the Rea-Gwenn and Travank wells are immature and were deposited in a much less subsiding area. They appear to be more similar to the Dorset section. As for the area south of Wytch Farm, 1-D Matoil modeling shows that the generation of oil or gas stopped at Levneg as soon as the process of inversion was initiated, which implies that, in this case, pre-Tertiary structural traps are more favorable targets than recently inverted structures.

Introduction

The Sea of Iroise represents the French part of the Western Approaches and is located to the west of the Paleozoic Armorican Massif in Brittany, to the east of the Bay of Biscay Cretaceous continental margin, to the south of the Mesozoic and Tertiary Celtic Sea basins, and to the west of the late Paleozoic-Mesozoic Channel basins (Fig. 1). An initial phase of petroleum exploration in the Sea of Iroise was completed by the middle of the 1980s with 12 exploration wells drilled in French waters. Two additional wells were drilled at the same time in the Channel basin and on the south Armorican shelf (Nautile and Penma, respectively). This chapter will be restricted to geochemical data from three of these wells: Levneg, Rea Gwenn, and Travank.

Although one oil show was reported from the Lennket well in lacustrine sandy Purbeckien facies, and one gas condensate show from the Krogen well in Callovian sandstones, no commercial accumulation has so far been encountered (Fig. 2). However, at the same time, drilling of about 50 wells in the southern Irish offshore led to the Kinsale Head and Ballycotton gas field discoveries, with reported reserves of 1.4 Tcf and 73 Bcf respectively. Also, the onshore English Wytch Farm oil field was developed and production was said to reach about 60000 bopd by the end of 1991 (Dranfield et al. 1987). Significant amounts of oil and gas may thus be present in the Sea of Iroise in still undisclosed structural or stratigraphic traps.

Studies presently undertaken benefit from an increased knowledge of the distribution of reservoir and source rocks from the available wells, direct observations and sampling of the Liassic source rock outcropping along the Dorset coast, and from a few recently acquired seismic lines of much better quality than the seismic previously available from the 1970s.

Geologic Framework

The geologic history of the Western Approaches and Channel basins may be discussed from released well data in both French and British waters, and from the SWAT-WAM regional seismic lines shot by the BIRPS-ECORS association (Evans 1990; Bois et al. 1991; Hamblin et al. 1992). Detailed geologic data are furthermore available from the Devon-Dorset outcrops and wells in southern

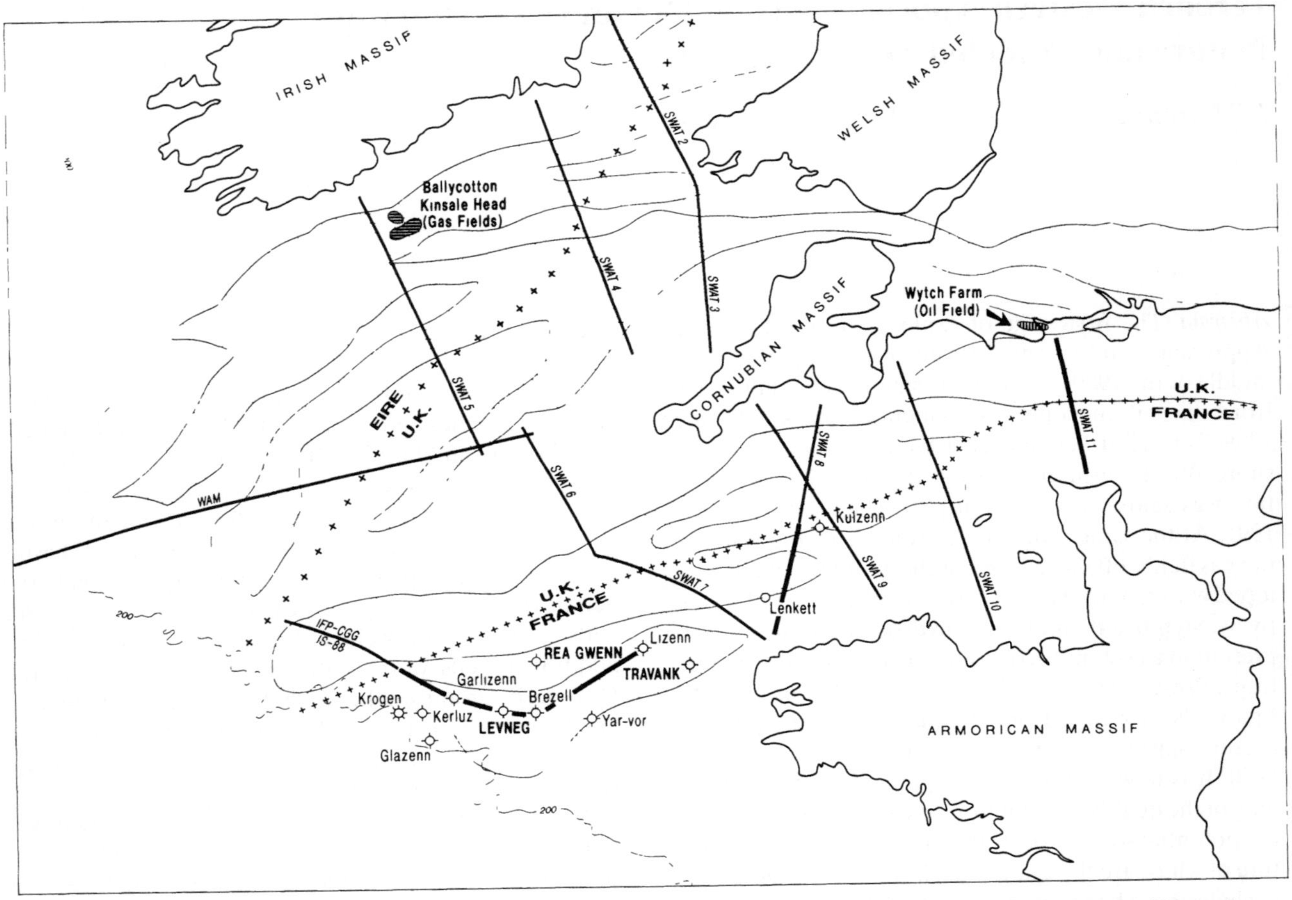

Fig. 1. Structural map of the Western Approaches with the location of the French wells, available regional seismic profiles, and main Mesozoic sedimentary basins

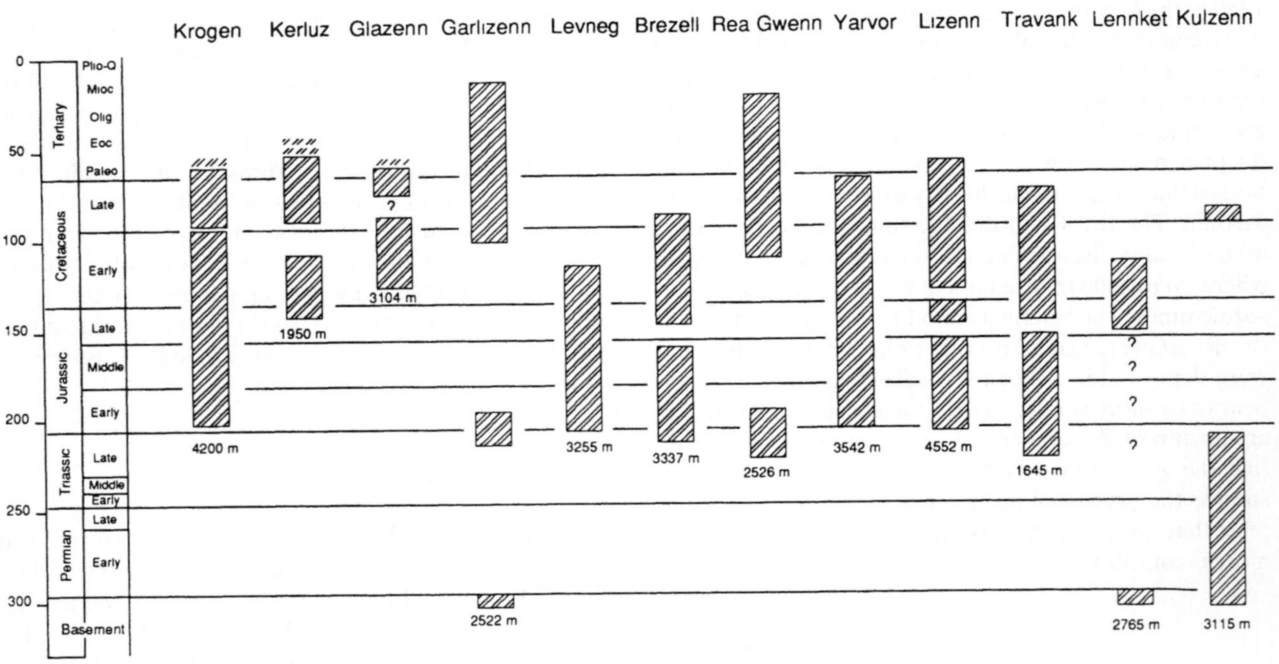

Fig. 2. Stratigraphic intervals encountered in the Iroise Sea wells

England (Stoneley 1982; House 1989; Jenkins and Senior 1991; Miles et al. 1993).

Stretching events causing subsidence occurred during Permian to early Cretaceous (Wealden) times (Fig. 3). These may be related first to the collapse of the thickened Paleozoic crust at the end of and just after the late stages of the Variscan orogeny, and subsequently to rifting events at the origin of the opening of the Tethyan and Atlantic Oceans in late Jurassic and early Cretaceous times. A major phase or regional uplift and subsequent erosion is documented in Aptian times and could be related to the "Austrian" E-W-orientated compressive event (Bénard et al. 1985; Gairaud and Le Théoff 1992). As a result, most of the early Cretaceous-Jurassic interval has been eroded in the British part of the Western Approaches, while this interval has fortunately been much better preserved in the French part. Regional subsidence and sedimentation resumed from Albian to Paleogene times, possibly in response to the cooling of the continental crust following Jurassic and early Cretaceous riftings. A major N-S-orientated compressive event occurred in late Paleogene times and induced locally severe basin inversion in the Western Approaches and Channel basins, as well as in the Wessex basin farther east (Shannon 1991; Chadwick 1993).

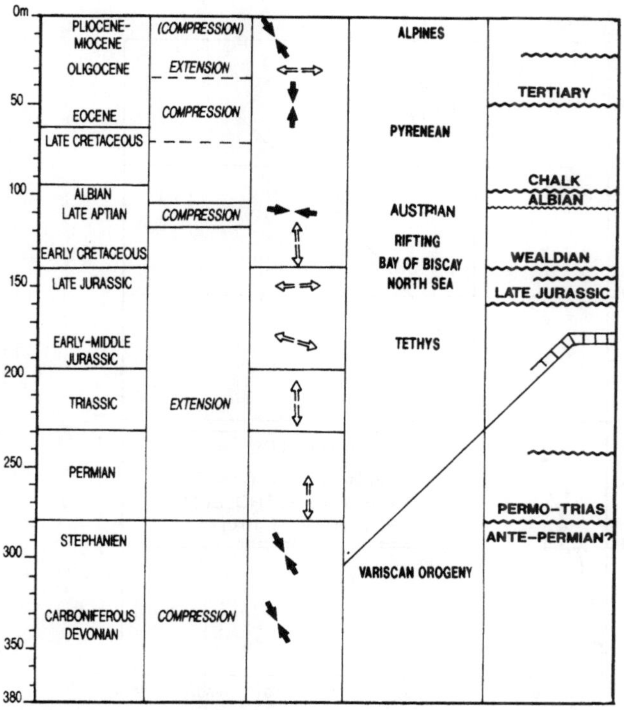

Fig. 3. Synthetic chart with the major tectonic events *From left to right* stages, prevailing stress regime, direction of maximal *(black arrows)* or minimal *(open arrows)* stresses, conventional names of related tectonic events, and main unconformities

Potential Source Rocks, Reservoirs, and Seals

According to data from the Wytch Farm oil field (Butler and Pullan 1990), the main source rock belongs to the marly early Lias, and could be of marine Type II, and/or a mixture of Type II and terrestrial Type III organic matter (Ebukanson and Kinghorn 1985, 1986). A similar origin is postulated for the Kinsale Head field (Colley et al. 1981). The peak of oil generation at Wytch Farm is generally believed to have occurred in late Cretaceous-early Tertiary times, i.e., before the Tertiary inversion (Butler and Pullan 1990). The migrating hydrocarbons were then able to be accumulated in structural traps such as preserved Jurassic tilted blocks or horsts (Selley and Stoneley 1987). Such accumulations have been preserved from any leakage as no further inversion occurred there. On the other hand, to the southwest of Wytch Farm, the Kimmeridge Bay producing well is located close to the top of a Tertiary inversion anticline. However, the fractured Dogger shales and carbonates have been producing without any decline for 25 years (2.5×10^6 bbl produced), suggesting some vertical leakage from a deeper reservoir or from an unknown trap (Selley and Stoneley 1987).

The two main Wytch Farm reservoirs are early-middle Triassic braid-plain and playa margin sandstones (Sherwood Fm.; Dranfield et al. 1987), and Toarcian shallow marine sandstones (Bridport Fm.; Penn et al. 1987). Seals are respectively provided by middle-late Triassic and early Dogger clays. A third reservoir has recently been found in Bathonian carbonates (Frome clay Fm.). In the Irish offshore region, the producing level at Kinsale Head is located in the Albian "Greensands" which are covered by the shaly Gault Clay Fm.

A few fair to good reservoirs have been encountered in the Sea of Iroise wells. They belong to the following time intervals: Permian (Kulzenn), Triassic (Kulzenn, Garlizenn, Rea-Gwenn), Callovian (Lizenn, Levneg, Yar-Vor) Purbeckian (Lenkett, Travank, Yar-Vor), Wealden, and Albian. Although they are widespread and of excellent quality, the two younger intervals generally do not have efficient seals at the location of the present wells.

This chapter will provide some data relating to the potential of Liassic shales from the Sea of Iroise with a major question being raised: are they, or have they been a significant source rock? We will first present a reference section sampled along the Dorset coast (sampling focused on the well-exposed Liassic cliffs surrounding Lyme Regis). Three Iroise Sea wells will then be described: Levneg 1, Rea-Gwenn 1, and Travank 1 bis. Rock Eval geochemical analysis has been performed on cuttings and includes the total organic carbon (TOC) content, petroleum potential (S2), maturity, and kinetics (Espitalié et al. 1985–1986). These parameters will then be used to calibrate some 1-D maturity modeling, south of Wytch Farm first and secondly at the site of the three Iroise Sea wells. The geometric constraints, i.e., sediment thicknesses on a virtual well south of Wytch Farm, and the amount of erosion following the Aptian and late Paleogene inversion, were obtained from the SWAT 8 and 11 seismic lines and from the CGG-IFP IS-88 profile.

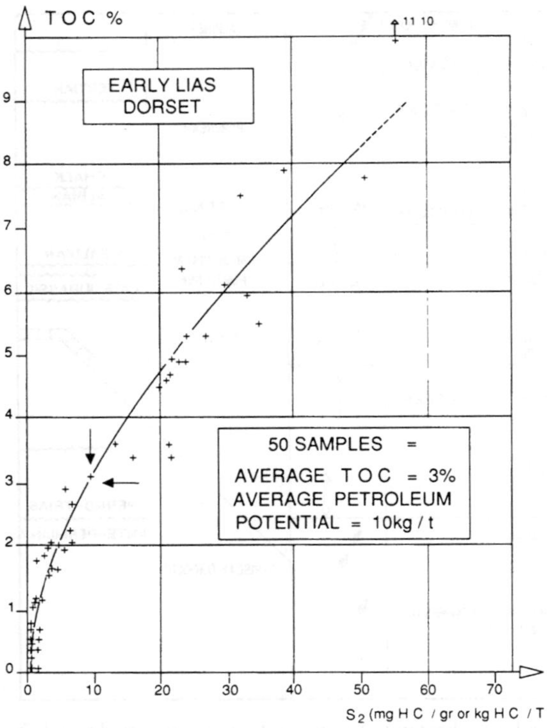

Fig. 4. Dorset early Lias source rock petroleum potential S2 versus TOC

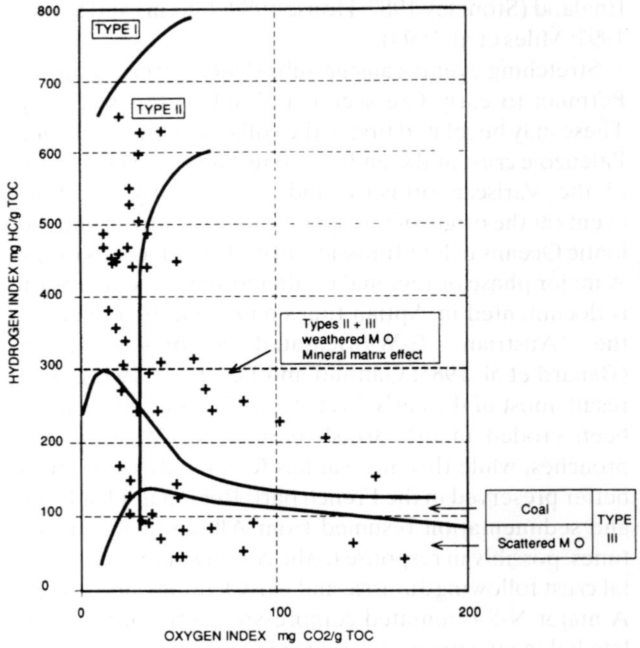

Fig. 5. Dorset early Lias section Hydrogen and Oxygen Index diagram

Table 1. Rockeval analysis from two selected sections of the Dorset Coast

L3 Section/Blue Lias Fm/Early Sinemurian					
Samples	S2	T_{max}	TOC	HI	OI
125596	0 2	420	0 3	61	258[b]
595	38 9	412	7 9	492	35[a]
594	0 6	421	0 7	84	56[b]
593	9 5	418	3 1	304	43[a]
592	56 0	411	11 1	502	30[a]
591	21 0	416	4 6	455	54[a]
590	8 6	422	2 7	314	58[a]
589	21 9	413	4 9	447	28[a]
588	32 9	412	7 5	437	34[a]
L2 Section/Blue Lias Fm/Early Hettangian					
Samples	S2	T_{max}	TOC	HI	OI
125917	0 4	413	0 8	47	48[b]
916	0 1	407	0 1	63	144[b]
915	21 7	414	4 7	462	21[a]
914	23 6	415	6 3	386	16[a]
913	0 2	409	0 3	52	85[b]
912	0 2	411	0 4	46	54[b]
911	1 3	419	1 2	102	25[b]
910	1 0	421	1 1	92	36[b]
909	0 9	420	1 0	95	34[b]
908	0 4	410	0 6	67	40[b]

Both sections are about 2 5 m thick The sampling interval is of 0 2 to 0 3 m, except for samples 125588 to 125589, where the sampling interval is of 0 05 m

[a] Family 1 good source rock

[b] Family 2 poor source rock

Dorset

The early Lias interval (Hettangian-Sinemurian-early Pliensbachian) is made up of either thick gray marls with fine interbedded carbonates or interbedded carbonates and dark marls. The total thickness is about 125 m (House 1989). The samples have been taken in the field, taking into account their apparent low weathering, but sampling was not biased to dark shales. The quality of this samples has been confirmed by the reasonably low OI of most of the samples (see Fig. 5).

Two selected field sections illustrate the potential but also the wide variability of the TOC content and petroleum potential of these rocks (Table 1). The sampled early Lias as a whole is an immature and good potential source rock (average TOC = 3% and average S2 = 10 kg/t, Fig. 4). The lowmaturity is in agreement with published organic maturity from estimated Ro% of 0.36 to 0.48 (Ebukanson and Kinghorn 1986). The interpretation of the HI/OI diagram (hydrogen and oxygen indices of the Rock-Eval pyrolysis, Fig. 5) allows two types of kerogen to be distinguished The first one, with HI up to 650, is obviously of marine Type II origin. The second one, with HI below 400 and OI above 50, could be either a mixture of Types II and III organic matter, or of degraded marine organic origin, or it could result from a mineral matrix effect altering the Rock-Eval results in the samples with low TOC content. Visual kerogen analyses were performed on four selected samples of Hettangian-Sinemurian age in order to obtain a better understanding of this organic matter composition (Ph Bertrand, unpubl. report). The TOC and HI of these samples were of 4.7, 2.1, 1.9, and 1.2 %, and 639, 303, 152, and 102 mg HC/g TOC respectively. They represent source rocks with a petroleum potential ranging from about 30 to 1 kg/t. A marine origin for all the samples is suggested by the very large proportion (90 to 95 %) of amorphous organic matter in all cases, and by the presence of plantonic algae such as acritarchs and Tasmanaceas. Continental fragments are also present, but the initial accurate ratio between marine and continental organic matter is impossible to determine because some of the amorphous organic matter could also be of terrestrial origin.

Iroise Sea

Levneg 1

The Levneg well was drilled in 1978 and was located on an inversion anticline as depicted in the recently acquired IFP-CGG IS 88 (Fig. 6). Below the thin Pleistocene cover, the youngest preserved sediments are of Barremian age. The total amount of erosion on top of the anticline is not acurately known, but it seems likely that it will be at least 1400 m according to the geometry and time thickness of the late Cretaceous and Tertiary strata as observed on the seismic. The Liassic interval is presently at depths of 1490 to 3053 m and is 1563 m thick (Fig. 7). Three main units may be distinguished: from 1490 to 2009 m, Aalenian and Toarcian dark marly and shaly deposits; from 2079 to 2736 m, Pliensbachian and late Sinemurian dark marls interbedded with grey marly limestones; and from 2736 to 3053 m, early Sinemurian and Hettangian grey marly limestones and shales. The TOC values for the Aalenian-Toarcian interval never exceed 0.9 %, and they are in the range of about 1 % for the Pliensbachian-Hettangian interval (Fig. 8). All samples are thermally mature to overmature so that the initial TOC content is unknown, but was certainly slightly higher at the origin. The T_{max} increase downhole to 520 °C. This suggests that a certain amount of Type III continental organic matter was present down to at least 2730 m (late Sinemurian). Below this, the very low residual petroleum potential (0.1 kg/t) rules out any hypothesis about the origin of the organic matter from chemical analysis.

Rea-Gwenn 1

The Rea-Gwenn well was drilled in 1981. We do not have any seismic section at the location of the well, but we believe that the Garlizenn well (Fig. 6) is located on a roughly similar geologic setting. The Liassic formation is presently located between 1809 and 2163 m and is thus only 354 m thick. The series is, however, restricted to the Hettangian and Sinemurian strata (interbedded gray mudstones and marls) as a result of Aptian erosion. This unconformity is overlain by early Albian sandstones, late Cretaceous chalks, and by a 1117 m-thick Paleogene and Neogene calcareous sequence (Fig. 7). No significant inversion tectonics has thus been active at this site. The Rock-Eval results (Fig. 9) show that this section is immature (T_{max} < 435 °C) and that two classes of organic facies may be defined. The first one is characterized by higher TOC and S2 values, up to 3.5 % and 15 kg/t respectively, with relatively high HI values of about 400. The second one shows lower TOC, S2 and HI values of about 1 %, 1 kg/t, and 100 respectively. Pyrogram shapes suggest a marine origin for Family 1 and a terrestrial origin or altered marine organic matter for Family 2 (Fig. 10).

Travank 1

This well was drilled in 1980 and is located in the vicinity of the Swat 8 seismic line (Fig. 11) on what appears to be a Jurassic tilted block preserved below the Aptian unconformity. The Liassic formation is encountered at depths from 927 to 1551 m and is thus 624 m thick (Fig. 7). Three main lithologies can be distinguished: from 927 to 1059 m, early Pliensbachian to early Aalenian silty gray marls with lignite-rich intervals; from 1059 m to 1191 m, late Sinemurian dark argillaceous mudstones; and from 1191 to 1551 m, early Sinemurian to Hettangian interbedded dark mudstones, argillaceous mudstones, and marls grading downhole to gray dolomites and mudstones. The whole section is immature (T_{max} < 435 °C) and four geochemical units can be distinguished (Fig. 12): from 900 to 1100 m, unit 1 is characterized by moderate TOC, S2 and HI values (1 %, 0.7 kg/t, and 60, respectively) and should be predominantly of terrestrial origin according to the pyrogram shapes. From 1100 to 1200 m, unit 2 is characterized by higher TOC, S2 and HI values, up to 2.9 %, 8 kg/t, and 280, respectively, and is very probably of marine origin. From 1200 to 1370 m, unit 3 is characterized by moderate TOC, S2 and HI values (1 %, 0.7 kg/t, and 150, respectively), but is stil possibly of marine origin; and from 1370 to 1550 m, unit 4 is characterized by low TOC, S2 and HI values, of about 0.2 %, 0.2 kg/t, and 70, respectively. A terrestrial origin, or altered marine organic matter, is again suggested by the pyrogram shape.

Maturity Modeling

The timing of hydrocarbon generation can be modeled from the previous geochemical data and integrated with burial histories from seismic profiles. The kinetics of organic matter transformation, the sedimentation versus erosion history, and the paleoheat flows first had to be defined.

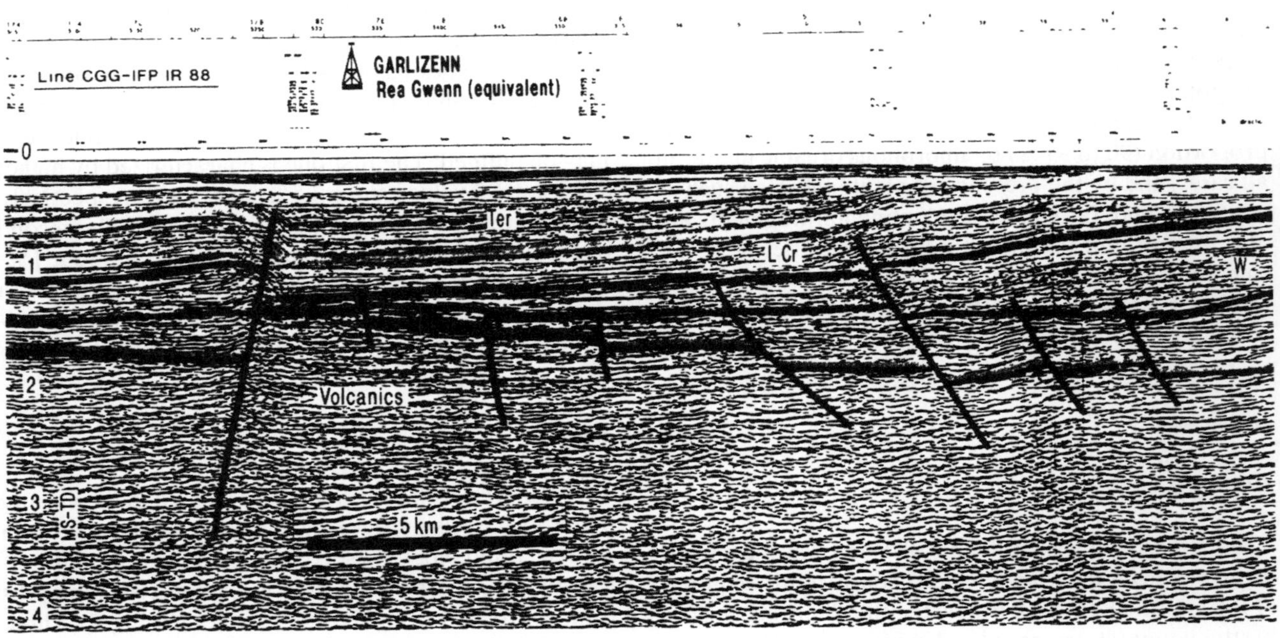

Fig. 6. IFP-CGG IS 88 seismic profile *Ter* Tertiary, *L Cr* late Cretaceous and Albian, *W* Wealden and Aptian, *Jur* Jurassic and Trias, *Paleo* Paleozoic basement, *Volcanics* Paleozoic volcanics

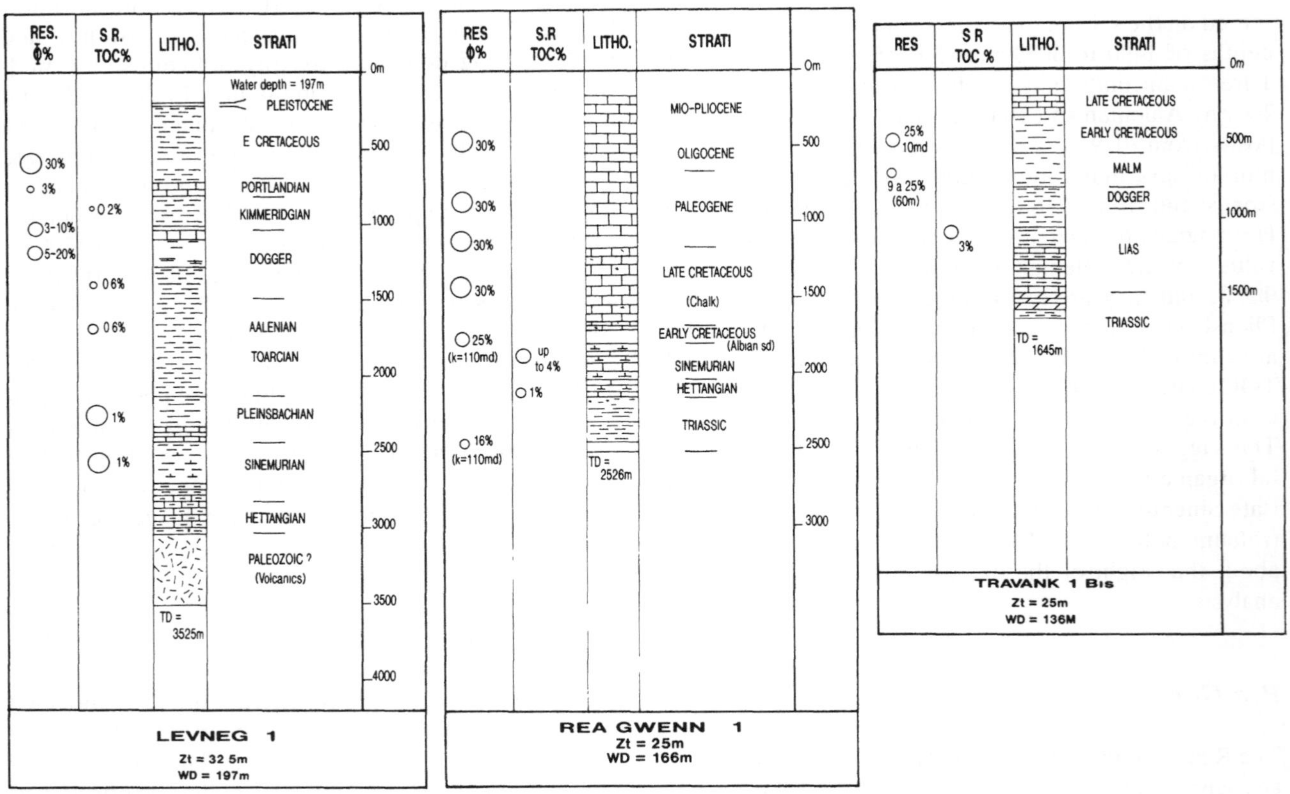

Fig. 7. Comparative stratigraphy of the Levneg, Rea Gwenn, and Travank wells

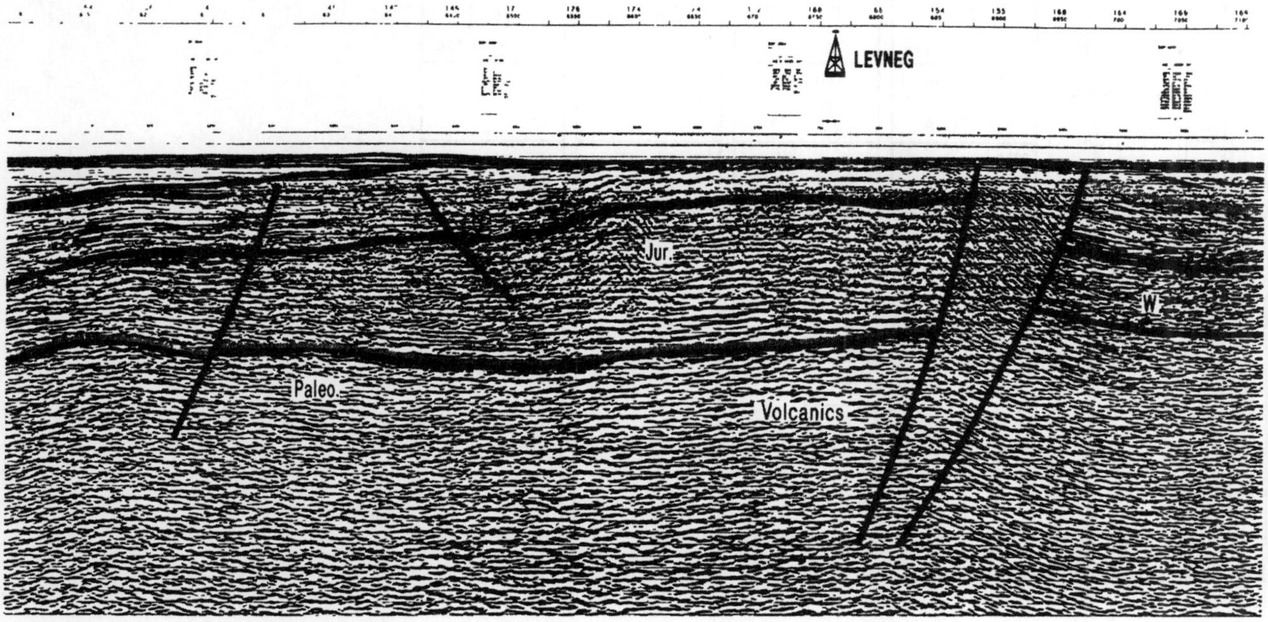

▲
Fig. 6 *(continued)* Legend see p 162

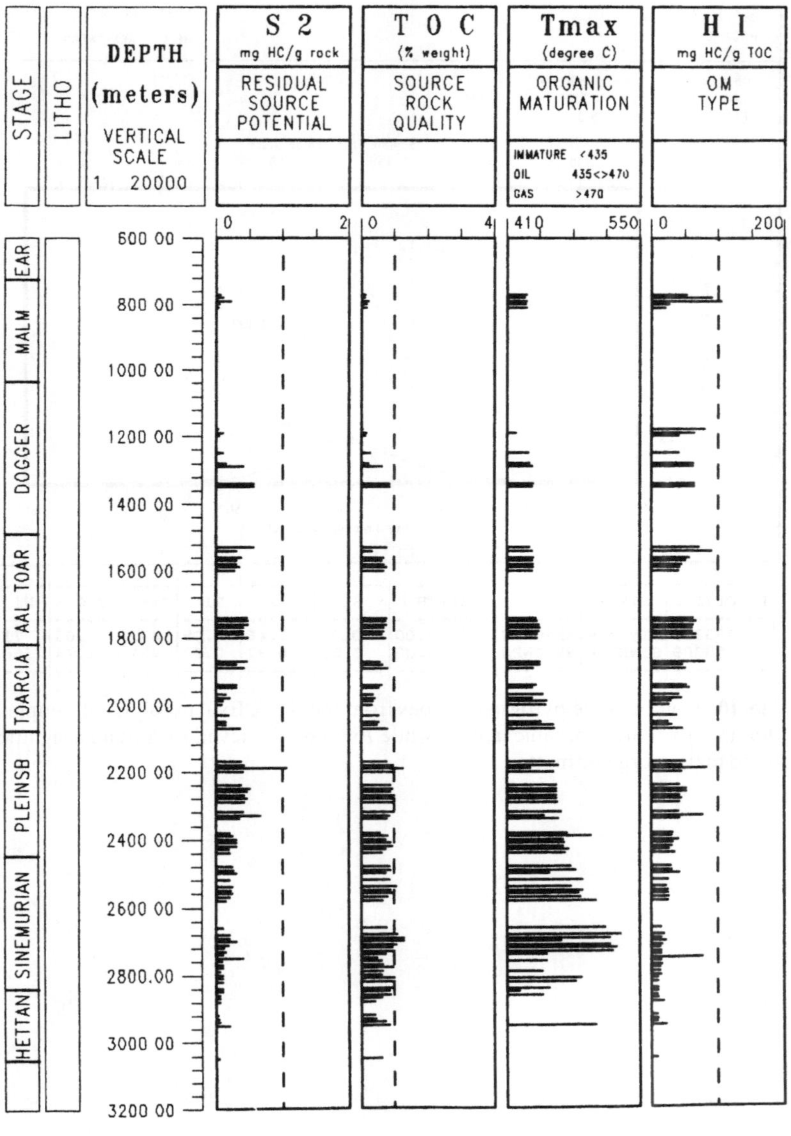

Fig. 8. Levneg well Rock Eval TOC results

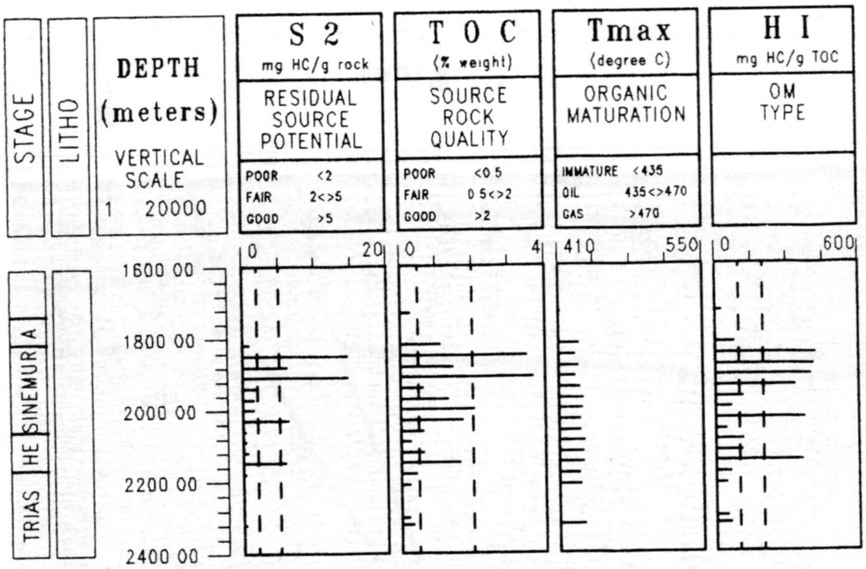

Fig. 9. Rea Gwenn well Rock Eval TOC results

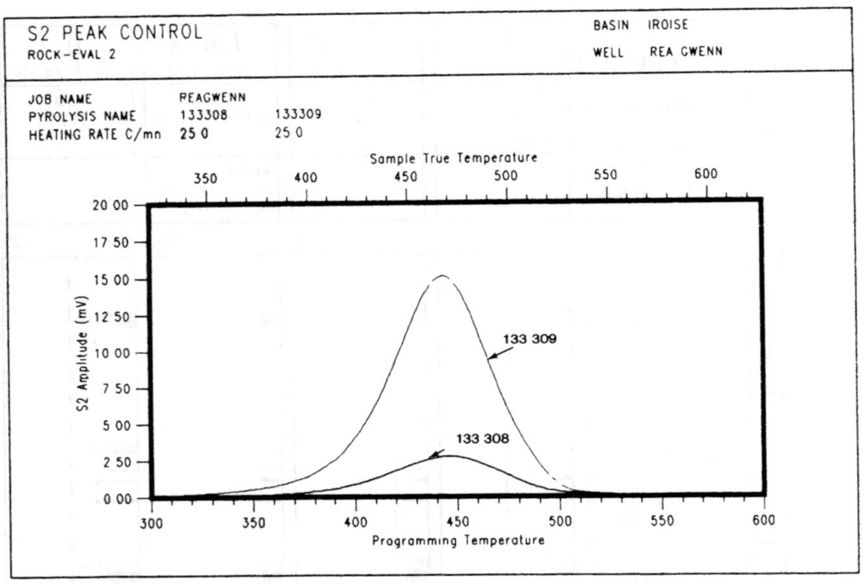

PYROLYSIS	STAGE	WELL	DEPTH -meters -	S1	S2	PI	Tmax	TOC	HI	OI
133308	SINEMUR	REAGWEN	2000.00	0.08	1.46	0.05	433	2.05	71	55
133309	SINEMUR	REAGWEN	2030.00	0.20	6.37	0.03	431	1.74	366	42

Fig. 10. Comparative pyrogram shapes from the Rock Eval pyrolysis of samples from the Rea Gwenn well *133309* is made of predominantly marine organic matter, while *133308* is believed to contain a significant amount of terrestrial organic matter and/or altered marine organic matter

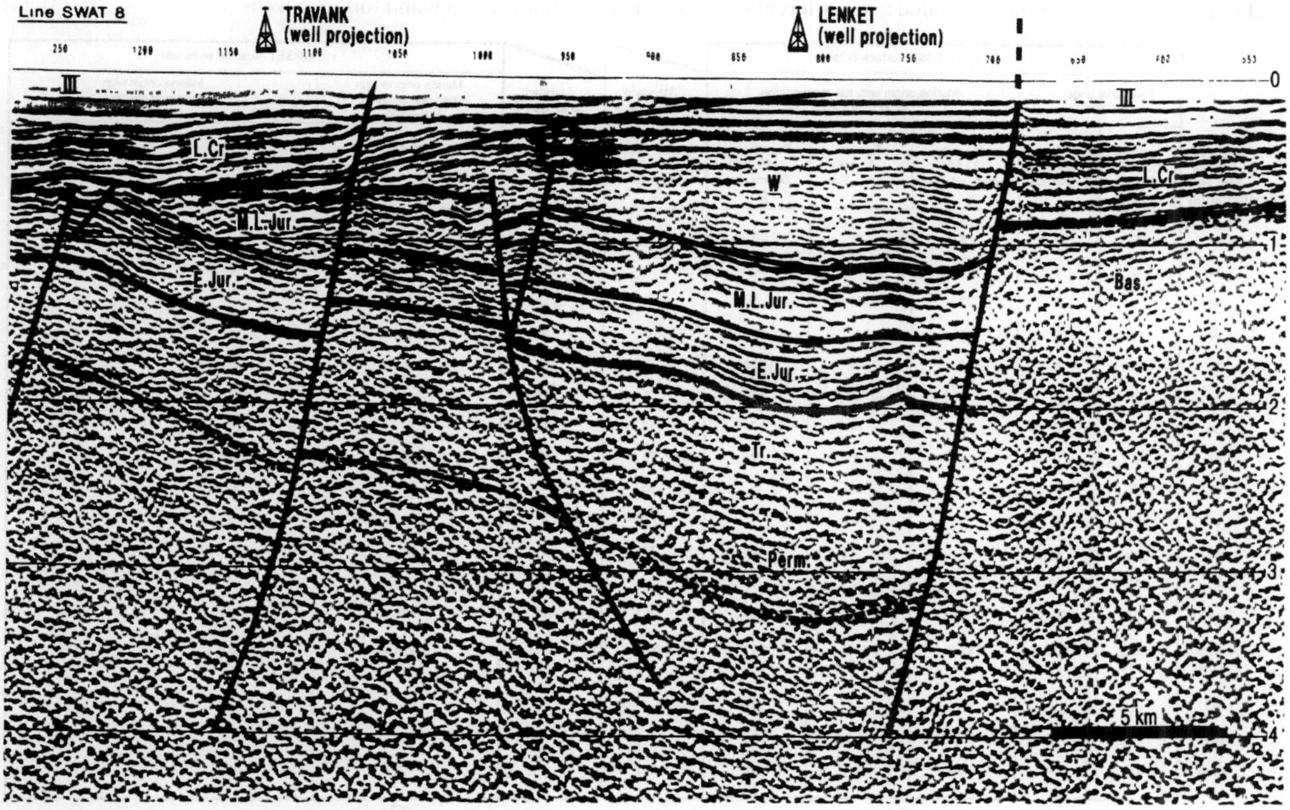

Fig. 11. SWAT 8 seismic section. *III* Tertiary, *L Cr* late Cretaceous and Albian, *W* Wealden and Aptian, *M L Jur* middle-late Jurassic, *E Jur* early Jurassic, *Tr* Trias, *Perm* Permian, *Bas* Paleozoic basement

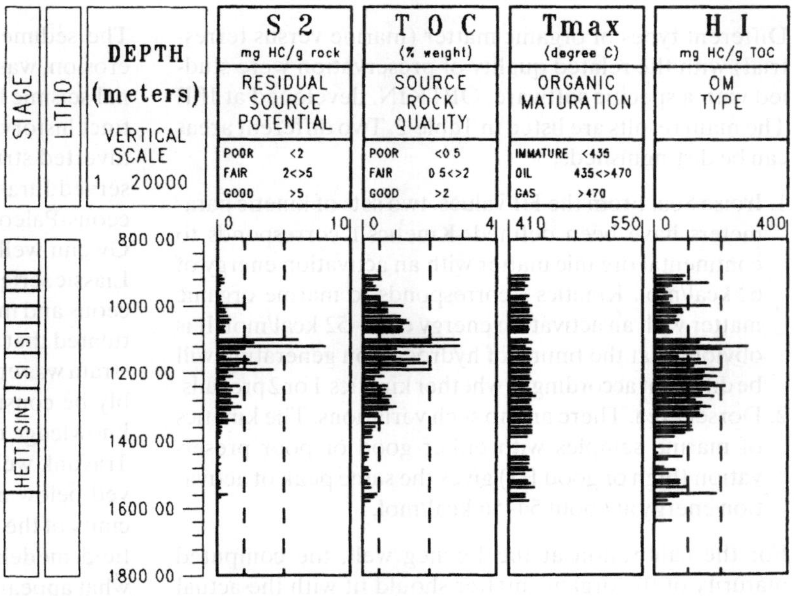

Fig. 12. Travank well Rock Eval TOC results

Table 2. Activation energies calculated from different samples from the Dorset outcrops and Ioise Sea wells

	IROISE (Subsurface-cuttings)							DORSET (outcrop samples)		
	Terrestrial origin HI ≈ 100		Marine origin with fair preservation (400 > HI > 200)			Standard O M Type 3	Standard O M type 2	Marine origine with bad preservation HI ≈ 100	Marine origin with good preservation (HI > 600)	
Activation Energie Kcal/mole	Levneg (132 665) Pleinsb	Rea-Gwenn (133 307) Sinemurian	Travank (133 066) Sinem Inf./Hett.	Travank (133 046) Sinem Sup	Rea-Gwenn (133 303) Sinemurian	North Sea Brent fn	Paris Basin Toarcian	125 911	125 905	125 906
Kinetic family	7	7	8	8	9	3	2	6	5	5
≤ 42	0	0	0	0	0	0		0	0	0
44	0	0	1	0	0	0	} 7	0	0	0
46	0	0	2	0	0	0		0	0	0
48	0	0	1	1	11	0	5	0	4	3
50	3	0	3	10	**252**	0	28	2	60	43
52	4	1	**93**	**125**	129	3	**273**	11	101	96
54	6	2	65	74	7	12	144	27	**245**	**223**
56	7	3	0	1	2	**75**	22	**46**	188	217
58	8	9	1	1	1	63	6	6	42	57
60	11	33	1	0	1	119	4	0	9	8
62	**14**	**43**	1	0	0	11	3	0	1	1
≥ 64	12	13	0	0	0	18		0	1	0
Total HI	65	104	168	212	403	201	492	106	652	648
Arrhenius Cte (/sec)	1 66E + 16	1 99 E + 16	4 95 E + 13	5 44 E + 13	1 53 E + 13	0 54 E + 15	0 90 E + 14	4 40 E + 14	3 28 E + 14	3 92 E + 14
S2	1 07	1 57	1 73	6 1	14 4	S2		1 26	34 2	30 3
TOC	1 28	1 33	1 02	2 9	3 5	TOC		1 24	5 4	4 7
Tmax	436	434	433	430	426	Tmax		419	420	414

□ (boxed values, shown in bold) Peak generation

Kinetic Parameters

Different types of organic matter (marine versus terrestrial) with the related quality of preservation were studied with a specific software, OPTKIN, developed at IFP. The main results are listed in Table 2. Two different areas can be distinguished:

1. Iroise Sea. From the HI values, two sets of kinetic parameters have been defined. Kinetics 1 corresponds to continental organic matter with an activation energy of 62 kcal/mol. Kinetics 2 corresponds to marine organic matter with an activation energy of 50–52 kcal/mol. It is obvious that the timing of hydrocarbon generation will be different according to whether kinetics 1 or 2 prevails.
2. Dorset area. There are no such variations. The kinetics of marine samples with either good or poor preservation (high or good HI) gives the same peak of activation energy at about 54–56 kcal/mol.

For the calibration at the Levneg well, the computed maturity of the organic matter should fit with the actual data. Two types of evolution were simulated: the first one with standard Type III organic matter (according to Rock-Eval procedure), and the second one with the actual Levneg organic matter. The best fit was obtained with the actual organic matter, for which hydrocarbon generation develops more slowly than for the standard Type III (although the present position of the top of the oil window remains the same (1780 m) in both cases).

Burial History

The sedimentation history, including the estimation of erosion, was deduced from the wells and available regional seismic lines. Each Iroise Sea well shows quite a distinct history. The Levneg well is located on top of a large inverted structure of late Paleogene age with thick preserved Jurassic series, but where at least 1400 m of Cretaceous-Paleogene sediments were eroded. The Rea-Gwenn well was drilled through thin and partly eroded Liassic series, unconformably overlain by thin late Cretaceous and moderately thick Tertiary strata. We have postulated that 500 m of middle Jurassic to early Cretaceous strata was eroded in Aptian times. This value must probably be considered as a minimum, as we have only poor knowledge on the actual thickness this interval had. The Travank well is located on a Jurassic tilted block preserved below a relatively thin Cretaceous cover. In the vicinity of the Dorset coast and south of the Wytch Farm oil field, modeling was performed on a virtual well located in what appear to have been the deepest part of the Channel Basin before the Tertiary inversion, as documented by the SWAT-11 profile (Bois et al. 1991).

Heat Flow History

A constant heat flow model using the present thermal data (obtained from downhole temperature measure-

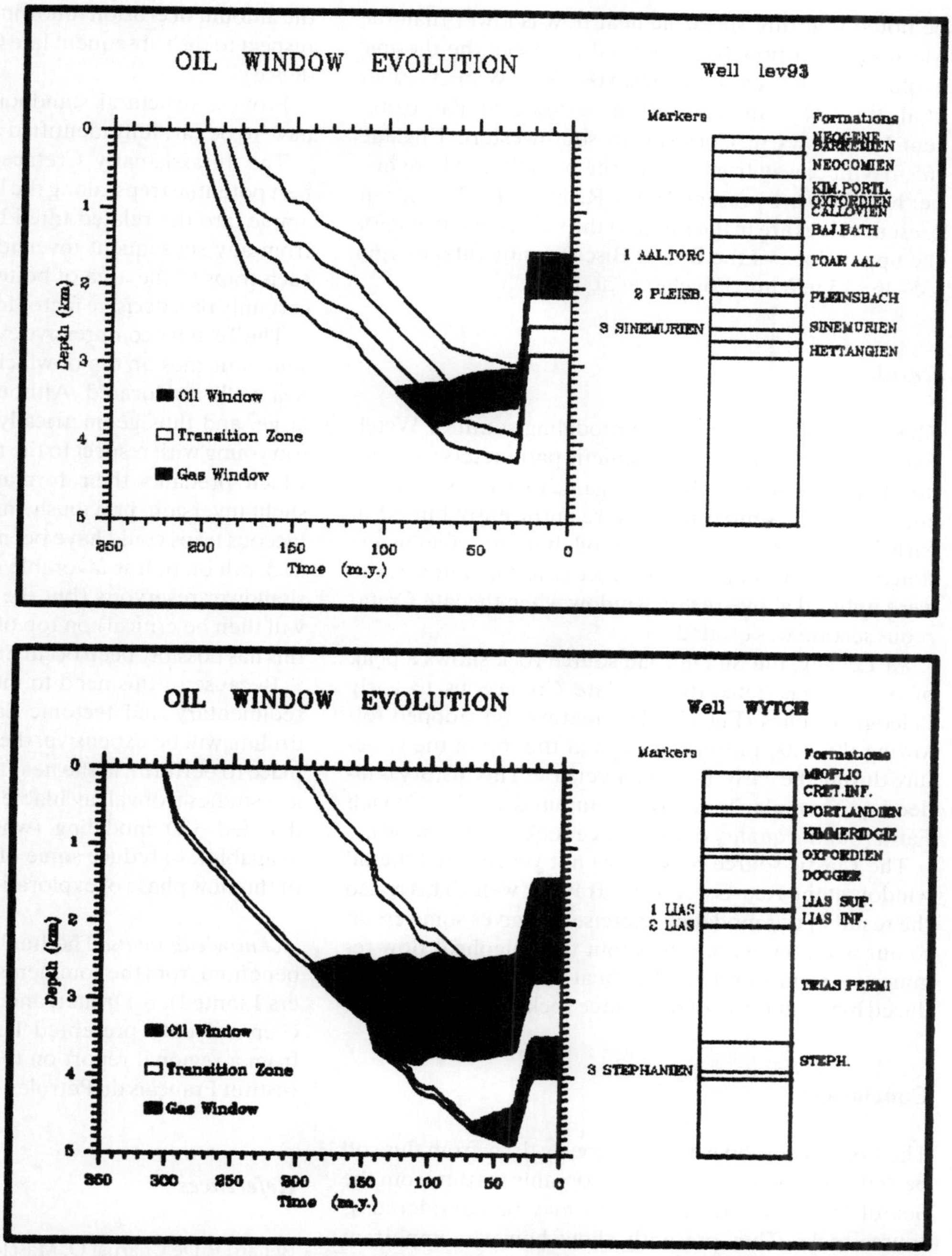

Fig. 13. Results of 1-D Matoil modeling at Levneg *(top)* and on a virtual well south of Wytch Farm *(bottom)*

ments in the wells) cannot explain the maturity observed for source rocks in a well such as Levneg. Higher heat flows thus probably prevailed in the past and have to be evaluated. This perilous exercise was tentatively done using the maturity data from Levneg. The subsidence history and, more particularly, the amount of erosion following the early Tertiary inversion were defined from the seismic data. Several heat flow histories can then be tested and the results of maturity computing can be compared with the actual data. Although a great number of compatible paleoheat-flow models can be produced, we finally favor one of them, i.e., a constant heat flow of 62 mW/m² in the Mesozoic followed by a gradual decrease to 41 mW/m² (present value according to calculations from downhole temperature measurements in a few Iroise Sea wells) in the Tertiary. A similar increase of the paleogeothermal gradient in Cretaceous times with respect to the present had already been postulated by Cornford et al. (1987) in an attempt to model the quantity of hydrocarbons generated in the vicinity of the Wytch Farm Oil Field. It is obvious that our choice is not related to any geodynamic model, but rather it should be taken as an equivalent paleoheat flow, making the modeling scenaros compatible with the actual maturity data. It should also be noted that this Mesozoic heat flow is lower than the Mesozoic heat flow recently used to model the thermic evolution of the near Paris Basin (80–85 mW/m^2, Gaulier et al., this Vol.) Our Mesozoic heat flow is similar to present heat flows in Cornwall in southwestern England (65 mW/m^2, away from the granite batholiths where higher heat flow are present; Willis-Richards 1993), and our present values are in the range of the values calculated for the upper slope of the Bay of Biscay continental margin (35–36±15 mW/m^2, Foucher et al. 1984)

Results

The result of the 1-D Matoil modeling south of Wytch Farm is shown in Fig. 13. The kinetic parameters were calibrated from marine Dorset organic matter. As expected, the Liassic source rocks were sufficiently buried in early Tertiary times to generate oil, but the broad inversion of the basin stopped this process, as the source rocks were uplifted above the oil window when the late Cretaceous section was eroded.

At Levneg, the Sinemurian source rock shows a peak of oil and gas generation in late Cretaceous to early Paleogene times (Fig. 13). The maturation stopped following the 1400-m-thick erosion at the top of the structure during the early Tertiary inversion. This history is indeed quite similar to the one computed south of Wytch Farm (but the quality of the source rock is not as good).

The Liassic source rocks have not yet reached the oil window at the Rea-Gwenn and Travank wells. This is also the result of our modeling exercise and gives some credit to our working hypothesis about the paleoheat flow regimes, as significantly higher heat flows would have induced hight maturity of the source rocks.

Conclusions

The first conclusion we can obviously draw from this still partial study of the early Liassic organic matter from the Sea of Iroise is that these rocks may be considered as source rocks. They presently show, however, important and quite rapid lateral changes in quantity and quality The thicker they are (such as at Levneg) does not necessarily mean the more attractive they are, as the kerogen could be, for a very significant proportion, of continental origin, and/or still of marine origine, but alterated. The initial TOC and petroleum potential are low as a result. In areas of moderate subsidence, i.e., away from the more subsiding troughs related to the early Jurassic rifting, the early Liassic interval is thinner, but the marine organic matter is only episodically (Travank and Rea-Gwenn wells, Dorset) diluted with continental material. As a result, the petroleum potential is good to excellent, especially in the late Sinemurian interval. The present diversity in maturity and timing of oil-gas generation is also very dependent on their burial history below the middle-upper Jurassic and lower Cretaceous sediments, and on the amount of erosion following the Aptian erosion (with respect to the subsequent late Cretaceous-Tertiary subsidence)

From a structural standpoint, two other events will have to be carefully identified on the seismic:

The Jurassic-early Cretaceous rifting events, as the best potential traps, along the lines of the Wytch Farm example, are the related tilted blocks or horsts preserved from any subsequent inversion. The close proximity of such traps to the area of better petroleum potential will certainly be a decisive factor for success

The Tertiary compressive event, with the related inversion anticlines on top of which more than half the Iroise Sea wells are located. Although such traps can be very large, and thus geometrically quite attractive, they are too young with respect to the main peak of oil generation which predates their formation. However, with only slight inversion, previously infilled Jurassic or early Cretaceous traps could have been partly or totally preserved at depth or, in less favorable cases, could escape toward shallower reservoirs (but the presence of effective seals will then be critical) on top of the inversion anticline (as this has possibly been occurring at Kimmeridge Bay).

Because of this need to integrate numerous complex sedimentary and tectonic data, and because any new drilling will be expensive, the Iroise Sea will be the right place to perform, in the near future, additional geochemical studies from all available wells, and in a second step, detailed 3-D modeling (when 3-D software becomes available), to reduce some of the uncertainties and risks of this new phase of exploration

Acknowledgments. The final version of this chapter has benefited from the comments of J. A. Miles (Oil Optimisers Limited), S. Drouet, and J. Espitalié (IFP). The geochemical data presented here are preliminary results from a regional report on the Sea of Iroise prepared by Institut Français du Pétrole.

References

Bénard F, De Charpal O, Mascle A, Trémolières P (1985) Mise en évidence d'une phase de serrage E-W au Crétacé inférieur en Europe de l'Ouest CRAS Paris 300, II, 15 765–768

Bois Ch, Gabriel O, Sibuet JC (eds) (1991) Les bassins sédimentaires de Mer Celtique et de Manche apport des profils seismiques SWAT Etude de la croûte terrestre par sismique profonde, Mer Celtique, Manche et ses Approches occidentales Profils SWAT et WAM Mém Soc Géol Fr 159

Butler M, Pullan CP (1990) Tertiary structues and hydrocarbon entrapments in the Weald Basin of southern England In Hardman RFP, Brooks J (eds) Tectonics events responsible for Britain's oil and gas reserves, Geol Soc Spec Publ 55 371–392

Chadwick RA (1993) Aspects of basin inversion in southern Britain J Geol Soc Lond 150 311–322

Colley MG, Mc Williams ASF, Myers RC (1981) Geology of the Kinsale Head Gas Field, Celtic Sea, Ireland In Petroleum geology of the Continental Shelf of North West Europe Institute of Petroleum, London, pp 504-510

Cornford Ch, Christie O, Endresen U, Jensen P, Mythr M-B (1987) Source-rock and seep oil maturity in Dorset, southern England Org Geochem 13, 1–3 399–409

Dranfield P, Begg SH, Carter RR (1987) Wytch Farm Oilfield Reservoir characterization of the Triassic Sherwood Sandstone for input to reservoir simulation studies In. Brooks J, Glenny KW (eds) Petroleum Geology of NW Europe Graham & Trotman, London, pp 149–160

Ebukanson EJ, Kinghorn RRF (1985) Kerogen facies in the major Jurassic mudrock formations of Southern England and the implication on the depositional environments of their precursors J Petrol Geol 8, 4· 435–462

Ebukanson EJ, Kinghorn RRF (1986) Maturity of organic matter in the Jurassic of Southern England and its relation to the burial history of the sediments J Petrol Geol 9, 3 259–280

Espitalié J, Deroo G, Marquis F (1985–1986) La pyrolyse Rock Eval et ses applications Rev IFP 40, 5 563–579, 6 755–784, 41, 1 73–89

Evans CDR (1990) The geology of the Western English Channel and its western approaches. BGS, U K offshore regional report, London, HMSO, 93 p

Foucher JP, Chenet PY, Montadert L, Roux JM (1984) Geothermal measurements during deep sea drilling project leg 80. In de Graciansky PCh, Poag CW et al (eds) Initial Reports of the Deep Sea Drilling Project volume LXXX, Washington U S Government Printing Office, pp 423–436

Gairaud H, Le Théoff B (1992) Geodynamic evolution of an intracratonic basin the Paris Basin EAPG Conf Paris, June 1982, Pap, pp 18–19

Hamblin RJO, Crosby A, Balson PS, Chadwick RA, Penn IE, Arthur MJ (1992) The geology of the English Channel. British Geological Survey, U K Offshore Regional Report, London, HMSO, 106 pp

House M (1989) Geology of the Dorset Coast Geologist' Association Guide, the Geologists' Association, London, 170 pp

Jenkins HC, Senior JM (1991) Geological evidence for intra-Jurassic faulting in the Wessex Basin and its margins J Geol Soc Lond 148, 2 245–260

Miles JA, Downes ChJ, Cook SE (1993) The fossile oil seep in Mupe Bay, Dorset· a myth investigated. Mar Pet Geol 10· 58–70

Penn IE, Chadwick RA, Holloway S, Roberts G, Pharaoh TC, Allsop JM, Hulbert AG, Burns IM (1987) Principal features of hydrocarbon prospectivity of the Wessex-Channel Basin. In Brooks J, Glenny KW (eds) Petroleum geology of NW Europe, Graham & Trotman, London, pp 109–118

Selley and Stoneley (1987) Petroleum habitat in South Dorset In Brooks J, Glenny KW (eds) Petroleum geology of NW Europe. Graham & Trotman, London, pp 139–148

Shannon PM (1991) Tectonic framework and petroleum potential of the Celtic Sea, Ireland. First Break 9 107–122

Stoneley R (1982) The structural development of the Wessex basin J Geol Soc 139; 4 545–554

Willis-Richards J (1993) Control on lithospheric strength, fracture opening and fluid movement in SW England In Parnell J, Ruffell AH, Moles NR Geofluids '93, Extended Abstr, pp 128–131

Aquitaine Basin, Bay of Biscay, Pyrenees

Deep Seismic Investigation in the Parentis Basin (Southwestern France)

C. Bois and O. Gariel

Abstract. The 300 km ECORS Bay of Biscay deep seismic profile was recorded to provide a better understanding of the development of the offshore part of the Parentis Basin. This basin appears in the seismic profile as a 100 km wide depression. Up to 15 km of Tertiary, Mesozoic, and possibly Paleozoic sediments rest on a very attenuated crust overlying a 15 km Moho uplift. After a gentle sagging of the basement in the Triassic-Jurassic, the basin was rifted in the early Cretaceous, essentially along the main Ibis Fault. Tectonic inversions in the early Albian and middle Eocene reactivated the faults and created anticlines and flexural depressions. After a lithospheric stretching in the early Cretaceous, further crustal thinning seems to have occurred beneath the basin through deep-seated processes involving metamorphism, magmatism, and/or delamination in the deep crust. Reflections occurring beneath the present Moho may be interpreted as the trace of a former lower crustal layering or a layered lower crust partly restored beneath the basin. Discrepancy between crustal stretching and attenuation and indications of crust-mantle interaction have also been found in deep seismic profiles shot across

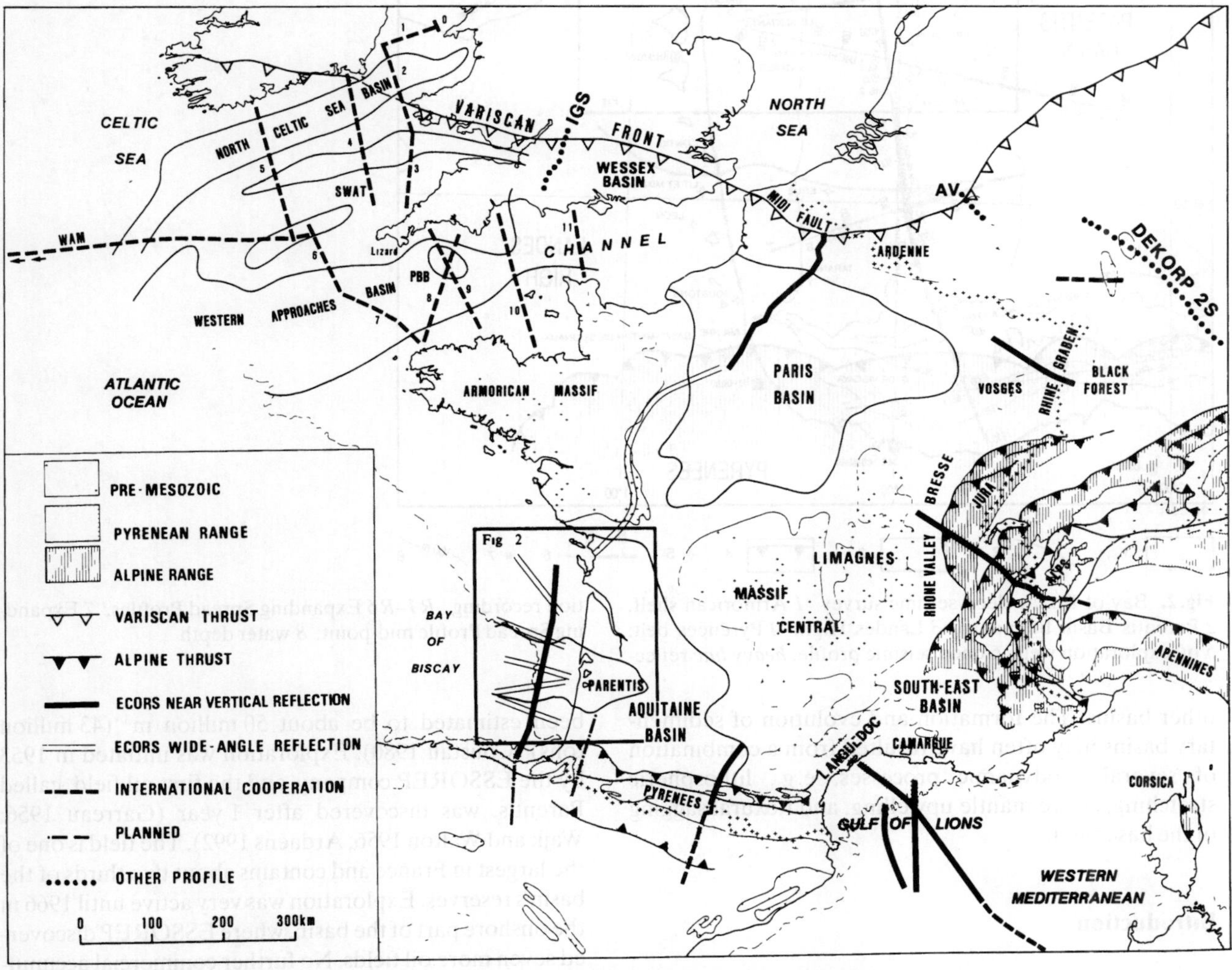

Fig. 1. Location map main structural features in France and adjacent areas and location of deep seismic profiles

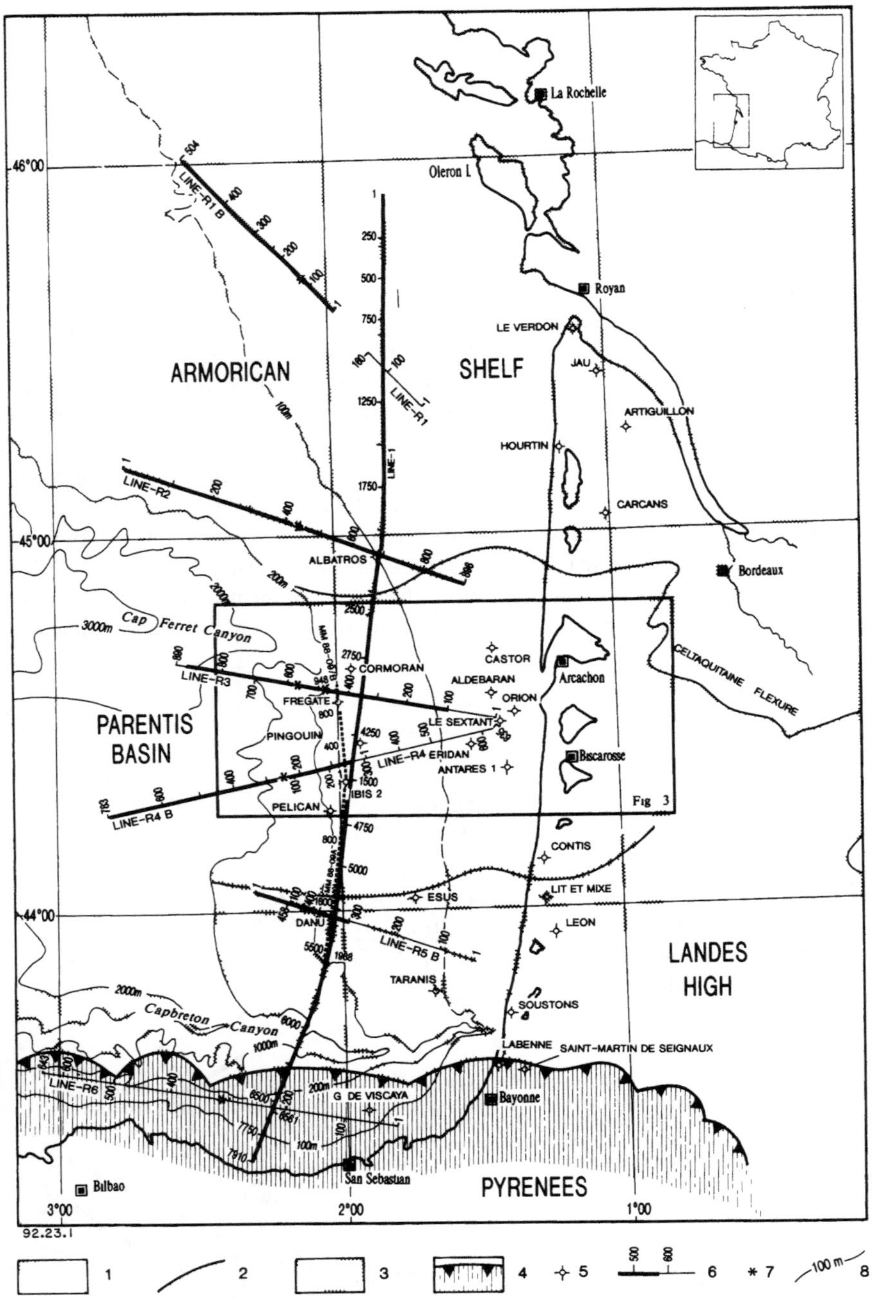

Fig. 2. Bay of Biscay deep seismic survey *1* Armorican shelf, *2* Parentis Basin boundary, *3* Landes High, *4* Pyrénées belt; *5* oil exploration well, *6* deep seismic profile, *heavy line* reflection recording, *R1–R6* Expanding Spread Profiles, *7* Expanding Spread Profile mid-point, *8* water depth

other basins. The formation and evolution of sedimentary basins may often have resulted from a combination of several geodynamic processes, e.g., lithospheric stretching, active mantle upwelling, and flexural sagging of the basement.

Introduction

The Parentis Basin, located in southwestern France (Fig. 1), is one of the main oil provinces in this country. The ultimate recovery of the oil pools discovered has been estimated to be about 50 million m^3 (43 million tons) (Coustau 1980). Exploration was initiated in 1953 by the ESSOREP company, and the first oil field, called Parentis, was discovered after 1 year (Garreau 1956; Wajk and Walton 1956; Ardaens 1992). The field is one of the largest in France and contains about two thirds of the basin's reserves. Exploration was very active until 1966 in the onshore part of the basin, where ESSOREP discovered seven more oil fields. No further commercial accumulation was discovered until 1991 in spite of a fairly active exploration pursued both onshore and offshore by four oil companies. Oil shows were tested offshore in only two

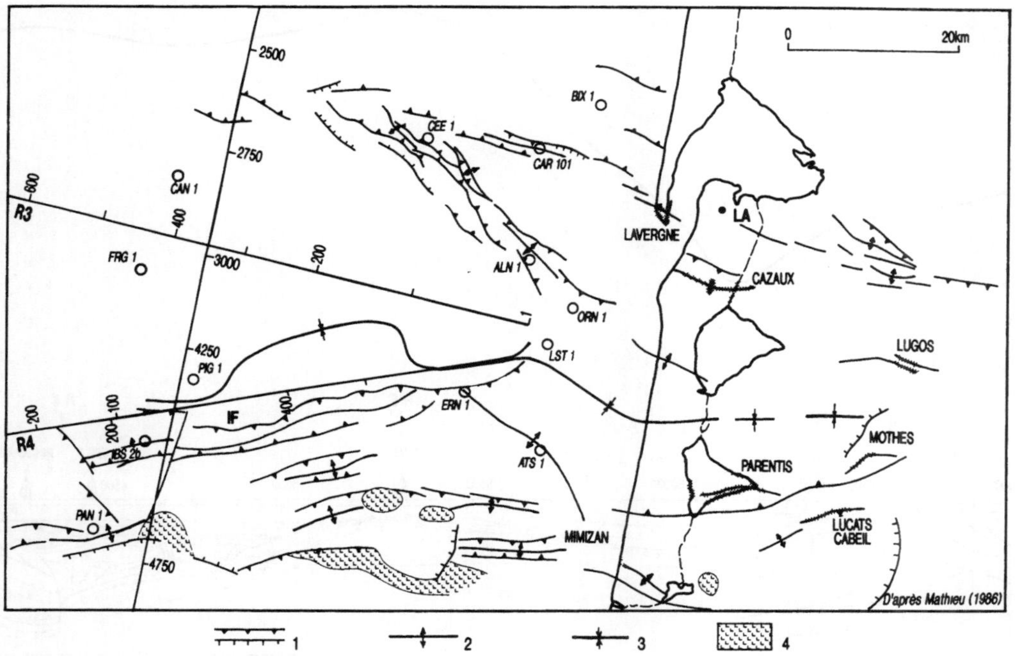

▲ **Fig. 3.** Main structural features of the Parentis Basin at the top of the Jurassic series (After Mathieu 1986) *1* Thrust and normal fault, *2* anticline, *3* syncline, *4* salt extrusion, *dark shade* oil field, portions of the Bay of Biscay profile and EPSs R3 and R4 are shown *IF* Ibis Fault, *LA* Les Arbousiers

structures (Ibis and Antares). A commercial accumulation was eventually discovered at the end of 1991 (Les Arbousiers), not far from old oil fields. To date, more than 130 exploration wells have been drilled, with 85% of them being located onshore.

A deep seismic investigation called the Bay of Biscay Survey was conducted in 1984 by the ECORS project with the co-operation of the REPSOL company (Pinet et al. 1987 a, b). The objective of this survey was to gain a better understanding of the formation of the offshore portion of the Parentis Basin. The purpose of this chapter is to show the main results of the survey and to draw tentative conclusions on the development of the basin.

Geologic Framework

The Parentis Basin covers about 7000 km^2 in Aquitaine and the adjacent continental shelf up to the 2000 m sea bottom contour (Fig. 2). The basin is bordered to the north and the south by the Armorican Shelf and the Landes High, respectively. These areas, where the basement is shallow, converge eastwards and make the basin a westward-opening embayment (Fig. 3), in the axial part of which up to 15 km of sediments may have been piled up. South of the Landes High, the basement deepens to 6000 m in the foreland basin of the Pyrénées thrust belt.

The thick sedimentary series of the Parentis Basin is essentially Mesozoic and Tertiary in age with a large part

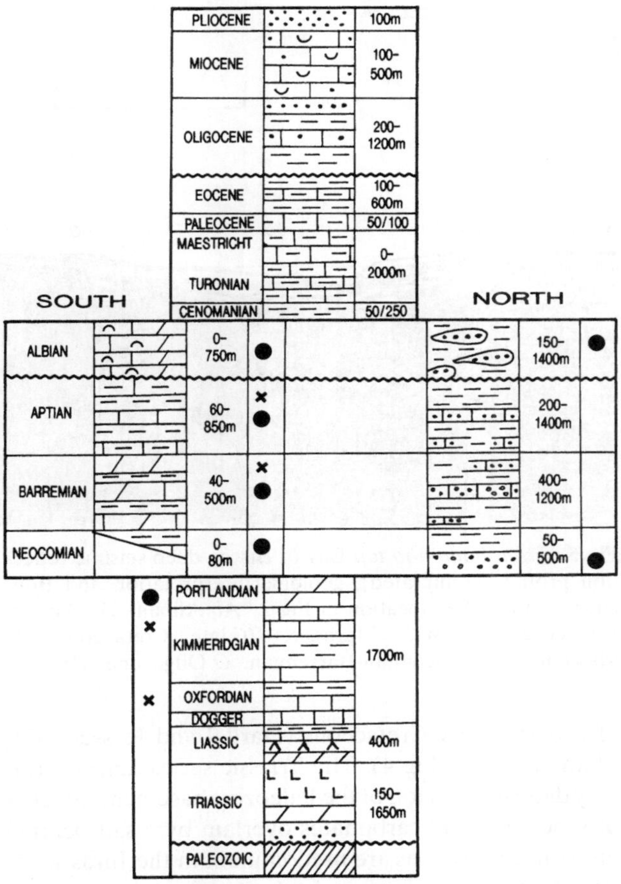

Fig. 4. Stratigraphic column of the Parentis Basin *Figures on the side of the lithological column* indicate thickness ranges (in m) *Black dot* Commercial oil reservoir; *X* potential source rock All discovered reserves were derived from late Jurassic source rocks

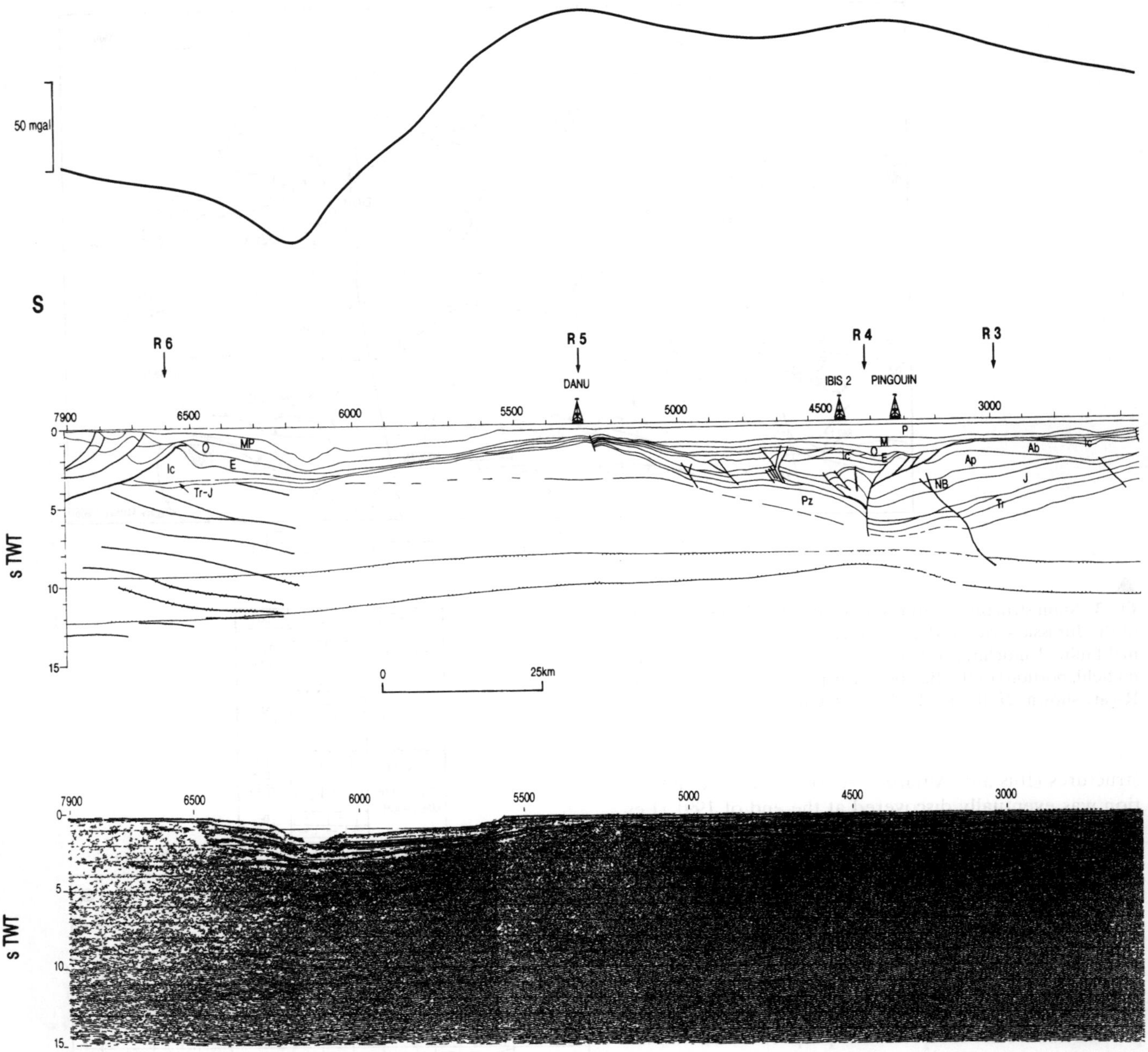

Fig. 5. *From bottom to top* Bay of Biscay deep seismic reflection profile (unmigrated), geologic interpretation and Bouguer anomaly See location in Fig 2 *Ab* Albian, *Ap* Aptian, *E* Paleocene-Eocene, *J* Jurassic; *IC* late Cretaceous, *M* Miocene, *NB* Neocomian-Barremian, *O* Oligocene, *PP* Plio-Quaternary, *Pz* Paleozoic sediment, *TR* Triassic, *Black* Triassic salt body, *shaded* high amplitude seismic layering in the lower crust, and in the upper crust in the northern part of the profile, *R 1–R 6* ESPs, *Danu to Albatros* oil exploration wells

of it being early Cretaceous (Dardel and Rosset 1971; Mathieu 1986) (Fig. 4). The Triassic series, unconformably deposited on top of the Paleozoic basement, consists of sandstone and carbonates overlain by a salt-bearing interval. Carbonates are predominent in the Jurassic series. Anhydrite occurs in the Lower Liassic series and marl in the Upper Jurassic series. The Lower Cretaceous series shows major differences in lithology between the northern and southern limbs of the basin. The Neocomian to Albian series is dominated in the south by platform carbonates with frequently dolomitized reefal bodies. Marl intercalations occur in the Baremian and Aptian. The Lower Cretaceous series passes northwards to shale intercalated with limestone and sandstone in the Barremian-Aptian formations and by sandstone and conglomerate in the Albian formations. A shallow marine and deltaic environment in the Barremian-Aptian and turbidites in the Albian controlled the deposition of clastic material derived from the Variscan basement of the Armorican massif. The Upper Cretaceous series consists of Cenomanian marl overlain by Turonian to Maestrichtian limestone. The Tertiary series includes Paleocene and Eocene limestone and marl, Oligocene marl and sandstone, Miocene marl and limestone, and Plio-

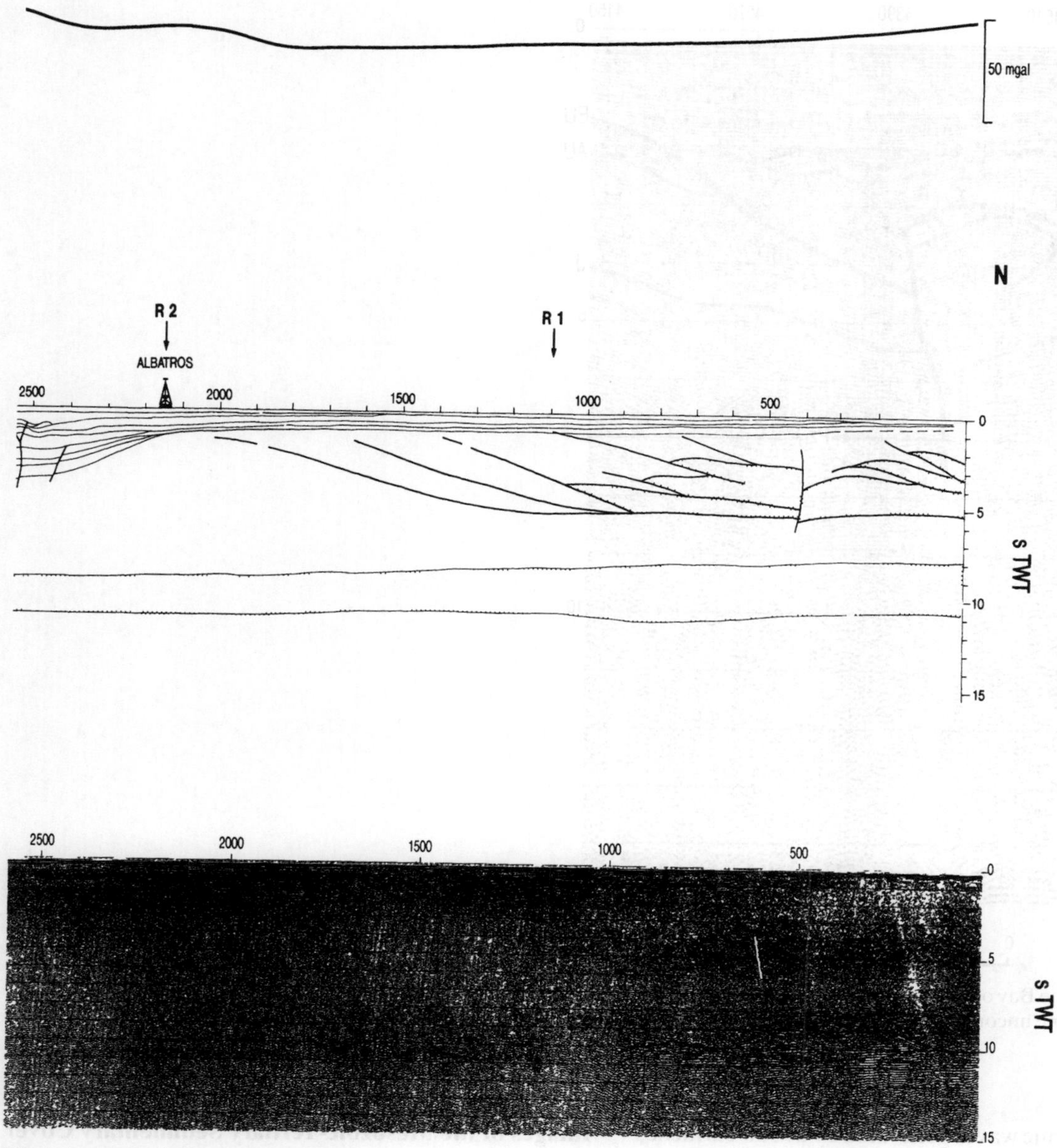

Fig. 5 *(continued)* Legend see p 176

Quaternary clastics. In addition to the basal Triassic unconformity, two major unconformities within the series bear witness to deformation events in the early Albian and middle Eocene. These deformations, together with the great subsidence, moved the Triassic salt whose extruded olistoliths were found in the Eocene series (Mathieu 1986).

Oil reservoirs are late Jurassic to Albian carbonates in the basin's southern limb, with the best production coming from vuggy and fractured dolomite intervals related to karsts (Fig. 2) (Séronie-Vivien et al. 1965; Choquin 1985; Martin 1986; Ardaens 1992). In the central part of the basin and its northern limb, reservoirs are Purbeckian, early Aptian and mainly Albian sandstone and conglomerates (Séronie-Vivien et al. 1965). The source rocks of the commercial accumulations are late Jurassic marl and carbonates (Espitalié and Drouet 1992).

The Bay of Biscay Deep Seismic Survey

The backbone of the ECORS of Biscay Survey was a deep seismic reflection profile running north-south over 300 km approximately parallel to the Aquitaine coast

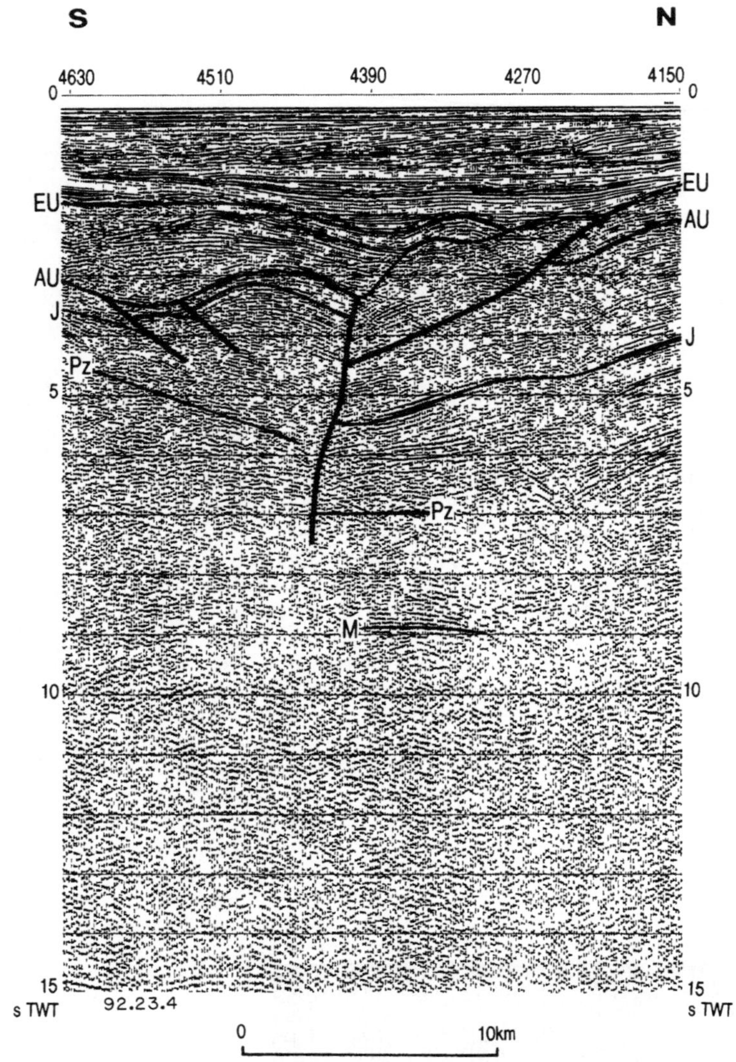

Fig. 6. Portion of the Bay of Biscay deep seismic profile (unmigrated) in the Ibis area See location in Figs 1 and 3 *AU* Albian unconformity, *EU* Eocene unconformity, *J* top Jurassic, *M* Moho, *Pz* possible top Paleozoic, *heavy line* fault

(Fig. 2) The seismic waves generated by a 8536 cu. inches airgun source were recorded by the shooting vessel down to 15 s and by a second vessel at a constant distance of 7.5 km (see Pinet et al. 1987 a, b, for acquisition parameters). In comparison to the vertical seismic reflection profile, the wide-aperture data show a notable enhancement of deep reflection continuity and amplitude. The greatly contributed to locating the bottom of the crust (Moho) Six expanding-spread profiles (ESP), with a maximum offset of 100 km, were also acquired at intervals along the near vertical reflection profile (Fig. 2) They provided the velocity distribution in the sediments and the basement, and established the location of the Moho characterized by a high-amplitude wide-angle reflection (Marillier et al. 1988). Vertical seismic reflections were also recorded along a large part of the ESPs.

Images of the Mesozoic-Tertiary Sedimentary Cover

The deep seismic reflection survey was designed to obtain a great penetration in the crust. Although the resolution is poorer than that of regular commercial seismic sections, the main features of the sedimentary series can be observed (Figs. 5 and 6). Formation boundaries were determined from exploration wells and a dense network of industrial seismic profiles. Four structural domains were crossed from north to south: the Armorican Shelf, the Parentis Basin, the Landes High, and the Cantabrian foreland in front of the Pyrénées belt (Figs. 2 and 5).

The Paleozoic basement forms a gently south-dipping monocline beneath the Armorican Shelf. Exposed some 60 km north of the end of the profile, basement was penetrated at 1800 m in the Albatros well on the northern hinge zone of the Parentis Basin The Tertiary and Upper Cretaceous series overlie a thin, eroded, Triassic-Jurassic series.

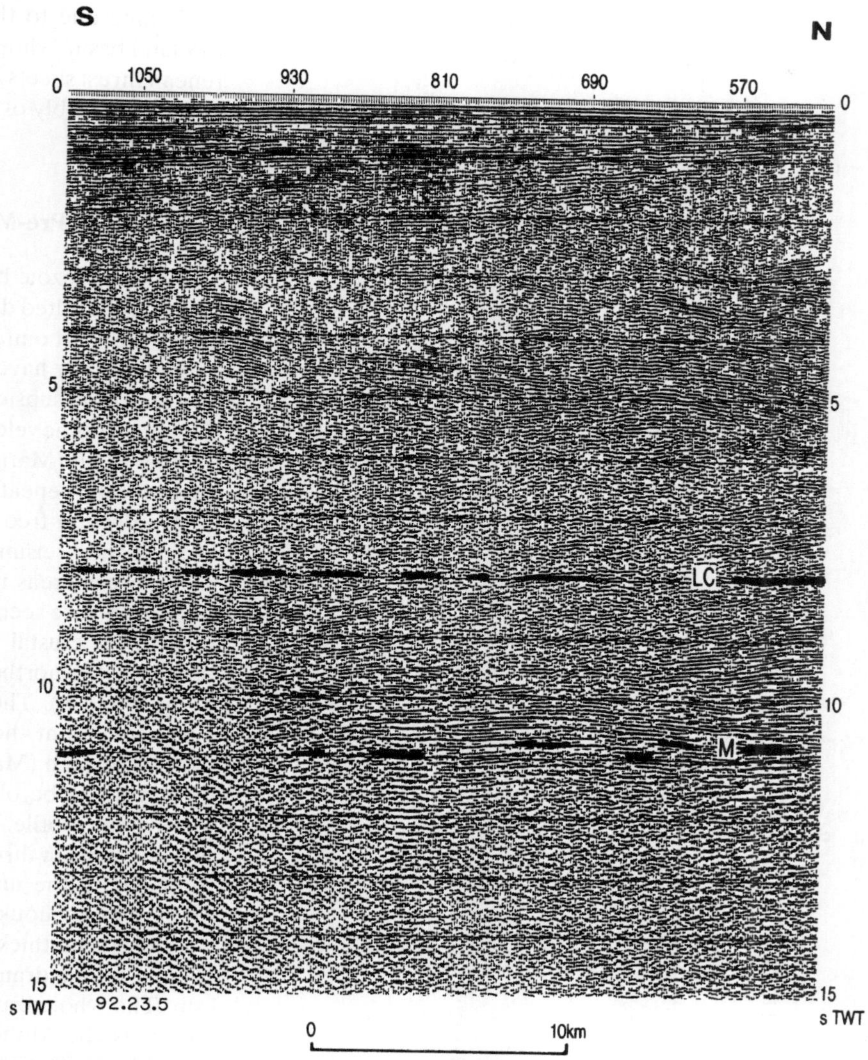

Fig. 7. Portion of the Bay of Biscay deep seismic profile (unmigrated) recorded on the Armorican shelf. See location in Figs. 2 and 5. *M* Moho, *LC* top of the lower crustal high-amplitude reflections

In the Parentis Basin, wide lateral variations in subsidence followed by tectonic inversions resulted in great lateral changes in the thickness of most of the series (Figs. 5 and 6). The Middle and Late Jurassic series is represented by low-frequency layered reflections bounded by reflections with higher amplitude that approximately correspond to the tops of the Liassic and Portlandian series. Another intervening high-amplitude marker represents Kimmeridgian marls. Weaker and shorter reflections beneath the Liassic top have been interpreted as Liassic and Triassic formations. A few areas with chaotic reflections have been tentatively interpreted as Triassic salt bodies.

The Lower Cretaceous series shows a thick set of reflections with lesser amplitude than in the Jurassic. The Barremian-Aptian boundary, determined from boreholes, can be fairly accurately traced. However, the early Albian unconformity is not always identifiable when it occurs between rocks of similar composition. Beneath this unconformity, the Aptian, Barremian-Neocomian, and even the Jurassic series were locally eroded, especially towards the basin edges. Above the unconformity, the Albian series is very irregularly distributed.

The Upper Cretaceous series conformably overlies the Albian series within the basin, but unconformably overlies Jurassic or Triassic rocks on the basin's rims. The Paleocene-Eocene series is imaged by high-amplitude reflections on the edges of the basin. It contains chaotic reflections regarded as Triassic salt bodies in the central part of the basin. The younger formations unconformably infill the remaining depression and largely overlap the basin's shoulders. High-frequency reflections outline channel and progradation deposits in this interval.

South of the Parentis Basin, the Landes High is overlain by a series that is thinner than 1500 m and consists of Triassic rocks without salt, Cretaceous to Oligocene carbonate and Plio-Quaternary clastics. Farther south, the Upper Cretaceous-Oligocene series shows a remarkable

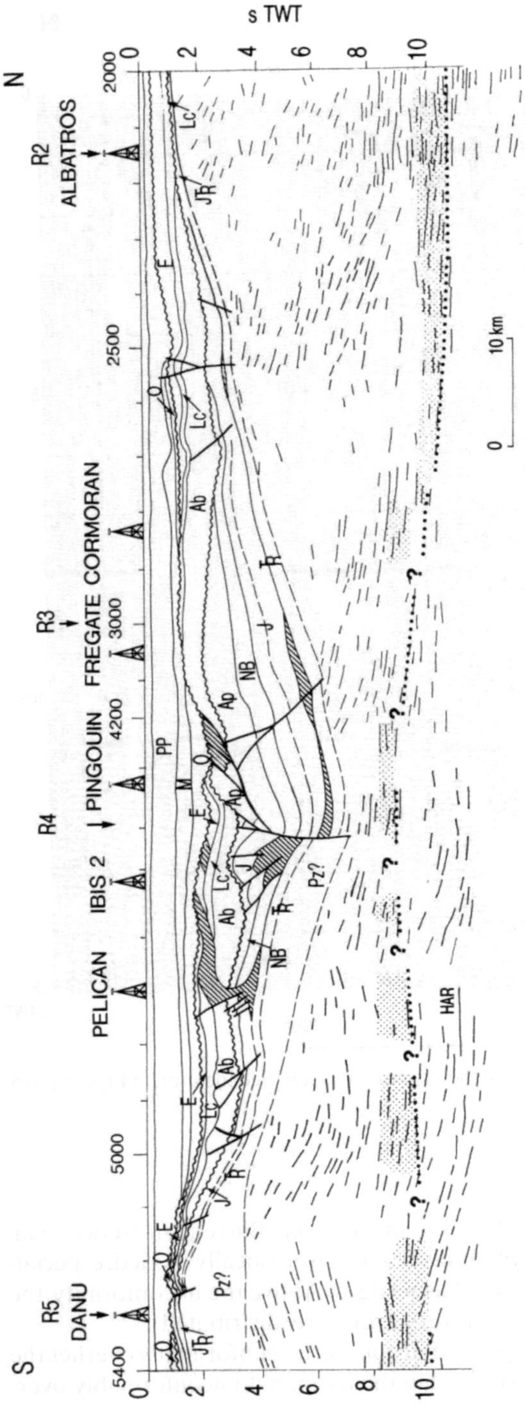

Fig. 8. Interpreted line drawing of the Bay of Biscay deep seismic profile across the Parentis Basin See location in Fig 2 Legends as in Fig 5 Salt bodies *oblique ruling*, *HAR* high amplitude reflection

thickening due to the development of the Cantabrian foreland basin, whose southern part was involved in Pyrenean thrust sheets (Fig 5) A thin Triassic-Liassic series is unconformably overlain by late Cretaceous marls and Eocene flysch

Image of the Pre-Mesozoic Basement

The pre-Mesozoic basement underlying the region was folded and faulted during the Variscan orogeny completed in the Carboniferous. Late Carboniferous-Permian sediments may have been deposited before the onset of the Mesozoic subsidence. This view is supported by the 5 2 km/s seismic velocity found down to 8 km beneath the Landes High (Marillier et al. 1988) and reflections that occur locally beneath the Mesozoic cover.

A reflection-free upper crust, a reflective layered lower crust and a seismically transparent upper mantle, similar to other areas in Europe (e.g., Blundell 1990; Bois et al. 1990) are seen in the Bay of Biscay profile (Fig. 7). The lower crustal seismic layering can clearly be observed in the northern and southern parts of the profile (Figs. 7 and 8). The Moho determined from ESPs has been located at the bottom of the layering, i.e., about 30–34 km deep (Marillier et al. 1988). North-dipping reflections can be observed at shallower depths at both ends of the profile. They have been interpreted as the image of Variscan thrust sheets, since the structures are not reflected in the unconformable Mesozoic cover (Pinet et al. 1987 a; Choukroune et al. 1990).

The crustal thickness and fabrics, as well as the Moho depth, show a dramatic change beneath the Parentis Basin. The whole crust is attenuated to less than 5–6 km, whereas the Moho is uplifted by 15 km and forms a roughly symmetrical bulge whose top is located beneath the deepest part of the basin (Marillier et al. 1988). Mantle upwelling is reflected by a gentle gravity high, despite the presence of a thick sedimentary pile (Lalaut et al. 1981; Marillier et al. 1988) (Fig. 5). Intrusion of dense material suggested by the high crustal velocity (6.3–6.7 km/s) beneath the basin (Marillier et al 1988) may also have played a part in the creation of the gravity high. The layered lower crust becomes very thin beneath the basin and disappears in places. The Moho reflection is faint. The Moho has been tied to the ESPs; however, subhorizontal and north-dipping reflections occur down to 11 s TWT beneath the basin and a high-amplitude marker occurs beneath the southern limb of the basin (Figs 8 and 10)

Structural Features

The depression of the Paleozoic basement beneath the Parentis Basin is about 100 km wide and 15 km deep (Figs. 5 and 8). A number of faults and anticlines can be observed in the sedimentary fill. Due to tectonic deformations, the structural setting of the Jurassic, Albian, and Upper Tertiary series are different.

The main structural feature affecting the Jurassic series is the Ibis Fault which appears to be more or less vertical with a displacement of 1.4 s TWT (about 3.5 km) (Figs. 5, 6, and 8). In the Lower Cretaceous series, the upper part of the fault shows a southerly dip and a reverse configuration. A south-verging overturned anticline with a probable salt core occurs south of the fault in the footwall block. Other faults of much less importance occur in several places. They are generally north-dipping with normal or reverse setting. The faults in the Pelican area are reverse and associated with a small structure with a probable salt core linked to a large salt extrusion (Fig. 8).

The Albian series occurs north and south of the Ibis area in two depressions (Figs. 5 and 8). A thrust anticline associated with salt extrusions affects the Pingouin area and oversteps a deeper fault. An anticlinal structure with a large salt extrusion occurs in the late Cretaceous-Eocene of the Pelican area. The Oligocene series gently drapes the underlying series deformed in the Albian and/or Eocene. It shows a broad swell above the southern limb of the basin and a terrace above its northern limb.

The prevalent northerly dip of the faults may have been inherited from the crustal Variscan fabrics. However, the shape of the faults cannot be properly determined deeper than the Jurassic series from the seismic profile. The only fault that clearly affects both the sediments and the basement is the Ibis Fault. Basement faults are suggested by reflection shifts. These faults cannot be clearly outlined and, in any case, they may be decoupled along salt layers from the faults in the sediments. The Moho and the layered lower crust are not clearly affected by any fault. The Ibis Fault may be rooted in the lower crust.

Besides the Parentis Basin, main structures can be observed in front of the Pyrénées belt. The basement was flexured in the Cantabrian foreland under the weight of the Pyrenean thrust sheets. The thrusts show increasing dip from north to south and may be rooted in a detachment located at the bottom of the sedimentary series beyond the end of the profile.

Geologic History

Although the early Albian and Eocene deformations in the Parentis basin were gentle, a reconstruction of the Mesozoic-Tertiary evolution of the basin is faced with three difficulties: (1) unknown amounts of rock were eroded after each compressional events, (2) Triassic salt bodies were deformed and displaced to an unknown extent during the geologic history, and (3) the very steep Ibis Fault may have undergone strike-slip movement making 2-D restoration questionable. The reconstruction in Fig. 9 has been prepared assuming that erosions were of small extent and material transfer outside of the section plane can be neglected.

Subsidence started as early as Triassic times with a gentle sagging of the basement. The irregular bottom of the series may have been caused by salt pillows or incipient diapirs. Salt rising in the Ibis area probably explains the thinning of the series towards that area (Fig. 6). However, the Ibis or other faults do not appear to have been very active during Jurassic times. Earlier fault-controlled rifting may have occurred, but this is not readily suggested by the seismic image. The subsidence rate increased in the late Triassic-early Liassic and the late Jurassic (Brunet 1993). Fault-controlled rifting essentially occurred in the early Cretaceous with the tilting of the northern limb of the basin along the Ibis Fault (Fig. 9). However, the southern limb was also somehow tilted basinwards.

The basin was uplifted and eroded in the early Albian, and its central part was gently deformed (Fig. 9). The Ibis Fault, which may have been a north-dipping normal fault during the early Cretaceous, was steepened and tilted northwards in the early Albian, with the formation of a subordinate northward thrust. Sinistral strike-slip movement possibly occurred. A normal setting can still be observed in the Jurassic series, whereas a reverse setting appears in the thick overlying Aptian series (Figs. 6 and 8). The cause of the Albian compression is controversal (Choukroune and Mattauer 1978; Bénard et al. 1985; Mathieu 1986; Brunet 1993). The orientation of the structures (Mathieu 1986) suggests a northeasterly compression (Fig. 3). The Albian-Upper Cretaceous series infilled flexural depressions created to the north and the south of the Ibis area. On the whole, the Parentis basinal area is characterized by a thick Lower Cretaceous series (Neocomian to Albian) that disappears outside of the basin (Fig. 9). In the late Cretaceous, a flexural foredeep developed in front of the Pyrénées.

The Pyrenean compression, which was active from the late Cretaceous to the middle Eocene with a N to N20° orientation, created a second deformation in the Parentis Basin. The Albian depocenters were shaped in broad wavy swells, and former faults and anticlines were reactivated (Fig. 9). The Ibis Fault may have undergone further sinistral strike-slipping. Salt extrusion and olistoliths were set in place in relation to the main faults. The Pyrenean compression produced a prominent shortening south of the Cantabrian foredeep and the stacking of thrust sheets (Fig. 5).

Formation of the Parentis Basin

The main purpose of the Bay of Biscay deep seismic experiment was to shed some light on the formation of the Parentis Basin. Several geodynamic models have been envisaged.

Pull-Apart Model

An overall ESE left-lateral shearing has often been postulated as having affected the whole Aquitaine lithosphere in the Albian-late Cretaceous and even Aptian times, and created the Parentis Basin by pull-apart (Choukroune and Mattauer 1978; Mathieu 1986). This deformation was associated with the well-known left-la-

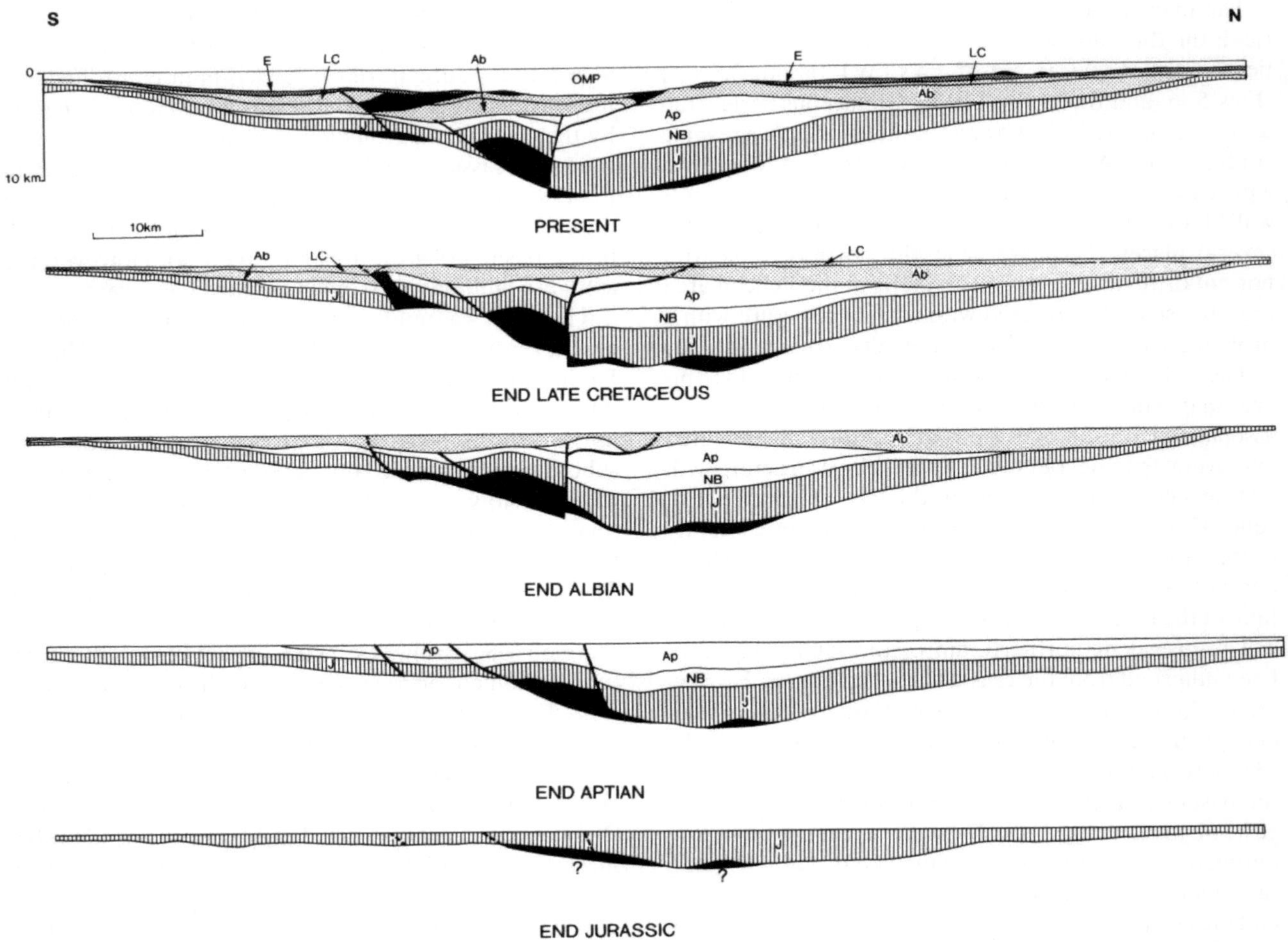

Fig. 9. Reconstruction of the Parentis Basin cross section through time Legends as in Fig 5

teral displacement of the Iberian plate along the North Pyrenean fault. However, the Bay of Biscay profile does not show any main strike-slip fault having controlled subsidence on the edges of the basin. The only possible strike-slip fault in the north (SP 2530) is quite minor (Fig. 8).

The Ibis Fault actually is the only major fault having controlled Cretaceous subsidence. From its geometry, it might be regarded as a flower structure. However, it is N 80°-oriented (Fig. 3) and should have been a normal fault during the above left-lateral deformation. To behave as a strike-slip fault, compression should have been northerly oriented, an orientation that is hardly supported by the regional strain pattern before Albian times. Therefore, we have preferred to interpret the Ibis Fault as an originally north-dipping normal fault which was steepened by both later deformations. Subordinate sinistral strike-slipping probably took place along the fault because of the NE- and N-oriented compressions related to these deformations; but this strike-slipping did not create the Parentis Basin, that was actually initiated at an earlier time The greater part of the sedimentary series was deposited from the Triassic till the Aptian when extension was approximately north-south-oriented and induced fairly straightforward stretching (Mathieu 1986) with possible subordinate dextral strike-slipping along the Ibis Fault

Lithospheric Stretching

Stretching in the Parentis Basin was mainly controlled by the Ibis Fault which forms the southern boundary of a broad half graben. However, this fault disappears laterally (Dardel and Rosset 1971; Mathieu 1986; Bois et al. 1994) suggesting that decoupling occurred during the rifting events along Triassic salt layers. Basement extension may not be directly related to the faults observed in the sediments

No clear low-angle detachment can be observed across the crust beneath the basin. The geometry of the basin and the underlying Moho uplift, and the distribution of the depocenters argue against the stretching models derived from the simple shear concept (e.g., Wernicke 1981)

Crustal attenuation affected both the upper and lower crust just as in a uniform extension model (McKenzie 1978) However, fault analysis has led to an average extension factor of 1.2 at the Jurassic level, whereas the crustal attenuation implies an average stretching factor of 1.6, with a maximum of 6 beneath the center of the basin (Pinet et al. 1987c; Pinet and Colletta 1990). Despite uncertainty as to extension estimate related to undetected faults, tectonic inversion, or oblique orientation of the profile with respect to the strain, we have assumed that the discrepancy between stretching and attenuation of the crust is real (Bois et al. 1990).

This discrepancy may have been accommodated by a major decoupling of the sedimentary cover along salt layers from its basement that was denuded on the basin's rims (Brunet 1994). Such a denudation might explain the lack of Jurassic rocks on the Landes High, where the Triassic is unconformably overlain by late Cretaceous carbonates; but the early Cretaceous rift series is also absent, and the missing formations may never have been deposited or have been eroded after the Albian deformation. On the northern rim of the basin, Jurassic rocks are present beneath the Lower Cretaceous series, and the denudation concept cannot be accepted

Crustal thinning may have been enhanced by creep of the deep crust; but such a creep should have been compensated for either by deep crustal thickening beneath the rims of the basin (Moretti and Pinet 1987) or by deep crustal denudation somewhere Available data fail to confirm this view.

Finally, we believe that the prominent crustal attenuation that occurs beneath the Parentis basin cannot be explained solely by mechanical stretching of the lithosphere Active mantle upwelling should also have occurred.

Deep-Seated Processes

Active rifting associated with a mantle plume implies an initial surface doming that can be evidenced by basement stripping beneath the rift sediments. There is no evidence to support such doming in the study area during the late Jurassic to the early Cretaceous (Mathieu 1986) Therefore, the early Cretaceous rifting was most probably initiated by passive lithospheric stretching. The discrepancy between stretching and attenuation of the crust suggests that the crust was further thinned by other processes such as magmatism, metamorphism, differentiation, and/or delamination. The high seismic velocity recorded in ESP R4 (Marillier et al. 1988) indicates that the crustal composition was deeply modified beneath the central part of the basin. Besides, the basinward tilt of the southern limb of the basin cannot be explained by the normal faults visible in the seismic section, and might be related to this late crustal attenuation.

Traces of the crustal attenuation might be present in the deep seismic section beneath the basin. A number of layered reflections occur down to 11 s TWT well beneath the present Moho determined from ESPs (Figs. 8 and 10). They are generally weak, but a higher amplitude one can be seen beneath the southern side of the mantle upwelling. This reflection approximately parallel to the basement top was first interpreted as the Moho (Pinet et al. 1987a, b). We have tentitatively interpreted it as a relict of a former Moho, whereas the overlying reflections may be relicts of a former layered lower crust. All these reflections are intersected by the present shallower Moho at an angle (Figs. 8 and 10).

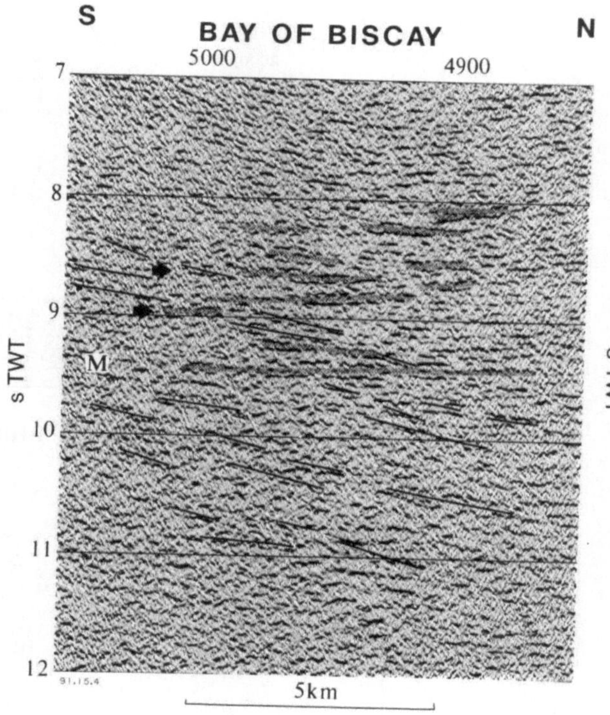

Fig. 10. Portion of the Bay of Biscay deep seismic profile See location in Figs 2 and 5 (Bois 1992) Gently dipping reflections in the mantle are intersected by the present Moho *(M)* and lower crustal layering *(arrows)* The Moho location has been derived from interpolation between ESPs R4 and R5 (Fig 8)

Comparison Between the Parentis Basin and Other European Rifts and Basins

Deep seismic profiles shot across Tertiary rifts such as the Rhine Graben also show a Moho bulge beneath the rift, resulting in a discrepancy between stretching and attenuation of the crust (Villemin et al. 1986; Bois 1992) (Fig. 11).

A similar discrepancy has also been observed in deep seismic profiles shot across Triassic-Jurassic basins such as the North Celtic Sea Basin (Bois et al. 1990; Dyment et al. 1990) (Fig. 11). This basin resulted in the Triassic-Jurassic from a gentle sagging of the basement with minor faulting. At present, a main difference between the two basins lies in the image of the layered lower crust. This unit, which is greatly attenuated beneath the Parentis Basin, does not show any attenuation beneath the North Celtic Sea Basin. It is probable that the layered lower crust was also attenuated in Triassic-Jurassic times beneath the North Celtic Sea Basin; but it was subsequently restored to its original thickness. The mode of

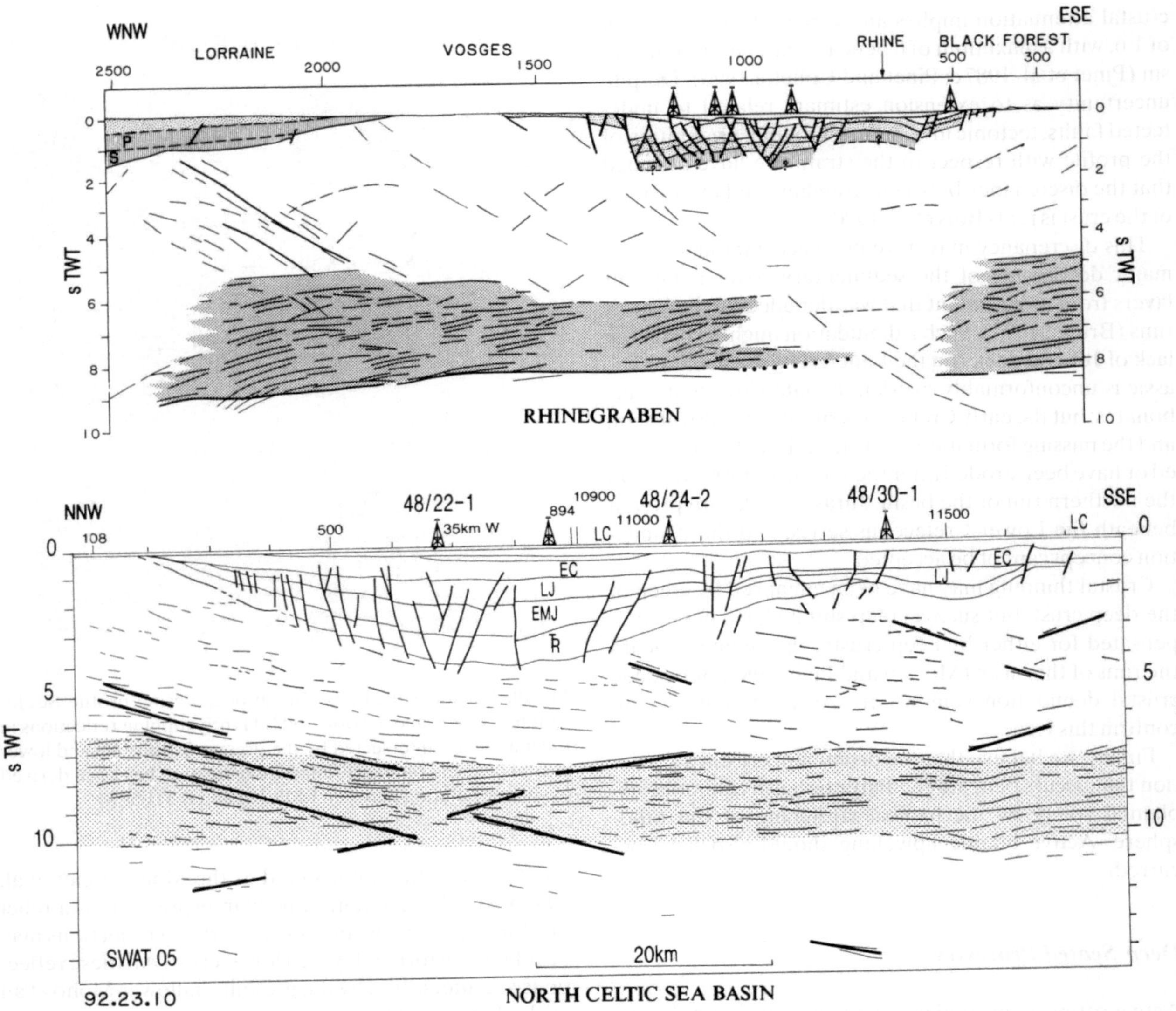

Fig. 11. Interpreted line drawing of deep seismic profiles across basins See location in Fig 1 *Above* Rhine Graben southern profile (Bois 1992) *1* Tertiary, *2* Mesozoic, *3* Paleozoic sediment, *4* layered lower crust, *5* main fault, *6* borehole, *dashed line* seismic diffraction *Below* SWAT 5 profile across the North Celtic Sea Basin (Bois 1992). *EC* early Cretaceous, *EMJ* early-middle Jurassic, *LC* late Cretaceous, *LJ* late Jurassic, *TR* Triassic, *shaded* layered lower crust, *heavy lines* major dipping reflections interpreted as traces of Paleozoic or older structures, oil exploration wells drilled near the profile have been shown

restoration has been debated: a new layered lower crust may have been emplaced by underplating or differentiation of the upper mantle (Bois et al. 1990; Dyment et al. 1990). Therefore, part of the reflections observed beneath the Parentis Basin's Moho might be regarded as the forerunners of the layered lower crust's restoration.

Conclusion

In the ECORS Bay of Biscay deep seismic section, the Parentis Basin appears to be a deep trough containing 15 km of Mesozoic and Tertiary sediments lying on a very attenuated crust overlying a prominent Moho upwelling The formation and evolution of the basin may be summarized as follows:

1. A gentle sagging of the basement occurred in the Triassic-Jurassic without any clear faulting. Subsidence increased greatly in the late Jurassic
2. A moderate lithospheric stretching in the early Cretaceous resulted in rifting essentially controlled by the Ibis Fault. This fault may have been rooted in the lower crust attenuated by ductile stretching Other faults may have controlled basement extension, but they were probably decoupled from the faults affecting the sedimentary infilling.
3. The crust is believed to have been further attenuated by active upwelling, as indicated by the discrepancy between the crustal attenuation and the stretching observed in the basin Lithospheric cooling controlled subsequent subsidence.

4. Tectonic deformation affected the basin in the early Albian and middle Eocene. Most of the faults kept a normal configuration at the Jurassic level, whereas anticlines and flexural depressions were created in Cretaceous layers and strike-sliping affected some faults.
5. The layered lower crust, which is prominent on both sides of the basin, was possibly partly restored beneath the central part of the basin as suggested by deep reflections occurring beneath the present Moho.

Discrepancies between the crustal attenuation and the stretching observed in the basin have been found in other basins and may indicate that deep-seated processes often contributed to basement subsidence. Traces of crust-mantle interaction may have been imaged by deep seismic profiles such as the ECORS Bay of Biscay line across the Parentis Basin. According to our interpretation, the formation and evolution of the Parentis Basin, and most probably of other basins, resulted from a combination of several processes, e.g., lithospheric stretching, active mantle upwelling, and flexural sagging.

Acknowledgments The ECORS deep seismic project is managed by the Institut Français du Pétrole, the Institut National des Sciences de l'Univers (CNRS), Elf Aquitaine, and the Institut Français pour la Recherche et l'Exploitation de la Mer. The detailed results of the ECORS Bay of Biscay deep seismic investigation including a copy of the seismic sections are being published in a Memoir by the Société Géologique de France (Bois et al. 1994). We thank L. Pages and an anonymous reviewer for their helpful comments.

References

Ardaens R (1992) Le champ de Parentis Réservoir carbonaté complexe et paléokarst Pétrole Tech Paris 372 34–42 (in French)

Bénard F, de Charpal A, Mascle A, Trémolières P (1985) Mise en évidence d'une phase de serrage Est-Ouest au Crétacé Inférieur en Europe de l'Ouest C R Acad Sci Paris II, 15 765–768 (in French)

Blundell DJ (1990) Seismic images of continental lithosphere J Geol Soc Lond 147 895–913

Bois C (1992) The evolution of the layered lower crust and the Moho through geological time in Western Europe contribution of deep seismic reflection profiles Terra Nova 4 99–108

Bois C (1993) Initiation and evolution of the Oligo-Miocene rift basins of southwestern Europe contribution of deep seismic reflection profiling Tectonophysics 226 227–252

Bois C, Gariel O, Pinet B (1990) ECORS deep seismic surveys across Paleozoic and Mesozoic basins in France and adjacent areas a review In Pinet B, Bois C (eds) The potential of deep seismic profiling for hydrocarbon exploration Technip, Paris, pp 381–401

Bois C, Pinet B, Gariel O (1994) The sedimentary cover along the ECORS Bay of Biscay deep seismic reflection profile A comparison between the Parentis basin and other European rifts and basins. In Bois C et al (eds) Deep seismic study of the Earth's crust The Bay of Biscay survey Mém Soc Géol Fr (in press)

Brunet MF (1994) Subsidence along the ECORS Bay of Biscay deep seismic profile In· Bois C et al (eds) Deep seismic study of the Earth's crust The Bay of Biscay survey Mém Soc Geol Fr (in press)

Choquin A (1985) La seconde vie d'un grand gisement Société Esso de recherche et d'exploitation pétrolière Ann Mines, Paris, March–April 1985 25–32 (in French)

Choukroune P, Mattauer M (1978) Tectonique des plaques et Pyrénées sur le fonctionnement de la faille transformante nord-pyrénéenne, une comparison avec les modèles actuels Bull Soc Géol Fr (7), XX 689–700 (in French)

Choukroune P, Pinet B, Roure F, Cazes M (1990) Major Hercynian thrust along the ECORS Pyrenees and Biscay lines Bull Soc Géol Fr (8), VI 313–320

Coustau H (1980) Les ressources en hydrocarbures du Bassin aquitain In Autran A, Dercourt J (eds) Evolutions géophysiques de la France Int Geol Congr Paris, Coll C7, Mém BRGM Orléans 107 331–334 (in French)

Dardel RA, Rosset R (1971) Histoire géologique et structurale du Basin de Parentis et son prolongement en mer In Debyser J et al (eds) Histoire structurale du Golfe de Gascogne Publ Inst Fr Pétrole, 22 Technip, Paris 1 IV.2 1–IV2 23 (in French)

Dyment J, Sibuet JC, Pinet B (1990) Deep structure of the Celtic Sea a discussion on the formation of basins Tectonophysics 173 435–444

Espitalié J, Drouet S (1992) Petroleum Generation and Accumulation in the Aquitaine Basin (France) In Spencer AM (ed) Generation, accumulation and production of Europe's hydrocarbons II Special publication of the European Association of Petroleum Geoscientists, 2 Springer, Berlin Heidelberg New York, pp 127–150

Garreau B (1956) Le gisement de Parentis Symp Sobre Y Yacimentos de Petrolio y Gas Int Geol Congr, Mexico 8 267–274 (in French)

Lalaut P, Sibuet JC, Williams C (1981) Présentation d'une carte gravimétrique de l'Atlantique Nord-Est C R Acad Sci Paris 292 597–600 (in French)

Marillier F, Tomassino A, Patriat P, Pinet B (1988) Deep structure of the Aquitaine shelf constraints from expanding spread profiles on the ECORS Bay of Biscay transect Mar Pet Geol 5, 1 65–74

Martin B (1986) Le renouveau du champ de Lugos Pétrole Tech Paris 327 20–23 (in French)

Mathieu C (1986) Histoire géologique du sous-bassin de Parentis Bull Cent Rech Elf Aquitaine (Prod) Pau 10 33–47 (in French)

McKenzie D (1978) Some remarks on the development of sedimentary basins Earth Planet Sci Lett 40 25–32

Moretti I, Pinet B (1987) Discrepancy between lower and upper crustal thinning In Beaumont C, Tankard AJ (eds) Sedimentary basins and basin-forming mechanisms Can Soc Petrol Geol Mem 12 233–239

Pinet B, Colletta B (1990) Probing into extensional sedimentary basins comparison of recent data and derivation of tentative models Tectonophysics 123 11–31

Pinet B, Montadert L, Curnelle R, Cazes M, Marillier F, Rolet J, Tomassino A, Galdeano A, Patriat P, Brunet MF, Olivet JL, Schaming M, Lefort JP, Arietta A, Riaza L (1987a) Crustal thinning in the Aquitaine shelf, Bay of Biscay, from deep seismic data Nature 325 513–516

Pinet B, Montadert L, ECORS Scientific Party (1987b) Deep seismic reflection and refraction profiling along the Aquitaine shelf (Bay of Biscay) Geophys JR Astron Soc 89 305–312

Pinet B, Montadert L, Mascle A, Cazes M, Bois C (1987c) New insights on the structure and the formation of sedimentary basins from deep seismic profiling in Western Europe In

Brooks J. Glennie KW (eds) Petroleum geology of North West Europe Graham & Trotman, London, 1, pp 11–31

Séronie-Vivien R, Sens J, Malmoutier G (1965) Contribution à l'étude des formations du Crétacé inférieur dans le Bassin de parentis (Aquitaine) Colloque sur le Crétacé inférieur, Lyon (1963) Bur Rech Geol Min, Mém 34 669–692 (in French)

Villemin T, Alvarez F, Angelier J (1986) The Rhine Graben extension, subsidence and shoulder uplift Tectonophysics 128 47–59

Wajk R, Walton G (1956) Geophysical history of the Parentis oil field, France Geophysics 21, 3 815–827

Wernicke B (1981) Low-angle normal faults in the Basin and Range province nappe tectonics in an extending orogen Nature 291 645–647

Subsidence in the Parentis Basin (Aquitaine, France): Implications of the Thermal Evolution

M. F. Brunet

Abstract. Subsidence analysis of the Aquitaine basin, using data of oil wells and seismic lines, allows the determination of some points whose history is representative of the sub-basins' evolution, with a multiphase rifting from the Paleozoic until the early Cretaceous. The deep crustal structure of the Parentis basin derived from ECORS survey data is interpreted in terms of crustal thinning history and lithospheric thermal evolution. It has been numerically modeled with a homogeneous thinning model, until a good fit with data observed The modeling gives an estimation of the maximum horizontal extension value along the section (around 40 km), when the crustal volume is preserved. Results of subsidence modeling, with the constraint of the present crustal thickness, show an excess of subsidence of 700 m at present, with regard to observed subsidence. Two processes may add their effects to contribute to the shallow observed subsidence: support of the basin by a nonzero flexural rigidity or a modification of the thermal regime. In this case, the lithosphere should be thinner than that predicted by a normal cooling; it would imply a supplementary thermal event or a process preventing the lithosphere to return to its initial thickness by cooling after the last rifting phase. The thermal modeling evolution is used for an estimation of conditions favorable to oil generation in the middle of the Parentis basin.

Introduction

The Aquitaine basin has been studied intensively because of its potential as a petroleum source region. This basin is divided into several basins: Parentis, Adour, Mirande, southern trough (Fig. 1); but these basins do not have a regular trough shape, they are themselves composed of several small sub-basins. The maximum accumulation of sediments in the middle of the Parentis and Adour basins is about 10–15 km. Recently, a program of deep seismic survey, ECORS Bay of Biscay (Pinet et al. 1987 a, b; Bois et al. 1993), was led across the marine part of the Parentis basin especially along a 300-km NS cross section. Oil wells, oil companies' seismic lines, and gravimetric data were compiled and added to the seismic reflection and refraction results of the ECORS cruise.

Lithospheric thinning is the main mechanism involved in the initiation and evolution of sedimentary basins established on continental crust in a divergent context. The lithospheric thinning leads to two phases of basement subsidence, "initial" during the rifting and "thermal" posteriously. During both phases, subsidence is amplified by isostatic compensation for sediment loading. The initial subsidence is due to the replacement by thinning of light continental-crust material with denser upper mantle material, passively upwelled. The thermal subsidence is caused by the cooling and thermal contraction of the hot mantle material. In fact, initial and thermal subsidences would be well separated if rifting was instantaneous As rifting phases have finite durations, a part of the heat is dissipated already during the rifting phase, which implies a superimposition of a portion of the thermal subsidence upon the initial subsidence.

The subsidence history of rifted basins or margins is recorded in the sediments deposited upon them. In the Aquitaine basin, a multiphase history of the subsidence has been generated by several phases of rifting. Numerical modeling of the lithospheric thinning can predict the evolution of subsidence and thermicity with the control of present data: lithostratigraphy and faciology, geophysical data such as crustal thickness and density, gravimetry, and thermicity.

Examples of Subsidence Curves in Some Aquitaine Sub-Basins

The observed subsidence registered in the sedimentary columns has two components: subsidence caused by the isostatic response to sediments and water loading, and tectonic subsidence due to thinning of continental crust and subsequent thermal cooling that would have occurred even without sediment loading. To examine the tectonic subsidence, the sediment-loading effects must have been removed from the total subsidence observed by the "backstripping" method. This technique was introduced by Perrier and Quiblier (1974) and improved by Steckler and Watts (1978). It was used to construct subsidence curves (Brunet 1989) with a local isostatic compensation after unloading the basement from the sediment effect. The local equilibrium (flexural rigidity = 0) is the easiest to implement, its use may be justified during the rifting phases, when the rigidity is supposed to be low, and in the middle of a basin. The regional compensation (nonzero flexural rigidity) implies knowing the history of the entire basin and making hypotheses on the value of the flexural

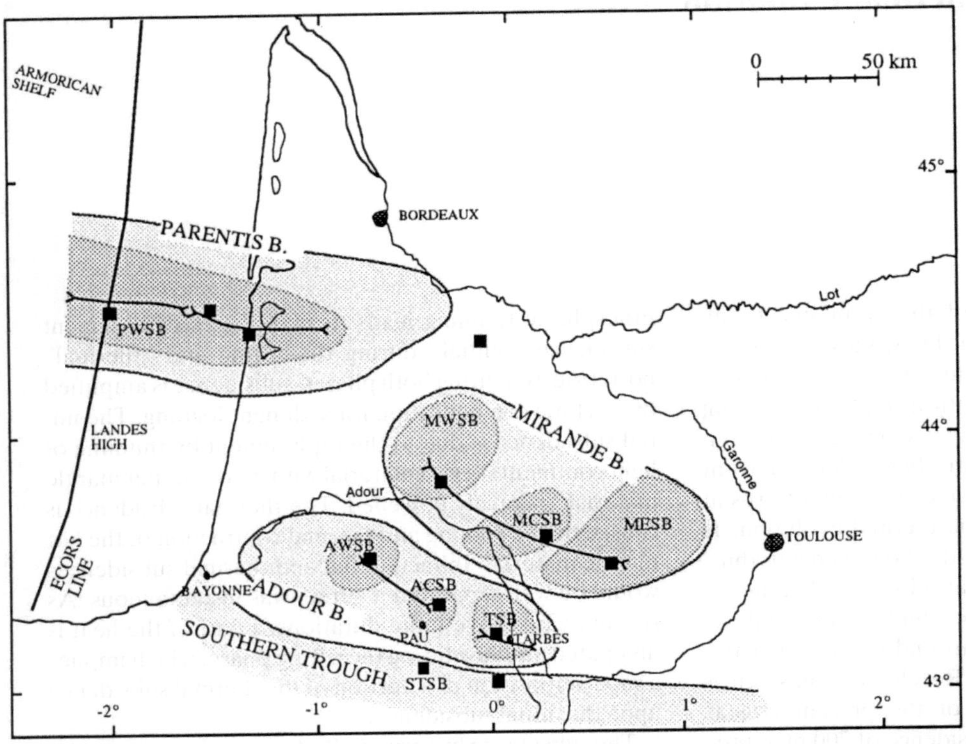

Fig. 1. Schematic representation of Aquitaine sub-basins *Light shaded zones* Basins, *shaded* sub-basins, centers of subsidence, *black squares* synthetic wells (Brunet 1991), *PWSB* Parentis west sub-basin, *MWSB, MCSB, MESB* Mirande west, center, and east sub-basins, *AWSB, ACSB, TSB* Adour west, center, and Tarbes sub-basins, *STSB* southern trough sub-basin), location of the N-S ECORS Biscay line

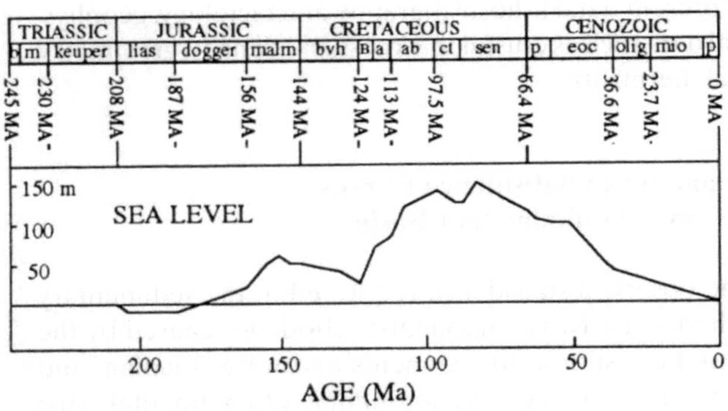

Fig. 2. Time scale (DNAG 1983) and sea-level variation curve used (Brunet 1991, 1993) Age abbreviations, *b* Buntsandstein, *m* muschelkalk, *bvh* Berriasian, Valanginian, Hauterivian, *B* Barremian, *a* Aptian, *ab* Albian, *ct* Cenomanian, Turonian, *sen* Senonian, *p* Paleocene, *eoc* Eocene, *ol* Oligocene, *mio* Miocene, *p* Plio-Quaternary

rigidity. The rigidity of the plate distributes the deformation through the basin beyond the application area of the load, near the margins of the basin, and induces the uplift of the edges by flexural bulging. A first attempt at regional isostatic modeling in the Parentis basin (Brunet 1991) seems to indicate low flexural rigidities

Biostratigraphic data, lithology, and porosity information are necessary for sediment compaction to be calculated and decompaction estimated during backstripping. According to the lithology, the laws of porosity versus

Table 1. Characteristics of the lithologic components used (After Brunet 1991 from data of oil wells and Becquey et al 1975)

Lithology	Φ 0%	Φ 2000	ρG
Clay	78	15	2 65–2 72
Calcareous clay	68	17	2 65
Marl	78	15	2 68–2 74
Limestone	60	9	2 71
Clayey limestone	60	9	2 66
Sand	48	10	2 65
Sandstone	45	6	2 65
Dolomitic limestone	30	5	2 79
Dolomite	15	5	2 86
Salt	0	0	2 17
Anhydrite	0	0	2 98

Φ 0, surface porosity, Φ 2000, porosity at depth 2000 m, ρG, grain density

Fig. 3. Subsidence curves of four synthetic wells representative of sub-basins (Brunet 1991) Location as in Fig 1 *For each well top (depth)* subsidence curves, *continuous line* tectonic subsidence curve under water *(upper curve)*, with local isostatic compensation, and basement deepening curve with sediments being decompacted *(lower curve)*, without correction for sea-level variations and water depths (++ nondecompacted sediments), *ornamented line* same curves with corrections, *middle (bathymetry)* water depths curve, *bottom (subsidence velocity)* diagram of tectonic subsidence velocity (under water), with corrections of sea-level and water depth variation ▶

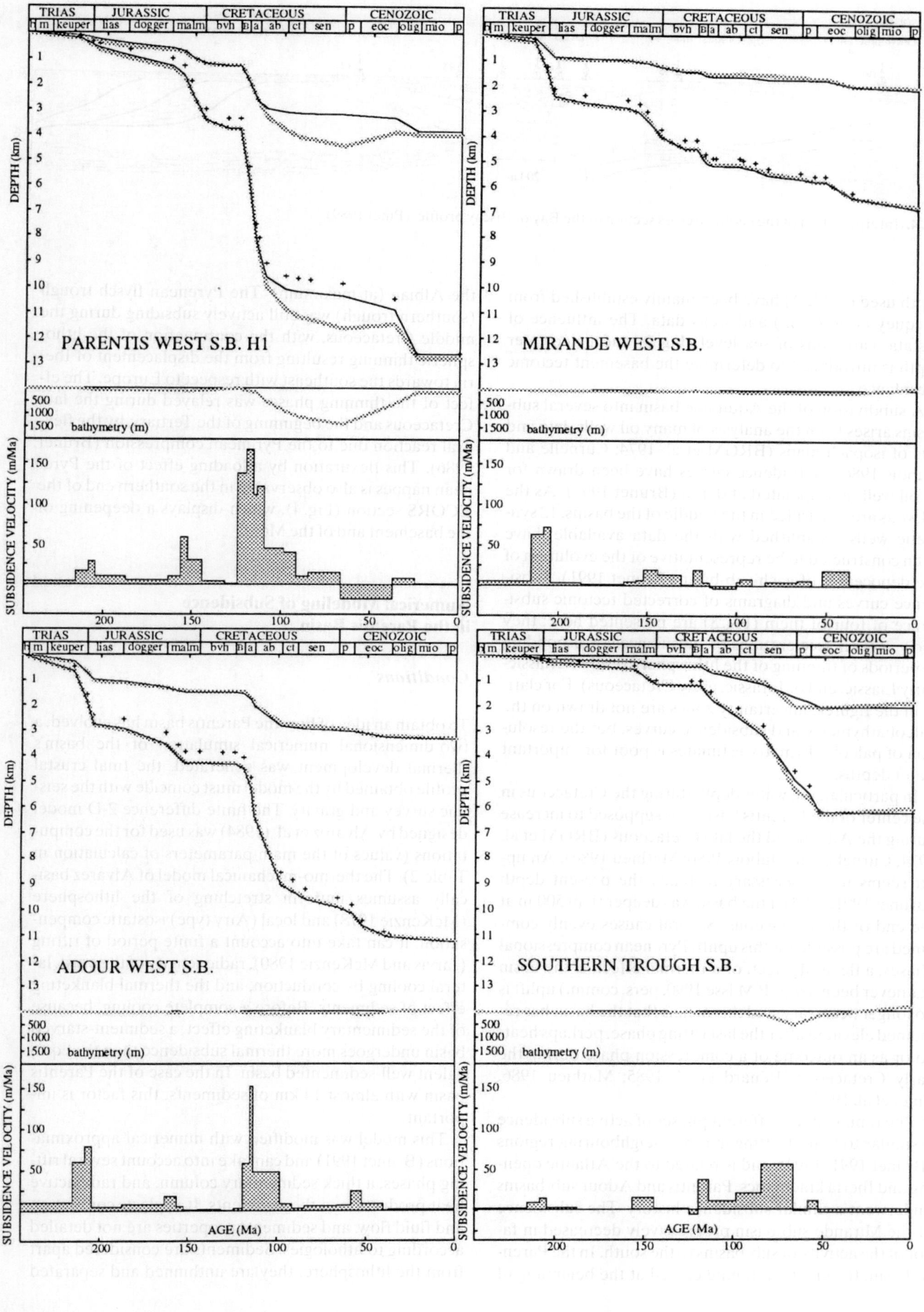
TRIAS
JURASSIC
CRETACEOUS
CENOZOIC
m keuper
lias
dogger
malm
bvh
a
ab
ct
sen
p
eoc
olig
mio
p
DEPTH (km)
PARENTIS WEST S.B. H1
MIRANDE WEST S.B.
ADOUR WEST S.B.
SOUTHERN TROUGH S.B.
bathymetry (m)
SUBSIDENCE VELOCITY (m/Ma)
AGE (Ma)

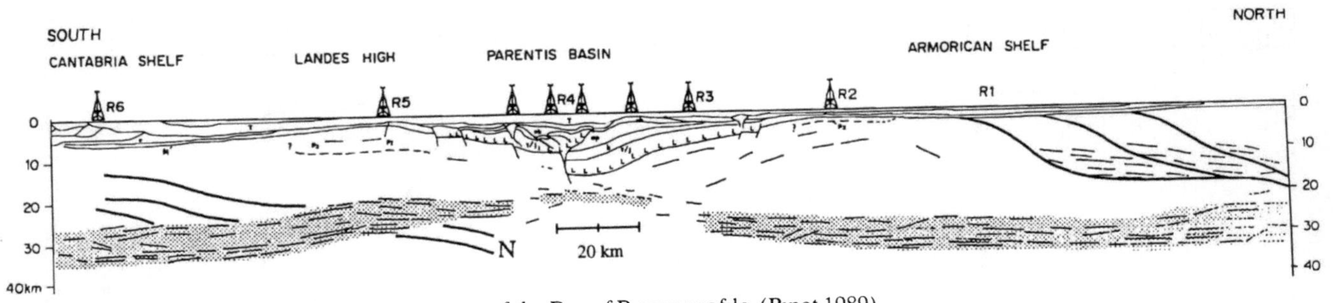

Fig. 4. Interpretation of the geologic cross section of the Bay of Biscay profile (Pinet 1989)

depth used (Table 1) have been mainly established from Becquey et al. (1975) and wells data. The influence of eustatic variations of sea level (Fig 2) and paleowater depth is introduced to determine the basement tectonic subsidence.

A subdivision of the Aquitaine basin into several sub-basins arises from the analysis of many oil wells data and sets of isopach maps (BRGM et al 1974; Curnelle and Dubois 1986). Subsidence curves have been drawn for 80 oil wells extrapolated at depth (Brunet 1991). As the oil wells are not drilled in the middle of the basins, 12 synthetic wells, established with the data available, have been constructed to be representative of the evolution of the depocenter of each sub-basin (Brunet 1991). Subsidence curves and diagrams of corrected tectonic subsidence of four of them (Fig. 3) are presented here; they show several accelerations of subsidence corresponding to periods of thinning of the lithosphere (end of Triassic-early Liassic, end of Jurassic, early Cretaceous). For clarity in the figures, uncertainty zones are not drawn on the paleobathymetry and subsidence curves, but the resolution of paleobathymetry estimates is poor for important water depths.

In particular, the water depth during the Cretaceous in the center of the Parentis basin was supposed to increase during the Albian and the late Cretaceous (BRGM et al. 1974; Curnelle and Dubois 1986; Mathieu 1986). An uplift seems to be necessary to reach the present depth (Brunet 1991, 1993) if the basin was deeper than 500 m at the end of the Cretaceous Several causes evenly combined are possible for this uplift: Pyrenean compressional phases, a flexural effect, or a thermal effect. If the basin has never been deep (P. Masse 1990, pers. comm.) uplift is no longer necessary, and this means that the basin has remained elevated after the last rifting phase, perhaps heated or as an outcome of a compression phase during the early Cretaceous (Bénard et al 1985; Mathieu 1986, Pinet et al. 1987b)

The timing of the different phases of active subsidence is similar to that of rifting in other neighbouring regions (Brunet 1991, 1993), and is related to the Atlantic opening and Iberia kinematics. Parentis and Adour sub-basins underwent a similar subsidence history The subsidence of the Mirande sub-basin progressively decreased in favor of the activity of sub-basins to the south. In the Parentis basin, the crustal thinning ceased at the beginning of the Albian (at maximum) The Pyrenean flysch trough (southern trough) was still actively subsiding during the middle Cretaceous, with the continuation of the lithospheric thinning resulting from the displacement of Iberia towards the southeast with respect to Europe. The effect of the thinning phases was relayed during the late Cretaceous and the beginning of the Tertiary, by the flexural reaction due to the Pyrenean compression (Brunet 1986). This flexuration by a loading effect of the Pyrenean nappes is also observable in the southern end of the ECORS section (Fig. 4), which displays a deepening of the basement and of the Moho

Numerical Modeling of Subsidence in the Parentis Basin

Conditions

To obtain an idea of how the Parentis basin has evolved, a two-dimensional numerical simulation of the basin's thermal development was generated, the final crustal profile obtained by the model must coincide with the seismic survey and gravity. The finite difference 2-D model designed by Alvarez et al (1984) was used for the computations (values of the main parameters of calculation in Table 2). The thermo-mechanical model of Alvarez basically assumes uniform stretching of the lithosphere (McKenzie 1978) and local (Airy type) isostatic compensation; it can take into account a finite period of rifting (Jarvis and McKenzie 1980), radioactivity of the crust, lateral cooling by conduction, and the thermal blanketing effect of sediments. Before a complete cooling, because of the sedimentary blanketing effect, a sediment-starved basin undergoes more thermal subsidence than the equivalent well-sedimented basin In the case of the Parentis basin with almost 14 km of sediments, this factor is important.

This model was modified with numerical approximations (Brunet 1991) and can take into account several rifting phases, a thick sedimentary column, and radioactive heat production in the sediments. It neglects convection and fluid flow, and sediments properties are not detailed according to lithologies. Sediments are considered apart from the lithosphere, they are unthinned and separated

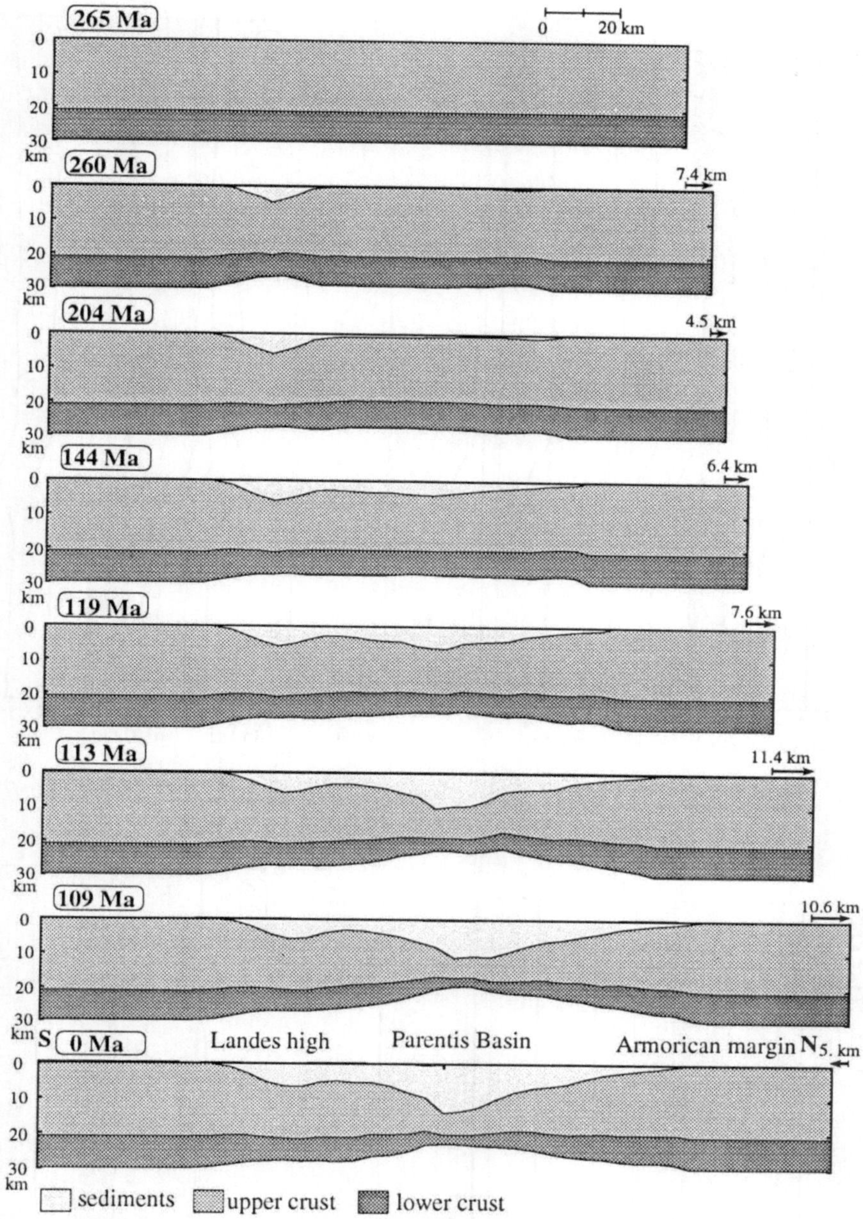

Fig. 5. Schematic evolution through time of the extended crustal section resulting from the computation of an homogeneous thinning model (After Brunet 1991) An image is given before any extension, at the end of each rifting phase (with the value of extension resulting from the thinning factors used in the computation), and at Present after the compression phases

from the crust by a decollement level in the late Triassic and/or early Liassic. Nevertheless, thermal disturbance may propagate laterally and vertically through the different levels (the grid used for the thermal computations is different from the one used for the extension). Another approximation is that the orientation of the section studied does not change for the different rifting phases and the possible variations of the extensional constraint orientation. In the real sedimentary column, the distribution of heat-producing elements is inhomogeneous with depth. We have estimated a mean value of 1.2 mW m^{-3} for the production of internal heat in the whole column of the synthetic well PWSB. This value has been calculated after mean values of radioactivity according to the lithologies (Becquey et al. 1975; Keen and Lewis 1982) and the percentages of these lithologies present in the column. The effect of heat production in the sediments is taken into account by a simple (1-D) conductive model in equilibrium (Rybach 1986).

The subsidence history control is obtained from the sedimentary record of eight oil wells (Brunet 1991, 1993) and a synthetic well issued from geological interpretation of the ECORS Bay of Biscay profile (Fig. 4). The study of the tectonic subsidence indicates six rifting phases for which the computation has been done (Figs. 5, 6): in the Paleozoic, the end of Triassic-early Liassic, late Jurassic, Barremian, early Aptian, and early Albian, with a rifting almost continuous but varying in intensity during the last

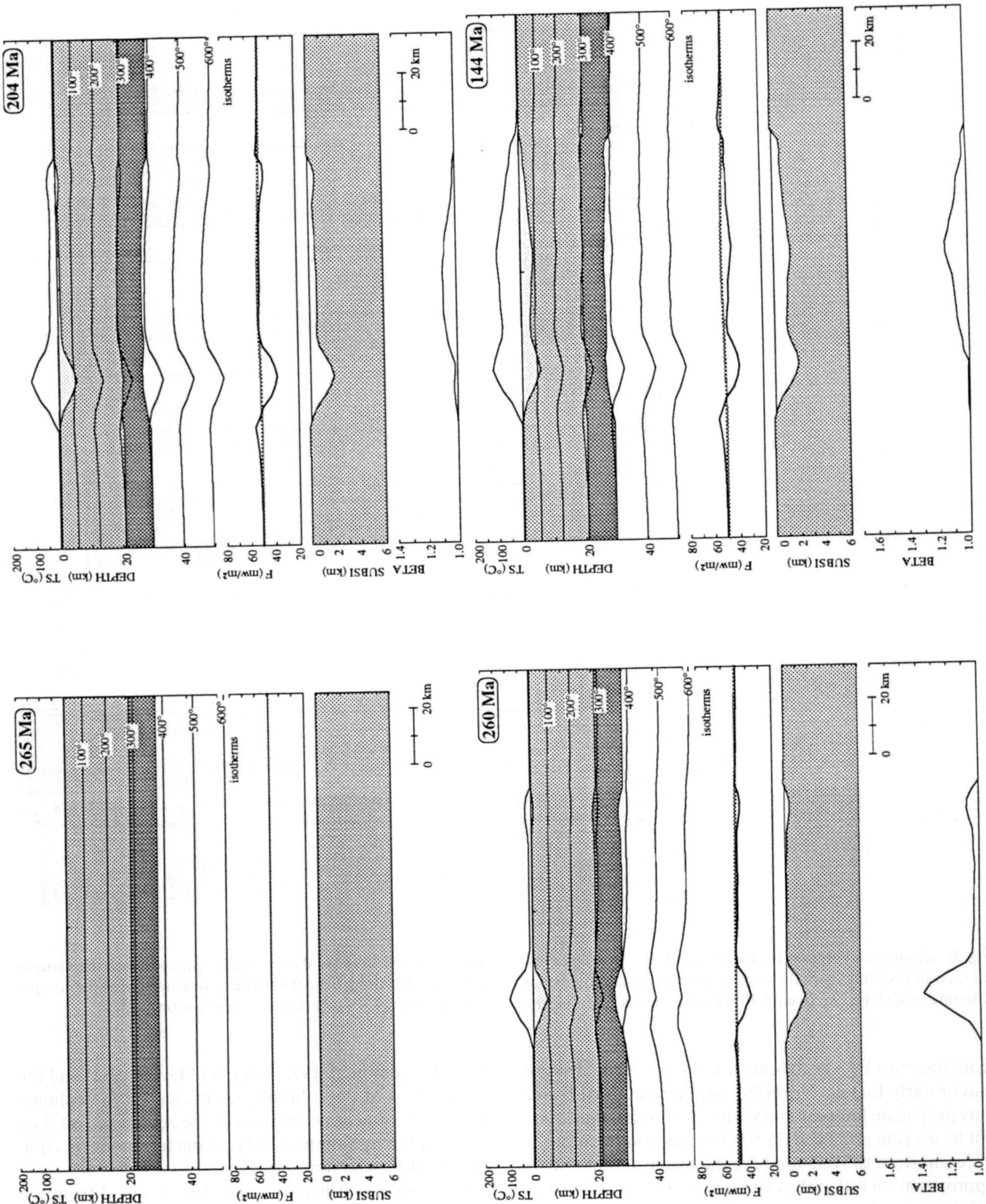

Fig. 6 a, b. Results of thermal modeling with a conservative model of homogeneous crustal thinning (After Brunet 1991) *265 Ma* before any extension. *260-204-144-119-113-109 Ma* end of each rifting phase. *0 Ma* stage corresponding to Present For each stage from the top *TS* temperature computed at the base of the sediments. *DEPTH* lithospheric profile for the first 50 km including sediments *(light shaded)*, upper crust *(shaded)*, lower crust *(heavy shaded)*, repartition of isotherms. *F* heat flow computed. *continuous line* at the top of the sediments. *figured line* without sediments. *SUBSI* corrected tectonic subsidence. *BETA* crustal thinning factor for each phase. *TOTAL BETA* total thinning factor for 0 MA. β *A* average beta along the thinned section

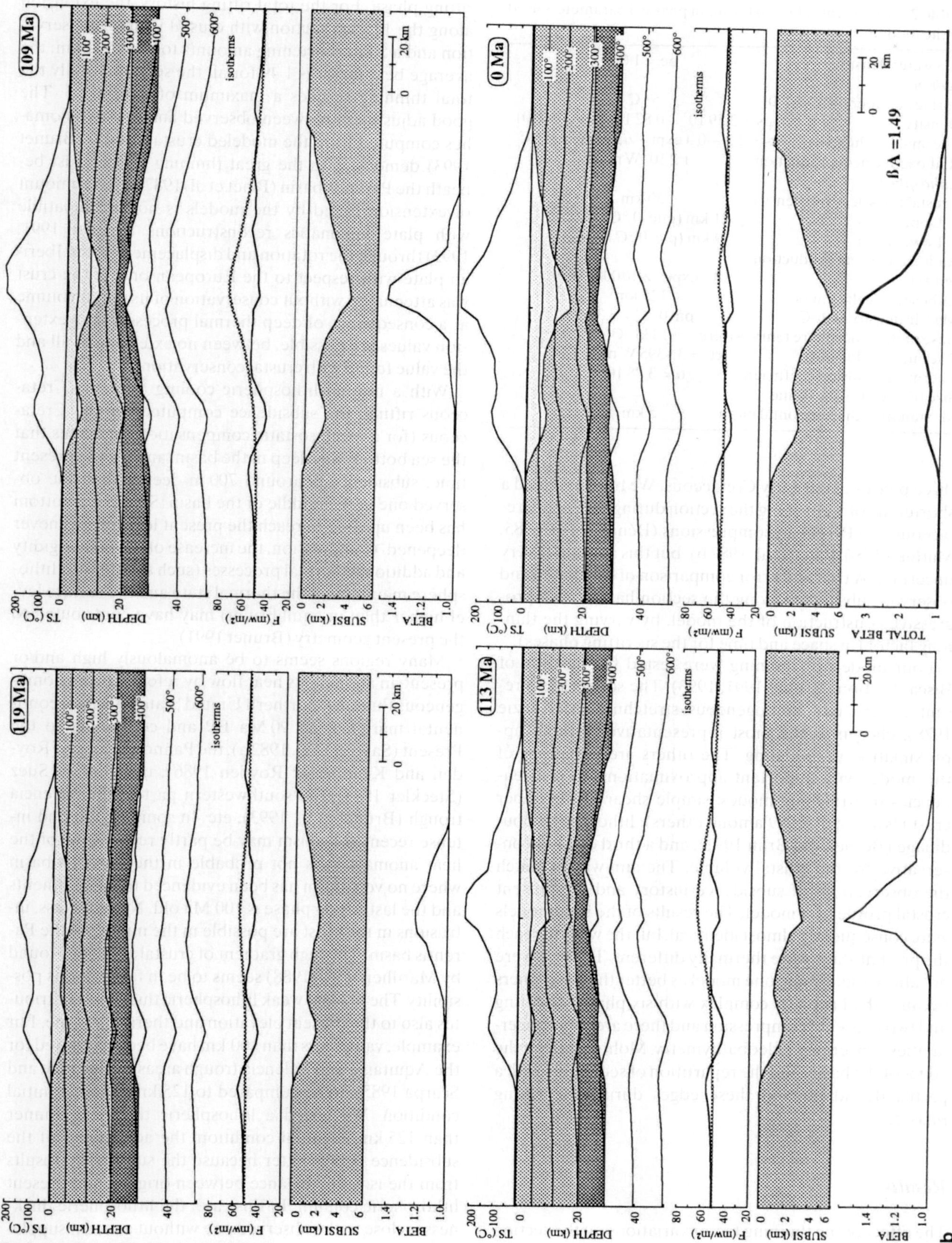

Fig. 6 *(continued)* Legend see p 192

Table 2. Numerical values of the main physical parameters used in the models

Sea water density	$\rho e = 1\ 03$
Sediments	
Surface temperature ($z = 0$)	5 °C
Density	$\rho s = 2\ 59\ 10^3 - 0\ 82\ 10^3 \exp(-z/750)$
Thermal conductivity	$\kappa s = 2\ 2 - 0\ 4 \exp(-z/750)$ W m^{-1} °C^{-1}
Radioactive heat production	$1\ 2\ 10^6$ W m^{-3}
Lithosphere	
Crustal thickness and density	30 km
Upper crust	21 km (ρuc 0 °C = 2 72)
Lower crust	9 km (ρlc 0 °C = 2 92)
Radioactive heat production (crust)	$3\ 10^{-6} \exp(-z/10000)$ W m^{-3}
Lithospheric thickness	125 km
Mantle density at 0 °C	ρm 0 °C = 3 35
Base of the lithosphere temperature	1333 °C
Thermal conductivity	$\kappa L = 3\ 1395$ W m^{-1} °C^{-1}
Thermal expansion coefficient	$\alpha = 3\ 28\ 10^{-5}$
Size of a vertical mesh and Maximum of the horizontal mesh	5 km

three phases of the early Cretaceous. We have assumed a shortening of 5 km along the section during the early Cretaceous and Pyrenean compressions (Bénard et al. 1985, Mathieu 1986; Pinet et al 1987b), but this amount is very uncertain. A trial and error comparison of calculated and observed subsidence along the section has allowed progressive construction of the model, by varying the thinning factor (in space and time for the six rifting phases).

Four models of thinning were tested for the Bay of Biscay section (Brunet 1991, 1993). The sole model presented here is a homogeneous stretching (McKenzie 1978), chosen as the most representative, without approximation in modeling The others are variations of this model with important approximations, to test tendencies of stretching modes: simple shear in the upper crust (Issler et al 1989 among others), lithospheric boudinage (Beslier and Brun 1991), and a third one nonconservative for the crustal volume. The aim was to match the observations of subsidence history and the present crustal profile by a model. The results of the four models were consequently almost identical, but the ways to reach the present stage were thermally different. Thus we were not able to determine one model as better than the others because the history is complex with six phases of rifting and two phases of compression and there are some uncertainties concerning paleobathymetry, Moho depth on the edges of the basin, and the repartition of sediments with a partial denudation on these edges during the rifting phases.

Results

The effect of the thinning factor variations on the section is shown for the six rifting phases and the model of homogeneous stretching (Fig. 5) Such a thinning is necessary to obtain a good fit between results of computations and observations of tectonic subsidence before and after each rifting phase. For the total rifting history, the extension along the 125-km section with crustal volume conservation and a 5 km shortening amounts to about 42 km: the average beta factor is 1 49 for all the section, locally the total thinning reaches a maximum of almost 3.5 The good adjustment between observed and gravity anomalies computed from the modeled crustal profile (Brunet 1993) demonstrates the great thinning of the crust beneath the Parentis basin (Pinet et al 1987b). The amount of extension found by the models is not incompatible with plate kinematics reconstructions (Brunet 1991, 1993) through the rotation and displacement of the Iberian plate with respect to the European one If the crust was attenuated without conservation of its initial volume as a consequence of deep thermal processes, any extension values are possible, between no extension at all and the value found with crustal conservation.

With a normal lithospheric cooling after the Cretaceous rifting, the subsidence computed for the Cretaceous (for a local isostatic compensation) indicates that the sea bottom was deep in the basin, and for the present time, subsidence is around 700 m deeper than the observed one in the middle of the basin So the sea bottom has been uplifted to reach the present level or was never deepened. Compression, the increase of the plate rigidity and additional thermal processes (such as a thinned lithosphere maintaining the thermal state and the surface level higher than in equilibrium) may have contributed to the present geometry (Brunet 1991)

Many regions seems to be anomalously high and/or present an anomalous heat flow by reference to a homogeneous thinning: northern United States Atlantic continental margin from 90 Ma BP and continuing to the Present (Sawyer et al. 1982a), the Pannonian basin (Royden and Keen 1980; Royden 1986), the Gulf of Suez (Steckler 1985), the southwestern part of the Valencia trough (Brunet et al. 1993), etc In some regions, an intense recent volcanism may be partly responsible of the heat anomaly. It is not probable in the Parentis basin where no volcanism has been evidenced in the sediments and the last rifting phase is 100 Ma old. Nevertheless, intrusions in the crust are possible in the middle of the Parentis basin. The high gradient of crustal velocities found by Marillier et al (1988) seems to be in favor of this possibility. The present weak lithospheric thickness contributes also to the present elevation and thermal regime. For example, values less than 100 km have been measured for the Aquitaine and Valencia trough areas (Calgagnile and Scarpa 1985), to be compared to 125 km taken as initial condition We tested a lithospheric thickness thinner than 125 km as initial condition; the adjustment of the subsidence is not better because the subsidence results from the isostatic balance between original and present lithospheric column In this case, the lithospheric thickness is close to the observed one without a recent supplementary modification of the thermal regime, but the tectonic subsidence is still too high

If the temperature has been elevated since the last rifting phase and not recently, the lithospheric thickness has been maintained thin (for example 100 km instead 125 km) during that time or crustal intrusions have influenced the thermal regime, then in this case the paleobathymetry has never been deep (Brunet 1991, 1993). If the thermal event is recent with an upwelling of asthenospheric material and thus of the isotherms in the lithosphere, the paleobathymetry may be important during the Cretaceous and decrease with the heating of the lithosphere.

Models and observations are very sensitive to the present thermal regime controlled by the presence of intrusions not cooled or the thickness of the lithosphere. The effect is an uprising of the isotherms in the lithosphere but with a wavelength different in the two cases.

Being not able at that time to simulate the presence of intrusions in the model, we chose to raise the isotherm corresponding to the base of the lithosphere during a very recent phase. This is equivalent to the modified extension model of Royden (1986) originally called non-uniform extension (Royden and Keen 1980; Sclater et al. 1980), allowing crustal thinning to occur almost independently of heating within the lithosphere by replacement of a large part of the mantle lithosphere by hot asthenosphere.

The model computed here for a modified thermal regime (Fig. 6, 0 Ma), provides more elevated temperatures at the base of the sediments, nearer to the observed ones, than a normal cooling. The present heat flow calculated is not too high because the sedimentary thickness is important. In effect, the heat flow observed is less elevated in the basin than in the surrounding areas where sediments are thinner and the crust (radioactive) thicker. By considering heat generation in the sediments (thickness of almost 14 km), and the thermal evolution of the lithosphere, the present temperature gradient (for all the sedimentary column) computed in the middle of the basin is around 21 °C km^{-1}, reaching 25 °C km^{-1} near the margins; measured values (Mathieu et al. 1985) in the oil wells of the Parentis zone are between 21 and 25 °C km^{-1}. The slight difference may be due to an underestimation of the radioactive heat production in the crust and in the sediments, to the absence in the thermal modeling of terms of fluid circulation and convection (possible variation of 40 °C, reported by Smith and Chapman (1983), in comparison with the conductive model), and to the absence of details in the sediments properties.

Another process, which is not exclusive of the thermal one, may have happened: this is the flexural support of the basin caused by the rigidity of the lithosphere (Brunet 1991, 1993). The recent model (Kooi et al. 1992) of levels of lithospheric necking acting on the state of flexure may also be qualitatively considered. According to this model, for the same amount of crustal thinning along the profile (which is an observation), the subsidence in the Parentis basin can be shallower than the one computed in local isostasy if the level of lithospheric necking is intermediate or shallow with a downward state of flexure (especially visible on the basin flanks flexurally downwarped). The depth of this level is controlled by the rheology of the lithosphere; it may be shallow for high extension rates, high geothermal gradient, and the presence of important crustal discontinuities (Kooi et al. 1992).

Modeling is nonunique, the isostatic compensation is certainly complex, and there exist other sets of values of the parameters taken into account and other modes of thinning which could satisfy the geological constraints used. Other constraints are necessary to improve the models.

Thermal Implications

The crust and upper mantle were heated during the rifting stages. Several studies have shown that the models describing the evolution of sedimentary basins are in a first approximation likely to predict the general thermal history of sediments (Keen 1979; Keen and Lewis 1982; Sawyer et al. 1982a, b, 1987). More refined models accounting for the measured properties of the sediments, convection, water circulation, hydrocarbon generation, and migration give obviously more reliable results (Smith and Chapman 1983; model combined with lithospheric evolution like the THEMIS model IFP: Doligez et al. 1986).

A cross section of the temporal variation in the isotherms depths across the basin was produced by the modeling of the Parentis basin. The thermal evolution of a particular stratigraphic unit can be traced as it undergoes burial, compaction, and heating with time. In a first approximation, integrating that time-temperature information allows estimation of hydrocarbon maturation. The maturation of organic material in the sediments and also its destruction depend upon two factors: time and temperature. Time-temperature models of thermal generation of oil have accounted for observations with the Time-Temperature Index TTI (Lopatin 1971; Waples 1980). The method is based on the observation that the rates of the first-order chemical reactions leading to the thermal maturation of organic matter doubles with each 10 °C rise in temperature. The easiest index to use is the thermal maturity index C of Royden et al. (1980).

$$C = \ln \int_0^t 2^{T/10} dt \text{ (T in °C, t in millions of years).}$$

Generation of oil begins at C = 10 and continues until C = 16. Between C = 16 and 19, oil is cracked to gas, and above 19, hydrocarbons are destroyed (Royden et al. 1980).

The results obtained for values of C at PWSB in the center of the Parentis basin are around 24 for the base of the sediments, 19 for the top of Jurassic, 15 for the top of Barremian, and 9 for the top of Albian (Fig. 7). According to the calibration of the index C (Royden et al. 1980) and to the thermal results of modeling, the Albian is at the beginning of oil generation in the middle of the basin, and older sediments are potentially source rocks. To be preserved without over-heating by deep elevated temperatures, hydrocarbons generated in strata older than

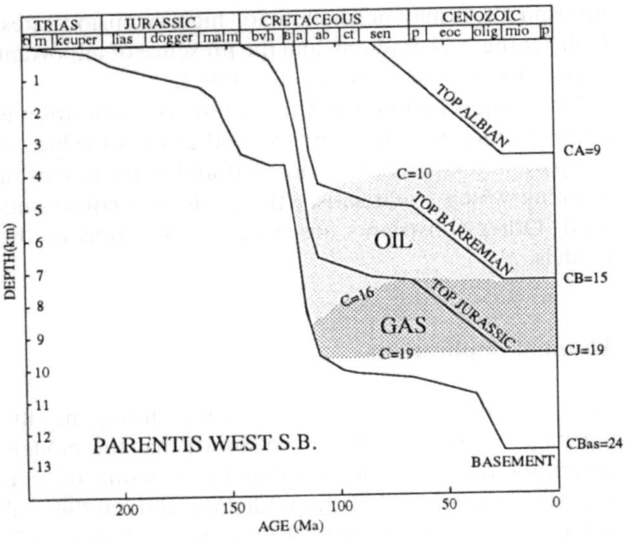

Fig. 7. Estimation of the generation of oil possibility in the middle of the Parentis basin derived from thermal modeling The *solid curves* show the depth history of several sedimentary horizons with the value of C calculated for each horizon *Shaded areas* are bounded by approximated constant C contours resulting from the model (C = 10, C = 16, C = 19, calibration after Royden et al 1980), delimiting zones of possible generation of oil and gas

Barremian must have migrated upward and laterally during tectonic phases, to be entrapped in adequate structures. Results should be extended to the basin, and especially to zones where a comparison is possible with maturation indicators of organic geochemistry, the marine upper Kimmeridgian (carbonated lithology and sapropelic facies) constituting a source rock, though relatively poor in organic matter (Claret et al 1981). If data of organic matter maturation through time are sufficiently precise, their comparison with the thermal evolution resulting from different models could help to choose between the different "scenarios" possible.

Conclusions

The Parentis and Adour sub-basins were created by the rifting and extensional events associated with the formation of the Atlantic, especially during the early Cretaceous, as a result of the large-scale rotation and displacement of the Iberian plate towards the European one.

Two-dimensional finite difference modeling of the Parentis basin defines a crustal structure that coincides with the seismic survey and gravity observations. In the case of a homogeneous thinning and thus stretching of the lithosphere, the whole section has been submitted to a horizontal extension of 42 km since the Paleozoic. This value is realistic only if the sediments of the basin were decoupled from the basement with a decollement level in the Triassic or early Liassic evaporites and shales, leading to a partial sedimentary denudation (during the rifting phases) of the borders of the basin. One assumption of the homogeneous thinning model is the necking of the crust and the lithosphere. In fact, due to the rheology of the lithosphere, the extension is more likely intermediate between simple and pure shear It is accommodated by faulting and half-graben formation in brittle levels of the lithosphere as the upper crust and upper mantle (for example Beslier and Brun 1991) and ductile deformation in ductile levels (lower crust and lower part of the lithospheric mantle). This leads to a complex deformation which produces differences in the subsidence of particular locations, but which does not affect a large-scale initial subsidence due to the crustal thinning all along a section or the long-term thermal subsidence. In basins where the amount of crustal thinning seems to be much greater than the factor of extension in surface measured from extensional faults, authors (Chenet et al. 1982; Moretti and Pinet 1987; Pinet et al. 1987 a, b; Pinet 1989) have proposed a nonconservative style of thinning caused by deep processes not well identified. In the case of a mode of crustal thinning nonconservative for the crustal volume, any amount of extension might have occurred between zero extension and the value of 42 km.

Nevertheless, the simple homogeneous thinning model provides good adjustments of many types of observations even if it is mechanically unrealistic ("paradox of stretching" evoked by Burrus 1989). In fact, this is due to the sensitivity of the subsidence essentially to the total crustal thickness and to the lithosperic thermal state Therefore, this model may be used in predicting other features as a minimum thermal evolution through time for the setting of a study. An elevation of temperatures in the mantle greater than that considered in a homogeneous thinning may disturb this evolution and add its effects, at present, to the basal thermal regime and to the state of isostatic compensation of the basin. In the Parentis basin, the great uncertainty about the paleowater depths during the end of lower Cretaceous and the upper Cretaceous prevent us from knowing the age of onset of a thermal anomaly and/or the action of a flexure.

The temperature-time histories of the sediments estimated from the subsidence may be useful in establishing the potential of a basin resulting from rifting or a rifted margin for petroleum generation, when data concerning sedimentary properties to oil/gas generation are poorly known. This method gives the image of a general time-thermal context in a subsiding basin, an important condition for the completion of hydrocarbon generation (Perrodon and Masse 1984).

Acknowledgements The author thanks ELF Aquitaine company for providing data, J.C Sibuet, an anonymous reviewer, and A. Mascle for their constructive remarks on the manuscript. This study was carried out under the financial support of the Institut National des Sciences de l'Univers Action d'accompagnement 87 ECORS 5 and 92 EC 08 and of ELF Aquitaine. Contribution no. 612 of Groupe d'Etude de la Marge continentale et de l'Océan, URA 718 CNRS.

References

Alvarez F, Virieux J, Le Pichon X (1984) Thermal consequences of lithosphere extension, the initial stretching phase Geophys J R Astr Soc 78: 389–411

Becquey M, De Charpal O, Lebreton F, Meynot C (1975) La compaction des roches carbonatées et argilo-carbonatées Variations de la porosité et de la vitesse du son avec la profondeur IFP-Elf Re Rep 22898 22

Bénard F, De Charpal O, Mascle A, Trémolières P (1985) Mise en évidence d'une phase de serrage est-ouest au Crétacé inférieur en Europe de l'Ouest C R Acad Sci Paris (II) 300 765–768

Beslier MO, Brun JP (1991) Boudinage de la lithosphère et formation des marges passives C R Acad Sci Paris (III) 313 951–958

Bois C, Gariel O, Lefort JP, Rolet J, Brunet MF, Masse P, Olivet JL (1993) Geologic contribution of the Bay of Biscay deep seismic survey a summary of the main scientific results, a discussion of the open questions and suggestions for further investigation In ECORS Bay of Biscay report Mém Soc Géol Fr (in press)

BRGM, Elf-Re, ESSO-REP, SNPA (1974) Géologie du Bassin d'Aquitaine BRGM Orléans, 26 pl

Brunet MF (1986) The influence of the Pyrenees evolution on adjacent basins Tectonophysics 129 343–354

Brunet MF (1989) Méthode d'étude quantitative de la subsidence In Assoc Sédim Fr (ed) Dynamique et méthodes d'étude des bassins sédimentaires Technip, Paris, pp 87–98

Brunet MF (1991) Subsidence et géodynamique du bassin d'Aquitaine Relations avec l'ouverture de l'Atlantique Thesis Doct Etat, Mém Sci Terre Univ P & M Curie, Paris, 91–8 288 pp

Brunet MF (1993) Subsidence along the ECORS Bay of Biscay deep seismic profile In ECORS Bay of Biscay report, Mém Soc Géol Fr (in press)

Brunet MF, Maillard A, Mauffret A (1993) Subsidence history of the Valencia trough Abs EUG VII meeting, Strasbourg, Terra Abstr 5 159

Burrus J (1989) Review of geodynamic models for extensional basins, the paradox of stretching in the Gulf of Lions (Northwest Mediterranean) Bull Soc Géol Fr (8), V 377–393

Calcagnile G, Scarpa R (1985) Deep structure of the European-Mediterranean area from seismological data Tectonophysics 118 93–111

Chenet PY, Montadert L, Gairaud H, Roberts D (1982) Extension ratio measurements on the Galicia, Portugal, and northern Biscay continental margins implications for evolutionary models of passive continental margins In Watkins JS, Drake C (eds) Studies in continental margin geology Am Assoc Pet Geol Mem 34 703–715

Claret J, Jardiné S, Robert P (1981) La diversité des roches mères pétrolières aspects géologiques et implications économiques à partir de quatre exemples Bull Cent Rech Expl Prod Elf-Aquitaine 5 383–417

Curnelle R, Dubois P (1986) Evolution mésozoique des grands bassins sédimentaires français (bassins de Paris, d'Aquitaine et du Sud-Est) Bull Soc Géol Fr (8), IV 529–546

DNAG (Decade of North American Geology) (1983) Geologic time scale Geological Society of America, Boulder, Co

Doligez B, Bessis F, Burrus J, Ungerer P, Chénet PY (1986) Integrated numerical simulation of the sedimentation heat transfer, hydrocarbon formation and fluid migration in a sedimentary basin the Themis model. In Burrus J (ed) Thermal modeling in sedimentary basins Technip, Paris, pp 173–195

Issler D, McQueen H, Beaumont C (1989) Thermal and isostatic consequences of simple shear extension of the continental lithosphere Earth Planet Sci Lett 91 341–358

Jarvis GT, McKenzie DP (1980) Sedimentary basin formation with finite extension rates Earth Planet Sci Lett 48 42–52

Keen CE (1979) Thermal history and subsidence of rifted continental margins evidence from wells on the Nova-Scotian and Labrador shelves Can J Earth Sci 16 505–522

Keen CE, Lewis T (1982) Measured radiogenic heat production in sediments from continental margin of eastern North America implications for petroleum generation Am Assoc Pet Geol Bull 66 1402–1407

Kooi H, Cloetingh S, Burrus J (1992) Lithospheric necking and regional isostasy at extensional basins 1 Subsidence and gravity modeling with an application to the Gulf of Lions margin (SE France) J Geophys Res 97 17553–17571

Lopatin NV (1971) Temperature and geologic time as factors in the coalification (in Russian) Akad Nauk SSSR Izv Ser Geol 3 95–106

Marillier F, Tomassino A, Patriat P, Pinet B (1988) Deep structure of the Aquitaine shelf constraints from expanding spread profiles on the ECORS Bay of Biscay transect Mar Petrol Geol 5 65–74

Mathieu C (1986) Histoire géologique du sous-bassin de Parentis Bull Cent Rech Expl Prod Elf-Aquitaine 10 33–47

Mathieu C, Modiano T, Mure E et al (1985) Bassin de Parentis synthèse géologique et géophysique, bilan pétrolier Elf-Aquitaine Conf Rep

McKenzie DP (1978) Some remarks on the development of sedimentary basins Earth Planet Sci Lett 40 25–32

Moretti I, Pinet B (1987) Discrepancy between lower and upper crustal thinning In Beaumont C, Tankard AJ (eds) Basin-forming mechanisms Can Soc Pet Geol Mem 12 233–239

Perrier P, Quiblier J (1974) Thickness changes in sedimentary layers during compaction history, methods for quantitative evaluation Am Assoc Pet Geol Bull 58 507–520

Perrodon A, Masse P (1984) Subsidence, sedimentation and petroleum systems J Petrol Geol 7 5–26

Pinet B (1989) Deep seismic profiling and sedimentary basins Bull Soc Géol Fr (8), IV 749–766

Pinet B, Montadert L, Curnelle R, Cazes M, Marillier F, Rolet J, Tomassino A, Galdeano A, Patriat P, Brunet MF, Olivet JL, Schaming M, Lefort JP, Arrieta A, Riaza C (1987a) Crustal thinning on the Aquitaine shelf, Bay of Biscay, from deep seismic data Nature 325 513–516

Pinet B, Montadert L, ECORS scientific party (1987b) Deep seismic reflection and refraction profiling along the Aquitaine shelf (Bay of Biscay) Geophys J R Astr Soc 89 305–312

Royden LH (1986) A simple method for analyzing subsidence and heat flow in extensional basins In Burrus J (ed) Thermal modeling in sedimentary basins Technip, Paris, pp 49–73

Royden LH, Keen CE (1980) Rifting process and thermal evolution of the continental margin of eastern Canada determined from subsidence curves Earth Planet Sci Lett 51 343–361

Royden LH, Sclater JG, Von Herzen RP (1980) Continental margin subsidence and heat flow important parameters in formation of petroleum hydrocarbons Am Assoc Pet Geol Bull 64 173–187

Rybach L (1986) Amount and significance of radioactive heat sources in sediments In Burrus J (ed) Thermal modeling in sedimentary basins Technip, Paris, pp 311–322

Sawyer DS, Swift BA, Sclater JG, Toksoz MN (1982a) Extensional model for the subsidence of the northern United States Atlantic continental margin Geology 10 134–140

Sawyer DS, Toksoz MN, Sclater JG, Swift BA (1982b) Thermal evolution of the Baltimore canyon trough and Georges Bank basin. In Watkins JS, Drake C (eds) Studies in continental margin geology Am Assoc Pet Geol Mem 34 743–762

Sawyer DS, Hsui AT, Toksoz MN (1987) Extension, subsidence and thermal evolution of the Los Angeles basin – a two-dimensional model Tectonophysics 10 15–32

Sclater JG, Royden L, Horvath F, Burchfield BC Semken S, Stegena L (1980) The formation of the intra-Carpathian basins as determined from subsidence data Earth Planet Sci Lett 51 139–162

Smith L, Chapman DS (1983) On the thermal effects of groundwater flow 1 Regional scale systems J Geophys Res 88 593–608

Steckler MS (1985) Uplift and extension at the Gulf of Suez indications of induced mantle convection Nature 317 135–139

Steckler MS, Watts AB (1978) Subsidence of the atlantic type continental margin off New York Earth Planet Sci Lett 41 1–13

Waples DG (1980) Time and temperature in petroleum formation application of Lopatin's method to petroleum exploration Am Assoc Pet Geol Bull 64 927–937

The Arzacq-Western Pyrenees ECORS Deep Seismic Profile

M. Daignières, M. Séguret, M. Specht, and ECORS Team

Abstact In late 1989, a 100-km-long vertical seismic reflection profile was recorded in the western Pyrenees from the French-Spanish border, at the limit between the Axial Zone and the Mendibelza Paleozoic massif, to the deep Arzacq domain of the Aquitaine basin.

The data was acquired with an explosive source and 24-fold coverage. Target objectives of this profile were firstly to document variations in crustal thickness and nature, in relation to the different structural position of the basement and asociated sedimentary cover, and secondly, to investigate possible effects of tectonic inversion during Pyrenean convergence.

The Iberian and European Mohos are identified at 14–15 s TWT and 9.5–10 5 s TWT, respectively, giving an overall structure similar to that in the east, as shown by a previous ECORS profile. Northward-directed thrusts (St Palais and Ste Suzanne) in the north Pyrenean zone, and the thickness of the Mesozoic sedimentary fill (more than 10 km) of the Arzacq basin are detailed. Northward-dipping reflectors (at 2–3 5 s) beneath the flysch basin of Mauléon are also detected but are still difficult to interpret (basement or Jurassic?)

Introduction

In October 1989, 5 years after the eastern Pyrenees ECORS profile (ECORS Pyrenees team 1988; Choukroune and ECORS Team 1989), a new vertical reflection profile was acquired in the western part of the chain, where knowledge of deep crustal structures, based mainly on old refraction and teleseismic studies from the early 1980s (Daignières et al. 1981, 1982; Gallart et al. 1981, 1985; Hirn et al. 1984) remained poor. Added to the Gulf of Biscay (Pinet et al. 1987) and Gulf of Lions (De Voogd et al. 1991) ECORS profile, this is the fourth deep vertical reflection profile through the Pyrenean range and its adjacent forelands

* The Arzacq Ecors Team is composed of members of the CNRS-INSU (1), Institut Francais du Pétrole (2) and E A P (3) (1) M Daignières, J Canérot, E Debroas, G Grandjean, A Hirn, J M Marthelot, D Rappin, M Séguret, (2) B Damotte, P Desegaulx, M Specht, (3) J Estival, M LeVot, A Morash, A Villien

Geological Setting

Classically, the main structural zones of the Pyrenees comprise, from north to south, the Aquitanian basin, the north Pyrenean zone, the axial zone and the south Pyrenean zone.

The Aquitanian foreland molasse sequence is overthrust from the south, by the north Pyrenean frontal thrust of the north Pyrenean zone, which carries Mesozoic flysch deposits and Triassic to Aptian carbonates cored by Hercynian basement of the north Pyrenean massifs.

The north Pyrenean zone is bounded to the south by the north Pyrenean fault zone trending east-west along the belt. This contains highly strained Mesozoic rocks (Choukroune 1976) metamorphosed under low pressure-high temperature conditions (Golberg et al. 1986) between 95 and 90 Ma (Albarède and Michard-Vitrac 1978; Montigny et al. 1986) and outcrops of mantle-derived rocks (Vielzeuf and Kornprobst 1983).

The axial zone, presently the highest part of the chain, comprises Hercynian rocks reworked during the Alpine orogeny by southward directed thrusts of the south Pyrenean zone (Séguret 1972; Camara and Klimowitz 1985, Labaume et al 1985).

In the western part of the range, the north Pyrenean fault is not clearly identified and the north Pyrenean zone is wider than in the eastern sector. The high-temperature metamorphism of Cretaceous age is absent and the axial zone is still covered by Cretaceous /Ecocene deposits. Borehole data and petroleum industry seismic profiles provide informations to constrain the thrust geometry and structure of the sedimantary cover in the western north Pyrenean zone where exploration has been active for the last 40 years (Curnelle et al. 1980, 1982). These industry data enabled the construction of geological cross sections (Seguret and Daignieres 1986; Velasque and Ducasse 1987) which were used to constrain the shallow part of the time section.

As previously mentioned, it is convenient to distinguish three principal areas. These are from north to south (Fig. 1):

1. *Arzacq Basin* This basin, with the Paleozoic substratum at a depth of 5000 m in the north and more than 10000 m in the south, is strongly asymmetric, and corresponds to a foreland basin during the Pyrenean orogeny at end-Cretaceous and Tertiary times, superim-

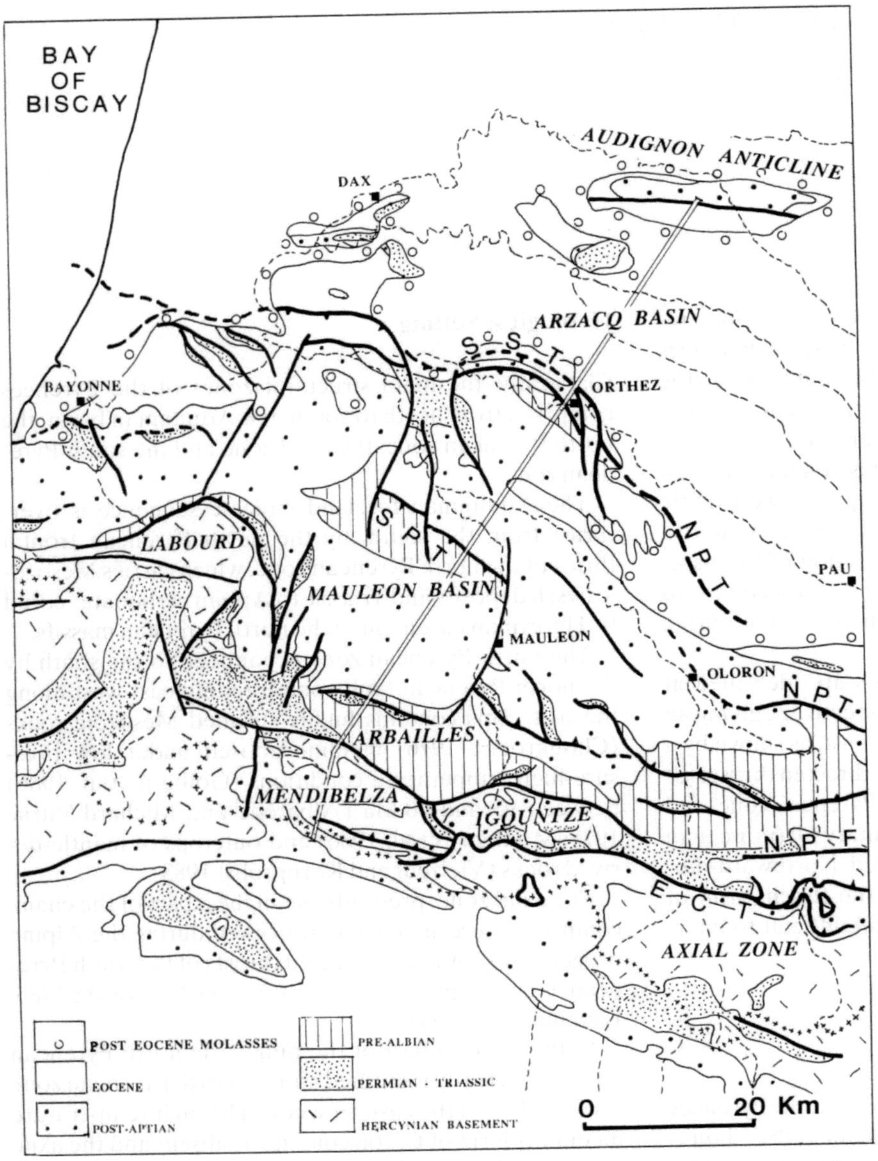

Fig. 1. Location map of the Arzacq-West Pyrenees ECORS profile *(double line)*, with main geological units and faults of the region *NPFT* North Pyrenean frontal thrust. *NPF* north Pyrenean fault. *SST* Sainte Suzanne thrust. *SPT* Saint Palais thrust. *ECT* Eaux Chaudes thrust

posed on an extensional basin that recorded a high rate of subsidence during the lower Cretaceous, accumulating more than 3500 m of marine sediments During the upper Cretaceous, it was characterized by a platform domain in the north which evolved into a flysch trough close the north Pyrenean frontal thrust. The northern basin margin is characterized by important halokinetic structures such as the ridge of Audignon, which was active during lower Cretaceous extension and later inverted during Pyrenean compression.

2. *North Pyrenean Zone* This zone is bounded to the north by the northward directed Sainte Suzanne thrust, a segment of the north Pyrenean frontal thrust The zone here essentially consits of the Mauléon basin divided into two tectonics units by the Saint Palais thrust, which involves Triassic evaporites as detachment level The Jurassic and lower Cretaceous series of the Mauléon basin thicken to the south and mainly correspond to shelf carbonate deposits. The Albian and upper Cretaceous series correspond to flysch sediments deposited in an extensional strike-slip setting during Albian times (Choukroune and Mattauer 1978) and a compressional setting during the upper Cretaceous (Curnelle et al. 1982). The southern part of the Arzacq basin is deformed in the footwall of the Sainte Suzanne Thrust.
The Arbailles and the Igountze-Mendibelza massifs form the southern part of the north Pyrenean zone The Arbailles massif, consisting of Jurassic-lower Cretaceous series, is bordered by thrust faults of limited importance The unconformities within the Jurassic to lower Cretaceous series are interpreted as resulting

from lower Cretaceous halokinetic deformation (Canerot 1989). The Igountze-Mendibelza massif is formed by Albian fan-delta/turbidite breccias and conglomerates directly onlapping the Paleozoic basement (Boirie and Souquet 1982). The massifs are thrust southward onto the Mesozoic cover of the axial zone, which obscures the western surface continuation of the north Pyrenean fault along this transect.

3. *Axial Zone.* Here, Cenomanian to Campanian shelf carbonates rest directly on Paleozoic rocks and are in turn overlain by Campanian-Maastrichtian slope mudstones, Paleocene carbonates, and Eocene turbidite deposits.

Deep Structure from Geophysics Prior to Deep Seismic Experiment

Geophysical information concerning the deep structure of the crust mainly comes from refraction studies (Daignières et al. 1981, 1982). In this western region, the middle and lower crust of the north Pyrenean zone appears to be highly heterogeneous with high velocities (6.7 to 6.8 km/s). Besides, the crust is 45 km thick south of the Mendibelza massif, but less than 30 km thick beneath the Mauléon basin.

Delays in teleseismic arrivals recorded by a temporary array (Hirn et al. 1984; Gallart et al. 1985) have shown the abruptness of the transition between two crustal thicknesses, a deepening of the Moho by 15 km over less than 10 km of horizontal distance, as occurs in the eastern part of the chain (Daignières et al. 1989).

This transition coincides with an EW zone, 70 km long and 20 km deep, of present-day seismicity following the western continuation of the north Pyrenean fault (Gagnepain-Beyneix 1987). The foci of these earthquakes are distributed on E-W-trending, north-dipping, left-stepping planes.

Gravity data show two negative Bouguer anomalies in the south and the north of the region, the first related to the crustal root and the second related to the Cretaceous sedimentary cover and the crustal flexure in Aquitaine. They surround a strong positive east-west elongated anomaly, up to 50 mGal, spanning the Mauléon basin and the Labourd granulitic massif. Its origin remains unclear and Mesozoic volcanism, basement thrusting, basic material in the middle and lower crust, and slices of lower crust or peridotites at intermediate depth are all possibilities (Zolnai 1971; Daignières et al. 1981).

Field Acquisition and Processing

The 98.4-km-long Arzacq-West Pyrenees profile (Fig. 1) was acquired between 09/14 and 10/09 1989. The source was explosive (30 to 10 kg per shot at 40 m depth) with a shot interval of 320 m. The recording system was a SERCEL SN 348, with 192 geophones groups (18 10 Hz geophones per group) spaced 80 m apart in a split spread. The nominal coverage is 24-fold, record length 20 s, and sampling interval 4 ms.

A routine processing sequence was applied (demultiplexing, amplitude recovery, deconvolution and CDP gathering) with static corrections calculated from the refracted first arrivals of every record. Velocities were picked by velocity scanning carried out on weathering corrected data. A band pass filter was applied to the whole stacked section, followed by amplitude equalization and a 0 to 20 s section was processed after an 8 ms resampling of the data (Fig. 2). Finally, the deep seismic profile processing was concluded with a wave-equation time migration effected up to 13 s (Fig. 3).

Description of the Time Sections

Several groups of reflectors are identified on the migrated time section (Fig. 3). They are broadly divided into a shallow group between 0 and 4 s TWT, and a deeper group between 8 and 11 s TWT.

Within the Arzacq Basin the reflectors A, B, C (Fig. 3) are identified from the Nassiet 101 well data, and correspond to the base of Molasse deposits, top of Paleocene (1100 ms, 1417 m), and base of upper Cretaceous (1600 ms, 2714 m) respectively. The reflectors labeled D probably correspond to the top Jurassic (3100 ms, 4205 m) and the reflectors in group E to the top Paleozoic basement (4200 ms).

The Sainte-Suzanne thrust is known from wells, in particular the Sainte Suzanne 101 (located 1.2 km SE of the section), which drilled Cretaceous and Jurassic rocks between 813 and 3109 m, before encountering a lower Cretaceous succession from 3109 to 5361 m. A lower, more minor, thrust is suspected in the footwall to the Sainte-Suzanne thrust on the basis of a discordance on migrated time section between north dipping reflectors (F, Fig. 3) than can be correlated to the Jurassic sequence of the Nassiet well, and horizontal reflectors (G, Fig. 3) in the footwall to the Sainte-Suzanne thrust. This discordance may correspond to the hanging wall and footwall ramp geometry of a south-dipping thrust.

The Orriule syncline in the northern Mauléon basin is reflected by a series of subparallel reflectors (H, Fig. 3) that display a geometry similar to the surface structure. The discordance between horizontal reflectors I, and north-dipping reflectors J, is interpreted as the footwall and hanging wall ramp geometry of a south-dipping thrust. The K group of south-dipping poor reflectors has no clear relation to surrounding reflectors.

The Saint-Palais thrust outcrops along the line as two branches and is inferred from the Saint-Palais well located 11 km NW of the section. Jurassic series were drilled between 1945 m and 2690 m, and a Liassic evaporite decollement level between 2690 m and 2900 m. The interruption of south-dipping to horizontal reflectors at 0.5–1 s (L, Fig. 3) may represent the hanging wall cut-off to the St-Palais thrust.

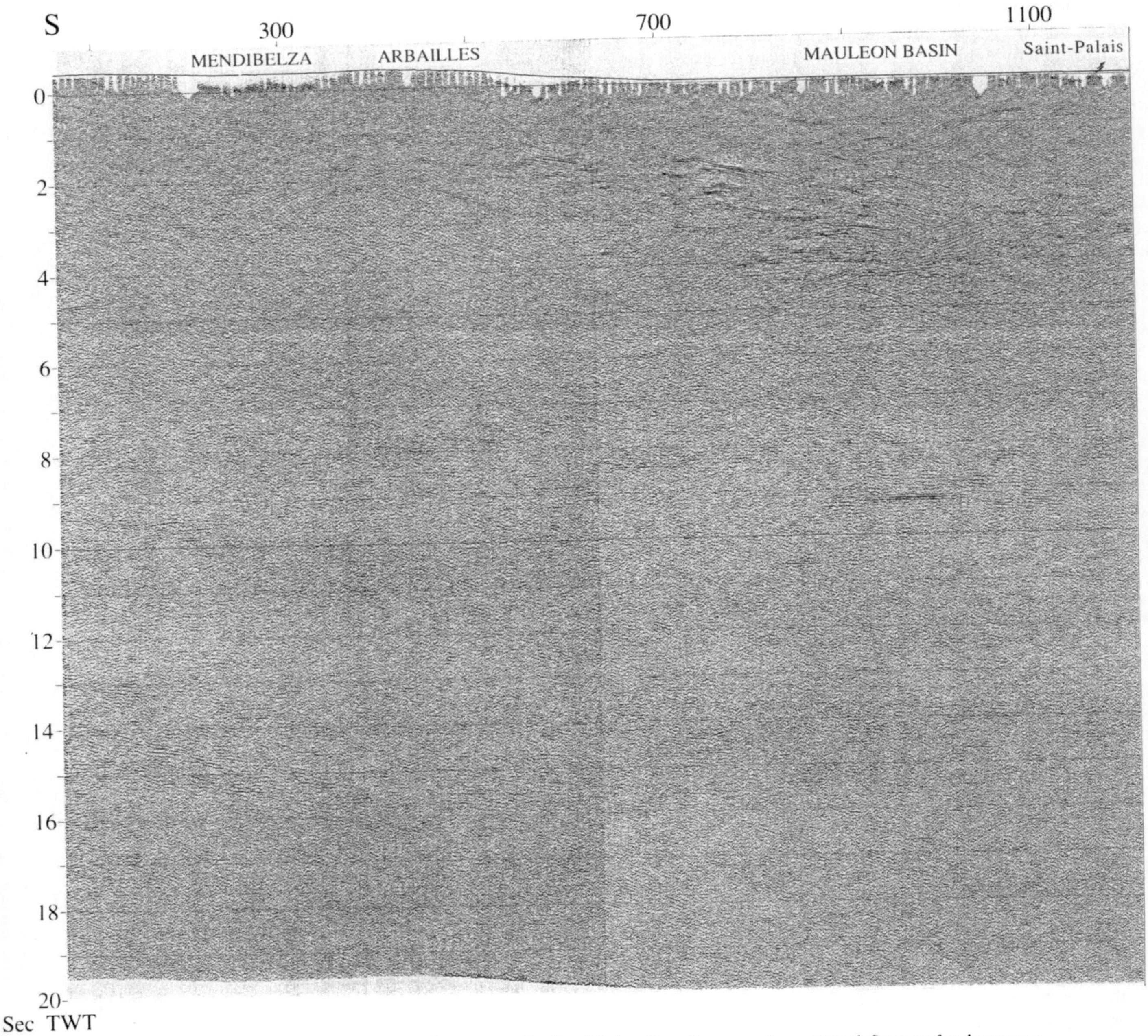

Fig. 2. The unmigrated Arzacq seismic reflection profile (0 to 20 s) with an 8 ms sampling interval See text for discussion

The north-dipping group (M–N, Fig. 3) of high amplitude and good continuity reflectors is either Jurassic or a reflective basement Their elevation and north dip may result from hanging wall deformation related to a south-dipping thrust ramp.

The structure of the Arbailles massif has been presented in the geological setting and no additional information is provided by the ECORS section.

The Mendibelza thrust is known from surface data (Larrau geological map 1970) Paleozoic rocks and Albian breccias (Boirie and Souquet 1982) of the Mendibelza massif are thrust southward over the "calschistes à Navarelles" formation, which corresponds to Maastrichtian slope facies overlying the "calcaires des canyons" formation which represents shelf deposits of Cenomanian-Santonian age that onlap the Paleozoic rocks of the axial zone Additional windows of Cretaceous slope facies that we have found in the bottom of some deeply incised valleys within the southern Mendibelza massif support the argument for a gentle dip of the principal thrust plane. These above-mentioned structures are not clearly imaged by the seismic profile

Concerning the structure of the deep crust and crust-mantle boundary, the most prominent feature is the absence of a clearly identified reflective lower crust. This is at variance with the east Pyrenees (Choukroune and ECORS Team 1989) and the Biscay (Pinet et al. 1987) ECORS profiles.

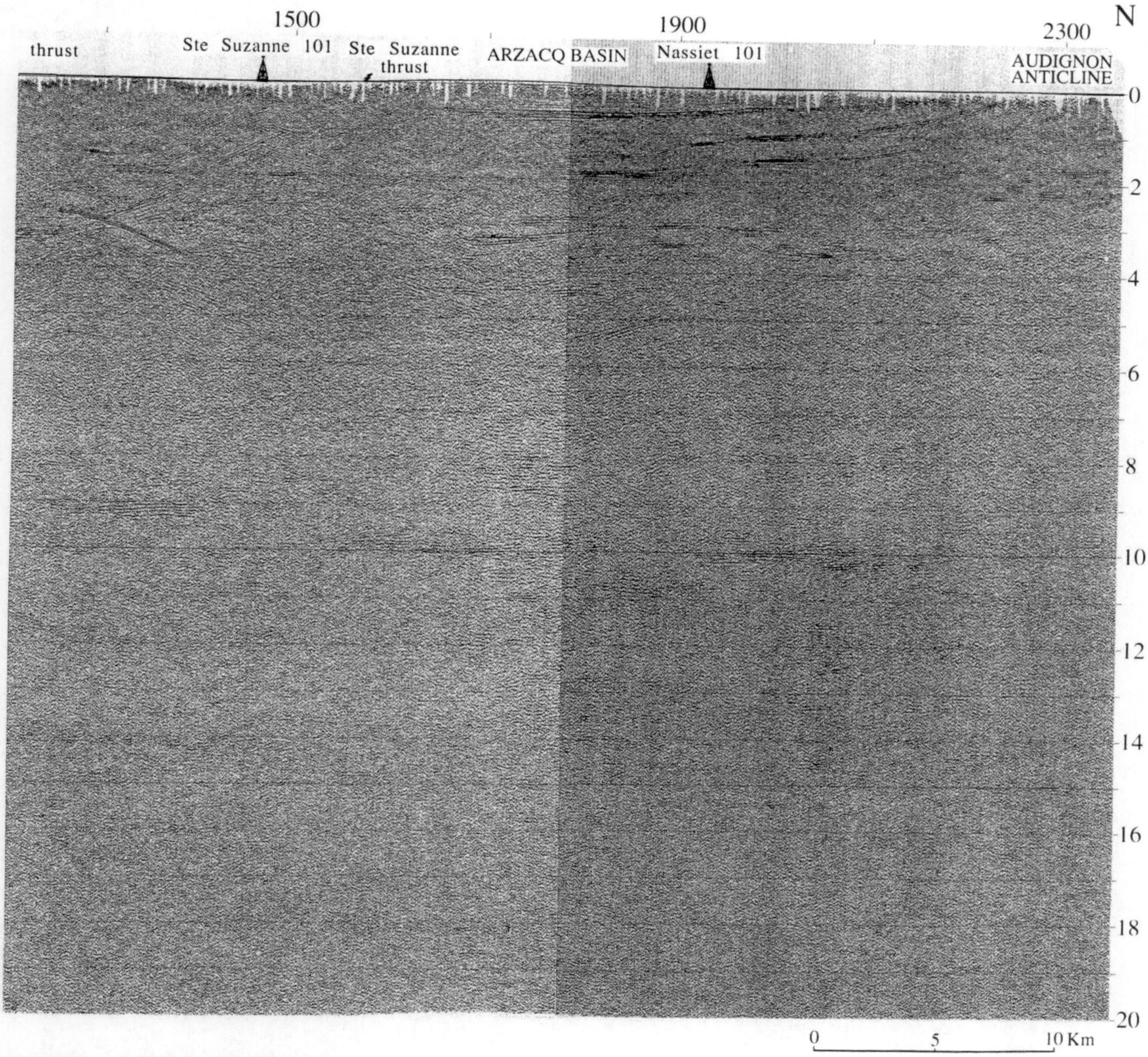

Fig. 2 *(continued)* Legend see p 202

On the unmigrated time section (Fig. 2) the crust-mantle boundary corresponds to a very clear reflector, steeply dipping towards the north from 13.8 s TWT at the southern limit of the section to 15.2 s TWT, 7 km further north. This is consistent with the deep position of the Moho identified from refraction to be around 40–45 km, slightly further south under the Pyrenean Axial Zone (Daignières et al. 1982). The quality of this reflection clearly demonstrates the northward dip of the deep crustal and Moho structures of Iberia.

The European Moho is clearly identified in the northern part of the profile at 10 s TWT. Beneath the Oriule syncline it is at 9 s TWT and some weak reflectors suggest a Moho at 8 s TWT south of the Saint-Palais thrust. However, taking into account lateral velocity variations of the sedimentary cover, one finds almost the same depth, 28.5 km, for the Moho from north to the southern limit of the European plate, as was obtained 12 years ago by refraction from the north Pyrenean zone in this part of the range (Daignières et al. 1982). The thickness of the pre-Triassic crust is therefore less than 6.5 s TWT beneath Arzacq and Mauléon basins (i.e., less than 20 km). Nevertheless, it is thicker than beneath the Parentis basin (Pinet et al. 1987). We will discuss in the next section the respective effects of crust stretching and thickening on final crust thickness.

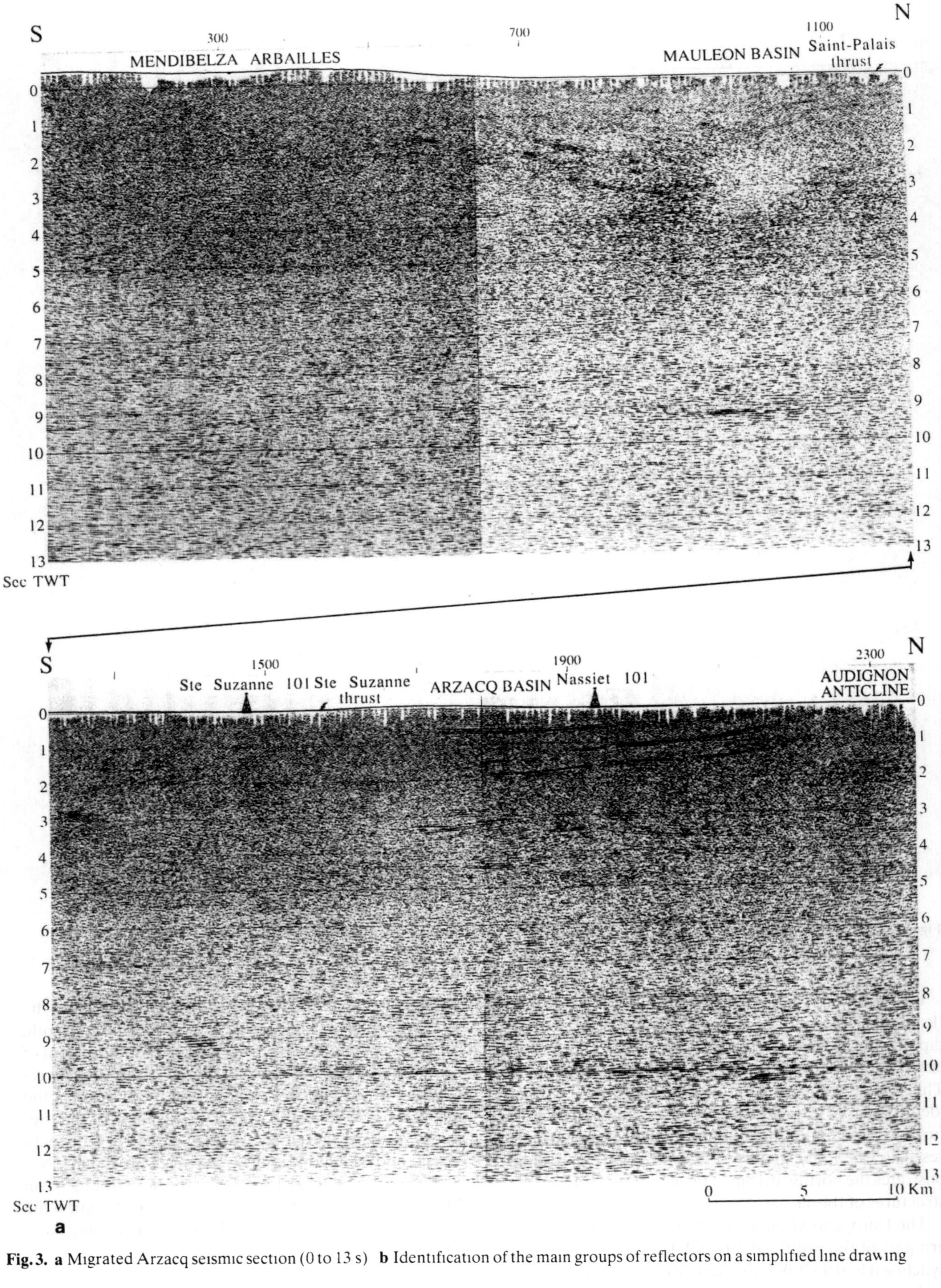

Fig. 3. **a** Migrated Arzacq seismic section (0 to 13 s) **b** Identification of the main groups of reflectors on a simplified line drawing

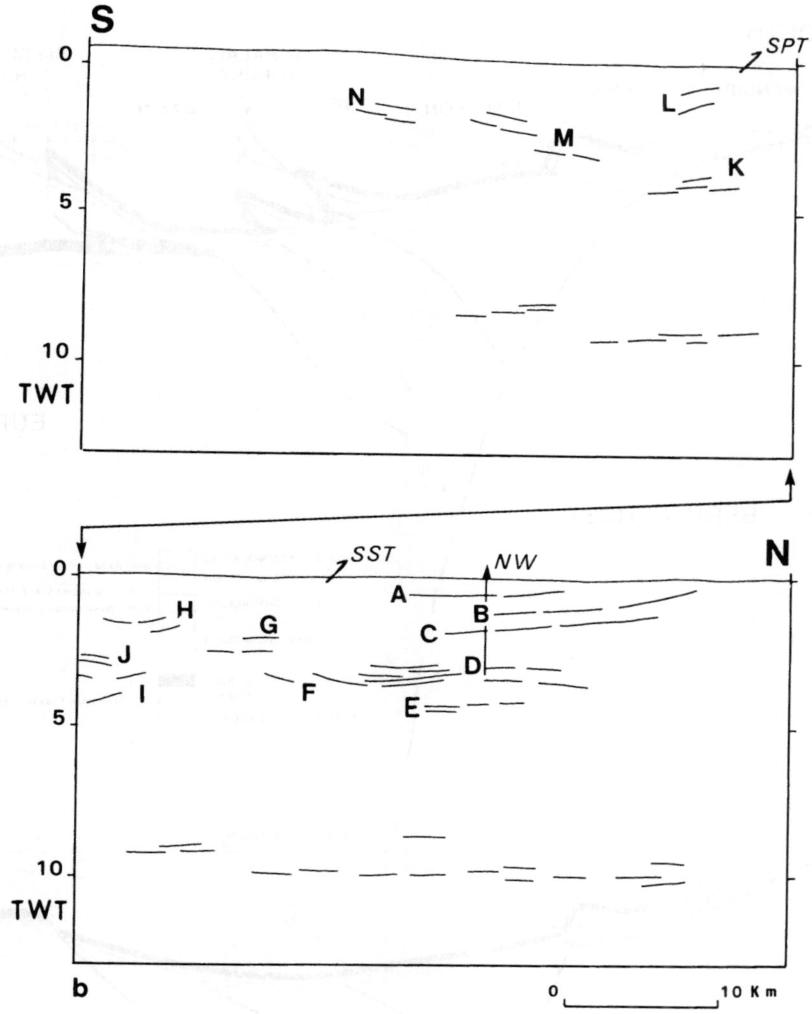

Fig. 3 *(continued)* Legend see p 204

Geological Interpretation of the Migrated Time Section

In order to obtain a depth section, a 2-D ray-based depth migration of a line drawing of the stack section was performed (Guillon 1991) using Raynaud's method (Raynaud 1988). The velocity model used for this depth migration results, for sedimentary cover, from the surface velocities calculated for the processing and available velocity measurements from wells, and for basement, from refraction data.

Using these constraints, as well as gravimetric and teleseismic data, a crustal scale structural cross section (Fig. 4a) was constructed. Only the macroscopic structures, such as large-scale thrusts and folds that can be resolved by seismic data, have been considered, and minor structures such as metric-decametric folds and the pervasive cleavage of the Mauléon basin are not included. Considering the weak deformation of the Jurassic carbonates observed on industrial seismic lines, it was assumed that the Mauléon Cretaceous flysch deposits are strongly folded by a decollement separating them from the underlying Jurassic sequence.

A tendatively balanced deformed state cross section (Fig. 4a) is restored to a pre-compressional template at end-Cretaceous time (Fig. 4b), just prior to the onset of the first shortening and inversion of the earlier extensional structures

According to extensive structural information reported in previous work, the thrust transport direction is close to N-S, and the ECORS profile has been located as far as possible from lateral ramps joining segments of the north Pyrenean thrust in order to avoid effects of strike-slip movements on the observed sturctures. In consequence, the section is drawn assuming a frontal plane strain-type inversion.

We have tried to obtain a geometrically viable solution for the bed length of base upper Cretaceous, Jurassic, top basement, and for the surface of pre-Triassic crust.

Both cover and basement structures are dominated by north vergent thrusts antithetic to the major south vergent Mendibelza thrust, along which the European crust is thrust over the Iberian.

The primary structures correspond to the basement tectonic wedging of the European crust overriding the Iberian one and the delamination of European cover sediments along the Triassic level. In relation to the tectonic

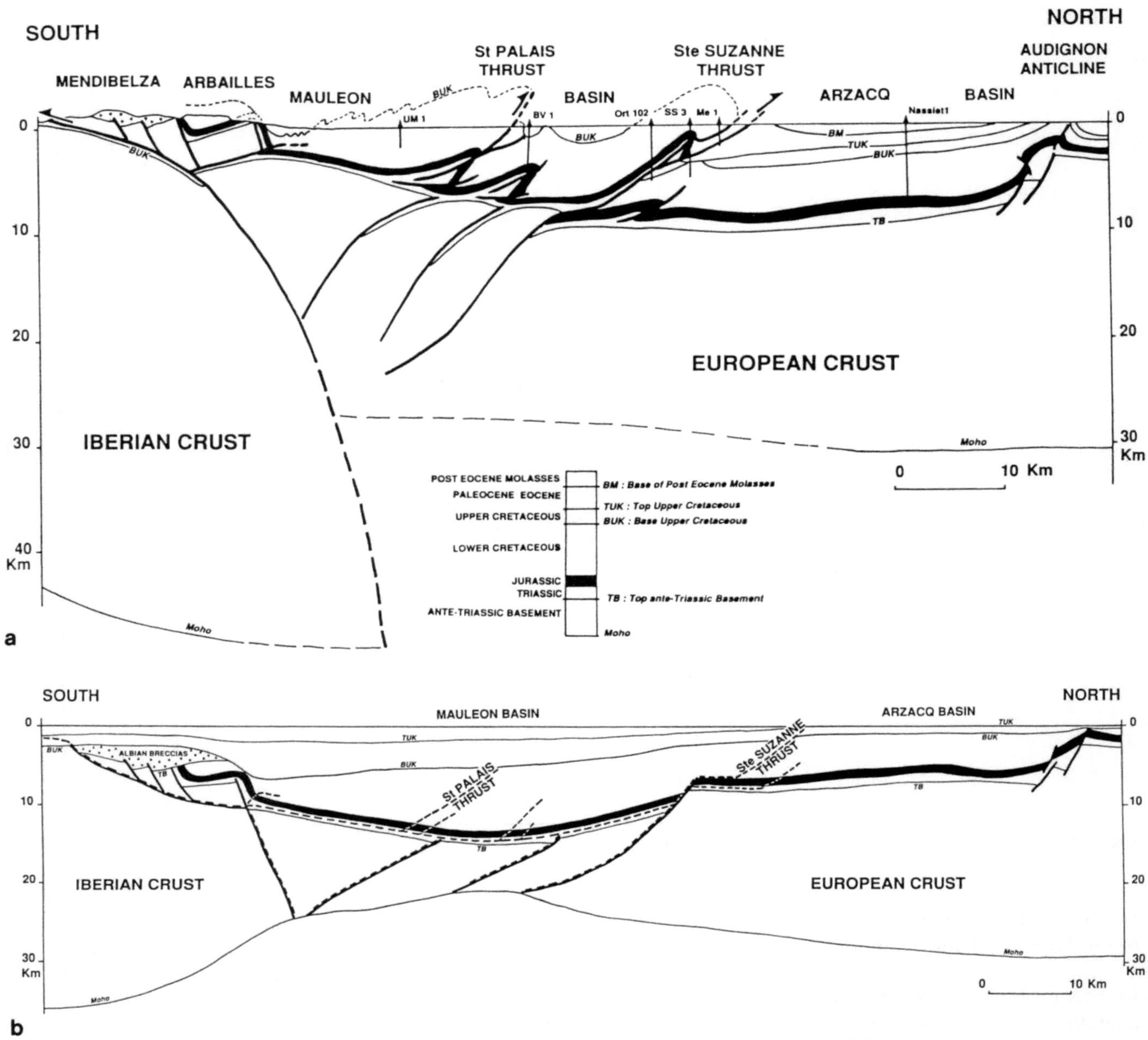

Fig. 4 a, b. Crustal-scale cross section **a** Deformed state cross section **b** Restored section at the end-Cretaceous time

wedging, the Paleozoic of the Arbailles massif is thought to be thrust northward over Triassic evaporites.

In addition to the Sainte Suzanne and Saint Palais thrusts, known from well data, a minor thrust is inferred in the footwall to the Sainte Suzanne thrust to explain the truncation of reflectors (F, G, Figure 3) The Bellevue thrust anticline in the footwall of the Saint Palais thrust was determined from the Bellevue well, which was still in progress at the time of writing.

In the proposed interpretation, the reflective markers M, N and J of Fig. 3 are interpreted to be basement or top of basement The location of the thrusts within the basement is inferred from the discordance between groups of reflectors (M and K, I and J, Fig. 3). The progressive shallowing of the top basement from the southern part of the Arzacq Basin to the axial zone is interpreted to result from the stacking of basement thrust sheets along presently relatively steep thrusts. It is postulated that these thrusts result from thrust reactivation and inversion of south-dipping normal faults inherited from Cretaceous left lateral strike-slip movements along the north Pyrenean zone.

As evidenced by previous structural data, the thickening of the Iberian crust results from both stacking of thrusts sheets and internal deformation (Camara and Klimovitz 1985, Labaume et al. 1985; Choukroune et al. 1989)

The restored section at the end-Cretaceous time (Fig. 4 b) infers an important extensional deformation related to Cretaceous strike-slip movement along the North Pyrenean Fault Zone. The Arbailles and Mendibelza blocks are interpreted to be detached on a gently north-dipping normal fault and the present north dip of the Jurassic, north of the Arbailles block, is interpreted as being related to Cretaceous extension due to the onlap of Cretaceous turbidites on previously tilted Jurassic strata. It is postulated that the Sainte Suzanne thrust results from the thrust reactivation of a Cretaceous normal fault to explain the presence of thicker and deeper water Cretaceous deposits in the hanging wall of the thrust. In a similar manner, the Audignon structure is interpreted to result from both salt diapirism localized by a Cretaceous normal fault within the basement and slight Pyrenean inversion from reactivation of the faults during shortening. Probably other normal faults offset the Jurassic deposits but are beyond the resolution of the seismic data.

To balance the surface of the pre-Triassic crust evidenced by the ECORS profile and refraction studies (Daignières et al 1982), the restored section infers an important crustal thinning especially in the central part of the Mauléon basin, where the crust was thinned up to 5 km. The subsidence history of Mauléon and Arzacq basins agrees with this interpretation (Grandjean 1992).

On the restored section, some geometric problems result from the left strike-slip movement along the north Pyrenean fault zone It is assumed that the low-angle detachment at the base of the Mendibelza-Arbailles massifs is not balanced in the plane of the section by reason of important lateral movement along the steep Iberian-European boundary.

The lack of Jurassic in the stratigraphic succession is not evidenced by surface or subsurface data in any location of the Mauléon basin. According to these data, Fig. 4b considers Jurassic to be present throughout the Mauléon basin. In consequence, the extension of Jurassic carbonates does not balance the thinning of the continental crust during Cretaceous extension. This geometric problem has not been solved.

The present thickness (15–20 km) of the continental crust beneath the Mauléon basin results from the thickening of the previously thinned crust. The total shortening of the north Pyrenean zone along this section line is estimated to be in the order of 40 km. At the basement level, most of this shortening concentrated in the area of previously thinned crust. Most of this basement shortening and later crustal thickening was probably concentrated in this area as a result of thermal weakening related to the extensional event that occurred 10 to 30 Ma before the shortening (Grandjean 1992).

Gravimetric data (Labourd Bouguer anomaly) and teleseismic delays over the same region imply high densities (at least 2.9) and high velocities (7 km/s) in the crustal block between 7 and 27 km depth beneath the Mauléon basin south of Ste Suzanne thrust (Grandjean 1992). This high-velocities, high-densities block is supposed to be lower crust or/and mantle-derived material. The granulite facies rocks of the Labourd massif (Fig. 1) may represent a higher unit of this block.

Conclusions

The Arzacq deep seismic profile is the fourth one across the intracontinental Pyrenean belt. It displays a geometry similar to the one already discovered for the eastern Pyrenees (Choukroune et al. 1989) that consists of (1) a strong Moho step 10–15 km high, (2) a northward vergence of basement and cover thrusts of the north Pyrenean zone and a southward vergence within the Iberian plate. However, the tectonic wedging is less, and the pre-Triassic crust remains presently thinner on the Arzacq transect in spite of an equivalent shortening rate. This results from higher thinning related to Cretaceous extension in the western Pyrenees. The crust thickness of the Parentis basin (Pinet et al. 1987) where, as in the Mauléon basin, 10 km of Cretaceous sediments accumulated, may represent the thickness of the crust beneath the Mauléon basin if it had not been thickened by later shortening.

Acknowledgments The authors wish to thank CNRS-INSU, IFP, and SNEA(P) for supporting the ECORS project. They also thank J. L. Mugnier and P. Gillcrist for helpful comments and condiderable improvement of the English.

References

Albarède F, Michard-Vitrac A (1978) Age and significance of the north Pyrenean metamorphism Earth Planet Sci Lett 40 327–332

Boirie JM, Souquet P (1982) Les poudingues de Mendibelza dépôts de cones sous-marins du rift Albien des Pyrénées Bull Cent Rech Expl Elf Aquitaine 6(2) 405–433

Camara P, Klimowith J (1985) Interpretacion geodinamica de la vestiente centro-occidental surpirenaica (Cuenca de Jaca-Tremp) Estud Geol 41 391–404

Canerot J (1989) Rifting éocrétacé et halocinèse sur la marge ibérique de Pyrénées occidentales (France), conséquences structurales Bull Cent Rech Expl Elf Aquitaine 13(1) 87–89

Choukroune P (1976) Structure et évolution tectonique de la Zone Nord Pyrénéenne Analyse de la déformation dans une partie de chaîne à schistosité sub-verticale Mém Soc Géol Fr LV(127) 116

Choukroune P, Mattauer M (1978) Tectonique des plaques et Pyrénées sur le fonctionnement de la faille transformante nord pyrénéenne, comparaison avec des modèles actuels Bull Soc Géol Fr 20 689–700

Choukroune P, Ecors team (1989) The ECORS Pyrenean deep seismic profile reflection data and overall structure of an orogenic belt Tectonics 8 23–39

Curnelle R, Dubois P, Seguin JC (1980) Le bassin d'Aquitaine, substratum ante-tertiaire et bordures mésozoiques les ressources naturelles, in Itinéraires géologiques Aquitaine-Languedoc-Pyrénées Bull Cent Rech Expl Elf Aquitaine Mem 3

Curnelle R, Dubois P, Seguin JC (1982) The Mesozoic-Tertiary evolution of the Aquitaine Basin. Philos Trans R Soc Lond A 305 63–84

Daignières M, Gallart J, Banda E (1981) Lateral variation of the crust in the North Pyrenean Zone Ann Geophys 37(3) 435–456

Daignières M, Gallart J, Banda E, Hirn A (1982) Implications of the seismic structure for the orogenic evolution of the Pyrenees range Earth Planet Sci Lett 57 88–110

Daignières M. De Cabissole B. Gallart J. Hirn A. Surinach E. Torne M (1989) Geophysical constraints on the deep structure along the Ecors Pyrenees line Tectonics 8(5) 1051–1058
De Voogd B. Nicolich R. Olivet JL. Fanucci F. Burrus J. Mauffret A. Pascal G. Argnani A. Auzende JM. Bernabini M. Bois C. Carmignani L. Fabbri. Finetti I. Galdeano A. Gorini CY. Labaume P. Lajat D. Patriat P. Pinet B. Ravat J. Ricci Lucchi F. Vernassa S (1991) First deep seismic reflection transect from the gulf of Lions to Sardinia (ECORS-CROP profiles in western Mediterranean) In Meissner R et al (eds) Continental lithosphere deep seismic reflections Am Geophys Union Geodyn Ser 22 265–274
ECORS Pyrenees team (1988) The ECORS deep reflection seismic survey across the Pyrenees Nature 331 508–511
Gagnepain-Beyneix J (1987) Etude expérimentale des tremblements de terre Exemple de la région d'Arette (France) Thèse Université Paris 7, 220 pp
Gallart J. Banda E. Daignières M (1981) Crustal structure of the Palaeozoic Axial Zone of the Pyrenees and transition to the North Pyrenean Zone Ann Géophys 37(3) 457–480
Gallart J. Daignieres M. Gagnepain-Beyneix J. Hirn A (1985) Relationship between deep structure and seismicity in the western Pyrenees Ann Geophys 3(2) 239–248
Golberg JM. Maluski H. Leyreloup A (1986) Petrological and age relationship between emplacement of magmatic breccia, alkaline magmatism and static metamorphism in the North Pyrenean Zone Tectonophysics 129 275–290
Grandjean G (1992) Mise en évidence des structures crustales dans une portion de chaine et de leur relation avec les bassins sédimentaires Application aux Pyrénées occidentales au travers du projet ECORS Arzacq-Pyrénées Thèse Université Montpellier II, 291 pp
Guillon D (1991) Approche des Pyrénées occidentales par sismique réflexion verticale et par sismologie locale DEA Université Montpellier, 48 pp
Hirn A. Poupinet G. Wittlinger G. Gallart J. Thouvenot F (1984) Pyrenees and Alps teleseismic prospecting of lithospheric contrasts Nature 308 531–533
Labaume P. Séguret M. Seyve C (1985) Evolution of a turbiditic foreland basin and analogy with an accretionary prism example of the Eocene south-Pyrenean basin Tectonics 4 661–685
Larau geological map (1970) Scale 1/50000 BRGM, Orléans
Montigny R. Azambre B. Rossy M. Thuizat R (1986) K-Ar study of Cretaceous magmatism and metamorphism from the Pyrenees age and length of rotation of the Iberian peninsula Tectonophysics 129 257–273
Pinet B. Montadert L. Ecors Scientific Party (1987) Deep seismic reflection and refraction profiling along the Aquitaine shelf (Bay of Biscay) Geophys J R Astr Soc 89 305–312
Raynaud B (1988) A 2-D ray-based depth migration method for deep seismic reflections Geophys J R Astr Soc 93 163–171
Séguret M (1972) Etude tectonique des nappes et séries décollées de la partie centrale du versant sud des Pyrénées caractère synsédimentaire, rôle de la compression et de la gravité Ser Geol Struct, USTL Montpellier 2, 155 pp
Séguret M. Daignières M (1986) Crustal scale balanced cross-sections of the Pyrenees, discussion Tectonophysics 129 303–318
Velasque PC, Ducasse L (1987) Geotraverse équilibrée dans les Pyrénées occidentales argument en faveur d'un rift intraplaque pour expliquer l'amincissement crustal crétacé C R Acad Sci 304, 2 383–386
Vielzeuf D. Kornprobst J (1983) Crustal splitting and the emplacement of Pyrenean lherzolites and granulites Earth Planet Sci Lett 67 87–96
Zolnai G (1971) Le front nord des Pyrénées occidentales In Histoire structurale du Golfe de Gascogne Technip, Paris IV(5) 1–10

A Reinterpretation of the Stress Regime of the Aquitaine Basin, Southwestern France, and Implications for Hydrocarbon Recovery

J. S. Bell and G. Caillet

Abstract. The logs and drilling records of exploration wells often can provide enough information to characterize the stress regime of a sedimentary basin, provided its geomechanical framework is understood. In the Aquitaine Basin, there are not many measurements for estimating subsurface in-situ stress magnitudes, but a significant number of wells have been logged suitably to provide stress directional data. Enough stratigraphic and structural information is available to assess the geomechanical response of the basin fill compared to that of the underlying basement. This study utilizes breakout ovalization analysis of 55 wells. The resulting horizontal principal stress orientations show no clearcut directional homogeneity. Around Bordeaux, σ_H exhibits a consistent NW-SE orientation. Towards the Pyrenees, the situation becomes less definitive. NE-SW σ_H orientations are widely present, but NW-SE σ_H orientations also occur. Highgrading the data does not improve the resolution.

This widespread lack of directional homogeneity, and the frequency with which ovalization was found to be aligned with well deviation, suggest that horizontal stress anisotropy is weakly developed in much of the Aquitaine Basin today. It is thought that this condition has arisen because most of the rocks in which stress orientations were measured are geomechanically detached from "basement" and cannot transmit a crustal stress signature. The detachment agent is believed to be early Mesozoic salt. As a result, the propagation directions of hydraulic fractures may vary considerably from place to place. It is also likely that preferred orientations for open fractures will not be consistent regionally.

Introduction

In most heavily drilled sedimentary basins, drilling and logging activities have yielded enough information to quantify the contemporary stress regime. Normally, there is sufficient documentation to show how stress magnitudes vary with depth and location, and to determine the orientations of the principal stresses. In principal, the methods used to map stress regimes are simple (Bell 1990a) and the results are dependent on the density of data available. However, they need to be interpreted in a geomechanical context.

To simplify the stress analysis, it is assumed that one principal stress is oriented vertically and therefore that the other two must be directed horizontally. This is a reasonable proposition, unless extreme factors such as significant surface relief render it questionable. With this generalization, good estimates can be made of the magnitudes, the vertical stresses, and the smaller horizontal principal stresses (magnitudes of the larger horizontal principal stresses usually are not measured), and accurate determinations can be made of the directions of all three principal stresses. There are, however, some provisos. For the most part, the measurements from which stress information can be derived were made for other purposes. They are not always suitable for precise inferences. Also, the most clearcut results are obtained from basins where there are significant differences in the magnitudes of the principal stresses. Where conditions are close to isotropic, the information tends to be less definitive.

This chapter reports the results of an investigation of the present-day stresses that are acting on sedimentary rocks in the Aquitaine Basin in southwest France. This basin contains France's largest oil and gas fields (Fig. 1). Production is centered in two sub-basins: the northern Parentis Basin and the southern Arzacq Basin (Winnock and Pontalier 1970; Winnock 1972; Mullan 1984). As in many onshore basins in Europe, production has declined in recent years and costly engineering procedures are being implemented, or being considered, to improve recovery. To be successful in these endeavors, it is necessary to have a detailed knowledge of the physical setting of the hydrocarbon pools and of their internal reservoir geometry. A key component of the physical setting is the contemporary stress regime, which describes the action of far field compressive forces on the reservoir rocks. In this chapter we examine the situation in the Aquitaine Basin to determine what can be documented there with respect to its stress regime, and we speculate about how this might have arisen and what it implies for hydrocarbon recovery.

Stress Magnitude Estimates for the Aquitaine Basin

Methods and Results

No subsurface measurements of in-situ stress magnitudes have been made in the Aquitaine Basin, so estimates were used. The vertical principal stress, σ_v, at a specific

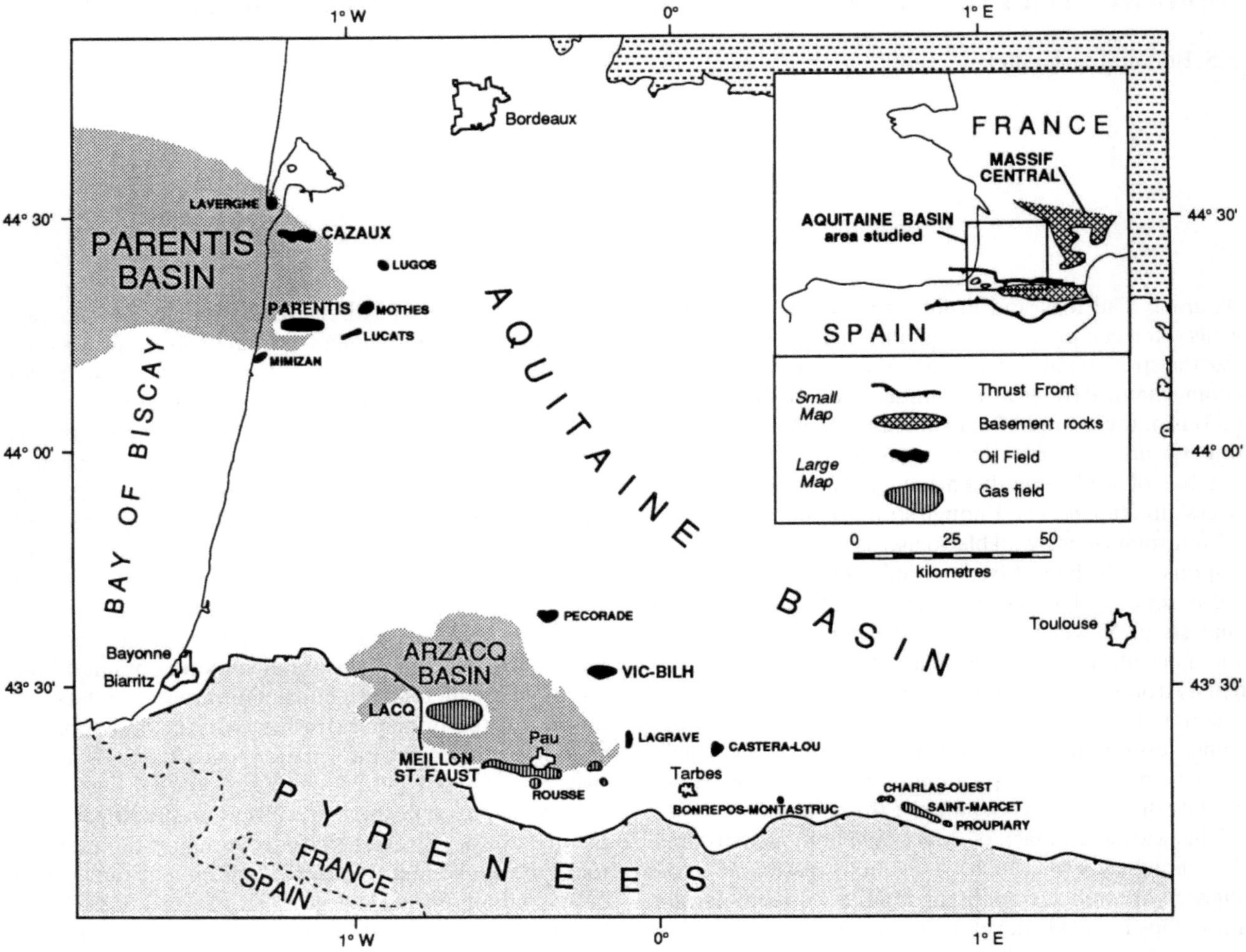

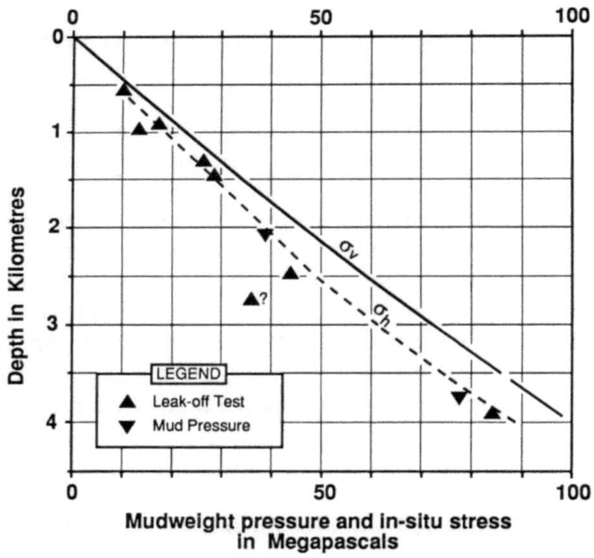

Fig. 2. Depth pressure relations of leak-off test pressures measured in five wells in the Aquitaine Basin, and elevated drilling mud pressures recorded in one well These pressures permit estimates of the magnitude of the smallest principal stress (in this case σ_h) to be made The leak-off test *labeled with a question mark* is one where a leak-off fracture may not have been initiated The vertical principal stress, σ_v, is estimated from density logs

▲ **Fig. 1.** Location map showing the oil and gas fields of the Aquitaine Basin and the area in which subsurface stress information was gathered

depth was equated with the overburden load, and the latter calculated by integrating a density log.

Leak-off tests in five wells give upper limits for the smallest principal stress magnitudes (Breckels and Van Eekelen 1981; Ervine and Bell 1987). High mudweights were used to drill the lower part of one well, without initiating any fractures. They suggest lower limits for smallest principal stress magnitudes The leak-off pressures and mudweights diagnose principal stress magnitudes that are less than σ_v, so they are presumed to refer to σ_h, the smaller horizontal stress (Fig. 2) It would appear that the magnitude of this principal stress amounts to approximately 80 % of the vertical stress in the Aquitaine Basin However, these limited data do not indicate the magnitude of σ_H, the larger horizontal principal stress, or show whether there is any significant horizontal stress anisotropy.

Stress Orientation Measurements in the Aquitaine Basin

Methods

Breakouts, or ovalized intervals in boreholes, have been shown to be sensitive indicators of far field stress trajectories (Bell and Gough 1979; Bell 1990a). Provided that significant horizontal stress anisotropy exists (Zoback et al. 1985; Maury 1993) and the well is nearly vertical in inclination (Mastin 1988), the long axis of a breakout (direction of ovalization) will be aligned parallel to σ_h and perpendicular to σ_H (Fig. 3). In many sedimentary basins, especially in North America, there is a notable consistency to the orientation of breakouts within individual wells and between neighboring wells (e.g., Bell and Babcock 1986; Plumb and Cox 1987; Dart and Zoback 1989), and the stress trajectories delineated by breakouts are confirmed by other indicators (Plumb and Hickman 1985; McLellan 1988; Zoback and Healy 1992). This agreement, and their directional homogeneity, argues strongly that breakouts can map horizontal stresses reliably and that, typically, sedimentary basins have regionally consistent directional stress signatures.

The term "breakout" was coined by Babcock (1978) to describe caving that results in lateral elliptical elongation of a borehole over discrete depth intervals. Breakouts were attributed to the action of anisotropic far field stresses by Bell and Gough (1979), and the term has since been widely used to describe caving that laterally elongates boreholes and is stress-initiated. Thus, the term "breakout" has become genetic. In this study, we encountered a large amount of borehole elongation that did not appear to be generated by far field stresses. To describe it, we employ the term "ovalisation", which was introduced by Janot et al. (1988). We use ovalisation to describe borehole intervals that are elongated elliptically, but where the initiating cause was not necessarily anisotropic stress. We use the term breakout where we are convinced that the elliptical caving was caused by localised shear fracturing of the borehole walls, promoted by differences in far field stress magnitudes.

Ovalization in the Aquitaine Basin was recorded by the calipers of four-arm dipmeter logs, originally run for structural information (Cox 1970). Their oriented caliper pad extension records show where borehole ovalization has occurred and document the orientation of the long axes of such intervals. The following criteria, evolved during studies by Babcock (1978), Fordjor et al. (1983); Springer (1987), and Dart and Zoback (1989), were applied to distinguish breakouts:

1. The dipmeter log must record systematic tool rotation above and below an ovalised interval.
2. Tool rotation must cease over the ovalised interval, so that there is a relatively constant azimuth in the range of +/− 10° for the No. 1 caliper arm.
3. One pair of calipers should record a diameter larger than hole gauge and the other pair should record a diameter which is not less than hole gauge. Ideally, the smaller caliper curve should remain straight, consistent with an intact borehole wall.
4. The larger diameter curve should record a vertically rapid increase in diameter at the top and bottom of the ovalised interval.
5. The borehole trajectory should be oriented within 5° of vertical.
6. The long axes of ovalized intervals should differ in their orientation by more than 30° from the well deviation direction. They should not be subparallel.

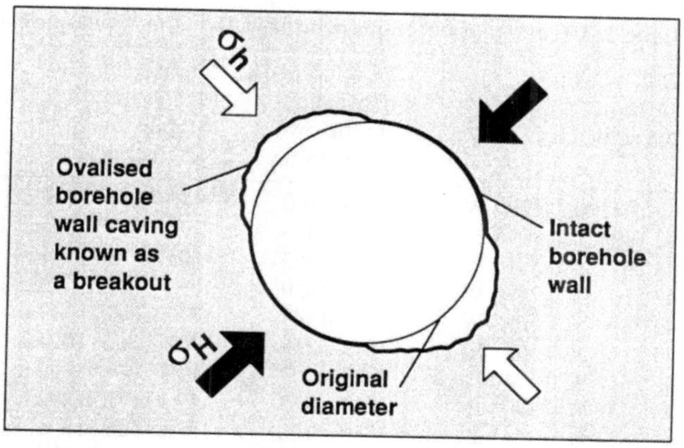

Fig. 3. Vertical view of an ovalized borehole looking down a well The far field stresses are shown on the assumption that their amplification around the borehole walls has led to shear fracturing, which has elongated the wellbore laterally Such stress-generated ovalizations are described as breakouts

If all these criteria were met, the ovalized intervals were considered to be stress-induced breakouts. Criterion 6 was developed on the basis of experience gained mapping stresses locally around producing fields in the Aquitaine Basin. It was especially important to apply because ovalized borehole intervals were widely present, and many of them were closely aligned with well deviation directions. It appeared that a considerable amount of borehole ovalization was caused by drill pipe wear on the walls of the holes, even if the wells were only mildly inclined. Dart and Zoback (1989) reported similar behavior in North American wells, but noted that it occurred typically in holes inclined at more than 5° from vertical. In the Aquitaine Basin, this was not the case: the long axes of ovalized zones commonly followed deviation over well intervals that were inclined within 1° or less of vertical. This made it difficult to identify ovalized zones (breakouts) that were unequivocally generated by unequal far field stresses.

Results

Fifty five wells were studied. They were selected on the basis of their geographic distribution and total depths, and because their trajectories were close to vertical. Ovalized zones that were not eliminated by the six criteria

Table 1. Breakout orientations from 55 wells in the Aquitaine Basin

Well location °Greenwich	Depth range of breakouts (m) below rotary table	Major population Mean azimuth (Stand deviation) Net thickness	Minor population Mean azimuth (Stand deviation) Net thickness	Inferred SHmax azimuth	Quality rating	Rotary table elevation (m)
43 194° N. 0 239° W	1620–3100	94 1° (20 7) 480 m	–	4 1°	C	46
43 160° N. 0 899° E	1165–1985	29 6° (6 1) 380 m	–	119 6°	A	7
43 161° N. 0 881° E	445–1453	110 0° (36 7) 131 m	9 4° (11 6) 85 m	20 0°	DD	48
43 210° N. 0 249° W	1800–5250	102 9° (20 8) > 300 m	–	12 9°	C	45
43 550° N. 1 578° W	555–1148	177 7° (13 2) 149 m	123 7° (2 1) 61 m	87 7°	CD	32 5
43 217° N. 0 582° E	2841–3259	94 5° (5 5) 1004 m	–	4 5°	C	43
43 261° N. 0 654° E	3620–4402	114 2° (13 5) 409 m	–	24 2°	B	7
43 243° N. 0 380° E.	1017–4423	49 1° (21 2) 492 m	170 0° (0 0) 4 m	139 1°	CE	93
43 265° N. 0 454° W	2550–5100	127 6° (26 3) n a	37 1° (11 7) n a	37 6°	C n a	41
43 454° N. 1 083° W	2460–4852	139 4° (19 8) 294 m	–	49 4°	B	5
43 257° N. 0 181° W	519–1967	142 6° (22 4) 327 m	–	52 6°	C	93
43 324° N. 0 165° E	3033–3131	2 3° (0 8) 9 m	–	92 3°	E	5
43 348° N. 0 169° E	1652–2839	27 0° (11 4) 83 m	159 0° (18 2) 10 m	117 0°	DE	5
43 126° N. 0 615° E	2812–4226	148 6° (12 4) 234 m	30 2° (9 8) 233 m	58 6°	BA	77
43 526° N. 0 094° W	837–3750	26 3° (6 0) 205 m	146 2° (26 8) 119 m	116 3°	AD	46
44 566° N. 0 922° W	110–140	67 0° (3 0) 30 m	–	157 0°	E	48
43 729° N. 0 859° W	2825–3458	146 8° (19 6) 191 m	–	56 8°	C	74
43 533° N. 0 265° W	1265–4224	25 1° (16 3) 185 m	102 0° (14 3) 46 m	115 1°	CE	48
44 602° N. 1 065° W	2329–2333	27 0° (0 0) 4 m	–	117 0°	E	48
43 406° N. 1 364° W	1817–3871	84 0° (7 3) 240 m	173 0° (3 0) 11 m	174 0°	AE	78
43 419° N. 0 646° W	117–275	9 9° (10 8) 126 m	–	99 9°	C	67
43 411° N. 0 632° W	23–108	32 5° (16 4) 55 m	–	122 5°	D	25
43 352° N. 0 106° W	2081–2273	114 5° (7 8) 25 m	–	24 5°	E	67
43 341° N. 0 234° E	785–2709	141 3° (14 9) 189 m	10 1° (18 9) 101 m	51 3°	CC	46
43 122° N. 0 416° E	317–3108	125 6° (17 1) 89 m	25 4° (14 3) 70 m	115 4°	DD	64
43 216° N. 0 632° W	398–3403	140 2° (19 5) 839 m	43 3° (13 3) 439 m	50 2°	BB	76
43 253° N. 0 514° W	1420–3850	71 2° (24 8) > 300 m	–	161 2°	C	41
44 841° N. 0 989° W	1827–1850	32 0° (3 0) 23 m	–	122 0°	E	55
43 782° N. 0 273° W	672–3745	55 4° (23 0) 225 m	121 1° (9 6) 107 m	145 4°	CC	78
44 478° N. 0 885° W	1011–1753	64 4° (8 8) 220 m	158 8° (7 8) 200 m	154 4°	AA	53
43 842° N. 0 361° W	2262–3795	106 7° (9 2) 247 m	–	16 7°	A	5
43 276° N. 0 342° W	2450–3200	24 9° (29 2) n a	–	114 9°	n a	47
44 150° N. 1 124° W	2267–2332	55 4° (6 6) 41 m	–	145 4°	E	38
43 119° N. 0 476° W	236–3325	53 6° (20 1) 548 m	–	143 6°	C	65
43 211° N. 0 370° E	820–1862	57 0° (15 0) 283 m	125 5° (24 4) 235 m	147 0°	BC	47
43 445° N. 0 961° E	3026–3047	154 9° (3 8) 13 m	–	64 9°	E	47
43 597° N. 0 692° W	1335–2876	120 3° (8 0) 213 m	49 7° (8 3) 52 m	30 3°	AD	93
43 756° N. 0 026° W	360–705	153 6° (15 0) 65 m	–	63 6°	D	34
43 309° N. 0 248° E	1624–4066	30 4° (6 7) 645 m	–	120 4°	A	48
43 114° N. 1 410° E	998–5545	92 4° (16 9) 1665 m	–	2 4°	B	92
43 355° N. 0 259° W	515–2652	134 8° (34 9) 330 m	–	44 8°	D	7
43 648° N. 0 371° W	1370–2823	162 2° (15 8) 68 m	–	72 2°	D	5
43 641° N. 0 371° W	2269–2819	151 9° (15 6) 126 m	–	61 9°	C	5
43 648° N. 0 353° W	1975–3143	159 5° (6 3) 199 m	–	69 5°	C	41
43 310° N. 0 049° W	2119–5659	128 8° (15 5) 593 m	59 2° (12 5) 56 m	38 8°	BD	43
43 586° N. 0 994° W	986–3261	146 3° (3 4) 214 m	–	56 3°	A	65
43 539° N. 1 240° W	55–145	142 7° (3 4) 23 m	–	52 7°	E	74
43 309° N. 0 568° E	4710–4980	58 5° (11 5) 113 m	–	148 5°	C	7
43 246° N. 0 167° W	2400–3650	97 0° (13 7) n a	8 5° (17 1) n a	7 0°	n a	46
43 509° N. 0 320° W	1855–5447	28 8° (10 9) 533 m	132 3° (25 6) 274 m	118 8°	AC	5
44 468° N. 0 990° W	2487–2490	82 0° (0 0) 3 m	–	172 0°	E	75
43 753° N. 0 661° W	2131–2768	99 9° (13 3) 25 m	–	9 9°	E	5
43 523° N. 0 212° W	970–3316	102 3° (15 1) 799 m	–	12 3°	B	47
43 516° N. 0 175° W	1014–2597	93 4° (21 2) 149 m	–	3 4°	D	65
43 520° N. 0 149° W	999–2607	122 2° (23 5) 233 m	–	32 2°	C	63

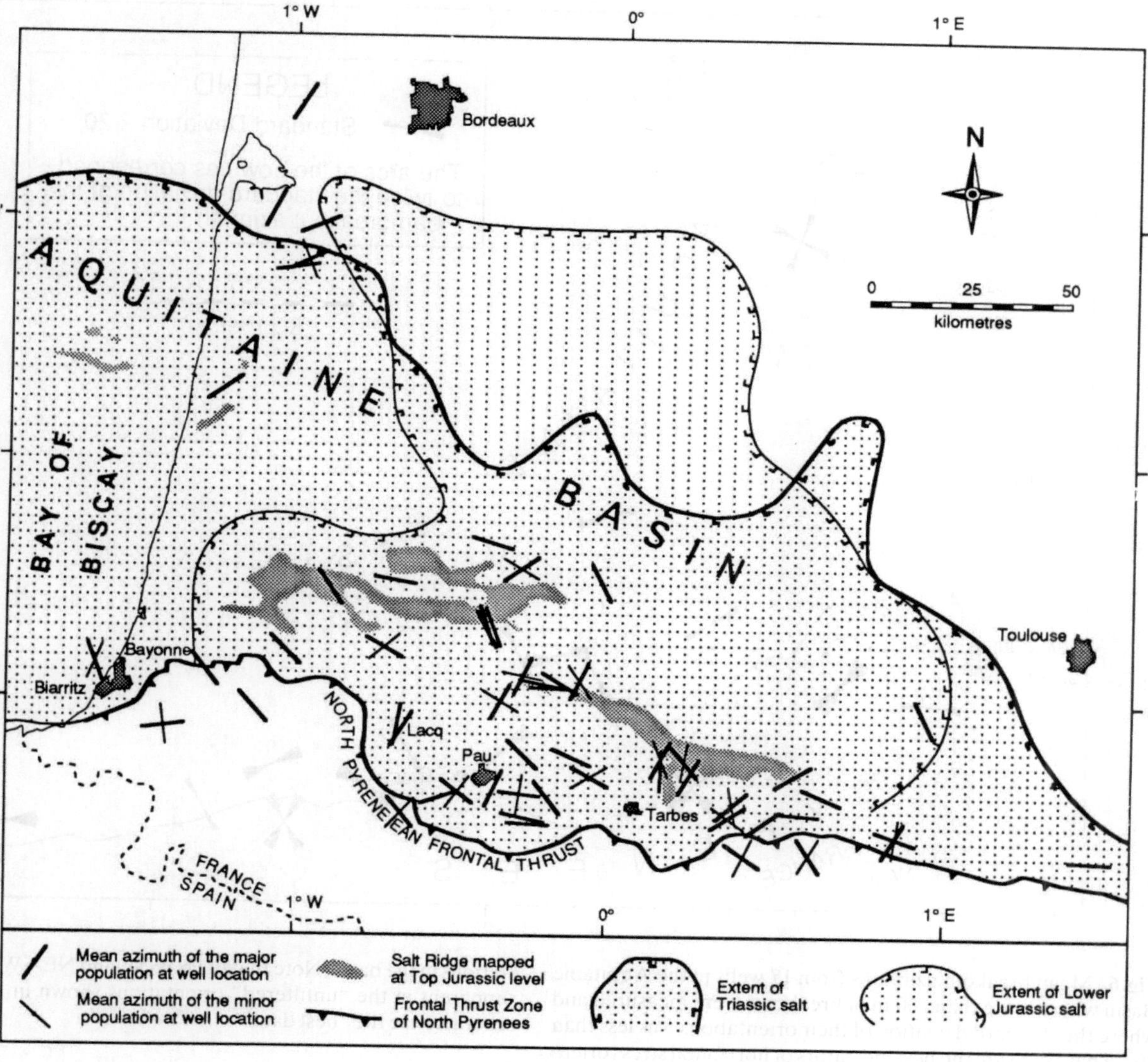

▲

Fig. 4. Mean breakout azimuths of 55 wells in the Aquitaine Basin Note that there is no clearcut directional consistency, but that there is a tendency for breakouts to be oriented NW-SE or NE-SW depending on location. Also shown are the extent of Triassic and early Jurassic halite deposition and the locations of diapiric ridges A distorted blanket of halite underlies much of the Aquitaine Basin

listed above were documented in the following manner. If the zone was less than 10 m long, a single measurement of the orientation of its long axis was made. For longer ovalized zones, orientation measurements were made at 10 m intervals and averaged. The orientations for each well were then histogrammed to identify populations. Mean population azimuths and standard deviations were later calculated for each well, with each individual breakout orientation being weighted according to its length (Mardia 1972).

The mean azimuths of the 55 wells are shown in Fig. 4 and listed in Table 1. They do not support a uniform configuration of horizontal stress orientations as has been found in other regions such as the Western Canadian Se-

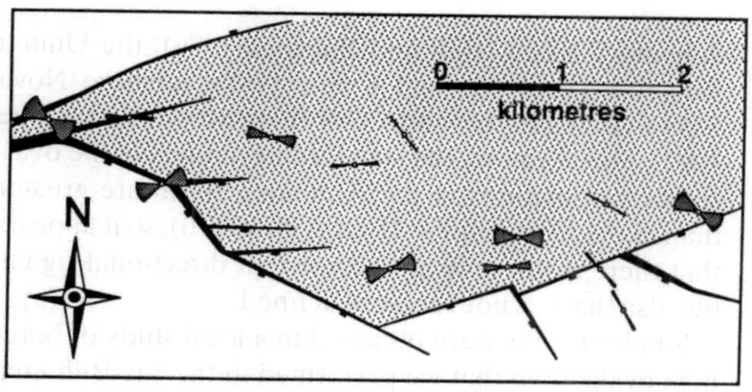

Fig. 5. Mean breakout orientations from 13 wells at the Vic-Bilh oil field in the Aquitaine Basin Measurements were made in unfaulted Tertiary and Cretaceous "cover" rocks, which overlie the fractured reservoirs that produce oil All the breakouts occur in well intervals that are inclined at less than 10° from vertical Note that the mean breakout azimuths show a degree of common alignment in a WNW-ESE direction The *speckled area* denotes the approximate extent of the Vic-Bilh structure beneath the unfaulted cover rocks. The faults shown are bounding fractures surrounding the Vic-Bilh structure They do not penetrate up into the "cover" rocks in which the breakouts were identified

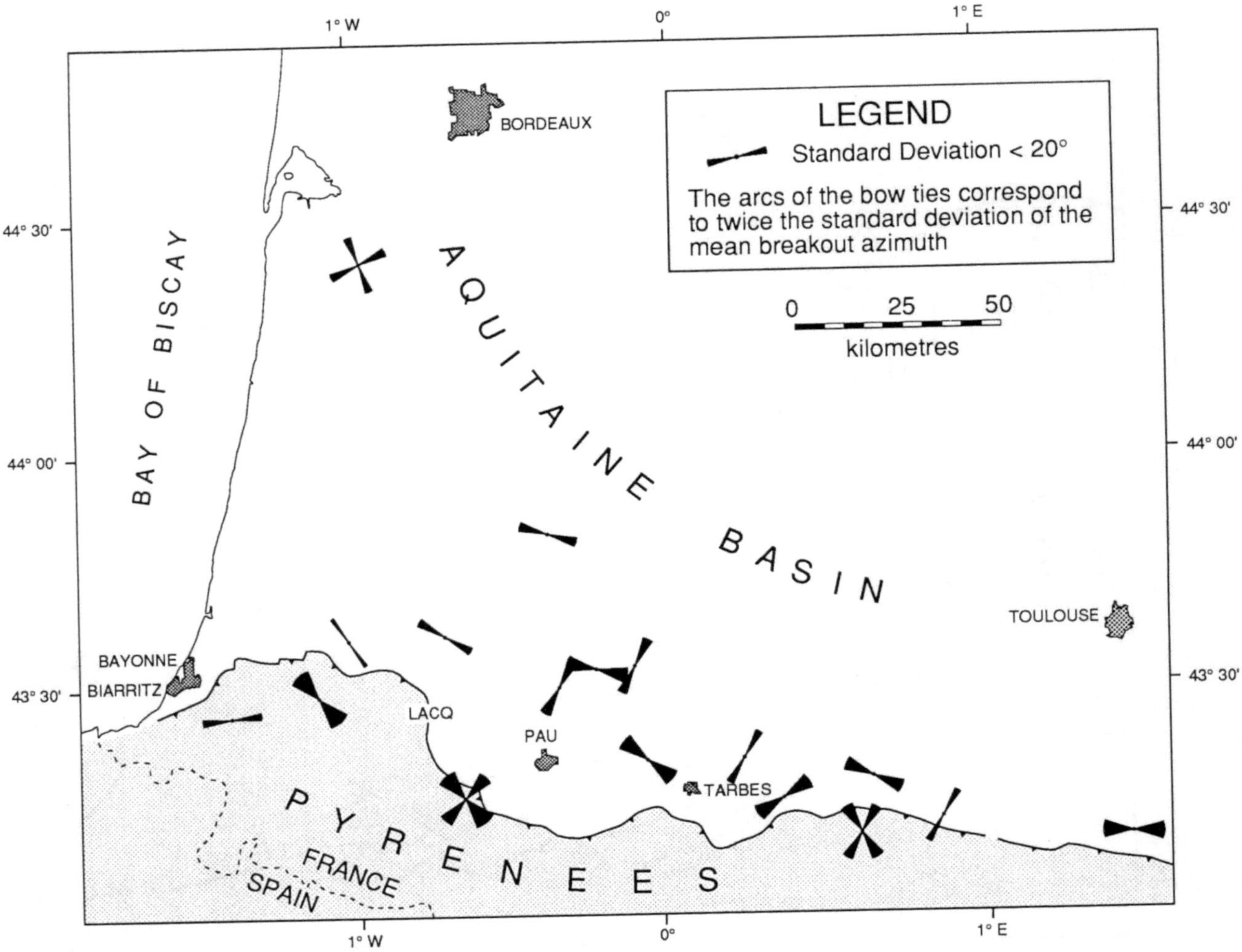

Fig. 6. Mean breakout azimuths from 17 wells in the Aquitaine Basin where more than 200 m of breakouts were measured and where the standard deviation of their orientations was less than 20° These data are the best indicators of horizontal stress orientations in the basin Note that the NW-SE and NE-SW azimuths, apparent in the "unfiltered" orientations shown in Fig 4, also characterize the "best data"

dimentary Basin (Bell and Babcock 1986), the United Kingdom (Brereton and Evans 1987), offshore Nova Scotia (Bell 1990b), and California (Mount and Suppe 1992). Furthermore, the standard deviations of the ovalization populations in the Aquitaine Basin are greater than in these other areas (Bell et al. 1992b), so it appears that there is not only an inconsistent directional signal, but also that it is not very well defined.

Similar results were obtained in a local study of borehole ovalization that we performed in the Vic-Bilh area (Fig. 1), where we examined 47 closely spaced exploration and production wells. Most of the Vic-Bilh wells were drilled at inclinations greater than 10° from vertical in their lower parts. These intervals contained ubiquitous ovalization parallel to the well deviation directions. In the less inclined intervals, ovalization was significant and over 80% parallelled the well deviation directions. This latter ovalization was attributed to drill bit and drill pipe wear on the borehole walls. Approximately 15% of the ovalization bore no obvious relationship to well trajectories and appeared to represent genuine breakouts generated by far field stresses. These ovalizations had a reasonably well-defined directional signature. Figure 5 portrays the measurements from 13 wells made in intervals where the vertical deviation was less than 10°. The inferred breakouts were measured over drill depths ranging between 924 and 2370 m in Tertiary and Cretaceous rocks. Three-D seismic coverage shows that these rocks are not offset by any major faults that pass close to the breakout intervals, so it is unlikely that in-situ stresses are reoriented locally by structural discontinuities (Bell et al 1992a). Although there is variation between wells, the mean azimuths of ovalization point to a WNW-ESE orientation for σ_h The lack of a well-developed directional consistency for breakouts is believed to be indicative of a weak signal. In other words, we suspect that horizontal stress anisotropy is not strongly developed in the Vic-Bilh area, but that there is enough inequality in the magnitude of the horizontal principal stresses for the boreholes to feel it locally and to respond by ovalizing themselves. However, it is clear that borehole trajectory is the cause of much more ovalization than the far field stress regime.

Experience at Vic-Bilh suggested that the regional breakout data should be high-graded, if possible. In an earlier study, we separated the breakouts measured at depths greater than 3000 m, on the assumption that the stress tensor signal would be stronger at depth and that the deeper breakouts might, therefore, better reflect the regional stress regime (Bell et al. 1992b). This exercise reduced the database to 30 wells, but it did not give rise to a more directionally coherent configuration. However, it did emphasize the NW-SE populations in the southwestern and southeasternmost areas, and the NE-SW signal in the south central region became more prominent. A second high-grading exercise involved considering boreholes with the best directional statistics, so we reduced the coverage to 17 wells, where the standard deviation of a breakout population was less than 20° and where the mean azimuth was averaged from more than 200 m of ovalized hole (Fig. 6). Again, the exercise emphasized the NE-SW and NW-SE populations and pointed towards a degree of directional orthogonality (Bell et al. 1992b).

After both high-grading exercises, the orientations still appeared to fit an orthogonal grid with NE-SW and NW-SE horizontal stress trajectories However, in the southern part of the basin, the larger and smaller horizontal stresses appeared to alternate locally.

Interpretation of the Stress Regime in the Aquitaine Basin

Any explanation for the breakout configuration in the Aquitaine Basin must account for two facts: (1) the disproportionately large amount of ovalization that is aligned with well deviation directions and (2) the apparent "orthogonality" of mean well azimuths for the genuine stress-induced breakouts.

The large amount of ovalization that parallels deviation, even in well intervals that are inclined at less than 1° from vertical, implies that pipe wear during drilling and drill bit abrasion during drillpipe entry have promoted most of the noncircular caving of the wells. Stresses appear to be insufficiently anisotropic to promote borehole elongation (Zoback et al. 1985) except in a few intervals, unlike in western Canada, for example, where that mechanism is regionally pervasive (Bell and Babcock 1986). However one regards these data, it is clear that borehole trajectory is much more influential in promoting ovalization than is in-situ stress anisotropy.

A lack of preferred horizontal stress trajectories was also observed around the Lacq field (Fig. 1). Limited oil production began in 1950, and major gas production started in 1963. Small to moderate-size earthquakes of magnitude 1.0 to 4.2 began to occur at Lacq in the early 1970s, about 10 years after the gas production began (Grasso and Witlinger 1990). The P-axis directions for the focal mechanisms of earthquakes that occurred during the 1976–1979 period show no obvious preferred orientation (Grasso and Feignier 1990), but are compatible with a radial configuration centered on the Lacq anticline. Therefore it is likely that the domal structure, and recent deformation caused by gas extraction and water injection, control the local stress trajectories (Grasso and Feignier 1990). These authors also point out that poroelastic modeling (Segall 1989) suggests that the stress changes associated with the gas production must be very small (maximum 1 MPa), although they are sufficient to trigger seismic activity. They emphasize that such small stress changes could not control stress trajectories in an active tectonic context, unless the regional deviatoric stress was extremely low (Grasso and Feignier 1990). This is precisely what is suggested by the regional ovalization and breakout data, that horizontal stress anisotropy is minimal in the Aquitaine Basin.

Thus, the stress regime for much of the Aquitaine Basin appears to be one where the largest principal stress is vertical and the two horizontal stresses are smaller and nearly coequal in magnitude.

Proposed Explanation

Why should the Aquitaine Basin be characterized by weak horizontal stress anisotropy? Other basins show significant magnitude differences between the two horizontal principal stresses (e.g., Western Canada and the St. Lawrence Platform, for summary data see Adams and Bell 1991). The answer may lie in the geomechanical setting. Most of the breakouts reported here were measured in wells that penetrate rocks which overlie a thick interval of Triassic halite and evaporites (Fig. 4). Up to 600 m of salt is present (Curnelle et al. 1982). Above the Triassic beds in the western part of the Aquitaine Basin are Lower Jurassic evaporites (Curnelle et al. 1982), again including halite in the succession (Fig. 4). It is suggested here that these basal Mesozoic evaporites geomechanically detach the sediments above them from the Paleozoic and Precambrian rocks below. In this model, the evaporites are acting as a surface of decollement and prevent the stress signature of the basement from being transmitted into the mid-Jurassic and younger beds. These latter units are decoupled and are effectively "floating" on the salt, particularly in the southern part of the basin where the salt is thickest, as indicated on the cross section through the Lacq area (Fig. 7).

On the other hand, according to this hypothesis, the most northerly wells located west and southwest of Bordeaux (Fig. 4) should be attached to basement and they should reflect its stress signature, because there is no salt separating them from pre-Mesozoic units. Interestingly, these particular wells all exhibit mean breakout azimuths between ENE-WSW and NE-SW, indicating a σ_H orientation of NNW-SSE to NW-SE (Figs. 4 and 6). This orientation of σ_H is compatible with the regional stress trajectories (Muller et al. 1992) in adjacent areas in western Europe (Fig. 8). What is not known is whether the horizontal stresses affecting these northerly wells exhibit greater anisotropy than those in the rest of the Aquitaine Basin. It would be worthwhile to determine this.

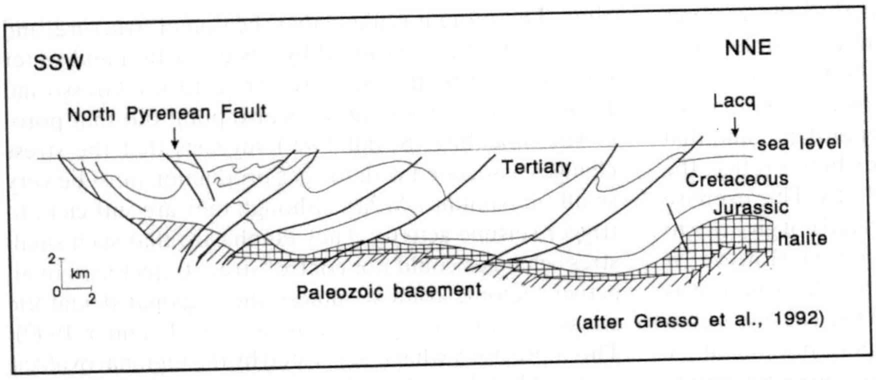

Fig. 7. Structural cross section from the Pyrenees to the Lacq gas field (After Grasso et al 1992) Note that Jurassic-Tertiary rocks are separated from older Paleozoic rocks by a layer of halite This halite layer is thought to geomechanically detach the rocks above it from those below it, so that the "basement" cannot transmit stress to the "cover"

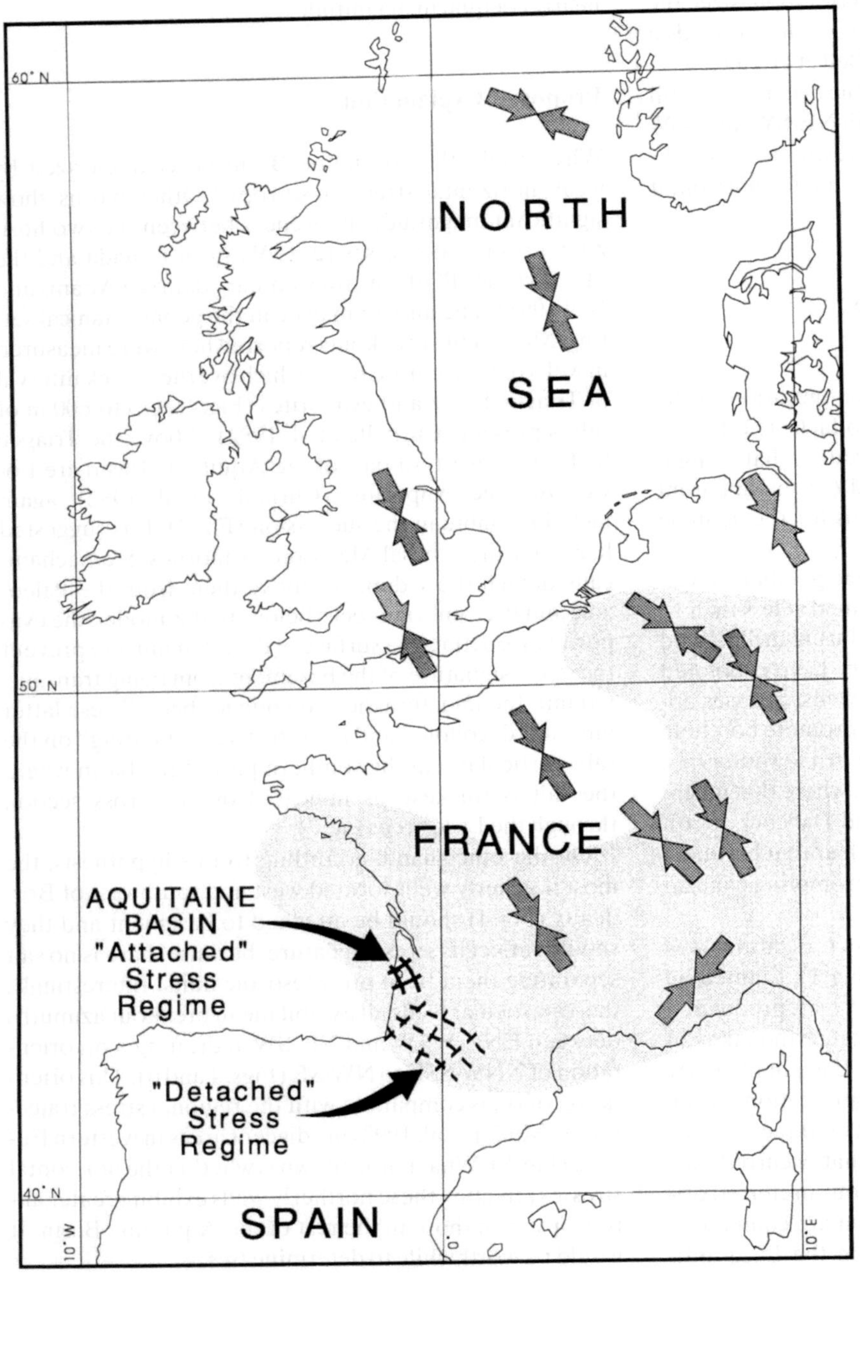

If this hypothesis of geomechanical detachment and minimal stress anisotropy is valid for most of the basin, how can we account for the apparent orthogonality of stress trajectories that was alluded to above? In an earlier paper (Bell et al. 1992b), we attributed the orthogonality to weakly developed stress anisotropy in the region and inferred that the trajectories reflected the regional crustal signal, albeit poorly. Previously (Bell et al. 1992a), we had suggested that the lithosphere beneath northwest Europe might be characterized by weak horizontal stress anisotropy, and that this reflected the relatively slow rate of absolute motion of the European plate (Minster and Jordan 1978).

We now believe that the apparent orthogonality of stress trajectories in the southern part of the Aquitaine Basin is likely to be structurally controlled. As Figs. 4 and 6 indicate, the mean axes of the breakouts, or σ_h, are either approximately parallel to the Pyrennean Mountain Front and to many of the salt ridges, or else they are more or less perpendicular to them. Breakout axes are also either approximately parallel or perpendicular to the regional structural dip of the basin. This parallelism is significant because bedding dip and faulting are likely to give rise to local extensional or compressional stress regimes. If the prevailing horizontal stresses exhibit minimal anisotropy, areal alternation of relative magnitudes could readily occur, as cartooned in Fig. 9. In fact, 19 wells have two populations of breakout orientations (Table 1), and this suggests that alternation of horizontal principal stress magnitudes occurs vertically as well as laterally in the Aquitaine Basin.

Discussion

We have gattered evidence to suggest that the southern part of the Aquitaine Basin is presently subject to a normal fault stress regime where $\sigma_v > \sigma_H \approx \sigma_h$. A normal fault stress regime is not anomalous for this part of Europe (Fig. 8), but minimal horizontal stress anisotropy may be. Muller et al. (1992) report that compressive stress regimes, with $\sigma_H > \sigma_v > \sigma_h$, exist in rocks attached to basement across much of Europe, but normal fault regimes have been reported in areas adjacent to the Aquitaine Basin in rocks that represent crystalline basement or are attached to it.

Muller et al. (1992) have shown that the deep crustal earthquakes in central and southern France reported by Nicolas et al. (1990) and Philip et al. (1993) yield normal fault focal mechanism solutions. Cornet and Burlet (1992) also report a normal fault regime, where $\sigma_v > \sigma_H > \sigma_h$, from stress measurements made in two wells located northeast of the Aquitaine Basin, in central France. Their results are based on a series of hydraulic fractures run between the surface and depths of 730 and 814 m in wells that penetrate crystalline rocks of the Massif Central (Fig. 1). The horizontal principal stress magnitudes were calculated on the assumption that preexisting fractures were opened. However, similar analyses of hydraulic fractures from two other wells in the Massif Central point to a strike-slip regime, where $\sigma_H > \sigma_v > \sigma_h$ (Cornet and Burlet 1992).

◀ **Fig. 8.** The stress regime of northwest Europe The *facing arrows* represent mean directions of σ_H, based on areal averages of five or more stress directional indicators (Muller et al 1992) They define a NNW-SSE/NW-SE directional signature for σ_H The *speckled area* represents a region where earthquake focal mechanisms and other stress indicators suggest that a normal fault regime ($\sigma_v > \sigma_H > \sigma_h$) is present Elsewhere, it is believed that a strike slip regime ($\sigma_H > \sigma_v > \sigma_h$) prevails The *grid lines* are speculative stress trajectories in the Aquitaine Basin The *solid lines* represent stress trajectories in rocks that are believed to be attached to basement The *dashed lines* represent poorly defined stress trajectories in rocks that are thought to be mechanically detached from basement In this region, it is believed that σ_h and σ_H may alternate locally due to minimal horizontal stress anisotropy in the area Note that the attached rocks exhibit regional stress trajectories

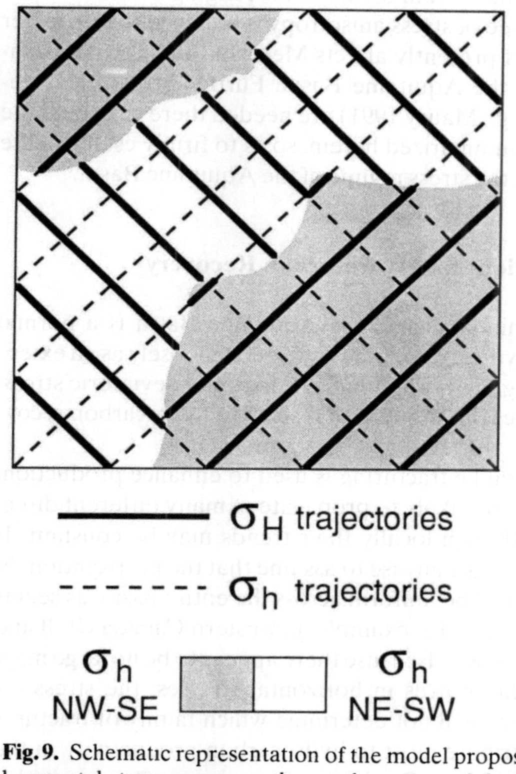

Fig. 9. Schematic representation of the model proposed for the horizontal stress regime in the southern part of the Aquitaine Basin, where the sediments are interpreted as being geomechanically detached from basement The character of the ovalization encountered in wellbores suggests that σ_h and σ_H are approximately equal in magnitude, and that one may replace the other depending on location The observed "orthogonality" of mean breakout azimuths suggests that principal stress trajectories respond to some structural control, such as overall basin geometry, but that the relative magnitudes vary locally As depicted, σ_h and σ_H alternate areally

There is geological evidence for recent normal faulting/extensional tectonics in the Massif Central of central France in the region where Cornet and Burlet made stress measurements by hydraulic fracturing. Pleistocene volcanism has been dated (Downes 1987), normal offsets of Quaternary basalt flows have been reported (Giot et al 1978) and striation analysis of post-Late Miocene faults suggests normal faulting with a mean azimuth of 078° for σ_h (Burg and Etchecopar 1980). These data and observations suggest that the basement of much of southern and central France may be subject to a normal faulting stress regime today.

The degree of horizontal stress anisotropy that exists in the cratonic basement is not clearly established Muller et al. (1992) report a significant number of NNW-SSE σ_H orientations on and around the Massif Central, so the principal horizontal stresses must vary in magnitude enough to generate well-defined stress trajectories. Cornet and Burlet have calculated σ_H magnitudes from their hydraulic fracture measurements in wells in the Massif Central, and obtain $\sigma_H:\sigma_h$ ratios ranging from 1.2 to 1.6. This degree of stress anisotropy is believed to be greater than what presently affects Mesozoic and Tertiary sediments in the Aquitaine Basin. Further stress measurements (e.g., Maury 1991) are needed there to extend the findings summarized herein, so as to firmly establish the nature of the stress regime of the Aquitaine Basin.

Implications for Hydrocarbon Recovery

In tectonic language, the Aquitaine Basin is a normal fault province, sometimes referred to loosely as an extensional regime, with minimal horizontal deviatoric stress. What does this imply with respect to hydrocarbon recovery?

If hydraulic fracturing is used to enhance production, fractures are likely to propagate in many different directions, although locally their trends may be constant. It would be most unwise to assume that the propagation directions will be uniform across the entire basin, as seems to be the case, for example, in western Canada (Bell and Babcock 1986). Because there appear to be no large magnitude differences in horizontal stresses, the stress regime alone will not determine which family of fractures will contribute most to hydrocarbon productivity in the Aquitaine Basin. Productive fractures may be oriented in any direction and factors other than the stress regime will control their transmissibility. However, all things being equal, the most productive fractures are likely to be those that are close to vertical, since dipping fractures will be more prone to close in response to vertical stress.

The proposed model of the stress regime of the Aquitaine Basin implies that, if future productivity enhancement of oil and gas fields is sought through horizontal drilling, information on the in-situ stresses will not provide the key information for guiding the boreholes, as it does in many basins Instead, detailed studies of reservoir fabric anisotropy will be mandatory for optimum results.

These implications are serious enough to have a major impact on the philosophy and practices of hydrocarbon extraction in the Aquitaine Basin, so it is essential to determine whether or not they are well founded What are lacking today are direct measurements of σ_H, the larger horizontal stress. Without them, it is impossible to verify the hypothesis of minimal stress anisotropy proposed herein We believe that the evidence we have compiled makes this conclusion almost inescapable, but it is not proved beyond all doubt, and it will not be until the complete stress tensor has been measured in some tens of subsurface locations. The only viable approaches for doing so at the present time involve using a combination of downhole techniques that can suffer from operational problems It is important to develop alternative technology to make such measurements as soon as possible

Acknowledgments This chapter has drawn upon many in-house studies conducted by Elf Aquitaine, particularly those of D. Fontanet, P Gauer, V. Maury, J-P. Richert, and C. Tourneret. We are most grateful to the management of Elf-Aquitaine for authorizing publication, and for assisting us to locate drilling data. The ideas presented here have evolved during discussions with R. Courel, G. Eisbacher, D. Grauls, N. Kessler, A. Le Marrec, V. Maury, L. Moen-Maurel, J-P. Petit, and J-P. Richert JSB thanks the Geological Survey of Canada for authorizing his Professional Development Leave with Elf-Aquitaine in 1990 and 1991, during which period the data reported here were generated. R. Deboaisne analyzed some dipmeter logs and advised us on interpretation M-C. Gellibert, C. Gout and R. Magalhaes-Gomez assisted with the reduction of the analytical data, and B. Cormier helped prepare the figures. The authors are most grateful to Alain Mascle for his encouragement in preparing the manuscript. An early draft was helpfully reviewed by P. J. McLellan and N A. Yassir. The observations and suggestions of two anonymous reviewers were extremely valuable and led the authors to revise their interpretation significantly. We remain, of course, solely responsible for the conclusions.

References

Adams J, Bell JS (1991) Crustal stresses in Canada In Slemmons DB, Engdahl ER, Zoback MD, Blackwell DD (eds) Neotectonics of North America Decade Map Vol 1, Geological Society of America, pp 367–386

Babcock EA (1978) Measurement of subsurface fractures from dipmeter logs Bull Am Assoc Pet Geol 62 1111–1126

Bell JS (1990a) Investigating stress regimes in sedimentary basins using information from oil industry wireline logs and drilling records In Hurst A, Lovell MA, Morton AC (eds) Geological applications of wireline logs Spec Publ Geol Soc Lond 48 305–325

Bell JS (1990b) The stress regime of the Scotian Shelf, offshore eastern Canada, to 6 kilometers depth and implications for rock mechanics and hydrocarbon migration In Maury V, Fourmaintraux D (eds) Rock at great depth, 3 Balkema, Rotterdam, pp 1243–1265

Bell JS, Babcock EA (1986) The stress regime of the Western Canadian Basin and implications for hydrocarbon production Bull Can Pet Geol 34 364–378

Bell JS, Gough DI (1979) Northeast-southwest compressive stress in Alberta evidence from oil wells Earth Planet Sci Lett 45 475–482

Bell JS, Caillet G, Adams J (1992a) Attempts to detect open fractures and non-sealing faults with dipmeter logs In Hurst A, Worthington PF, Griffiths CM (eds) Geological applications of wireline logs II Spec Publ 65 Geological Society of London, London, pp 211–220

Bell JS, Caillet G, Le Marrec A (1992b) The present-day subsurface stress regime of the southwestern part of the Aquitaine Basin, France, as indicated by oil well data J Struct Geol 14 1019–1032

Breckels IM, Van Eekelen HAM (1981) Relations between horizontal stress and depth in sedimentary basins Paper SPE 10336, 56th Annual Fall Technical Conference Society of Petroleum Engineers of AIME, San Antonio, Texas, October 5–7, 1981

Brereton NR, Evans CJ (1987) Rock stress orientations in the United Kingdom from borehole breakouts Rep RG 87/14, British Geological Survey, 36 pp, Keyworth, Notts, UK

Burg JP, Etchecopar A (1980) Détermination des systèmes de contraintes liés à la tectonique cassante au cœur du Massif Central français la région de Brioude (Haut Allier) C R Hebd Seances Acad Sci Serie D 290 397–400

Cornet FH, Burlet D (1992) Stress field determinations in France by hydraulic tests in boreholes J Geophys Res 97 11,829–11,849

Cox JW (1970) The high resolution dipmeter reveals dip-related and borehole and formation characteristics 11th Annu Logging Symp SPWLA, Los Angeles, May 1970

Curnelle R, Dubois P, Seguin JC (1982) The Mesozoic-Tertiary evolution of the Aquitaine Basin Philos Trans R Soc Lond A305 63–84

Dart RL, Zoback MD (1989) Wellbore breakout stress analysis within Central and Eastern Continental United States Log Analyst, Jan–Feb 1989 12–23

Downes H (1987) Tertiary and Quaternary volcanism in the Massif Central, France In Fitton JG, Upton BG (eds) Alkaline igneous rocks Geological Society of London, London, pp 517–530

Ervine WB, Bell JS (1987) Subsurface in-situ stress magnitudes from oil-well drilling records. an example from the Venture area, offshore eastern Canada Can J Earth Sci 24 1748–1759

Fordjor CK, Bell JS, Gough DI (1983) Breakouts in Alberta and stress in the North American plate Can J Earth Sci 20 1445–1455

Giot D, Clozier, L, Fleury R (1978) Manifestations tectoniques récentes (quaternaires) en Limagne d'Allier (Massif central, Puy de Dome, Allier) Bull Bur Rech Geol Min 2 149–155

Grasso J-R, Feignier B (1990) Seismicity induced by gas production II Lithology correlated events, induced stresses and deformation. Pure Appl Geophys 134 427–450

Grasso J-R, Wittlinger G (1990) Ten years of seismic monitoring over a gas field Bull Seismol Soc Am 80 450–473

Grasso J-R, Gratier J-P, Gamond JF (1992) Stress transfer and seismic instabilities in the upper crust example of the western Pyrenees. J Struct Geol 14 915–924

Grellet B, Combes P, Philip H, Mohammadioun B, Haessler H (1991) Methodology of seismotectonic zoning in the intraplate low seismicity domain, example of France and surrounding areas In. Volume 2, Proceedings of the Fourth International Conference on Seismic Zonation Borchert RD, Shah HC (eds) Earthquake Engineering Research Institute, Oakland, California, USA, pp 35–42

Janot P, Gauer P, Gross E (1988) Orientation de la contrainte tectonique dans l'Europe de l'ouest à partir des ovalisations de trous de forages Rev Inst Fr Pétrole 43 517–521

Klein RJ, Barr MV (1987) Regional state of stress in western Europe In Stephansson O (ed) Rock Stress and Rock Stress Measurement Proc Int Symp, Stockholm, Balkema, Rotterdam, pp 33–44

Mardia L (1972) Statistics of directional data probability and mathematical statistics Academic Press, London, 527 pp

Mastin L (1988) Effect of borehole deviation on breakout orientations J Geophys Res 93 9187–9195

Maury V (1991) The role of rock mechanics in oil and gas exploration Nature Suppl 350 8–10

Maury V (1993) An overview of tunnel, underground excavation and borehole collapse mechanisms Chapter 14, v 4, Comprehensive rock engineering, principles, practice and projects Hudson JA (ed-in-chief) Pergamon, pp 369–412

McLellan PJ (1988) In situ stress prediction and measurement by hydraulic fracturing, Wapiti, Alberta J Can Pet Technol 27, 2 85–95

Minster JB, Jordan TH (1978) Present-day plate motions J Geophys Res 83 5331–5354

Mount VS, Suppe J (1992) Present-day stress orientations adjacent to active strike-slip faults California and Sumatra J Geophys Res 97 11,995–12,013

Mullan HS (1984) Deep gas potential of the Aquitaine Basin, France Am Assoc Pet Geol Bull 68 1857–1869

Muller B, Zoback ML, Fuchs K, Mastin L, Gregersen S, Pavoni N, Stephansson O, Ljunggren C (1992) Regional patterns of tectonic stress in Europe J Geophys Res 97 11,783–11,803

Nicolas M, Santoir JP, Delpech PY (1990) Intraplate seismicity new seismotectonic data in western Europe Tectonophysics 179 367–380

Philip H, Grellet B, Combes P, Haessler H (1993) Methodology of seismotectonic zoning in an intraplate low seismicity domain – example of France and surrounding areas In Davidovici VE (ed) Recent advances in earthquake engineering and structural dynamics French Association of Earthquake Engineering, Paris (in press)

Plumb RA, Cox JW (1987) Stress distributions in eastern North America to 4 5 km from borehole elongation measurements J Geophys Res 92 4805–4816

Plumb RA, Hickman SH (1985) Stress-induced borehole elongation a comparison between the four-arm dipmeter and the borehole televiewer in the Auburn geothermal well J Geophys Res 90 5513–5522

Segall P (1989) Earthquakes triggered by fluid extraction Geology 17 942–946

Springer JE (1987) Stress orientations from wellbore breakouts in the Coalinga region Tectonics 6 667–676

Winnock E (1972) Exposé succinct de l'evolution paléogéologique de l'Aquitaine Bull Soc Géol Fr 7, 15 5–12

Winnock E, Pontalier Y (1970) Lacq gas field, France In Halbouty MT (ed) Geology of giant petroleum fields Mem 14, American Association of Petroleum Geologists, Ulsa, OK, pp 370–387

Zoback MD, Healy JH (1992) In-situ stress measurements to 3 5 km depth in the Cajon Pass scientific research borehole implications for the mechanics of crustal faulting J Geophys Res 97 5039–5057

Zoback MD, Moos D, Mastin L, Anderson R (1985) Wellbore breakouts and in-situ stress Geophys Res 90 5523–5530

South-East Basin, Gulf of Lions, Western Alps

Structure of the Gulf of Lions (Northwestern Mediterranean Sea): A Review

C. Gorini, A. Mauffret, P. Guennoc, and A. Le Marrec

Abstract. After a brief review of previous work, mainly related to the results of the deep penetration (ECORS) seismic lines, we present a picture of the Gulf of Lions tectonic framework based on a dense grid of seismic industrial lines. This region is divided into two zones: a northeastern one where narrow grabens are oriented NE-SW and a southwestern area where large and deep depressions are limited by faults and horsts which trend N-S. The onshore geology indicates that the second zone has been formed during the Late Eocene epoch in relation to the western European graben system, whereas the NE-SW horsts and grabens are Oligocene in age and are connected to the opening of the northwestern Mediterranean basin. Beneath the Tertiary several deep reflections may represent a southward extension of Mesozoic rocks of the southeast basin and some Eocene thrusts. The thrusting surfaces have been reactivated during the subsequent Oligocene rifting. The structure of the Gulf of Lions is replaced in a general geodynamic framework.

Introduction

The sedimentary cover of the Gulf of Lions continental shelf is formed by a large Pliocene-Quaternary progradational body fed by the Rhône river, a Miocene layer more or less eroded during the Messinian desiccation and Oligocene-Early Miocene beds that filled the deep grabens formed during the rifting of this passive margin of the Western Mediterranean basin. This area has been intensively surveyed by oil companies which completed 35 000 km of seismic profiles and 11 wells. From a geological point of view, the interest for this zone has been reactivated by the acquisition of deep penetration seismic profiles (ECORS lines). The interpretation of these profiles is aided by the dense grid of industrial lines, and a detailed picture of the extensional features is now available. We have been able to draw a fine three-dimensional picture of the extensional features. We present briefly in this chapter the main results already published or in press and then highlight some new aspects of the rifting based upon the detailed mapping.

Geological Setting

Beneath the thick (about 3 km) Miocene to Quaternary sedimentary cover the Oligocene-Early Miocene extensional features are evident. They mainly trend NE-SW and are in strike with the main faults of the onshore zone: Cévennes, Nimes, and Aix-Durance faults (Fig. 1) which bound some important onland grabens (Arthaud et al. 1981). However, these faults are preexistent to the rifting which led to the opening of the Western Mediterranean basin, and the structural framework of the Gulf of Lions results from a long and complex tectonic history.

The NE-SW fault system formed during a Late Hercynian strike-slip faulting (Arthaud and Matte 1975). The Villefort fault, which limits the Carboniferous Cévennes basin (Debrand-Passard and Courbouleix 1984), belongs to the conjugate system. We will show the influence of the southeastern extension (Gard fault, Deroin et al. 1992; Arlesian fault zone) of this fault during the subsequent geological history.

The Triassic evaporitic beds, overlying a thin cover of sandstones and mudstones, are thick and formed a decollement level during the Middle Jurassic stretching, but a deeper level of decollement cut through the Hercynian basement (Roure et al. 1992).

A large extension, limited by the Cévennes and Durance faults, occurred during the Middle Jurassic epoch in relation to the formation of the Tethys and the Central Atlantic ocean (Baudrimont and Dubois 1977; Curnelle and Dubois 1986). Several kilometers of Jurassic black shales and carbonate have been accumulated in the southeast basin. For example, a map (Fig. 2) of the Late Oxfordian marine sediments (Debrand-Passard and Courbouleix 1984) shows the main southeast basin where the depocenter is 400 m thick. Two troughs, opened towards the Gulf of Lions, are separated by the Camargue Mesozoic high. This high is limited by the SE extension of the Villefort fault, named the Arlesian fault, and the Nimes fault.

During Albian time, the Gulf of Lions region underwent a large tectonic event related to the opening of the Bay of Biscay and the North Atlantic Ocean and the coeval rotation of the Iberian plate (Olivet et al. 1984; Philip et al. 1987; Chorowicz and Mekarnia 1992; Hibsh et al. 1992). During this epoch, Iberia, Corsica, and Sardinia formed a single plate; consequently, the North Pyrenean fault, which limits the Iberian plate, may have an east-

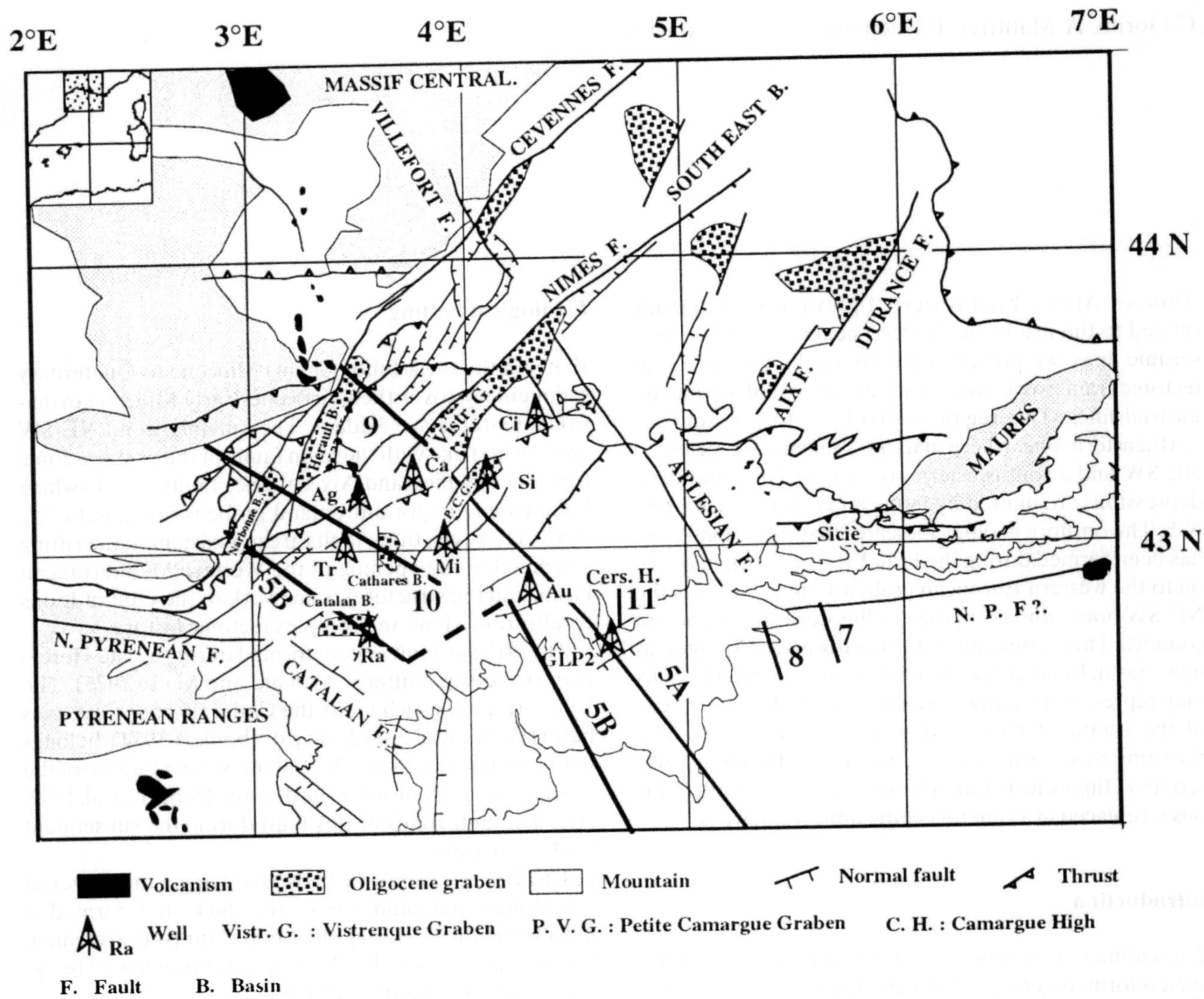

Fig. 1. Geological framework of the Gulf of Lions, location map of the seismic profiles and cross sections shown in following figures Figure 9 is a detail of Fig 5 a and Fig 10 a detail of Fig 5 b *N P F* North Pyrenean Fault

Fig. 2. Mesozoic southeast basin and the main troughs of the European rift system Note the Camargue high which divided the Mesozoic basin into two parts *Inset* Europen rift system (After Bergerat 1982) ▶

ward extension located south of Provence (Arthaud and Mattauer 1972; Olivet et al. 1984, Olivet 1987).

During the Late Cretaceous and the Early Paleogene, the Iberia plate converged towards Europe, and the formation of the Pyrenean Ranges resulted from a collision which culminated during the Early Eocene (Lutetian) The Eocene compressional orogeny is well known (Arthaud and Mattauer 1972; Arthaud and Seguret 1981) in the Gulf of Lions region and can be extended towards the east as far as Provence (Sicié thrust Figs 1 and 2) We will provide some offshore evidence of Eocene thrusting, but the compressional tectonics is clearly less intense than in the Pyrenees. On land, the Mesozoic cover is thrust towards the north and the NE-SW fault system (Cévennes, Nimes, and Durance-Aix faults) are reactivated with a left-lateral strike-slip component (Arthaud and Mattauer 1972; Arthaud and Seguret 1981, Bergerat 1982).

After the Lutetian last compressional event, Europe was affected by a widespread extension. Most of the authors attribute an Oligocene age to this rifting event. However, the first occurrence of extension can be as old as Late Eocene (Ludian) (Triat and Truc 1983). A geological cross section of a Paleogene graben (Fig. 3; Debrand-Passard and Courbouleix 1984) shows three unconformities: the oldest is related to the Pyrenean orogeny, although the picture does not indicate the folding which should affect the pre-Lutetian layers, the middle unconformity is placed between the Ludian beds and a Stampian (Early Oligocene) layer, the youngest one seals the extensional faults and is overlain by the undisturbed Miocene cover. The Late Eocene rifting trends N-S and can be followed (Fig. 2, inset) from the north (Rhenan trough) to the south (Catalan margin). The Oligocene rifts result from the extensional opening of the NE-SW fault system that previously had a strike-slip component (Bergerat 1982; Triat and Truc 1983): Alès and Hérault grabens bounded by the Cévennes fault; Vistrenque trough limited by the Nimes fault: Aix and Manosque ba-

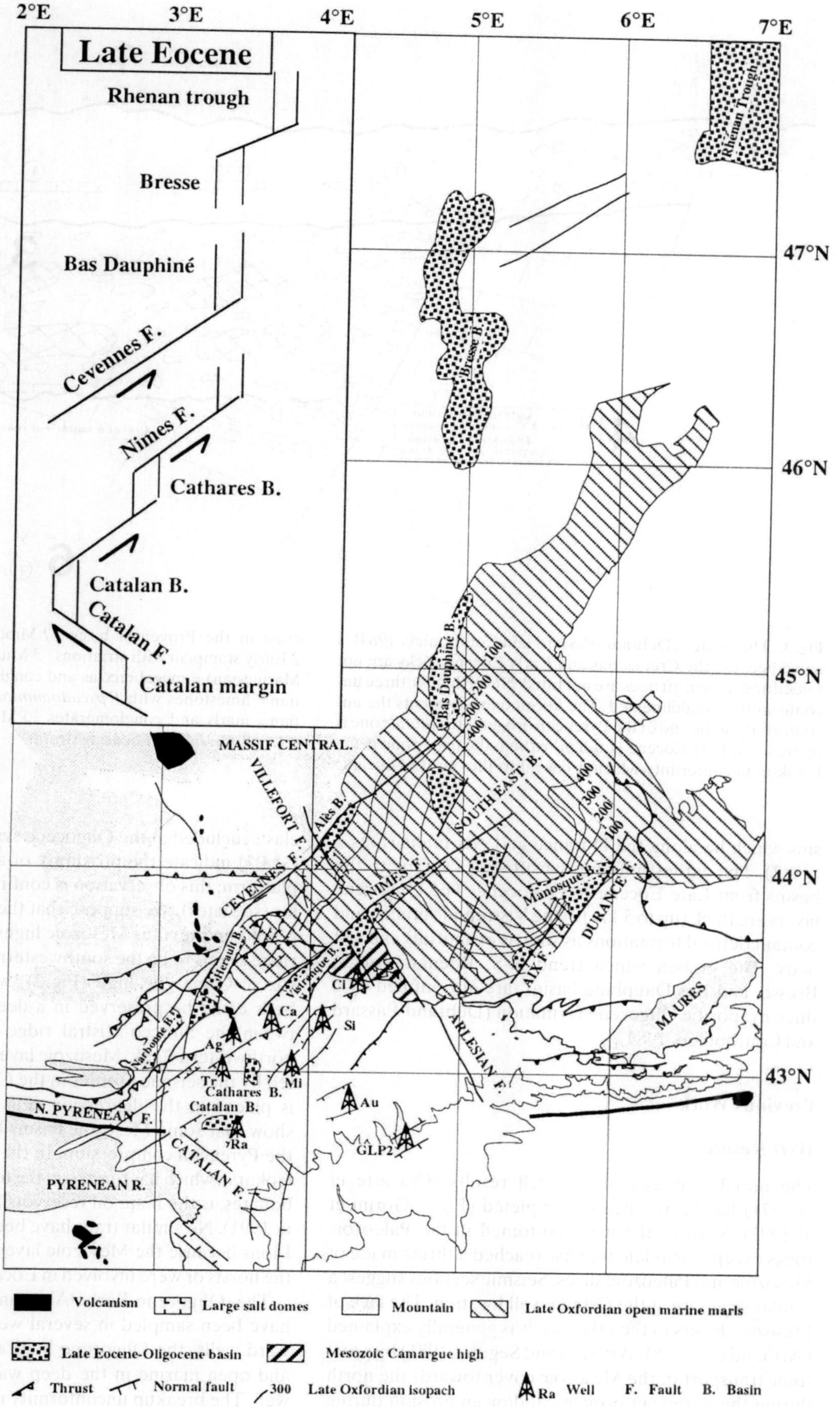
2°E
3°E
4°E
5°E
6°E
7°E
47°N
46°N
45°N
44°N
43°N
Late Eocene
Rhenan trough
Bresse
Bas Dauphiné
Cevennes F.
Nimes F.
Cathares B.
Catalan B.
Catalan F.
Catalan margin
Rhenan Trough
Bresse B.
Bas Dauphiné B.
100
200
300
400
SOUTH EAST B.
MASSIF CENTRAL.
VILLEFORT F.
Alès B.
CEVENNES F.
NIMES F.
Manosque B.
DURANCE F.
AIX F.
MAURES
ARLESIAN F.
Herault B.
Vistrenque B.
Narbonne B.
Cl
Ca
Si
Ag
Tr
Mi
Cathares B.
Catalan B.
Ra
Au
GLP2
N. PYRENEAN F.
CATALAN F.
PYRENEAN R.
Volcanism
Large salt domes
Mountain
Late Oxfordian open marine marls
Late Eocene-Oligocene basin
Mesozoic Camargue high
Thrust
Normal fault
300 Late Oxfordian isopach
Ra Well
F. Fault
B. Basin

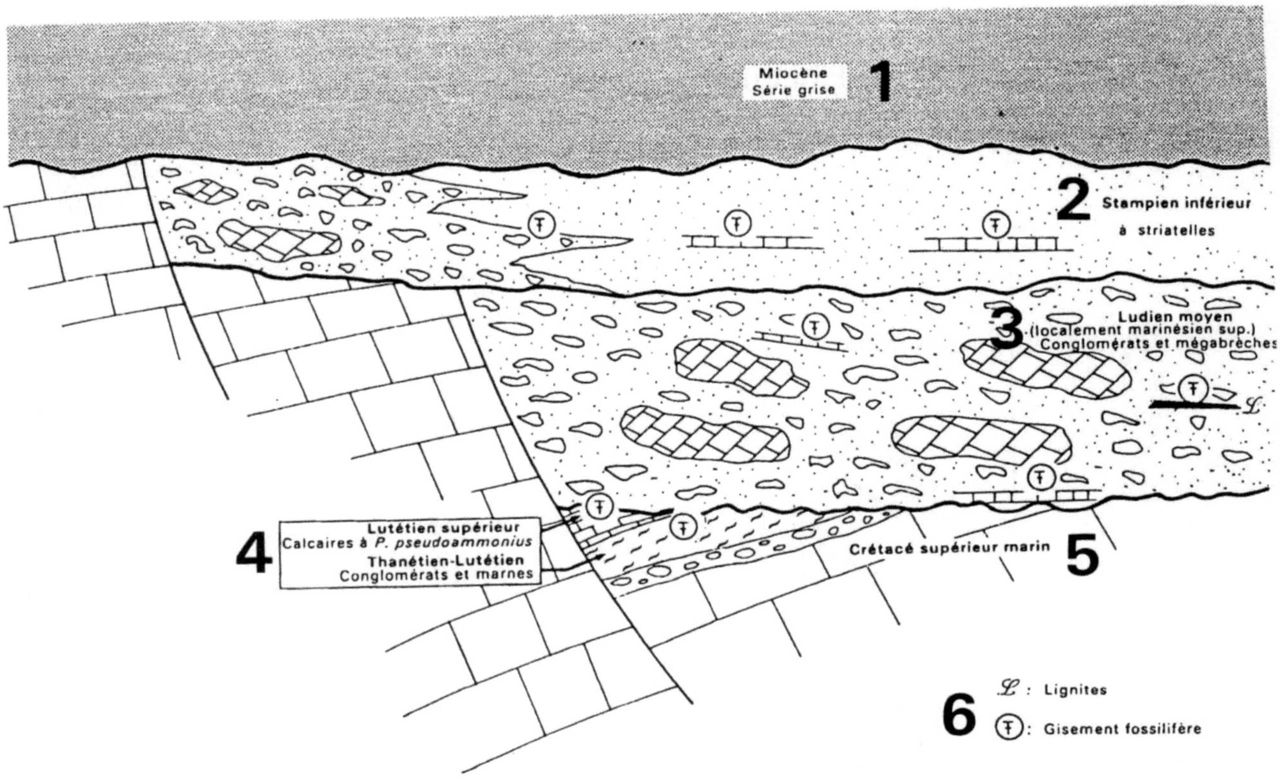

Fig. 3. This sketch (Debrand-Passard and Courbouleix 1984) is naive because the Cretaceous and Early Eocene rocks are not folded and the synrift beds are not tilted Nevertheless, three unconformities are delineated The *lowest* one represents the unconformity at the end of the Pyrenean orogeny, the *middle* one is related to a Late Eocene (Ludian) rifting, the *upper* one is the breakup unconformity which is coeval to the onset of oceanic crust in the Provençal basin *1* Miocene – gray formation *2* Early stampian with striations *3* Middle Ludian (locally late Marinesian) = megabreccias and conglomerates *4* Late Lutetian = limestones with *P pseudoammonius*, Thanetian – Lutetian = marls and conglomerates *5* Marine Late Cretaceous *6 L* lignites, *F* Fossil-bearing strata

sins which lie along the Durance and Aix faults (Figs. 1 and 2). The earliest infilling synrift series vary with the basins from Late Eocene to Oligocene. The Oligocene layers are thick (up to 5 km) in the Vistrenque graben and contain detrital formations as well as evaporites and salt beds. The graben which trend N-S, Rhenan trough, Bresse, and Bas Dauphiné basins, are also infilled by a thick evaporitic Oligocene formation (Debrand-Passard and Courbouleix 1984).

Previous Work

Well Results

The initial synthesis of the well results (Cravatte et al. 1974) has been recently completed (Fig. 4; Gorini et al. 1993). Most of the wells bottomed in the Paleozoic rocks except Cicindèle that has reached a thrust made of Mesozoic and Paleozoic slices. Seismic sections suggest a similar structure at the Calmar well location. The lack of Mesozoic layers in the other wells is generally explained (Arthaud et al. 1981; Arthaud and Seguret 1981) by a tectonic transport of the Mesozoic cover towards the north during the Pyrenean orogeny and/or an erosion during the subsequent Oligocene rifting. However, the Jurassic clasts included in the Oligocene synrift beds (Autan and GLP2) indicate the proximity of a Mesozoic carbonate platform; this observation is confirmed by seismic profiles (see later). We suppose that the Oligocene horsts are placed on previous Mesozoic highs. The Sirocco-Mistral ridge seems to be the southwestern extension of the Mesozoic Camargue high (Fig. 2), whereas the Mesozoic beds could be preserved in a deep graben located between the Sirocco-Mistral ridge and the Tramontane northwestern flank Mesozoic layers are, indeed, suggested by the seismic profiles in the Calmar well area which is placed in the depression (Fig. 4). However, we will show that some Mesozoic basins have been inverted by the Pyrenean compression. In the Valencia trough a paleokarst, which has formed at the top of the Mesozoic carbonates, is the main oil reservoir (Clavell and Berastegui 1991). No similar traps have been found in the Gulf of Lions because the Mesozoic layers are absent on top of the horsts or were involved in Eocene thrusting.

The Oligocene-Early Aquitanian synrift sediments have been sampled in several wells. In the distant seaward wells, the Oligocene beds are evaporitic (Autan) and open marine in the deep water GLP2 exploration well The breakup unconformity is dated exactly as middle Aquitanian in age (Cravatte et al. 1974).

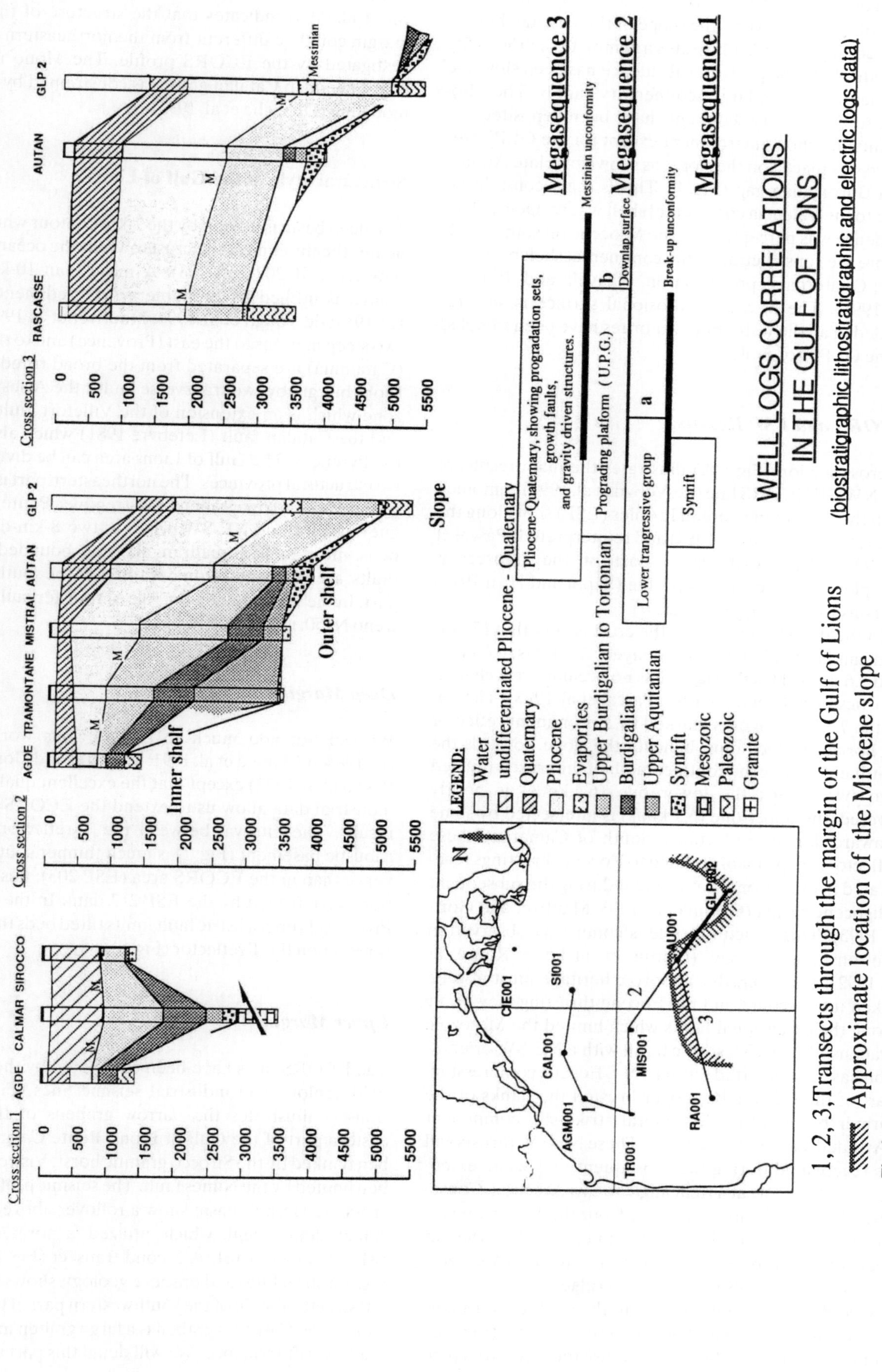

Fig. 4. Main results of the industrial wells (Gorini et al 1993)

A complete description of the post-rift evolution of the Gulf of Lions is beyond the scope of this chapter, but the transect Agde-GLP2 illustrates a complete section of the margin where the post-rift subsidence has been slow and compensated by a large sedimentary supply. The Miocene to Quaternary sediments have been deposited in a continental shelf environment except for the GLP2 area which was placed on the slope, as shown by a late Aquitanian-Burdigalian gap (Fig. 4). This is the second difference to the Valencia continental shelf where open marine influence was perceptible in the Miocene oil source beds before the construction of the continental shelf by a Plio-cene-Quaternary progradation (Clavell and Berastegui 1991). The Messinian erosional surface is well recorded in all the wells, but evaporites have been sampled in the GLP2 borehole.

ECORS and ESP Results

A cross section (Fig. 5 A) illustrates the main results of the NW-SE ECORS line (de Voogd et al. 1991) combined with the Expanding Spread Profiles (ESP) shot along the line (Le Douaran et al. 1984) and reinterpreted (Pascal et al. 1993). The northeastern part of the transect is completed by a refraction profile (Sapin and Hirn 1974) and onshore geology (see below).

In the deep basin domain the crust is very thin (5 km) and can be divided into three layers: 7.3 km/s, 6.6 km/s, and 5.6 km/s. This thin layer has no oceanic crust characteristics (Mauffret et al. 1993; Pascal et al. 1993). The top of the 7.3 km/s layer is marked by a prominent reflector (T horizon) which dips, beneath the slope, towards the continent. Below the continental shelf the crust is divided into two layers: the lower one (6.4 km/s) is poorly layered, the upper one (6.2 km/s) is intersected by some seaward-dipping reflections. South of Camargue these reflectors are probably related to Eocene thrustings reactivated in extensional detachment during the subsequent Oligocene rifting (Gorini et al. 1993; Mauffret and Gorini 1993). This structural style is commonly observed in the onshore geology (Gorini et al. 1991; Roure et al. 1992). However, the Mesozoic heritage must also be taken into account and the Eocene thrusting may occur along the extensional faults which limited the Mesozoic grabens. When a Mesozoic basin with a NE-SW orientation has been inverted during a N-S Eocene compression, part of the infilling beds overthrusted the flanks of the former basin with a left-lateral strike-slip component (Arthaud and Mattauer 1972). These NE-SW thrusts will be reactivated during the subsequent Oligocene extension which occurs at a right angle to this direction. Consequently, some highs may result from the Eocene inversion of the Mesozoic basin (Deroin et al. 1992), whereas some Paleozoic ridges could be the remnants of Mesozoic horsts (Camargue-Sirocco-Mistral ridge).

Geophysical data are scarce in the southwestern part of the region and we offset the cross section (Fig. 5 B) from Rascasse to Autan well because there is no deep information in the basin, and the free air gravity map (Pascal et al. 1993) indicates that the structure of the deep margin could be different from the northeastern one investigated by the ECORS profile. The Moho rise beneath the deep Catalan graben is constrained by gravity modeling (Debeglia et al. 1993).

Structural Style of the Gulf of Lions

The deep basin is limited by the 7-km contour which delineates the thinning of the crust towards the ocean (Fig. 6, between ESP 202 and 203). A more than 10-km-deep trough is infilled by probable synrift sediments (Burrus 1984; de Voogd et al. 1991; Mauffret et al. 1993). The two steep margins to the east (Provence) and to the south (Catalonia) are separated from the broad rifted Gulf of Lions margin by two transverse faults: the Arlesian fault zone which is an extension of the Villefort fault (Fig. 1) and the Catalan fault (Lefebvre 1981) which also limits the Pyrenees. The Gulf of Lions area can be divided into two structural provinces. The northeastern part is formed by several narrow grabens (Vistrenque, Petite Camargue) which trend NE-SW, whereas two 8-km-deep grabens, more or less quadrangular and bounded by N-S faults, are the main tectonic features of the southwestern part. In the whole region the role of transfer faults, which trend N150, is prominent.

Deep Margin

We can not add much to the previous works (Burrus 1984; de Voogd et al. 1991; Mauffret and Gorini 1993; Pascal et al. 1993) except that the excellent quality of the industrial data allow us to extend the ECORS and ESP results. The interval between the T reflector and the acoustic basement (Fig 7) is much thinner south of Provence than in the ECORS area (ESP 203); this observation is confirmed by the ESP 212 data. In the southern Provençal zone, a listric fault limits tilted beds that directly rest upon the T reflector (Fig. 8).

Upper Margin

The ECORS lines have been completed by the onshore field geology and industrial seismic lines (Fig. 9). The transect illustrates the narrow grabens of the northeastern part of the Gulf of Lions: Petite Camargue graben flanked by the Sirocco granitic horst, Vistrenque graben limited by the Nimes fault. The seismic profiles which cross the Hérault basin show a rollover above an extensional detachment which utilized a former Eocene (Montpellier) thrust. A second transect (Fig. 10), based on industrial lines and onshore geology, shows the different structural style of the southwestern part of the Gulf of Lions. The Cathares graben is a large graben infilled by a thick synrift sequence. We will detail this part in the next

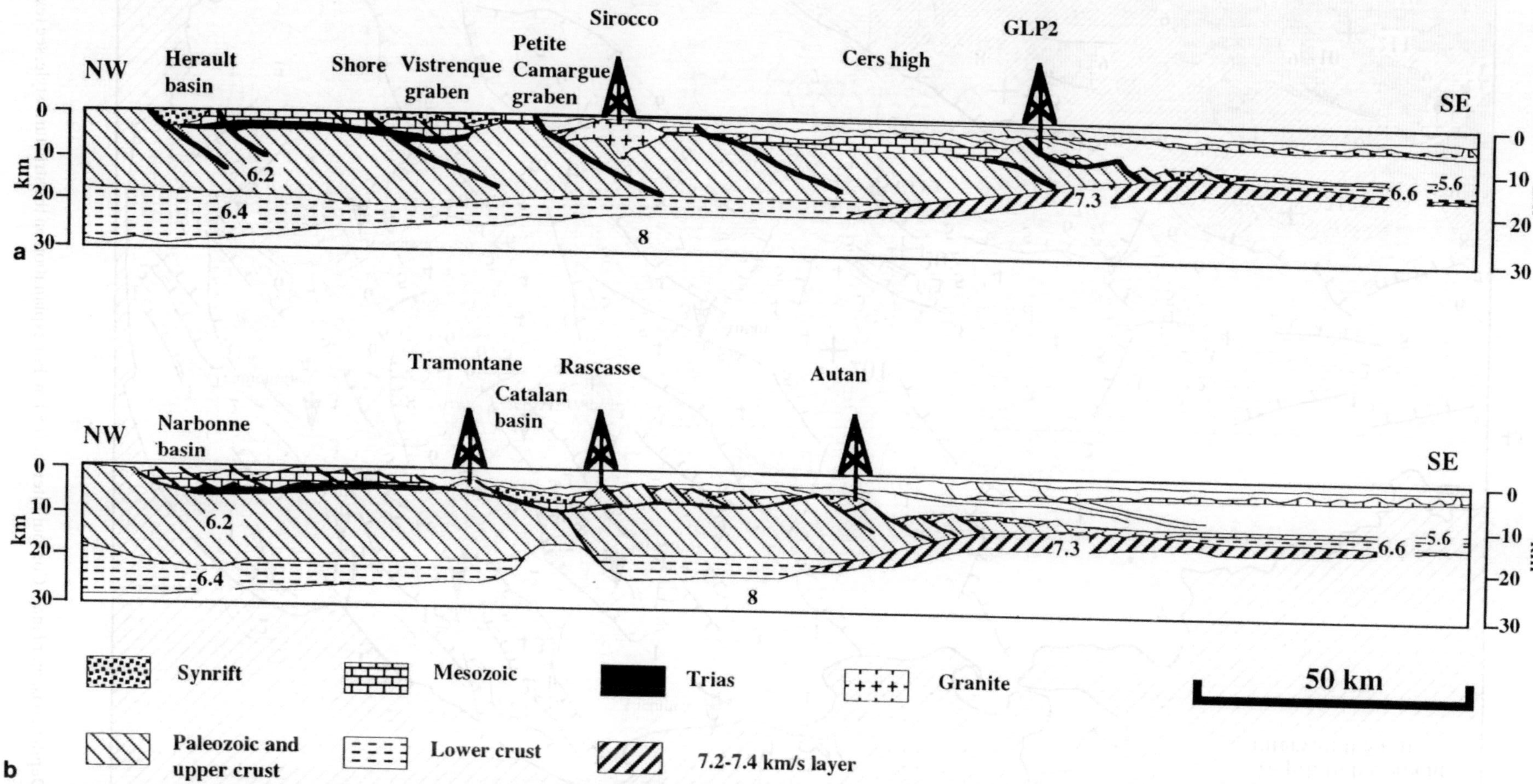

Fig. 5 a, b. Two transects of the Gulf of Lions (After Gorini 1993) The position is indicated in Fig 1 The transect is offset from Rascasse to Autan (see Fig 1) because we have no deep seismic information from the basin The northwestern parts of both sections are detailed in Figs 9 and 10 Note the difference between the two transects The northeastern transect shows several narrow grabens (Vistrenque, Petite Camargue), whereas the large Catalan basin is the main tectonic feature of the southwestern transect

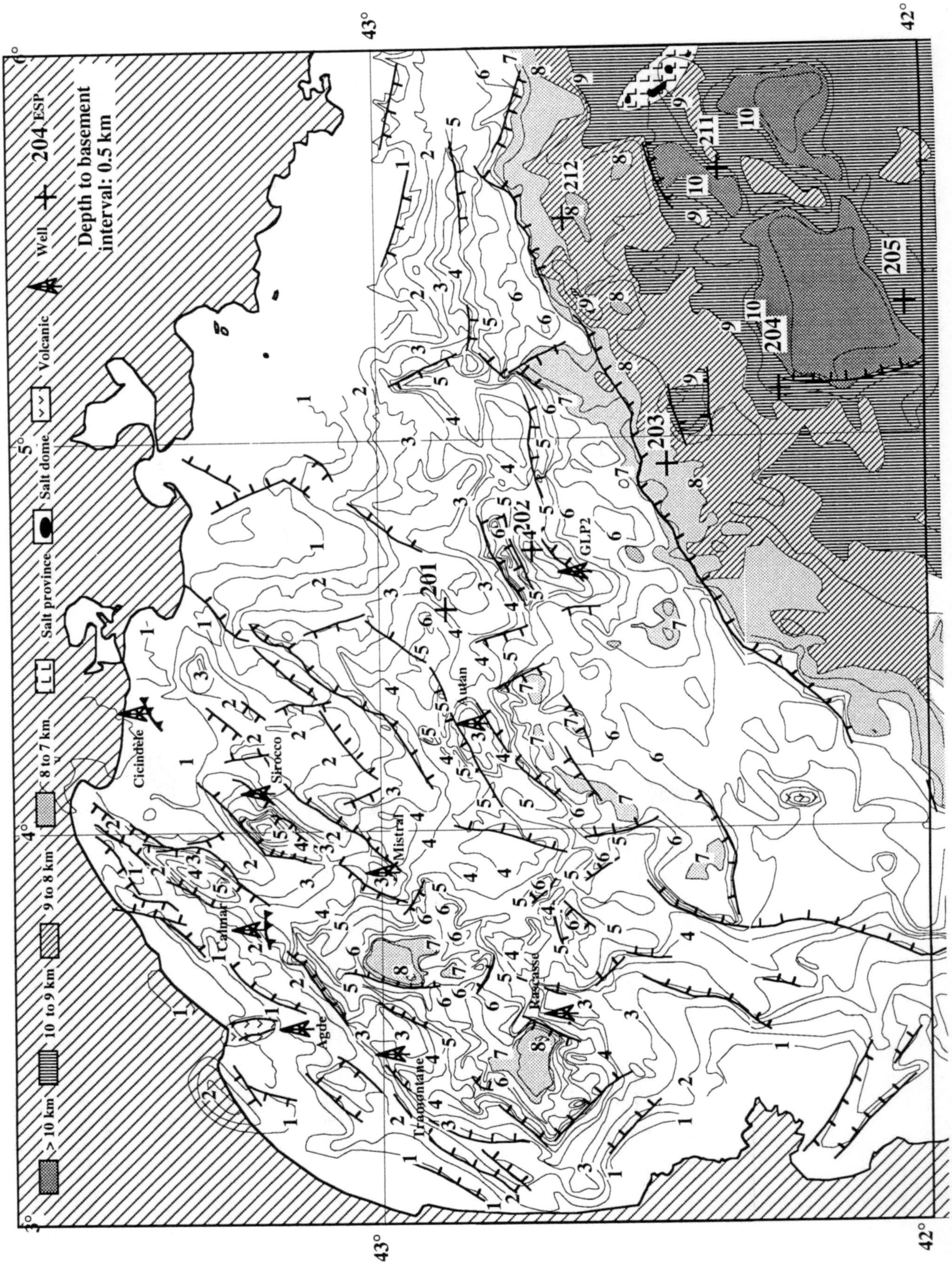

Fig. 6. Depth to basement map Contour interval 0.5 km For explanation on the structural style, see text

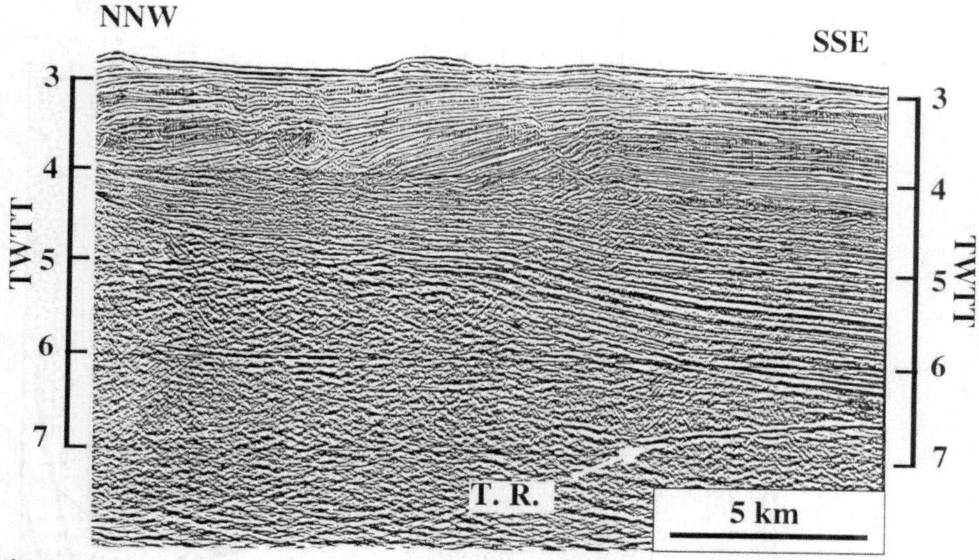

Fig. 7. *T R* T reflector Enlarged portion of a profile presented in Gorini (1993) The location of the profile is indicated in Fig 1 Note the thin interval between the T horizon and the acoustic basement This observation is confirmed by the results of ESP 212 (Pascal et al 1993, location of the ESP 212 in Fig 6)

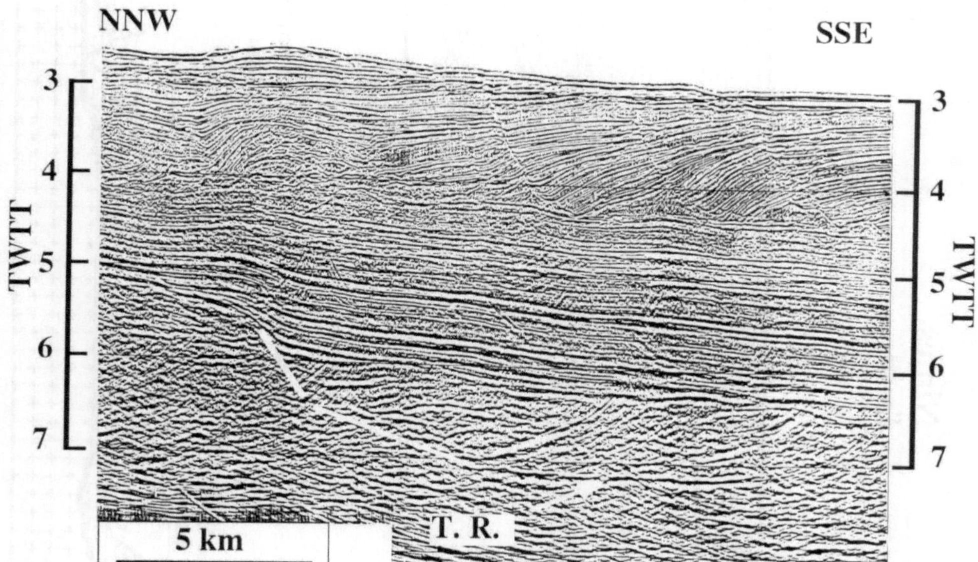

Fig. 8. *T R* T reflector Enlarged portion of a profile presented in Gorini (1993). The location of the profile is indicated in Fig 1 Note a listric fault which merges in the T horizon, and the tilted synrift layers

section. Although the two transects show a different structure below the continental shelf, the onshore Narbonne basin has, nevertheless, the same tectonic evolution as the Hérault basin and the Eocene ramps of the Corbières nappes were reactivated during the Oligocene extension (Gorini et al. 1991). On the inner continental shelf, off the Narbonne basin, the seismic profiles show several internal reflections beneath the acoustic basement, which is placed at the top of the pre-rift rocks. These internal reflections are correlated to folded and thrusted Mesozoic terranes and sometimes to the thrusts themselves. The same types of reflection have been identified north of the GLP2 well (DR: deep reflection, Fig. 11) and, in this case, the folding is not obvious. A previous seismic analysis (Mauffret and Gorini 1993) shows the large extension of these presumed Mesozoic carbonate rocks from north of the GLP2 well up to the Camargue shore. This region was probably a sub-basin connected to the southeast Mesozoic basin. The frequent occurrence of Mesozoic pebbles in the synrifts beds sampled at Autan and the GLP2 wells supports this hypothesis.

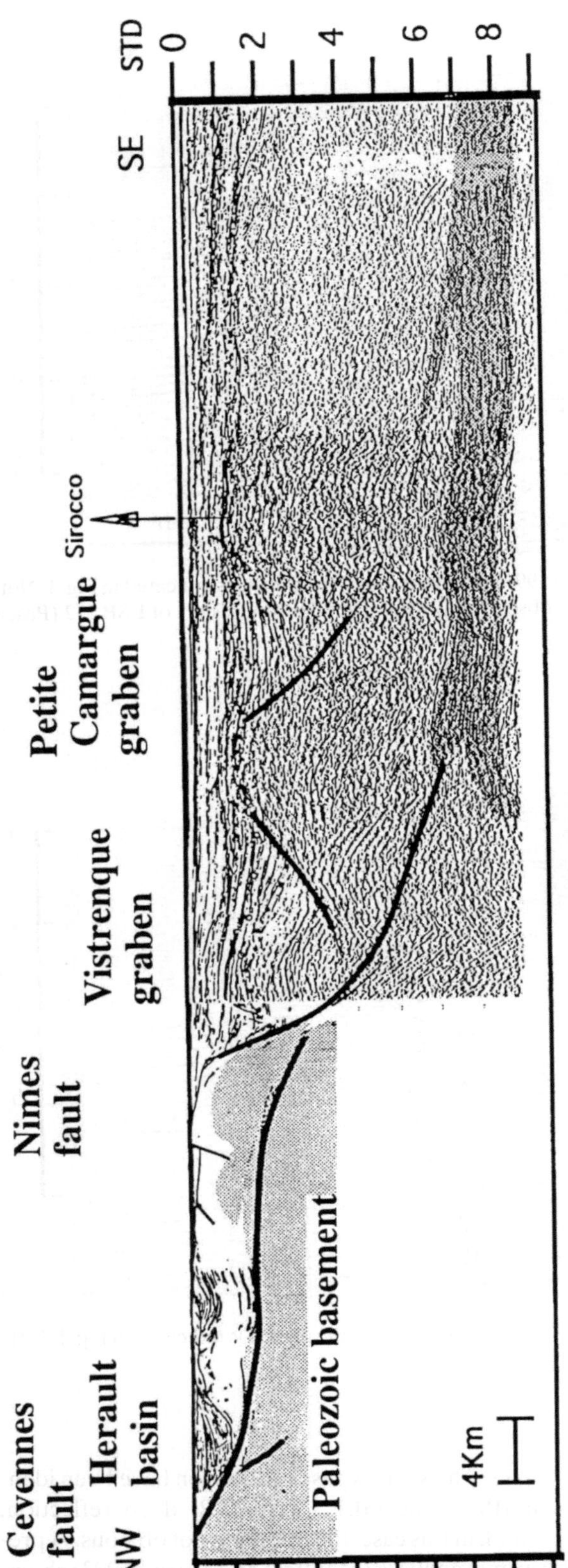

Fig. 9. Composite cross section that completes the ECORS seismic line (Gorini 1993) The location of the transect is indicated in Fig 1 Note the rollover in the Herault basin and the reactivation in the extensional decollement of the Montpellier thrust

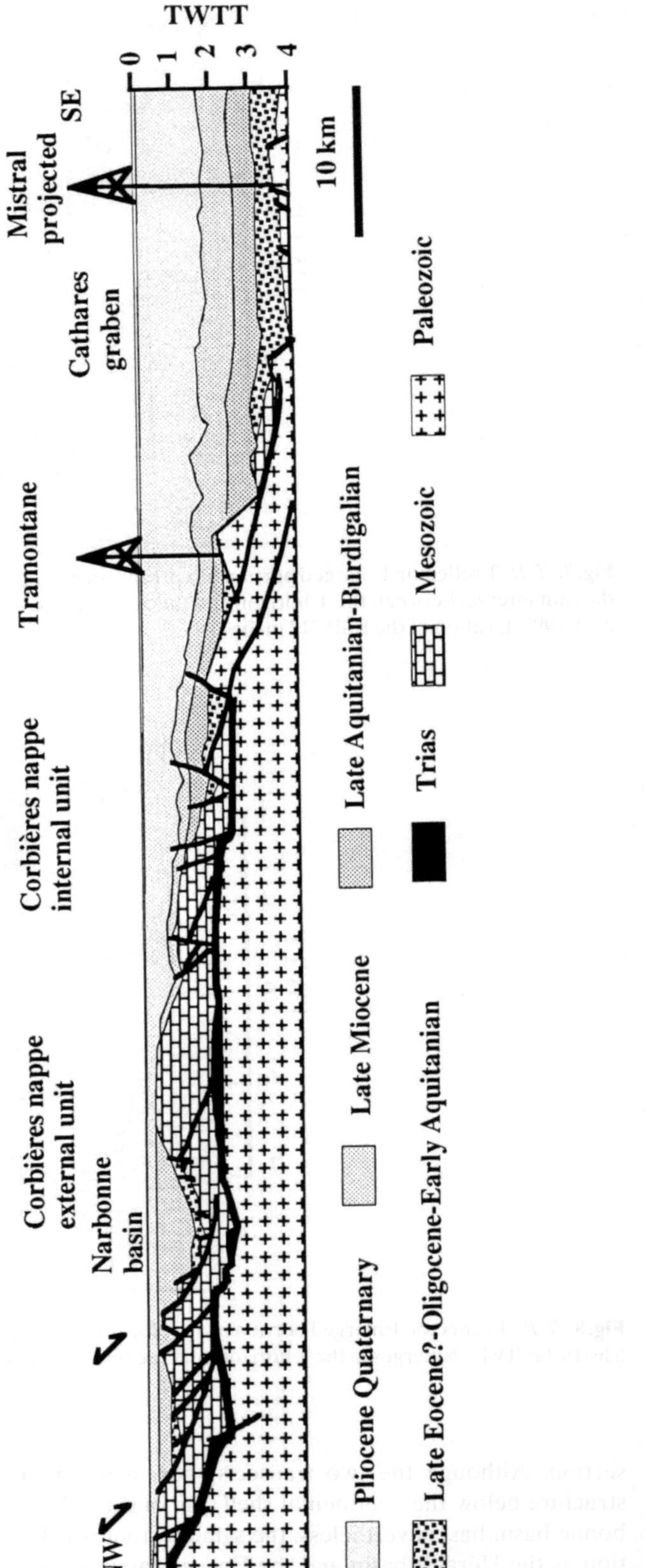

Fig. 10. Composite cross section including the onshore geology (After Gorini 1993) The location of the transect is indicated in Fig 1 Note the reactivation of the Corbières base nappe during the formation of the Narbonne basin The structure of Mistral horst and Cathares graben is detailed later

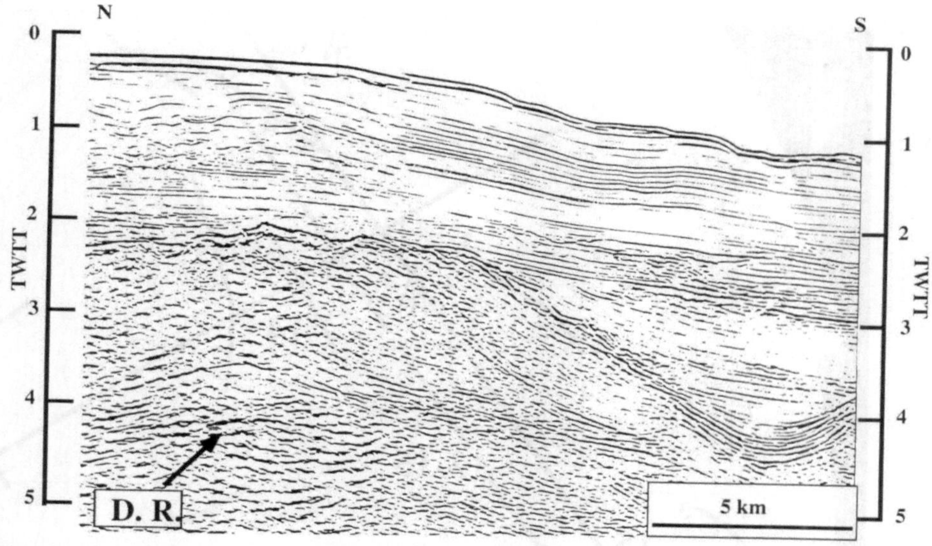

Fig. 11. N-S ECORS profile *D R* Deep reflection. The location of the profile is indicated in Figs 1 and 20 The Cers high, north of the GLP2 well, is probably made up of Mesozoic rocks

Detailed Study of the Southwestern Part of the Gulf of Lions

Tramontane Structure and Cathares Basin

The detailed depth to basement map and the Oligocene bird's-eye view (Fig. 12) show that the Tramontane Paleozoic block slid from the slopes which trend NE-SW; we can observe the scar which resulted from this collapse. A previous study (Mauffret and Gennesseaux 1989) of this area showed a prominent reflector beneath the Paleozoic Tramontane structure. This deep reflection (Fig. 13) was interpreted as the seismic signature of an intra-Paleozoic thrust formed during the Pyrenean orogeny. This deep reflection is visible on several seismic lines (Fig. 14). In order to understand the relationships between the Tramontane structure and the Cathares basin and to determine the true dip of the possible Eocene thrust, we transform the seismic lines in depth sections (Fig. 15) and draw a depth map to the deep reflector with 6 km/s for the interval velocity between the acoustic basement and the reflector (Fig. 16). The dip line (Fig. 15 A) shows that the Tramontane slide is not directly related to the deep reflector, but a shallow horizon could be a decollement level. This shallow decollement may merge with the base of the 8-km-deep Cathares basin, but the lower reflector is deeper than 10 km. These observations do not change significantly with a velocity interval below 6 km/s. On the other hand, the deep reflector rises towards the scar of the Tramontane slide. Northwest of the scar depression a internal reflection is correlated with the base of the Mesozoic carbonate rocks or a thrusting involving these Mesozoic layers. A simple slide of the Tramontane block from the scar to its present situation should occur along a detachment dipping towards the southeast, but the depth map of the deep reflector (Fig. 16) shows a southward component. Therefore, we conclude that at least two detachment levels exist beneath the Paleozoic Tramontane slide and the deeper detachment does not have a direct relationship with the Tramontane collapse. In fact, we may suppose that two successive collapses occurred: a first giant Paleozoic slide formed the scar, then the top of this block was scalped and the present situation of the Tramontane block resulted from this second foundering. The strike lines (Fig. 15 B, C) show the northern rise of the deep reflector, then the deep reflector merges with the acoustic basement but some deeper reflections appear. This zone is near the Calmar well (Fig. 6), where a Mesozoic thrust is suggested by the seismic profiles. The intra-Paleozoic reflections may represent the roots of this thrusting, but the Calmar well is placed close to the bottom of the Oligocene graben (Fig. 6), whereas Tramontane was drilled on the flank and this situation may explain the presence or the lack of Mesozoic rocks, respectively, involved by the Eocene thrusting if we suppose that the Oligocene graben was a former Mesozoic depression. The depth map of the deep reflector (Fig. 16) shows that the surface has a trend almost parallel to the NE-SW extensional faults, but a southward-dipping trend is also perceptible. This surface is probably a good image of an Eocene thrusting towards the north with a left-lateral strike-slip component. If a clear correlation exists between the NE-SW extensional faults and the trend of the deep surface, on the other hand, the N-S trend of the fault that bounds the Cathares graben is completely different.

Rascasse Structure and the Catalan Graben

The Catalan graben is an 8-km-deep depression bounded by the Rascasse high which trends N-S (Fig. 12). In a previous interpretation (Fig. 17) the presence of intra-Paleozoic reflections was suspected (Mauffret and Gennes-

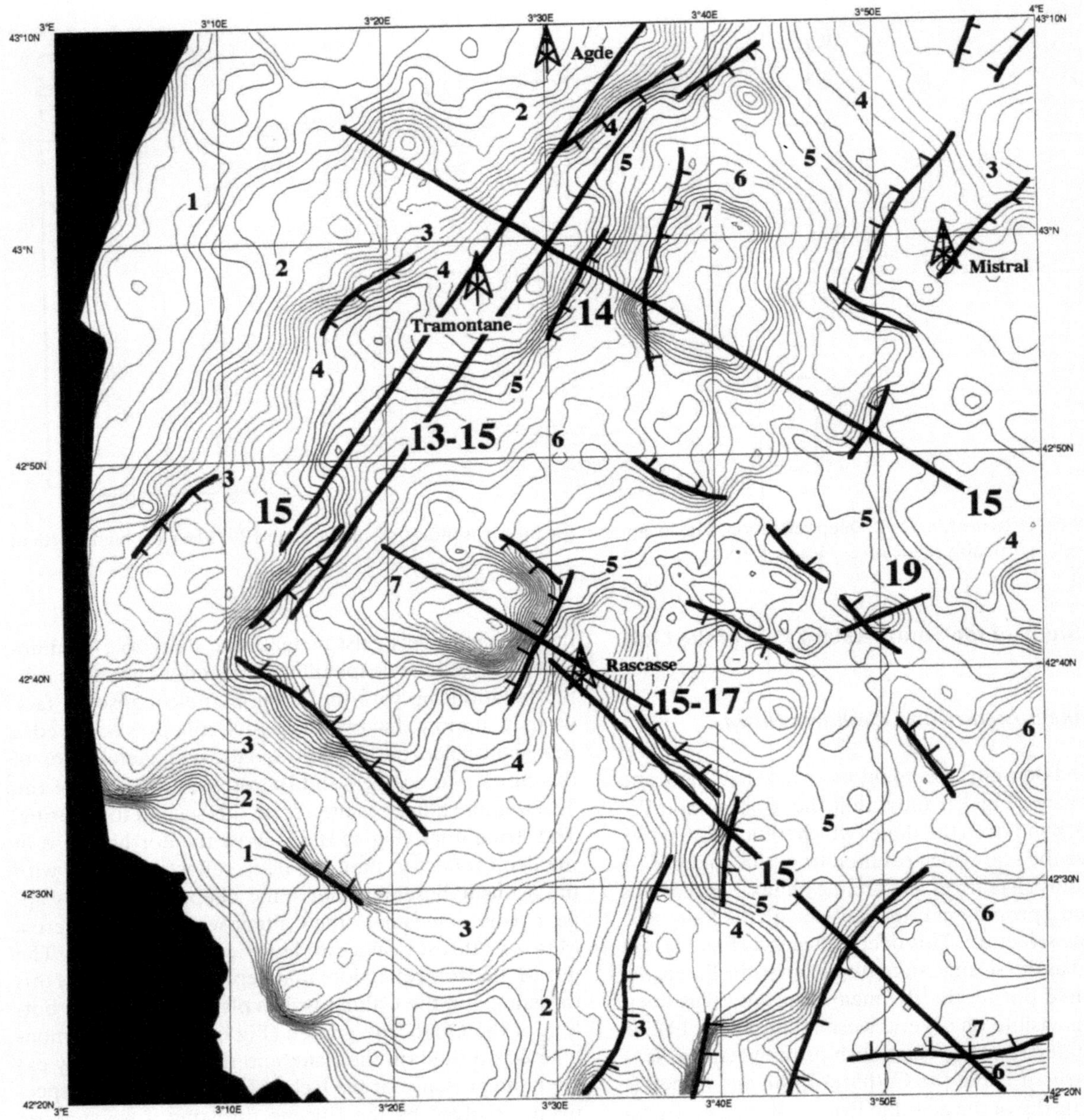

3°E
3°10E
3°20E
3°30E
3°40E
3°50E
4°E
43°10N
43°N
42°50N
42°40N
42°30N
42°20N
Agde
Tramontane
Mistral
Rascasse
13-15
14
15
15-17
19

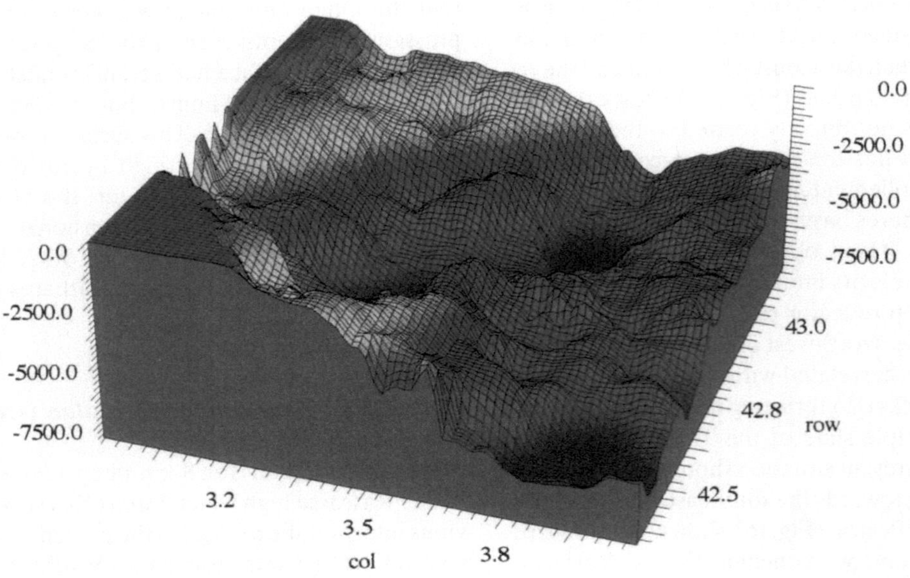

0.0
-2500.0
-5000.0
-7500.0
43.0
42.8
42.5
row
3.2
3.5
3.8
col

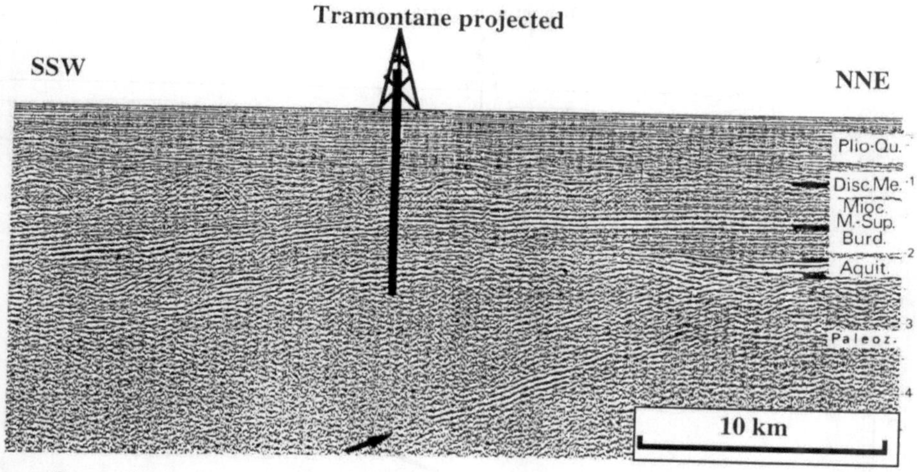

Fig. 13. Seismic profile showing a deep reflection beneath the Tramontane Paleozoic block (Mauffret and Gennesseaux 1989) The location of the profile is indicated in Figs 12 and 16 The deep reflection is probably the trace of an Eocene thrust Note that the line is in strike with the Tramontane structure

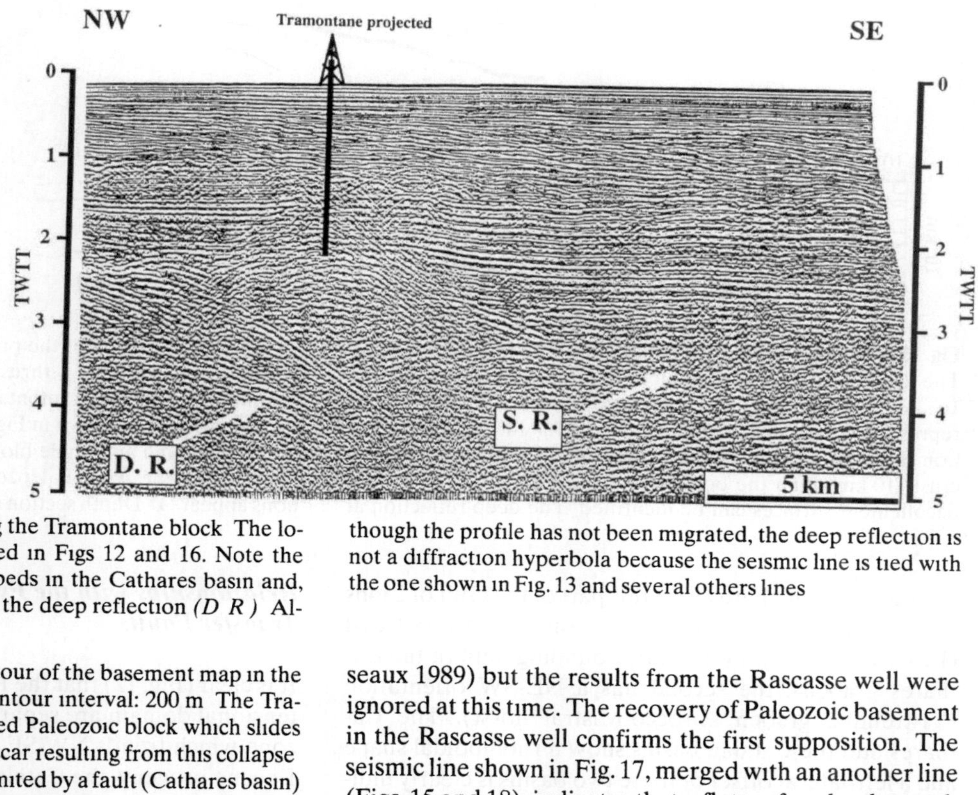

Fig. 14. Seismic profile crossing the Tramontane block The location of the profile is indicated in Figs 12 and 16. Note the thickness of the synrift *(S R)* beds in the Cathares basin and, beneath the Tramontane horst, the deep reflection *(D R)* Although the profile has not been migrated, the deep reflection is not a diffraction hyperbola because the seismic line is tied with the one shown in Fig. 13 and several others lines

◀ **Fig. 12.** *Above* Automatic contour of the basement map in the Rascasse-Tramontane area. Contour interval: 200 m The Tramontane well is placed on a tilted Paleozoic block which slides from the upper slope. Note the scar resulting from this collapse The two large depressions are limited by a fault (Cathares basin) and the Rascasse horst (Catalan basin) which trend N-S. The same orientation is observed along the Catalan margin The Pyrenean Range is limited by a transverse structure (Catalan fault, Lefebvre 1981) *Below* Oligocene bird's-eye view from the southwest. Note the collapsed Tramontane block and the offset of the two basins

seaux 1989) but the results from the Rascasse well were ignored at this time. The recovery of Paleozoic basement in the Rascasse well confirms the first supposition. The seismic line shown in Fig. 17, merged with an another line (Figs. 15 and 18), indicates that a flat surface lies beneath tilted blocks. Although the surface and the bottom graben are situated at about the same level, there is no obvious correlation between the formation of graben and the deep reflector. The depth map (Fig. 16) indicates that seismic lines are parallel to the slope, which apparently dips towards the northeast, but we do not have enough seismic control to draw a picture as good as in the Tramontane area. The flat surface is probably an intra-Paleozoic decollement level, which is the base of tilted blocks transported during the extension (Fig. 5). The true dip of the flat surface (northeast or southeast) is not determined.

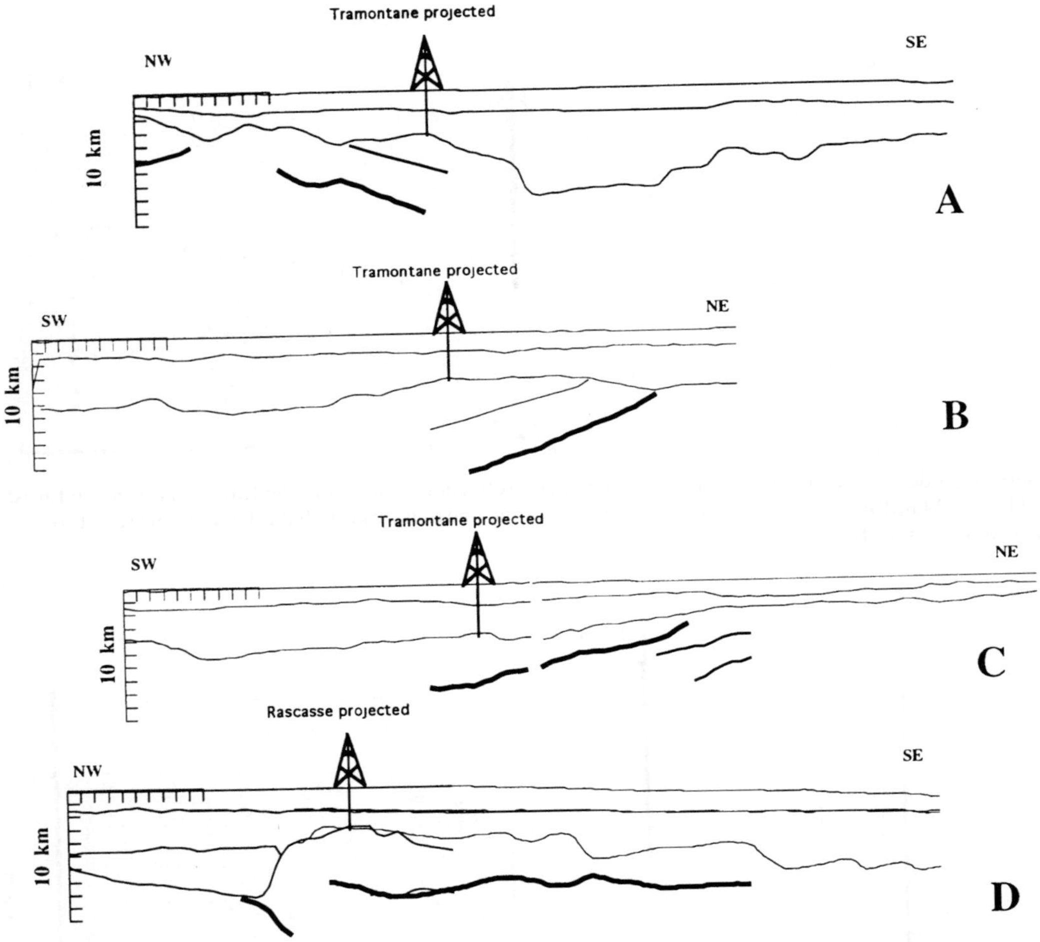

Fig. 15 A-D. Depth sections in the Tramontane-Rascasse area The location of the depth sections are indicated in Figs 12 and 16 **A** Cross section of the Tramontane block The central part represents the seismic profile shown in Fig 14 The dip reflection dips towards the Cathares basin but goes deeper on the crust (10 km) than the bottom of the depression (8 km) Two decollement surfaces can be identified The deep reflection at the northwestern end of the profile corresponds to the base of the Mesozoic cover or a thrust involving this cover **B** Strike seismic line along the Tramontane block This a depth section of the seismic profile shown in Fig 13 **C** Strike seismic line along the Tramontane and Agde blocks The deep reflection is shallow beneath the Agde Paleozoic well but some deeper reflections appear **D** Depth section of the Rascasse area (see later)

In conclusion, there are in this part of the Gulf of Lions two directions of rifting: the first one has a N-S trend (Rascasse horst and westward-bounding fault of the Cathares graben), the second has a NE-SW orientation (Vistrenque graben, Sirocco-Mistral horst). The Cathares and the Catalan basins show a rhomboidal shape and a left-lateral offset of these two basins is clearly indicated by the depth to basement map (Fig. 12). These depressions belong to the European rift system (Fig. 2), which was probably active with a N-S orientation during the Late Eocene (Ludian, Triat and Truc 1983; Debrand-Passard and Courbouleix 1984) just after the Pyrenean orogeny. For a second time, the Oligocene grabens have been opened along the former transpressional structures (Cévennes fault system) which trend NE-SW

Relationships with the Pyrenees and Transfer Faults

It is clear (Fig. 12) that the Pyrenees and the offshore extensional domain are separated by a large transfer fault zone which trends NW-SE (Lefebvre 1981). During the Eocene orogeny, the Catalan fault transmitted the compressional stress from the Pyrenees to Calabria (Rehault et al. 1984, Olivet 1987). During the Oligocene extension, this fault zone connected the rifts of the Gulf of Lions to the Catalan margin, which is affected by N-S normal faults. The Cretaceous plate boundary between Iberia and Europe, the North Pyrenean fault, has been reactivated in the central Pyrenean Range, and a large step in the Moho depth (Daignières et al. 1982) indicates an underthrusting of the Iberian plate (Choukroune and Ecors Team 1989). However, this step disappears in the eastern part of the Pyrenees and a thinning of the crust is perceptible beneath the Mediterranean coast (Gallart et al 1980) Consequently, the North Pyrenean

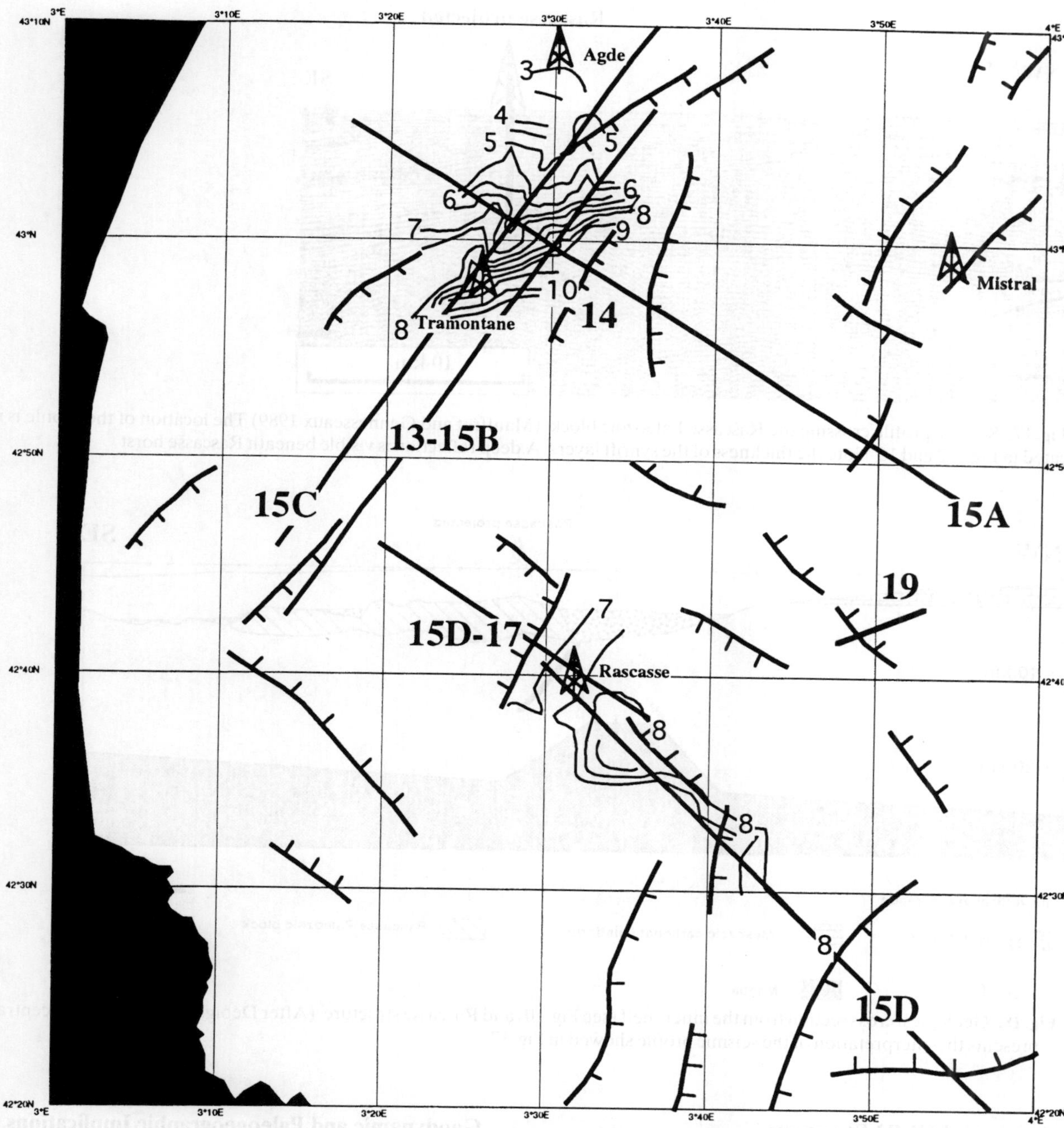

Fig. 16. Depth to the deep reflection Contour interval: 0 5 km Note the dip towards the southeast beneath the Tramontane well This deep reflection is probably the seismic image of on Eocene thrusting surface with a strike-slip component along the NE-SW direction The deep reflection beneath the Rascasse (see later) is apparently dipping towards the northeast but the seismic control is poor The *numbers* refer to the position of the seismic profiles shown in Figs 13, 14, 15, 17 and 19

fault is probably weakly remobilized during the Eocene compression and should run straight across (Fig. 1) the Gulf of Lions (Arthaud and Mattauer 1972; Arthaud and Seguret 1981; Olivet 1987). The North Pyrenean fault can be followed up to the shore at 42°40N (Fig. 1) and may limit the Catalan basin to the south (Fig. 12). A seismic expression of the fault has not been identified, either on the ECORS profiles or on the industrial seismic lines except south of Provence, where some weak seismic evidence of opposite vergence has been described (Mauffret and Gorini 1993). However, from the Rascasse basin to the Provence margin, a succession of troughs, south of the Autan well, north of the GL2 well, is slightly offset by apparent right-lateral strike-slip faults which strike NW-SE (Fig. 6). Consequently, we assume that the possible extension of the North Pyrenean fault has been offset by a transfer fault during the Oligocene extension.

These transfer faults are particularly evident on the basement map (Fig. 12). The Mistral ridge is abruptly interrupted by one of these faults. A seismic profile (Fig. 19), which crosses a transfer fault, shows a typical half-graben, indicating a transtensive motion.

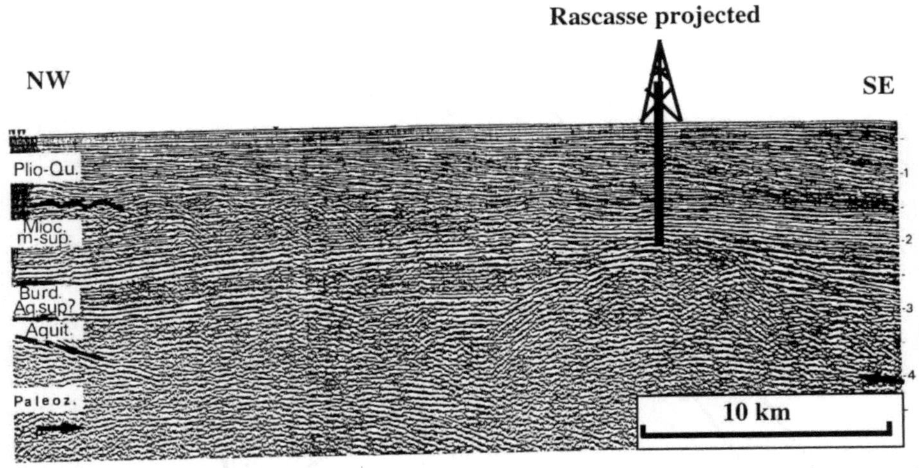

Fig. 17. Seismic profile crossing the Rascasse Paleozoic block (Mauffret and Gennesseaux 1989) The location of the profile is indicated in Figs 12 and 16 Note the thickness of the synrift layers A deep reflection is visible beneath Rascasse horst

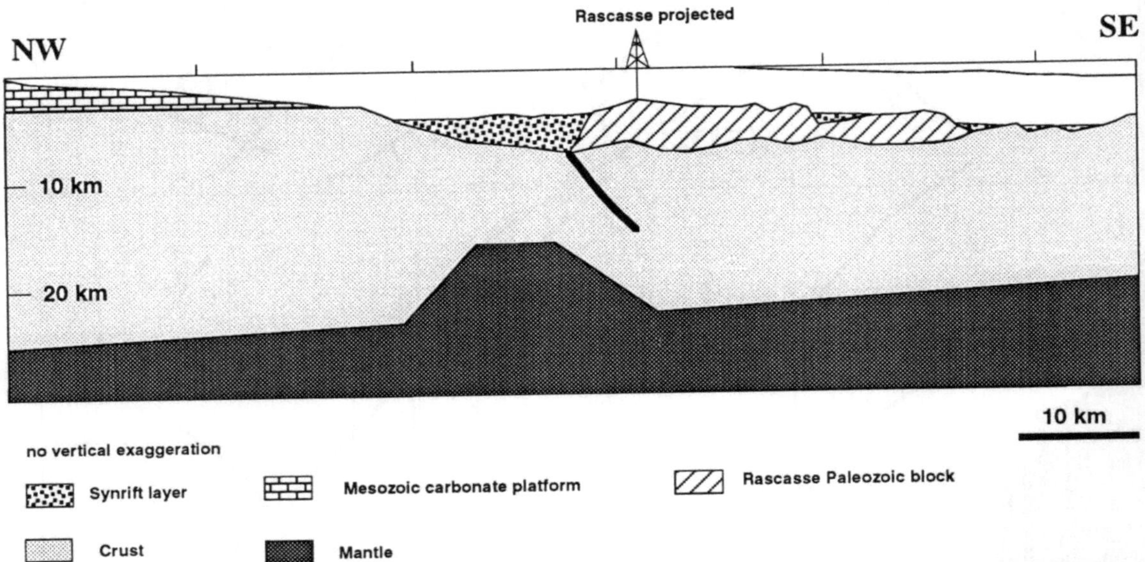

Fig. 18. Geological cross section from the inner shelf (see Fig 10) and Rascasse structure (After Debiglia et al 1993) The central part represents the interpretation of the seismic profile showed in Fig 17

Autan and GLP2 Structures

Again, the role of the transfer faults is particularly visible in the Autan-GLP2 area (Fig. 20) and the two structures are clearly right laterally offset, although the two features do not have the same structural expression because the seaward slope of the Autan structure is steeper than that of the GLP2 ridge. The right-lateral offset of a transfer fault could be apparent and this type of fault may delimit two different tectonic provinces. Abundant literature describe the GLP2 structure (de Voogd et al. 1991; Gorini et al. 1993), but the Autan structure is less well known. The seaward slope is steep but not affected by prominent normal faulting. Nevertheless, the Autan high is limited by a northward-facing extensional fault (Fig 21). Deep reflections are visible beneath the Autan Paleozoic structure but the seismic control is too weak to determine a surface

Geodynamic and Paleogeographic Implications

The initial position of the Sardinia and Corsica blocks (Fig. 22) is based on the works of Olivet (1987) and Gueguen et al. (1993). In this kinematic model, the thick crusts of Sardinia and the Gulf of Lions are adjacent. A complete description of the structural implications of this model (Gueguen et al. 1993; Mauffret et al. 1993) is beyond the scope of this chapter, but we can note that the first crack occurred with a N-S orientation between Sardinia and Catalonia, and this first rift is connected to the large basins of the Gulf of Lions by the Catalan transfer fault. This reconstruction has some important paleogeographic implications. Oligocene marine sediments are known in the Betics, Mallorca (Azema et al. 1974), in the canyons south of Provence (Labarbarie 1979) and GLP2 (Gorini et al. 1993). Oligocene evaporites with marine in-

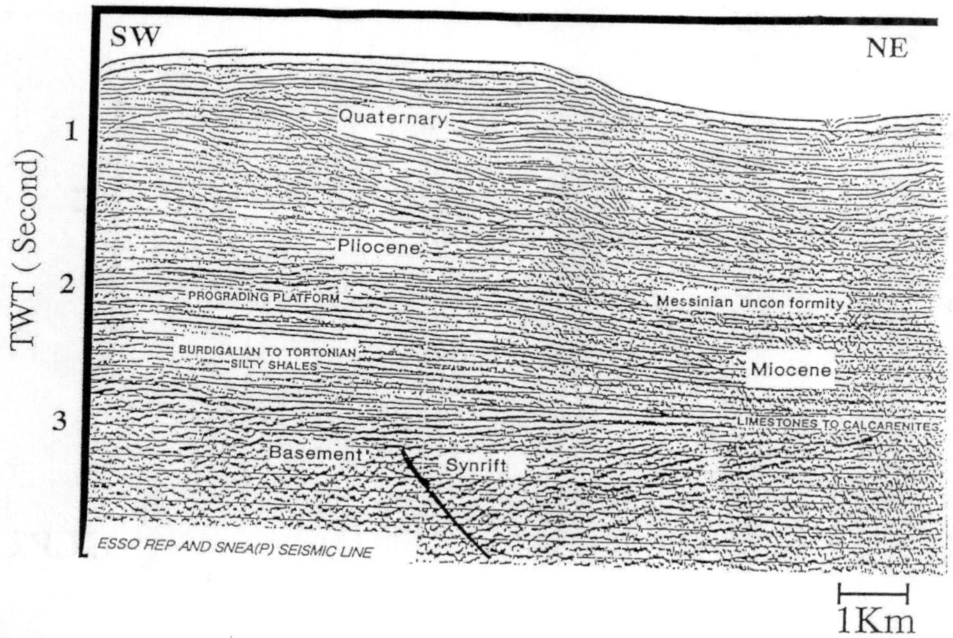

Fig. 19. Seismic profile crossing a transfer fault (Gorini 1993) The location of the profile is indicated in Figs 12 and 16 There is no evidence of a strike-slip fault but the extension is clear in the half-graben which trends NW-SE

fluence are described in the Vistrenque graben, the Marseille-Aix basin (Nury 1988), the Catalan graben (Bartrina et al. 1992), and have been sampled in the Autan well (Cravatte et al. 1974). The Cathares and Catalan basins, which were probably formed before evaporite deposition, were probably an arm of the Oligocene sea (Fig. 23).

Summary and Conclusions

This chapter shows that the deep penetration seismic lines (ECORS) are a powerful tool, however, a three-dimensional map using the industrial seismic lines is necessary to understand the complex structures. The present structural framework is the result of a long geological history which began during the Late Paleozoic with the formation of a conjugate system of faults. These faults, mainly oriented NE-SW, limited the Mesozoic rifts but the influence of the NW-SE faults can also be detected. These Mesozoic basins may be extended to the Gulf of Lions with the shape of a glove finger. However, this assumption is based on seismic evidence and most of the wells bottomed to the Paleozoic basement. These Paleozoic ridges could be the remnants of Mesozoic horsts. We cannot discern the influence of the Middle Cretaceous rotation of the Iberian plate but an eastern extension of the North Pyrenean fault, right-laterally offset by Oligocene transfer faults, has been proposed. This proposition is based on the existence of deep and narrow troughs lying in a E-W direction but we cannot discern a seismic signature of such a fault cutting through the deep crust. The Eocene compressional deformation is evident and several deep reflections in the Mesozoic and the Paleozoic rocks can be related to Eocene thrusts. The Mesozoic basins have been inverted and the thrusts are oriented E-W but also NE-SW with a left-lateral strike-slip component. Detailed mapping of the thrusts surfaces indicated that they are deeply rooted in the crust. Nevertheless, the compressional deformation is not as intense as in the Pyrenean Range. The Middle Eocene compression has been immediately followed by a Late Eocene rifting which trends N-S and is related to the European rift system. The deep Cathares and the Catalan basins have a rhomboidal shape and are connected by a left-lateral strike-slip fault. The major transfer Catalan fault, oriented NW-SE, transmitted the extensional motion from the Gulf of Lions to the Catalan margin where the first separation of Sardinia and Iberia occurred. After this initial N-S rifting a second extension, with a NE-SW orientation, reactivated the previous thrusting surfaces. This second extensional event concerns mainly the northeastern part of the Gulf of Lions and preceded the opening of the Provençal basin. The role of the transfer faulting during this rifting is important. The Valencia trough and the Gulf of Lions region underwent the same kind of extensional tectonics during the Late Paleogene and the Early Miocene, but the first zone is oil-productive, whereas the wells from the Gulf of Lions are dry. The Mesozoic history is the same but the carbonate platform formed in this epoch has been disturbed, eroded, and transported towards the north during the Pyrenean orogeny in the second case, whereas the Eocene compressional tectonics is minor in the Valencia Trough. After the rifting the basement was deep beneath the Valencia continental shelf and a high heat flow favors the generation of oil and gas. In the Gulf of Lions the first generation of N-S rifts aborted rapidly and the subsidence which followed the second NE-SW rifting was compensated by a large progradational sedimentation in a cold environment.

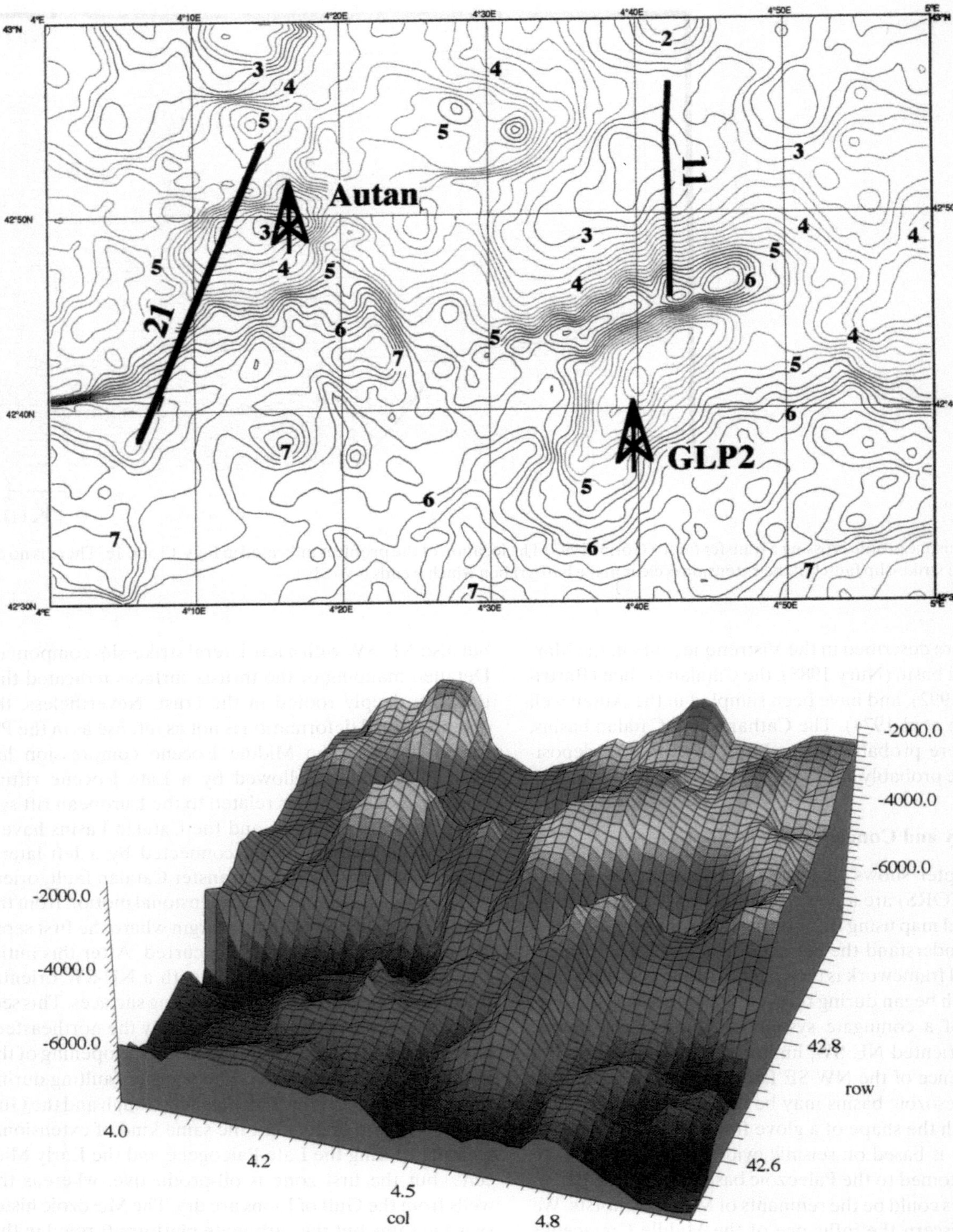

Fig. 20. *Above* Automatic contour map of the depth to basement in the Autan GLP2 area Contour interval 200 m Note the deep depressions south of the Autan well and north of the GLP2 horst The two drilled structures are separated by several transfer faults *Below* Oligocene bird's-eye view of the Autan-GLP2 area Note the Mesozoic Cers high illustrated in Fig 11

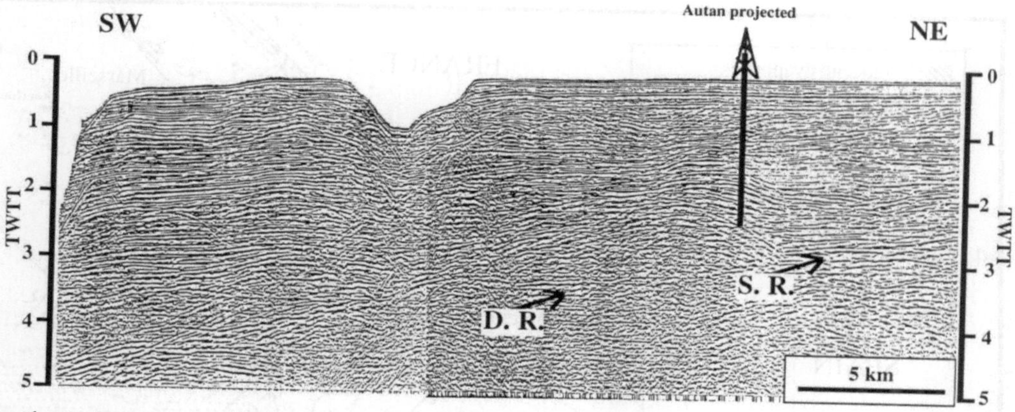

Fig. 21. Seismic line crossing the Autan structure The location of the profile is indicated in Fig 20 *S R* Synrift layer, *D R* deep reflection beneath the Autan Paleozoic horst

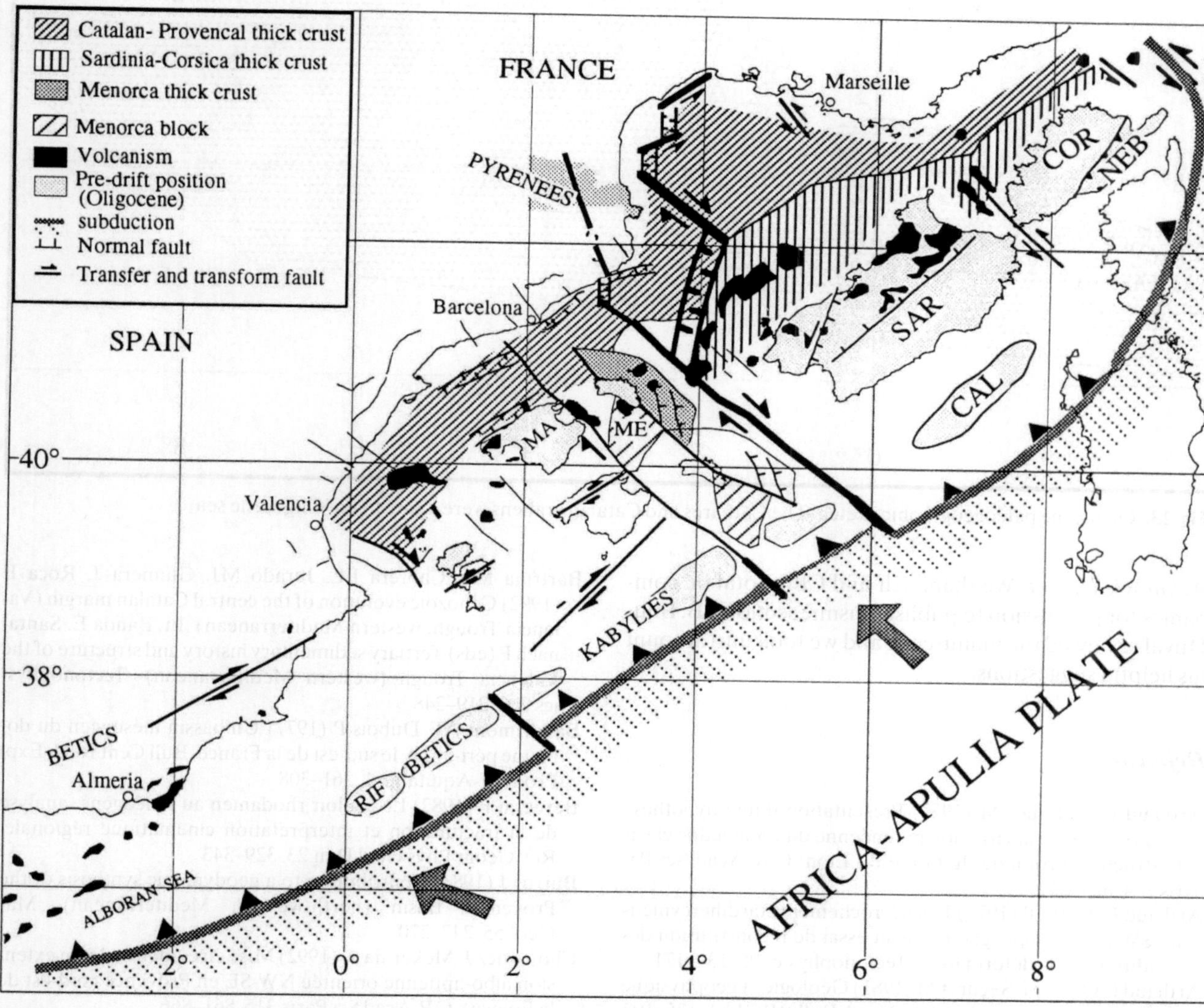

Fig. 22. Geodynamic reconstruction of the northwestern Mediterranean basin The first separation (Late Eocene?) occurred with a N-S orientation between Sardinia and Catalonia This rift is related to the Gulf of Lions basins via the Catalan fault

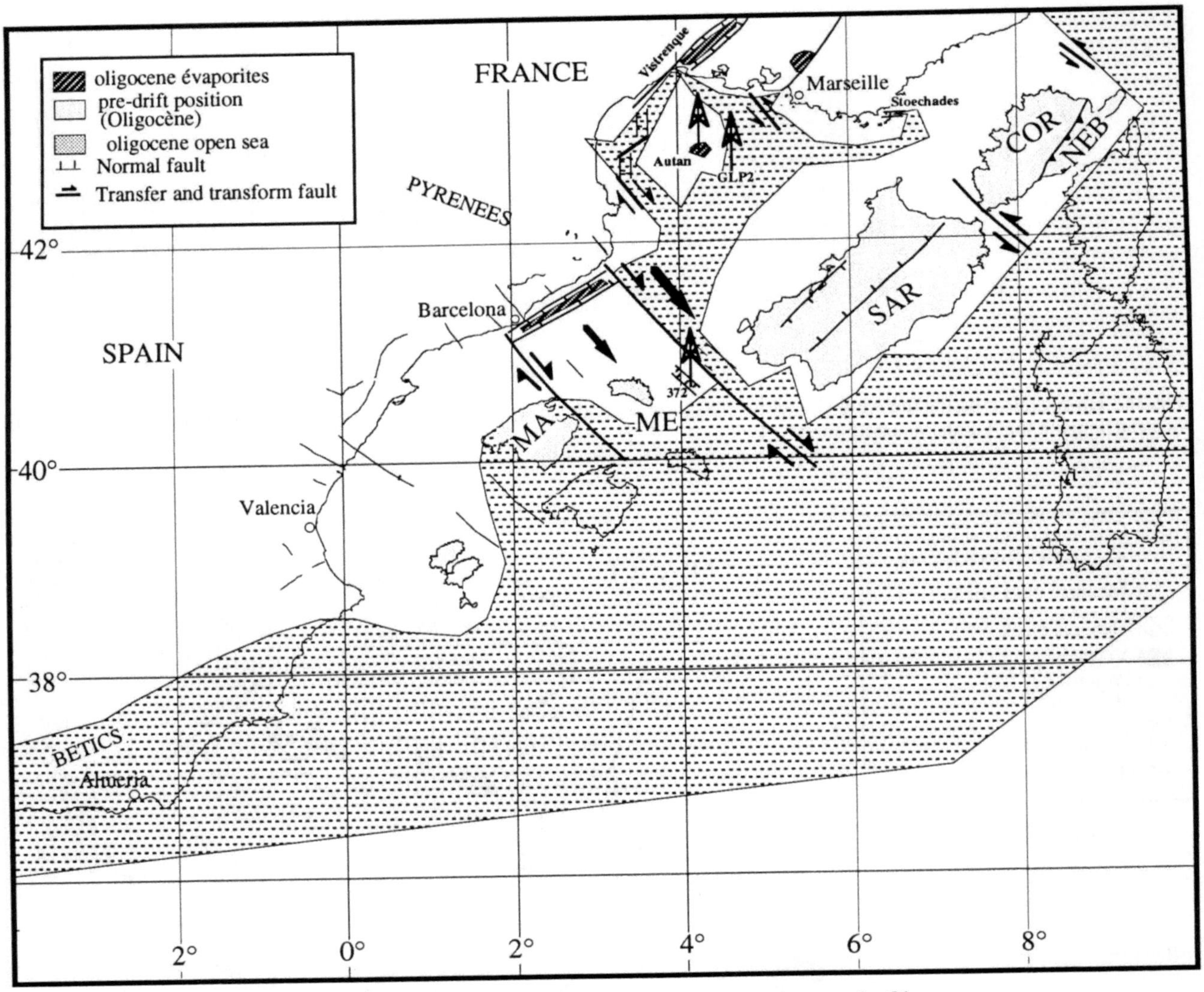

Fig. 23. Oligocene paleogeographic sketch The Cathares and Catalan grabens were a gate to the Oligocene sea

Acknowledgment. We thank Elf and CFP-Total oil companies for permission to publish seismic sections. B. Biju-Duval reviewed the manuscript and we took into account his helpful suggestions.

References

Arthaud F. Mattauer M (1972) Présentation d'une hypothèse sur la génèse de la virgation pyrénéenne du Languedoc et sur la structure profonde du Golfe du Lion C R Acad Sci Paris 274 524–527

Arthaud F. Matte P (1975) Les décrochements tardihercyniens du SW de l'Europe, géométrie et essai de reconstitution des conditions de la déformation Tectonophysics 25 139–171

Arthaud F. Ogier M. Seguret M (1981) Géologie et géophysique du Golfe du Lion et sa bordure Nord Bull BRGM 1 175–193

Arthaud F. Seguret M (1981) Les structures pyrénéennes du Languedoc et du Golfe du Lion (Sud de la France) Bull Soc Géol Fr 7 51–63

Azema J. Bourrouilh R. Champetier Y. Fourcade E. Rangheard Y (1974) Rapports stratigraphiques, paleogéographiques et structuraux entre la Chaîne Ibérique, les Cordillères Bétiques et les Baléares Bull Soc Géol Fr 16 140–160

Bartrina MT. Cabrera LL. Jurado MJ. Guimera J. Roca E (1992) Cenozoic evolution of the central Catalan margin (Valencia Trough, western Mediterranean) In Banda E. Santanach P (eds) Tertiary sedimentary history and structure of the Valencia Trough (western Mediterranean) Tectonophysics 203 219–248

Baudrimont AF. Dubois P (1977) Un bassin mésogéen du domaine péri-alpin le sud-est de la France Bull Cent Rech Expl Prod Elf-Aquitaine 1 261–308

Bergerat F (1982) Le couloir rhodanien au Paleogène analyse de la fracturation et interprétation cinématique régionale Rev Géogr Phys Géol Dyn 23 329–343

Burrus J (1984) Contribution to a geodynamic synthesis of the Provencal Basin (north-western Mediterranean) Mar Geol 55 247–270

Chorowicz J. Mekarnia A (1992) Mise en evidence d'une extension albo-aptienne orientée NW-SE en Provence (sud-est de la France) C R Acad Sci Paris 315 861–866

Choukroune P. Ecors Team (1989) The ECORS pyrenean deep seismic data and overall structure of an orogenic belt Tectonics 8 23–39

Clavell E. Berastegui X (1991) Petroleum geology of the Gulf of Valencia In Spencer AM (ed) Generation, accumulation and production of Europe's hydrocarbon Oxford University Press, Oxford, pp 355–368

Cravatte J, Dufaure P, Prim M, Rouaix S (1974) Les sondages du Golfe du Lion stratigraphie et sédimentologie Not Mém CFP 2 209–274

Curnelle R, Dubois P (1986) Evolution mésozoique des grands bassins sédimentaires français bassin de Paris, d'Aquitaine et du sud-est Bull Soc Géol Fr 8 526–546

Daignières M, Gallart J, Banda E, Hirn A (1982) Implications of the seismic structure for the orogenic evolution of the Pyrenean range Earth Planet Sci Lett 57 88–100

Debeglia N, Gorini C, Guennoc P, Le Marrec A, Mauffret A (1993) Oligocene rifting in the Gulf of Lions, geophysical contribution to a new structural scheme Terra Abstr 5 277

Debrand-Passard S, Courbouleix S (1984) Synthèse géologique du sud-est de la France Mém BRGM 126 pp

Deroin JP, Chorowicz J, Curnelle R (1992) Utilisation de l'imagerie Spot en géologie régionale Contribution à l'analyse de la tectonique post-hercynienne du Bas-Languedoc méditerranéen (France) CR Acad Sci Paris 314 511–517

de Voogd B, Nicolich R, Olivet JL, Fanucci F, Burrus J, Mauffret A, Pascal G, Argnani A, Auzende JM, Bernabini M, Bois C, Carmignani L, Fabbri A, Finetti I, Galdeano A, Gorini CY, Labaume P, Lajat D, Patriat P, Pinet B, Ravat J, Ricci Lucchi F, Vernassa S (1991) First deep seismic reflection transect from the gulf of Lions to Sardinia (Ecors-Crop profiles in western Mediterranean) In Continental lithosphere deep seismic reflections Am Geophys Union, pp 265–274

Gallart J, Daignières M, Banda E, Surinach E, Hirn A (1980) The eastern pyrenean domain lateral variations at crust-mantle level Ann Geophys 36 141–157

Gorini C (1993) Géodynamique d'une marge passive le Golfe du Lion (Méditerranée occidentale) Thèse de l'Université de Toulouse

Gorini C, Viallard P, Deramond J (1991) Modèle d'inversion négative la tectonique extensive post-nappe du fossé de Narbonne-Sigean (Corbières, sud de la France) CR Acad Sci Paris 312 1013–1019

Gorini C, Le Marrec A, Mauffret A (1993) Contribution to the structural and sedimentary history of the Gulf of Lions (western Mediterranean), from the ECORS profiles, industrial seismic profiles and well data Bull Soc Géol Fr 164 353–363

Gueguen E, Olivet JL, Rehault JP (1993) Kinematic model in the western Mediterranean new constraints Terra Abstr 5 167

Hibsh C, Kandel D, Montenat C, Ott d'Estevou P (1992) Evénement crétacés dans la partie mériodonale du bassin subalpin (massif Ventoux-Lure et partie orientale de l'arc de Castellane, SE France) Implications géodynamiques Bull Soc Géol Fr 163 146–158

Labarbarie M (1979) Etude structurale des canyons sous-marins de la marge des Maures (Var) Thèse de 3° cycle, Paris 6, 53 pp

Le Douaran S, Burrus J, Avedik F (1984) Deep structure of the north-western Mediterranean basin results of a two-ship seismic survey Mar Geol 55 325–345

Lefebvre D (1981) Evolution morphologique et structurale du Golfe du Lion Essai de traitement statistique des données Thèse de 3° cycle, Paris 6, 163 pp

Mauffret A, Gennesseaux M (1989) Compression, décrochement et distension sur le pourtour méditerranéen nord-occidental CR Acad Sci Paris 308 961–967

Mauffret A, Gorini C (1993) Structural styles of the Camargue area and western Provençal basin (southeastern France), geodynamic consequences Tectonics (submitted)

Mauffret A, Pascal G, Maillard A, Gorini C (1993) Structure of the deep northwestern Mediterranean basin J Petrol Geol (submitted)

Nury D (1988) L'Oligocène de Provence méridionale Doc BRGM 163 1–410

Olivet JL (1987) L'origine du bassin nord-occidental de la Méditerranée du point de vue de la cinématique des plaques In Burrus J, Olivet JL (eds) Profils ECORS Golfe du Lion rapport d'implantation IFP, Paris, 35 941-1, pp 10–49

Olivet JL, Beuzart P, Auzende JM, Bonnin J (1984) Cinématique de l'Atlantique Nord et Central Cnexo 54 107

Pascal G, Mauffret A, Patriat P (1993) The ocean-continent boundary in the Gulf of Lions from analysis of expanding spread profiles and gravity modelling Geophy J Int 113 701–726

Philip J, Masse PJL, Machhour L (1987) L'évolution paléogéographique et structurale du front de chevauchement nord-toulonnais (Basse-Provence occidentale, France) Bull Soc Géol Fr 8 541–550

Rehault JP, Boillot G, Mauffret A (1984) The western Mediterranean basin, geological evolution Mar Geol 55 447–477

Roure F, Brun JP, Colletta B, Van den Driessche J (1992) Geometry and kinematics of extensional structures in the Alpine foreland basin of southeastern France J Struct Geol 14 503–519

Sapin M, Hirn A (1974) Results of explosion seismology in the southern Rhône valley Ann Geophys 30 181–202

Triat JM, Truc G (1983) Le rôle des failles N 50 dans la sédimentation des temps méso et cénozoiques et dans l'évolution tectonique du bassin du sud-est (France) Bull Cent Rech Expl Prod Elf-Aquitaine 7 425–432

Multiphase Extensional Structures, Fault Reactivation, and Petroleum Plays in the Alpine Foreland Basin of Southeastern France

F. Roure, J. P. Brun, B. Colletta, and R. Vially

Abstract. Jurassic and Oligocene extensional structures are still well preserved in the compressional basin of southeastern France. Due to the occurrence of numerous potential decollement levels (i. e., the Carboniferous coal measures, the Triassic evaporites, the Jurassic blackshales, or the Neocomian marls), these structures frequently independently involve the crystalline basement and the sedimentary cover, thus illustrating various structural styles that are clearly imaged in seismic profiles.

Local Paleogene and Neogene tectonic inversions have modified the early extensional structures, especially at the boundary between Mesozoic basinal and platform domains. A large wedging of the shallow Cretaceous platform sequences thus occurred over a deep indenter made of Mesozoic basinal sequences, and is best imaged along La Lance-Valence transect in the Rhône Valley. Alternatively, wedging of the Mesozoic cover relates to the tectonic inversion of a buried Paleozoic basin along the Pic-St Loup profile in Languedoc. A similar backthrusting occurs in the thin-skinned cover units of the Jura Mtns near Oyonnax, and probably also relates to the recent inversion of a deeper Paleozoic basin. The relative timing of surface backthrusting and deeper basin inversion was analyzed by means of various techniques. Intermediate geometries of these complex triangle structures are thus discussed by using cross-section balancing techniques, and are then compared with analogue models.

In Neogene times, a collapse of the Pyrenean edifice occurred synchronously with the opening of the Gulf of Lions. During the building up of this new Mediterranean continental margin, the late Cretaceous to Paleogene sole-thrust of the tectonic pile was reactivated as an extensional detachment. Distensive piggy-back basins developed locally above the flats of this multiphase decollement surface north of La Clape Mtn. The development and intermediate geometries of these post-compressional collapse basins are also discussed with reference to analogue models.

Besides its complex architecture and partial inversion, which make this area a frontier zone for exploration, the basin of southeastern France still preserves locally unaltered Jurassic extensional structures, either (1) near the Ardèche margin, away from the areas reactivated during the Oligocene extension or Neogene inversions; (2) in the autochthon, along the Drôme River beneath the Vercors allochthon or in the Gabian area beneath the St Chinian allochthon; (3) inside reactivated blocks, in between the sole and roof thrusts of cover duplexes. Early generated oils have hopefully been preserved from secondary migration or cracking in these relict Jurassic structures.

Alternatively, during the Alpine thrust emplacement (Vercors), the overburial of the external parts of the basin led to a new episode of maturation. Similarly, the thick sedimentary infill of Oligocene (Bresse, Valence, Alès) or Neogene (Camargue, southern Languedoc) basins and the contemporaneous and local increase of the heat flow may have contributed to the generation of new oils, either from Tertiary source rocks or from deeper Mesozoic or Paleozoic horizons. In these latter cases, post-Jurassic traps should also be prospected.

Introduction

The petroleum potential of tectonically complex areas relates directly to the compatibility between the timing of the deformation (age of the traps), the burial history of the source rocks, the age of expulsion and migration of the hydrocarbons, and the availability of migration pathways between source areas (kitchen) and potential traps.

Usually in foreland areas, extensional structures relate to early rifting episodes (passive margin stage of the basin) and thus predate the compressive deformations (flexural stage or later episodes of inversion and thrusting). The maximum burial of the source rocks can be reached during either the syn-rift tectonic subsidence or the post-rift thermal subsidence of the margin prior to the deformation of the basin, or else during the compression itself, then resulting either from a rapid syntectonic sedimentation in the foredeep and piggy-back basins, or from the overthrusting of allochthonous tectonic units (tectonic overburial).

The case of the Alpine foreland basin in southeastern France is taken here as a typical example to describe a tectonically complex area for which the petroleum potential still requires better evaluation. Effectively, the occurrence of various decollement horizons in the Mesozoic sedimentary cover or even locally in the Paleozoic substratum induced complex geometries, either thin-skinned or basement-involved, during the successive Jurassic, early Cretaceous, and Oligo-Aquitanian extensional episodes that affected the basin. Later, Pyrenean (late Cretaceous to Eocene) or Alpine (Neogene) compressions led to the local inversion of these early extensional structures (basement fault reactivation) and to the formation of new thin-skinned structures (detachment of the sedimentary cover), thus largely modifying both the geometry of the basin and the burial of the potential source rocks.

Cross-section balancing techniques and sand-box analogue experiments were then used to examine the intermediate geometries of major structures, and to propose a coherent explanation for the building up of traps

Finally, new petroleum plays and exploration trends are proposed for this frontier area, in the light of the results of our structural modeling.

Regional Geologic and Geophysical Background

Detailed geologic descriptions of the basin are found elsewhere (Boillot et al. 1984, Debrand-Passard et al. 1984; Curnelle and Dubois 1986; de Graciansky et al. 1988; Roure et al. 1992) We shall here focus only on its major structures or events, which induced its present architecture.

General Geologic Setting and Major Episodes of Deformation

The basin of southeastern France is a remnant of the Tethyan paleomargin (Lemoine and Trumpy 1987). Its initial configuration and sedimentary infill mainly result from an early Jurassic (Liassic) episode of rifting (Elmi 1983, 1984; Giot et al. 1991 a, b), and the whole architecture of the basin has been only slightly modified by later Oligocene extension (southern extent of the west European rift system; Bois et al. 1991; Bois 1993) or the late Cretaceous-Paleogene and Neogene episodes of inversion (Roure et al. 1992).

Eastward from the Saône and Rhône rivers, north-trending folds and thrusts of the Jura and Vercors Mtns (Fig. 1) relate to the Neogene Alpine deformations. The eastern limit of the basin consists effectively of the Penninic thrust front, which marks the western extent of the inner Alps.

Between the Vercors Mtns and Provence, but still east of the Rhône valley, the Diois and Baronnies areas outline a complex pattern of either north-, east- or southeast-trending folds, which result from the superposition of both Pyrenean (late Cretaceous to Eocene) and Alpine (Neogene) deformations. Alpine fault reactivations are also known to affect the southeast- to east-trending La Lance and Ventoux-Lure thrust folds (Villeger and Andrieux 1987, Ford 1993), as well as the Alpilles and Luberon trends (Fig. 1), which effectively mark the southernmost extent of the Neogene compression toward the Mediterranean

In the south, east-trending folds and thrusts in southern Languedoc (St Chinian arc, Montpellier and Pic St Loup folds) and Provence (Ste. Victoire, Ste. Baume) (Fig 1) relate solely to late Cretaceous to Eocene Pyrenean deformations (Arthaud and Mattauer 1969; Tempier 1987)

Superimposed on these early compressive structures, the Oligocene rifting and Neogene opening of the Gulf of Lions have drasically modified the previous organization (Burrus 1984, 1989, Boillot et al. 1984; Gorini 1989; Gorini et al 1991, 1993) The new Mediterranean margin is thus a composite of (1) E-W segments, which are still parallel to the Pyrenean structures, and (2) SW-NE segments, which on the contrary are parallel to the Jurassic or Oligocene trends.

Sedimentary Infill of the Basin and Major Decollement Levels

Thick Carboniferous and Permian sequences crop out locally at the periphery of the basin, either along its western border (Lodève Permian basin and Alès or St Etienne late Carboniferous basins in the Massif Central), its southeastern corner (Permian basins from Provence) or its eastern border (Carboniferous basins from the External Crystalline Massifs, i.e., La Mure) (Fig 1; Mascle 1990)

The exact extent of the Paleozoic strata beneath the Mesozoic infill of the basin is only partially known. Some exploration wells have reached either the Permian or the Carboniferous sequences, and deep sedimentary horizons are frequently imaged on the seismic profiles. However, as will be stipulated below, the role of the Paleozoic sequences is crucial in both (1) the structuration of the basin itself (Carboniferous coal measures frequently acted as a decollement level during the Jurassic and/or Oligocene episodes of extension; Roure et al. 1992; Bonijoly et al. 1993) and (2) its petroleum potential. One of the recognized source rocks of the area consists of Autunian (early Permian) strata (Mascle 1990; Mascle et al 1993, this Vol.), whereas other potential sources consist of Stephanian coal measures.

Despite the fact that the Triassic sequence is relatively homogeneous across the whole basin, especially with the constant occurrence of evaporites, salt-related structures (diapirism) are known to occur only between the La Lance and Ventoux trends (Montmirail, Propiac and Condorcet diapirs, Fig. 1; Martinod 1988; Dardeau et al. 1990) or eastward (Laragne, External Crystalline Massifs, and Castellane arc; Mascle et al. 1988). Undoubtedly, these weak Triassic horizons acted as the major decollement level during both the Jurassic and Oligocene episodes of extension, and the Pyrenean and Alpine compressions (Petit et al. 1973; Roure et al. 1988, 1992).

Both north-south and east-west thickness and facies changes occur in the overlying Jurassic and Cretaceous sequences, and help to separate various sub-basins for which the structural evolution has been relatively distinct. In the north (Vercors Mtns) and in the south (La Lance and Ventoux Mtns, Provence area), platform carbonates (mainly Barremian to Aptian-Urgonian limestone) form a thick (1 to 2 km thick) and rigid body that

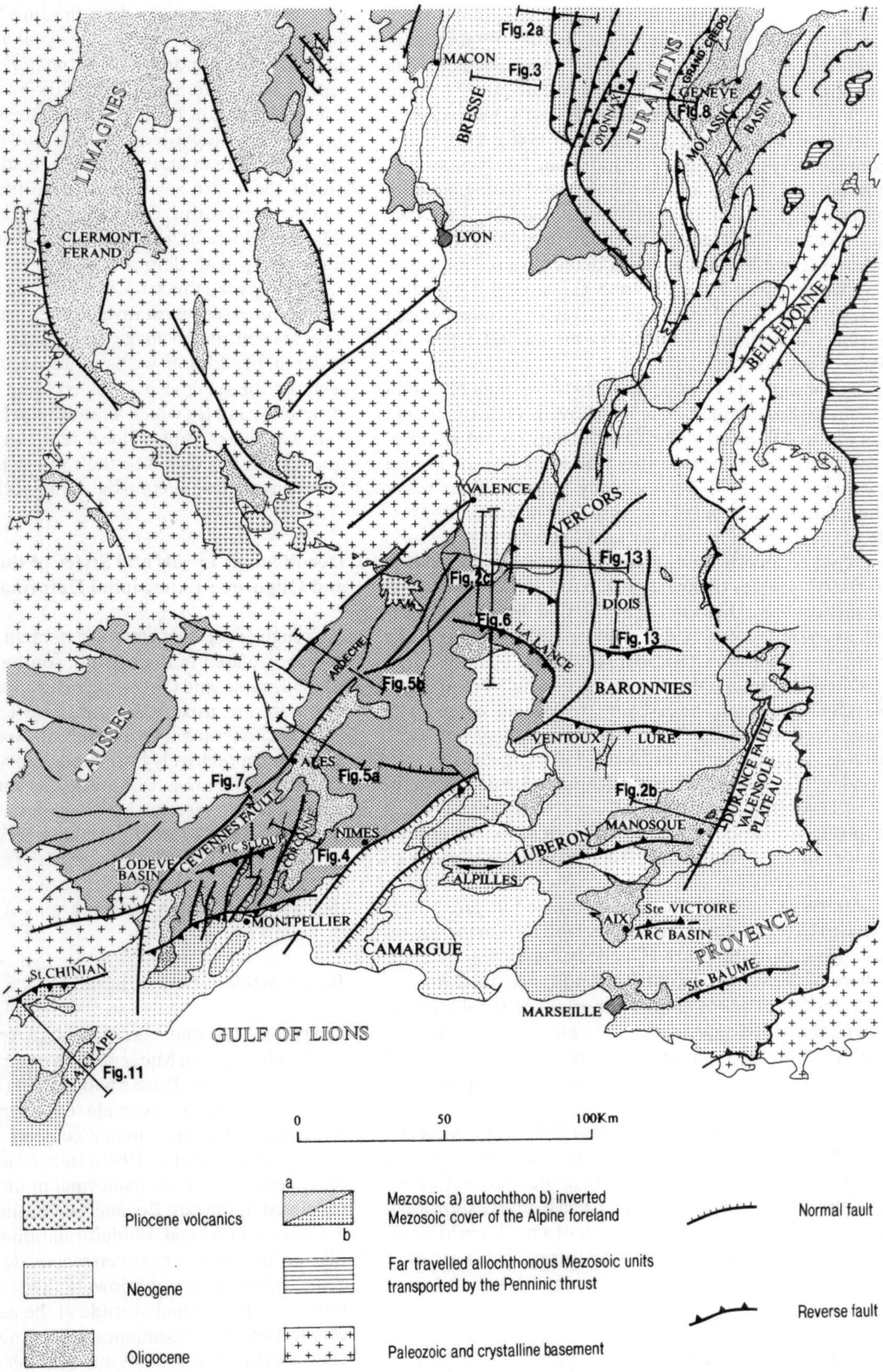

Fig. 1. Structural map of the Alpine foreland basin of southeastern France and location of the seismic profiles or geologic cross-sections discussed in the text

overlies a relatively thin Jurassic sequence. In between, i.e., in the Diois and Baronnies areas, the Cretaceous sequence is, on the contrary, made of pelagic beds that lie on top of a thick disharmonic Jurassic mud pile (5 to 6 km thick Liassic to Oxfordian blackshales) (Debrand-Passard et al. 1984, Curnelle and Dubois 1986). This lateral contrast between the thick disharmonic levels of the Vocontian basin (Diois and Baronnies) and the rigid carbonates of the surrounding platforms (Vercors, La Lance and Ventoux) has frequently led to complex triangle geometries (Ford 1993) in which both the Lower Cretaceous (Neocomian) and the Jurassic (Liassic or Oxfordian marls) sequences have been activated as decollement surfaces in addition to the above-mentioned Triassic or Paleozoic ones.

The ubiquitous Jurassic blackshales (mainly Pliensbachian-Toarcian-Aalenian and Callovian-Oxfordian) and very local Albian gray marls are the only potential Mesozoic source rocks in the basin of southeastern France (Mascle et al. 1993). They usually acquired their maximum burial at the end of the episode of thermal subsidence of the Tethyan margin, by mid-Cretaceous times, and were subsequently uplifted during the late Cretaceous to Tertiary episodes of inversion. Up to 2 km of Cretaceous sequences were thus effectively eroded during the Eocene, even on the poorly deformed western margin of the basin (Ardèche area; Giot et al. 1991 a, b).

Oligocene sediments, locally exceeding 2 km in thickness in Camargue, have been trapped in extensional basins related to the West European rift system (i.e., Bresse, Valence, and Alès basins, Fig. 1), leading to a secondary burial episode for the deeper horizons (Bergerat et al 1990; Blanc et al. 1991). Coeval Paleogene flexural deposits are also known to have affected the burial history of Mesozoic source rocks farther east in the Alps (Vially, this Vol.).

Effects of the Neogene Alpine foreland flexure are restricted to the Molasse basin in the north, or to the Valensole Plateau in the south (Guellec et al. 1990 a, b) Elsewhere in the basin of southeastern France, the Miocene sedimentation was of limited importance (mainly restricted to elongated synclines in the east, and to the Rhône valley at the deformation front), probably due to a still active inversion process in this part of the Alpine foreland.

However, coeval and important Oligocene-Miocene and Plio-Quaternary sedimentation also occurred away from the Alpine deformation front in the south in Camargue and along the Mediterranean coast, in an extensional context related to the development of a neoformed passive margin (Burrus 1984, 1989; Gorini 1989; Gorini et al. 1991, 1993)

Geophysical Background

Besides a single deep seismic reflection profile which crosses the northern part of the basin from the Massif Central to the External Crystalline Massifs (ECORS profile, across the Bresse graben, the Jura Mtns and the Molasse basin, Fig 1, Bergerat et al 1990, Guellec et al. 1990 a, b), hundreds of kilometers of conventional seismic lines cover most parts of the basin west of the Alpine deformation front (i e , the Bresse, Alès, and Valence basins, the Ardèche margin, and the southern Languedoc, Roure et at. 1988, 1992). In addition, numerous profiles have also been shot in deformed portions of the bordering platform, i.e., in the Jura and Vercors Mtns (platform domain), in Provence, or even in the Diois and Baronnies (highly deformed basinal domain).

We have used this set of seismic data here to select surface features representative for the Jurassic and Oligocene extension, and to image the regional extensional styles at depth. In addition, we have prepared depth-converted sections for specific profiles crossing typical inversion structures, and for which a calibration by wells or surface data was also possible These depth sections have subsequently been restored to their intermediate geometries by means of cross-section balancing techniques.

Decollement Levels and Styles of the Jurassic and Oligocene Extensional Deformations

Although for a long time a basement involvement has been postulated in the basin of southeastern France for all the Jurassic and Oligocene extensional structures, surface data are usually not sufficient to determine their deep architecture. Frequently, surface faults delineate areas with a relatively thin sedimentary sequence (hanging wall) from areas with thick deposits (footwall). However, their connection with deeper basement structures is not always direct and evident, since numerous potential detachment levels do occur in the sedimentary cover and usually help to transmit the surface deformation to the basement in other distal portions of the basin.

Basement-Involved Structures

Despite the fact that its surface trace is presently partially masked by the late Miocene Jura overthrust, the eastern border fault of the Bresse half graben (Figs. 1, 2 a) is probably the simplest example of extensional basement structure in the area under consideration (Chauve et al. 1988; Bergerat et al. 1990). Imaged in the ECORS and other seismic profiles, its normal motion was effectively restricted to the late Eocene-Oligocene and induced the deposition of a thick syndeformational sedimentary infill. As this fault has not been reactivated during the Neogene deformations, its downfaulted compartment still preserves the original attitude of the beds and outlines a classic drag of the sedimentary strata near the fault plane (Fig. 2 a, Hamblin 1965, White et al. 1986).

Other basement-involved faults in the basin of southeastern France usually show a more complex, multiphase evolution.

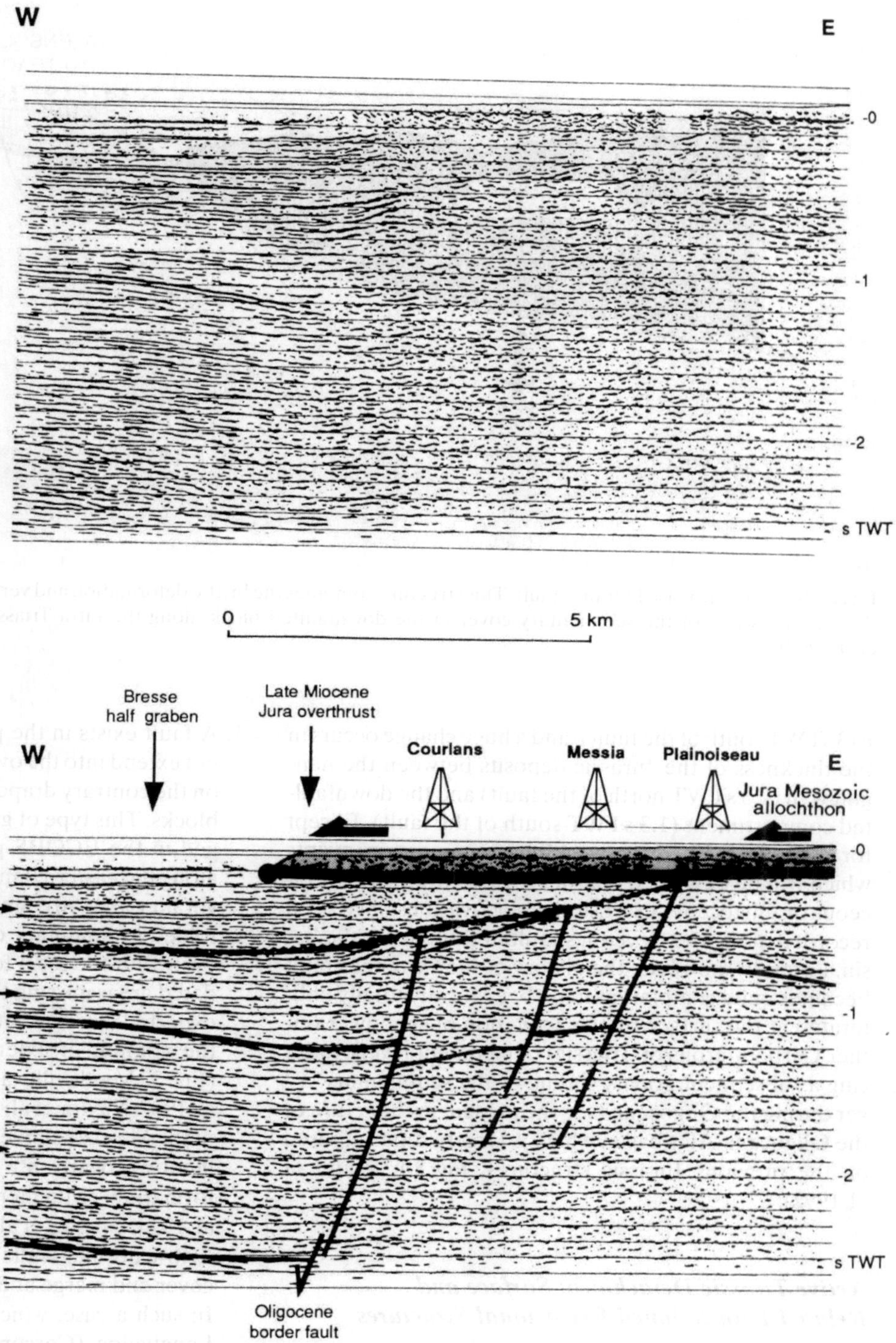

Fig. 2 a–c. Examples of basement-involved extensional structures **a** The Oligocene border fault of the Bresse half graben Note the drag of the beds in the downfaulted block **b, c** see pp 250, 251

In the east, the initial Jurassic geometry of the Durance fault was thus constantly modified during subsequent Cretaceous or Oligocene extension, or Eocene and Miocene compression (inversions) (Roure et al. 1992). Moreover, despite most of the vertical motion observed at the surface along this structure was balanced at depth by an equal displacement of the basement, part of the cover deformation was also absorbed laterally by the activation of a Triassic decollement level in the downfaulted block (Fig. 2 b).

In the west, the famous Cévennes fault itself (Arthaud and Mattauer 1969; Arthaud and Matte 1974) preserves its initial Jurassic configuration almost unmodified south of Valence (Fig. 1). There, as shown in Fig. 2 c, the basement is effectively largely offset by the fault at depth (at 2 sTWT below the sea level north of the fault, compared

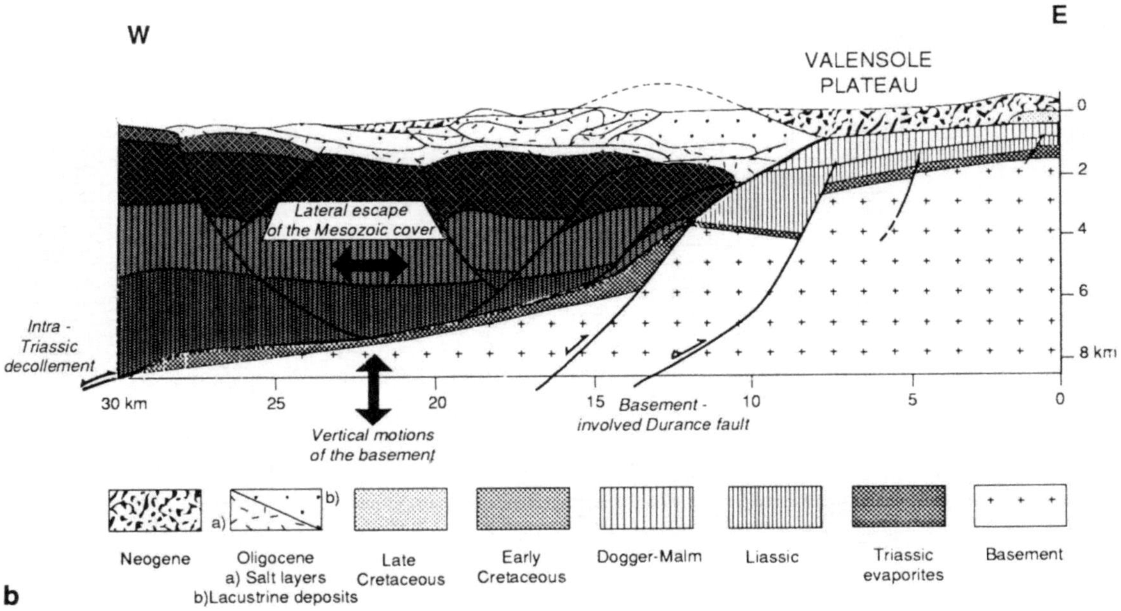

Fig. 2. b The multiphase Durance fault This structure combines the brittle deformation and vertical movement of the basement with the lateral escape of the sedimentary cover in the downfaulted block along the intra-Triassic decollement level (After Roure et al 1992)

to 3 sTWT south of the fault), and a huge change occurs in the thickness of the Jurassic deposits between the hangingwall (0.3 sTWT north of the fault) and the downfaulted compartment (1.3 sTWT south of the fault). Except for the slight Eocene and Miocene inversion events which will be described below, no other younger (Cretaceous or Oligocene) extensional reactivations have been recorded along the Cévennes fault in this part of the basin. Laterally, however, the history of the Cévennes fault becomes quite complex, as it has been locally totally obliterated at the surface by the activation of shallower (either Carboniferous or Triassic) detachment levels, inducing a subsequent gravity sliding of the sedimentary cover during the Oligocene (see the present architecture of the Oligocene Alès basin, which is largely superimposed on the previous Jurassic structures; Fig. 5 a; Roure et al. 1992).

Active Triassic Detachment Surface and Related Thin-Skinned Extensional Structures

Except for near its borders, the Mesozoic sedimentary sequences of the basin of southeast France are entirely detached from their pre-Triassic substratum along the Triassic evaporite horizon. This total decoupling between the cover and the basement accounts for most of the extensional structures observed throughout the area. No direct connection can be drawn between the surface and deep structures, and two extreme configurations are frequent:

1. A fault exists in the pre-Triassic substratum but does not extend into the overlying sedimentary cover, which on the contrary drapes gently over the basement-tilted blocks. This type of gravitational structure is well imaged in the ECORS profile across the Bresse graben along the eastern flank of the Cormoz high (Fig 3; Bergerat et al. 1990). Since shortening was induced in the folded sedimentary cover during this gravity gliding, even with the local formation of reverse faults, denudated areas missing their initial sedimentary cover are expected to occur laterally to balance, at the surface, the amount of extension observed in the basement (Fig. 3; Vendeville and Cobbold 1987; Vendeville et al. 1987; Naylor et al. 1994). In the Bresse example, part of the Mesozoic cover of the Jura was effectively translated westward into the Oligocene graben during its formation (Mugnier and Viallon 1986; Bergerat et al. 1990; Guellec et al. 1990 a, b).
2. Listric normal faults are restricted to the sedimentary cover and merge at depth with the Triassic evaporites. In such a case, which is best imaged in the southern Languedoc (Corconne fault and related Oligocene structures, Fig. 4; Roure et al. 1988), the top of the underlying substratum remains undeformed, even for very large (more than 1 km) vertical offsets in the sedimentary cover. Clearly, the amount of extension measured at the surface is partially balanced at depth by other basement faults in distal downhill portions of the basin, but also by the occurrence of denudated areas (zones with a bald basement, devoided of its initial sedimentary cover) uphill. At a regional scale effectively, the whole process can be interpreted as a gravity sliding of the sedimentary cover toward the center of the basin, above a gently dipping substratum

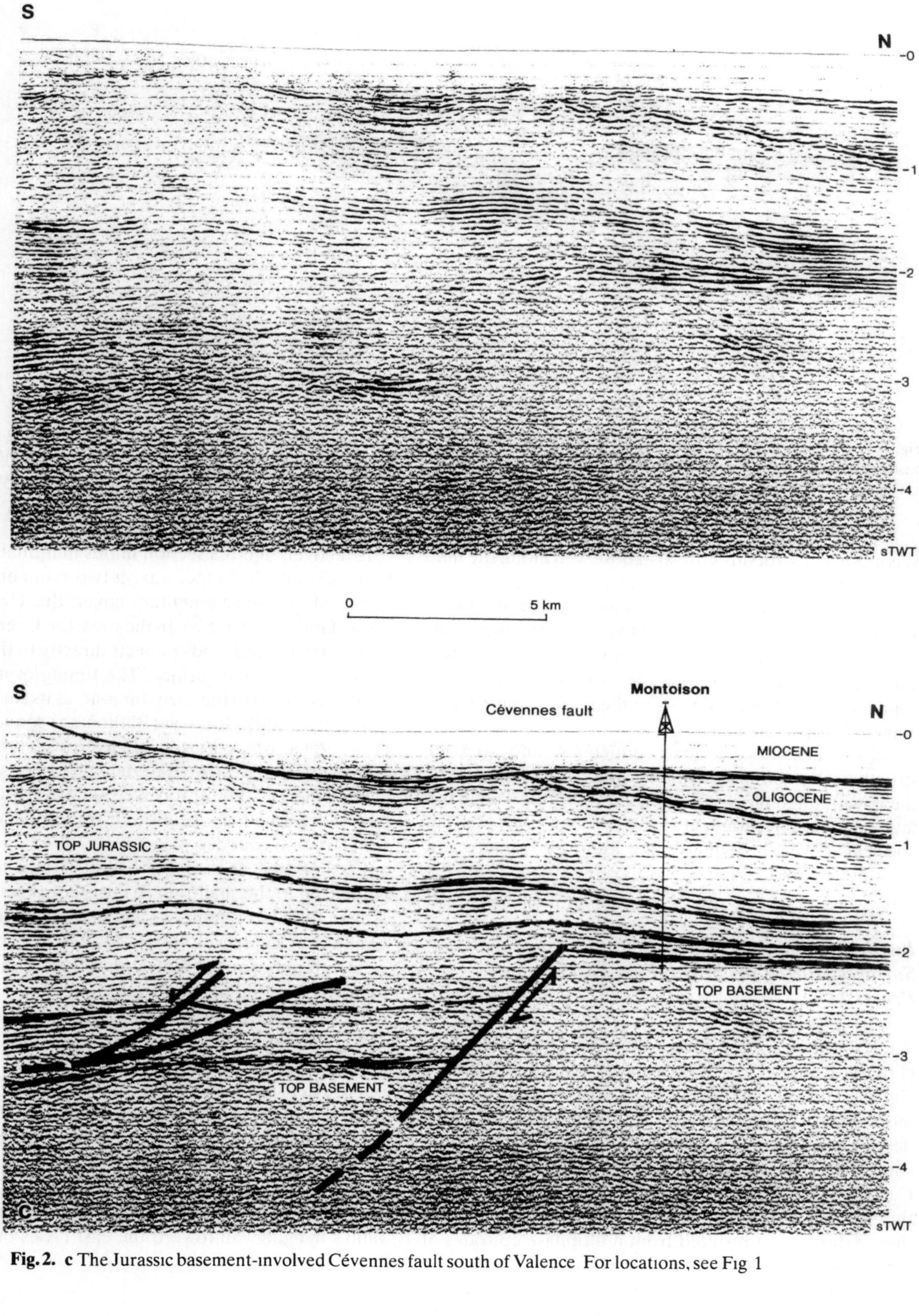

Fig. 2. c The Jurassic basement-involved Cévennes fault south of Valence For locations, see Fig 1

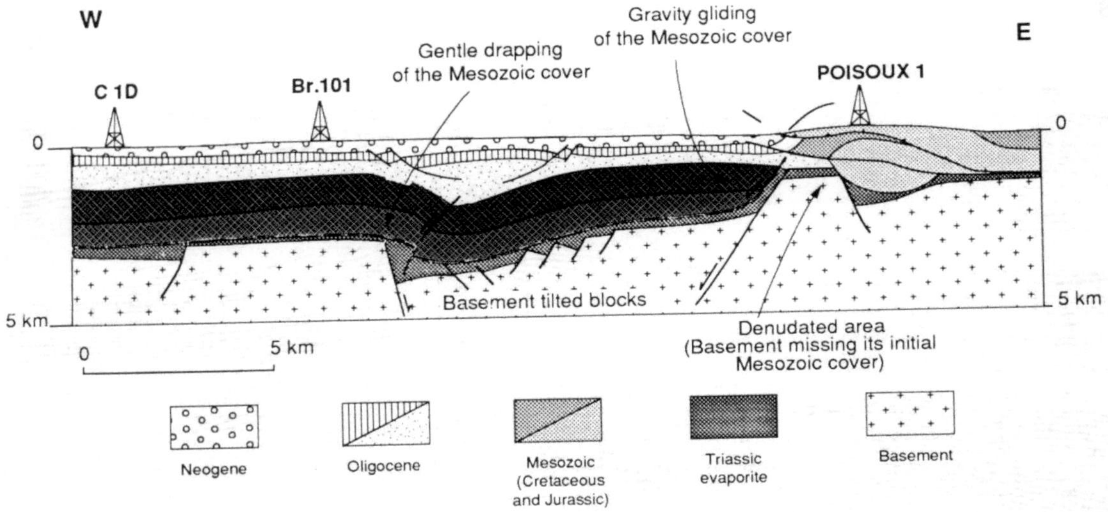

Fig. 3. Gentle draping of the Mesozoic sedimentary cover over a basement block in the Bresse graben (ECORS profile, location in Fig 1, Bergerat et al 1990) Note the activation of the Triassic evaporites as a decollement level during the Oligocene extension, and the total decoupling between the sedimentary cover (thin-skinned units) and the basement (tilted blocks)

Second-order structures are also observed along the major fault planes and account for vertical changes in the lithostratigraphic column. Refractions of the fault planes are obvious at the boundary between brittle and weak horizons, and steeply dipping ramps (up to 60°) in the early Liassic, middle Jurassic, or Tithonian carbonates connect with flat detachments in the late Liassic, Oxfordian, or Neocomian marls.

Drag synclines, secondary anticlines, and rollover structures are common features, which accommodate in the downfaulted blocks the irregularities observed in the fault surfaces (Bernoulli et al. 1981; Gibbs 1987; McClay and Ellis 1987 a, b; Williams and Vann 1987; Faure and Chermette 1989).

Incidence of Deeper Carboniferous Detachment Surfaces During the Multiphase Extensional Processes

As described earlier beneath the Alès Oligocene basin (Fig. 5 a; Roure et al. 1992), the coal measures in buried Carboniferous basins were locally activated as a decollement surface during the post-Triassic extensional episodes In this example, however, surface constraints and subsurface data are not sufficent to specify whether the Carboniferous horizons were mobilized solely during either the Jurassic or the Oligocene extension, or whether their deformation resulted from multiphase episodes of extension.

More recently, a new processing of seismic reflection lines across the Ardèche margin has recognized the northwestward extent of this Carboniferous detachment as far north as Balazuc (Bonijoly et al. 1993). Unlike in the Alès basin, surface constraints are sufficient there to relate the timing of activation of this intra-Carboniferous detachment to the Jurassic rifting event.

Effectively, a cross section in this marginal part of the basin of southeast France images two major normal faults in the Mesozoic sedimentary cover, the Uzer and Ardèche faults (Figs. 1, 5 b) In the west, the Uzer fault becomes listric at depth and connects directly to the Carboniferous decollement surface. The timing of its deformation is restricted to the early Jurassic, as its surface trace is sealed by Middle Jurassic (Bathonian) sequences (Giot et al. 1991 a, b). A Liassic vertical offset of more than 1 km is attested to in the Balazuc well for the sedimentary cover, whereas the reprocessed seimic profile images a still undeformed crystalline substratum beneath the intra-Carboniferous decollement.

Farther east, the Ardèche fault is also listric but is connected to the shallower Triassic decollement surface. Unlike the Uzer fault, its surface trace cuts through Cretaceous sequences, thus indicating a probable Oligocene age here for the activation of the Triassic detachment.

By comparing the Ardèche and Alès cross-sections and eventually more distal transects like the one in Fig 2 c across the Cévennes fault, a regional 3D balance of the extension can be attempted. More probably, the intra-Carboniferous detachment was mobilized solely during the Liassic rifting at the same time as the Cévennes fault. Subsequent Oligocene extension only locally reactivated the shallower Triassic decollement, inducing a gravity sliding of the Mesozoic cover of the Cévennes fault's hangingwall toward the east (Alès basin section, Fig. 5 a).

Paleogene and Neogene Basin Inversions

Small-scale fault reactivations, as well as regional basin inversions, are obvious over all the Alpine foreland (Ziegler 1983; de Graciansky et al 1988, Coward and

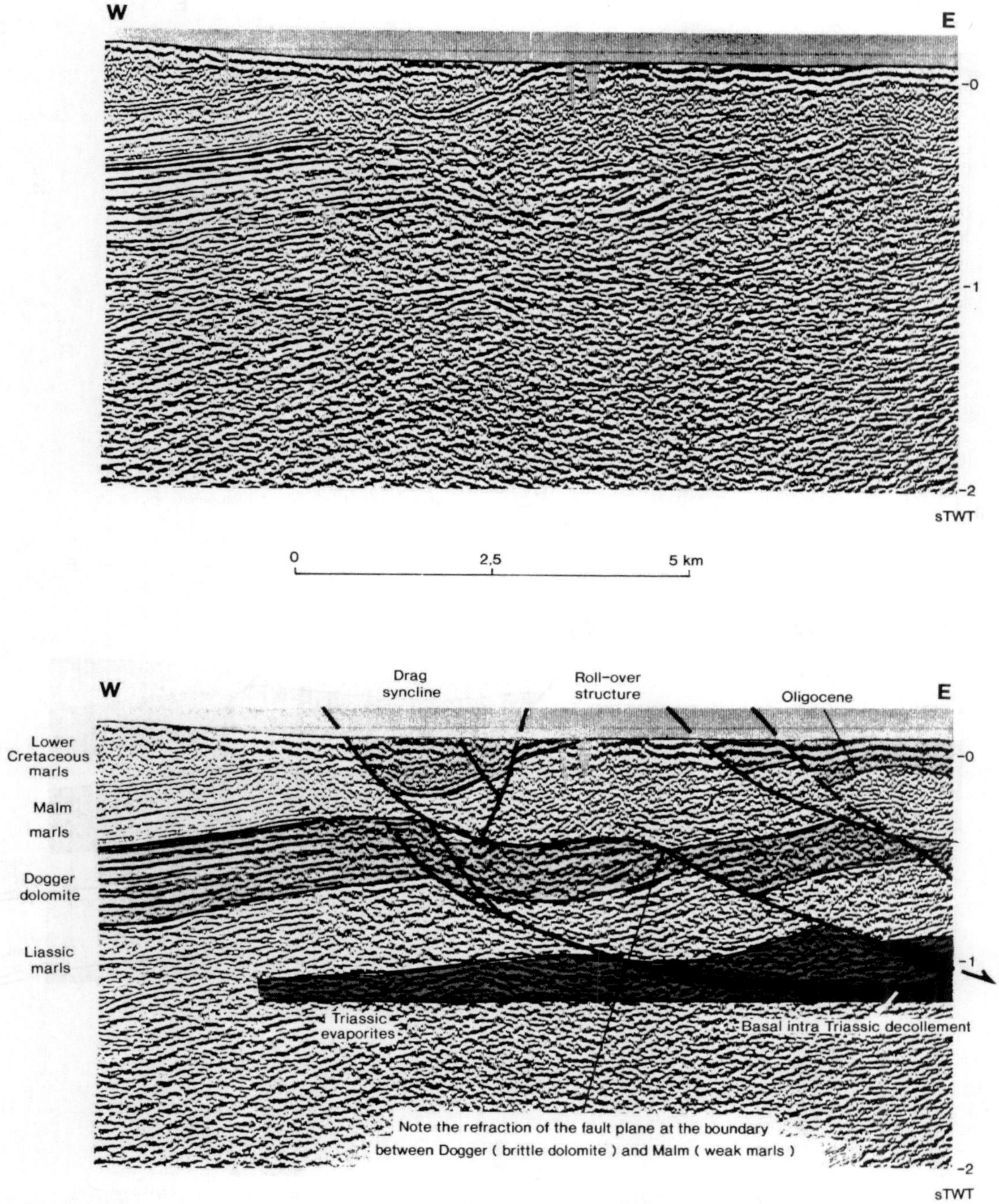

Fig. 4. Oligocene listric faults near Corconne (location in Fig 1) Note the activation of the Triassic evaporites as the main decollement level Shallower refractions of the fault plane are also observed at the boundary between brittle and ductile horizons (i e , the boundary between early Liassic limestones and late Liassic marls, or the boundary between the Tithonian limestone and the early Cretaceous marly limestones, Roure et al 1988)

Dietrich 1989), especially in the basin of southeastern France, and have been described elsewhere (External Crystalline Massifs; Gillcrist et al. 1987, Mugnier et al. 1990; Ste Victoire; Biberon 1988; Vercors Mtns, Butler 1989; Roure et al. 1992; Durance fault and Manosque basin, Roure et al. 1992; Alès basin, Roy and Trémolières 1992).

In all these published examples, however, the thick inverted sequences of the basins (footwall) were largely thrust over the reduced stratigraphic pile of their paleomargins (hangingwall).

In the following examples, the surface configuration of the inverted structures is quite different, just because it is the sedimentary cover of the hangingwall (usually the platform domain surrounding the basins) which is defor-

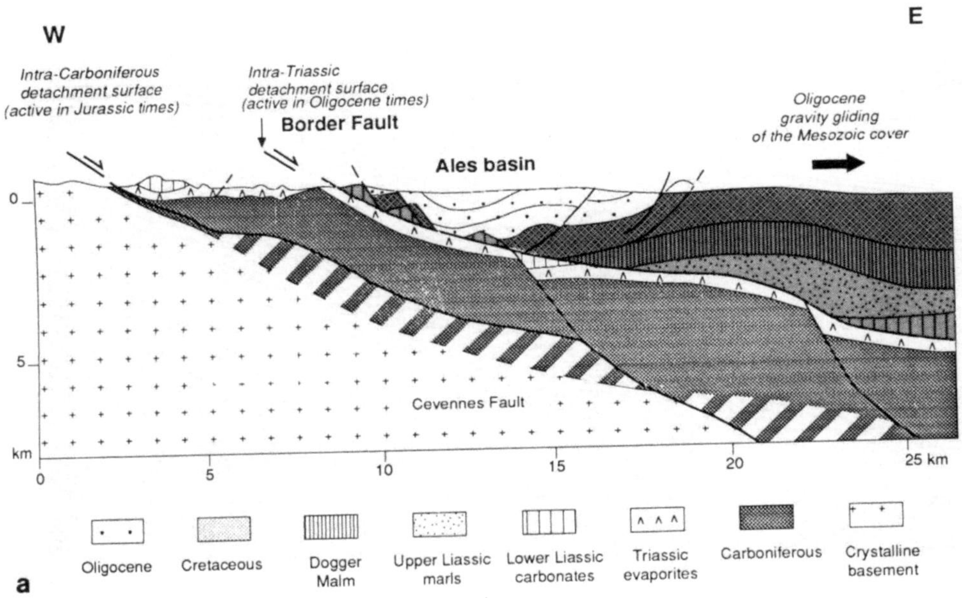

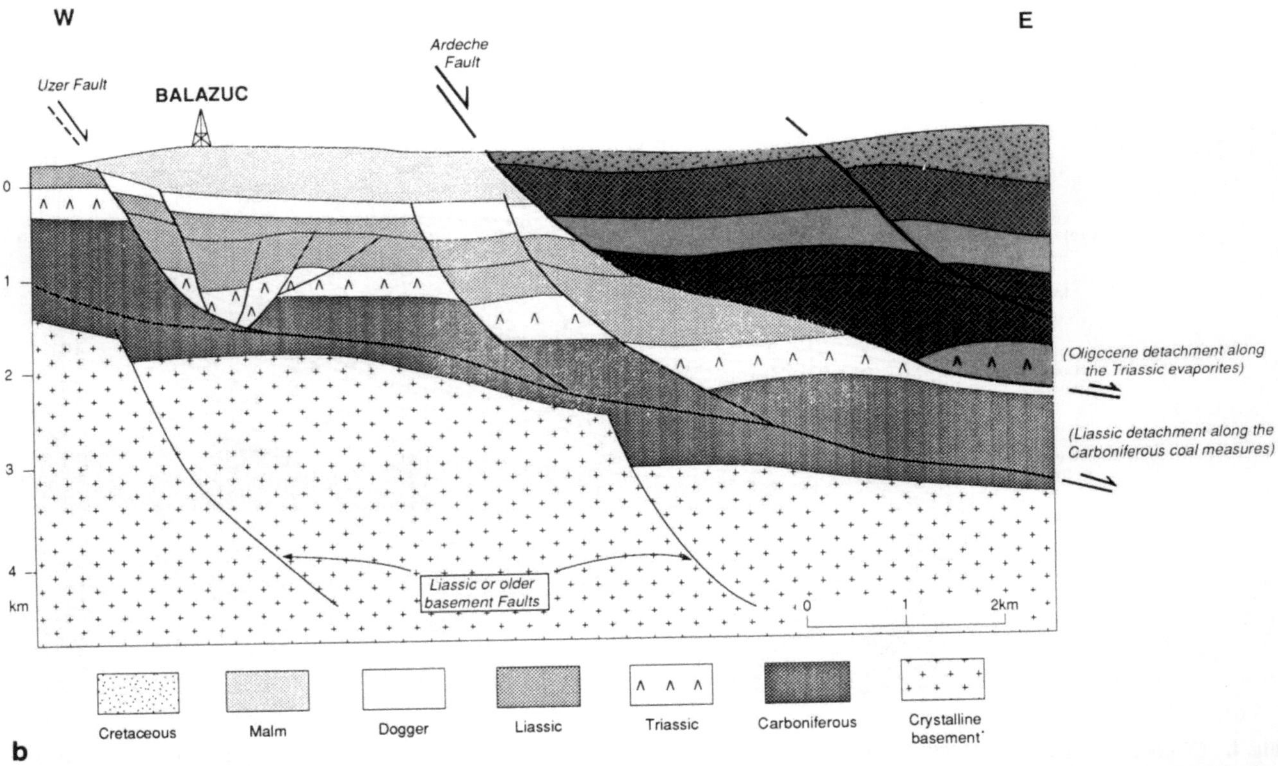

Fig. 5 a, b. Intra-Carboniferous detachment and multiphase extensional structures **a** The Alès basin section Note the local superimposition of the Triassic and Carboniferous detachment surfaces A downhill gravitary gliding of the Mesozoic cover of the hanging-wall of the Cévennes fault occurred during the Oligocene (After Roure et al 1992) **b** The Ardèche margin section (see location in Fig 1, after Bonijoly et al 1993) Note the listric nature of the Uzer fault, which individualizes a thick early Jurassic basin, but becomes listric in the Carboniferous horizons The Uzer fault and this intra-Carboniferous decollement were not reactivated in Oligocene times, unlike the Ardeche fault in the east, which merges at depth with the Triassic evaporites and, near the surface, cuts the Cretaceous beds

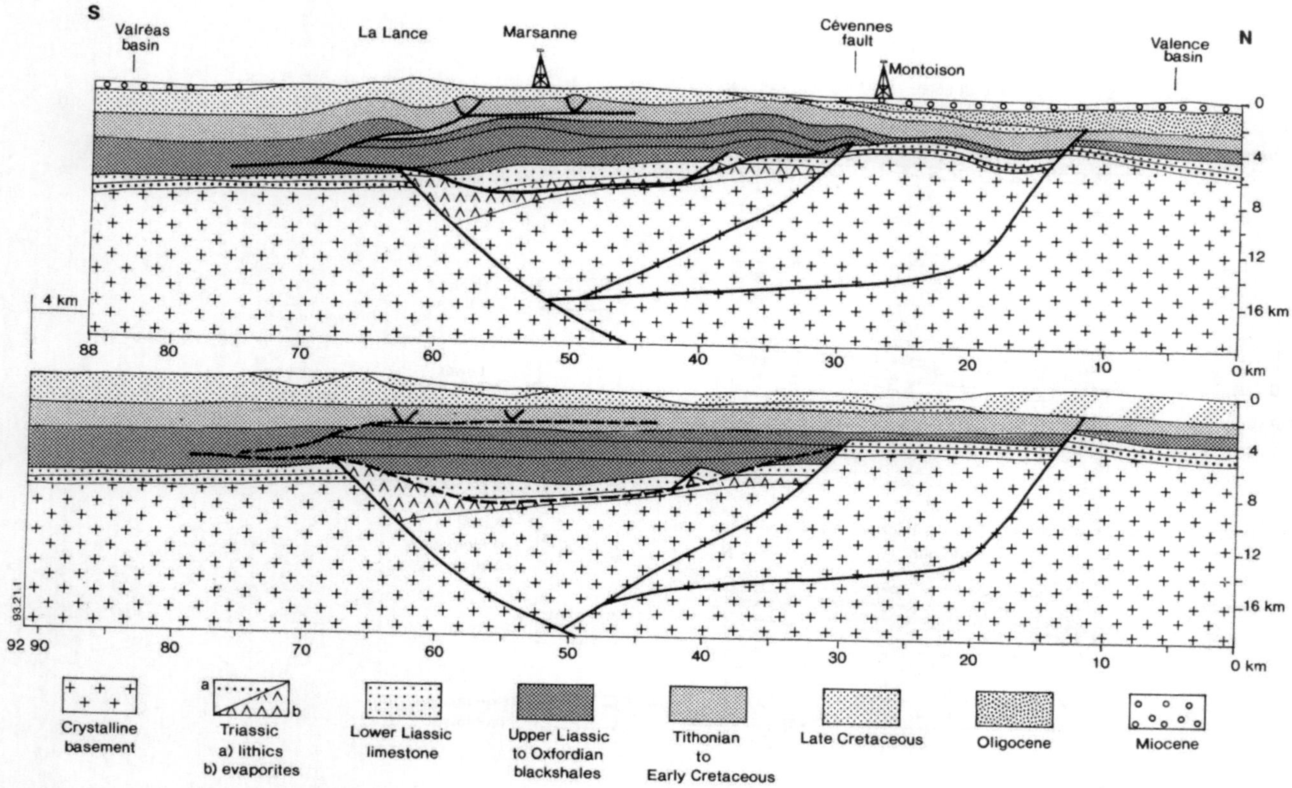

Fig. 6. *Above* North-south depth-converted cross-section along the Rhône valley from Valence to Valréas (see location in Fig 1 and the seismic profile in Fig 9) Note the Neogene inversion of the Mesozoic basinal sequences, and the resulting wedging of the Cretaceous platform sequences in the south (La Lance overthrust) *Below* Tentative restorations of the same inverted structures in their pre-compressive geometries Transport direction was parallel to the section only during the Paleogene and early Neogene, which effectively constitute the main episodes of inversion However, the Jurassic extension was oblique (about 45°) to the section, making these restorations unsuitable for measuring of the Jurassic extension

med and thrust basinwards. At depth, the shallow thrusts connect directly to deep high angle basement faults, thus building up a classic triangle zone.

A reference section for such unusual structures is given by the Ventoux-Lure overthrust (Fig. 1; Villeger and Andrieux 1987; Ford 1993), where the Urgonian reefal limestone largely overrides the Jurassic mud pile of the Baronnies sub-basin. In the nearby La Lance structure (Figs. 1, 6), regional seismic lines are used in addition to surface data to constrain the deep architecture of such a complex feature.

Similar structures in Provence (Ste Victoire) and Languedoc (Pic St Loup, Fig. 7), or even in the Jura Mtns (the Oyonnax backthrust and the Grand Crêt d'Eau overthrust, Fig. 8) are also discussed in terms of basin inversions, even though they do not relate to major thickness variations in the Mesozoic sedimentary cover. Effectively, as indicated below, all these structures are best described as basement-involved triangle zones, which remobilize the border faults of deeply buried Paleozoic basins.

Balanced cross sections are also confronted with sandbox analogue models to determine the intermediate geometries and the relative timing of deformation in these complex triangle structures.

Regional Examples

Only examples with sufficient subsurface constraints (exploration wells and seismic profiles) are described and discussed here.

La Lance-Valence Transect. A north-south depth-converted cross-section along the Rhône valley has been compiled from Valence to Valréas along a regional seismic line (Figs. 1, 6, and 9). This transect cross the Cévennes fault in the north, and the La Lance overthrust in the south, two structures that respectively mark the northern and southern limits of the thick Jurassic blackshale basin. Thick Cretaceous platform facies on the contrary are restricted to the La Lance allochthon, and pass northward into more pelagic basinal facies.

Basin inversion is evidenced by the attitude of the Jurassic sequence, which is presently at a higher elevation in the basin than in its previous margins. Similarly, the Neogene regional depocenters are located away from and on both sides of the inverted structure in the Valreas and Valence basins, in areas that initially belonged to the bordering platform domains.

The northern border fault of the Jurassic basin (the northeast-trending Cévennes fault) is clearly imaged on the seismic, while its southern counterpart is presently

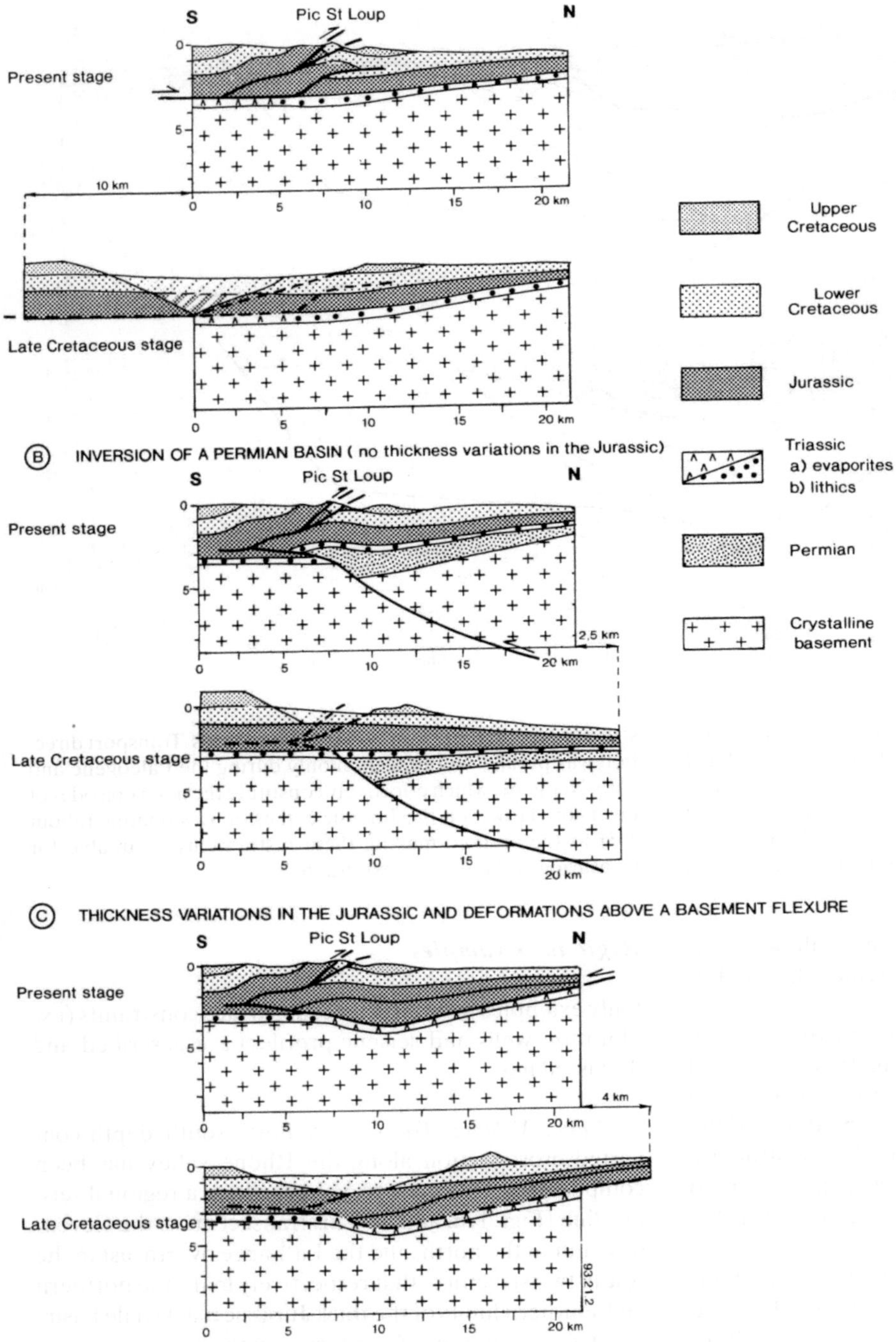

Fig. 7 A–C. North-south cross section of Pic St Loup (see location on Fig 1) Note the northward overthrust of the platform sequences (Pic St Loup) over the basinal domain (St Martin de Londres basin) For location see Fig 1 **A** Previous model favoring thin-skinned deformations (Roure et al 1988) This model does not account for the required balance between basement and cover shortenings A minimum amount of 10 km of basement shortening needs to be found elsewhere in the south **B** Model involving a triangle zone and the wedging of the platform sequences by the inversion of a Permian basin **C** Model involving a gentle flexure in the basement, and a progressive thickening in the Mesozoic sedimentary cover (similar to geometries proposed for the Ste Victoire overthrust by Biberon 1988) Note that this model requires a minimum of 4 km of basement shortening north of the St Martin de Londres basin

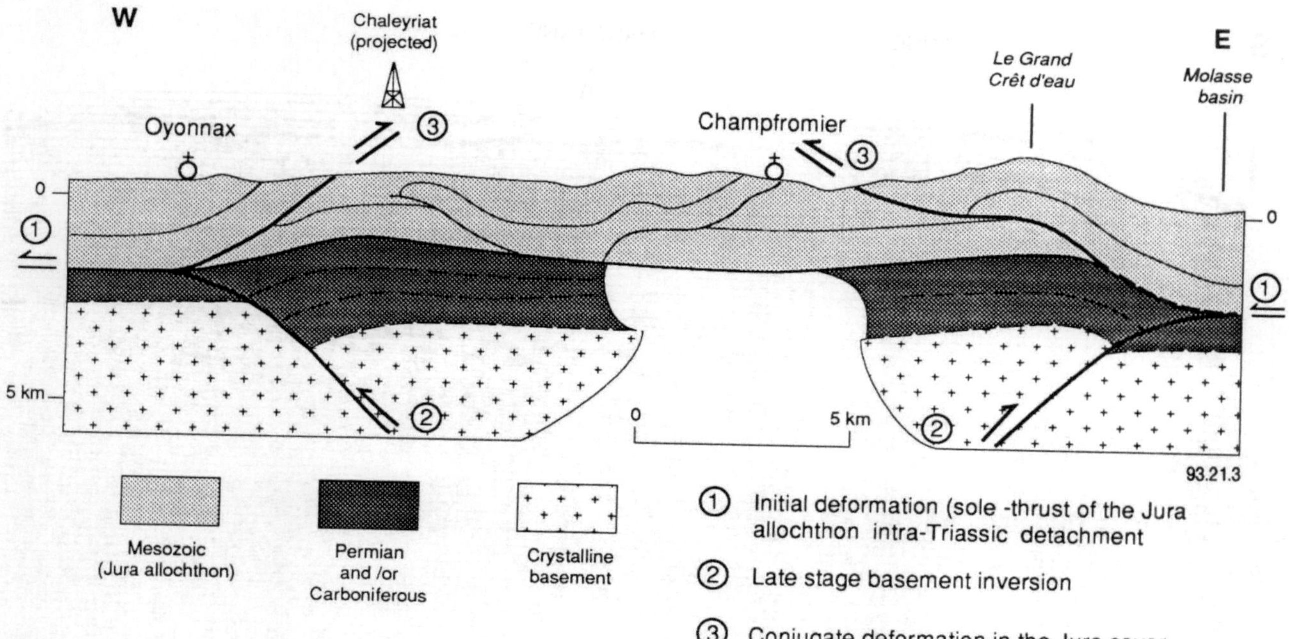

Fig. 8. The Oyonnax backthrust and its possible relation to the late stage inversion of a deeper Paleozoic basin (After Guellec et al 1990a, b and Philippe, this Vol)

masked by the surficial north-verging overthrust of La Lance Mtn. According to gravimetric modeling (Martinod 1988) and to the occurrence of diapiric structures beneath the Dieulefit syncline, however (Figs. 1, 6), a northwest-trending-northeast-facing normal fault is also assumed to limit the Jurassic basin in the south and to offset its basement.

A restoration of the section to its precompressive (Cretaceous) geometry has been attempted in order to compare the respective amounts of Cenozoic shortening in the sedimentary cover and in the basement (Fig. 6b). Despite detailed observations, the seismic data attest to two successive episodes of inversion (the northern margin of the basin was inverted first in Paleogene times prior to the deposition of the Oligocene infill of the Valence basin, whereas the La Lance structure mainly results from the Miocene inversion episode). No attempt has yet been made to differentiate the respective amounts of Pyrenean and Alpine shortenings. Similarly, due to the orientation of the profile, which parallels only the transport direction of these Eocene and early Neogene inversion episodes, no balance of the section has been made for older stages, nor has any attempt been made to measure the exact amount of Jurassic extension.

Clearly, during the Alpine inversion episode most of the basement shortening (about 4 km) was balanced at the surface by the large overthrust of the La Lance platform unit over the basinal domain, with the thick Jurassic sedimentary infill of the basin acting as an indenter, thus progressively decoupling (wedging) the reduced bordering sequences along shallow intra-Jurassic or intra-Neocomian decollement horizons. Part of the basement shortening was also accommodated in the basin itself by the reactivation of Triassic evaporites.

The Pic St Loup Triangle Structure. This north-south profile successively crosses the north-verging Pyrenean (late Cretaceous to lowermost Oligocene) overthrust of the Pic St Loup and the St Martin de Londres basin (Figs. 1, 7; Roure et al. 1988).

For a long time, a basement structure has been inferred beneath the Pic St Loup and is attested to by the occurrence of an E-W lineament still imaged for a long distance east of the Corconne fault, far behind the surficial termination of the cover overthrust (Chorowicz et al. 1991).

As in the previous example, surface geology images a reduced stratigraphic sequence of platform affinities in the south (the Pic St Loup unit, outlining numerous erosional events and bauxitic horizons) which is thrust northward over a more basinal domain (the St Martin de Londres basin). However, these slight thickness variations in the Mesozoic sedimentary cover are not sufficient to account for the geometry observed. On the contrary, subsurface data attest to the occurrence of Paleozoic (Permian pro parte) sedimentary strata north of the Pic St Loup unit beneath the St Martin de Londres basin.

Previous models favoring a relative isopacity of the Mesozoic sedimentary cover and a thin-skinned mode of deformation (Fig. 7a; Roure et al. 1988) do not account for a necessary balance between cover and basement shortening (10 km of shortening in the sedimentary cover, to be balanced somewhere in the south by a cryptic but equal basement shortening).

Another discussed solution (Fig. 7C) is similar to models proposed earlier for the Ste Victoire architecture in Provence (Biberon 1988). There, the basement is only slightly flexed toward the north beneath the Pic St Loup unit, but accounts for a progressive deepening of the substratum and a coeval thickening of the Jurassic sedimen-

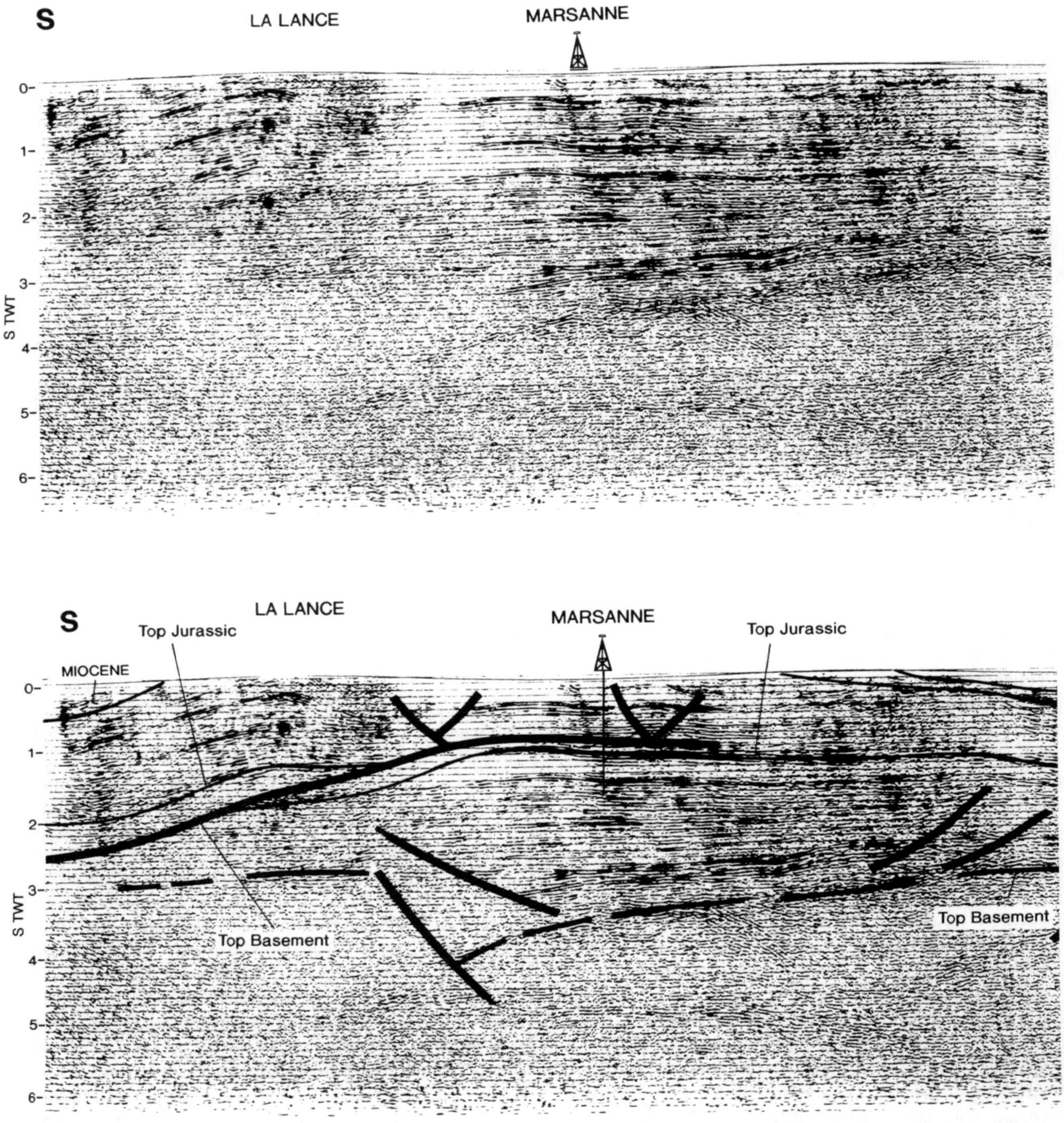

Fig. 9. North-south seismic profile along the Rhône Valley from Valence to Valréas A depth-converted section and a tentative restoration of the same profile to its precompressive geometry are shown in Fig 6 and discussed in the text

tary cover. In such a case, the triangle zone required to connect the basement flexure and the shallow backthrust involves only the basal Triassic decollement of the St Martin de Londres basin and the north verging Pic St Loup overthrust. As a result, basement shortening (around 4 km) still needs to be found elsewhere, i.e., along the northern border of the St Martin de Londres basin, which is unlikely according to the general northward transport direction of the Pyrenean thin-skinned structures in the Languedoc region. Despite their poor quality, however, the seismic data beneath the Pic St Loup overthrust enables us to propose a coherent solution to account for both the basement and cover shortening, with the occurrence of a triangle zone in the sedimentary cover directly connected at depth with a basement involved structure. Effectively, seismic data account for a relatively shallow location of the Triassic-Permian interface beneath the St Martin de Londres basin,

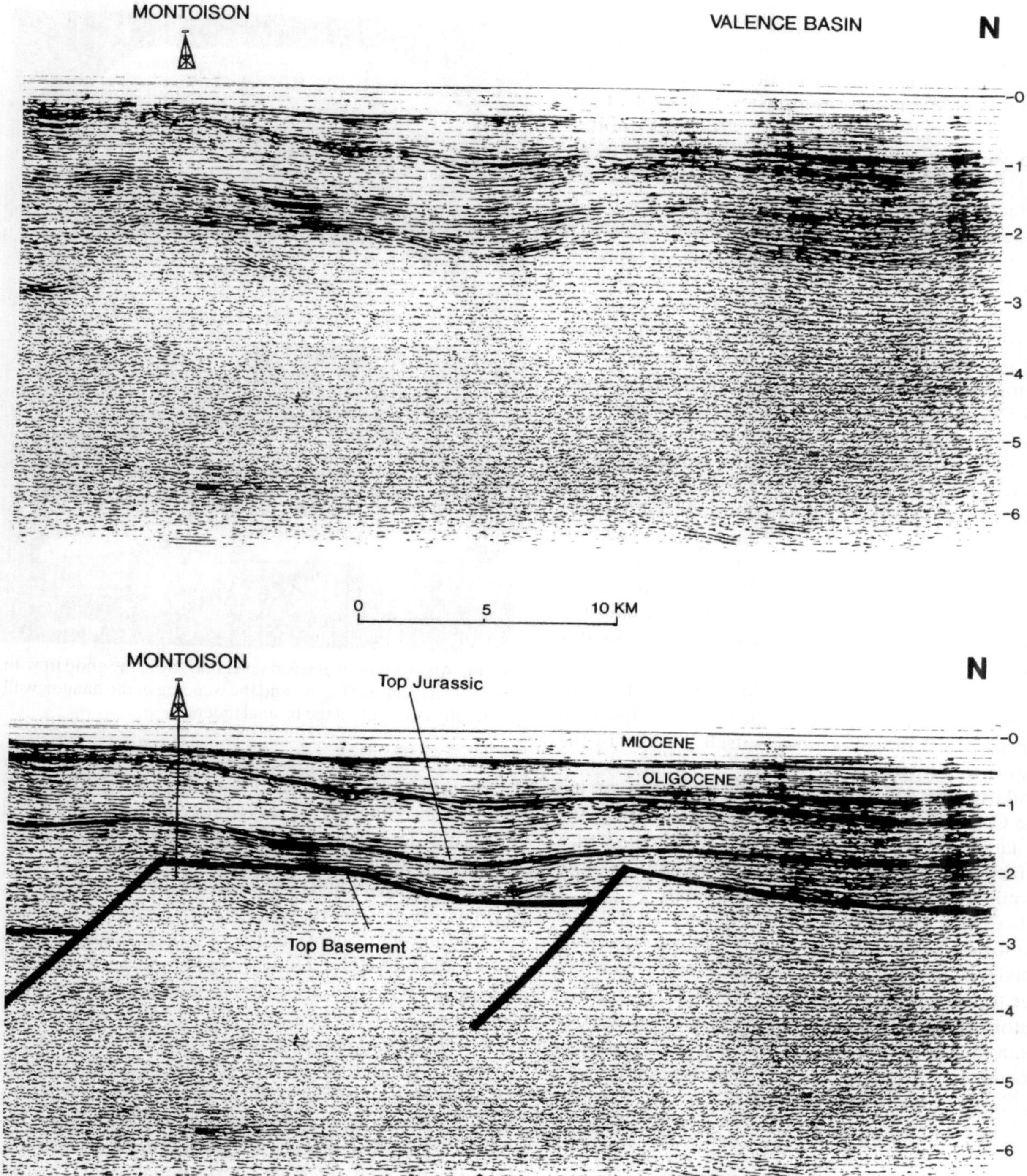

Fig. 9 *(continued)*. Legend see p. 258

compared to the deeper intra-Triassic detachment that underlies the Pic St Loup unit in the south. In addition, deep seismic reflectors are also imaged beneath St Martin de Londres, attesting to a thick Paleozoic sequence north of the Pic St Loup overthrust.

In this favored solution, a north-facing basement fault effectively underlies the Pic St Loup structure, as suggested earlier on the basis of remote sensing studies (Chorowicz et al. 1991), and wedging and backthrusting of the reduced platform domain are connected directly to the Eocene inversion of the border fault of this basin. The 2.5 km of local basement shortening (reactivation and inversion of the Permian normal fault) are sufficient to balance the overlying deformation observed in the sedimentary cover.

On a regional scale, this reactivated Permian structure is best compared to those of the nearby Lodève basin (Fig. 1), where a coeval north-facing normal fault is exposed at the surface and bounds a Permian basin in the south. Similar structures in Provence, where thickness variations in the Mesozoic sedimentary cover are hardly defined and probably remain insufficient to explain the overall geometry of the inverted structures (north-verging Ste Baume or south-verging Ste Victoire overthrusts), probably also relate to the reactivation of buried Paleozoic basins (thick Permian strata are effectively known to occur at depth beneath the Arc basin, precisely between the conjugated Ste Victoire and Ste Baume structures: Fig 1; Biberon 1988)

The Oyonnax Backthrust and Its Relations to the Inversion of a Deeper Paleozoic Basin Despite the fact that the Jura structural context is totally different, the ECORS deep seismic profile images a quite similar triangle geometry near Oyonnax, with an east-verging surface backthrust in the sedimentary cover being closely associated with a deeper inverted Paleozoic basin (reactivation of an east-facing basement-involved normal fault; Fig. 8; Guellec et al 1990 a, b; Philippe, this Vol.).

Because no synorogenic sediments are preserved in this area, it still remains unclear whether the basement fault inversion occurred synchronously or postdates the backthrusting in the sedimentary cover. In any case, this inversion of a buried Paleozoic basin postdates the regional westward emplacement of the Mesozoic Jura allochthon over the intra-Triassic detachment, since this basal sole thrust of the Jura allochthon is itself refolded by the deep inverted structure.

Similarly, part of the surficial deformations of the Grand Crêt d'Eau can be interpreted as the result of the same late-stage basement inversion (reactivation of a west-facing normal fault), since it is also underlain by an abnormally high basement-involved structure.

Between the Oyonnax backthrust and the Grand Crêt d'Eau structure, recent exploration wells have effectively reached Paleozoic sedimentary strata, which remain on a higher topographic level than the nearby Triassic basal sandstones beneath the Oyonnax syncline in the west, or beneath the Molassic basin in the east, thus attesting to the regional incidence of this late stage basement inversion (Philippe, this. Vol.).

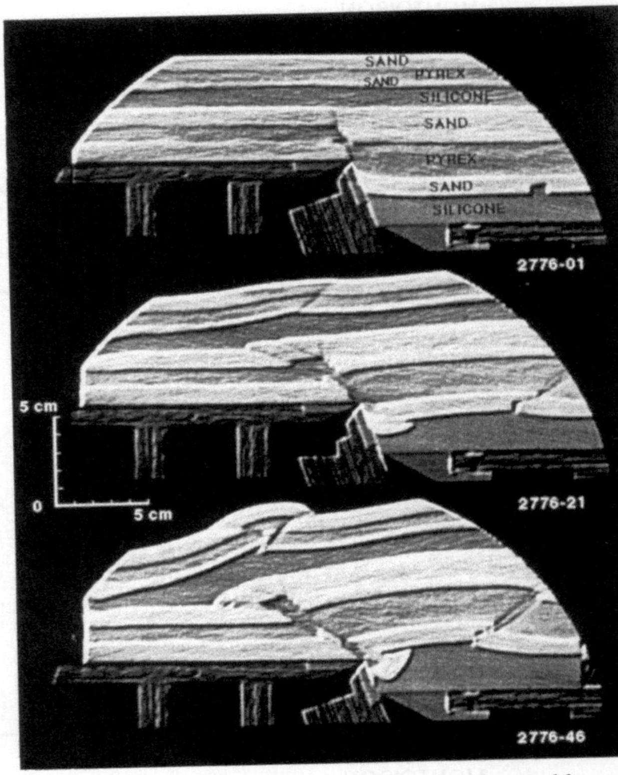

Fig. 10. Analog model of basin inversion Note the good fit with the geologic section (Fig 6), and the wegding of the hangingwall in the left, as a result of the basinal indenter

Sandbox Experiments of Basin Inversion

The present geometry of the above-mentioned examples of triangle zones related to basin inversion is relatively well calibrated by surface and subsurface data. However, numerous questions remain unsolved regarding (1) the relative timing of the cover backthrusting compared to the deeper basement inversion (La Lance and Oyonnax examples, Figs. 6, 8), or (2) the exact geometry of the basement beneath the cover backthrust (La Lance and Pic St Loup examples, Figs. 6, 7). To solve these problems, sandbox analogue experiments of basin inversion have been performed and compared with both the present and intermediate geometries of the sections discussed.

The experiment was carried out in a normal gravity field using both viscous and brittle material. Sand and Pyrex powders simulated brittle rocks, while silicone putty (Dow Corning PDMS SGM36) simulated natural rocks with viscous behavior such as salt, shales, or marls. Taking into account that a length/scale ratio of 1 : 25 000 and a time equivalent such as 1 h in an experiment is about 3 Ma in nature, natural rocks simulated by the silicone putty (viscosity = 5×10^4 Pa s) should have an equivalent viscosity of 10^{20} Pa s.

The deformation box was composed of two rigid basal plates separated by a steeply dipping normal fault (Fig. 10). The model was 30 cm wide. In a first step the right-hand-side plate was progressively lowered to simulate a downfaulted block. The downfaulted compartment was progressively filled with various materials. The first material deposited was silicone putty, which was deposited only in the downthrown part of the model. While normal dip slip proceeded along the normal fault, two layers of sand and one layer of Pyrex powder were deposited on both sides of the fault. After extensional movement, a 1-cm-thick layer of silicone and three layers of brittle material (sand and Pyrex) were deposited. The silicone layer and the uppermost brittle layers thus sealed the normal fault that developed in the lowermost brittle part of the model.

The inversion tectonic process was simulated by moving up the basal plate on the right-hand-side and by shortening the model with the right-hand-side vertical wall (shortening velocity = 1 cm/h; total amount of shortening = 4.5 cm).

Before inversion, the model displayed a clearly steeply inclined (73°) normal fault in the lowermost brittle series with a thick downfaulted block (9.5 cm) and a thinner upthrown block (6 cm) (Fig. 10, image 2776-01).

When inversion tectonics proceeded, the previous normal fault was not reactivated, but it was cut by a new left-verging low-ange (30°) thrust fault producing a typical short-cut geometry on both sides of the fault.

After a shortening of 2 cm (Fig. 10, image 2776-21) a low amplitude anticline developed in the uppermost layers. This anticline is associated with a right-verging thrust. The uppermost layer of silicone induced a complete decoupling between the two brittle units. Moreover, the vergences of lower and upper thrust are opposite, producing a structure comparable to a passive roof-duplex.

For greater amounts of shortening (Fig. 10, image 2777-46), the triangle geometry was accentuated. The downthrown brittle block acted as an indenter that was forced in the ductile horizon. Shortening in the uppermost brittle unit was accommodated independently of the lowermost structures. The old normal fault was passively transported but not reactivated.

The experiment (Fig. 10) is thus in good agreement with the surface observations made along the La Lance and Pic St Loup sections. The reactivation of a normal fault in the rigid basement (the two rigid basal plates of the model) can effectively induce neoformed thin-skinned and triangle compressive structures in the sedimentary cover, by activating successively its various weak horizons.

Short-cuts are predicted at depth by the model, both in the hangingwall and downthrown compartments, with the initial normal fault being preserved and passively transported in the indenter. However, detailed observations are still lacking in the mentioned examples to confirm the generality of this mode of deformation, which has effectively been proved elsewhere (Vially et al. 1994).

Neogene Extensional Reactivation of Thrust Faults

Extensional basins related to the collapse of an earlier orogen have been extensively described in various places around the world (Coward et al. 1987; Howards and John 1987; Dewey 1988; Coward and Trudgill 1989; Roure et al. 1991).

In such a context, the previous sole thrust of the compressive allochthon is usually reactivated as a normal detachment surface. Simultaneously, syn-deformation sediments progressively infill distensive piggy-back basins, which develop at the top of the allochthon above the flats of the basal decollement.

As for compressive piggy-back basins (Roure et al. 1991, 1993, Zoetemeijer et al. 1992), a migration of the post-compressional depocenters is progressively observed during this gravity sliding of the compressional edifice, and helps to constrain the restoration of the intermediate geometries of the structure.

To determine the timing of the deformation and estimate the incidence of such a multiphase deformation on the burial history of potential source rocks, a regional geologic cross section has been selected in the basin of south-eastern France and restored to its pre-extensional geometry Present and intermediate geometries have subsequently been compared with analogue models to take into account the attitude of the syndeformation sediments in the reconstructions.

The St Chinian-La Clape Negative Inversion

Figure 11 images a north-south geologic section in the southern Languedoc region. From the Massif Central to the Mediterranean, it successively crosses the St Chinian Pyrenean (late Cretaceous to Eocene) north-verging fold and thrust belt in the north, the Miocene Narbonne basin, and the La Clape antiformal structure in the south. Surface geology helps to constrain the deep architecture of the St Chinian tectonic edifice, whose sole thrust merges at the surface in the late Cretaceous to Eocene synflexural foredeep deposits of the autochthon, but reaches progressively deeper stratigraphic horizons southwards (Roure et al. 1988).

Effectively, with the exception of the two northernmost tectonic units, which involve solely synorogenic deposits, all the other tectonic slices of the St Chinian allochthon also contain Jurassic and Triassic sequences, thus attesting to the activation of the Triassic evaporites as the main decollement level for the Pyrenean deformations (Roure et al. 1988).

A south-facing high-angle normal fault (the Quarante fault) borders, in the south, the Mesozoic outcrops of the St Chinian arc and separates the early deformed Pyrenean structures from the post-nappe Miocene infill of the Narbonne basin. Seismic profiles cross this basin between the Quarante fault and the La Clape antiforme (Gorini 1989; Gorini et al. 1991), thus helping to trace the southern extent of the basal detachment, which eventually even reaches the pre-Triassic substratum farther south.

According to its surface geometry (arcuate pattern on Landsat imagery), the Quarante fault appears to be listric and to connect at depth with the basal Triassic evaporites (sole-thrust of the Pyrenean edifice) which was thus reactivated as a normal detachment during the Miocene (episode of opening of the Gulf of Lions).

In such an extensional context, the La Clape antiformal structure, which involves not only the Mesozoic sequence but also the Miocene sedimentary infill of the Narbonne basin, is more logically interpreted as a roll-over structure developed during the positive inversion of the Pyrenean sole-thrust (Gorini 1989; Gorini et al. 1991).

Restoration of the section to its pre-Oligocene configuration attests for 4.5 km of Oligocene to Miocene extension in the sedimentary cover, which are only partially

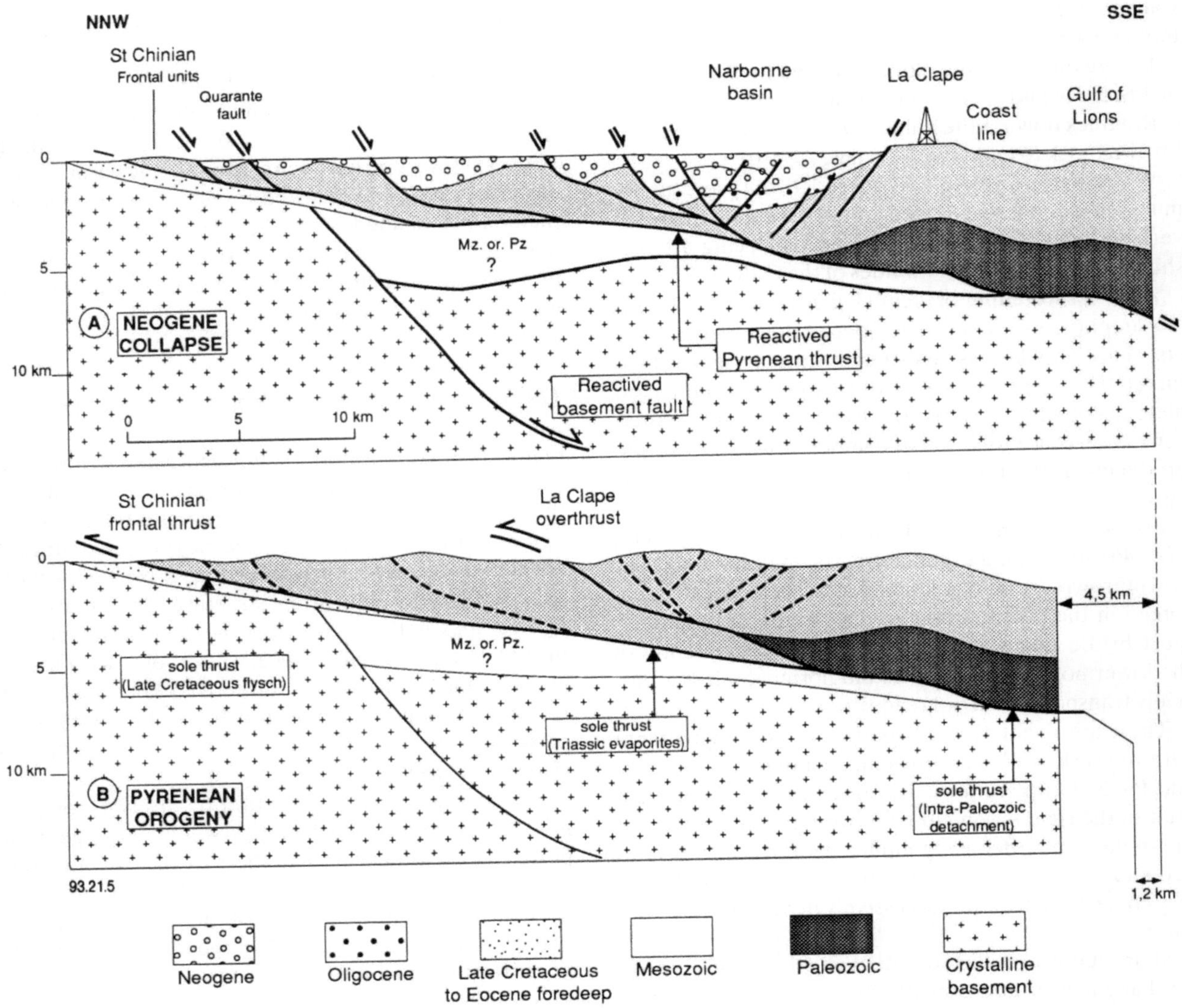

Fig. 11 A, B. Geologic cross section from the St-Chinian fold-and-thrust belt to the Mediterranean (see location in Fig 1) Note the extensional reactivation of the sole-thrust of the Paleogene tectonic pile This collapse of the Pyrenean edifice occurred in Neogene times, in relation to the opening of the Gulf of Lions, inducing the development of Neogene depocenters (extensional piggy-back basins) above the flats of the basal decollement

balanced by 1.2 km of measured extension in the underlying basement (Fig. 11 a, b). Additional basement remobilization is, however, expected to occur further south, offshore, in the Gulf of Lions (Burrus 1989; Gorini et al. 1993).

Sandbox Experiments of Collapse or Flat-Ramp Detachments

The St Chinian-La Clape geologic section has been restored to its pre-Miocene geometry in order to determine its shape at the final stage of the Pyrenean deformations, prior to the collapse of the compressive tectonic pile.

In addition, the present and intermediate geometries of the section have been confronted with analog experiments where syn-deformation sediments are trapped in extensional piggy-back basins. By so doing, it has been possible to determine the genetic relations between the La Clape rollover structure, the internal architecture of the Miocene infill of the Narbonne basin, and the Miocene reactivation of the sole-thrust of the Pyrenean edifice as a normal detachment.

Figure 12 shows two characteristic deformation patterns obtained in laboratory models of flat-ramp extensional systems (Roure et al. 1992). The models are made of sand with a frictional angle of 30° like most rocks in the upper crust (Byerlee 1978). They are built above a rigid base with a flat-ramp geometry. A thin plastic sheet intercalated at the bottom of the sand layer is pulled from the right to obtain the deformation. The insert at the bottom of Fig. 12 a and b illustrates the principles of the experiments and shows the basic pattern of the faults that are produced in each of them. The basin formed above the ramp is progressively filled with fresh sand during extension.

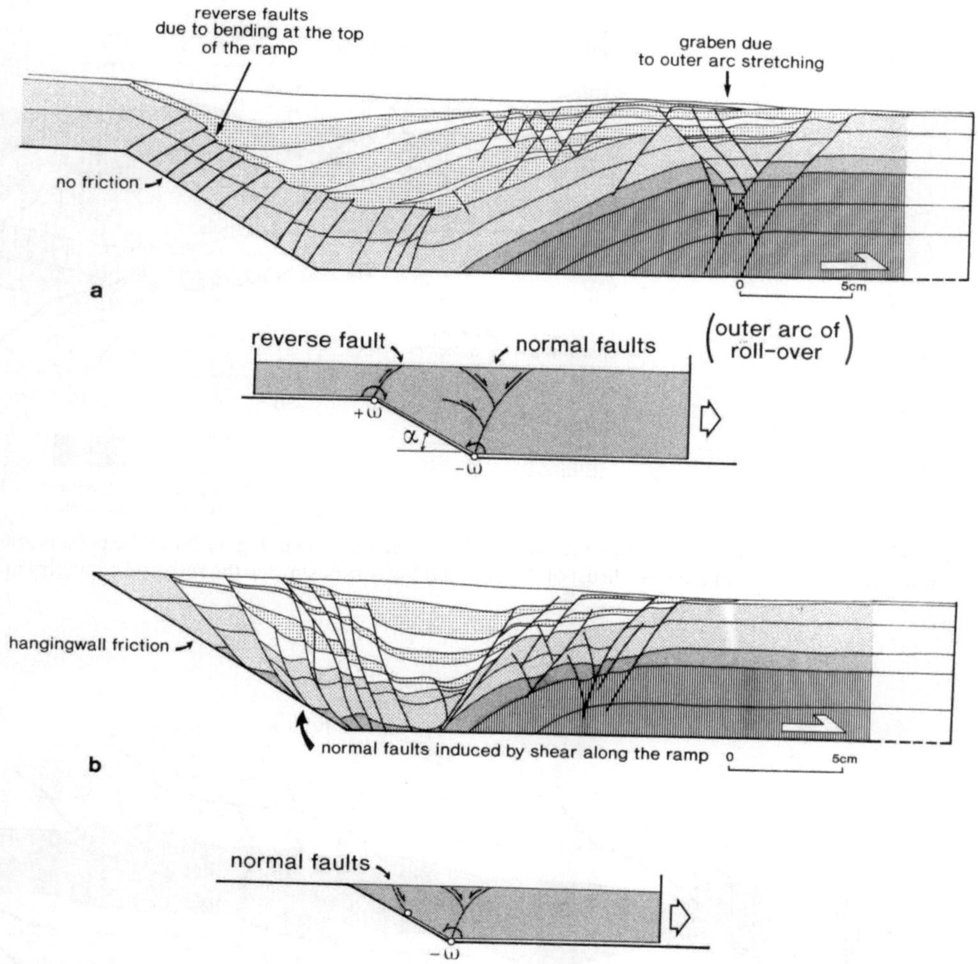

Fig. 12 a, b. Analogue model of collapse and basin development above a reactivated flat detachment (extensional piggyback basins, Roure et al 1992)

In Fig 12a, the hangingwall has a perfect slip without friction along both flat and ramp. During its displacement to the right, the hangingwall undergoes flexuration at the upper (+ w) and lower (− w) extremities of the ramps, which causes two different modes of faulting. Steeply dipping reverse faults with a small offset develop sequentially in the upper part of the ramp to accommodate the flexuration + w. The flexuration − w at the bottom of the ramp produces the rollover anticline and a series of Y-shaped normal fault patterns whose dominant faults dip toward the ramp. During the entire experiment the depocenter of the induced sedimentary basin is located near the upper extremity of the ramp.

In Fig. 12b the hangingwall also has a perfect slip along the flat but exhibits friction along the ramp. Steeply dipping normal faults are induced by shear along the ramp. These faults migrate toward the bottom of the ramp during extension, giving a migration of the basin depocenter. By reference to the overall structure, the basin depocenter starts near the upper extremity of the ramp and migrates with time toward the rollover anticline.

A comparison of these two experiments with the La Clape section leads to the following remarks. The normal faults observed in the NW limb of the La Clape anticline dip toward the Tertiary sedimentary basin. This observation, which can be compared directly to the experiments, sustains the interpretation of the La Clape structure as a rollover anticline associated with a ramp located to the NW and dipping toward the anticline. In the NW of the basin, normal faults dip toward the SE, i.e., in the same direction as the inferred ramp. This pattern is comparable to the experiment in Fig. 12b, and suggests that movement along the ramp was characterized by frictional slip. This conclusion would be in agreement with an interpretation of the ramp as an inherited thrust fault reactivated in extension during the Tertiary.

From the surface point of view, this mode of post-compressional extension looks very similar to classic extensional basins developed above a relatively flat horizontal surface or in assymetric rifted margins (Wernicke and Burchfiel 1982; Gibbs 1987; McClay and Ellis 1987a, b; Vendeville et al. 1987). However, the burial and thermal history of the sediments underlying the basal detachment will differ greatly in the case of a simple or multiphase extension (i.e., the Alès basin; Fig. 5a), or in the case of post-nappe collapse.

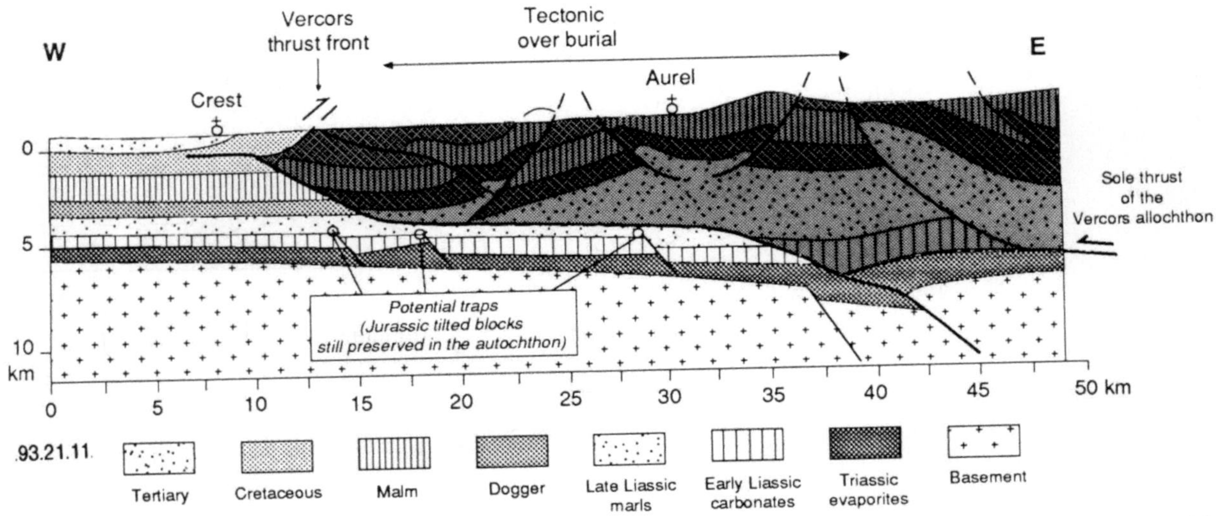

Fig. 13. Geologic cross section along the Drôme River (see location in Fig 1) Note the preservation of early Jurassic tilted blocks in the autochthon, beneath the sole-thrust of the inverted Vercors basin See the restored geometry in Fig 15 a (After Roure et al 1992)

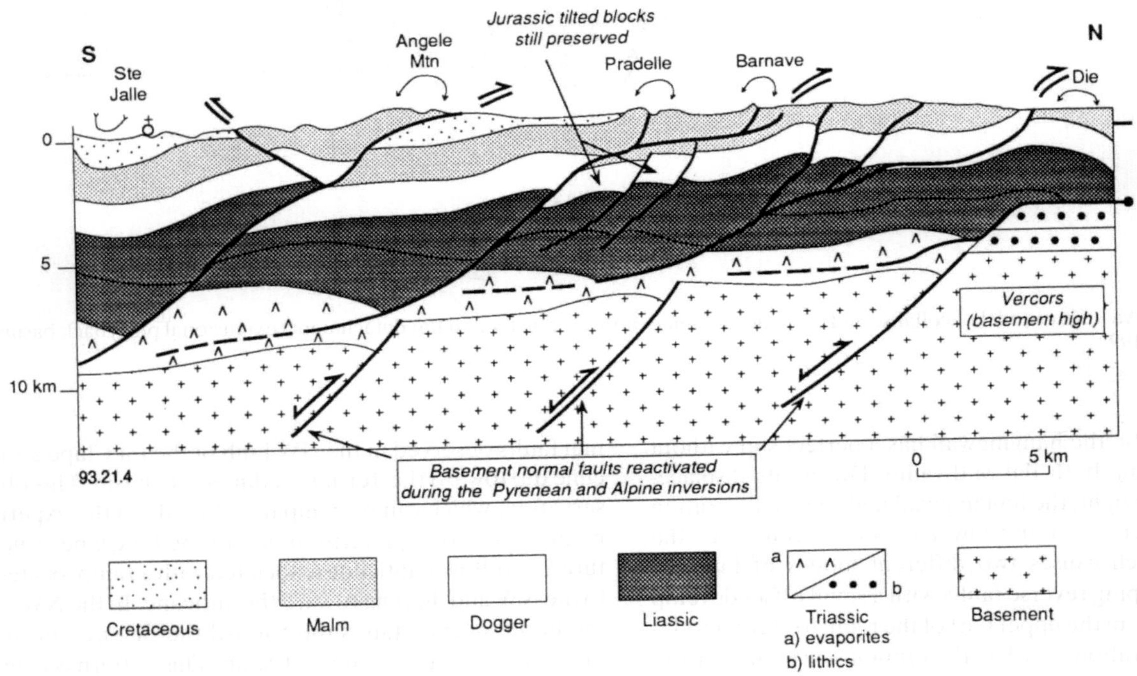

Fig. 14. North-south geologic section across the Diois area outlining normal faults and tilted blocks still preserved in the Dogger limestones beneath a shallow decollement surface in the Oxfordian blackshales (see location in Fig 1)

Multiphase Structures and Petroleum Plays

Major reservoirs are located in the basal Triassic sandstone beneath the evaporites, or in the early Liassic carbonates. Additional valuable reservoirs are found locally in the Oligocene sequences.

In contrast, potential source rocks are distributed in various portions of the lithostratigraphic column (i.e., the Stephanian coal and the Autunian oil shales, the Toarcian and Oxfordian blackshales, the Albian-Cenomanian gray marls, and the Oligocene bituminous shales; Mascle et al. 1993, this Vol.), implying distinct burial and maturation histories for each of them As a result, the petroleum potential of the basin of southeastern France is directly related to the compatibility between the age of the traps and the age of expulsion and migration of the oil and gas, which may differ greatly from one source rock to another.

As mentioned earlier, maximum burial was usually obtained at the end of the thermal subsidence of the passive

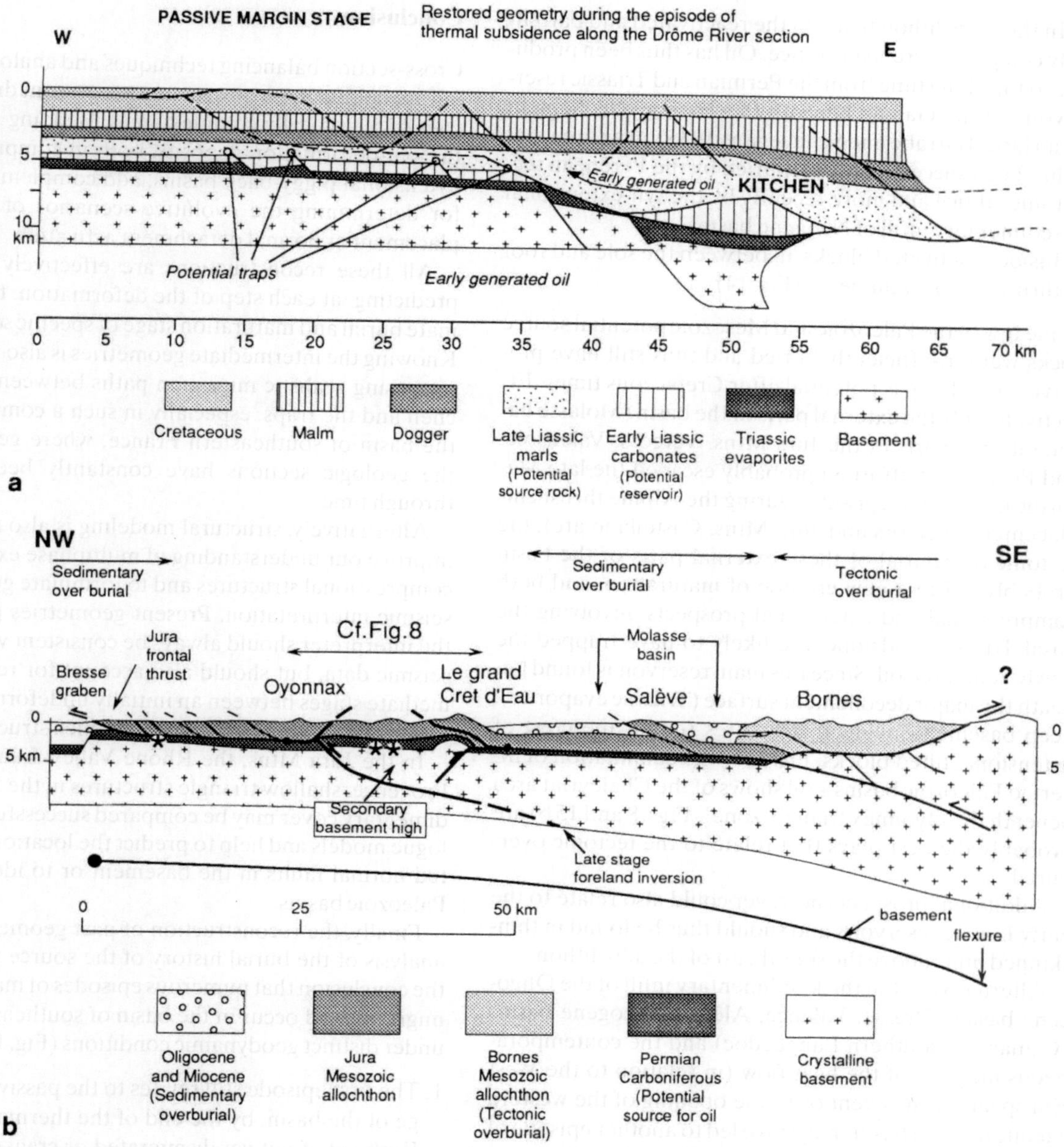

Fig. 15 a, b. Schematic diagrams outlining the connection between the geodynamic settings and the petroleum habitats **a** Passive-margin traps and early generated oil (cf Fig 13, this Drôme River section is restored here into its Cretaceous pre-compressive geometry). **b** Late-stage compressive traps and tectonic or synflexural overburial (cf ECORS transect across the Molasse Basin and the inner Jura Mtns) * Initial area of trapping = first migration along the flexural trend, ** late stage trapping = migration toward a secondary high

margin, prior to the late Cretaceous to Eocene and Neogene structural inversions. Clearly, the thickest and central parts of the basin were totally mature or even already in the gas window a long time before the late stage compressive deformations (more than 10 km of sediments were accumulated in the deepest parts of the basin). Oil generated early hopefully could have been preserved from secondary migration or cracking in the traps that relate to the early extensional events, mainly early Jurassic, but also locally as young as the late Jurassic (Baronnies, de Graciansky et al. 1988) or even the early Cretaceous (Manosque basin, Roure et al. 1992).

The search for undeformed early extensional traps would thus essentially concern areas located away from the Oligocene extension or Pyrenean and Alpine inversions:

1. Along the western and southern borders of the basin, i.e., near the Ardèche margin where Jurassic tilted blocks are still well preserved (Giot et al. 1991 a, b) and where oil shows have been observed in some wells in Hettangian limestone or Triassic sandsone.
2. In the autochthon along the Drôme River and northwards beneath the Vercors allochthon (Fig. 13).

3 In the autochthon beneath the reactivated and partially collapsed Pyrenean edifice. Oil has thus been produced for a long time from the Permian and Triassic reservoirs of the Gabian beneath the Triassic decollement surface (Barrabé and Schneegans 1935), in an area that has been successively overburied by the Pyrenean tectonic edifice and more recently by the syn-extensional sedimentary infill of Neogene basins.
4. Inside reactivated blocks, in between the sole and roof thrusts of cover duplexes (Fig 14).

Some few of the Paleozoic and Mesozoic potential source rocks were insufficiently buried and thus still have preserved an oil or gas potential after Cretaceous times Effectively, only the external parts of the basin (Molasse basin, internal parts of the Jura Mtns. Vercors. Valensole, and Provence platforms) probably escaped the late Mesozoic maturation episode. During the Alpine thrust emplacement (Vercors and Jura Mtns. Castellane arc), the tectonic overburial of these external parts of the basin probably led to a new episode of maturation, and both compressional and extensional prospects, involving the basal Triassic sandstone, are likely to have trapped the newly generated oil Since this main reservoir is found beneath the major decollement surface (Triassic evaporite), deep basement-involved structures (either the crest of extensional tilted blocks, Fig. 3; or the culmination of inverted Paleozoic basins – oil shows of the Chaleyriat area beneath the Oyonnax triangle zone – Figs. 8 and 15 b) are probably the best plays that relate to the tectonic overburial.

Additional prospects however could also relate to the early Liassic reservoirs and should thus be found in thin-skinned units above the sole-thrust of the allochthon.

Alternatively, the thick sedimentary infill of the Oligocene basins (Bresse, Valence, Alès) or Neogene basins (Camargue, southern Languedoc) and the contemporaneous increase of the heat flow (in relation to the West European rifting event or to the opening of the western Mediterranean basin) may have led to another episode of oil generation (Blanc et al. 1991). Both the Oligocene source rocks found at the bottom of the sedimentary infill of these new extensional basins, or the deeper Paleozoic or Mesozoic horizons that escaped early maturation are likely to have generated and expelled additional oil and gas during these recent extensional episodes Numerous oil shows have indeed been encountered in the Valence or Alès basins.

As in the case of tectonic overburial, the plays related to this sedimentary overburial are mainly related to post-Jurassic traps. In particular, since some of the potential source rocks are located in these neoformed basins, Oligocene and Neogene reservoirs involved in recent extensional structures should also be prospected.

Conclusions

Cross-section balancing techniques and analogue modeling are helpful in reconstructing the intermediate geometries of tectonically complex areas Bedding attitudes of the syn-deformation deposits, both in compressional or extensional piggy-back basins, add complementary data for determining the evolutive scenarios of thrust emplacement or normal detachment activation

All these reconstructions are effectively crucial for predicting, at each step of the deformation, the approximate burial and maturation stage of specific source rocks. Knowing the intermediate geometries is also required for proposing realistic migration paths between the oil kitchen and the traps, especially in such a complex area as the basin of southeastern France, where geometries of the geologic sections have constantly been modified through time

Alternatively, structural modeling is also necessary to improve our understanding of multiphase extensional or compressional structures and to formulate guidelines for seismic interpretation. Present geometries proposed by the interpreter should always be consistent with well and seismic data, but should also account for realistic intermediate stages between an initially undeformed configuration and progressively more complex structures.

In the Jura Mtns, the Rhône Valley, Languedoc, and Provence, shallow triangle structures in the Mesozoic sedimentary cover may be compared successfully with analogue models and help to predict the location of reactivated normal faults in the basement or to identify buried Paleozoic basins.

Finally, the reconstruction of past geometries and the analysis of the burial history of the source rocks lead to the conclusion that numerous episodes of maturation and migration did occur in the basin of southeastern France, under distinct geodynamic conditions (Fig. 15):

1. The first episode still relates to the passive margin stage of the basin, by the end of the thermal subsidence. Early oil, if not yet dismigrated or cracked, should be found in still undeformed Jurassic structures in the external parts of the basin (Fig. 15 a).
2 Subsequent episodes result from the Pyrenean or Alpine tectonic overburial. Oil generated during these periods may be found in compressional structures (Fig. 15 b)
3 The last episodes result from a sedimentary overburial due to the thick terrigenous infill of neoformed (post-compressional) Oligocene and/or Neogene extensional basins (Figs 5 a and 11).

Acknowledgments A. Mascle provided helpful comments. Seismic data were obtained from EAP/ESSO/BP/CFP/IFP/FSH (Rhône Valley, Diois, and Bresse profiles), CFP (Languedoc), the ECORS, and GPF Programs. J F. Ballard contributed to two of the experiments.

References

Arthaud F, Mattauer M (1969) Sur les décrochements NE-SO sénestres contemporains des plis pyrénéens du Languedoc CR Soc Geol Fr 8 290–291

Arthaud F, Matte P (1974) Les décrochements tardi-hercıniens du sud-ouest de l'Europe Géométrie et essai de reconstitution de la déformation Tectonophysics 25 139–171

Barrabé ML, Schneegans MD (1935) Rapport d'ensemble sur le gisement pétrolifère de Gabian (Hérault) Ann Office Nat Combustibles Liquides 10, 4 595–665, 10, 5 819–897

Bergerat F, Mugnier JL, Guellec S, Truffert C, Cazes M, Damotte B, Roure F (1990) Extensional tectonics and subsidence of the Bresse basin: an interpretation from ECORS data In Roure F, Heitzmann P, Polino R (eds) Deep structure of the Alps Mém Soc Geol Fr 156, Mem Soc Geol Suisse 1, Vol Spec Soc Geol Ital 1 145–156

Bernoulli D, Davis GA, Montadert L (1981) Listric normal faults Oceanologica Acta 4 Proc 26th Int Geol Congr, Geology of Continental Margins Symp, Paris, Colloque C3, pp 87–101

Biberon B (1988) Mécanismes et évolution de chevauchements à vergences opposées Exemple des structures de la Ste Victoire (Provence) PhD thesis, Grenoble University, 190 pp

Blanc G, Doligez B, Lajat D, Mascle A (1991) Evaluation du potentiel pétrolier des formations paléozoiques de la Bresse et de sa bordure jurassienne Bull Soc Geol Fr 162 409–422

Boillot G, Montadert L, Lemoine M, Biju-Duval B (1984) Les marges continentales actuelles et fossiles autour de la France Masson Paris

Bois C (1993) Orogenic belts and sedimentary basins Thoughts on crustal evolution suggested by deep seismic reflection images Bull Soc Geol Fr 164 327–342

Bois C, ECORS Scientific party (1991) Post orogenic evolution of the European crust studied from ECORS deep seismic profiles In Meissner R et al (eds) Continental lithosphere deep seismic reflection AGU Geodyn Ser 22 59–68

Bonijoly D, Roure F, Mignot A, Perrin J, Le Strat P (1993) The Ardeche passive margin consequence of a detachment within the Stephanian. EUG VII, Strasburg, Terra Abstr 675

Burrus J (1984) Contribution to a geodynamic synthesis of the Provençal basin (northwestern Mediterranean) Mar Geol 55, 34 247–270

Burrus J (1989) Review of geodynamic models for extensional basins the paradox of stretching in the Gulf of Lions (northwest Mediterranean) Bull Soc Geol Fr 8, V 377–393

Butler RWH (198) Thrust evolution within previously rifted regions an example from the Vercors, French subalpine chains Mem Soc Geol Ital 1987, 38 5–18

Byerlee JD (1978) Friction of rocks Geoph Appl 116 615–626

Chauve P, Martin J, Petitjean E, Sequeiros F (1988) Le chevauchement du Jura sur la Bresse. Données nouvelles et réinterprétation des sondages Bull Soc Géol 8 861–870

Chorowicz J, Deroin JP, Gesse G, Huger J, Becue B, Curnelle R, Perrin G, Ronfola D (1991) A methodology for the use of SPOT imagery in oil exploration. Example of the Bas-Languedoc exploration permits area (France) Int J Remote Sensing 12, 100 2087–2108

Coward M, Dietrich D (1989) Alpine tectonics, an overview In Coward MP, Dietrich D, Park RG (eds) Alpine tectonics Geol Soc Spec Publ 45 1–25

Coward MP, Trudgill B (1989) Basin development and basement of the Celtic Sea basin (SW Britain) Bull Soc Geol Fr 8, V· 423–436

Coward MP, Dewey JF, Hancock PL (1987) Continental extensional tectonics Geol Soc Spec Publ 28 139–152

Curnelle R, Dubois P (1986) Evolution mésozoique des grands bassins français (bassin de Paris, d'Aquitaine et du Sud-Est). Bull Soc Geol Fr 8 529–546

Dardeau G, Fortwengler D, de Gracianski PC, Jacquin T, Marchand D, Martinod J (1990) Halocinèse et jeu de blocs dans les Baronnies diapirs de Propiac, Montaulieu et Condorcet (Drôme, France) Bull Cent Rech Expl Prod Elf-France 14 111–159

Debrand-Passard S, Courbouleix S, Lienhardt MJ (eds) (1984) Synthése géologique du sud-est de la France Mém Bur Rech Geol Min 125

de Graciansky PC, Dardeau G, Lemoine M, Tricart P (1988) De la distension à la compression l'inversion structurale dans les Alpes Bull Soc Géol Fr 8, IV 779–785

Dewey JF (1988) Extensional collapse of orogens Tctonics 7 1123–1139

Elmi S (1983) La structure du sud-est de la France une approche à partir de la bordure vivaro-cévenole du Massif Central CR Acad Sci Paris 296 1615–1619

Elmi S (1984) Tectonique et sédimentation In Synthèse géologique du sud-est de la France Mem Bur Rech Géol Min 125 166–175

Faure JL, Chermette JC (1989) Deformation of tilted blocks, consequences on block geometry and extension measurements Bull Soc Geol Fr 8, V 461–476

Ford M (1993) The geometry of a deformed basin-platform transition the Ventoux-Lure fault zone, SE France In Peri-Tethyan platforms, IFP/Eurotethys Research Conference, Arles, Abstr

Gibbs AD (1987) Development of extension and mixed-mode sedimentary basins In Coward MP, Dewey JF, Hancock PL (eds) Continental extensional tectonics Geol Soc Spec Publ 28 19–33

Gillcrist R, Coward M, Mugnier JL (1987) Structural inversion and its controls examples from the Alpine foreland and the French Alps Geodin Acta 1 5–34

Giot D, Roure F, Elmi S, Lajat D, Steinberg M (1991a) Géométrie et dynamique d'une marge continentale mésozoique, Ardèche, France CR Acad Sci Paris 312 747–754

Giot D, Roure F, Dromart G, Elmi S, Naville C, Perrin J, Steinberg M (1991b) Dynamique d'une marge passive jurassique les principales étapes du rejeu de la faille d'Uzer révélées par le forage de Balazuc 1 (Ardèche, France) CR Acad Sci Paris 313 1463–1469

Gorini C (1989) Inversion structurale négative la tectonique oligo-miocène dans le fossé de Narbonne Sigean (Aude, France) DEA report, Toulouse Univ , 80 pp (unpublished)

Gorini C, Viallard P, Deramond J (1991) Modèle d'inversion structurale négative la tectonique extensive post-nappe du fossé de Narbonne-Sigean (Corbières, sud de la France) CR Acad Sci Paris 312, II 1013–1019

Gorini C, Le Marrec A, Mauffret A (1993) Contribution to the structural and sedimentary history of the Gulf of Lions (western Mediterranean), from the ECORS profiles, industrial seismic profiles and wells data Bull Soc Geol Fr 164 353–363

Guellec S, Lajat D, Mascle A, Roure F, Tardy M (1990a) Deep seismic profiling and petroleum potential in the western Alps constraints with ECORS data, balanced cross-sections and hydrocarbon modeling In Pinet B, Bois C (eds) The Potential of deep seismic profiling for hydrocarbon exploration. IFP Explor and Produc Research Conf, Arles Technip, Paris, pp 425–438

Guellec S, Mugnier JL, Tardy M, Roure F (1990b) Neogene evolution of the western Alpine foreland in the light of ECORS data and balanced cross-sections In Roure F, Heitzmann P, Polini R (eds) Deep structure of the Alps Mém Soc Geol Fr 156, Mem Soc Geol Suisse 1, Vol Spec Soc Geol Ital 1

Hamblin WK (1965) Origin of "reverse drag" on the downthrown side of normal faults GSA Bull 76 1145–1164
Howards KA, John BE (1987) Crustal extension along a rooted system of imbricate low-angle faults Colorado extensional corridor, California and Arizona In Coward MP, Dewey JF, Hancock PL (eds) Continental extensional tectonics Spec Publ Geol Soc Lond 28 299–311
Lemoine M, Trumpy R (1987) Pre-oceanic rifting in the Alps Tectonophysics, Amsterdam 133 305–320
Martinod J (1988) Compression alpine, extension mésozoique et diapirisme dans les Baronnies (Drôme) IFP Rep 36523 (unpublished)
Mascle A (1990) Géologie pétrolière des bassins permiens français Comparaison avec les bassins permiens du nord de l'Europe In Potential économique des bassins permiens français Chron Rech Min 499 69–86
Mascle A, Jimenez MC, Duval P, Biju-Duval B, Trémolières P, Arnaud H, Carrio E (1993) Field trip to the western Alps and their foreland in France EAEG, Paris, 1992, Guidebook
Mascle G, Arnaud H, Dardeau G, Debelmas J, Dubois P, Gidon M, de Graciansky PC, Kerckhove C, Lemoine M (1988) Salt tectonics, Tethyan rifting and Alpine folding in the French Alps Bull Soc Geol Fr 747–758
McClay KR (1989) Analogue models of inversion tectonics In Cooper MA, Williams GD (eds) Inversion tectonics Spec Publ Geol Soc Lond 44 41–62
McClay KR, Ellis PG (1987 a) Analogue models of extensional fault geometries In Coward MP, Dewey JF, Hancock P (eds) Continental extensional tectonics Spec Publ Geol Soc Lond 28 109–125
McClay KR, Ellis PG (1987 b) Geometry of extensional fault systems developed in experimental models Geology 15 341–344
Mugnier JL, Vialon P (1986) Deformation and displacement of the Jura cover on its basement J Struct Geol Oxford 8 373–387
Mugnier JL, Guellec S, Ménard G, Roure F, Tardy M, Viallon P (1990) Crustal balanced corrs-sections through the external Alps deduced from the ECORS profile In Roure F, Heitzmann P, Polino R (eds) Deep structure of the Alps Mem Soc Geol Fr 156 15–28
Naylor MA, Larroque JM, Gauthier BM (1994) Understanding extensional tectonics, insights from sand box models In Roure F et al (eds) Geodynamic evolution of sedimentary basins Moscow Symp Proc (in press)
Petit JP, Bousquet JC, Mattei J (1973) Glissement synsédimentaire et troncature basale de blocs hettangiens du bord sud du Causse du Larzac entre Arboras et Salces (Languedoc) CR Acad Sci 227
Roure F, Séguret M, Villien A (1988) Structural styles of the Pyrenees a view from seismic reflexion to surface studies Guide-book Field-Trip 3 Am Assoc Pet Geol Mediterranean basins conference, Nice, 140 pp
Roure F, Casero P, Vially R (1991) Growth processes and melange formation in the southern Apennines accretionary wedge Earth Planet Sci Lett 102 395–412
Roure F, Brun, JP, Colletta B, Van Den Driessche J (1992) Geometry and kinematics of extensional structures in the Alpine foreland of southeastern France J Struct Geol 14 503–519
Roure F, Carnevali JO, Gou Y, Subieta T (1994) Geometry and kinematics of the north Monagas thrust belt (Venezuela) Mar Pet Geol (in press)
Rowan MG, Kligfield R (1989) Cross-section restoration and balancing as aid to seismic interpretation in extensional terranes Am Assoc Pet Geol Bull 73 955–966
Roy JP, Trémolières P (1992) L'inversion du demi-graben oligocène d'Alès (France) données de terrain et modélisation CR Acad Sci Paris 355, II 1777–1781
Séranne M, Chauvet A, Séguret M, Brunel M (1989) Tectonics of the Devonian collapse basins of western Norway Bull Soc Geol Fr 8, V 489–499
Tempier C (1987) Modèle nouveau de mise en place des structures provençales Bull Soc Geol Fr 533–540
Vendeville B, Cobbold PR (1987) Glissements gravitaires synsédimentaires et failles normales listriques modèles expérimentaux CR Acad Sci Paris 305, II 1313–1319
Vendeville B, Cobbold PR, Davy P, Brun JP, Choukroune P (1987) Physical models of extensional tectonics at various scales In Coward MP, Dewey JF, Hancock P (eds) Continental extensional tectonics Spec Publ Geol Soc Lond 28 95–107
Vially R, Haddadi N, Boudjema A, Letouzey J, Desforges G, Askri H, Bénard F (1994) Basin inversion along the north African plate margin, Algerian Atlas In IFP-Peri-Tethys Research Conference, Peri-Tethyan platforms, Arles meeting, Proc (in press)
Villeger M, Andrieux J (1987) Phases tectoniques post-éocènes et structuration polyphasée du panneau de couverture nord-provençal (Alpes externes méridionales) Bull Soc Géol Fr 8, III 147–156
Wernicke B, Burchfiel BC (1982) Modes of extensional tectonics J Struct Geol 4 105–115
White NJ, Jackson JA, McKenzie DP (1986) The relationships between the geometry of normal faults and that of sedimentary layers in their hanging wall J Struct Geol 8 897–909
Williams G, Vann I (1987) The geometry of listric normal faults and deformation in their hanging walls J Struct Geol 9 789–795
Ziegler PA (1983) Inverted basins in the Alpine foreland In Seismic expression of structural styles – a picture and work atlas Am Assoc Pet Geol Stud Geol 15 3 3-3–3 3-12
Zoetemeijer R, Sassi W, Roure F, Cloetingh S (1992) Stratigraphic and kinematic modeling of thrust evolution, northern Apennines, Italy Geology 20 1035–1038

Thrust Propagation and Syntectonic Sedimentation in the Savoy Tertiary Molasse Basin (Alpine Foreland)

E. Deville, E. Blanc, M. Tardy, C. Beck, M. Cousin, and G. Ménard

Abstract. The Savoy Tertiary molasse basin is part of the Neogene foreland thrust belt of the Alps that was overlain by Tertiary syntectonic sediments during the polyphase development of the Alpine foreland. The analysis of these stratigraphic sequences is helpful for understanding the dynamics of the area because it provides efficient constraints for restoring the tectonic subsidence and the progressive deformation history of the basin. During Rupelian to Aquitanian times prograding sedimentary sequences were deposited, while passive flexuring of the foreland developed. This was probably induced by the tectonic overload created by the Oligocene thrust sheets in the inner parts of the Alps. From the Rupelian to the Chattian, this basin developed with high progradation rates (6 mm/year) and high subsidence rates (> 0.2 mm/year) in the eastern parts of the basin, whereas during the Aquitanian, progradation was low (< 1 mm/ year) but subsidence remained high (> 0.15 mm/year). Afterwards, this early passive foreland flexural basin was inverted and became involved in the Neogene foreland thrust system during the Miocene thrust propagation. During the Burdigalian to the Serravallian, this led to the development of sparce syn-thrust infillings with heterogeneous deposition rates. This notably highlights the important migration of the Alpine front in this area during Burdigalian times. The homogenized strain on the whole molasse basin in Savoy during the Burdigalian-Langhian was around 2 mm/year. Taking into account the proposed evolution, hydrocarbon maturation and migration are also considered in estimating the petroleum potential of the area.

Introduction

The outer parts of the western Alps form a classic foreland fold-thrust belt. On the Savoy transect, this belt encompasses, from west to east, the Jura mountains, the Savoy Tertiary molasse basin and the Bornes and Bauges massifs of the Subalpine chains (Fig. 1). The sediments in these regions form a thin thrust wedge that was detached from its initial basement during Tertiary-late Alpine collision stages (Winnock 1961; Laubscher 1965; Mugnier 1984; Mugnier and Ménard 1986; Guellec et al. 1990 a, b). This orogenic front accommodates, within the sedimentary cover of the Alpine foreland, the last frontal few tens of kilometers of the several hundred kilometers of NW-directed shortening recorded in the crustal/lithospheric wedge of the Alps (Tardy et al. 1990). As such, the molasse basin constitutes simply the part of the foreland thrust system of the Alps that has been overlain by syntectonic Tertiary deposits.

Because the Tertiary sedimentation occurred above the thrust system of the detached cover wedge and during tectonic shortening, these syntectonic sediments constitute precious markers that have recorded the history of the Alpine foreland tectonic movements. These sediments are good indicators of each step in the progressive deformation because they have fossilized each tectonic increment that occurred in the thrust system. This chapter proposes a structural overview of this molasse basin and describes the relationships between tectonics and sedimentation that occurred during the Oligocene-Neogene development of the Alpine foreland. Hydrocarbon maturation and migration problems are also discussed.

Geologic Background

The Savoy molasse basin extends from the Geneva area to the Chambéry area between the Subalpine frontal thrust and the Jura mountains (Figs. 1, 2). The surface geology shows several N-S or NE-SW Mesozoic ramp anticlines constituted by Jurassic and Lower Cretaceous formations that emerge from the Tertiary molasse (Fig. 2). These ramp anticlines are bounded by several NW-SE-trending transverse faults (Cruseilles fault, Vuache fault, Droisy fault, Culoz fault) that acted as transfer-faults during thrust tectonics, and that have also influenced the Tertiary sedimentation areas.

Lithostratigraphy

The Mesozoic series consists of platform carbonates, very similar to those of the Jura mountains. Facies changes with respect to the Jura appeared in the Late Jurassic, where coral reef carbonates are well developed in the western part of the basin, and they are replaced by hemipelagic limestones in the eastern part of the basin. A change also appears in the Valanginian to Berriasian deposits, where platform progradation facies developed in

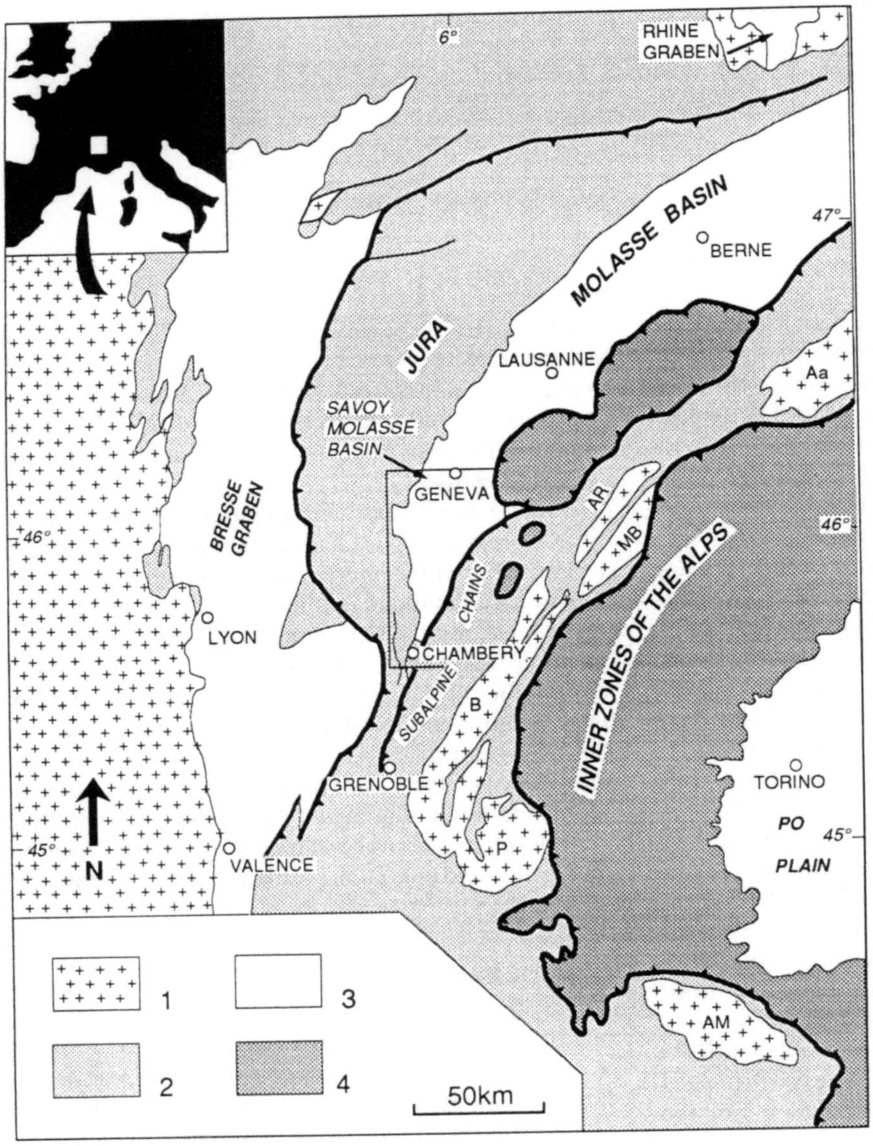

Fig. 1. Structural framework of the Savoy molasse basin within the western Alps overthrust foreland *1* Crystalline basement, *2* Mesozoic cover, *3* Tertiary, *4* inner zones of the Alps, *Aa* Aar, *AR* Aiguilles Rouges, *B* Belledonne, *M* Mercantour, *MB* Mont Blanc, *P* Pelvoux

the eastern part of the basin and in the Subalpine massifs (Arnaud and Arnaud-Vanneau 1989; Arnaud-Vanneau and Arnaud 1991).

The Tertiary sediments of the Alpine molasse basin are lithostratigraphically subdivided into four groups (Rigassi 1957; Lamiraux 1977; Homewood and Allen 1981; Jung 1982; Paolillo 1984; Homewood et al. 1986), as shown in Fig. 3

1. The Lower Marine Molasse (Untere Meeresmolasse of the German authors: UMM) consists of Rupelian shallow marine sandstones and conglomerates (Doriaz Sandstone), marine and brackish shales (Meletta Marls), and interbedded shales and well-bedded shallow marine sandstones that correspond to turbidites and littoral sheet sandstones (Déserts Sandstone). The Lower Marine Molasse was deposited in a narrow, shallow, brackish, and badly aerated marine gulf. In Savoy, it is preserved in the Bauges and Bornes massifs, and it is also present in the northeastern part of the molasse basin (east of the Salève anticline)
2. The Lower "Freshwater" Molasse (Untere Suss-wassermolasse· USM) includes at the base (Chattian) freshwater carbonates, red silty, and sandy shales with paleosoils (red molasse) and typical fluviatile channelized thinning-upward cycles. Locally, gypsiferous marls yielded ostracodes indicating brackish water. The upper part of the Lower "Freshwater" Molasse (Aquitanian) is mainly composed of subarkosic sandstones that correspond to the growth of sandy fans rich in Alpine detrital metamorphic minerals which constitute erosion products from the inland metamorphic belt of the Alps. Detailed biostratigraphic control is well constrained in the Lower "Freshwater" Molasse

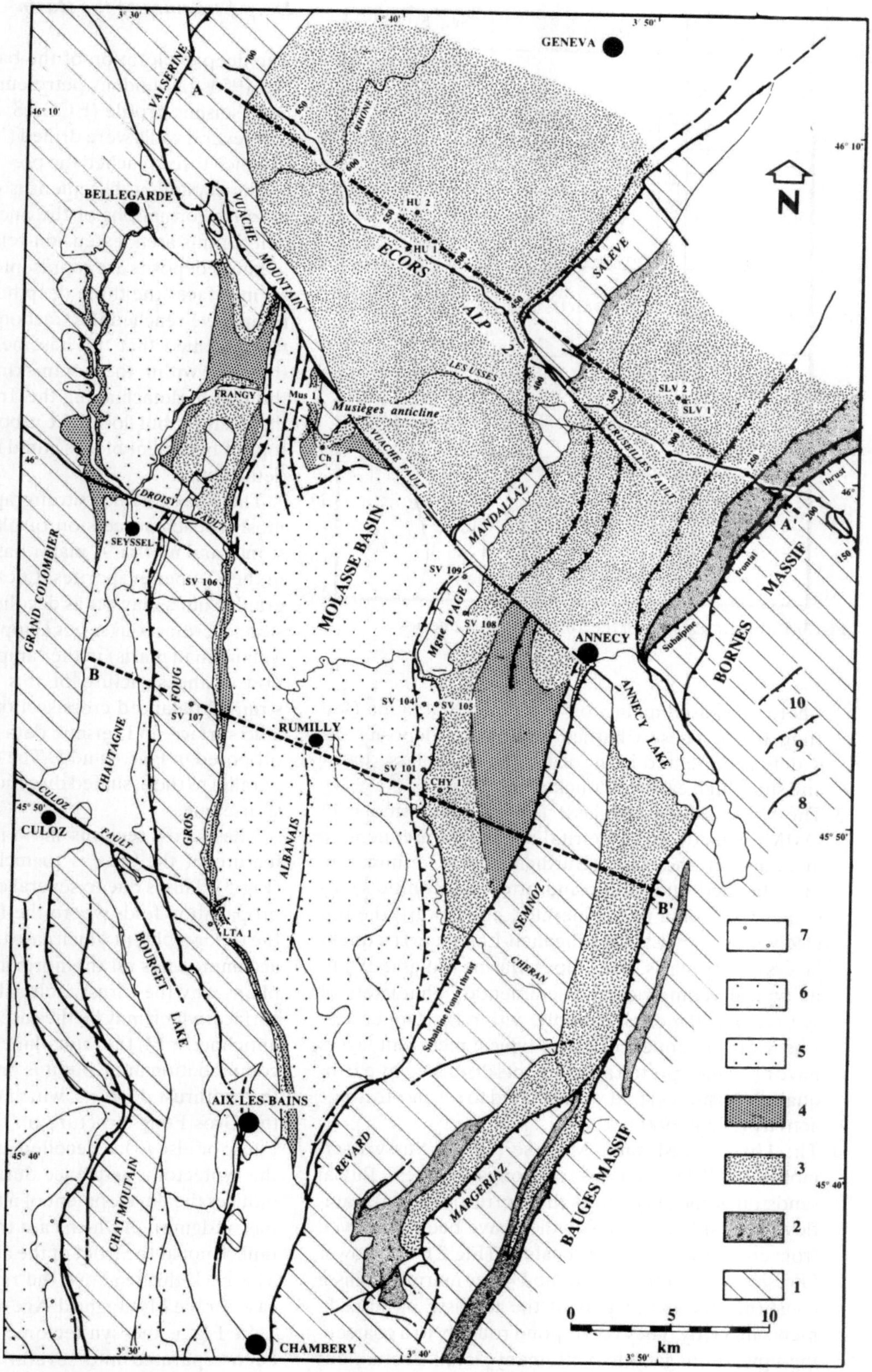

Fig. 2. Structural sketch map of the Savoy molasse basin and location of the ECORS ALP2 deep seismic profile, and of the cross sections in Figs. 4 and 5. *1* Jura and Subalpine Mesozoic cover; *2* Rupelian, *3* Chattian, *4* Aquitanian; *5* Burdigalian, *6* Langhian-Serravallian; *7* Tortonian, *8* Thrust fault, *9* Blind-thrust, *AA'* location of the cross section of Fig 4, *BB'* location of the cross section of Fig. 5

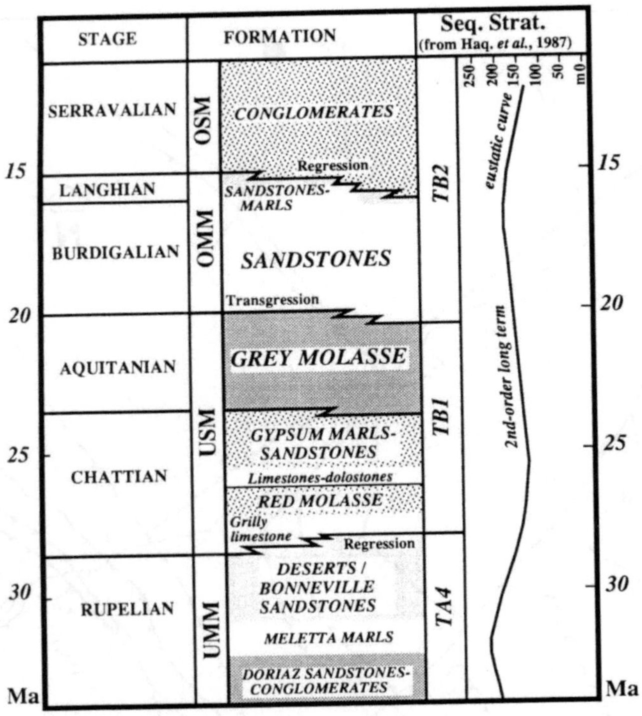

Fig. 3. Lithostratigraphy of the Tertiary molasse series

where it is based mainly on micromammal fauna (see Jung 1982). These Chattian-Aquitanian deposits are widely represented in all of the Savoy molasse basin and show high eastward thickening.

3. The Upper Marine Molasse (Obere Meeresmolasse: OMM) is made up of detrital marine parasequences, including at the base (Burdigalian) predominantly thick-bedded feldspathic and often glauconitic sandstones with some conglomerates, and toward the top (Helvetian, e. g., Late Burdigalian-Langhian) frequent layers of silty marls. The Upper Marine Molasse exhibits typical geometries and sedimentary characters of syn-tectonic deposits. Notably, within the Upper Marine Molasse sequences, some typical recurrent layers have been interpreted as sediments disturbed by earthquakes (seismites) that were related to seismo-tectonic activity (Blanc 1991).
4. The Upper Freshwater Molasse (Obere Susswassermolasse: OSM) consists of conglomerates, fluvial sandstones and silty shales that correspond to streamflow alluvial fans. In Savoy they have been preserved from erosion only in a few localities (Fig. 2). The Lower "Freshwater" Molasse and the Upper marine molasse constitute the major part of the molasse basin sediments in Savoy. They correspond roughly to a coarsening-upward megacycle, with many detail fluctuations and regional variations. Drainage occurred essentially from the northeast (perpendicularly to the main tectonic shortening direction), from the erosion of the ante-Miocene Alpine nappes.

Deep Structure of the Basin

The deep exploration of the basin started as early as in the 1950s Abundant petroleum seismic profiles and a deep seismic profile (ECORS ALP 2; Fig. 4) were shot, and several wells were drilled (Michel and Caillon 1957), of which three reached the pre-Triassic basement.

The crystalline basement is constituted by the metamorphic formations of the ancient Hercynian orogenic belt. In this area, extension-related Carboniferous sedimentary deposits are represented at the basement-Triassic interface (the three deep holes in the Savoy molasse basin have reached the Carboniferous). Conventionally, it is assumed that the basement is not involved in the thrusting within these thin-skinned structures which are thought to detach along the Triassic evaporites decollement layer; but locally, Carboniferous deposits and/or tops of basement blocks could have been involved in the detachment.

The Mesozoic lithostratigraphic column shows an almost isopach succession of platform carbonates (limestones and marls) A major basal decollement is located in the Keuper evaporites. Locally, also, weak lithostratigraphic units may act as decollement levels within Jurassic and Cretaceous series (Upper Liassic, Oxfordian, and Valanginian marls) in the ramp anticlines. As an illustration of the structure of this molasse basin, well-constrained balanced cross sections, constructed from reliable surface and seismic data and borehole results, are proposed in Figs. 4 and 5. The cross sections were drawn parallel to the assumed direction of maximum horizontal shortening.

These cross sections show, notably, that a widespread feature of the area is to include several blind-thrusts (Fig. 5). This is due to several causes, either (1) to thrust-propagation folds where the fault plane does not reach the topography, in which case the thrust-propagation fold accommodates all the horizontal shortening, which explains why the thrust fault is blind at the topographical surface, which may be the case for parts of the Gros Foug anticline; or (2) the fault plane reaches the deposition accommodation area, but it is progressively overlapped by the synthrust deposits, which is the case for the north of the Gros Foug structure on the cross section shown in Fig. 5; or else (3), a decollement layer is activated within the syntectonic sequence during thrust tectonics, which induces the development of a triangle zone (intercutaneous wedging), in which case the emergence of the thrust fault is not at the front of the active border, but it appears as a back-thrust above the ramp anticline, which is the case for the Montagne d'Age anticline (Fig 5)

In Fig. 5, the syn-tectonic sedimentary bodies have been separated into several deformed modules corresponding to deformed syntectonic sedimentary sequences (Fig. 5). Only supercycles (second-order sequences) have been distinguished. The sequence boundaries (borders of each module) have been defined by different types of segments, (1) conformable segments (isochrons), (2) erosional segments, (3) onlap segments,

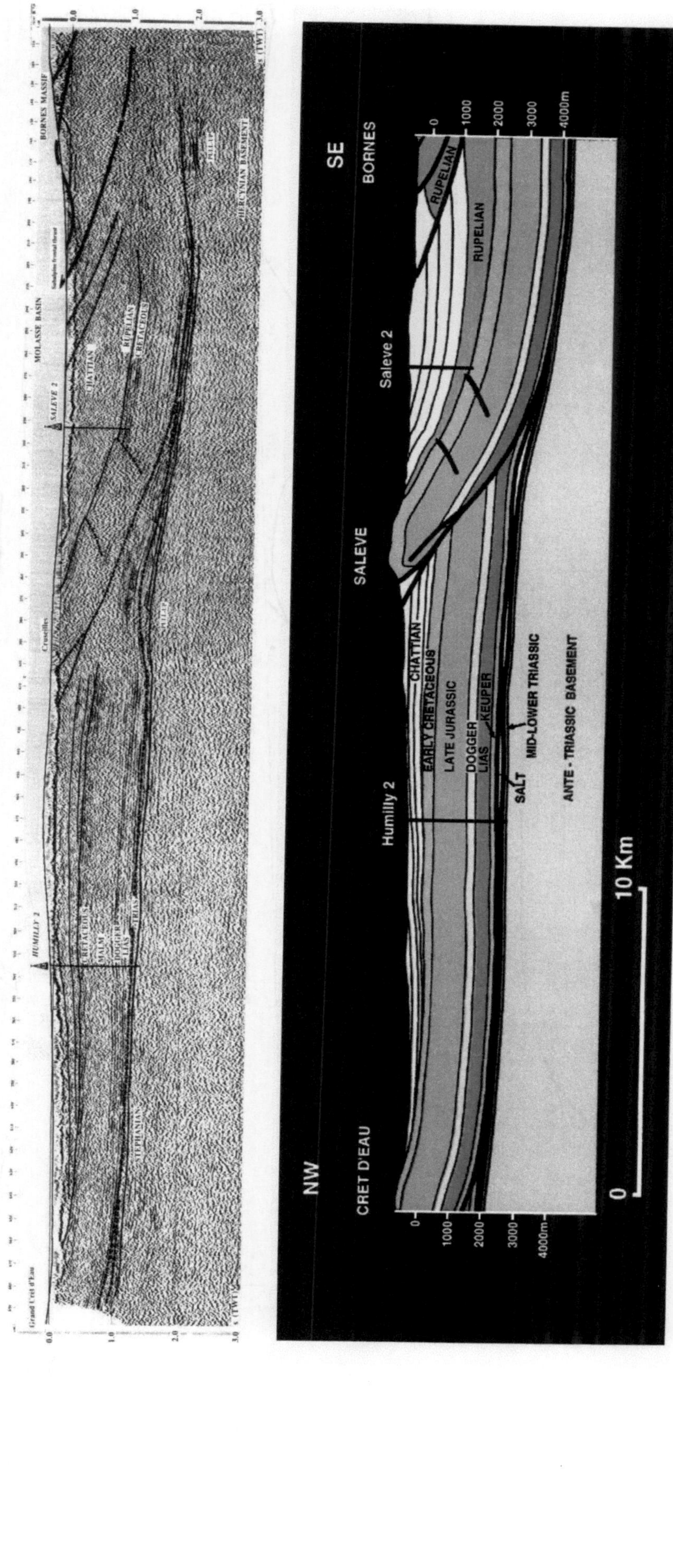

Fig. 4. The ECORS ALPES2 deep seismic profile and its interpretation

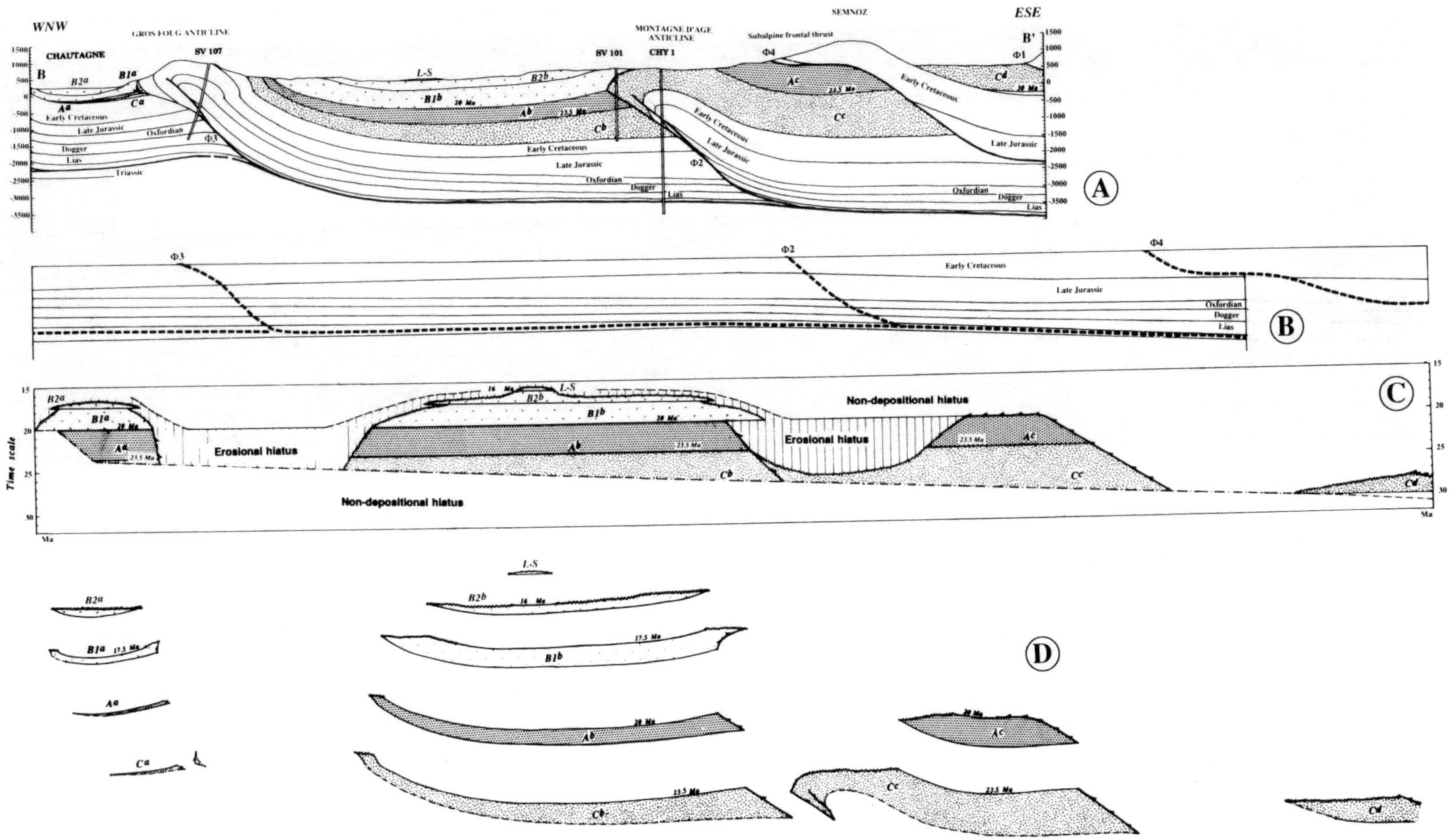

Fig. 5 A–D. Structure of the syntectonic sequences of the Savoy molasse basin **A** A cross section through the Savoy molasse basin south of the Vuache fault (*Ca to Cd* Chattian sequences, *Aa to Ac* Aquitanian sequences, *B1a* and *B1b* lower Burdigalian sequences, *B2a* and *B2b* upper Burdigalian sequences, *L-S* Langhian-Serravallian sequences) **B** Restoration of the section prior to the Tertiary sedimentation **C** Chronostratigraphic chart **D** Distinction of the syntectonic sequences

(4) downlap segments, (5) tectonic truncations (note that the same boundary between two sequences may be defined by different types of segments on the border of each module). This also enables us to define the nondepositional hiatus and erosional hiatus that are illustrated on a chronostratigraphic section (Fig. 5). Some reference chronostratigraphic surfaces (isochrons), for which the stratigraphic age is suitably constrained over the whole transect, were selected, and each was assigned a numerical age (Figs. 3, 5). For this, we used the recent correlative chart of the European Oligocene and Miocene from Berger (1992). The actual deformed geometry of the reference isochron surfaces is shown in Fig. 5.

Geodynamics

In the syntectonic sedimentary sequences of the Savoy molasse basin, the Chattian-Aquitanian Lower "Freshwater" Molasse exhibit characters of inverted flexural sequences, notably with a regular thinning toward the foreland after restoration, whereas the Burdigalian-Serravalian Upper Marine Molasse shows characters of syn-thrust sedimentary sequences (growth strata made of discontinuous sedimentary infillings of variable longitudinal extension, confined between ramp anticlines; see Specht et al. 1991; Suppe et al. 1991; Zoetemeijer and Sassi 1991; Zoetemeijer et al. 1992).

The stucture of the syntectonic sedimentary sequences enables us to reconstruct the thrust sequence that occurred in the basin. For the evolution of the thrust sequence we propose the following succession:

1. An early Chattian shortening responsible for the development of the Upper Oligocene orogenic thrust front (that probably corresponds to the present eastern Bauges frontal thrust, Φ 1 in Fig. 5). Indeed, it is necessary to invoke a Chattian tectonic setting for the eastern Bauges over the easternmost part of the Chattian lithostratigraphic unit to give to this sequence a kinematically possible structure.
2. A Burdigalian piggyback thrust sequence, responsible for the development of the main thrust-tectonics increments recorded within the central molasse basin anticlines (Montagne d'Age, Gros Foug; Figs. 5, 7).
3. A late out-of-sequence thrusting responsible for the final setting of the Subalpine front that overthrust the Late Burdigalian-Langhian molasse in Savoy (Doudoux et al. 1982; Fig. 2).

In Fig. 6, we propose a tectonic restoration attempt of the Chattian-Aquitanian continental basin. Each step of restoration was obtained by the unfolding of the successive modules, from younger to older. The restoration mode was chosen so that the uppermost conformable segment of each retro-tectonic step is restored in compatible state with the sedimentary environment during deposition. The conformable segments (isochron stratigraphic surfaces) have been simply restored horizontally, because the syntectonic molasse sediments in the area always

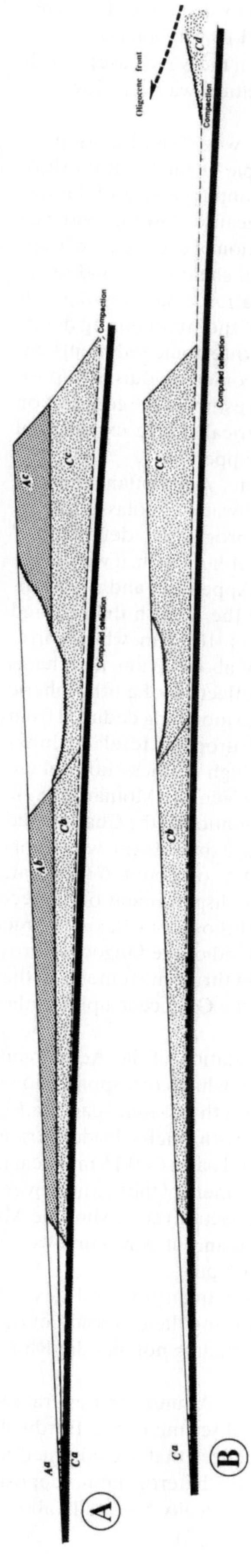

Fig. 6 A, B. Attempt to restore of the Chattian-Aquitanian continental basin **A** Late Aquitanian stage **B** Late Chattian stage

have characters either of lagoon to shallow-water lacustrine alternances, or of intertidal to very shallow-water subtidal marine alternances (upon the wave base), which implies that the angle of deposition was very low (assumed to be ≤ 1 or 2°).

The retro-deformation mode was chosen as isopach folding, with layer-parallel simple shear (flexural-slip), and with no shearing in the horizontal portions of the segment. This mode was used because natural evidence shows that the internal deformation of each sequence appears to be mostly layer-parallel shear or second-order faulting. This mode of deformation conserves the volume, the length and thickness of the strata during deformation. Decompaction of the syntectonic sediments enables us to restore the initial geometries during deposition and before deformation. To estimate the compaction during burial we used an empirical simple exponential porosity decay with depth (see Appendix).

The Late Chattian and Late Aquitanian restored geometries of the lower "freshwater" molasse can be used as a tool to construct the progressive deflection of the European foreland lithosphere. As such, if we consider a simple elastic model (see Appendix) and if we estimate the flexural rigidity from the fit with the restored deflection, we obtained about 8×10^{22} Nm, which corresponds to an elastic thickness of about 23 km, and that is compatible with the flexural deflection of a lithospheric plate. This result is equivalent to modeling deduced from the actual deflection of the European foreland lithosphere and corresponds to a high-stiffness lithosphere (Karner and Watts 1983; Lyon-Caen and Molnard 1989).

The high sedimentary progradation of the Chattian red molasse and gypsum marls (more than 30 km, which corresponds to progradation rates of about 6 mm/year, Fig. 6) can be related to a large displacement of the tectonic overload that created the lithospheric flexure (large overthrusting toward the foreland of the Oligocene orogenic front). The early Chattian thrust movements in the Alpine front can be related to the Oligocene uplift of the inner zones of the Alps.

The low sedimentary progradation of the Aquitanian grey molasse (1 to 2 km in Savoy, which corresponds, however, to lower progradation rates than 1 mm/year; Fig. 6), associated during the same time with a high subsidence rate in the eastern parts of the flexural basin (> 0.15 mm/year), can be regarded as a low displacement of the tectonic overload that creates the tectonic deflection (early Miocene Alpine nappes) during the Aquitanian, but in the same way, as a thickening of this tectonic overload.

Note that the Chattian-Aquitanian foreland flexural basin of the Alps developed during the extension of the west-European rift system (including notably the Rhine and Bresse grabens, Fig 1).

The inversion of the Chattian-Aquitanian flexural sequences explains the structural setting of the Burdigalian-Langhian syn-thrust sequences that are confined to only some small basins with very heterogeneous deposition rates (Fig. 7) and that can locally be totally absent (northeast of the Vuache fault; Fig. 2).

The inversion is associated with subsidence characterized by the transgression of the Miocene sea. As such, there is no evident correlation here between the Alpine Miocene marine transgression and the usually assumed global eustatic cycles of sea-level fluctuation (the Aquitanian-Burdigalian boundary classically corresponds to a marine regression episode, Haq et al 1987) In the Alpine foreland, the relative sea-level changes are related first to tectonic events, whereas the global eustatic fluctuations appear as second-order phenomena with respect to tectonics. As such, we propose that the sea-level fluctuations recorded within the third-order marine Burdigalian-Langhian parasequences (Blanc 1991) correspond to regional modifications of the kinematics of fault propagation in the Alpine foreland and not to global sea-level fluctuations. This could explain why each sequence appears to be related to second-order imbricate development in the ramp-anticlines as has been described in overthrust belts (Specht et al. 1991). Nevertheless, the second-order global fluctuations of sea level can easily be recognized, because the marine sequences in the area correspond to periods of global sea-level highstands, whereas the continental sequences correspond to periods of global sea-level lowstands (Fig. 3). The structure of these Burdigalian-Langhian syn-thrust sequences is compatible with homogenized strain rates over the whole molasse basin of about 2 mm/year. The Burdigalian-Langhian tectonics could be related to the initial uplift of the eastern units of the External Crystalline massifs (Belledonne massif; Ménard 1988).

The late Miocene to recent generalized uplift of the Alpine foreland remains a poorly understood phenomena (uplift from the Langhian below sea level to the present mean altitudes ranging between 500–800 m, e.g., much higher than the mean altitude of coeval sediments in the Bresse basin; Bergerat et al 1990). This could be related to the late active tectonic deformation of the foreland. Indeed, a possible interpretation could be the present development, at depth, of a new crustal imbrication that is suggested by the ECORS deep seismic results where a thickening of the lower layered crust appears under the molasse basin (Guellec et al. 1990a). The crustal thickening could be responsible for an isostatic uplift of the foreland. Note also that an active strong uplift is currently taking place in the southeastern parts of the Jura (Fourniguet 1977).

Hydrocarbon Maturation and Migration

The main potential hydrocarbon source rocks in the basin and adjacent areas are distributed in three lithostratigraphic units:

1. The Stephanian that has coal characters (type III organic matter, with TOC > 50%, see Blanc et al. 1991).
2. The Toarcian that has organic matter of marine origin (type II); samples taken from several wells in the adjacent area and from outcrops have TOC that range from 1 to 10%.

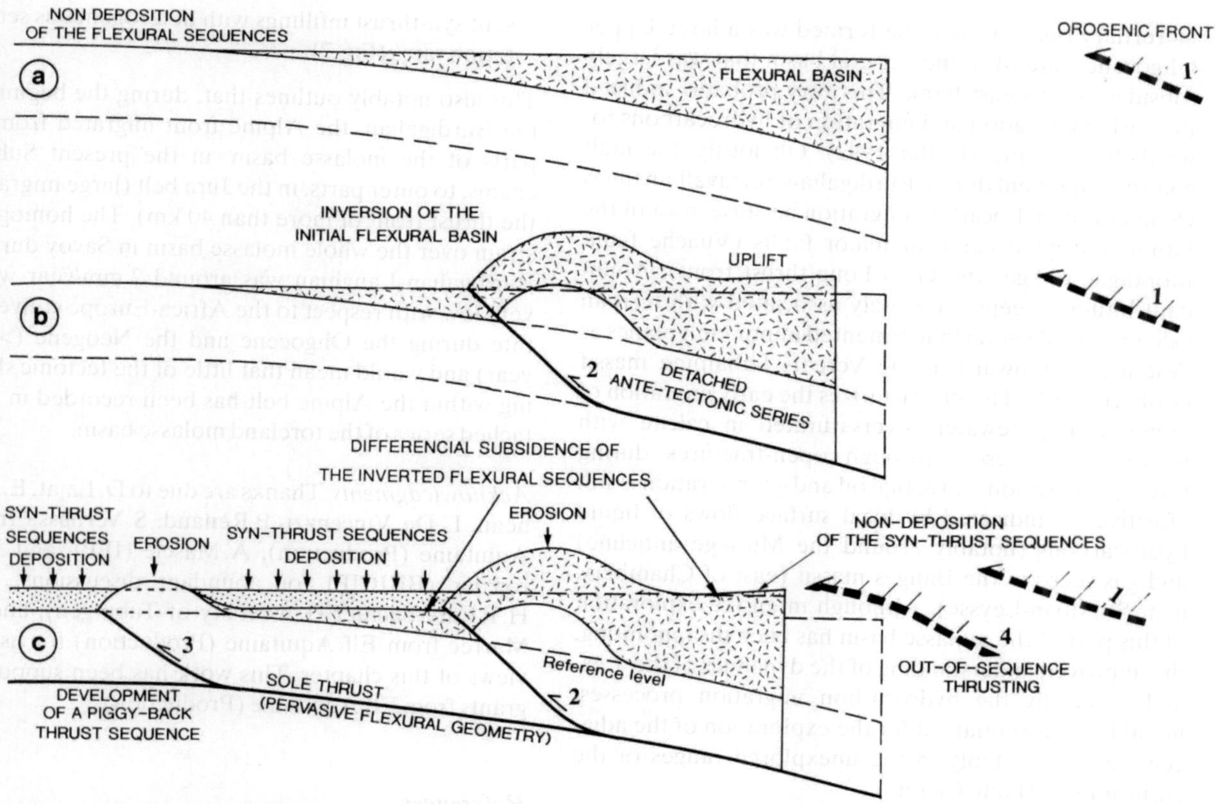

Fig. 7 a–c. Simplified evolution of the molasse basin development **a** Development of the Chattian-Aquitanian passive flexural basin **b** Inversion of the early flexural basin at the beginning of the Burdigalian transgression **c** Progressive development of Burdigalian-Serravalian piggyback basins and late out-of-sequence overthrusting of the Subalpine units

3. Locally, the Oligocene that includes a few layers with organic matter of types II and III in the molasse basin and adjacent Subalpine massifs (Gorin et al. 1989; Schegg 1993).

In the area, rocks with good reservoir properties that are in a good structural position to receive hydrocarbon migration are present at the bottom of the Mesozoic series. The most promising reservoirs are located in the fluviatile sandstones of the Early Triassic (Buntsandstein formation). They are locally absent in the south of the basin (la Tailla 1). Toward the east, in the Subalpine chains, they are enriched by carbonate horizons and their reservoir properties decrease. These sandstones are overlain by the limestones and dolostones of the Middle Triassic that also have good reservoir properties locally. They are in turn overlain by the Upper Triassic evaporites (Keuper) that constitute a potential efficient seal, notably in the northern part of the basin where thick salt accumulations are present (see Humilly 2). Furthermore, sandstone from the Rhetian and carbonates from Middle Triassic, Lias, and Dogger may also locally have reservoir qualities. In the eastern part of the basin and in the Subalpine chains, the Upper Jurassic limestone exhibits platform progradation facies that may locally have good porosity and permeability. Good reservoirs are present in the Eocene, Oligocene, and Aquitanian sandstones, at the bottom of the Tertiary molasse series, where they are interbedded with marl and clay layers. Their reservoir properties decrease in the eastern part of the basin, probably because of compaction processes related to their deeper burial toward the Subalpine front. These sandstones form westward-prograding onlaps (Fig. 4) and, as such, they locally constitute good potential stratigraphic traps. They have provided good oil shows in several wells (Humilly 2, Salève 2).

According to the proposed burial evolution and the thermal history of the area indicating relatively low heat flow densities during the Mesozoic (actual values are estimated around 70 mW m^{-2}; Lucazeau and Vasseur 1989), the Liassic source rock was probably buried, in the deep parts of the basin, in oil-window conditions during the Aquitanian-Burdigalian, and the Stephanian source rock probably reached the oil-window conditions as early as the Cretaceous and was buried in the gas-window during the Tertiary subsidence. Note also that in the east of the molasse basin, in the Bornes Subalpine massif, the Brizon 1 well (Charollais and Jamet 1990) has shown that rocks located under 700 m are actually at the beginning of the oil-window and that the true oil-window really starts below 1000 m. At depth, the bottom of the Mesozoic series is overmature.

Moreover, abundant surface bitumen and oil seeps indicate the maturation of source rocks and the migration of hydrocarbons within the thrust belt (Zweidler 1985). The timing between thrusting and hydrocarbon maturation is essential. Most of the structural traps were formed in Tertiary times. First to be formed was a large Upper Oligocene-Late Miocene flexural basin that was largely subsident on its east flank This very probably induced the early maturation and migration of hydrocarbons toward the foreland (Butler 1992). Obviously, the fault zone development during Burdigalian-Serravallian times channelized hydrocarbon migration because most of the bitumen seeps occur near major faults (Vuache fault, Montagne d'Age, and Gros Foug thrust front). Abundant bitumen seeps are widely associated with ferroan calcite crystallisation that cemented minor fault planes as described southward in the Vercors Subalpine massif (Roberts 1992). This characterizes the early migration of mineralized porewaters oversaturated in calcite with hydrocarbon present through open-fractures during thrust propagation. An actual oil and gas migration is still effective, as indicated by local surface flows of liquid hydrocarbons (notably around the Musiège anticline) and gas seeps in the Bauges massif (east of Chambéry, near St-Alban-Leysse). Although intensive exploration of this part of the molasse basin has been disappointing, the improved understanding of the dynamics of the area, and especially the hydrocarbon migration processes, might be an essential aid for the exploration of the adjacent regions, notably in the unexplored ranges of the eastern Jura (Haute Chaîne).

Conclusion

On the Savoy transect, the molasse basin appears to be a part of the Alpine foreland thrust system that was overlain by syntectonic sediments during the Tertiary development of the foreland.

As such, the western French part of the molasse basin of the European Alpine foreland differs from its eastern German and Austrian part, which has remained a passive foreland basin in its outer part until the present and was not inverted and displaced during the Miocene (see Bachmann et al. 1982, 1987; Wagner et al. 1986).

The proposed evolution of the Savoy molasse basin may be summarized as follows:

1. From the Rupelian to the Aquitanian a passive foreland flexural basin developed, in which sedimentation started in a relatively shallow marine environment in this outer part of the Alps and evolved during the Chattian and the Aquitanian into continental deposits (mostly lacustrine sedimentation). From the Rupelian to the Chattian, this basin developed with high progradation rates (6 mm/year) and high subsidence rates (> 0.2 mm/year) in the eastern parts of the basin, whereas during the Aquitanian, progradation was low (< 1 mm/year) but subsidence remained high (> 0.15 mm/year);
2. During the Burdigalian-Langhian, the inversion of the Oligocene-Aquitanian passive flexural foredeep occurred, and this part of the foreland was then involved in the Neogene Alpine foreland thrust system. This led notably to the development of sparce and locally absent syn-thrust infillings with heterogeneous sedimentation rates (Fig. 7).

This also notably outlines that, during the beginning of the Burdigalian, the Alpine front migrated from inner parts of the molasse basin, in the present Subalpine chains, to outer parts, in the Jura belt (large migration of the thrust front of more than 40 km) The homogenized strain over the whole molasse basin in Savoy during the Burdigalian-Langhian was around 2 mm/year which is very low with respect to the Africa-Europe convergence rate during the Oligocene and the Neogene (> 5 mm/year) and would mean that little of the tectonic shortening within the Alpine belt has been recorded in the detached series of the foreland molasse basin.

Acknowledgments Thanks are due to D. Lajat, E. Ducorneau, L. De Vincenzzi, P. Renaud, S. Vernassa from Elf Aquitaine (Production), A. Mascle (IFP), and J. F. Deronzier (BEICIP) for abundant discussions, and to H. P. Luterbachter (University of Tubingen), and A. Le Marrec from Elf Aquitaine (Production) for useful reviews of this chapter. This work has been supported by grants from Elf Aquitaine (Production)

References

Arnaud H, Arnaud-Vanneau A (1989) Séquences de dépôt et variations du niveau relatif de la mer au Barrémien et à l'Aptien inférieur dans les massifs subalpins septentrionaux et le Jura (SE de la France) Bull Soc Géol Fr 8, V 651–660

Arnaud-Vanneau A, Arnaud H (1991) Sédimentation et variations relatives du niveau de la mer sur les plate-formes carbonatés du Berriasien-Valanginien et du Barrémien dans les massifs subalpins septentrionaux et le Jura (Sud-Est de la France) Bull Soc Géol Fr 162, 3 535–545

Bachmann GH, Dorh G, Muller M (1982) Exploration in a classic thrust belt and its foreland Bavarian Alps, Germany AAPG Bull 62, 12 2529–2542

Bachmann GH, Muller M, Weggen K (1987) Evolution of the molasse basin (Germany, Switzerland) Tectonophysics 137 77–92

Berger JP (1992) Correlative chart of the European Oligocene and Miocene application to the Swiss molasse basin Eclogae Geol Helv 85, 3 573–609

Bergerat F, Mugnier JL, Guellec S, Truffert C, Cazes M, Damotte M, Roure F (1990) Extensional tectonics and subsidence of the Bresse Basin a view from ECORS data In Roure F, Heitzman P, Polino R (eds) Deep structure of the Alps Mém Soc Géol Fr 156, Mém Soc Géol Suisse 1, Vol Spec Soc Geol Ital 1 145–156

Blanc E (1991) Evolution sédimentaire syntectonique au front d'une chaîne de collision en environnement littoral Le cas du Miocène (molasse marine) de la région de Chambéry, France Rapp DEA, Université de Savoie, Chambéry (unpublished)

Blanc G, Doligez B, Lajat D, Mascle A (1991) Evaluation du potential pétrolier des Formations paléozoiques de la Bresse et de sa bordure jurassienne, France Bull Soc Géol Fr 162, 2 409–422

Butler RW (1992) Hydrocarbon maturation and tectonic loading in the Western Alpine foreland thrust belt In England WA, Fleet AJ (eds) Petroleum migration Geol Soc Spec Publ 59 227–244

Charollais J, Jamet M (1990) Principaux résultats géologiques du forage de Brizon 1 (BZN 1), Haute-Savoie, France In Roure F, Heitzman P, Polino R (eds) Deep structure of the Alps Mém Soc Géol Fr 156, Mém Soc Géol Suisse 1, Vol Spec Soc Geol Ital 1 185–202

Doudoux B, Mercier B, Tardy M (1982) Une interprétation nouvelle de la structure des massifs subalpins savoyards (Alpes occidentales) nappes de charriage oligocènes et déformations superposées C R Acad Sci Paris 295 63–68

Fourniguet J (1977) Mise en évidence de mouvements néotectoniques actuels verticaux dans le SE de la France par comparaison de nivellements successifs Rapp Interne BRGM, Orléans

Gorin G, Gulocar F, Cornialey Y (1989) Organic geochemistry, maturity, palynofacies and paleoenvironment of Upper Kimmeridgian and lower Tertiary organic-rich samples in the Southern Jura (Ain, France) and Subalpine massifs (Haute-Savoie, France) Eclogae Geol Helv 82/2 491–515

Guellec S, Mugnier JL, Tardy M, Roure F (1990a) Neogene evolution of the western Alpine foreland in the light of ECORS data and balanced cross-section In Roure F, Heitzman P, Polino R (eds) Deep structure of the Alps Mém Soc Géol Fr 156, Mém Soc Géol Suisse 1, Vol Spec Soc Geol Ital 1 165–184

Guellec S, Lajat D, Mascle A, Roure F, Tardy M (1990b) Deep seismic profiling and petroleum potential in the western Alps constraints with ECORS data, balanced cross-sections and hydrocarbon modeling In Pinet B, Bois C (eds) The potential of deep seismic profiling for hydrocarbon exploration, IFP exploration and production research conference, Technip, Paris, pp 425–438

Haq BU, Hardenbol J, Vail PR (1987) The chronology of fluctuating sea level since the Triassic Sciences 235 1156–1167

Homewood P, Allen P (1981) Wave-, tide- and current-controlled sandbodies of Miocene Molasse, western Switzerland AAPG Bull 65, 12 2534–2545

Homewood P, Allen PA, William GD (1986) Dynamics of the Molasse Basin of western Switzerland Spec Publ Int Assoc Sediment 8, Foreland Basins 199–217

Jung P (ed) (1982) Nouveaux résultats biostratigraphiques dans le bassin molassique, depuis le Vorarlberg jusqu'en haute Savoie Doc Lab Géol Lyon HS 7

Karner GD, Watts AB (1983) Gravity anomalies and flexure of the lithosphere J Geophys Res 88, B12 449–477

Lamiraux C (1977) Géologie du Miocène des chaînons jurassiens méridionaux et du bas Dauphiné nord oriental entre Chambéry et la tour du Pin, Thèse 3eme cycle, Université de Savoie, Chambéry (unpublished)

Laubscher HP (1965) Ein kinematisches Modell der Jurafaltung Eclogae Geol Helv 58 232–318

Lucazeau F, Vasseur G (1989) Heat flow density data from France and surrounding margins Tectonophysics 164: 251–258

Lyon-Caen H, Molnar P (1989) Constraints on the deep structure and dynamic processes beneath the Alps and adjacent regions from the analysis of gravity anomalies. Geophys J Int 99 19–32

Ménard G (1988) Structure et cinématique d'une chaîne de collision les Alpes occidentales et centrales Thèse d'état, Université de Grenoble, 268 pp

Michel P, Caillon G (1957) Quelques résultats des sondages exécutés en Savoie par la Régie Autonome des Pétroles Bull Soc Géol Fr 7, 7 995–1008

Mugnier JL (1984) Déplacement et déformation dans l'avant-pays d'une chaîne de collision Thèse 3ème cycle, Université de Grenoble, 163 pp

Mugnier JL, Ménard G (1986) Le développement du bassin molassique suisse et l'évolution des Alpes externes un modèle cinématique Bull Cent Rech Expl Prod Elf-Aquitaine 10, 1 167–180

Paolillo JP (1984) Essai de synthèse des caractéristiques géologiques et géotechniques des molasses périalpines savoyardes Thèse 3ème cycle, Université de Grenoble (unpublished)

Roberts G (1992) Structural controls on fluid migration through the Rencurel thrust zone, Vercors, French sub-Alpine chains In England WA, Fleet AJ (eds) Petroleum migration Geol Soc Spec Publ 59 245–262

Rigassi D (1957) Le Tertiaire de la région genevoise et savoisienne Bull Ver Schweiz Petrol Geol Ing 24, 66 19–34

Schegg R (1993) Thermal maturity and history of sediments in the North Alpine foreland basin (Switzerland, France) Publications du département de Géologie et Paléontologie, Université de Genève 15, 194 pp

Specht M, Deramond J, Souquet P (1991) Relations tectonique sédimentation dans les bassins d'avant-pays utilisation des surfaces stratigraphiques isochrones comme marqueurs de la déformation Bull Soc Géol Fr 162 553–562

Suppe J, Chou GT, Hook SC (1991) Rates of folding and faulting determined from growth strata In MacClay KR (ed) Thrust tectonics Chapmann and Hall, London, pp 105–122

Tardy M, Deville E, Fudral S, Guellec S, Ménard G, Thouvenot F, Vialon P (1990) Interprétation structurale des données du profil de sismique réflexion profonde ECORS-CROP Alpes entre le front pennique et la ligne du Canavese (Alpes occidentales) In Roure F, Heitzman P, Polino R (eds) Deep structure of the Alps Mém Soc Géol Fr 156, Mém Soc Géol Suisse 1, Vol Spec Soc Geol Ital 1 217–226

Wagner K, Kuchelkorn K, Hiltmann W (1986) Neue Ergebnisse zur alpinen Gebirgsbildung Oberosterreichs aus der Bohrung Oberhofen 1-Stratigraphie, Fazies, Maturital and Tektonik Erdol-erdgas-Kohle 102, 1 12–19

Winnock E (1961) Résultats géologiques du forage Risoux 1 Bull Ver Schweiz Petrol Geol Ing 28, 74 17–26

Zoetemeijer R, Sassi W (1991) 2D reconstitution of thrust evolution using the fault-bend method In McClay KR (ed) Thrust tectonics Chapman and Hall, London, pp 133–140

Zoetemeijer R, Sassi W, Roure F, Cloeting S (1992) Stratigraphic and kinematics modeling of thrust evolution, northern Apennines, Italy Geology 20 1035–1038

Zweidler D (1985) Genèse des gisements d'asphalte des formations de la Pierre Jaune de Neuchâtel et des calcaires urgoniens du Jura (Jura neuchatelois et nord-vaudois) Thèse de l'Université de Neuchâtel

Appendix

Flexural Model of the Lithosphere

Flexural models of lithospheric plates suppose that the lithosphere is represented as a semi-infinite elastic layered rigid plate overlying the asthenosphere considered as an inviscid fluid. We used the following equation that describes the deflection $\omega(x)$ of an elastic plate initially at an uniform depth d°:

$$\omega(x) = d^\circ + e^{-x/\lambda}\,[A \cos(x/\lambda) + B \sin(x/\lambda)]$$

with, $\lambda = (4\Gamma / \Delta\rho\, g)^{1/4}$,

where A and B are parameters depending on the tectonic stresses applied on the foreland lithosphere and on the lithostatic stresses created by the overloading responsible for the lithosphere flexure (tectonic thrust load and sedimentary filling load in the basin). $\Delta\rho$ is the difference between the surrounding densities (ρm mean density of the upper mantle, ρs mean density of the sediments), g the gravitational acceleration, and Γ the flexural rigidity Γ is given by

$$\Gamma = \frac{EH^3}{12(1 - \vartheta)^2} \quad .$$

where H is the elastic thickness of the foreland lithosphere, E the Young modulus, and ϑ the Poisson ratio. The flexural wavelength, $2\pi\lambda$, is independent of the value of the tectonic load.

Decompaction Model

To estimate compaction during burial we used the following empirical equation assuming a simple exponential porosity decay with depth:

$$\Phi = \Phi^\circ e^{-cz},$$

where Φ is the porosity of the buried sediments, Φ° the porosity of the unburied sediments, c an empirical parameter depending on the lithology, and z the thickness of the overburden

The compaction at depth z is

$$\alpha = \Phi^\circ (1 - e^{-cz}),$$

and the calculated compaction thickness of a sedimentary pile (n layers) is given by

$$C_t = \sum_{1}^{n} \int_{z_{i-1}}^{z_i} \Phi^\circ (1 - e^{-c_i z})\, dz$$

For the uplifted syntectonic sediments thrust tectonics, we have assumed that the compaction is not reversible and remains constant with the value of maximum overburden. This supposes that compaction occurred during subsidence and that the overall volume of the syntectonic sediments remains constant during thrust-tectonics deformation. For this application we used the following mean values, $c = 5 \times 10^{-4}\ m^{-1}$ $\Phi^\circ = 0.05$ for the lower "freshwater" molasse and $c = 5.4 \times 10^{-4}\ m^{-1}$ $\Phi^\circ = 0.1$ for the upper marine molasse, which are adjusted to match the measurements on surface and borehole samples

The Southern French Alps Paleogene Basin: Subsidence Modeling and Geodynamic Implications

R. Vially

Abstract. This chapter is based on detailed field surveys performed in the 1980s by the Ecole Nationale Supérieure du Pétrole et des Moteurs (ENSPM) and the Institut Français du Pétrole (IFP) in the southern French Alps Paleogene basin. In this basin, the so-called Priabonian trilogy (Nummulitic Limestone–Blue Marls–Annot Sandstone) is well developed and represents (with the uppermost olistostrome) a complete sedimentary cycle.

Understanding gained about the sedimentary infilling of the basin enables us to reconstruct the major outlines of the sedimentary thickness, the paleotopography of the basin, as well as the paleobathymetries of these deposits. These parameters are indispensable for computing the subsidence of a basin.

We can affirm that the hypothesis of a flexural origin (internal tectonic overloading) is plausible from the mechanical and geological standpoints. A new schematic reconstruction of the paleogene basin is proposed in which the "cordilleras" are the emerged part of more internal overthrusting structures.

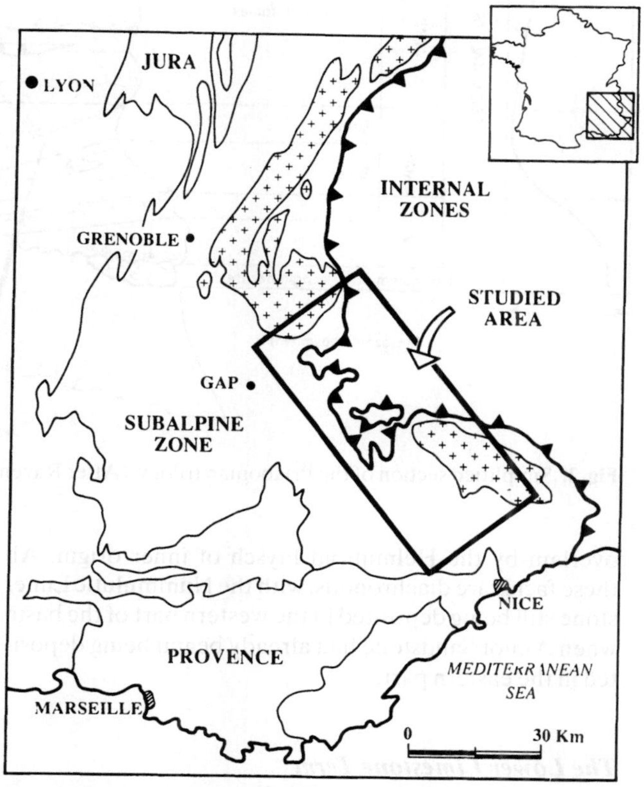

Fig. 1. Location of the area studied

Introduction

The Eocene-Oligocene marine Nummulitic basin in the southern French Alps (Fig. 1) was greatly deformed and eroded during Mio-Pliocene tectonic events (Alpine phase) so that outcrops are now restricted to a few synclinal areas. In its present situation, the southern French Alps Paleogene basin stretches for 160 km from north to south and 80 km from east to west (the eastern boundary of the basin is not known), but it continues farther north (Aiguilles d'Arves flysch near the Galibier Pass, Taveyannaz sandstone in Savoie and Switzerland; Sinclair 1992).

An examination of its infilling, however, enables its architecture to be reconstructed, and a modeling approach to its subsidence enables us to propose a geodynamic interpretation of the development of the front of the Alpine thrust-belt in the Paleogene.

Numerous authors (Tricart 1980; Lemoine 1984; Pairis 1987 and others) have connected the development of this basin with the Paleogene compressive events but nobody has tested the mechanical plausibility of such a hypothesis. Other authors (Ravenne et al. 1987) have proposed linking the formation of the basin with the E-W distension phase of the Oligocene. The subsidence study is a helpful method for determining the possible geodynamic origin of this basin.

Sedimentary Infilling

The sedimentary infilling includes the three conventional terms of the Priabonian trilogy as defined by Boussac in 1912, to which we must add a unit known as the Schistes à Blocs (a sedimentary melange) gullying the top of the Annot sandstone (cf. Fig. 2). This series as a whole forms a veritable sedimentary cycle starting with an abrupt deepening (nummulitic limestone and blue marls), continuing with progressive infilling by silico-clastic turbiditic deposits (Annot sandstone in the broad sense), with definitive infilling becoming effective with the gravity emplacement of the Autapie olistostrome. The Autapie olistostrome is made up of the Schistes à Blocs as a sole

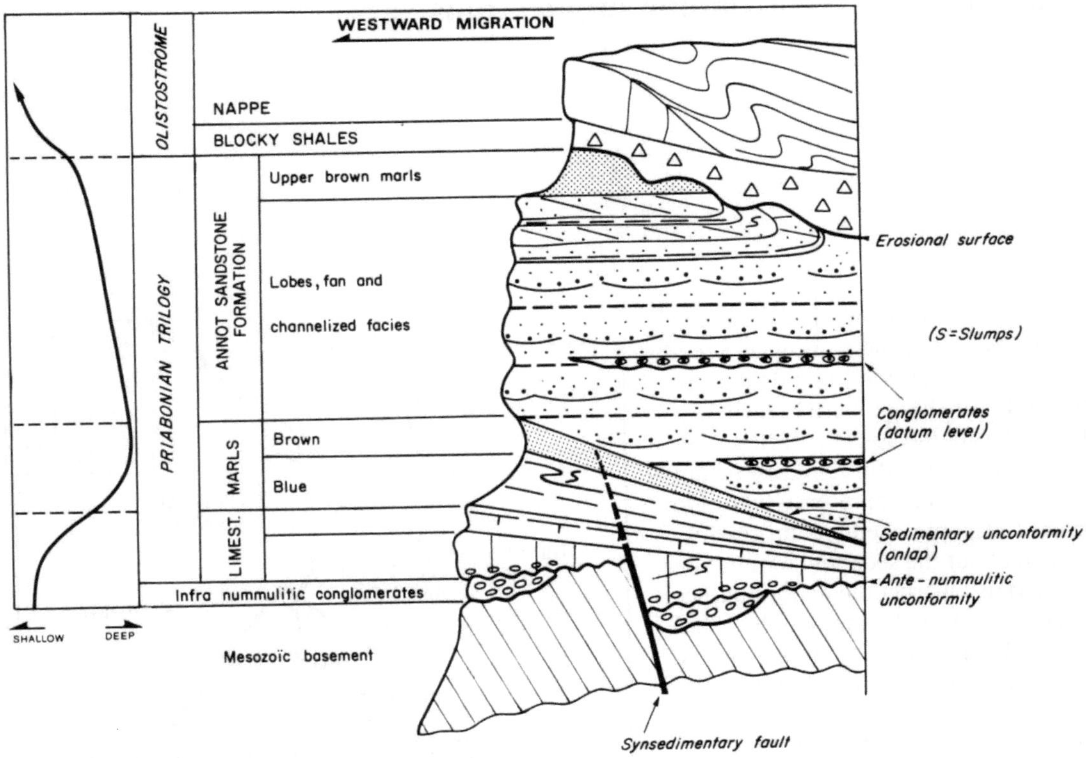

Fig. 2. Simplified section of the Priabonian trilogy (After Ravenne et al 1987)

overlain by the Helmintoid Flysch of inner origin. All these facies are diachronous, with the Nummulutic Limestone still being deposited in the western part of the basin when Annot Sandstone had already begun being deposited in the eastern part.

The Lower Limestone Term (Nummulitic Limestone, Fig. 3*)*

The shoreline migrated from east to west from the Lutetian to the Priabonian (Campredon 1977) on an already structured Mesozoic series more than 5 km thick as a result of the Pyreneo-Provençal N-S shortening (Upper Cretaceous or Paleocene). The distribution and extension of the transgression facies are governed directly by the morphology of the basin floor inherited from the tectonic structure of the underlying sediments. Note should also be taken of the set of synsedimentary accidents that are generally normal faults. Whatever the location of these faults in the basin, they were active only during the sedimentation of the Lower Limestone and the Intermediate Term.

Transgression facies were deposited at the bottom, with briny or deltaic lacustrine formations locally reworking the fluviatile facies deposited during the previous period of emersion. Above, we find littoral marine series with numerous biodetrital limestones with benthic microfauna (nummulites, discocyclines) associated with facies containing algae and polyps.

The Intermediate Term (Blue Marls)

The transgression layers gradually pass into intermediate facies in a deeper and quiet marine environment. The Blue Marls (Fig. 3) contain benthic and pelagic foraminifers and are more or less calcareous. At the top fine detrital horizons and submarine sliding structures (slumps, collapse faults) are generally present. Due to the deepening of the basin, the depocenters are located on the edges of the basin and at the foot of synsedimentary faults, while the center of the basin was in vacuity. This was the period of maximum subsidence, when the bathymetry of the deposits probably increased from 100 m at the bootom of the series to 900 m at the top (Mougin 1978).

The Sandstone Term (Annot Sandstones s. l.)

The Annot Sandstone deposit (Fig. 4) corresponds to the sudden influx of material mainly coming from the Corsican-Sardinian massif (Stanley 1961; Jean 1985) and locally from the Pelvoux range. This sudden influx seems to have resulted from the uplifting of the Corsican-Sardinian massif in relation with the rifting of the Ligurian Sea (Burrus 1984)

The first coarse detrital influxes appear in the eastern part of the basin at the bottom of the Priabonian (Campredon 1977). They are laterally accompanied by a fine "detrital cloud" mixed with the marly sedimentation at Annot (P15 or NP18 micropaleontological zone). In the Annot region, the coarse detrital material comes only at

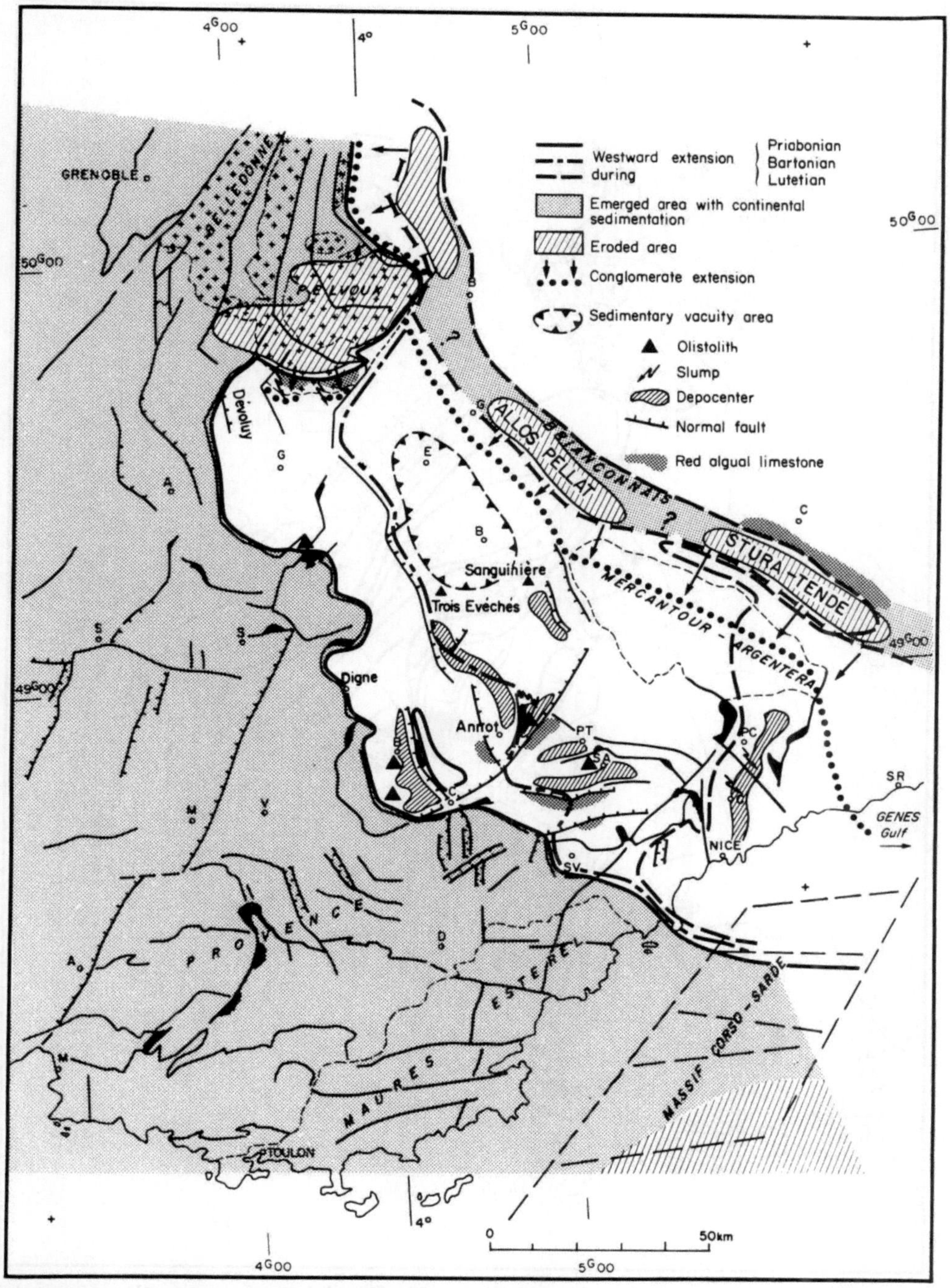

Fig. 3. Paleogeographic map of the Nummulitic Limestone and Blue Marls deposits (Ravenne et al 1987)

the end of the P16 zone (Laval and Albussaidi 1984). This facies migration corresponds to the progressive infilling of the basin from east (Contes/Peira Cava) to west (Annot) by south-north distributaries. The contact of sandstones with the underlying series is a east-to-west onlap that is clearly visible at Peira-Cava (Fig. 5 a) and at Annot and from north to south along the entire Trois Evéchés massif. The mean unconformity value is about 3°. The mapping of this unconformity associated with the correlation of the sandstone sequences proposed by Inglis et al. (1981) enables the topography of the Annot Sandstone basin to be reconstructed.

This supply of material to this basin came mainly from the Corsican-Sardinian continent via several S-N distributaries. Two distributaries have been effectively located in the Contes-Peira Cava region (Stanley 1975; Conort and Odishou 1977) and the Annot region (Stanley 1961; Laval and Albussaidi 1984; Jean 1985). Most of the sandstone term was deposited by gravity processes ranging from debris flows to more or less diluted turbidites (Stanley 1961, 1978; Normak 1978; Walker 1978; Ravenne and Beghin 1983; Ravenne et al. 1987). Figure 4 shows:

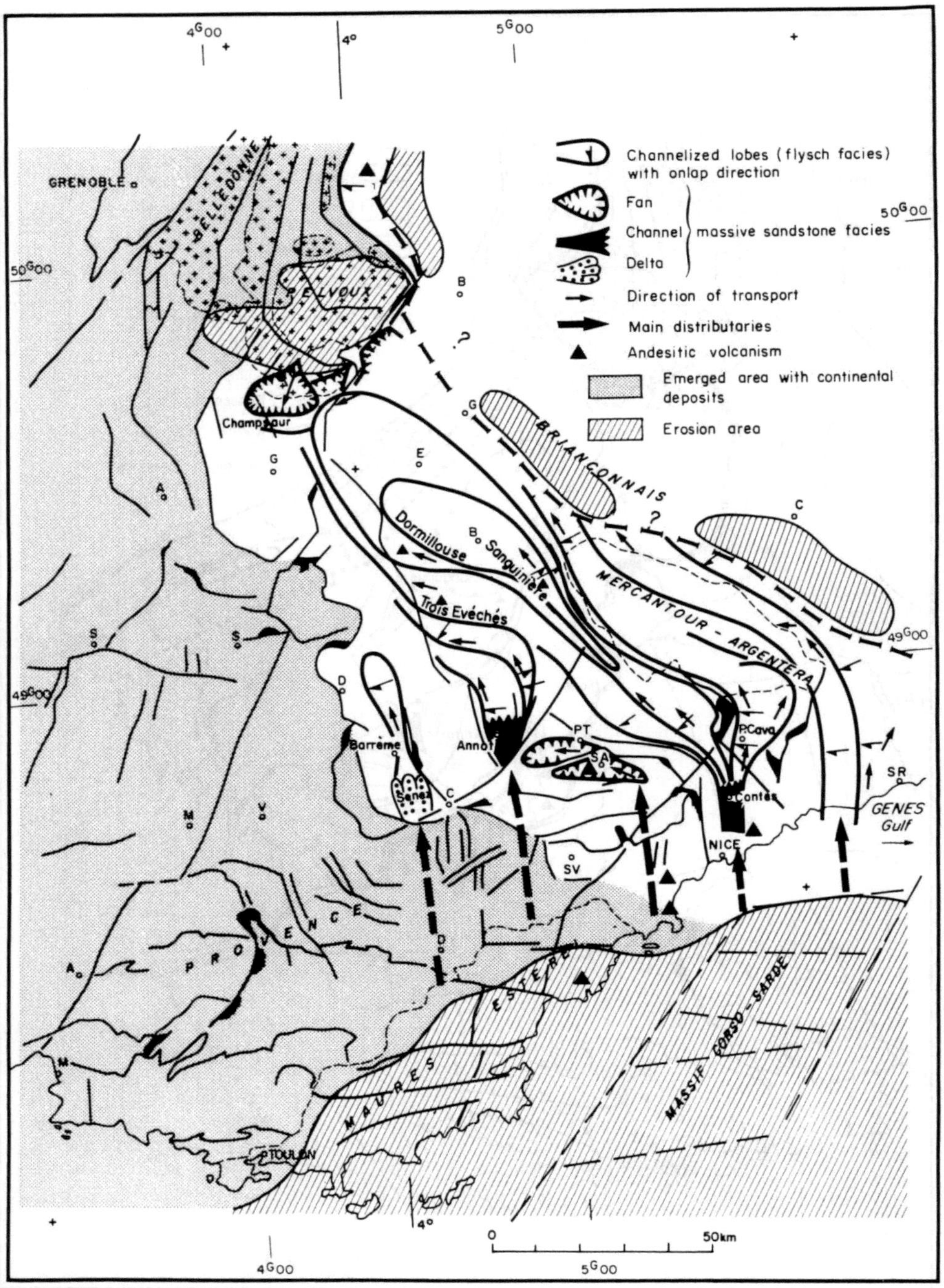

Fig. 4. Paleogeographic map of the deposition area of the Grès d'Annot formation (Ravenne et al 1987)

- Massive sands corresponding to the main distributaries from the Provençal border as well as to very low-efficiency transport fans (Champsaur Sandstone) implanted on the side of the Pelvoux range (Butin-Kiener and Calatayud 1983):
- Alternating sequences of more or less coarse sands and silts corresponding to the "lobes" of fan that remain highly channeled by the morphology of the basin.

These turbiditic deposits progressively infill the topography of the basin inherited during the emplacement of the Blue Marls.

Schistes à Blocs Sedimentation and the Autapie Olistostrome

The top of the Annot Sandstone series is affected by a large erosion surface as well as an abrupt change in sedimentation (Inglis et al. 1981; Wazi et al. 1985; Vially and Ravenne 1990) in relation to the gravity emplacement of the Autapie olistostrome.

A detailed analysis of this erosion surface shows that the transition between Annot Sandstone and Schistes à Blocs may be more continuous than appears (Fig. 6). The Annot Sandstone is less coarse and less channeling to-

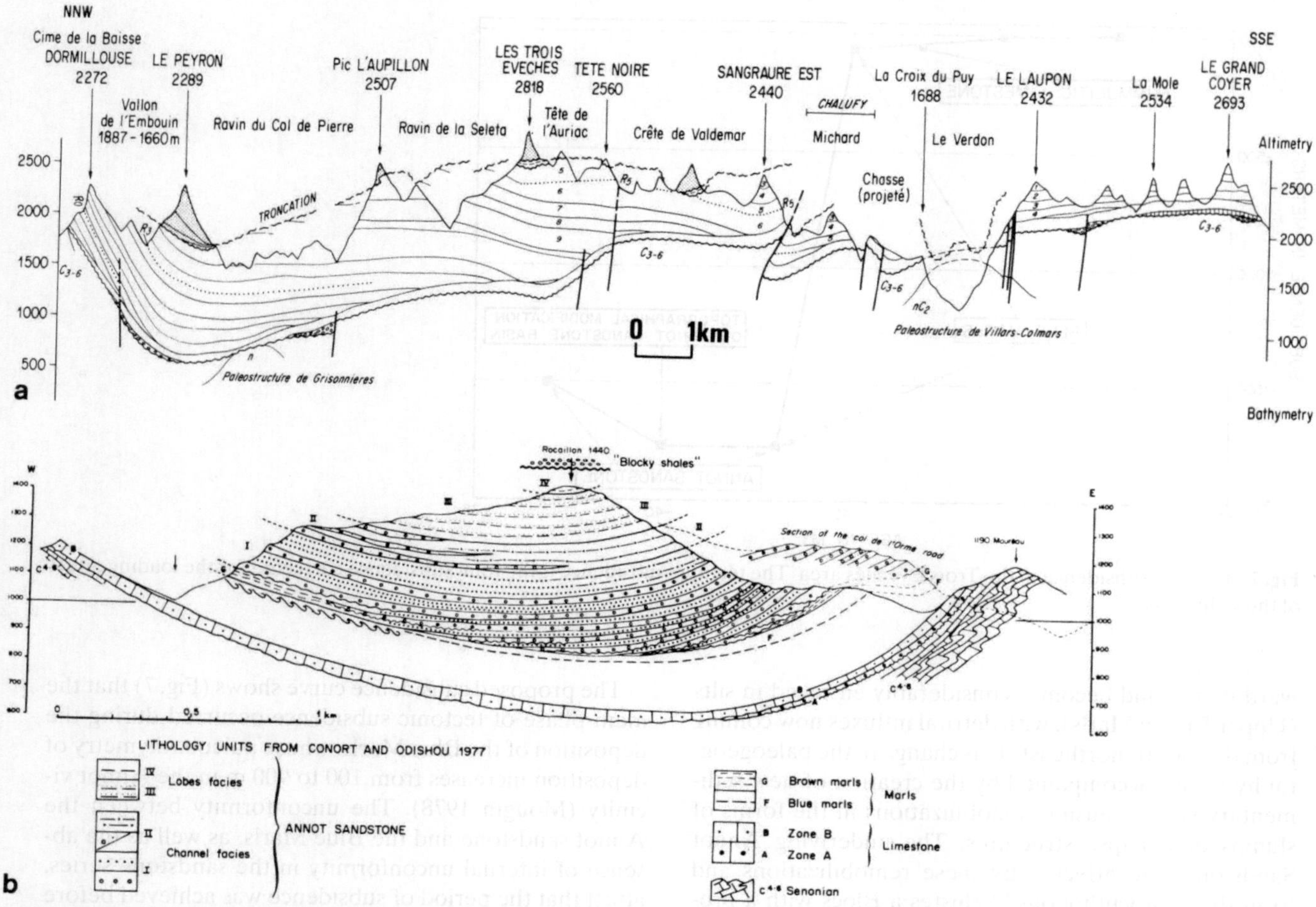

Fig. 5. a NW cross section through the nummulitic basin from Dormillouse to Grand Coyer (45 km), compiled from Le Varlet and Roy (1983) and Inglis et al (1981) **b** E-W cross section through the Peira Cava syncline (After Conort and Odishou 1977)

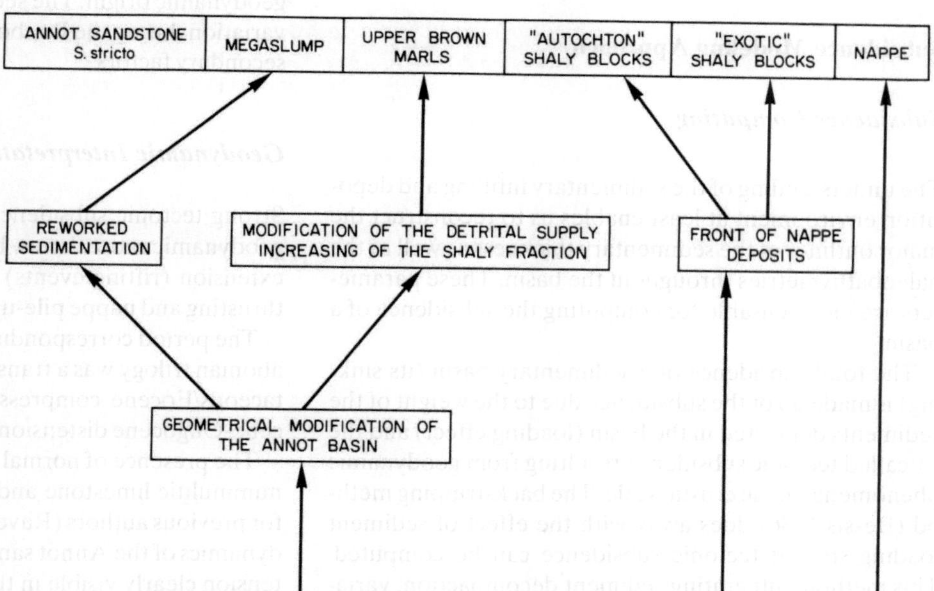

Fig. 6. The uppermost sedimentation of the Annot sandstone formation

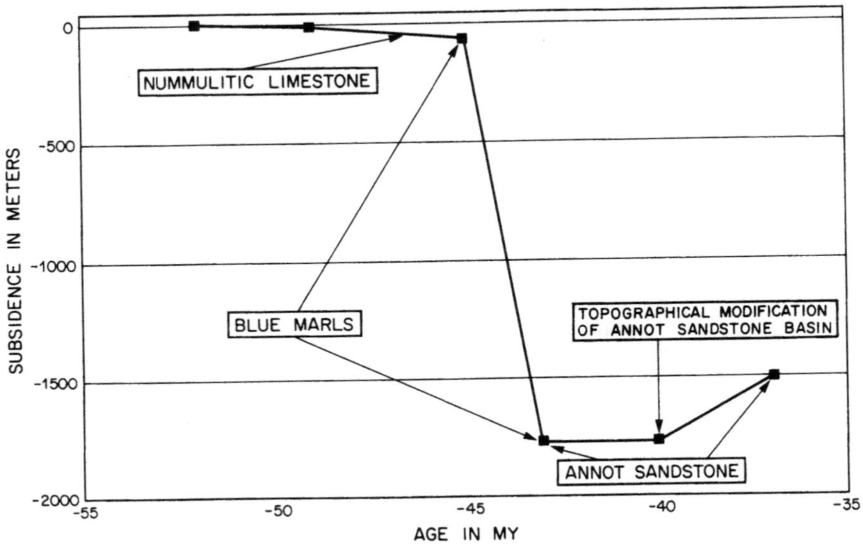

Fig. 7. Tectonic subsidence in the Trois Evéchés area The tectonic subsidence is the total subsidence corrected for the loading effect of the sediments

ward its top and becomes considerably enriched in silts (Upper Brown Marls), with detrital influxes now coming from the east or northeast. This change in the paleogeography is also accompanied by the creation of new sedimentary slopes causing remobilizations in the forms of slumps or collapse structures. The underlying Annot Sandstones are affected by these remobilizations and form the "autochthonous" Schistes à Blocs with a progressive enriching of "allochthonous" blocks of inner origin (Helminthoid Flysch) anouncing the arrival of the Autapie olistostrome.

Subsidence Modeling Approach

Subsidence Computing

The understanding of the sedimentary infilling and deposition environment at least enables us to reconstruct the major outlines of the sedimentary thickness as well as the paleobathymetries throughout the basin. These parameters are indispensable for computing the subsidence of a basin.

The total subsidence of a sedimentary basin (its sinking) is made up of the subsidence due to the weight of the sediments deposited in the basin (loading effect) and the so-called tectonic subsidence resulting from geodynamic phenomena on the crustal scale. The backstripping method (Bessis 1986) does away with the effect of sediment loading so that tectonic subsidence can be computed. This method, integrating sediment decompaction, variations in the paleobathymetry of deposits, and global eustatic variations, does not underlie any geodynamic hypothesis. We have only to choose the model of isostatic compensation. Isostasy refers to the compensation in depth for crustal loads and relief disparities.

The proposed subsidence curve shows (Fig. 7) that the main phase of tectonic subsidence occurred during the deposition of the Blue Marls, whose paleobathymetry of deposition increases from 100 to 900 m in the Annot vicinity (Mougin 1978). The unconformity between the Annot sandstone and the Blue Marls, as well as the absence of internal unconformity in the sandstone series, attest that the period of subsidence was achieved before the deposition of the thick sandstone series that infill the basin. The coincidence between tectonic subsidence and water depth (and not with sediment thicknesses and loading) shows that the main cause of the basin formation is of geodynamic origin. The sediment weight and the eustatic variation during the Priabonian (Haq et al. 1987) are only secondary factors

Geodynamic Interpretation of Subsidence (Fig. 8)

Strong tectonic subsidence does not imply any specific geodynamic cause. It may be due either to a strong crustal extension (rifting events) or to tectonic loading due to thrusting and nappe pile-up in the thrustbelt.

The period corresponding to the deposition of the Priabonian trilogy was a transition period between late Cretaceous/Eocene compression phases and the Late Eocene-Oligocene distensions.

The presence of normal faults during the deposition of nummulitic limestone and Blue Marls was an argument for previous authors (Ravenne et al. 1987) to link the geodynamics of the Annot sandstone with the Oligocene distension clearly visible in the autochthon (Camargue and Manosque basins).

However, the spatial distribution of the Paleogene flysch facies all around the Western Alps (from Switzerland to Corsica), Eastern Alps, Carpathians, Caucasus and Algerian Alpine structure in an identical structural

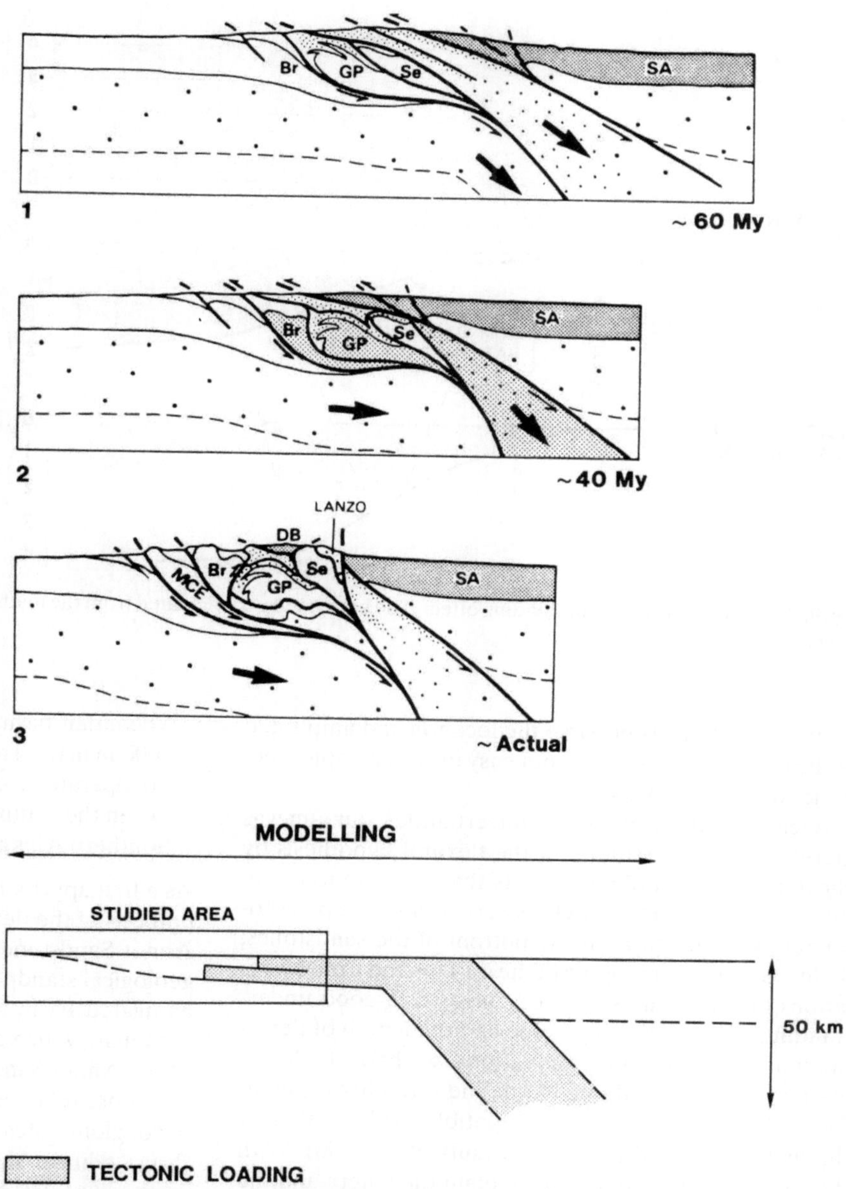

Fig. 8. Geological context of the western Alps during the Tertiary The flexural modeling has taken into account the loading effect of the *dotted area*

position suggests a purely Alpine origin. The two hypotheses will be tested.

Subsidence Modeling

The Distension Hypothesis. McKenzie (1978) proposed a uniform stretching model to explain tectonic subsidence. In this model, initial and post-rift tectonic subsidence depend on the stretching factor of the lithosphere. This model predicts the theoretical subsidence curves for various values of stretching.

In applying this model for the Southern French Alps Paleogene basin, we need a stretching factor of between 3 and 4 to explain such a tectonic subsidence. This value, that might correspond to a partial oceanization of the initial continental crust, is not compatible with the geological setting. Hence, we reject the distension hypothesis.

The Flexural Hypothesis. We have chosen a flexure model in which the vertical shear stress due to tectonic, sedimentary, and water loading is transmitted by the flexural rigidity of the lithosphere so that deformations also occur in the region adjacent to the loading area. In foreland flexural basins, numerous investigations have shown that the flexural behavior of the continental lithosphere can be modeled as the flexure of an elastic plate (Haxby et al. 1976; Watts and Ryan 1976; Steckler and Watts 1978; Beaumont et al. 1982; Desegaulx 1990). To perform this modeling, access must be gained to numerous mechanical parameters concerning the lithosphere (elastic thickness, Poisson's ratio, Young's modulus). We also have to

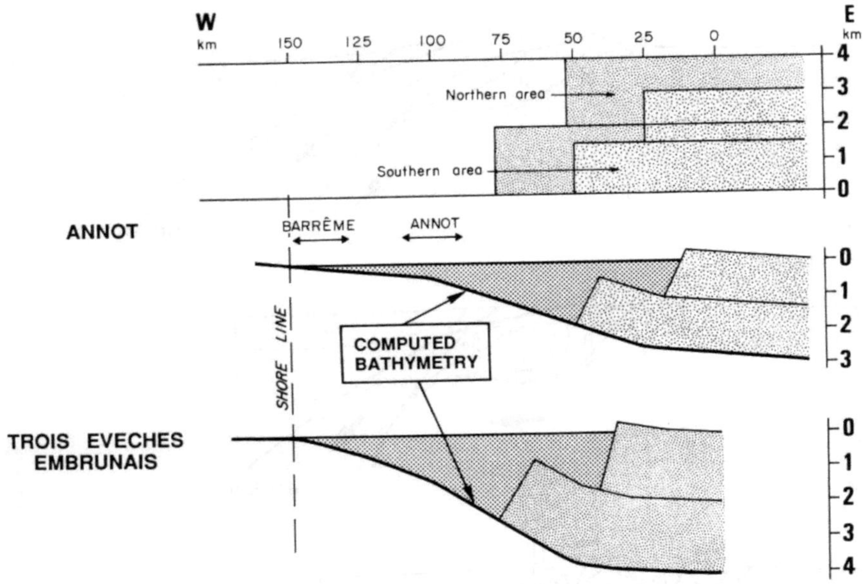

Fig. 9. Modeling of the tectonic loading effect The bathymetry is calculated from the loading deflection corrected for the thickness of the nappes

input as carefully as possible the location and amplitude of the overloading, which is not easy in such complex tectonic zones as the Alps.

Because of these numerous uncertainties, our aim was merely to test the validity of the flexural hypothesis by checking the flexural parameters that we were to use in the model to reflect the field observations. The paleotopography of the basin at the bottom of the sandstones, deduced from the mapping of the on-laps and from correlations of sandstone sequences, gives us a good understanding of the deflection value and the length of deformation. To reflect our observations, we had to take an elastic thickness of about 17 km and a rigidity of about 2.4×10^{22} Nm, which are compatible with a slightly thinned continental crust (Desegaulx et al. 1990). With this set of parameters we can explain the length and the depth of the basin along a regional cross section from Barrême to Peira Cava.

Once the modeling parameters had been set, it was then possible to compute the deflection over the entire basin and to compare the results with our data. The computed bathymetry is the result of the computed deflection of the plate corrected for the assumed thickness of the Sub-Briançonnais and Briançonnais units (Fig. 9). In this first modeling iteration we obtained an elongated north-south flexural basin that does not fit well with our data. In the next iterations with the method of trial and error, we have to modify some parameters to take field data into consideration better (Fig. 10).

- To assume an uplift of the Corsican-Sardinian block to explain the closure of the basin in the south which is an acceptable geologic hypothesis.
- To assume a decrease in the thickness of the elastic plate in the northward direction, which could indicate a thinner continental plate in the prolongation of the Vocontian basin. The maximum bathymetry is about 2000 m in the Trois Evéchés and Sanguinière area
- To operate a complete mechanical uncoupling between the southern edge of the Pelvoux range and the Southern Alps basin (Meollion fault?).

As a first approximation, we can thus affirm that the hypothesis of the flexural origin of the nummulitic basin of Annot Sandstone is plausible from the mechanical and geological standpoints. We have obtained a north-south elongated basin bordered in the south by an emerged continent, which could have provided the detrital supply of the Annot Sandstone. To the north, the Pelvoux massif, whose relative position is not known, is also a source of conglomerates and sandstones. The top of the internal thrust edifice, slightly emerged, does not constitute a great source of detrital material. Episodically and perhaps in relation with tectonic events, regional mass flows of internal origins can be recognized (Inglis et al. 1981; Jean 1985).

It is important to note that the wavelength and deflection for the same overloading is directly dependent on the flexural rigidity.

- A low flexural rigidity (stretched continental crust) produced a narrow but deep flexural basin in which we can expect to fing flysch type deposits.
- A high flexural rigidity (normal continental crust) produced a large but shallow flexural basin in which molasse type deposits can be deposited.

So, in a typical collision belt where two stretched passive margins come into collision, the first flexural stage is flysch-type (Paleogene flysch in the Alps). When the deformation migrates to the less or nonstretched external domain, the same overloading will produce a molasse-type flexural basin (Neogene molasse in the Alps).

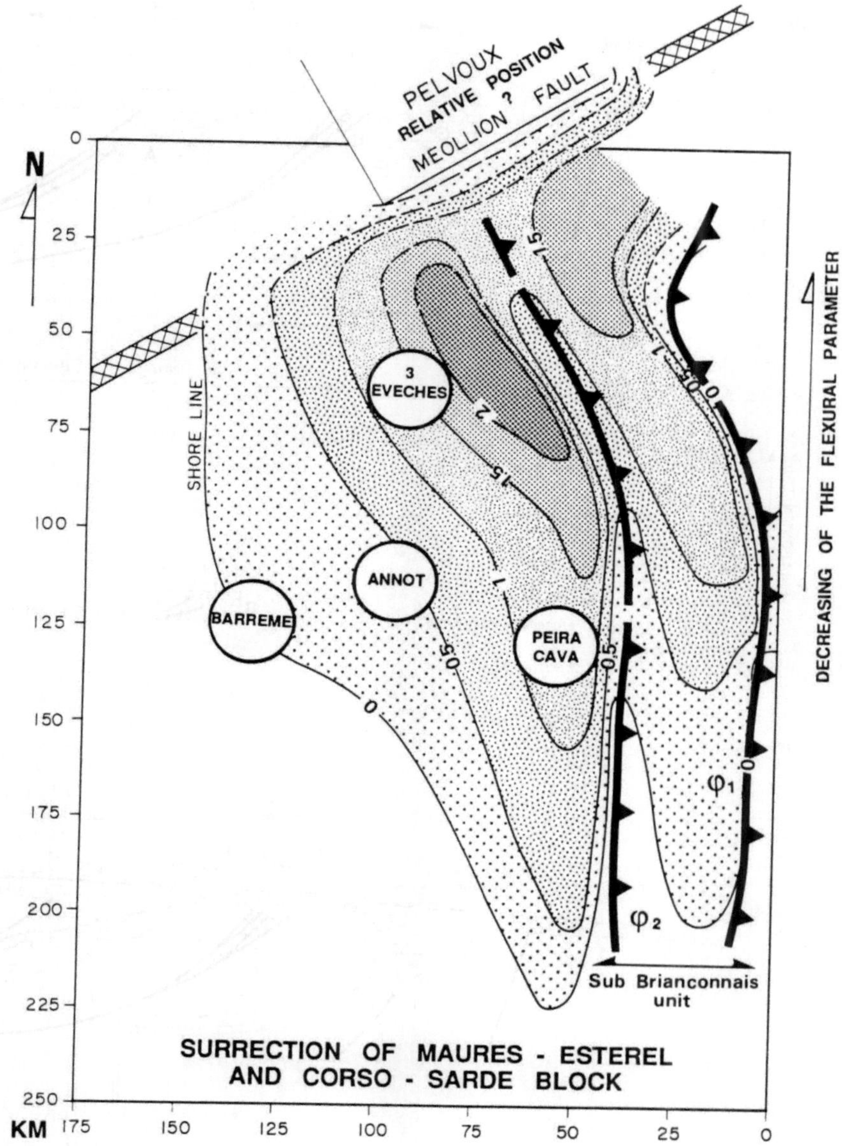

Fig. 10. Computed bathymetry during the beginning of Annot sandstone deposition Deflection corrected for the thickness of the sub-Briançonnais (Φ 1) and Briançonnais (Φ 2) nappes

Geodynamics Conclusions

By integrating regional data as well as the subsidence modeling, we can propose a schematic reconstruction of the evolution of the Paleogene basin in the southern French Alps (Fig. 11).

In relation to thrusting and loading in the internal zones, a flexural basin formed at the front of the Paleogene Alpine thrust belt and migrated progressively to the west. In an initial phase, flexuring caused normal faults that affected the sedimentation of nummulitic limestone and Blue Marls (Pairis 1987; Ravenne et al. 1987).

These faults accommodate but do not explain the very great subsidence of the basin. They stopped being active before the deposition of the Annot Sandstones, whatever their location in the basin may have been. They are thus diachronous. The Annot Sandstone, for which the material is essentially of southern origin, infills the paleotopography inherited from the flexure.

The Stura-Tende and Allos-Pellat cordilleras can be interpreted as partly emerged parts of internal overthrusting structures. The overthrusting nature of the sub-Briançonnais unit in the middle Eocene is revealed by the Saint-Clément complex and the Pinfol series, which have been interpreted by Pairis and Kerckhove (1987, 1988) as a small accretion paleoprism. If we accept this hypothesis, the unconformable position of the medio-Eocene flysches (lateral equivalent of the Annot Sandstones) on the sub-briançonnais unit of the Pellat (Blanc 1987 a) can possibly be explained as a small piggyback basin.

At the beginning of the Priabonian, the flexuring and overthrusting stopped and only an uplifting of the entire range, in relation to an isostatic readjustment, made possible the departure of the internal olistostrome, which

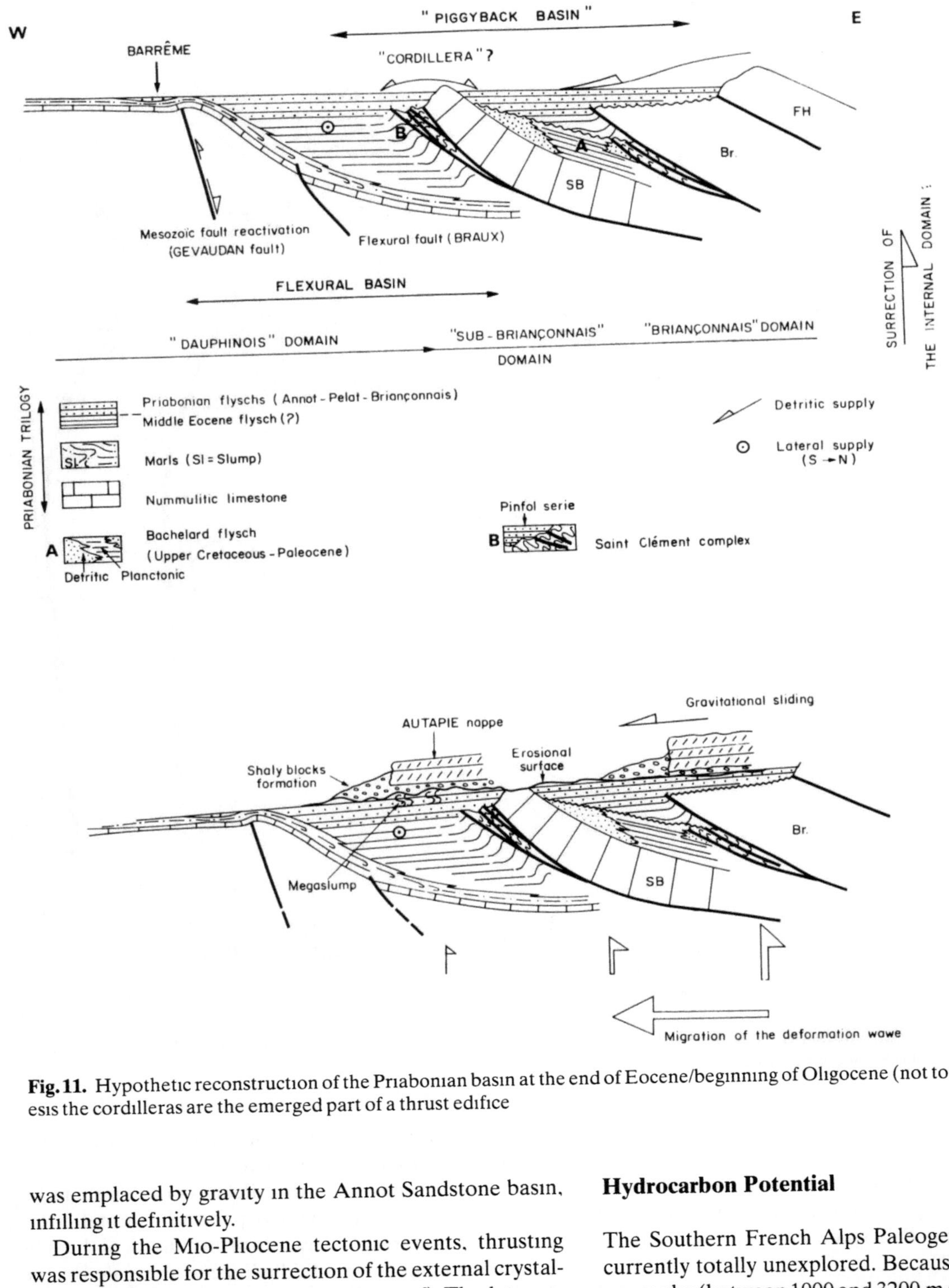

Fig. 11. Hypothetic reconstruction of the Priabonian basin at the end of Eocene/beginning of Oligocene (not to scale) In this hypothesis the cordilleras are the emerged part of a thrust edifice

was emplaced by gravity in the Annot Sandstone basin, infilling it definitively.

During the Mio-Pliocene tectonic events, thrusting was responsible for the surrection of the external crystalline massif (Argentera-Mercantour massif). This late surrection explains why the external crystalline massif does not provide detrital material for the Annot Sandstone (Ivaldi 1974; Jean 1985). The front of deformation propagated to the west and the Digne nappe was emplaced over the sub-Alpine "autochthon" domain passively carying the Southern French Alps Paleogene basin.

Hydrocarbon Potential

The Southern French Alps Paleogene basin is an area currently totally unexplored. Because of the marked topography (between 1000 and 3200 m altitude), no seismic survey has been carried out, and the structural complexity has already frightened petroleum explorationists! In addition no petroleum system[1] has been found in this basin.

[1] *Petroleum system* (Demaison and Huizingua 1991) "a 'petroleum system' is a dynamic, petroleum-generating and -concentrating system It requires the timely convergence of various geologic elements (e g , mature source rock, expulsion, accumulation) essential to the formation of petroleum deposits

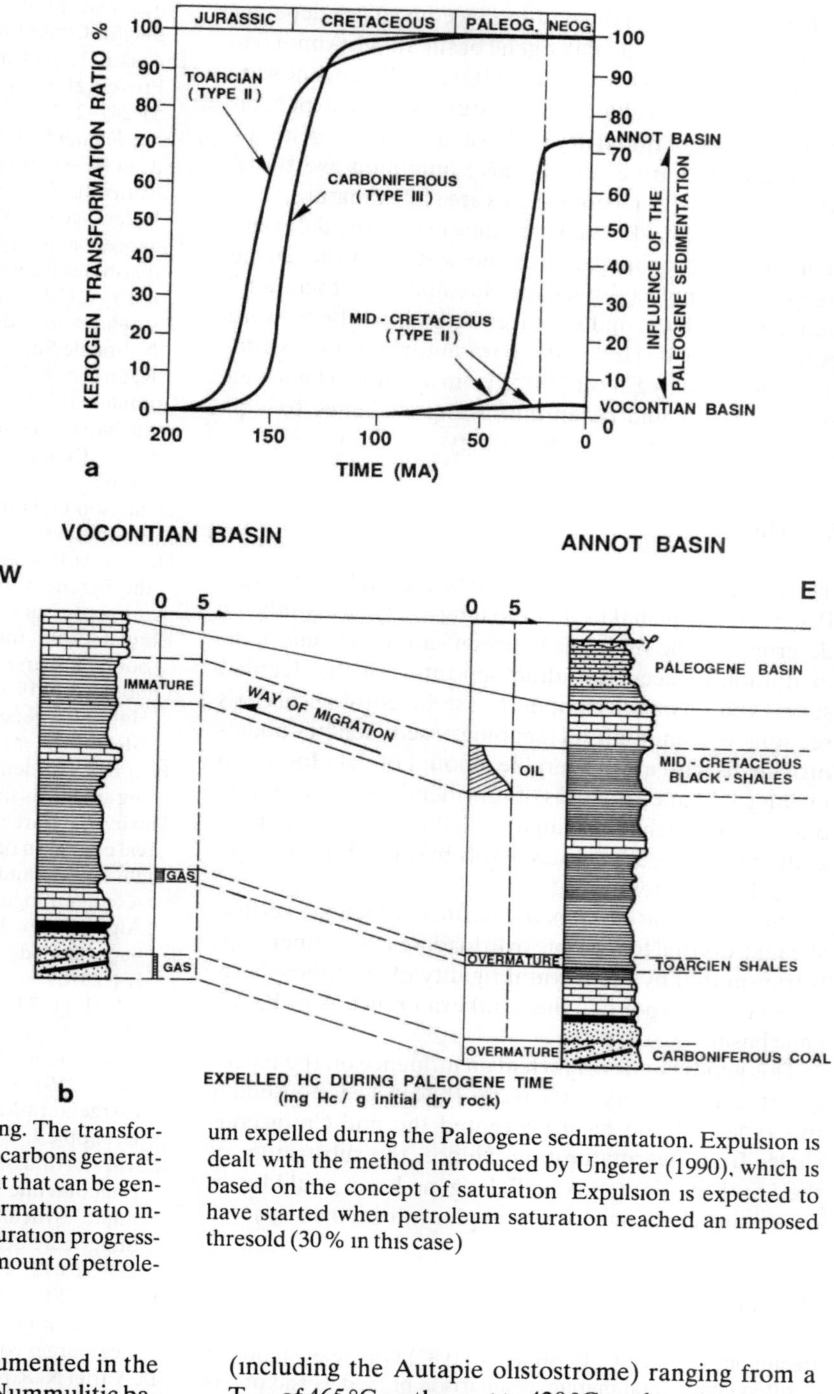

Fig. 12 a, b. Maturation and expulsion modeling. The transformation ratio is the ratio of the amount of hydrocarbons generated by primary cracking to the maximum amount that can be generated during complete evolution The transformation ratio increases from 0 to 100 % when burying and maturation progresses The expelled petroleum is the cumulative amount of petroleum expelled during the Paleogene sedimentation. Expulsion is dealt with the method introduced by Ungerer (1990), which is based on the concept of saturation Expulsion is expected to have started when petroleum saturation reached an imposed thresold (30 % in this case)

Numerous source rocks have been documented in the Paleozoic and Mesozoic basement of the Nummulitic basin.

- Stephanian coal (Type III) (Guiomar 1990) with a very high maturation level (PR = 3.5 and T_{max} near 580°)
- Liassic marls (Type II) (Tribovillard 1989) with a residual TOC ranging up to 3 % in spite of an overmaturation level.
- Mid-Cretaceous shales (Type II) (Crumière 1991; Bréheret and Crumière 1987; Bréheret, this. Vol.). Numerous "black shale" facies from Aptian to Cenomano-Turonian times contain organic-rich episodes with COT up to 2 %. The maturation of these levels depends directly on the thickness of the Paleogene series (including the Autapie olistostrome) ranging from a T_{max} of 465 °C on the east to 420 °C on the west.

Nevertheless, the formation of a Paleogene foreland with a deposit of more than 2000 m of sediments (including the Autapie olistostrome) has influenced the maturation of the potential source rocks of the underlying Mesozoic and Paleozoic series. By using a 1-D numerical simulation program (GENEX), which integrates subsidence, thermal reconstitution, maturity and compositional petroleum generation and expulsion (Fig. 12), it is possible to evaluate the maturity of the potential source rocks.

The maturation of the Stephanian coals and Liassic marls occurred during late Jurassic and early Cretaceous time and was totally completed during the Paleogene.

The maturation of the mid-Cretaceous shales depends on the thickness of the Paleogene basin In the Annot basin, the deposition of more than 1500 m of Paleogene sediments was responsible for the maturation and expulsion (about 5 mg/g of initial dry rock) of oil. This oil, generated during flexuring, could have migrated westward where the mid-Cretaceous shales are still immature.

During the Mio-Pliocene tectonic events the deformation front was propagated to the west, and the Digne nappe was emplaced over the Subalpine "autochthon" domain in which mid-Cretaceous hydrocarbons were perhaps trapped Hence the para-autochthonous structures, deduced by Elliott (1985) from a regional balanced cross section could be an interesting challenge for the economic interest of Alpine geology.

Conclusions

The quality of the outcrops in the Southern French Alps Paleogene basin makes them an interesting case study for determining the relations between subsidence and sedimentation of deep turbiditic deposits. All the detailed surveys based on the compiling of sedimentological cross sections, correlation, and mapping of sedimentary bodies provide us with a considerable amount of data for use in making a detailed analysis of subsidence. This pluridisciplinary approach to the analysis led to the reliable integration of the geodynamics of this basin within the Alpine geologic context

The hypothesis of a flexural basin, in which the vertical shear stress due to tectonic overloading in the inner Alps is transmitted by the flexural rigidity of the lithosphere, seems to correspond to the Southern French Alps Paleogene basin.

This geodynamic origin had an influence on the Paleogene thermal history of the basin; the low heat flow under this type of basin have prevented the mid-Cretaceous shales from overmaturation. Hence, the substratum of the Southern French Alps Paleogene basin could be an interesting challenge for new exploration in the Alps

References

Beaumont C, Keen C, Boutilier R (1982) On the evolution of rifted continental margin comparison of models and observations for the Nova Scotia margin Geophys J R Astron Sci 70 667–715

Bessis F (1986) Some remarks on subsidence study of sedimentary basins application to the Gulf of Lion margin (W Mediterranean) Mar Pet Geol 3, 1 37–63

Blanc C, Jean S, Kerckhove C, Pairis JL, Perriaux J (1987a) Les flyschs éocènes des unités sub-briançonnaises méridionales des nappes de l'Ubaye Géol Alp Mém hors-série 13 357–371

Blanc C, Pairis JL, Kerckhove C, Perriaux J (1987b) La formation du flysch du Bachelard (Néocrétacé-Paléocène) dans l'Unité du Pelat Géol Alp Mém hors-série 13 273–283

Boussac J (1912) Etude stratigraphique sur le Nummulitique alpin Mémoire de la Carte Géologique de France

Breheret JG, Crumiere JP (1987) Episodes à accumulation de matière organique dans les faciès pélagiques du bassin vocontien (SE France) au Crétacé moyen Livre des résumés p 80/81, Congr Français Sédimentologie, 1987

Burrus J (1984) Contribution to a geodynamic synthesis of the Provençal basin (North western mediterranean) Mar Geol 55 247–267

Butin-Kiener M, Calatayud P (1983) Analyse sédimentologique de la série nummulitique du Champsaur Diplôme de l'Ecole Nationale Supérieure des Pétroles et des moteurs (ENSPM) Référence IFP 31 437

Campredon R (1977) Les formations paléogènes des Alpes maritimes franco-italiennes Mém hors-série Soc Géol Fr 9

Conort A, Odishou A (1977) Etude de l'éventail sédimentaire sub-aquatique de la région de Peira Cava Diplôme de l'Ecole Nationale Supérieure des Pétroles et des moteurs (ENSPM) Référence IFP 26 362

Crumiere JP (1991) Les potentialités pétroligènes des dépôts du bassin vocontien oriental (Sud-Est de la France) à la limite Cénomano-Turonien Thèse 3° cycle Université de Lyon

Demaison G, Huizingua B (1991) Genetic classification of petroleum systems AAPG Bull 75, 10 1626–1643

Desegaulx P, Roure F, Villien A (1990) Structural evolution of the Pyrenees tectonic inheritance and flexural behavior in the continental crust Tectonophysics 182, 3/4 211–225

Elliot T (1985) International Symposium of foreland basin, Fribourg Excursion guidebook

Guiomar M (1990) Le Carbonifère des chaînes subalpines de Haute-Provence (France) dans son cadre structural Doc BRGM 188

Haq BU, Hardenbol J, Vail PR (1987) Chronology of fluctuating sea levels since the Triassic Science 235 1156–1167

Haxby WF, Turcotte DL, Bird JM (1976) Thermal and mechanical evolution of the Michigan basin Tectonophysics 36 57–75

Inglis I, Lepvraud A, Mousset E, Salim A, Vially R (1981) Etude sédimentologique des grès d'Annot (région de Colmars les Alpes et du col de la Cayolle) Diplôme de l'Ecole Nationale Supérieure des Pétroles et des moteurs (ENSPM) Référence IFP 29765

Ivaldi JP (1974) Origine du matériel détritique des séries des "Grès d'Annot" d'après les données de la thermoluminescence Géol Alpine 50 75–98

Jean S (1985) Les grès d'Annot au NW du massif de l'Argentera-Mercantour Thèse 3° cycle Université de Grenoble

Laval A, Albussaidi S (1984) Nouvelles observations de la série priabonienne Evolution latérale en relation avec la tectonique (synclinal d'Annot) Diplôme de l'Ecole Nationale Supérieure des Pétroles et des moteurs (ENSPM) Référence IFP 32 677

Lemoine M (1984) La marge occidentale de la Téthys Ligure In Boilot G, Montadert L, Lemoine M, Biju-Duval B (eds) Les marges continentales Masson, Paris, pp 157–251

Le Varlet X, Roy JP (1983) Etude de la série priabonienne de la région de la vallée de l'Ubaye, les Trois Evéchés Diplôme de l'Ecole Nationale Supérieure des Pétroles et des moteurs (ENSPM) Référence IFP 32 428

Mc Kenzie D (1978) Some remarks on the development of sedimentary basins Earth Planet Sci Lett 40 25–32

Mougin F (1978) Contribution à l'étude des sédiments tertiaires dans la partie orientale du synclinal d'Annot Thèse 3° cycle, Grenoble

Normak WR (1978) Fan valleys, channels and depositional lobes on modern submarine fans characters for recognition of sandy environments AAPG Bull 62, 6 912–931

Pairis JL (1987) Les domaines de sédimentation paléogènes dans le Sud-Est français Géol Alp Mém hors-série 13

Pairis JL, Kerckhove C (1987) Le flysch de Saint-Clément (Haut-Embrunais) un paléoprisme d'accrétion nummulitique dans la zone sub-briançonnaise Géol Alp Mém hors-série 13: 283–299

Pairis JL, Kerckhove C (1988) La cordillère du lac d'Allos Géol Alp Mém hors-série 14 155–163

Ravenne C, Beghin P (1983) Apport des expériences en canal à l'interprétation sédimentologique des dépôts de cônes sous-marins Rev Inst Fr Pétrole 38, 3 279–297

Ravenne C, Vially R, Riche P, Tremolieres P (1987) Sédimentation et tectonique dans le bassin marin Eocène supérieur-Oligocène des Alpes du sud Rev Inst Fr Pétrole 42, 5 529–553

Sinclair HD (1992) Turbidite sedimentation during alpine thrusting the Taveyannaz sandstone of eastern Switzerland Sedimentology 39 837–856

Stanley D (1961) Etude sédimentologique des grès d'Annot et de leurs équivalents latéraux Thèse Grenoble et rapport IFP 6 821

Stanley D (1975) Submarine canyon and slope sedimentation (Grès d'Annot) in the French maritime Alps In 9° Congr Int Sédimentologie, Nice

Stanley D (1978) Coarse sediments transport by mass-flow and turbidity current processes and downslope transformations in Annot Sandstone canyon-fan valley system In Stanley D, Kelling M (eds) Sedimentation in submarine canyons, fans and trenches Dowden, Hutchinson and Ross, Stroudsberg, Pennsylvania, 359 pp

Steckler MS, Watts AB (1978) The Gulf of Lion subsidence of a young continental margin Nature 187 425–429

Tribovillard NP (1989) Contrôle de la sédimentation marneuse en milieu pélagique semi-anoxique Exemple dans le Mésozoique de SE de la France et de l'Atlantique Doc Lab Géol Lyon 109

Tricart P (1980) Tectoniques superposées dans les Alpes Occidentales au sud du Pelvoux Evolution structurale d'une chaine en collision These d'état, Strasbourg, 407 pp

Ungerer P (1990) State of the art of research in kinetic modelling of oil formation and expulsion Adv Organic chem 1989 Org Chem 16, 1–3 1–25

Vially R, Ravenne C (1990) Sedimentation and geodynamics of the grès d'Annot foreland basin (southern part of the western Alps, France) AAPG Ann Convention, San Francisco, AAPG Bull 74, 5 785

Walker RG (1978) Deep-water sandstone facies and ancient submarine fans models for exploration for stratigraphic traps AAPG Bull 62, 6 932–966

Watts AB, Ryan WBR (1976) Flexure of the lithosphere at continental margins Tectonophysics 36 25–44

Wazi N, Gigot P, Lanteaume M (1985) Découverte de "Schistes à Blocs" au sommet de la série paléogène du synclinal de Peira Cava (Alpes maritimes) conséquence sur l'évolution tectono-sédimentaire Bull Soc Géol Fr I,4 491–494

The Mid-Cretaceous Organic-Rich Sediments from the Vocontian Zone of the French Southeast Basin

J.-G. Bréhéret

Abstract. During Cretaceous time, the Vocontian basin, which is a part of the French "Bassin Subalpin Méridional", was characterized by a pelagic sedimentation more or less diluted by fine siliciclastic supplies. In the series, several intervals display organic-rich sediments which are distributed from the Upper Hauterivian to the Lower Turonian.

At least, certain organic-rich horizons or intervals appear to constitute local records of widespread oceanographical phenomena. These have been called oceanic anoxic events (OAE) by Schlanger and Jenkyns (1976), and this term has subsequently been used by a number of authors. However, the concept of OAE suffers from a certain vagueness as, for instance, it applies to relatively short-lived episodes (as for the Cenomanian-Turonian event) as well as to periods of more than 20 million years as for Aptian-Albian. More recently, a distinction was made of "subevents" (OASEs) in this last stratigraphical interval (Arthur et al. 1990). The accuracy of this terminology is discussed, thanks to the detailed study of the Aptian-Albian Marnes Bleues Formation.

This Marnes Bleues Formation displays a variety of organic-rich sediments. They are recognized as belonging to four types according to their fabric (essentially degree of lamination) and to their organic content (as total organic carbon or TOC), following the example of Bralower and Thierstein (1984). They correspond to class 1 to class 4, from the poorest (< 0.7% TOC) to the richest ones (usually > 2.5% and up to 8% TOC). The arrangement of the organic-rich layers in the series is not at random, but corresponds to an organization into a hierarchy. The study of this organization relies on the identification of sedimentary sequences at different scales.

The Marnes Bleues are composed, for the most part, of the stacking of sequences of different orders which are superimposed. A banding is mainly distinguished thanks to a weak contrast in color (from light to dark gray). It presents a very high frequency, and the couplet "dark-light" may be referred to elementary sequences (Ferry 1991). This banding is not always detectable and in several intervals it displays a complex pattern. Most of dark layers belong to class 1 and class 2. Less order sequences are depicted, which very usually group five elementary sequences. These bundles occur very regularly and are particularly well identifiable. Nearly all of them are easily correlated on sections all over the basin. The analogy of these sequences with the genetic sequences described by Cross (1988) and by Guillocheau (1991) in other contexts is strong enough to suggest that they may be another expression of the same kind of sedimentary units generated by fluctuations in relative sea level. In several intervals (as the Lower Albian), these sequences are punctuated by black shales belonging to class 3 and often grouped by two or three. The crushing majority of organic-rich beds occur in such a way, and thus are typically periodites. The frequency (estimated within the limits of the stratigraphical resolution) of these sets of five elementary couplets suggests a possible consequence of variations in Earth's orbital eccentricity. The close study of the position of organic-rich beds within the genetic sequences allows one to distinguish two types. A first type, "condensed", occurs more or less in phase with carbonate-rich layers. It is interpreted as formed during lowerings in detrital supply, corresponding to episodes of small rises in relative see level and/or dry climatic phases. A second type, "fed", is in opposite phase with carbonate-rich layers. It is better explained by small falls in relative sea level and/or wet climatic phases. In any case, these black-shale periodites correspond to a normal recurrent sedimentation which lasts during million years. As a whole, they represent the consequence of long-term trends in global climatic and oceanographic parameters. However, at present, no accurate inter-basinal correlations have been made for individual beds.

On the other hand, several horizons are particularly noteworthy firstly because of their high organic content, secondly because they break the regular stacking of the basic sedimentation. They are composed of thin sequences with contrasted facies which record large fluctuations in oceanographical conditions. These beds are rare and aperiodic, and they constitute key beds at the scale of the Vocontian basin. Furthermore, these markers are recognized also in different basins at regional or even at global scale. They may be viewed as records of real events. They come within the framework of low-order sequences (third or second order according to the schemes of Vail et al. 1977). They correspond to the condensation of a whole genetic sequence or even of several, and are the result of episodes of major flooding. Their genesis was only possible because of the propensity for oceanic water to suboxia or anoxia.

Introduction

The huge accumulation of black shales in the mid-Cretaceous marine strata of most parts of the world has been emphasized by numerous studies since their early discovery in deep sea sediments (Lancelot et al. 1972). As underlined by Schlanger and Jenkyns (1976), Jenkyns (1980), and Arthur et al. (1990), these organic carbon-rich shales appeared to be confined to particular stratigraphic horizons. The so-called oceanic anoxic events (Schlanger and Jenkyns 1976) correspond to periods during which the oceanographic conditions on the global scale were favorable to the trapping of organic matter in the sediments. As recalled by Arthur et al. (1990, p. 78), "the term 'OAE' was originally defined to signify time envelopes during which black shale deposition was particularly prevalent." As such, "any period of time can be classified as an event depending on the applied stratigraphic resolution." So two "OAEs" were initially distinguished in the Cretaceous: the first (OAE 1) during the Barremian-Aptian-Albian interval, with OAE 2 corresponding to the Cenomanian-Turonian boundary. Subsequently, a third OAE was recognized in the Coniacian-Santonian interval (Arthur and Schlanger 1979, Jenkyns 1980). However, an ambiguity concerning the proper sense of the term OAE, because of large differences in duration, brought Arthur et al. (1990, pp 78–88) to distinguish in the first OAE (about 20 million years) three "subevents": OASEs 1 A, 1 B, and 1 C, respectively placed with- in the Early Aptian, the Early Albian, and the Late Albian. Nevertheless, a clear definition of the subevents is not given by the authors. In particular, there is no defined distinction in the hierarchical position of the Albian subevents and the Cenomanian-Turonian event.

Furthermore, Funnell (1987) discusses quite accurately the anoxic event concept from both the anoxic and event points of view. The reality of anoxia is, moreover, questioned by Pedersen and Calvert (1990) and Calvert et al. (1992). The arguments concerning these two subjects will not be repeated here. On the one hand, Tyson and Pearson (1991) have very well exposed the problems of anoxia and oxygen deficiency, and, on the other, the detailed biostratigraphical study of Bralower (1988), for instance, points to some small discrepancies in spatiotemporal distribution of oxygen depletion phenomenons for the well-known Cenomanian-Turonian event itself. Whatever some authors may have advanced (e. g., Pedersen and Calvert 1990, for instance) many indications (lithological, paleontological, geochemical) argue for the repeated occurrences of phenomena of at least severe oxygen depletion, if not true anoxia, in the mid-Cretaceous oceans. According to a number of workers (Demaison 1991; De Boer 1991), the organic richness of the sediment is above all a question of preservation which would be enhanced thanks to the oxygen depletion, at least for certain types of organic matter (Emerson and Hedges 1988). Mid-Cretaceous black shales investigated by Bralower and Thierstein (1984), for instance, have been shown to correspond to a rather low primary production recorded because of sluggish deep-water circulation.

If, as Pedersen and Calvert (1990) suggested, the organic matter content of black shales represents in all cases a consequence of a high productivity, it would rather be more accurate to speak of productivity events instead of anoxic events. However, we may suspect that high productivity was not a strict prerequisite. So it can be supposed that the conditions of primary production for all the organic-rich deposits have varied in wide proportions. Furthermore, a consequence of high production is generally a strong oxygen depletion of deep waters, leading to the rarefaction or absence of macrobenthic life and the preservation of original lamination (as recalled by Demaison 1990). In this way, it seems less wrong a priori to speak of anoxic events than of productivity events, although an expression referring to the trapping of the organic matter in the sediment would be more suitable.

However, all detailed studies demonstrate the existence of a patchy distribution or of a cyclicity expressed at different scales. So it is difficult to admit the persistence of suboxia or anoxia over a large time span, in the same way that "anoxic conditions existed throughout the global oceanic water column or even that organic carbon burial took place in every marine section during the event" cannot be attested as specified by Arthur et al. (1987, p. 401). These authors rather consider the term OAE as "a useful way of referring to time-bounded envelopes of particularly, perhaps more globally, widespread deposition of org.C-rich sediment in marine environments." If we have to know if short anoxic episodes occurred on a regional or a global scale, correlations of high resolution have to be established between sites, which remains to be done for practically all the recognized organic-rich horizons.

In fact, the cyclicity affecting most of the black-shale series is widely recognized (see, for instance, De Boer 1991). So what events with respect to cycles? Do we consider as events "all that happens" according to the definition of a French dictionary. This was the option taken by Weissert et al. (1979) in their paper entitled *Cyclic Anoxic Events in the Early Cretaceous Tethys Ocean*. According to Bralower (1988, p. 312): "Any period can be classified as an event, depending on the stratigraphic solution applied." As Seilacher recalls (1991, p. 222): "Cycles and events are relative terms that defy rigorous definition." However, "Cycles are described as sinuous curves with certain periodicities and more or less uniform amplitudes. [...] Events, in contrast, are not only aperiodic, they also occur in varying magnitudes, the larger the rarer."

In the following, using the example of the Vocontian basin, we shall try to distinguish what can be called events of organic matter trapping and to see what their significance may be.

Previous studies have shown that the small Vocontian area displays a succession of organic-matter rich horizons distributed from the Hauterivian up to the Turonian (Bréhéret 1983, 1984, 1988; Deroo et al. 1988; Ferry 1988, 1989 b; Bréhéret and Crumière 1989, Crumière 1989,

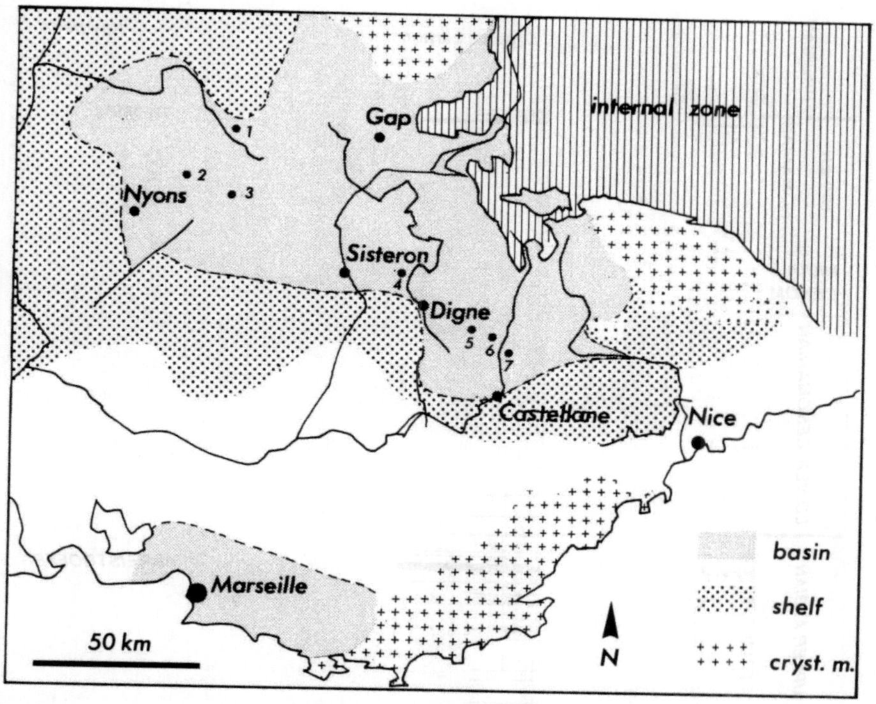

Fig. 1. Location map of the Vocontian pelagic zone of the French Southeastern Basin *Shaded area* is the pelagic zone, *dotted area* is the shelf zone Sections illustrated on Figs 6 to 14 are indicated as follows *1* Serre-Chaitieu, *2* Pré-Guittard, *3* Palluel, *4* Gaubert, *5* Les Sauzeries, *6* Les Ravels, *7* Les Briers

Magniez-Jannin 1991, 1992). The position of this basin imparts to it the role of a marginal sea in the Tethyan realm, so it can be regarded as a catchment basin receiving a noticeable terrigenous detrital input but largely submitted to the pelagic influence. In this way, it displays a series which is moderately expanded and then able to record, in spite of the local imprint, events of more general oceanographic significance. Among the variety of black shales recognized, our goal is to establish their relative importance and their potential for recording oceanographic events. After having presented the regional setting and the stratigraphic distribution of organic matter-rich beds, we focus our attention on the Aptian-Albian interval.

Material and Methods

The stratigraphic distribution of Cretaceous organic matter-rich beds from the Vocontian basin is presented here. Some data are taken from the literature (as concerns Hauterivian to Barremian and Cenomanian to Turonian intervals). However, as an example, we focus our attention on the Aptian-Albian interval upon which a detailed study is in progress. The stratigraphy of this interval rests upon works among which the most important are those of Paquier (1900), Jacob (1907), Fromaget and Varcin (1954) and Thomel (1960, 1964) for the ammonites, Moullade (1966), then Fries (1986) and Fries et al. (1986) for the foraminifera, and Thierstein (1973) for the calcareous nannofossils. In addition, the discoveries of numerous ammonites (as pyritic molds and particularly as imprints, i. e., crushed aragonitic shells) and also bivalves (inoceramids) during the fieldwork of the author have led to the improvement of the stratigraphical frame. Planktonic foraminifera have also been studied. In this respect, some data can already be found in Bréhéret et al. (1986; 1989), and a synthesis will be published in collaboration with M. Delamette. The stratigraphic boundaries (as they are indicated on the figures) are based on the ammonites which have been studied by M. Delamette (Geneva).

Numerous sections have been studied in the Aptian-Albian series, and most of them have been drawn bed by bed. Thus, precise correlations have been established between sections across the Vocontian basin at the scale of high frequency sequences (elementary sequences and genetic sequences, i. e., couplets and bundles of couplets; see below). Analysis of organic matter from bulk samples has been performed on a Rock-Eval device in the IFP Geochemical Laboratory of Rueil-Malmaison (France) thanks to J. Espitalié. Several parameters have been obtained (see Espitalié et al. 1985–1986 for theoretical and technical considerations), among them: total organic carbon content (TOC, expressed as a %), hydrogen index (HI, in mg HC/g TOC), oxygen index (OI, in mg CO_2/g TOC), maximum temperature of pyrolysis (T_{max}, as °C). Geochemical data (major and trace elements) have been obtained by quantometry in the CRPG (CNRS) Laboratory at Nancy (France). Many data have been provided by H. Brumsack (University of Gottingen, FRG). Carbonate content was calculated by calcimetry using a Bernard calcimeter.

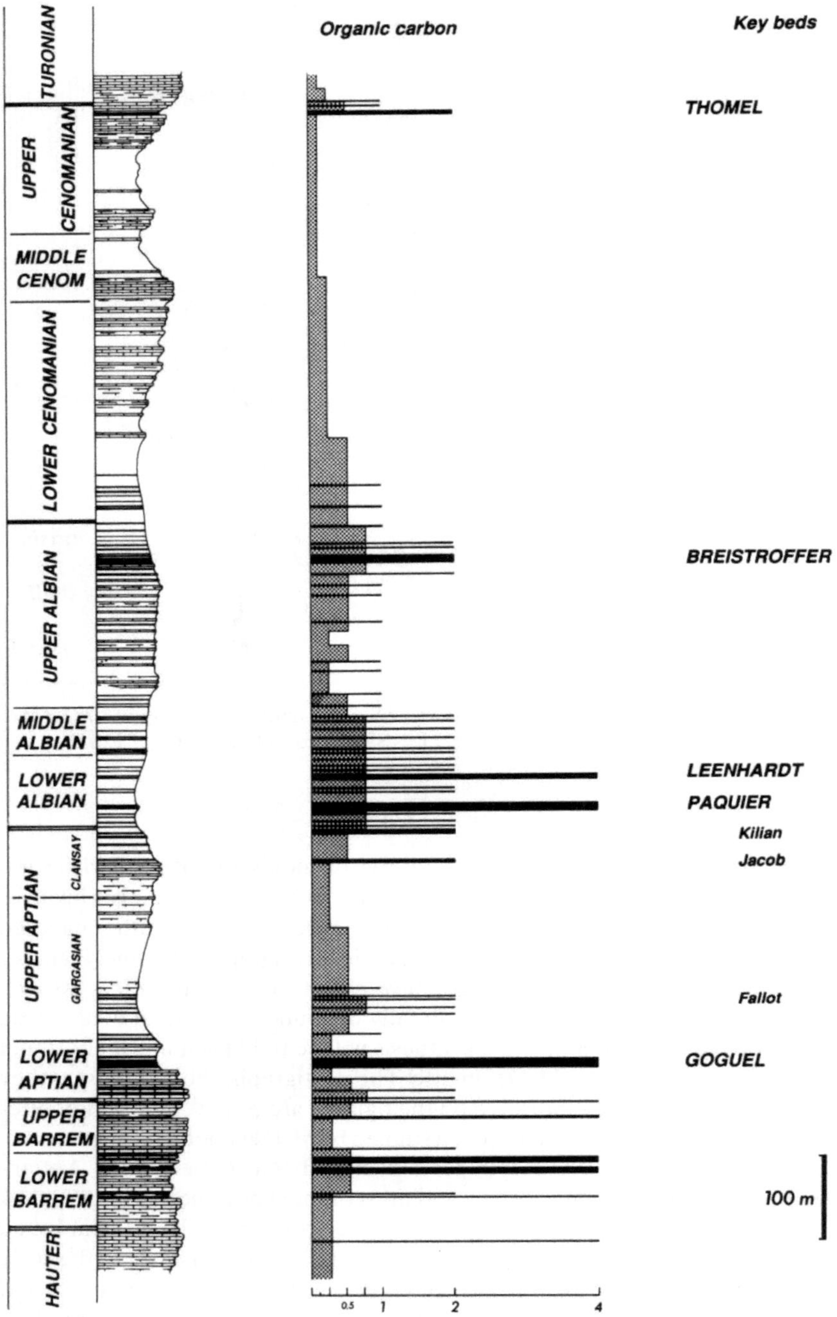

Fig. 2. A comprehensive section of the mid-Cretaceous (Hauterivian to Turonian) pelagic sediments in the Vocontian zone with the simplified organic carbon record *Horizontal bars* represent organic content of black shales, *short bars* about 1 %, *medium bars* about 2 %, *long bars* 3–4 % TOC The *thickness of the dotted area* represents the mean TOC of enclosing bioturbated marls (0 2, 0 3, 0 5, 0 7 %)

Regional Setting and Outline of the Basinal Series from Hauterivian to Turonian

In the southeast of France, the Vocontian "trough", which was named by Paquier (1900), constitutes a pelagic part of the Bassin Subalpin Méridional (Fig. 1). This area is enclosed between the Vercors massif, the river Rhône, the Ventoux-Lure axis, and the Nice-Castellane Arch. By mid-Cretaceous time, it corresponded to the western part of the northern margin of the Alpine-Tethys sea. The Vocontian area was surrounded by coral-rudistid Urgonian carbonate platforms developed from the Barremian to the Early Aptian (Arnaud-Vanneau et al. 1984). Here we give a brief outline of the basinal series.

Lithology (Fig. 2)

In the Early Cretaceous, the Hauterivian series (about 60 m) are characterized by a very regular alternation of micritic limestones and marls (see Ferry 1984, 1989a, 1991) and show a relatively constant thicknesses. Such a pattern continues in the Lower Barremian (60 m), but in its upper part the limestone beds thicken at the expense of marls. Except for a small interval called the Vire à Heteroceras, in the Upper Barremian, the limestone predominance (on about 70 m) is a typical feature up to the lower part of the Lower Aptian. This is particularly true for the base of the Lower Aptian. Marls may be reduced to joints between thick limestone beds. The pelagic limestones (with the exception of resedimented ones) are mainly constituted of nannoconids, which are rare in marly interbeds (Noel 1968). These limestones are completely bioturbated. On the contrary, certain marl interbeds are laminated.

A sharp change occurs in the upper part of the Lower Aptian: here the lime-dominated sediment gives way to a relatively homogeneous (although banded throughout) marly series almost completely devoid of pure limestone beds for more than 700 m. This series has been called Marnes Bleues by Flandrin (1963). The calcereous component increases progressively in the Upper Albian. The Cenomanian (about 400 m) consists of a succession of bundles of small calcareous beds scattered in the marls. The last marly interval in the Upper Cenomanian corresponds to the Niveau Thomel, a black-shale interval described by Crumière (1989). Then the more limey Turonian sequence (200 m) is composed of a tight alternation of limestones and marls.

Resedimentation

Abundant resedimentation characteristic of the basins structuration occurs throughout the stratigraphic interval evoked here (Ferry 1984). These gravity deposits are particularly important in the Lower Cretaceous basin submitted to extensional dynamics (Ferry 1989a). They are composed either of platform (turbidites) or of pelagic-hemipelagic material (slumps, mudflows, debris flows). Up to the Lower Aptian, the gravity deposits stemming from platforms are composed of calciturbidites; they are particularly abundant in the Barremian and early Aptian interval. Their disposition expresses the resedimentation of bioclastic sands of the peripheral Urgonian facies (Ferry 1984, 1989a). With the so-called drowning of the Urgonian platform during the Early Aptian, the products issued from it vanish abruptly and are replaced by siliciclastic supplies. Frequent turbidites recycle sands composed of quartz, mica, sporadic glaucony, and rare bioclasts. Slumps are abundant in the Aptian. Most of them come from the siliciclastic shelf and are composed of heterolithic facies (mixed mud and sand, many of them being tempestites). Pelagic slumps are sporadic and mainly found in the late Upper Aptian (Clansayesian). Slumps are very rare in the Albian (with the exception of the Vraconian) and the number of turbidites disminishes greatly. However, several thick ones (grainflows originating from shelf dunes and sandwaves) are observed in the Middle Albian and the base of the Upper Albian. A certain resumption of turbidites and slumps occurs in the Upper Cenomanian (Ferry et al. 1988).

Stratigraphic Distribution of Organic Matter-Rich Beds

Because of the dilution by carbonate (of primary sediment and of diagenetic origin) the total organic carbon content of limestone beds (> 75 % $CaCO_3$) is systematically lower than 0.2 %. So the data which lead to the establishment of the vertical distribution of organic matter-rich beds (Fig. 2) are essentially based on the total organic carbon (TOC) content of marls (i. e., "interbeds" in the intervals of limestone-marl alternations).

According to Magniez-Jannin (1991, 1992), in the Upper Hauterivian (*Balearis* and *Angulicostata* zones), a small interval, I, is identified thanks to some interbeds where the TOC, content is clearly higher (2.5 %) than the background (0.3 to 0.5 %). A second interval, II, more clearly expressed, develops in the upper part of the Lower Barremian (*Nicklesi* to *Caillaudi* zones); in several black shales the TOC content is over 4 %. A third one, III, is situated in the middle part of the Upper Barremian (*Feraudi* zone). A last interval, IV, that displays organic carbon-rich layers is found in the uppermost Barremian-base of the Lower Aptian. First indications concerning these organic-rich intervals were given earlier by Bréhéret (1984) and Deroo et al. (1988).

The most prominent is the 2nd interval (Magniez-Jannin 1992), several horizons of which display the characteristic paper-shale lithology. A first episode of organic enrichment is recorded in the middle part of the Lower Barremian, at the base of the *Compressissima* zone, and a second in the upper part of the Lower Barremian (*Moutoniceras* zone). This latter may tentatively be correlated with an organic carbon-rich interval displaying the same paper-shale lithology in several European sites: the Hauptblatterton (*Aulacoteuthis* zone) in the Lower Saxony basin (Kemper and Zimmerle 1978; Mutterlose and Harding 1987), the Munk Marl bed (from the Tuxen Formation) in the Danish Central Trough (Jensen et al. 1986; Jensen and Buchardt 1987) and the Lower B beds (LB1B–LB2C) of the Speeton Clay Formation of Yorkshire (Rawson and Mutterlose 1983). It may also be correlated, but there is more doubt, with a black-shale interval described in the Lombardian basin (cf. Vialli 1949; Weissert et al. 1979). The possible correlations with several European sites let us think that this organic-rich interval could represent an event on a regional scale.

The 3rd interval could be coeval with the Blatterton beds of the *brunsvicensis* and lower *germanica* zones of the Lower Saxony basin (Mutterlose and Harding 1987). As for the Vocontian black shales, they occur as dispersed horizons interbedded in the host sediment, which is composed of normal shales instead of limestones.

In the Central North Atlantic, as summarized by de Graciansky et al. (1987), the Hauterivian-Barremian-

Lower Aptian interval, which corresponds to the upper part of the Blake Bahama Formation, consists of the alternation of bioturbated white limestones and laminated black shales. This pattern strongly recalls that depicted in the Vocontian basin and north and central Italy.

Just above the top of the limestone-marl alternation of the Lowermost Aptian rests an organic matter-rich bed known as the Niveau Goguel (Bréhéret 1988), with the characteristics of a paper shale (about 4 m). TOC values range up to 4%. This bed belongs to the *Deshayesi* zone (Lower Aptian = Bedoulian; *G. blowi* Foram. zone). It is correlated with the well-known Livello Selli found in Umbria (Wezel 1985; Coccioni et al. 1987, 1992), and with the Fischschiefer of the Lower Saxony basin (Kemper and Zimmerle 1978, Gaida et al. 1981; Mutterlose 1992). It seems to have a very widespread distribution, since such a coeval organic-rich layer has been recognized even in the Pacific Ocean by Sliter (1989, 1992). It is recognized as the OASE 1A by Arthur et al. (1987).

The next organic matter-rich interval (Niveau Fallot, about 20 m thick) is located in the base of the Upper Aptian (Gargasian). Its TOC content is moderate (up to 2.5%). Possible correspondance to other European sites has not yet been established.

In the upper part of the Upper Aptian (Clansayesian) occurs the Niveau Jacob (Bréhéret 1983), 2.50 m thick, with numerous ammonites (*Jacobi* zone). Its TOC content is also moderate; it reaches 2.5%. An equivalent may occur in the Umbrian area (cf. Premoli-Silva et al. 1989a).

The next interval is about 150 m thick and concerns the whole Lower and Middle Albian. Several tens of horizons display a TOC content above 2%. The organic content of enclosing marls and shales is itself between 0.7 and 1% (Bréhéret et al. 1986). This set may be correlated with the Argiloso marrone et Marnoso verde units of the Umbrian Scisti a fucoidi. Two layers are particularly outstanding in this interval. The Niveau Paquier (Bréhéret 1983), very rich in ammonites (*L. tardefurcata*) with characteristics of paper shales and a TOC content generally above 3% (up to 8%), is a key bed recognized in different sites of the Alpine sea (Bréhéret 1985, 1988; Arthur et al. 1990) and which extends to Iran (Alavi-Naini 1972). It may correspond (?) to one horizon of the Livello Monte Nerone (Coccioni et al. 1989) in the Umbrian Scisti a fucoidi. The second level, called Niveau Leenhardt, about 30 m above Niveau Paquier, has a TOC content of 3%. It represents another key bed at the basin scale. It may tentatively be correlated with the Livello Urbino (?) of the Umbrian Scisti a fucoidi.

In the depocenter, the base of the Upper Albian (base of *Inflatum* zone; *B. breggiensis* Foram. zone) displays several layers moderately enriched in organic carbon (up to 2% TOC). However, in most of the basin, this interval is missing or condensed, and is expressed as a glaucony-rich layer. Recurrent black shales of the same age occur in the Scisti a fucoidi (Herbert and Fischer 1986; Pratt and King 1986; Premoli-Silva et al. 1989a). However, there is no major event in organic matter trapping such as recorded in Australia with the oil shales of the Toolebuc formation (Ozimic 1988).

Several moderately C_{org}-enriched layers are recognized in the upper part of the *Inflatum* zone and the lower part of the *Dispar* zone, i.e., the base of "Vraconian". The upper part of this substage shows another interval with frequent well-marked organic-rich layers, it is called the Niveau Breistroffer (Bréhéret 1988), and corresponds to the extent of the *P. buxtorfi* foram subzone. The TOC is moderate (up to 2%). This interval may be correlated with other sites across the world: in Tunisia (Talbi 1986), in the US-Canadian Western Interior, in the Central Pacific, and in Central Australia (see Arthur et al. 1990 for a review).

The last interval which displays a marked enrichment in organic content (up to 2% TOC) is the Niveau Thomel (Crumière et al. 1990a), characteristic of the upper part of the Upper Cenomanian and basal Turonian. The quasi-global extent of contemporary black shales does not need to be emphasized there, given the very abundant literature published on this subject.

A Hierarchy in the Nature and Sequential Position of Organic Matter-Rich Beds in the Aptian-Albian Interval

A closer inspection of the mid-Cretaceous series of the Vocontian area shows some differences between the black-shale beds relative to their stratigraphical position. These elements make it possible to determine what can be considered as an event in comparison with what cannot.

The Nature of Black Shales

Different kinds of sedimentary rocks are recognized in the series according to the proportions of their organic-carbon content (TOC), their lime content, and their clay content. This may be displayed on a ternary plot (Fig. 3A–B), the end-members of which are CaO, AL_2O_3 and TOC (× 10). A first axis appears as evident (Fig. 4A, B): it corresponds to dilution by lime from the limestones (above 75% $CaCO_3$ and very poor in TOC) to clays and marls (below 50% $CaCO_3$, relatively enriched in TOC). However, among clays and marls, another axis is characterized; it corresponds to organic enrichment. Four classes are distinguished according to these three components and mainly the TOC; they are closely linked to lithological characteristics and to paleontological parameters. They correspond to those distinguished by Bralower and Thierstein (1984).

1. Bioturbated marls and shales, light to dark, blocky, with a TOC < 0.7%.
2. Dark homogeneous marls and shales, bioturbations rarely visible and often limited to chondrites, blocky to flaggy or platy, with a TOC between 0.7 and 1.0%.

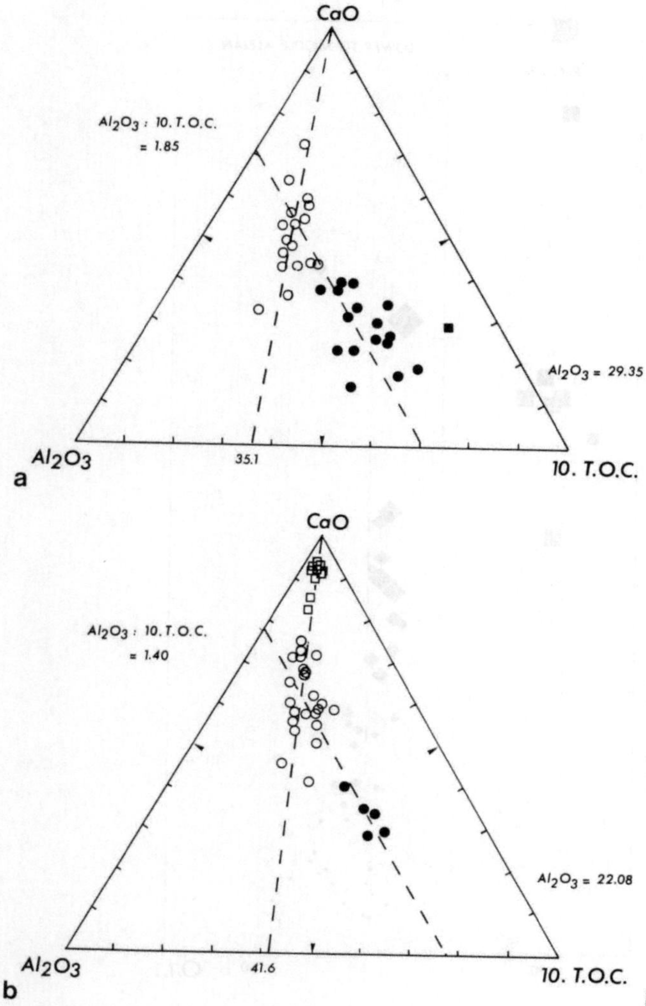

Fig. 3 a, b. Ternary plot of geochemical data (CaO-Al_2O_3-10 TOC) **a** From samples of Upper Aptian to Lower and Middle Albian (section of Pré-Guittard) **b** From samples of Middle Albian and the lower part of the Upper Albian (section of Bert) *Open squares* Limestones, *open circles* marls and shales of class 1 and 2, *black circles* marls and shales of class 3 and 4, *black squares* niveau Paquier

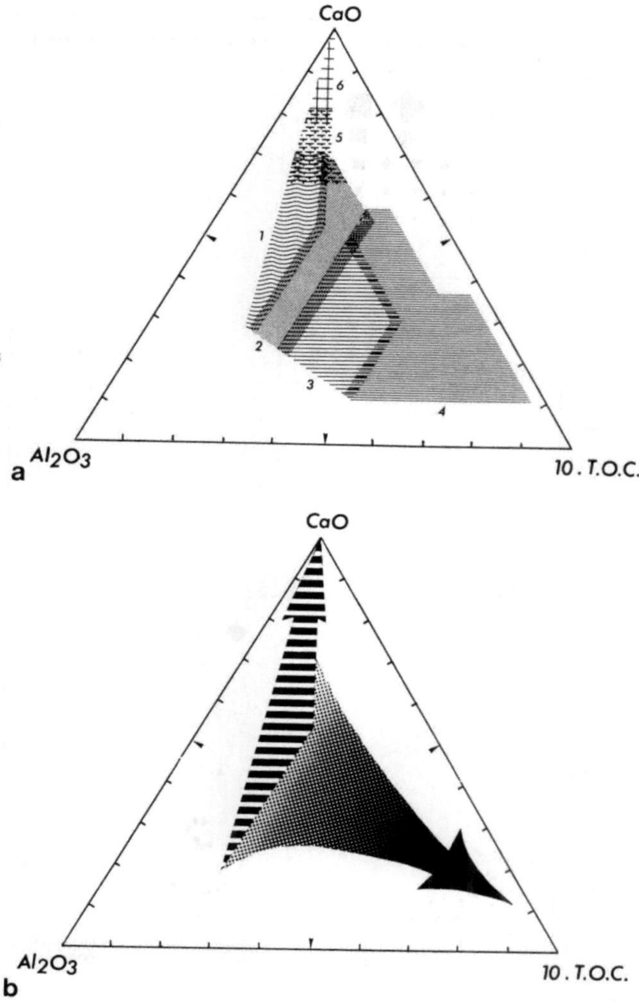

Fig. 4 a, b. Ternary plot of geochemical data (CaO-Al_2O_3-10 TOC) **a** The different fields *1* Bioturbated marls and shales, *2* homogeneous dark marls and shales, *3* faintly laminated dark marls and shales, *4* dark paper shales, *5* limey beds, *6* limestones **b** The *two arrows* corresponding to the dilution by CaO and the enrichment in TOC

3. Dark faintly laminated marls and shales, with practically no bioturbation visible, usually platy to fissile, with a TOC up to 2.5%.
4. Dark paper shales displaying a conspicuous even lamination, with a TOC up to 8%.

The analyses of organic matter shows that the enrichment in TOC is closely linked to the quality of preservation of organic matter, particularly the labile one, as has been observed in many other instances. This shows well on a Van Krevelen-type diagram (OI/HI plot, cf. Espitalié et al. 1985–1986). The section which represents Upper Aptian-Lower to Middle Albian outcropping on the Col de Pré-Guittard, in the western part of the basin, provides a good example of this (Fig. 5). The organic matter contained in the series is clearly immature, before the oil window, with a T_{max} between 415 and 435 °C (Bréhéret et al. 1986, Figs. 7, 9). In organic matter-poor rocks, HI is below 50 mg and OI is usually lower than 100 mg; the organic matter is mainly residual: oxydized marine material or recycled terrestrial material. An increase in TOC is correlated with a higher HI, and in moderately enriched sediments, dinoflagellates may be abundant. However, the richest samples are characterized more by a high content in hydrogenous amorphous organic matter (Bréhéret 1988), which adds to a background of refractory organic matter, as is testified by palynofacies. This fact is interpreted as the result of a strong oxygen depletion (suboxia) leading to less efficient microbial degradation of the sedimented planktonic matter. The disappearance or scarcity of benthic organisms (Mollusca, Forams) as well as the absence of bioturbation are good evidence (Bréhéret et al. 1986; Bréhéret and Delamette 1989).

The most elementary observations show that the bioturbated facies (shales, marls as well as limestones) are very widespread; these are the "normal marls" (although

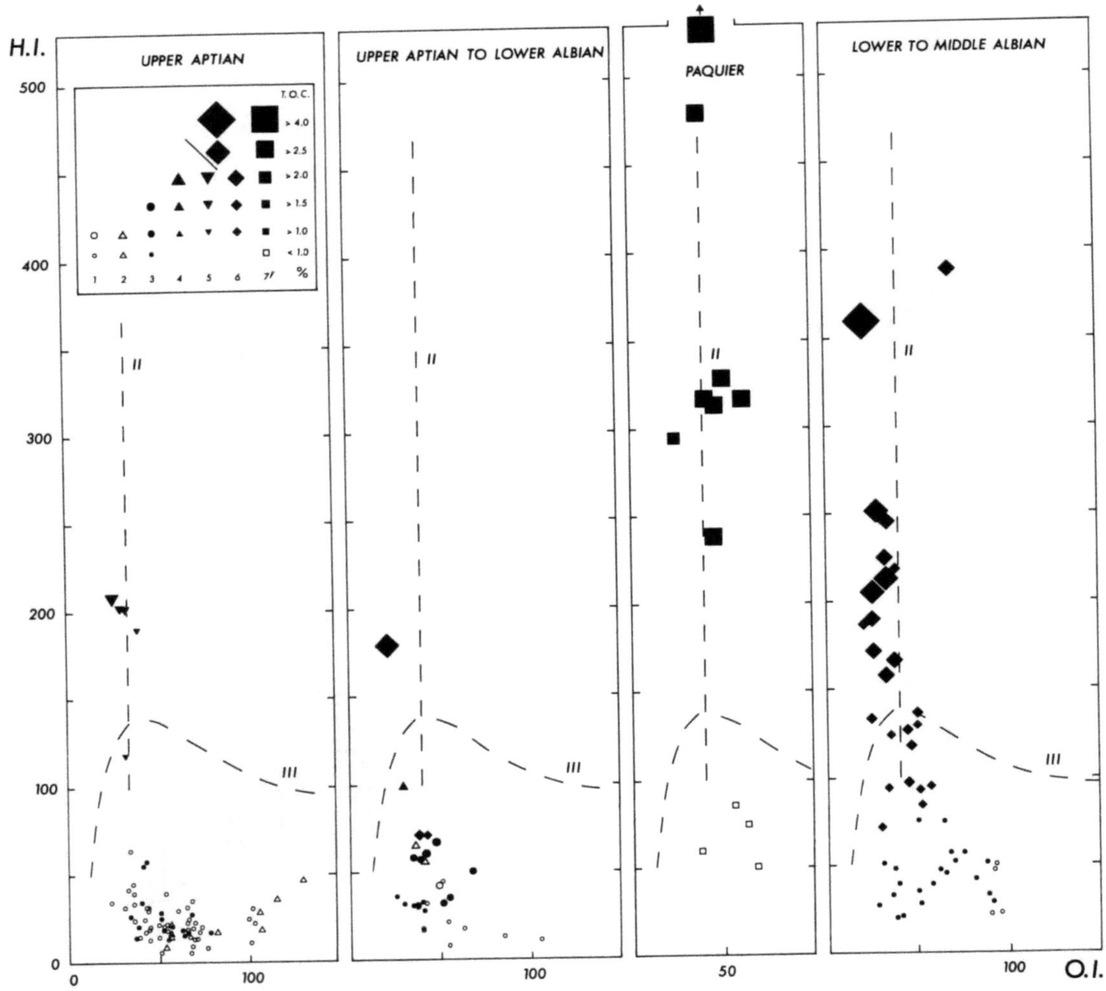

Fig. 5. Types of organic matter of sediments from Upper Aptian and Lower-Middle Albian from the Pré-Guittard section. Samples are plotted on four OI/HI Van Krevelen type diagrams. Each one corresponds to a part of the section. *1* Upper Aptian with mostly bioturbated marls. *2* Top of Aptian and base of Lower Albian formed by bioturbated marls and the first levels of black shales periodites. *3* Lower Albian Niveau Paquier formed of paper shales. *4* upper part of Lower Albian and base of Middle Albian corresponding to the main interval of black-shale periodites. The meaning of the symbols is as follows: *1* bioturbated marls. *2* bioturbated limey beds. *3* homogeneous dark marls. *4* black shale of the Niveau Kilian. *5* black shale of the Niveau Jacob. *6* black shales of the Lower to Middle Albian periodites including the Niveau Leenhardt. *7* black shales of the Niveau Paquier (*open squares* correspond to intercalated bioturbated horizons). The stratigraphic position of the key beds is given in Fig. 2

in many cases ichnoforms are indistinct, the sediment being mottled). The dark homogeneous marls are rather frequent, especially in the basal part of the Upper Aptian (Gargasian) and in the Lower to Middle Albian. The low-grade laminated marls and shales are sporadic (about 5 % of the series considered here); usually they occur in a cyclic pattern. The high grade paper-shales are exceptional (< 1 % of the series). Geochemical logs of the type sections (Serre Chaitieu, Gaubert, Pré-Guittard, and Col de Palluel) are shown in Figs. 6 to 9.

The Position of Black Shales

The whole Marnes Bleues series investigated appears to be formed by the alternation of two types of layers with a thickness varying from one to several decimeters: either couplets of darker and lighter bands in marly intervals (Fig. 10) or – more rarely – limestone-marl alternation. These couplets correspond to elementary sequences according to the terminology of Ferry (1991, pp 56–57, Fig. 9), who places them at the level of the sixth order (or the fifth order following Vail et al. 1991).

These elementary sequences are themselves clustered in fours or more frequently fives. These vertically stacked units (normally 1 to 5 m thick) are often quite conspicuous and generally easily correlated all over the basin whatever the lithology, provided it is pelagic. The striking steadiness in the succession of these small units is highly suggestive that it consists of the record of a periodic signal of allocyclic origin, as is noted for a number of other Mesozoic intervals in the Vocontian basin (Ferry 1991). Furthermore, the bundling of five couplets at this scale appears extremely widespread in geological series, as emphasized by

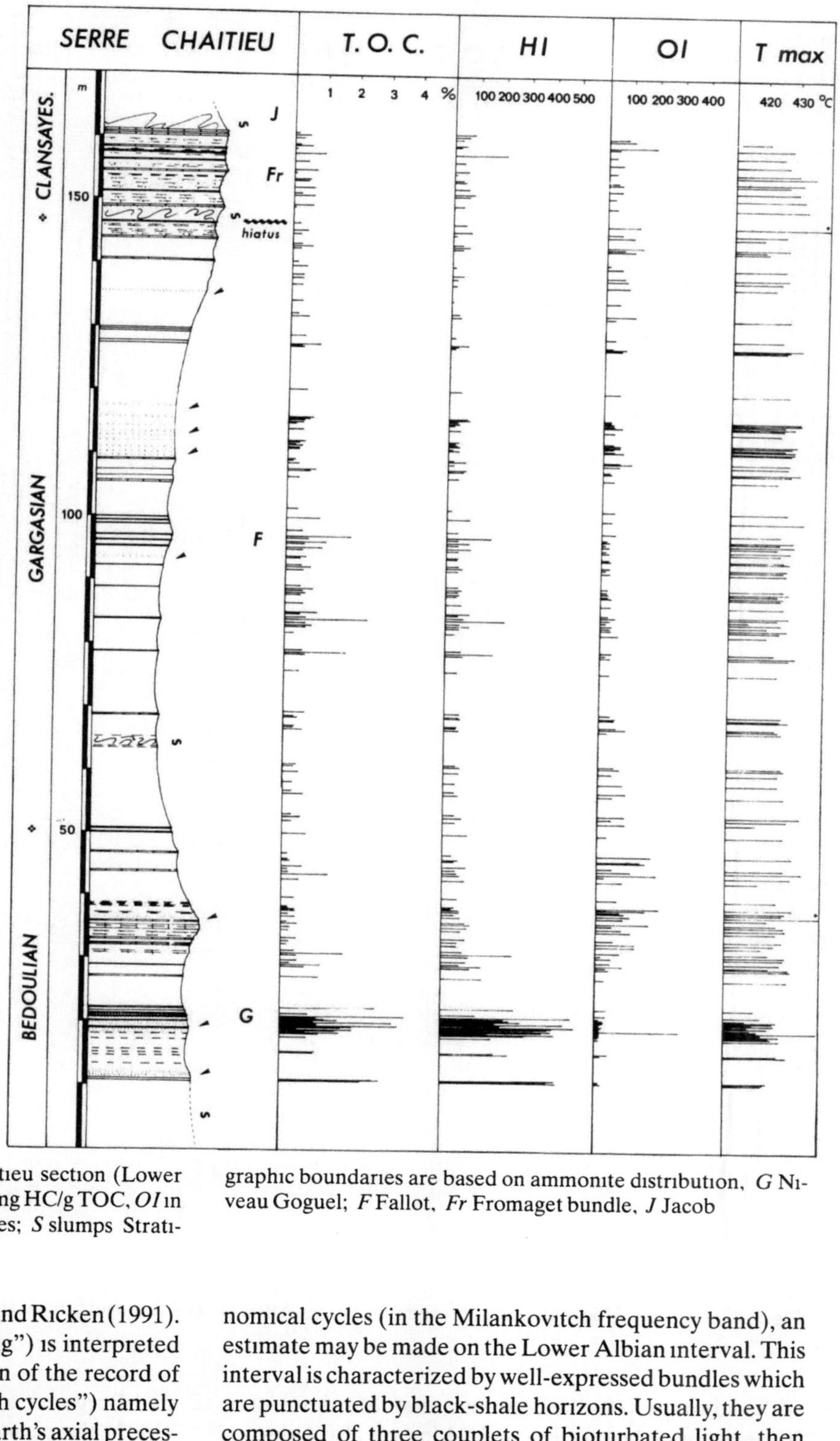

Fig. 6. Geochemical log of the Serre Chaitieu section (Lower and Upper Aptian) *TOC* in percent; *HI* in mg HC/g TOC, *OI* in mg CO_2/g TOC; T_{max} in °C; *arrows* turbidites; *S* slumps Stratigraphic boundaries are based on ammonite distribution, *G* Niveau Goguel; *F* Fallot, *Fr* Fromaget bundle, *J* Jacob

Fischer and Bottjer (1991) and Einsele and Ricken (1991). This bundling ("Schwarzacher bundling") is interpreted by these authors as the superimposition of the record of two astronomical cycles ("Milankovitch cycles") namely the short orbital eccentricity and the Earth's axial precession. Their periodicities (quasi-periodicities) are respectively of ca. 100 000 and ca. 21 000 (for more details see, for instance, Berger et al. 1989, 1992, and Fischer 1991). The orbital eccentricity would be the prominent feature of many series according to Fischer (1991) and Premoli-Silva et al. (1989b), this parameter acting upon climatic and oceanographic parameters and relative sea level. In order to see if the time corresponding to the deposition of the sequences is compatible with the quasi-periods of the astronomical cycles (in the Milankovitch frequency band), an estimate may be made on the Lower Albian interval. This interval is characterized by well-expressed bundles which are punctuated by black-shale horizons. Usually, they are composed of three couplets of bioturbated light, then dark, layers and two couplets, each formed by a bioturbated dark layer followed by a laminated dark one (class 3). At least 32 bundles occur in this Lower Albian interval. However, they may be more numerous owing to the eventualities of hiatuses and amalgamations of beds because of condensation. On the other hand, the numerical dating of the Lower Albian boundaries is rather imprecise. According to the well-documented syntheses of Kennedy and Odin (1982) and Odin and Hunziker (1982), the duration

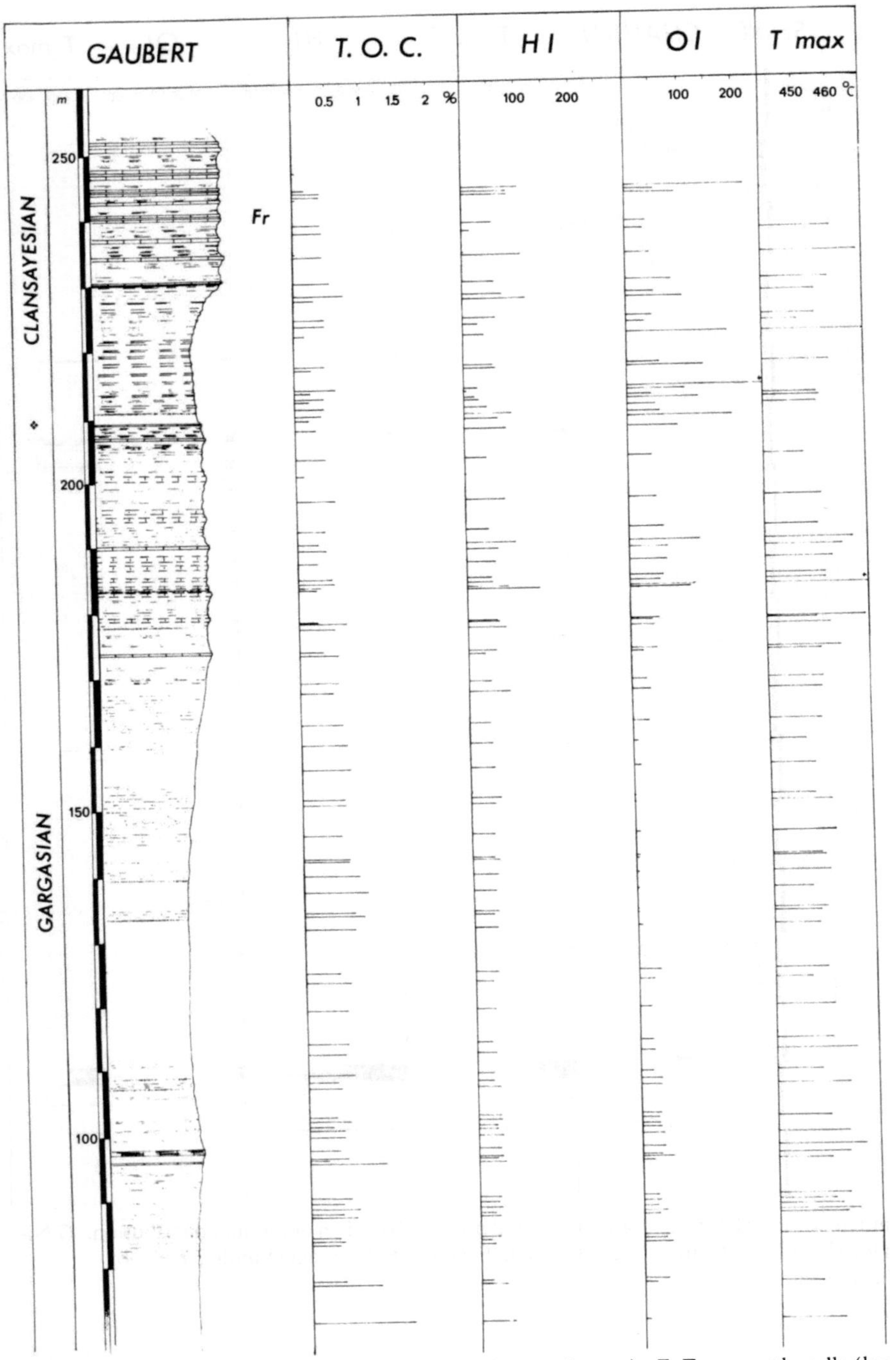

Fig. 7. Geochemical log of the Gaubert section (Upper Aptian) *Fr* Fromaget bundle (legend as in Fig 6)

of this substage would be between 3 and 8 Ma: the most reliable estimate being close to 5.5 Ma (see also Odin and Odin 1990). If this estimate is taken into account, this brings the average time covered by one bundle close to 172 Ka. However, given the degree of uncertainty, the margin is comprised between 90–190 Ka. These values fall in the Milankovitch frequency band, and may be tentatively related to the record of the short period of the Earth's orbital eccentricity, the mode of which is centered around 100 Ka (in fact, more precisely two modes are recognized around 95 and 123 Ka according to Berger 1978 and Berger et al. 1992). As calculated by Berger et al. (1992), this period of eccentricity was highly stable during the last billion years. So a causal connection with this orbital parameter is assumed here for these sets of five couplets The couplets themselves are assumed to correspond to the record of the Earth's axial precession, the duration of which is close to 21 Ka. Although more precise time constraints are needed, the hierarchical organization of the superimposed cycles is highly suggestive of the record of astronomical periods (see Fischer and Bottjer 1991, p. 1067).

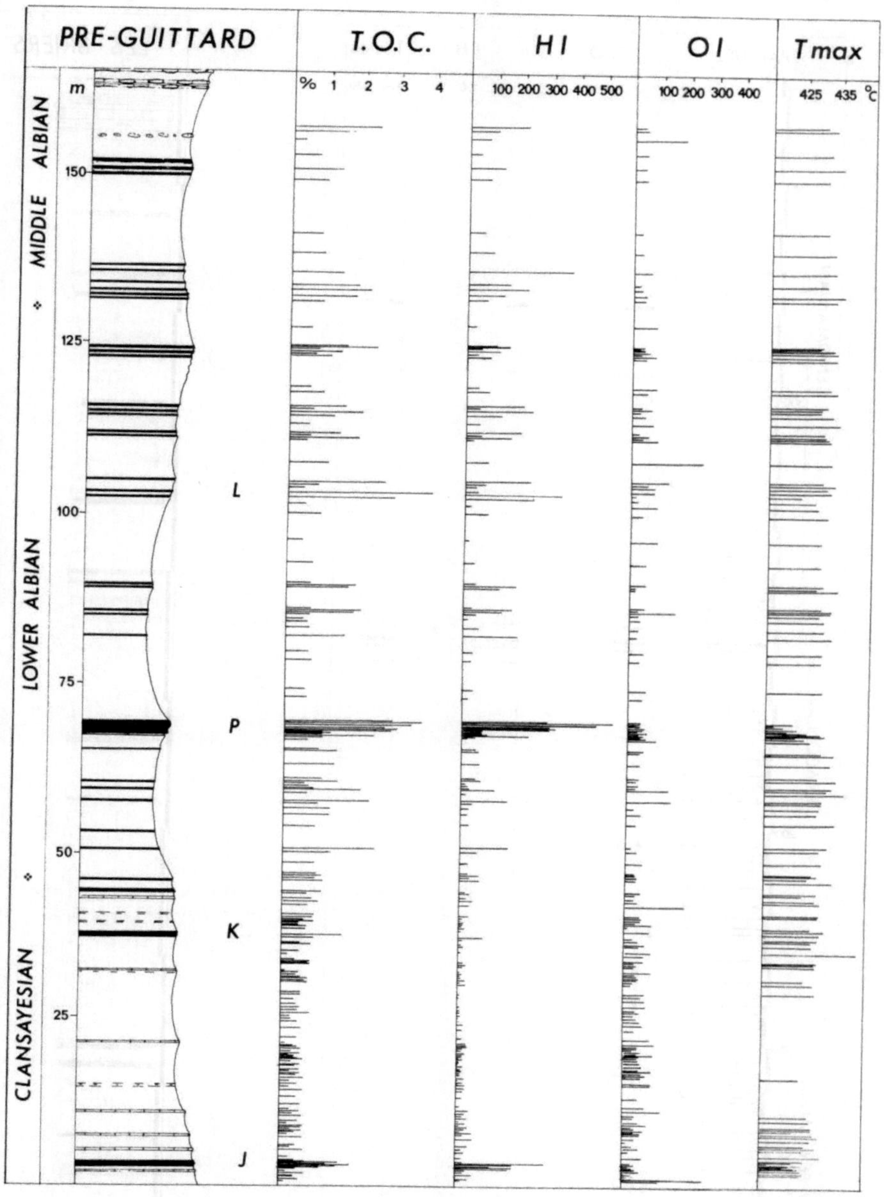

Fig. 8. Geochemical log of the Pré-Guittard section (Upper Aptian and Lower to Middle Albian) *J* Niveau Jacob, *K* Kilian, *P* Paquier, *L* Leenhardt (legend as in Fig 6)

The bundles of five couplets constitute the fundamental building blocks of most of the stratigraphic intervals of the Marnes Bleues. These vertically stacked units are constituted of only several types of basic motifs (Bréhéret, in prep.). They are particularly conspicuous on the outcrops and are easily correlated all over the basin (as it is the case for the Lower Albian). As such these pelagic sequences may be compared to the genetic sequences of strata initially described by Busch (1971), the concept of which was developed by Cross (1988), then Guillocheau (1991) and Homewood et al. (1992). They are considered to be sedimentary responses to changes in relative sea-level fluctuations (derived from the interaction of eustatic fluctuations, tectonic movement, and sediment supply). As emphasized by Guillocheau (1991), a number of observations recorded in the literature suggest that they are a common feature of numerous sedimentary environments, from coastal (Guillocheau et al. 1989) to open-marine ones (Van Buchem and McCave 1990).

As has been suggested by Cross (1988) and Ferry (1991), such sedimentary packets correspond to the parasequences of Van Wagoner et al. (1988) and Mitchum and Van Wagoner (1991) which they consider to be of fourth order. According to the hierarchy used by Kauffman (1986) and followed by Ferry (1991) and Fulthorpe (1991), they would be of fifth order.

As mentioned above, these units of five couplets are quite conspicuous in the Lower to Middle Albian for example (Figs. 10, 11). In this interval, the series is regularly punctuated by small bundles of two or sometimes three black-shale layers belonging to class 3, with a moderate organic-matter enrichment. A close examination shows

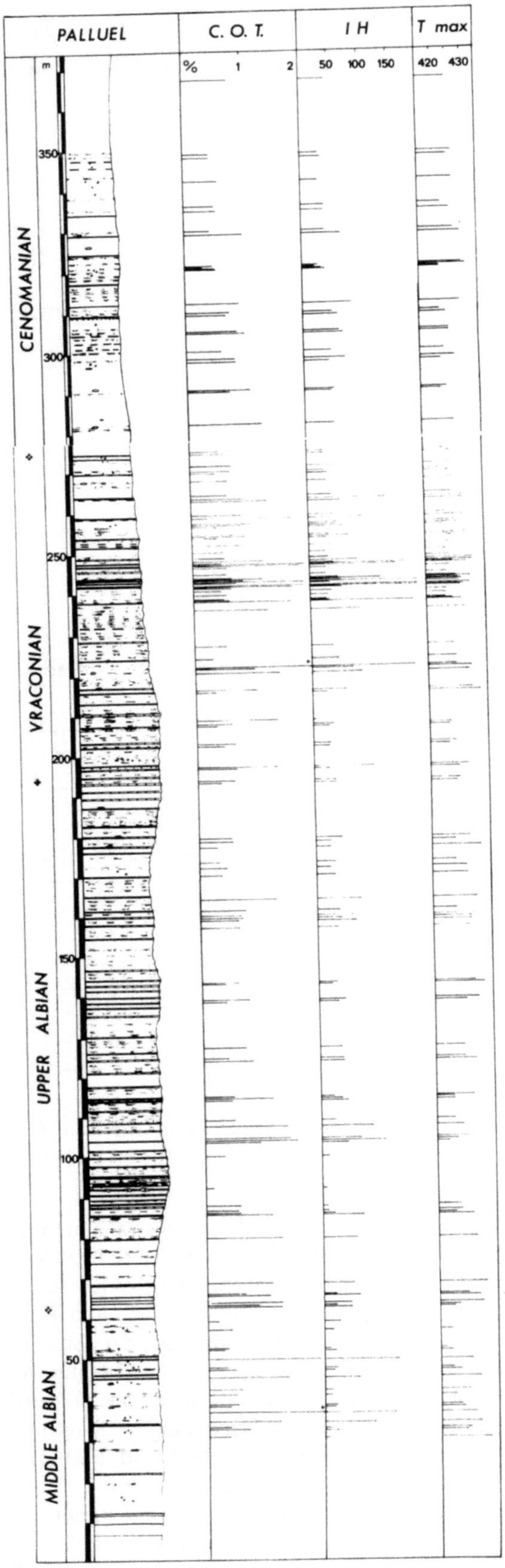

Fig. 9. Geochemical log of the Palluel section (Middle Albian to Basal Cenomanian). *BR* Niveau Breistroffer (legend as in Fig. 6)

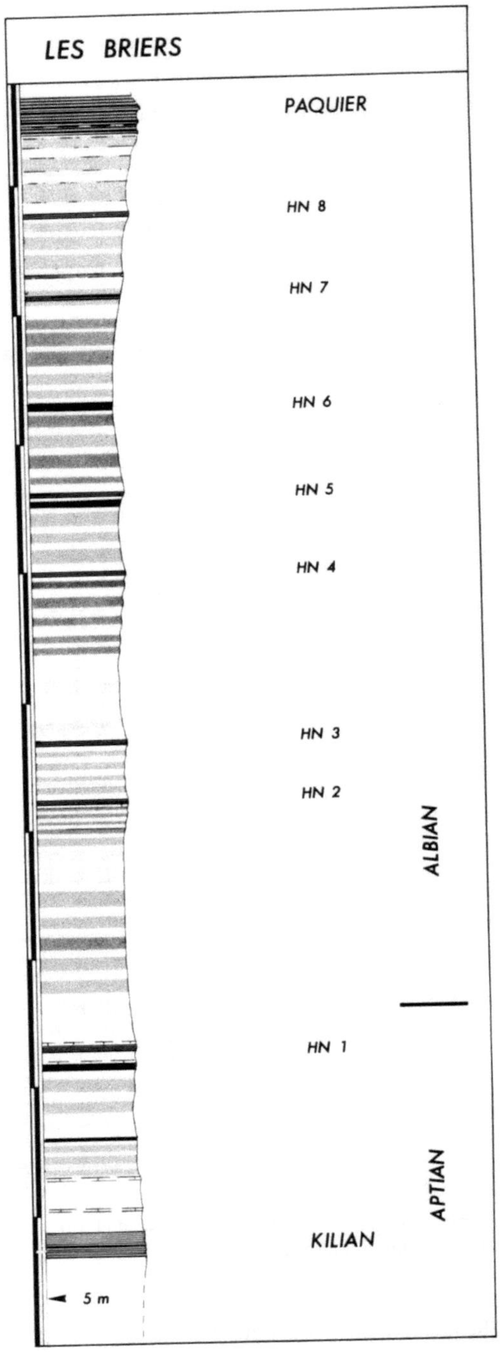

Fig. 10. Section of Les Briers (near Saint-André-Les-Alpes, Eastern part of the Vocontian zone) displaying the upper part of bioturbated marls of Upper Aptian (Clansayesian) age and the lower part of the Lower Albian, from the Niveau Kilian to the Niveau Paquier. The bundles of fives couplets (genetic sequences) are obvious. Sixteen are distinguished between the Kilian and the Paquier. The laminated organic-rich layers which end a lot of these sequences are noted HN ("horizons noirs", here *HN 1 to 8*.)

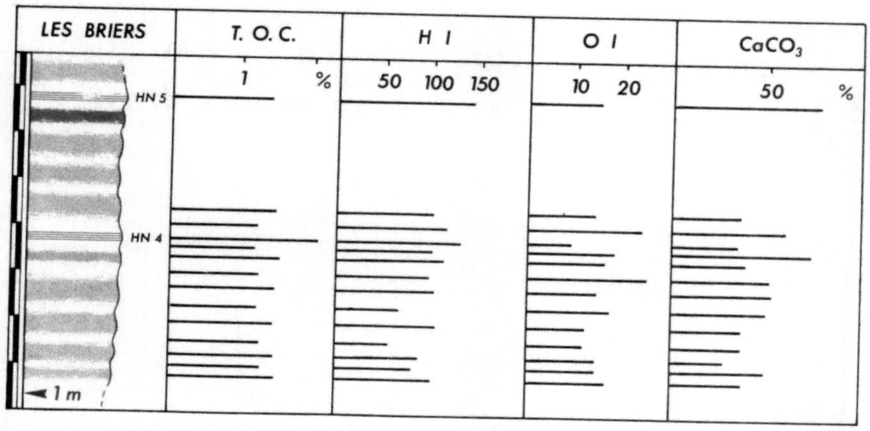

Fig. 11. Organic parameters (organic carbon, hydrogen index, and oxygen index) and carbonate content of two genetic sequences (respectively ended by HN 4 and HN 5) from the Lower Albian Les Briers section (see Fig. 10)

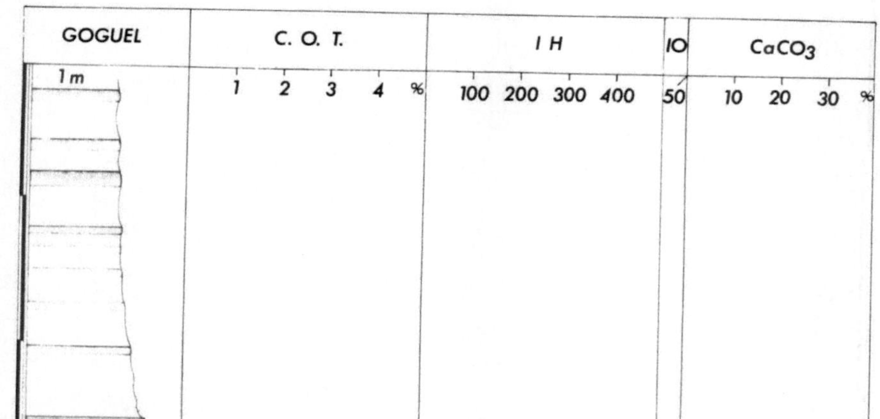

Fig. 12. The Niveau Goguel (Lower Aptian) in the Les Sauzeries section (eastern part of the Vocontian zone). Horizons with *horizontal lines* and noted *Go 1* to *Go 6* are paper shales. They mark the top of small-sized genetic sequences

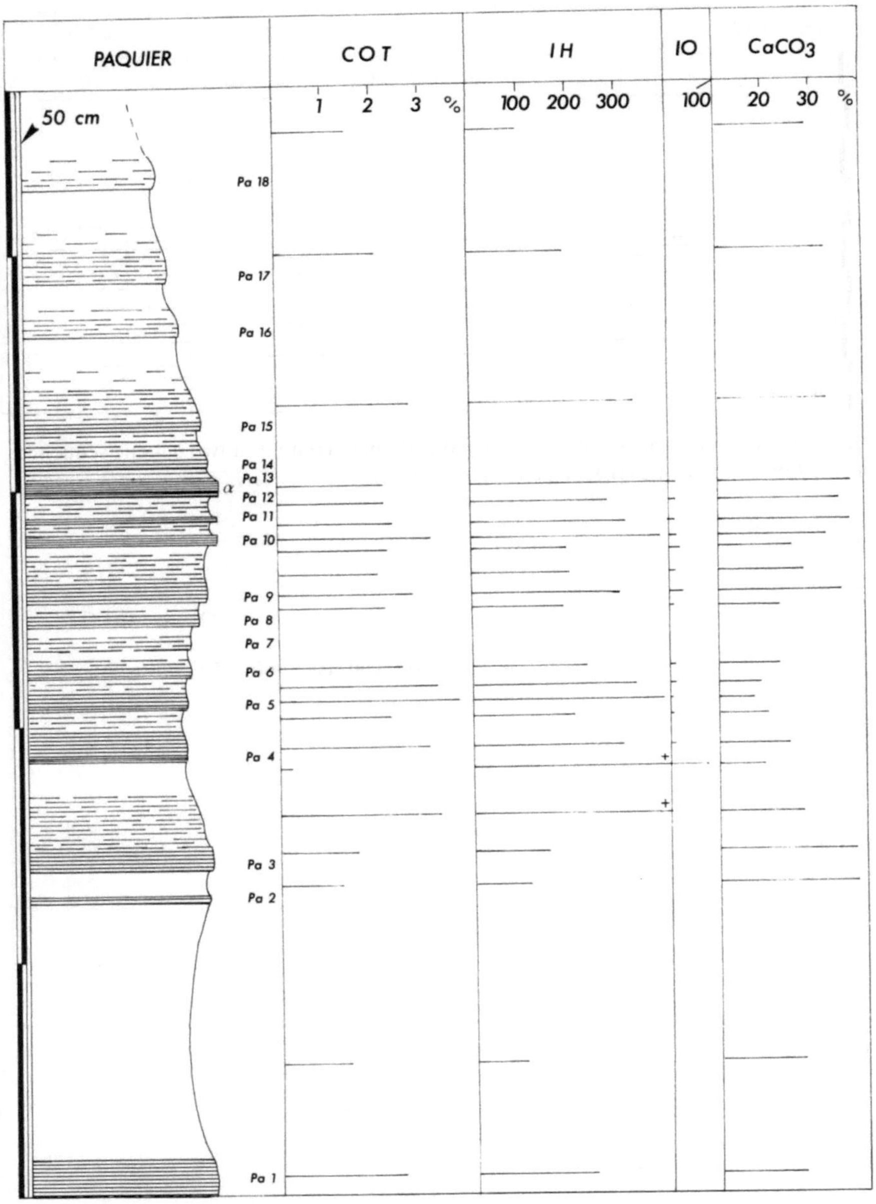

Fig. 13. The Niveau Paquier (Lower Albian) in the Ravel Section (near Hyèges, eastern part of the Vocontian zone). Eighteen small-sized sequences *(Pa 1* to *Pa 18)* considered as elementary are distinguished. They are correlated all over the Vocontian basin. *Horizontal lines* paper shales, *broken horizontal lines* moderately fissil shales, *white areas* homogeneous or weakly bioturbated shales

that they correspond generally to two couplets each formed of a bioturbated layer followed by a dark laminated layer. A complete genetic sequence is composed of these two couplets plus three couplets of marls and shales mainly of class 2. Not surprisingly, this pattern of black-shale occurrence appears similar, at first glance, to those observed in other series, particularly that shown in the Albian sequence of the Scisti a fucoidi (Herbert and Fischer 1986) or, for much more recent sediments, in the Mediterranean Quaternary sapropel sequence (Rossignol-Strick 1987; Hilgen 1991). Trapping of organic matter is efficient whenever the circumstances are favorable. As such, much of the low grade black shales (class 3) may be considered as periodites (following the terminology of Einsele 1982, Einsele et al. 1991). As stated above, the C_{org}-rich black shales result rather from an enhanced preservation of organic matter when bottom waters are impoverished in dissolved oxygen. During the greater part of the mid-Cretaceous, in most basinal marine environments, the depletion in dissolved oxygen giving way to organic matter-rich black-shale desposition, if not effective, was at least latent and recurrent. Periods on the scale of 10^6 years characterized by the permanence of normal oxygenation appear rather exceptionally. This may be due to a variety of factors acting simultaneously, as mentioned by many authors (see review in de Boer 1991). Nevertheless, and above all, a subsequent question to what has already been discussed still exists: is suboxia or

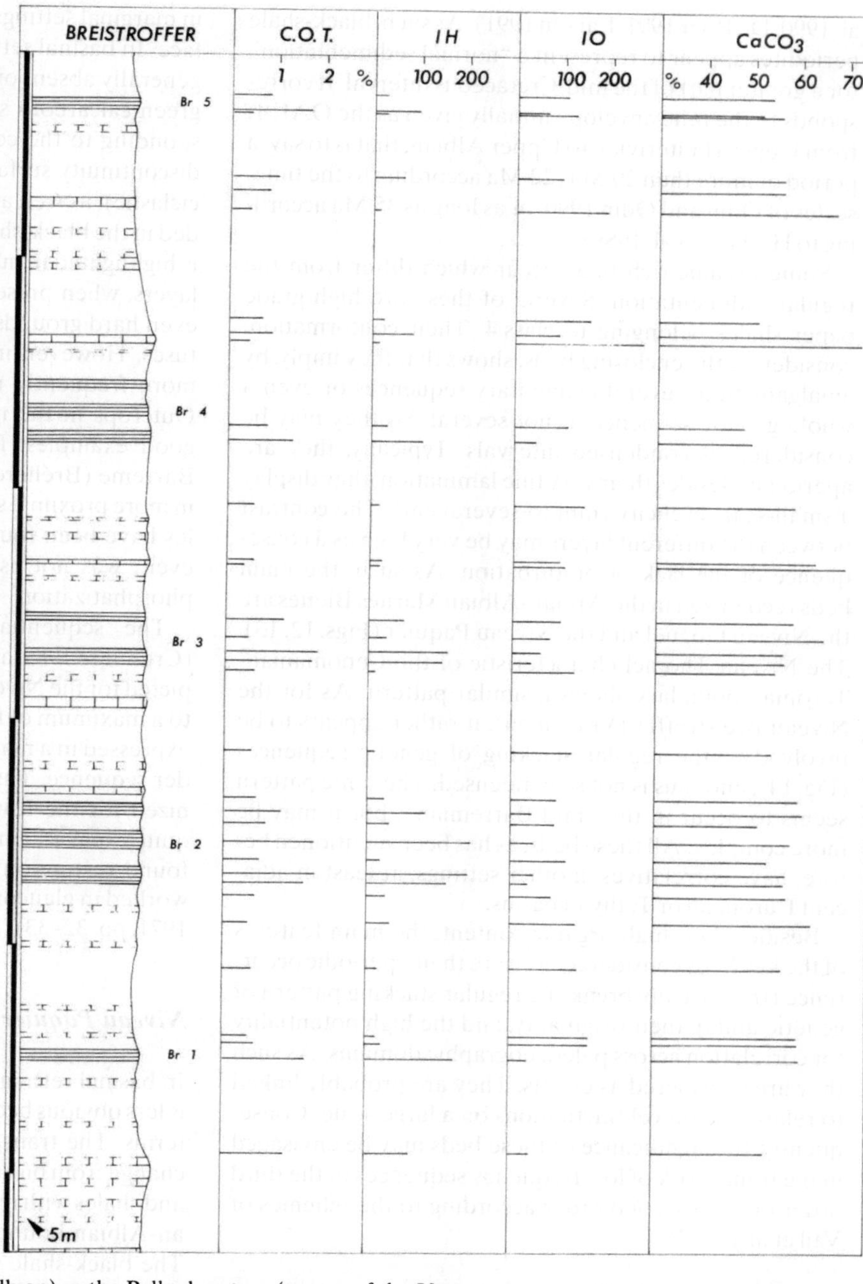

Fig. 14. The niveau Breistroffer (Upper Albian) in the Palluel section (centre of the Vocontian zone) The laminated *(horizontal lines)* layers, noted here *Br 1* to *Br 5*, are inscribed in genetic sequences between limey beds

anoxia of Cretaceous oceans (as well as other periods) driven by enhanced production due to high fertility of oceanic surface waters (Jenkyns 1980), or is it primarily due to a previously established density stratification, climatically and tectonically induced (Bralower and Thierstein 1984)? In spite of indications of high organic production for some intervals at certain sites (i. e., Eastern Atlantic), it would appear that the major factor is oxygen depletion which favors the preservation of organic matter. Anyway, among all the factors implied in the accumulation of organic matter, it can be seen that they are of one of two kinds of process: (1) the geodynamic ones (the fit of the continents, leading to the existence of a broad east-west Tethyan seaway that has been broken up: Ricou 1987), a huge CO_2 degassing by intense volcanism inducing a greenhouse effect; an important rise in sea level from the Aptian to the Turonian ("Cenomanian transgression") induced by a high rate of sea-floor spreading; the very high mean global temperatures of oceans and the atmosphere inferred from many studies, thus lowering the oxygen solubility in marine water and enhancing evaporation or precipitation depending on the case; the lack of large-scale continental glaciers diminishing the oceanic circulation) and (2) the astronomical ones (Earth's orbital variations) which interfere with the foregoing parameters (among an extensive literature see, for instance: Arthur et al. 1990, De Boer 1991, Larson 1991). As such, black-shale periodites appear to represent a "normal sedimentation" for a greater part of the mid-Cretaceous interval. It corresponds to the time envelope initially given as the OAE 1, from Upper Hauterivian to Upper Albian, that is to say a period of more than 20 Ma (22 Ma according to the time-scales of Odin and Odin 1990, or as long as 35 Ma according to Harland et al. 1989).

Some organic-rich beds occur which differ from the regular sedimentation. Several of these are high-grade paper shales belonging to class 4. Their conformation, considering the enclosing beds, shows that they imply, by amalgamation, several elementary sequences or even a whole genetic sequence, if not several. So they may be considered as condensed intervals. Typically, they are aperiodic. Besides their very fine lamination, they display a small-scale cyclicity (mm to several cm). The contrast between the different layers may be very high as a consequence of the lack of bioturbation. As such, the main beds recognized in the Aptian-Albian Marnes Bleues are the Niveau Goguel and the Niveau Paquier (Figs. 12, 13). The Niveau Thomel characteristic of the Cenomanian-Turonian boundary shows a similar pattern. As for the Niveau Breistroffer (Vraconian), it rather appears to be involved in the regular stacking of genetic sequences (Fig. 14), and thus is not so condensed. The same pattern seems to occur in the "mid-Barremian", but it may be more complex. All these beds, as has been mentioned before, have correlatives in other settings, at least in adjacent European or Tethyan basins.

Besides their high organic content, the main features of the key beds considered here are their aperiodic occurrence (they usually break the regular stacking pattern of genetic units), their originality, and the high potentiality for correlation across paleogeographic domains. As such they are considered as events. They are probably linked to relative sea-level fluctuations on a large scale. Consequently, the significance of these beds may be envisaged in the framework of low frequency sequences of the third order or even second order according to the schemes of Vail et al. (1977).

Sequential Position of Black-Shale Key Beds of the Aptian and Albian Marnes Bleues

Taking into account the position of these black-shale key beds according to the series, each of them (including "mid-Barremian" and Breistroffer) appears to correspond to a major transgressive pulse, or maximum flooding. Each case is examined here.

Niveau Goguel

This black-shale key bed is located above an important surface of discontinuity. In deep basinal settings where the sedimentary record is more or less complete, it represents a conformity that evolves towards an unconformity in marginal settings. It coincides with a transgressive surface. In basinal settings the transgressive systems tract is generally absent or formed by very thin (several meters) green calcareous shales, the black-shale key bed corresponding to the condensed section usually rests on the discontinuity surface. Many turbidites (the first are calciclastic), as well as mud- and debris flows, are interbedded in the black-shale interval. The stratigraphic position is highlighted in marginal settings where the organic-rich layers, when preserved, usually rest on firm grounds or even hard grounds, thus expressing the existence of hiatuses. However, in that context, these black shales are more frequently replaced by highly glauconitic marls. Outcrops in the north of the Castellane Arch provide good examples, for instance Montblanc, la Beaume, Barrême (Bréhéret 1991). No record of this event exists in more proximal settings. No reworked phosphate nodules have been found, however, it seems that this anoxic event was not associated with an important phase of phosphatization.

The sequential position of the Niveau Thomel (Crumière 1989; and see below) is quite similar to that depicted for the Niveau Goguel. Both appear to correspond to a maximum of flooding following a major sea level fall expressed in a major discontinuity, initiating a second order sequence. This disposition appears also to be recognized for the Lower Barremian event although with a quite weaker amplitude. Here phosphate nodules are found resting on the hiatal surface on the shelf as reworked in glauconitic marls and limestones (see Cotillon 1971, pp. 32–33).

Niveau Paquier

In basinal settings the disposition of the Niveau Paquier is less obvious because of a lack of contrasting lithological terms. The transgressive phase appears to begin with a change from bioturbated marly facies towards dark marls and shales with interbedded black shales, below the Aptian-Albian boundary, at the close of the Niveau Kilian. The black-shale periodites of the lowermost Albian represent a transgressive systems tract (as is shown by geometrical considerations), whereas the Paquier corresponds to a maximum of flooding. This fact is testified by the opening of thresholds having allowed faunal migration (ammonites of the genus *Leymeriella*) from the European realm to the Vocontian domain (Delamette in Bréhéret et al. 1986). It corresponds precisely to the time when the Anglo-Paris basin itself was invaded by the *Leymeriella* fauna (Regularis zone) due to the opening of the London-Brabant threshold (Amedro 1992). This fauna is represented by reworked phosphatic molds which are the only remnants of the condensed interval. In the Vocontian area, the development of a relatively thick transgressive systems tract (up to 40 m in the deponcenters) was then possible due to a favorable accommodation potential caused by a suitable subsidence of the basin center accompanied by a progressive rise in sea level. The onset of the transgressive phase induced a restriction which may be at the origin of the drastic crisis of planktonic foraminifera: *Ticinella* and the normal-sized forms of *Hedbergella* disappeared. Only poor populations composed of minute *Hedbergella* are observed (Caron, in Bréhéret et al. 1986). However, this crisis probably has a global significance as suggested by Caron (ibid.), Leckie (1989), and Sliter (1992). As has been shown by Bréhéret (1991), the Niveau Paquier as preserved in the southern part of the Vocontian area (Sisteron area) is strongly glauconitized. This Lower Albian layer rests upon an unconformity surface (dated of Upper Aptian: from upper Clansayesian down to the basal Clansayesian substage). This arrangement is somewhat analogous to that of the Niveau Goguel or Thomel; however, in the marginal settings, the potentiality for preservation of this key bed are better. Phosphate lag deposits are particularly rare in proximal settings (indices of phosphatization are quite rare). However, several sites distributed from the Ardèche (West) to the south of the Castellane Arch (southeast) yield very rare reworked ammonite steinkerns coeval with the Paquier event (see data provided by Cotillon 1971, for instance). Furthermore, phosphate lag deposits or reworked lag deposits are known on the northern margin, in the Vercors, in the Helvetic domain (Delamette 1989) and, as already mentioned, they are also known from the Anglo-Paris basin.

The Niveau Paquier is overlain by highstand marls and shales with low-grade black-shale periodites, but a new rise in relative sea level occurs with the onset of the second ammonite zone *(Mammillatum)* of the Lower Albian. It does not follow any noticeable fall in sea level, as evidence of breaks in sedimentation are rare (Sisteron) in what is considered to be part of the upper part of the *Tardefurcata* zone. A new maximum of flooding is attained in the *Mammillatum* zone with the black-shale bed called Niveau Leenhardt, some horizons of which are paper shales of class 4. The record of this layer on the margins is even better than that of the Paquier. Indices of glauconitization and phosphatization, such as ammonite steinkerns, in marginal settings are relatively frequent. They are also found on adjacent shelves. Higher up, the interval, rich in black-shale periodites (class 3), appears to be a highstand systems tract. It is well recorded in marginal settings.

Niveau Breistroffer

The sequential position of the Niveau Breistroffer is less clear, although it may be equivalent. The onset of the *Dispar* zone (Vraconian) is marked by a set of laminated horizons (low grade black shales of class 3) that may correspond to a new rise in relative sea level. They are interbedded between evenly bedded limey beds in the depocenter. These beds are nodulized in the marginal settings where the series is thinned (Bréhéret 1992b); this is the "faisceau michoide" (Bréhéret and Delamette 1987). This bundle resting on a discontinuity surface in marginal settings may be interpreted as a transgressive systems tract. It pinches out shelfwards. The set of organic-rich layers corresponding to the Niveau Breistroffer begins at the top of this bundle. As mentioned earlier, it does not appear to be as condensed as the Paquier or Goguel. Rather, the black-shale beds are inscribed in the stacked genetic sequences. Although it corresponds to a maximum in flooding, it is rather thick because of a sufficient sediment supply with an available accommodation. The Niveau Breistroffer itself may rest upon the transgressive surface in more marginal settings.

A certain analogy appears in the cases illustrated above. These black-shale key beds correspond to major flooding events, and they follow important breaks in sedimentation; either they rest upon a transgressive surface (even in basinal settings), as is the case for the Goguel (and the Thomel, see Crumière et al. 1990b), or they surmount a transgressive systems tract, as illustrated by the Paquier and the Breistroffer.

This situation shows that each main black-shale key bed closely follows a discontinuity surface evolving towards an unconformity landward, and expressing a major break in sedimentation. Given the fact that they are transgressive, these beds may be lacking on marginal sites due to a weak accommodation potential and to erosion by current activity. Furthermore, anoxia or suboxia being rather restricted to the deep basin center, diagenetic facies such as glaucony draperies and phosphates are to be found in the proximal sites, when they are preserved (Bréhéret 1991, 1992a). On the shelf, in the best cases, they are represented only by phosphatic lags (or reworked lags). These represent concretions around burrows or invertebrate (sponges, Brachiopoda, Mollusca) steinkerns exhumed after the restriction event, when the circulation is reactivated. However, the most prominent black-shale beds have not been associated to important phosphatization in the close shelf area.

Considerations on the Transgressive Position of Black-Shale Key Beds

In the North Atlantic series, Sigal (in de Graciansky et al. 1982; Muller et al. 1983, 1984) for example, showed that there is a gap, which he called événement E1 (Initially the evidence of a widespread hiatus by Sigal gave rise to its being identified as an "event"; it appears important to distinguish this "événement E1", a hiatus, from the "OAE 1", or more precisely "OASE 1 a subevent" of Arthur et al. 1990, the effective sedimentary record of an anoxia), in the Lower to Upper Aptian (at the base of the black shales of the Hatteras Formation). It is recognized now, after Sigal (in Muller et al. 1983, 1984 and de Graciansky et al. 1987), that two major breaks occur, the one affecting the boundary between Lower and Upper Aptian: upper part of the Bedoulian-lower part of the Gargasian (given as E0 in de Graciansky et al. 1987, and E1a in Robaszynski 1989); the other at the boundary between Aptian and Albian: upper part of Clansayesian-lower part of Albian (given as E1 in de Graciansky et al. 1987, and E1b in Robaszynski 1989). No clear evidence of a major organic-rich interval has been shown until now in the North Atlantic basin in the upper part of Lower Aptian. However, the core 11 of site 391 (Western Atlantic), which is characterized by black shales of class 3 and 4, according to the data of Herbin and Deroo (1982) could be an equivalent (taken into account the biostratigraphical data presented by Muller et al. (1983, 1984). As for the Lower Albian, a peak in the preservation of organic matter is noted for the North Atlantic by Arthur et al. (1990).

In the Anglo-Paris basin (Aube area, stratotype), the basal Lower Albian is represented by a phosphatic lag coeval with the Paquier event (with ammonites of the *Regularis* zone) (see Destombes 1980; Amedro 1992). This lag rests on the Sables Verts Formation; these green sands attributed to the lowermost Albian by Rat et al. (1980) are now thought to belong to the Upper Aptian (Clansayesian) by Amedro (1992). Everywhere in the Anglo-Paris basin, if it is present, the Lowermost Albian is characterized by reworked phosphatic ammonite steinkerns testifying to the subequent winnowing of a condensed section. Thus the sequential position apparently matches well with that observed for the Paquier in the neighboring Vocontian basin.

Sigal (in de Graciansky et al. 1982) showed the existence of a second gap corresponding to the Mid- to Upper Cenomanian that he called événement E2. According to other authors, this time span would be represented by a thin "condensed" layer as in ODP site 641 A (Thurow et al. 1988). However, Bralower (1988) has shown that such a hiatus is widespread on European shelves. It is also emphasized by Robaszynski (1989), for whom it represents the Mid-Cenomanian nonsequence.

This pattern is also depicted in the Vocontian basin, as shown by Crumière et al. (1990a, Fig. 8). In the basinal situation (Vergons), on the scar left by the hiatus as a firm ground rests the Thomel organic-rich layer. The break affects all of the upper part of the *cushmani* Foram zone, then the Thomel level is dated as the *archaeocretacea* Foram zone. In shelf setting of the Domaine Provençal, the hiatus is more pronounced and no black shale is preserved. As mentioned above, this significance of the Cenomanian-Turonian Boundary Event (noted as CTBE by Thurow and Kuhnt 1986) as the maximum of a transgression is testified in many settings. It is interpreted by many authors as the record of a global event and, for instance, the Niveau Thomel is correlated with the Livello Bonarelli from Italy.

The similarity in the sequential position of this Upper Cenomanian Livello Bonarelli and the Lower Aptian Livello Selli from central Italy is emphasized by Coccioni et al. (1991). Because of the parallelism in the evolution of the environment, the authors suggest analogous sedimentary and paleoceanographic conditions for both.

In fact, more generally, this pattern is quite stricking and recognized by numerous workers in so many cases (see Hallam and Bradshaw 1979; Wignal and Hallam 1991, for instance): a high-grade organic-rich bed, with a condensed character follows a major break in sedimentation, the black-shale-roofed discontinuities, so called by Baird and Brett (1991). Examples may be found in many other contexts such as, for example, the Upper Devonian-Lower Mississipian black shales of North America (Baird and Brett 1991), or the Lower Toarcian black shales of Europe (Jenkyns 1985; Wignall 1991; Wignal and Hallam 1991). Numerous authors stress the starvation induced by the rise in sea level and the subsequent transgressive pulse. Among them, Wignal and Hallam (1991) emphasize the role of the deepening of the basin center reinforced by the lack of sediment input.

This analogy of their sequential context and their organic matter richness does not signify that these beds are highly comparable in all respects. In fact, as emphasized by Bréhéret and Delamette (1989), each of these beds of the Vocontian mid-Cretaceous interval presents its own characteristics, particularly concerning their faunal content. This is due to different geological and oceanographical circumstances accompanying and leading to suboxia or anoxia.

An additional observation may be made in the deep depocenters, where the series is near enough complete. These striking layers (at least in the mid-Cretaceous interval) are generally interbedded amid intervals themselves punctuated by black-shale periodites (of lower grade and shorter duration) climatically modulated, that may occur below and above. In marginal settings, the basal black-shale periodites are obviously missing, the organic-rich key bed rests on the unconformity surface, or is itself even lacking; but upper black-shale periodites (corresponding to a higher sea level) are more often observed. The case has been noticed, for instance, by Coccioni et al. (1991) and Premoli-Silva (1991) for the Bonarelli Level below which thin black-shale layers occur in several sites of the Umbria-Marchean Basin: "Marker nero" in the "lower critical interval". This interval corresponds to a transgressive systems tract.

Organic-rich key beds appear here, indicating what may be called oceanic anoxic events, a record of extraordinary conditions arising while the general marine environment was already prone to suboxia or anoxia. Authors still disagree on the reasons for these anoxic events which occur during periods of pronounced rises in sea level.

Considerations on the Condensed Character of the Black-Shale Key Beds

As the enclosing sediments, the organic-rich key beds like Goguel and Paquier are not homogeneous; they are characterized by a cyclicity organized into a hierarchy. Small-scale sequences (cm to dm) are generally grouped by five in bundles of several decimeters. Six are counted in the Goguel (Fig. 12) and 3.5 (i.e., 18 high-frequency sequences noted Pa 1–Pa 18) in the Paquier (Fig. 13) on the best outcrops. A close examination showed a transition between the elementary sequences of the enclosing sediments towards the small-scale sequences of the key bed and vice versa. This is especially true for the Paquier (Fig. 13). There, the elementary sequences become progressively thinner, reaching a minimum at the level of the "horizon α" (Bréhéret 1983), which is composed of two sequences Pa 12 and Pa 13, and then progressively thicken. All these elementary sequences may be correlated all over the Vocontian basin. Their relative thinness is interpreted here as consequence of a decrease in detrital supply. This gives these horizons a condensed character.

A comparable short-scale cyclicity has been noticed for other organic-rich intervals, particularly for the Cenomanian-Turonian boundary, i.e., the Niveau Thomel in the Vocontian basin (Crumière et al. 1990a), and the DSDP site 603 by Herbin et al. (1987); other examples are given by Schlanger et al. (1987), notably the Oued Bahloul (Tunisia), and the Bridge Creek Limestone (USA).

A similar pattern depicted for the Livello Selli (coeval with the Niveau Goguel: Early Aptian) in Umbria has brought Herbert (1992) to interpret the small bundles as the record of the 100-Ka Earth's eccentricity cycles. Given a number of 10 to 12 bundles for the different sections, he estimates the total duration of the event at approximately 1 to 1.2 Ma. This author considers that this estimate based on cyclostratigraphy displays a higher precision than other obtained by extrapolating sedimentation rates from the intervals above or below the horizon. According to such a mode of calculation, the duration of the event would be between 200 (thickness referred to Maiolica sedimentation rate) and 500 Ka (Aptian marly limestones sedimentation rate). To give a comparison, Jenkyns (1985) for the Early Toarcian Event and Arthur et al. (1987) for the Cenomanian-Turonian Event estimated that each one lasted approximately half a million years.

If the mode of calculation of Herbert based on the number of sequences is applied to the Goguel and Paquier Event, their duration would be respectively about 600 Ka and 360 Ka.

The Position of Black-Shale Periodites Within Genetic Sequences

The genetic sequences are recognized on the outcrops according to the lithology which expresses the result of the fluctuations of four parameters: detrital supply, primary production of calcareous and noncalcareous organisms in surface water, conditions of oxygenation in the bottom water and in the topmost sediment, and early diagenesis. As a result of the distance from shore, and the relative depth of the depositional environment, the sedimentation process is essentially a calm setting, with the exception of several episodes of resedimentation (slumps or turbidites) generally well identifiable. It ensues that the contrast between the different lithological types which make up the genetic sequences is weak. Their characteristics are, in fact, basically restricted to the interplay of two parameters: planktonic production (calcareous/noncalcareous)-bottom water oxygenation. As these genetic sequences probably are the result of an orbital forcing, as has been mentioned above, they may be mainly the response to two kinds of factors: climatic-oceanographic parameters (essentially temperature, chemistry, and oceanic circulation) and sea level. Among other things, this last factor is linked to temperature. Probably high-frequency fluctuations of at least several meters have occurred (in the absence of an ice cap) according to the steric expansion of the water as a result of temperature changes (see Paskoff 1987; Revelle 1990 in Plint et al. 1992; Roemmich 1992). Climatically controlled changes in the volume of groundwater stored on continents represents another possibility (with a comparable order of magnitude) which is notably evoked by Hay and Leslie (1990) in Revelle (1990) and Jacobs and Sahagian (1993). Moreover, as recalled by Einsele and Ricken (1991, p. 613) "increasing evidence indicates that such high-frequency sea-level variations occurred throughout the Phanerozoic" with an amplitude between 1 and 10 m (see Strasser 1988, for example). In this respect, Grotsch (1992) has described small cycles on the mid-Cretaceous (Barremian to Albian) carbonate platform of the Gavrovo zone (Greece); these cycles represent the consequence of high-frequency sea-level oscillations. We are then brought to examine the position of the black-shale layers within the genetic sequence in order to see how their formation may be explained as regards the variations of the two factors mentioned above, i.e., climatic-oceanographic parameters and sea-level fluctuations.

Remarks At the scale of the genetic sequence, in the bathyal zone, the concept of accommodation potential (see Cross 1988) cannot be used, bacause of the meaningless changes in the bathymetric range (and the relatively low sedimentation rate). On the other hand, even small fluctuations in relative sea level may cause shifts in the sediment parting, so the pelagic environment may be alternately starved and fed. However, the wetter or drier character of the prevailing climate on continental areas may produce similar effects on alluviation. These two kinds of phenomena may combine.

In the Marnes Bleues, the black-shale periodites (mainly of class 3) appear to be distributed according to two patterns in the genetic sequences.

Condensed Type

The first type, and the most characteristic (Figs. 11, 15), is that yielded by the Lower to Middle Albian interval (see also Bréhéret and Delamette 1989, Fig. 2). There, the black-shale layers usually of class 3 (between 2 and 2.5 % TOC) and rarely of class 4 (> 2.5 %) are sometimes isolated (as is usually the case for the lower Lower Albian) but often occur grouped in twos or even threes: "doublets" and "triplets" (as shown in the upper Lower Albian and the lower Middle Albian). They are inscribed

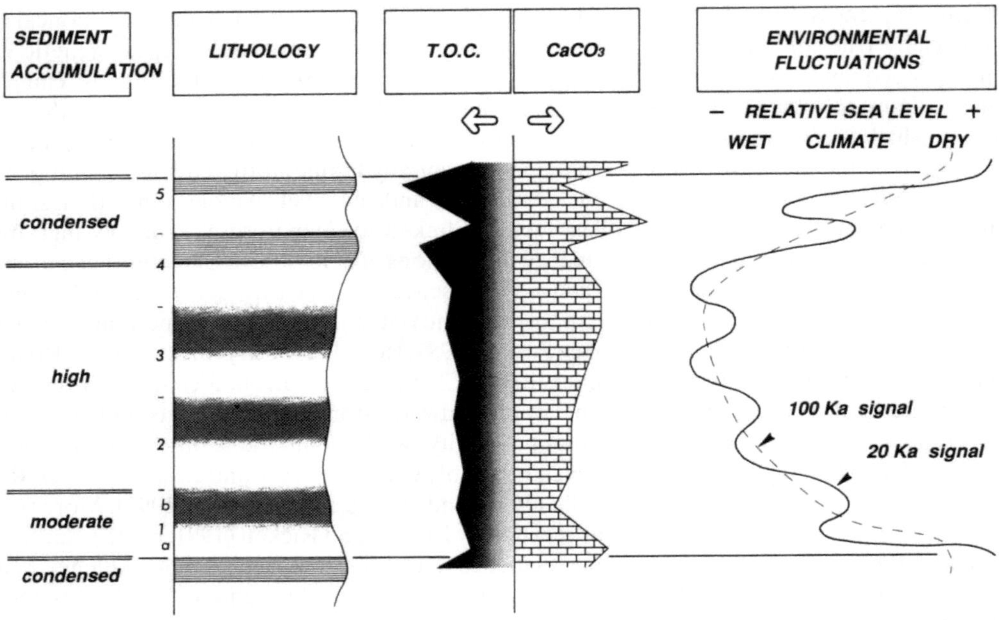

Fig. 15. Model of the "condensed type" black shale as inscribed in genetic sequences (bundles of five couplets) of the Lower and Middle Albian interval

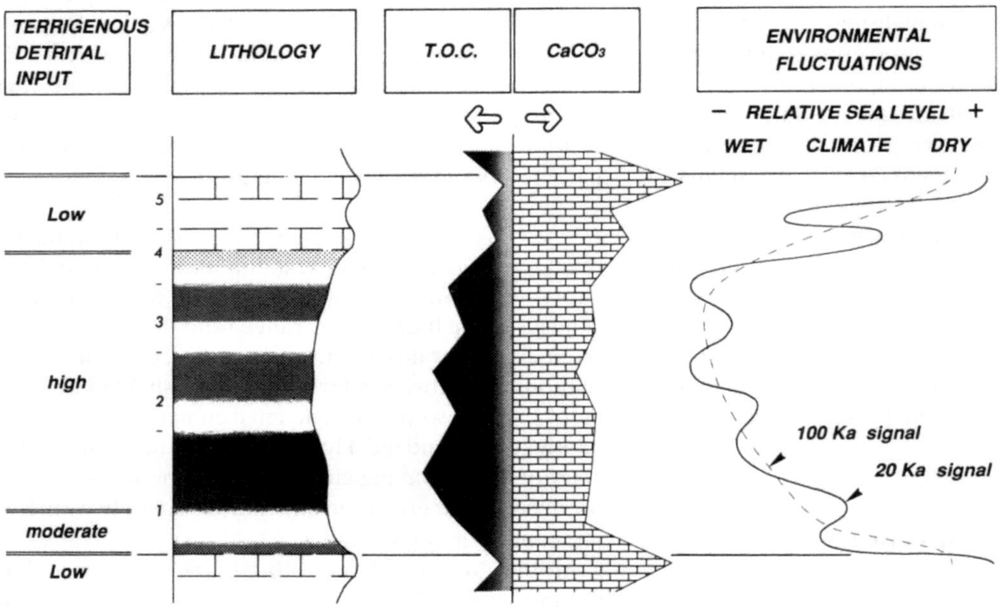

Fig. 16. Model of the "fed" type black shale as inscribed in genetic sequences of the Upper Aptian or the Upper Albian

in units otherwise formed solely of more or less dark marls or shales of class 2 and sometimes of class 1, displaying a subtile banding. Given the dark color and the relative homogeneity of the sediment, the only difference between the layers often lies in the more or less pronounced anisotropy of the material and the fissility which ensues because of alteration due to outcropping It must be noted that from one genetic sequence to another, the laminated characteristic of the black-shale layers is more or less pronounced. Furthermore, the contrast between black shales and the enclosing beds may be so slight that they become indistinct: such a pattern may be repeated on two or more successive genetic units

The carbonate content varies in small proportions. Depending on the interval considered, the carbonate contrast is generally lower than 15% between richest and poorest layers (10 to 25% in lowermost Albian, 20 to 35% in the basal Middle Albian for instance).

The most typical succession is as follows (Fig. 15). Three basic cycles (elementary sequences or couplets, noted 1 to 3) each composed of one layer (a) rather light and bioturbated and of one layer (b) rather dark, often more fissile but homogeneous, where bioturbations generally are not easily detectable. The last two basic cycles (4 to 5) finishing the genetic sequence are then each composed of one rather light more or less bioturbated layer (4 a and 5 a) which is followed by a laminated black-shale layer (4 b and 5 b). If the first bioturbated layer (4 a) of this set of couplets displays a thickness comparable to that of layers 1 a to 3 a, the set of the last three layers of the genetic sequence, including 4 b, 5 a, and 5 b generally display a distinctly reduced thickness. In some cases it may correspond to half of the thickness of the underlying layers. The most calcareous layers are adjacent to the black-shale layers (4 b and 5 b). So the maximum contrast in lithology is found in cycles 4 and 5. Given the homogeneity of the lithology, such a reduced thickness accompanied with a higher carbonate content may be attributable to a weaker detrital sediment supply for cycles 4 and 5, and from a stratigraphical point of view to a relatively condensed interval. Referring to the sequential models given, for example by Cross (1988), this interval may be interpreted as recording maxima in flooding which occur during a relative sea-level rise. The effect of a rather dry climate may be an alternative explanation, although both phenomena may combine.

Fed Type

The second type of genetic sequence including black shales (Fig. 16) is typical in the late Upper Aptian (Clansayesian) interval and in the late Upper Albian (Vraconian) interval. There, the mean value in carbonate is generally higher than for the Lower to Middle Albian, as is the contrast between the richest and the poorest layers in carbonate: it may be of 25 to 30 %. In addition to black shales, the marly sequences are punctuated by limestone beds or limey marl beds that are more or less indurated. The base of the genetic sequence is considered to be formed by the succession of three basic cycles (elementary sequences), each one composed of a bioturbated, rather light marl layer and a dark, weakly bioturbated, homogeneous or laminated layer that usually represents a black shale of class 2 but that may be of class 3 (with up to 2.5 % TOC) as is the case for the Niveau Jacob in the Clansayesian, or the Breistroffer in the Vraconian. In some examples (they can be seen particularly in the lowermost Upper Aptian = basal part of Gargasian), the dark layers may be split into two dark layers separated by a slightly lighter one, thus yielding a complex pattern which recalls, to some extent, that depicted by Hilgen (1987, his "quadruplets" in Fig. 4) in the Pliocene of the Sicilian Trubi section. However, in that case, the subdivision concerns the light marly layers instead of the dark ones. The upper part of the genetic sequence is composed of two basic cycles, each one made of a bioturbated marl layer and a bioturbated limey bed. The limey bed is generally thinner than the marl layer. Likewise, the limey beds are thinner than the black-shale layers of the lower part of the section. Therefore they may be related to a weakening of the detrital supply and may constitute a condensed interval. Such a pattern suggests for the black-shale layers a meaning different from that in the first case. Perhaps it records a restriction in the water column which is linked to a relative lowstand in sea level. In this respect it can be observed that the organic content is rather lower than for the black shales of the condensed type. Alternatively, these black shales may correspond to the start of a rather humid climate which entails a high continental runoff. This leads to the expansion of low-salinity surface water and the subsequent hydrologic more or less stable stratification. In this respect, several "fed" black shales of the uppermost Aptian are relatively rich in continental organic matter (as macro- and microdebris), as in the Niveau Jacob.

Remarks

Instead of a simple univocal pattern, two kinds of periodic black shales are therefore distinguished here according to their position within the genetic sequences. This brings their interpretation as being the consequence of quasi-opposite conditions of genesis. In the first case, black shales are in phase with detritism lowerings (condensation) giving way to an increase in calcareous production (generating limestones in other circumstances). In the second case, black shales are out of phase with these episodes.

The fed type would be similar to that depicted in the Upper Albian interval (*T. praeticinensis* zone) by Herbert and Fischer (1986). These are low-grade source rocks (< 1.5 % TOC) according to the data of Pratt and King (1986, Table 1, Fig. 1). An oxygen-deficient benthic environment is supposed, and it is inferred that marine productivity was moderate to low (ibid, p. 519). This is also the conclusion of Premoli-Silva et al. (1989 a, p. 230). However, according to Pratt and King (ibid.), most of the fine detrital components correspond to eolian dusts. In fact, as suggested by these authors, these rhythmic black shales would represent the consequences of a salinity stratification linked to an arid climate, strong winds, and stagnant benthic conditions. As regards to comparable black shales (according to the sequential position) in the Vocontian area, the hypothesis presented here is different, as they are rather considered to be fed black shales deposited during either periods of relative lowstands in sea level, or periods of humid climates, or both.

The deposition of black shales during periods of humid climate/lowstand would then be opposed to the deposition of limey beds during periods of dry climate/highstand.

The condensed type (first evoked case) shows a close association of black shales with more limey layers and suggests a deposition during drier climates and/or of maxima in flooding. An enhanced planktonic productivity is not to be disregarded.

To verify these interpretations would require following the genetic sequences in proximal environments. However, the conditions are not favorable, principally because of amalgamation due to the very weak accommodation potential Usually, the series correlated with the sequences depicted in basin settings are composed of an alternation of black-shale layers and bioturbated glauconitic marls (see, for example, Bréhéret 1991). Phosphate nodules are common in both facies, either pristine (cf. Follmi 1989) or reworked, as is generally the case (almost exclusively in the glauconitic marls) The top of the black-shale layers is always bioturbated, and usually presents a "smallpox pattern" yielded by the crowded *Chondrites* networks. In more proximal sites, black shales are missing. In certain cases, the small sequences are punctuated by omission surfaces characterized by *Thalassinoides* burrows and covered by highly glauconitic marls sometimes rich in ammonites (i.e., for the equivalent of the Breistroffer at Consegudes).

Conclusions

The mid-Cretaceous sediments of the Vocontian area display several intervals with organic-rich layers which represent the local consequence of global oceanographical trends (OAEs in the broad sense). Detailed study of the Marnes Bleues series has allowed the recognition of a hierarchy in these deposits. Most of the black-shale beds are periodites inscribed in high-frequency cycles as a consequence of fluctuations in climatic and oceanographic parameters, probably astronomically brought about. Among these black shales, two types may be differentiated according to their sequential position considering the carbonate-rich layers. The first type, "condensed", is more or less in phase with carbonate-rich layers and is deposited during lowering in detrital input interpreted as a result of transgressive episodes and/or dry climatic phases The second type, "fed", is an opposite phase, and rather corresponds to falls in relative sea level and/or to wet climate phases.

Several organic-rich beds break this regular pattern, and are composed of amalgamated genetic sequences. They represent key beds at the scale of the Vocontian area and, as is the case of the Goguel and the Paquier, are local records of very widespread "anoxic events". Each one displays its own characteristics, but a certain similarity emerges in their sequential position as, following a transgressive trend, they illustrate maxima in floodings depicted at the scale of third- and second-order sequences.

It follows that a single depositional model is rather hard to imagine in order to explain all types of black shales depicted here

In any case, the lithological heterogeneities in the black-shale beds, even at the millimetric scale, express high contrasts in detrital input, plankton production, and organic matter preservation. They were deposited under the influence of a variable intensity in suboxia induced by a continuously fluctuating oxygenation level of sea water.

Acknowledgments Thanks are expressed to J. Espitalié (Geochemical Laboratory of the Institut Français du Pétrole, Paris) for the use of the Rock-Eval device and for the numerous constructive discussions. H.J Brumsack (Univ. Oldenburg) is acknowledged for having provided several geochemical data Fruitful discussions have been highly appreciated with G. Busson and D. Noel (Muséum National d'Histoire Naturelle, Paris), M. Delamette (Genève), J R. Disnar (Univ Orléans), P. Cotillon, J P. Crumière and S. Ferry (Univ. Lyon), F Guillocheau (Univ. Rennes), P. Le Strat (Bureau Recherches Géologiques & Minières, Orléans), and also with participants to the Apticore-Albicore Workshop held in Perugia (Italy), particularly I. Premoli-Silva, E. Erba (Univ. Milano), A Fischer (Univ. Los Angeles), and T. Herbert (Scripps Inst. of Oceanogr., La Jolla). Thanks also go to A Mascle, who initiated this publication, and to J. P. Herbin (IFP, Paris) and W. Kuhnt (Univ. Tubingen), who reviewed the manuscript and gave many valuable suggestions R. Williams, who improved the English manuscript, is greatly acknowledged. The help of the Laboratoire de Géologie de l'Université François Rabelais (Tours) has been greatly appreciated

References

Alavi-Naini (1972) Etude géologique de la région de Djam Geol Surv Iran, Téhéran, Rep 23 288

Amedro F (1992) L'Albien du bassin anglo-parisien ammonites, zonation phylétique, séquences Bull Cent Rech Expl Prod Elf-Aquitaine 16, 1 187–233

Arnaud-Vanneau A, Arnaud H, Masse JP (1984) Faciès urgoniens du Sud-Est de la France In Debrand-Passard S (ed) Synthèse géologique de Sud-Est de la France, vol 1, stratigraphie et paléogéographie Mém BRGM 125 335–338

Arthur MA, Schlanger SO (1979) Cretaceous "oceanic anoxic events" as causal factors in development of reef-reservoir giant oil fields Bull Am Assoc Pet Geol 63 870–885

Arthur MA, Schlanger SO, Jenkyns HC (1987) The Cenomanian-Turonian Oceanic Anoxic Event, II Paleoceanographic controls on organic-matter production and preservation In Brooks J, Fleet AJ (eds) Marine Petroleum Source Rocks Geol Soc Spec Publ 26 401–420

Arthur MA, Jenkyns HC, Brumsack HJ, Schlanger SO (1990) Stratigraphy, geochemistry, and paleoceanography of organic carbon-rich Cretaceous sequences In Ginsburg RN, Beaudoin B (eds) Cretaceous resources, events and rhythms NATO ASI series, Series C Mathematical and physical sciences, vol 304 Kluwer, pp 75–119

Baird GC, Brett CE (1991) Submarine erosion on the anoxic sea floor stratinomic, palaeoenvironmental, and temporal significance of reworked pyrite-bone deposits In Tyson RV, Pearson TH (eds) Modern and ancient continental shelf anoxia Geol Soc Spec Publ 58 233–257

Berger A (1978) Long term variations of daily insolation and Quaternary climatic changes J Atmos Sci 35, 2 2362–2367

Berger A, Loutre MF, Dehant V (1989) Pre-Quaternary Milankovitch frequencies Nature 342 133

Berger A, Loutre MF, Laskar J (1992) Stability of the astronomical frequencies over the Earth's history for paleoclimate studies Science 255 560–566

Bralower TJ (1988) Calcareous nannofossil biostratigraphy and assemblages of the Cenomanian-Turonian boundary interval

implications for the origin and timing of oceanic anoxia Paleoceanography 3, 3 275–316
Bralower TJ, Thierstein HR (1984) Low productivity and slow deep-water circulation in mid-Cretaceous oceans Geology 12 614–618
Bréhéret J-G (1983) Sur des niveaux de black shales dans l'Albien inférieur et moyen du domaine vocontien (SE de la France) étude de nannofaciès et signification des paléoenvironnements Bull Mus Natl Hist Nat Paris 5, 1, C, 1 113–159
Bréhéret J-G (1984) Nannofacies of laminated horizons in Upper Barremian-Lower Aptian of the Central North Atlantic and the occidental Tethys indications of anoxic recurrent episodes 5th IAS European Meeting, Marseille, Abstr
Bréhéret J-G (1985) Indices d'un événement anoxique étendu à la Téthys alpine, à l'Albien inférieur (événement Paquier) CR Acad Sci Paris 300, II, 8 355–358
Bréhéret J-G (1988) Episodes de sédimentation riche en matière organique dans les marnes bleues d'âge aptien et albien de la partie pélagique du bassin vocontien Bull Soc Géol Fr 8, 4 349–356
Bréhéret J-G (1991) Glauconitization episodes in marginal settings as echoes of mid-Cretaceous anoxic events in the Vocontian basin In Tyson RV, Pearson TH (eds) Modern and ancient continental shelf anoxia Geol Soc Spec Publ 58 415–425
Bréhéret J-G (1992a) Organic-rich beds and early diagenetic products in the mid-Cretaceous series of the Vocontian basin, their meaning in the sedimentary dynamics Sequence stratigraphy of European basins CNRS-IFP, Dijon, May 18–20, 1992, Abstr, pp 284–285
Bréhéret J-G (1992b) Lithologic and diagenetic expression of the thinning of the Marnes bleues series (Vocontian basin) towards the shelf, the role played by short term pauses in sedimentation Platform margins Int Symp IAS, ASF, CNRS, Univ Grenoble (France), Chichilianne, May 6–8, 1992, Abstr, pp 23–24
Bréhéret J-G, Crumière J-P (1989) Organic-rich episodes in the mid-Cretaceous (Aptian to Turonian) pelagic realm of the Vocontian basin (SE France) In Cotillon P (ed) Les événements de la partie moyenne du Crétacé (Aptien à Turonien) Geobios Lyon Mém Spéc 11 205–209
Bréhéret J-G, Delamette M (1987) Séquences de dépôt et couches riches en matiére organique (CRMO) dans les marnes bleues aptiennes et albiennes du bassin vocontien In Ferry S, Rubino JL (eds) Eustatisme et séquences de dépôt dans le Crétacé du Sud-Est de la France, livret-guide de l'excursion du Groupe Français du Crétacé en fosse vocontienne (1987), GeoTrope, Lyon 1 83–93
Bréhéret J-G, Delamette M (1989) Faunal fluctuations related to oceanographical changes in the vocontian basin (S-E France) during Aptian-Albian time In Cotillon P (ed) Les événements de la partie moyenne du Crétacé (Aptien à Turonien) Geobios Lyon Mém Spéc 11 267–277
Bréhéret J-G, Caron M, Delamette M (1986) Niveaux riches en matière organique dans l'Albien vocontien, quelques caractères du paléoenvironnement, essai d'interprétation génétique Doc Bur Rech Géol Min 110 141–191
Burollet PF (1989) Table ronde "Evénéments téthysiens" au cours de la partie moyenne du Crétacé Compte rendu In Cotillon P (ed) Les événements de la partie moyenne du Crétacé (Aptien à Turonien) Geobios Lyon Mém Spéc 11 309–310
Busch DA (1971) Genetic units in delta prospecting Bull Am Assoc Pet Geol 55 1137–1154
Calvert SE, Bustin RM, Pedersen TF (1992) Lack of evidence for enhanced preservation of sedimentary organic matter in the oxygen minimum zone of the Gulf of California Geology, 20 757–760
Coccioni R, Nesci O, Tramontana M, Wezel FC, Moretti E (1987) Descrizione di un livello-guida "radiolaritico-bituminoso-ittiolitico" alla base delle Marne a fucoidi nell" Appennino umbro-marchigiano Boll Soc Geol Ital 106 183–192
Coccioni R, Franchi R, Nesci O, Perilli N, Wezel FC, Battistini F (1989) Stratigrafia, micropaleontologia e mineralogia delle Marne a Fucoidi (Aptiano inferiore – Albiano superiore) delle sezioni di Poggio le Guaine e del Fiume Bosso (Appennino umbro-marchigiano) Atti 2° Conv Fossili, Evoluzione, Ambiente, Pergola, 25–30 october 1987, 25 pp
Coccioni R, Erba E, Premoli-Silva I (1991) Litho- and biostratigraphy of the Livello Bonarelli close to the Cenomanian/Turonian boundary (Umbria-Marche Apennine, Italy) and possible paleoceanographic significance Colloque international sur les événements de la limite Cénomanien-Turonien (mai 1991) Géol Alp Mém hs 17 25–26
Coccioni R, Erba E, Premoli-Silva I (1992) Barremian-Aptian calcareous plankton biostratigraphy from the Gorgo Cerbara section (Marche, Central Italy) and implications for plankton evolution Cretaceous Res 13 517–537
Cotillon P (1971) Le Crétacé inférieur de l'arc subalpin de Castellane entre l'Asse et le Var Mém BRGM 68 313
Cross T (1988) Controls on coal distribution in transgressive-regressive cycles, Upper Cretaceous, Western Interior, USA In Sea-level changes An integrated approach SEPM Spec Publ 42 371–380
Crumière J-P (1989) Crise anoxique à la limite Cénomanien-Turonien dans le bassin subalpin oriental (Sud-Est de la France), relation avec l'eustatisme In Cotillon P (ed) Les événements de la partie moyenne du Crétacé (Aptien à Turonien) Geobios Lyon Mém Spéc 11 189–203
Crumière J-P, Crumière-Airaud C, Espitalié J (1990a) Préservation cyclique de la matière organique amorphe des sédiments au passage Cénomanien-Turonien dans le bassin vocontien (Sud-Est France) Contrôles paléo-océanographiques Bull Soc Géol Fr (8), VI, 3 469–478
Crumière J-P, Crumière-Airaud C, Espitalié J, Cotillon P (1990b) Global and regional controls on potential source-rock deposition and preservation the Cenomanian-Turonian oceanic anoxic event (CTOAE) on the European Tethyan Margin (southeastern France) In Huc AY (ed) Deposition of organic facies AAPG Stud Geol 30, Tulsa 107–118
de Boer PL (1991) Pelagic black shale-carbonate rhythms orbital forcing and oceanographic response In: Einsele G, Ricken W, Seilacher A (eds) Cycles and events in stratigraphy Springer, Berlin Heidelberg New York, pp 63–78
de Graciansky PC, Brosse E, Deroo G, Herbin JP, Montadert L, Muller C, Sigal J, Schaaf A (1982) Les formations d'âge Crétacé de l'Atlantique Nord et leur matière organique paléogéographie et milieux de dépôt Rev Inst Fr Pét 37, 3 275–337
de Graciansky PC, Brosse E, Deroo G, Herbin JP, Montadert L, Muller C, Sigal J, Schaaf A (1987) Organic-rich sediments and palaeoenvironmental reconstructions of the Cretaceous North Atlantic In Brooks J, Fleet AJ (eds) Marine petroleum source rocks Geol Soc Spec Publ 26 317–344
Demaison G (1991) Anoxia vs. productivity what controls the formation of organic-carbon-rich sediments and sedimentary rocks? Discussion Am Assoc Pet Geol Bull 75, 3 499
Deroo G, Bréhéret J-G, Crumière J-P (1988) Exemples d'études géochimiques dans un bassin à partir des échantillons d'affleurement L'exemple du "Crétacé moyen" du bassin vocontien (Sud-Est de la France) Sémin Géochimie de la matière organique IFP-ENSPM Formation Industrie, Rueil-Malmaison, 17 pp 49 fig HT

Destombes P (1980) Les ammonites de l'Albien inférieur et moyen dans le stratotype de l'Albien gisements, paléontologie, biozonation In Rat P et al (ed) L'Albien de l'Aube, les stratotypes français CNRS 5 51–194
Einsele G (1982) Limestone-marl cycles (periodites) diagnosis, significance, causes – a review In Einsele G Seilacher A (eds) Cyclic and event stratification Springer, Berlin Heidelberg New York, pp 8–53
Einsele G, Ricken W (1991) Larger cycles and sequences Introductory remarks In Einsele G, Ricken W, Seilacher A (eds) Cycle and events in stratigraphy Springer, Berlin Heidelberg New York, pp 611–616
Einsele G, Ricken W, Seilacher A (1991) Cycles and events in stratigraphy – basic concepts and terms In Einsele G, Ricken W, Seilacher A (eds) Cycles and events in stratigraphy Springer, Berlin Heidelberg New York, pp 1–19
Emerson S, Hedges JI (1988) Processes controlling the organic carbon content of open ocean sediments Paleoceanography 3, 5 621–634
Espitalié J, Deroo G, Marquis F (1985–1986) La pyrolyse Rock-Eval et ses applications Rev IFP Paris 40, 5 563–579, 40, 6 755–784, 41, 1: 73–89
Ferry S (1984) Apports détritiques dans le bassin vocontien In Debrand-Passard S (ed) Synthèse géologique du Sud-Est de la France, chapitre Crétacé inférieur Mém BRGM 125 332–334
Ferry S (1988) Contrôle eustatique de la resédimentation calcaire en fosse vocontienne (Mésozoique, S-E de la France) In Ferry S, Rubino JL (eds) Eustatisme et séquences de dépôt dans le Crétacé du Sud-Est de la France (livret-guide de l'excursion du Groupe Français du Crétacé en fosse vocontienne, mai 1988) GeoTrope Lyon 1 40–55
Ferry S (1989 a) Post field-trip comments In Ferry S, Rubino JL (eds) Mesozoic eustacy record on western Tethyan margins Post-meeting field-trip in the Vocontian Trough Publ Assoc Sédiment Fr 12 121–140
Ferry S (1989 b) Systems tract analysis in alternating basinal carbonate series free of gravity reworkings Eastern Vocontian Trough, stop 19 b Angles section Barremian stratotype In Ferry S, Rubino JL (eds) Mesozoic eustacy record on western Tethyan margins Post-meeting field-trip in the Vocontian Trough Publ Assoc Sédiment Fr 12 113–118
Ferry S (1991) Une alternative au modèle de stratigraphie séquentielle d'Exxon la modulation tectono-climatique des cycles orbitaux Géol Alp Mém hs 18 Enregistrement de l'eustatisme sur les marges ouest-téthysiennes au Mésozoique (colloque Association des Géologues du Sud-Est, Lyon, novembre 1989) pp 47–99
Ferry S, Rubino JL, Crumière JP (1988) La régression eustatique fini-cénomanienne dans le Sud-Est de la France In Ferry S, Rubino JL (eds) Eustatisme et séquences de dépôt dans le Crétacé du Sud-Est de la France (livret-guide de l'excursion du Groupe Français du Crétacé en fosse vocontienne, mai 1988) GeoTrope Lyon 1 108–121
Fischer AG (1991) Orbital cyclicity in Mesozoic strata In Einsele G, Ricken W, Seilacher A (eds) Cycles and events in stratigraphy Springer, Berlin Heidelberg New York, pp 48–62
Fischer AG, Bottjer DJ (1991) Orbital forcing and sedimentary sequences J Sed Petrol 61, 7 1063–1069
Flandrin J (1963) Remarques stratigraphiques, paléontologiques et structurales sur la région de Séderon Bull Serv Carte Géol Fr 272 815–845
Follmi KB (1989) Evolution of the mid-Cretaceous triad Platform carbonates, phosphatic sediments, and pelagic carbonates along the northern Tethys margin Lecture Notes in Earth Sciences, 23, Springer, Berlin Heidelberg New York, 153 pp
Friès G (1986) Dynamique du bassin subalpin méridional de l'Aptien au Cénomanien Thèse Doct Etat Univ Paris VI 370 pp
Friès G, Bizon G, Bizon J-J Beaudoin B (1986) Séquences sédimentaires et évolution de la microfaune de l'Aptien au Cénomanien l'exemple du Sud-Est de la France Bull Cent Rech Expl Prod Elf-Aquitaine 10, 2 393–403
Fromaget J, Varcin P (1954) Nouvelle contribution à la connaissance du Crétacé moyen dans la fosse vocontienne C R Acad Sci Paris 238 1048–1050
Fullthorpe CS (1991) Geological controls on seismic sequence resolution Geology 19 61–65
Funnell BM (1987) Anoxic non-events, alternative explanations In Brooks J, Fleet AJ (eds) Marine petroleum source rocks Geol Soc Spec Publ 26 421–422
Gaida KH, Gedenk R, Kemper E, Michaelis W, Scheuch R, Schmitz HH, Zimmerle W (1981) Lithologische, mineralogische und organisch-geochemische Untersuchungen an Tonsteinen und Tonmergelsteinen der Unterkreide Nordwestdeutschlands (unter besonderer Berucksichtigung der Schwarzschiefer) Geol Jahrb A 58 15–47
Grotsch J (1992) Stacking pattern of shallowing-upward cycles on a mid-Cretaceous open ocean carbonate platform (Gavrovo zone, NW-Greece) Sequence stratigraphy of European basins CNRS-IFP, Dijon, May 18–20, 1992, Abstr, pp 416–417
Guillocheau F (1991) Modalités d'empilement des séquences génétiques dans un bassin de plate-forme (Dévonien armoricain) nature et distorsion des différents ordres de séquences de dépôts emboîtées Bull Cent Rech Expl Prod Elf-Aquitaine 15, 2 383–410
Guillocheau F, Lejay A, Homewood P, Hoffert M (1989) Effet de variations du niveau relatif de la mer à très haute fréquence sur l'agencement de corps sédimentaires côtiers tidaux importance relative des contrôles auto- et allocycliques C R Acad Sci Paris, 309, sér II 1047–1053
Hallam A, Bradshaw MJ (1979) Bituminous shales and oolitic ironstones as indicators of transgressions and regressions J Geol Soc Lond 136 157–164
Haq BU, Hardenbol J, Vail PR (1987) Chronology of fluctuating sea levels since the Triassic Science 235 1156–1166
Harland WB, Armstrong RL, Craig LE, Smith AG, Smith DG (1989) A geologic timescale Cambridge University Press, Cambridge
Herbert TD (1992) Paleomagnetic calibration of Milankovitch cyclicity in Lower Cretaceous sediments Earth Planet Sci Lett 112 15–28
Herbert TD, Fischer AG (1986) Milankovitch climatic origin of mid-Cretaceous black shale rhythms in central Italy Nature 321 739–743
Herbin JP, Deroo G (1982) Sédimentologie de la matière organique dans les formations du Mésozoique de l'Altlantique Nord Bull Soc Géol Fr (7), 24, 3 497–510
Herbin JP, Masure E, Roucaché J (1987) Cretaceous formations from the lower continental rise off Cape Hatteras organic geochemistry, dinoflagellate cysts, and the Cenomanian/Turonian boundary event at sites 603 (Leg 93) and 105 (Leg 11) In Van Hinte JE, Wise SW Jr et al (eds) Initial Reports of the Deep Sea Drilling Project, vol 93 US Government Printing Office, Washington, pp 1139–1162
Hilgen FJ (1987) Sedimentary rhythms and high-resolution chronostratigraphic correlations in the Mediterranean Pliocene News P Stratigr 17, 2 109–127
Hilgen FJ (1991) Astronomical calibration of Gauss to Matuyama sapropels in the Mediterranean and implication for the geomagnetic polarity time scale Earth Planet Sci Lett 104 226–244
Homewood P, Guillocheau F, Eschard R, Cross TA (1992) Corrélations haute résolution et stratigraphie génétique une

démarche intégrée Bull Centres Rech Explor-Prod Elf Aquitaine, 16, 2 357–381
Jacob Ch (1907) Etudes paléontologiques et stratigraphiques sur la partie moyenne des terrains crétacés dans les Alpes françaises et les régions voisines Thèse, Paris, 314 pp
Jacobs DK, Sahagian DL (1993) Climate-induced fluctuations in sea level during non-glacial times Nature 361 710–712
Jenkyns HC (1980) Cretaceous anoxic events from continents to oceans Geol Soc Lond J 137 171–188
Jenkyns HC (1985) The early Toarcian and Cenomanian-Turonian anoxic events in Europe comparisons and contrasts Geol Rundsch 74, 3 505–518
Jensen TF, Buchardt B (1987) Sedimentology and geochemistry of the organic carbon-rich Lower Cretaceous Sola Formation (Barremian-Albian), Danish North Sea In Brooks J, Glennie K (eds) Petroleum geology of north west Europe Graham & Trotman, London, pp 431–440
Jensen TF, Holm L, Frandsen N, Michelsen O (1986) Jurassic-Lower Cretaceous lithostratigraphic nomenclature for the Danish Central Trough Dan Geol Unders Afh Raekke 1, 12 65
Kauffman EG (1986) High-resolution event stratigraphy regional and global Cretaceous bioevents In Walliser OH (ed) Global bioevents a critical approach Lecture Notes in Earth Sciences Springer, Berlin Heidelberg New York, pp 279–335
Kemper E, Zimmerle W (1978) Die anoxischen Sedimente der praoberaptischen Unterkreide NW-Deutschlands und ihr palaogeographischer Rahmen Geol Jahrb A, 45 3–41
Kennedy WJ, Odin GS (1982) The Jurassic and Cretaceous time scale in 1981 In Odin GS (ed) Numerical dating in stratigraphy John Wiley, New York, pp 557–592
Lancelot Y, Hathaway JC, Hollister CD (1972) Lithology of sediments from the western North Atlantic, Leg 11, DSDP In Hollister CD, Ewing JI et al (eds) Initial Reports of the Deep Sea Drilling Project, 11 US Government Printing Office, Washington DC, pp 901–949
Larson RL (1991) Geological consequences of superplumes Geology 19 963–966
Leckie RM (1989) A paleoceanographic model for the early evolutionary history of planktonic foraminifera Palaeogeogr Palaeoclimatol Palaeoecol 73 107–138
Magniez-Jannin F (1991) Renouvellements de foraminifères et séquences de dépôt dans le Crétacé inférieur du Bassin vocontien Bull Soc Géol Fr 162, 5 887–895
Magniez-Jannin F (1992) Foraminifers tolerance to dysaerobic conditions within early Cretaceous depositional sequences of the Vocontian basin (S E France) Sequence stratigraphy of European basins CNRS-IFP, Dijon, May 18–20 1992, Abstr, pp 284–285
Moullade M (1966) Etude stratigraphique et micropaléontologique du Crétacé inférieur de la fosse vocontienne. Thèse, Lyon, Doc Lab Géol Fac Sci Lyon 15 369
Muller C, Schaaf A, Sigal J (1983) Biochronostratigraphie des formations d'âge Crétacé dans les forages du DSDP dans l'Océan Atlantique Nord 1ère partie Rev Inst Fr Pétrole 38, 6 683–708
Muller C, Schaaf A, Sigal J (1984) Biochronostratigraphie des formations d'âge Crétacé dans les forages du DSDP dans L'Océan Atlantique Nord 2ème partie Rev Inst Fr Pétrole 39, 1 3–23
Mutterlose J (1992) Migration and evolution patterns of floras and faunas in marine Early Cretaceous sediments of NW Europe. Palaeogeogr Palaeoclimatol Palaeoecol 94 261–282
Mutterlose J, Harding I (1987) Phytoplankton from the anoxic sediments of the Barremian (Lower Cretaceous) of North-West-Germany. Abh Geol B-A, Wien, 39 177–215
Noel D (1968) Nature et genèse des alternances de marnes et de calcaires du Barrémien supérieur d'Angles (Fosse vocontienne, Basses Alpes) C R Acad Sci Paris 266, D 1223–1225
Odin GS, Hunziker JC (1982) Radiometric dating of the Albian-Cenomanian boundary In Odin GS (ed) Numerical dating in stratigraphy John Wiley, New York, pp 537–556
Odin GS, Odin C (1990) Echelle numérique des temps géologiques Géochronique, Paris 35 12–21
Ozimic S (1988) Palaeoenvironmental and geological controls on the quality of the Toolebuc Formation oil shale, Queensland Rec Bur Min Resour Geol Geophys 42 108–116
Paquier V (1900) Recherches géologiques dans le Diois et les Baronnies orientales Thesis, Grenoble, 402 pp
Paskoff R (1987) Les variations du niveau de la mer Recherche 18, 191 1010–1019
Pedersen TF, Calvert SE (1990) Anoxia vs productivity what controls the formation of organic-carbon-rich sediments and sedimentary rocks? Am Assoc Pet Geol Bull 74 454–466
Plint AG, Eyles N, Eyles CH, Walker RG (1992) Control of sea level change In Walker RG, James NP (eds) Facies models Response to sea level change Geological Association of Canada, Waterloo, pp 15–25
Pratt LM, King JD (1986) Variable marine productivity and high eolian input recorded by rhythmic black shales in mid-Cretaceous pelagic deposits from central Italy Paleoceanography 1, 4 507–522
Premoli-Silva I (1991) The latest Cenomanian Anoxic Event an update Colloque international sur les événements de la limite Cénomanien-Turonien (mai 1991) Géol Alp Mém hs 17 109
Premoli-Silva I, Erba E, Tornaghi ME (1989 a) Paleoenvironmental signals and changes in surface fertility in mid-Cretaceous Corg-rich pelagic facies of the fucoid marls (central Italy) In Cotillon P (ed) Les événements de la partie moyenne du Crétacé (Aptian à Turonien) Geobios Lyon Mém Spéc 11 225–236
Premoli-Silva I, Ripepe M, Tornaghi ME (1989 b) Planktonic foraminiferal distribution record productivity cycles evidence from the Aptian-Albian Piobbico core (central Italy) Terra Nova 1 443–448
Rat P et al (1980) L'Albien de l'Aube, les stratotypes français CNRS 5 446
Rawson PF, Mutterlose J (1983) Stratigraphy of the Lower B and basal Cement Beds (Barremian) of the Speeton Clay, Yorkshire, England Proc Geol Assoc 94 (2) 133–146
Renz O (1979) Ammonoidea from the Lower Cretaceous of hole 402 A in the bay of Biscay, DSDP leg 48 In Montadert L, Roberts DG et al (eds) Initial Reports of the Deep Sea Drilling Project, 48 US Government Printing Office, Washington, pp 641–645
Revelle R (ed) (1990) Sea level change National Research Council, studies in geophysics National Academy Press, Washington, DC, 234 pp
Ricou LE (1987) The Tethyan oceanic gates a tectonic approach to major sedimentary changes within Tethys Geodyn Acta 1, 4/5 225–232
Robaszynski F (1989) Evénements à l'échelle globale pendant la partie moyenne du Crétacé Compte-rendu de la table ronde tenue au terme du Colloque du Groupe Français du Crétacé à Lyon le 28 octobre 1987 In Cotillon P (ed) Les événements de la partie moyenne du Crétacé (Aptien à Turonien) Géobios Mém Spec 11 311–319
Roemmich D (1992) Ocean warming and sea level rise along the southwest US coast Science 257 373–375
Rossignol-Strick M (1987) Variations de l'insolation, formation des sapropèles de méditerranées au quaternaires Implications géologiques pour les black-shales du Crétacé. In Trichet J (ed) Géologie de la matière organique (colloque Orléans, mars 1985) Mém Soc Géol Fr NS 151 7–26
Schlanger SO, Jenkyns HC (1976) Cretaceous oceanic anoxic events causes and consequences. Geol Mijnbouw 55 179–184

Schlanger SO, Arthur MA, Jenkyns HC, Scholle PA (1987) The Cenomanian-Turonian Oceanic Anoxic Event. I Stratigraphy and distribution of organic-carbon-rich beds and the marine ∂ 13 C excursion In Brooks J, Fleet AJ (eds) Marine petroleum source rocks Geol Soc Spec Publ 26 371–399
Seilacher A (1991) Events and their signatures – an overview In Einsele G, Ricken W, Seilacher A (eds) Cycles and events in stratigraphy Springer, Berlin Heidelberg New York, pp 222–226
Sliter WV (1989) Aptian anoxia in the Pacific basin Geology 17 909–912
Sliter WV (1992) Cretaceous planktonic foraminiferal biostratigraphy and paleoceanographic events in the Pacific ocean with emphasis on indurated sediment In Ishizaki K, Saito T (eds) Centenary of Japanese Micropaleontology pp 281–299
Strasser A (1988) Shallowing-upward sequences in Purbeckian peritidal carbonates (lowermost Cretaceous, Swiss and French Jura Mountains) Sedimentology 35 369–383
Talbi R (1986) Intérêt pétrolier et potentialité bitumineuse des faciès d'âge Vraconien Application à la prospection des schistes bitumineux Mém DEA, Faculté des Sciences Tunis, 95 pp
Thierstein HR (1973) Lower Cretaceous calcareous nannoplankton biostratigraphy Abh Geol B-A 29 52 pp
Thomel G (1960) Observations sur l'Aptien et sur l'Albien des environs de Puget-Théniers (Alpes-Maritimes) Bull Soc Géol Fr Paris (7), II 87–93
Thomel G (1964) Les zones d'Ammonites de l'Aptien des Basses-Alpes C R Acad Sci Paris 258, (9) 4308–4310
Thurow J, Kuhnt W (1986) Mid-Cretaceous of the Gibraltar Arch area In Summerhayes CP, Shackleton NJ (eds) North Atlantic palaeooceanography Spec Publ Geol Soc Lond 21 423–444
Thurow J, Moullade M, Brumsack HJ, Masure E, Taugourdeau-Lantz J, Dunham K (1988) The Cenomanian/Turonian boundary event (CTBE) at hole 641 A, ODP leg 103 (compared with the CTBE interval at site 398) In Boillot G, Winterer EL et al (eds) Proceedings of the Ocean Drilling Program, Scientific results, 103 587–634
Thurow J, BCCP-group (1992) Cycles and sequences in the boreal Lower Cretaceous Sequence stratigraphy of European basins. CNRS-IFP, Dijon, May 18–20, 1992, Abstr. pp 284–285
Tyson RV, Pearson TH (1991) Modern and ancient continental shelf anoxia an overview In Tyson RV, Pearson TH (eds) Modern and ancient continental shelf anoxia Geol Soc Spec Publ 58 1–24
Vail PR, Audemard A, Bowman SA, Eisner PN, Perez-Cruz C (1991) The stratigraphic signatures of tectonics, eustacy and sedimentology – an overview In Einsele G, Ricken W, Seilacher A (eds) Cycles and events in stratigraphy Springer, Berlin Heidelberg New York, pp 617–659
Vail PR, Mitchum RM, Thompson S III (1977) Seismic stratigraphy and global changes of sea level, part 4 global cycles of related changes of sea level In Payton CE (ed) Seismic stratigraphy – application to hydrocarbon exploration Am Assoc Pet Geol Mem 26 83–98
Van Buchem FSP, McCave IN (1990) Cyclic sedimentation patterns in lower Lias mudstones of Yorkshire (GB) Terra Nova 1 461–467
Van Wagoner JC, Posamentier HW, Michum RM, Vail PR, Sarg JF, Loutit JS, Hardenbol J (1988) An overview of the fundamentals of sequence stratigraphy and key definitions In Wilgus C, Hastings B, Ross C, Posamentier H, Van Wagoner J, Kendall LG (eds) Sea-level changes, an integrated approach Soc Econ Paleontol Mineral Spec Publ 42 39–45
Vialli V (1949) Nuova fauna ad Ammoniti der Barremiano superiore Lombardo Atti Soc Ital Sci Nat Mus Civ Stor Nat Milano 88, 1–2 35–64
Weissert H, McKenzie J, Hochulie P (1979) Cyclic anoxic events in the Early Cretaceous Tethys Ocean Geology 7 147–151
Wezel CF (1985) Facies anossiche ed episodi geotettonici globali Giorn Geol (3), 47, 1–2 281–286
Wignall PB (1991) Model for transgressive black shales? Geology 19 167–170
Wignall PB, Hallam A (1991) Biofacies, stratigraphic distribution and depositional models of British onshore Jurassic black shales In Tyson RV, Pearson TH (eds) Modern and ancient continental shelf anoxia Geol Soc Spec Publ 58 291–309

Reservoir Aspects in Relationship with Sequence Boundaries in Carbonate Platforms

A. Arnaud-Vanneau and E. Carrio-Schaffhauser

Three types of reservoir associated with diagenetic anomaly structures overlain by three important sequence boundaries age recognized in the Lower Cretaceous carbonate platforms of the Jura and the subalpine chains (Fig. 1; Arnaud-Vanneau et al. 1991). These reservoirs, several meters thick (5 to 15 m), present a conformable stratigraphic habit and are located in the inner platform. Their appearance was due to an aerial exposition, but they developed later during diagenetic evolution.

The first of these diagenetic structures, corresponding to root molds, appears during the Upper Berriasian (Sb B5) of the Chartreuse massif (La Chambotte area). These decimetric molds show a succession of centripetal cementations with, in places, dolomitization (Fig. 2 A). After an external pedogenetic rim, probably bacterial in origin, a first cementation develops, several millimeters thick, made up of micritic infillings. This first stage of infilling seems to be due to a weak retraction of roots inducing muddy infiltrations. The following type of infilling shows a coarse bioclastic limestone interpreted as tempestites according to the included foraminifera that usually appear in more external parts of the platform. So, this type of sediment fills a more important void due to a strong retraction of roots. The last stage of cementation, made up of sparitic calcite, reveals a complete disappearance of the roots. By isotopic analysis (Arnaud-Vanneau 1980), these cements have been interpreted as a late burial stage of diagenesis. A central cavity is always preserved.

In this case, the initial porosity is attributed to biological activity, but the actual voids are due only to a lack of cementation. This type of reservoir exists in plurikilometric areas, behind the Berriasian platform margin. It reveals, at this period, the development of mangroves, under tropical conditions, during a brief, weak sea level fall.

In the Vercors massif (locality of Gorges du Nant), under a major sequence boundary during the middle Barremian (Sb B4), paleokarstification and important dolomitization-dedolomitization processes mark another type of reservoir structure. In fined grainstones with keystone vugs (beach record), a thick dolomitized level appears (up to 7 m), which appears to correspond to a mixed zone at the transition between marine and meteoric waters, according to its stratigraphical pattern, thickness, occurrence with a compatible geological setting (beach position) and association with meteoric vadose fabrics as karst structures. In the dolomitized matrix, interconnected karstic microcavities (mm to cm in size) with bedded infillings can be observed.

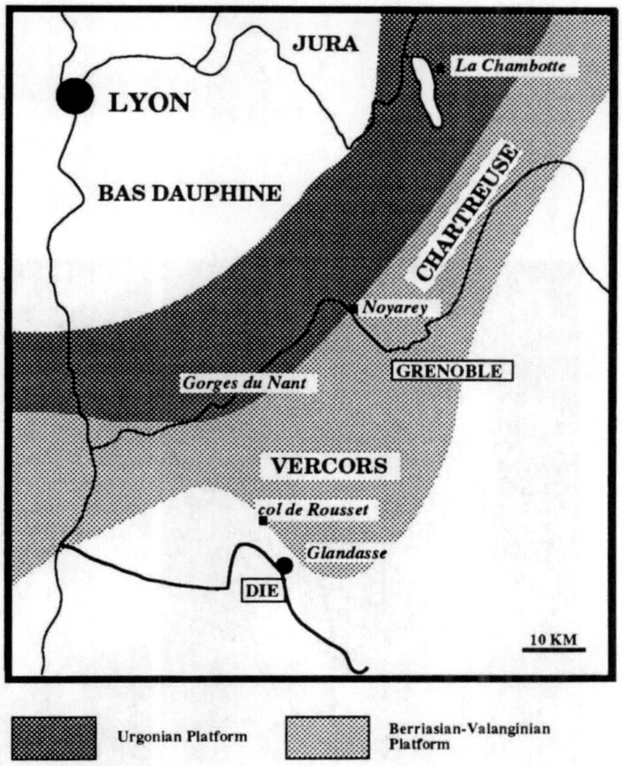

Fig. 1. Location map of the Vercors and Chartreuse massifs showing limits of the Berriaso-Valanginian and Urgonian platforms

Towards the top of this sequence, these tiny vugs increase in size and grade to karstic megacavities (meters to plurimeters in size) with Turonian infillings dated by a bryozoan fauna. Therefore, the meteoric dissolution activity begins at the middle Barremian and develops later in the Turonian period (Arnaud-Vanneau 1980).

This kind of reservoir is present in the inner part of the platform, along a decakilometric belt parallel to the platform margin. This megaporous diagenetic anomaly reveals a fall in sea level, in a relatively humid climate owning to the karst development and the lack of associated evaporitic sequences.

The last of these reservoir structures, corresponding to two stratigraphical chalky layers, appears in the Vercors

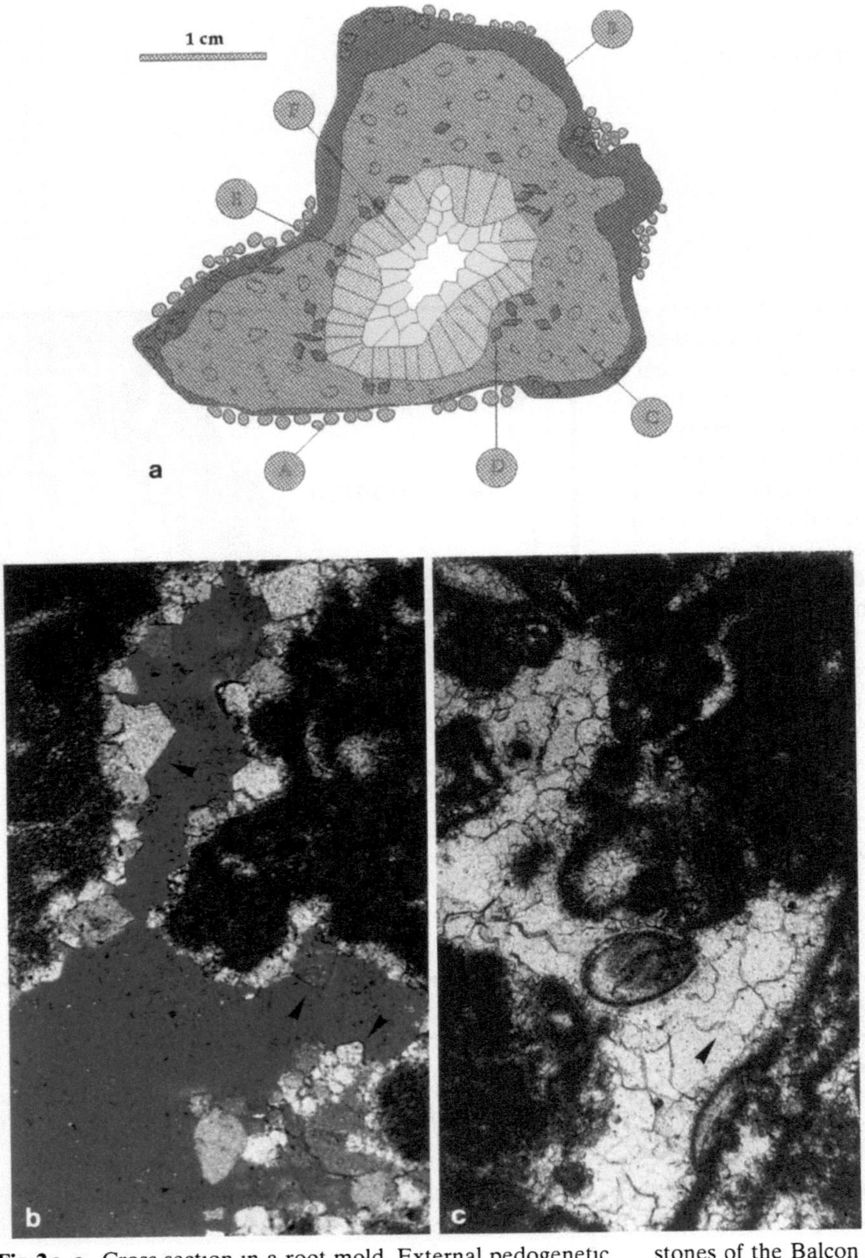

Fig. 2 a–c. Cross section in a root mold External pedogenetic rim, probably bacterial in origin, first stage of infilling made up of micritic sediments, coarse bioclastic limestones of the second stage of infilling interpreted as tempestites, dolomitization, *E* and *F* last stages of cementation made up of late burial calcite Variations in the diagenetic evolution of the Urgonian limestones of the Balcon des Ecouges Partial cementation of the porous levels A primary porosity is observed according to the preserved shape of the calcite crystals *(arrowheads)*, the cementation is complete in the indurated levels ending with mosaic calcite *(arrowheads)*

massif, at the locality called Balcon des Ecouges, at the Barremian-Aptian boundary (SbA1), between the Upper Barremian late highstand and the first Aptian transgressive levels. This sequence boundary is associated with a general sea level fall inducing local emersions (presence of charophytes). The parasequences of this late highstand system tract are made up of biosparitic limestones with micritized oncolites and small rudists They show two porous levels (N = 16%) interbedded with indurated, nonporous limestones (less than 1%) The only difference between these two types of limestones with the same microfacies consists in variations in their diagenetic evolution. In the porous carbonates, the cementation is incomplete and consists of a calcitic rim fringing the micritized oncolites. In the indurated limestones, the cementation finishes with mosaic calcite (Fig. 2 B, C). The type of cementation is the same in the intergranular, porous network and in the voids left by dissolution inside the shells According to the preserved shape of the calcite crystals (without dissolution marks) forming the partial cement in the porous deposits, this important porosity (μm to mm) can be regarded as primary and due to a lack of cementation associated with a weak compaction. In places, the presence of interpreted, asymmetric cements allows these deposits to be correlated to a vadose meteoric zone. Moreover, in the indurated limestones, the mosaic cement marks the phreatic zone.

The presence of two superimposed vadose and phreatic zones in succession could be due to the superposition of two aquifer systems. A local, floating meteoric water lens overlies a permanent meteoric aquifer as observed today in islands like Bermuda, where rainfalls are strongly seasonal, and the presence of the floating lens thus tends to be cyclic (Moore 1989). This type of reservoir is restricted to small areas (around 10 km^2) of the inner platform, near the margin.

Even if these reservoirs are linked to major sequence boundaries, their nature and aspect are closely connected to the paleoclimatic and paleogeographic situation. In the Jura and the subalpine chains, environments change from equatorial-tropical Berriasian to arid Aptian, from mangroves (indicated by root molds) to a subdesertic zone (unlithified sediments).

References

Arnaud-Vanneau A (1980) Micropaléontologie, paléoécologie et sédimentologie d'une plate-forme carbonatée de la marge passive de la Téthys. l'Urgonien du Vercors septentrional et de la Chartreuse (Alpes occidentales) Géol Alp 1: 267 pp

Arnaud-Vanneau A, Arnaud H, Beck Ch, Carrio-Schaffhauser E, Cousin M, Dumont Th (1991) Le concept de stratigraphie séquentielle applications dans le domaine des plates-formes carbonatées et de leurs marges CNRS, Ecole Thématique, Grenoble-Chambéry, 36 pp

Moore CH (1989) Carbonate diagenesis and porosity Elsevier, Amsterdam, 339 pp

Eastern and Northern France

Transfer Zone in the Southern Jura Thrust Belt (Eastern France): Geometry, Development, and Comparison with Analogue Modeling Experiments

Y. Philippe

Abstract. The Jura thrust belt constitutes the arcuate leading edge of the French Alps, which was developed during the Neogene and forms a conventional thin-skinned fold-and-thrust belt, whose general orientation is NNE-SSW. It consists of a thin Mesozoic cover displaced towards the NW above a regional sole thrust located in the Upper Triassic evaporitic layers, during the Late Miocene – Early Pliocene Alpine phase. During the Late Eocene and Oligocene, this area was involved in a significant extensional tectonic phase which created basement topography that has been preserved below the decollement level. These basement highs contitute potential traps, as shown by recent boreholes where oil has been tested in the southern part of the belt. Reservoirs are in the sandstones of the Buntsandstein Formation (Early Triassic) and dolomites of the Muschelkalk Formation (Middle Triassic), and potential source rocks are believed to be coal measures of Stephanian or Autunian age. In the southern part of the belt, the frontal thrust of the Jura belt is laterally linked to the frontal thrusts of the Chartreuse massif, with an offset of up to about 55 km. This area is interpreted as a complex transfer zone, which accommodates the southern termination of the belt. Balanced serial cross sections and microtectonic analysis of this zone reveal a fold-and-thrust geometry without any significant strike-slip motion along external thrusts. This particular geometry is due to a local N 50–60 ° shortening, which is strongly oblique with respect to the regional shortening of the southern Jura belt (N 100 °), and to the Chartreuse massif shortening (N 110 °). We interpret this structural setup as the result of the conjugated effect of three parameters: (1) the lateral disappearance of the basal decollement level (main parameter), (2) the influence of thickness variations in the cover between the internal and external part of the belt (indentation), and (3) the local tectonic control by previous normal faults in the basement. According to these data, analogue modeling experiments were performed and analyzed by computerized X-ray tomography in order to validate our hypothesis and to constrain the boundary conditions that produced this transfer zone and controlled its evolution.

Introduction

For many years, numerous works in structural geology dealing with fold-and-thrust belts have made it possible to identify the different deformation styles in compressive regime and to describe the geometry of structures associated with these deformations, on several scales (e.g., Bally et al. 1966; Dahlstrom 1970; Boyer and Elliott 1982). In the light of modern structural geology concepts, these studies have more particularly demonstrated the role of decollement levels located at the base and/or within sedimentary covers involved in thrust tectonics. External or foreland structural zones in mountain belts generally display series of folds and imbricate thrusts implying a sedimentary cover displaced above a passive basement by means of a weak stratigraphical level (salt, evaporites, shales, etc.). In a zone that underwent homogeneous shortening, the structures generated must persist along the strike of the area. As a matter of fact, ramps and associated folds are not cylindrical but have along-strike ends, which are oblique (or transverse) with respect to the transport direction (Thomas 1990). These anomalous tectonic features, which can be observed on all scales, are particularly obvious when they affect frontal parts of fold-and-thrust belts. They are therefore expressed in map view by a strike offset of the frontal thrust.

The case of the southern termination of the Jura foreland thrust belt provides a spectacular example of a transverse zone in front of the French Alps (Fig. 1). In the southern part of the belt, the External Jura thrusts are laterally linked to the frontal thrusts of the Chartreuse subalpine massif, with a lateral offset of up to about 55 km. Therefore this area can be interpreted as a transfer zone which accommodates the southern termination of the belt.

Regional Geological Setting

The Jura is an arcuate-shaped thin-skinned fold-and-thrust belt which constitutes the most external part of the western Alps (Fig. 1). It consists of a Mesozoic cover folded and displaced towards the northwest during the Mio-Pliocene Alpine phase, above a regional detachment level hosted in the Upper Triassic salt layers. This belt is flanked by two main structural zones: to the west and north, the Bresse and Rhine grabens, and to the southeast, the Molasse basin. The Bresse and Rhine grabens take part in the Oligocene West-European rift system (Maury and Varet 1980), whereas the Molasse basin corresponds to an Oligo-Miocene synorogenic foreland

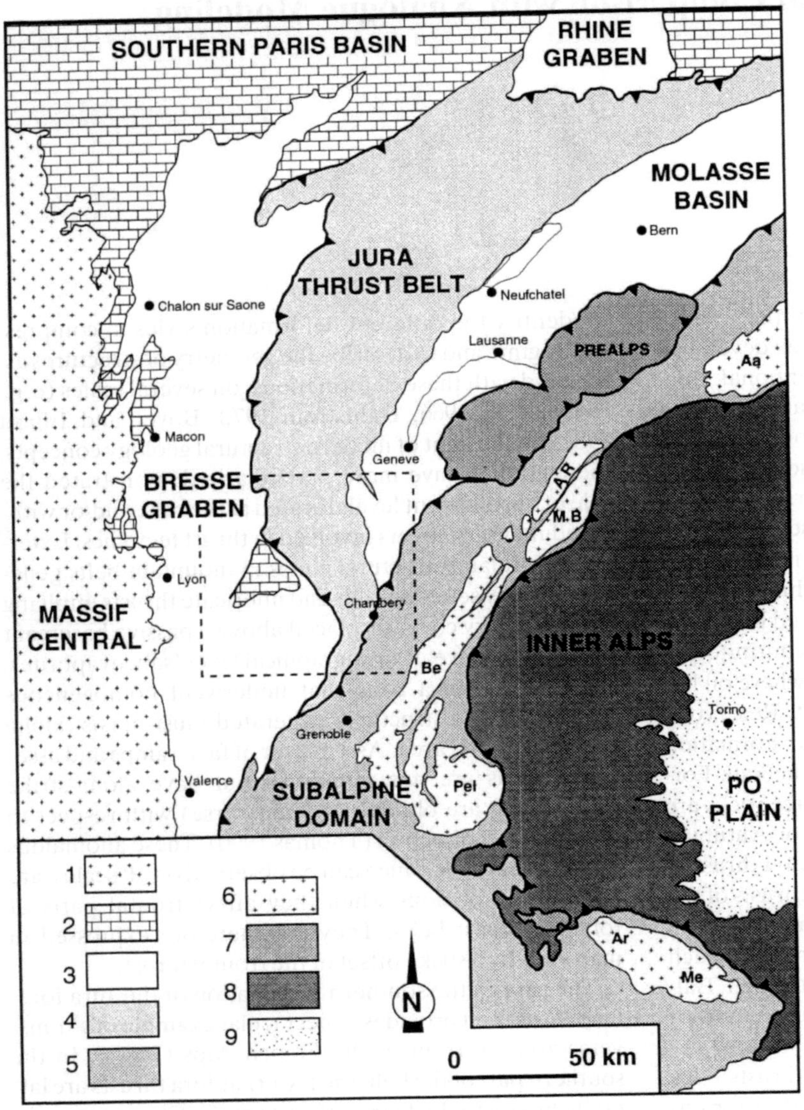

Fig. 1. Structural map of the western Alpine Belt and its foreland in France *1* Paleozoic basement, *2* autochtonous Mesozoic cover, *3* Tertiary, *4* folded Mesozoic cover of the Jura thrust belt, *5, 6* Subalpine domain (*5* Mesozoic cover, *6* Paleozoic basement), *7* Prealpine klippen, *8* Inner Alps, *9* Cenozoic fill of the Po plain *Aa* Aar massif, *Ar* Argentière massif, *AR* Aiguilles Rouges massif, *Be* Belledonne massif, *Me* Mercantour massif, *MB* Mont Blanc massif, *Pel* Pelvoux massif The *dotted square* indicates the boundaries of the study area

trough associated with the continental collision and crustal thrusting in the Alps (Mugnier and Ménard 1986; Deville et al., this Vol.). The Jura mountain proper is divided into two major structural zones (Chauve et al. 1980):

1. To the southeast, the Internal Jura or Haute-Chaîne, which is characterized by large deformations generally corresponding to box-shaped folds (Buxtorf 1916; Laubscher 1977; Mitra and Namson 1989), and sometimes considerable overthrustings (e.g., the Risoux Nappe, see Winnock 1961; Bitterli 1972).
2. To the northwest, the External Jura in which we can distinguish two subzones displaying opposite structural styles:
   - The Plateaux zone striking between the internal zone and the Faisceaux jurassiens, corresponding to slightly deformed tabular rocks.
   - The Faisceaux Zone on the western part of the belt, on the contrary affected by strong deformation characterized by a succession of numerous small-scale imbricate thrusts and strike-slip structures.

The belt is bounded in the west by a major frontal thrust which places Mesozoic cover rocks on top of Tertiary sediments of the Bresse graben, as demonstrated by several boreholes (Enay 1982; Chauve et al. 1988) or field observations (Philippe 1991).

To the east, the Jura sedimentary cover underlies the Molasse basin, but outcrops locally by means of ramp anticlines (e.g., the Salève anticline)

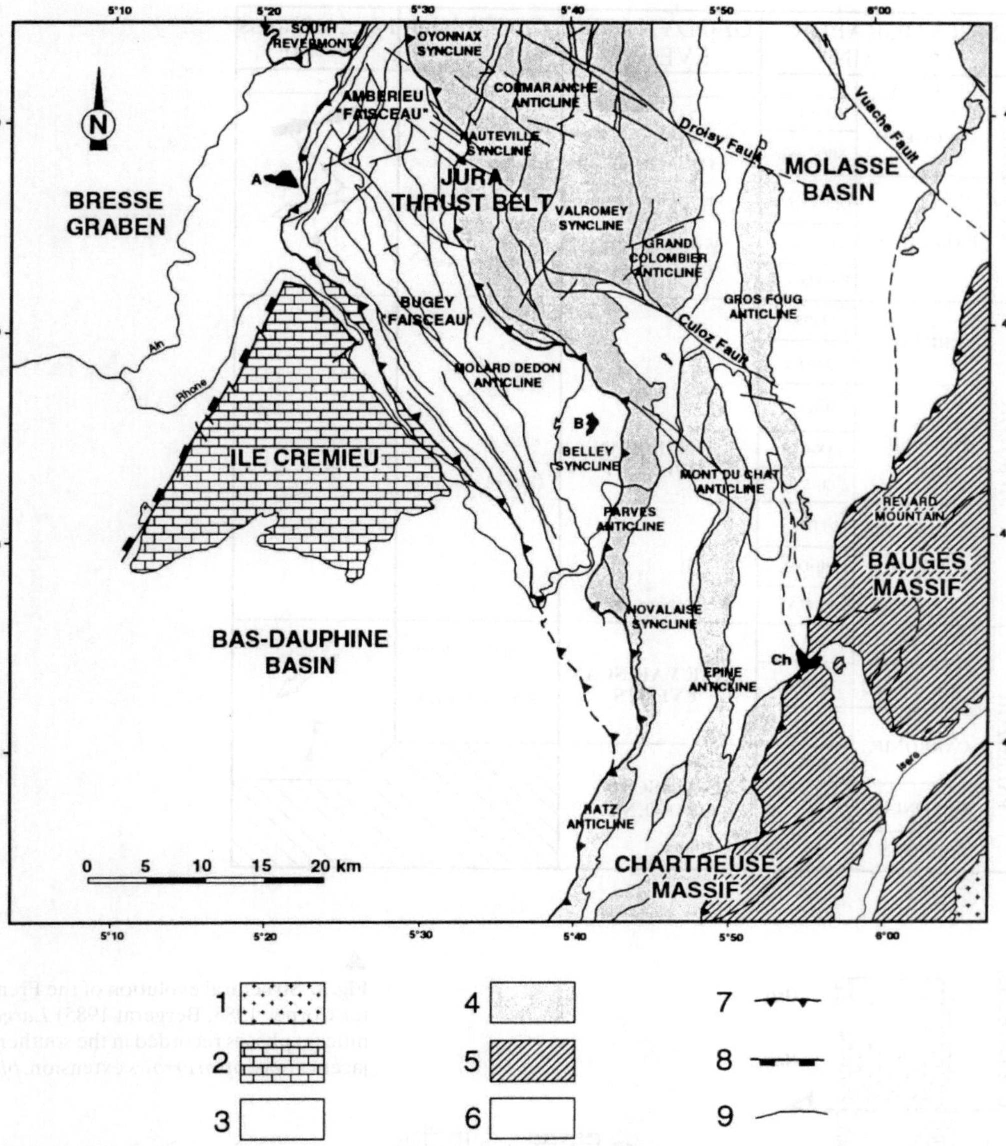

Fig. 2. Structural map of the southern Jura fold-and-thrust belt and adjacent areas *1* Paleozoic basement, *2* autochtonous Mesozoic cover, *3, 4* folded Mesozoic cover of the Jura mountain belt (*3* External Jura, *4* Internal Jura), *5* Mesozoic cover of the Subalpine Domain, *6* Cenozoic; *7* major thrusts; *8* normal faults; *9* minor thrusts or strike-slip faults, *A* Ambérieu-en-Bugey, *B* Belley; *Ch* Chambery

The Study Area

Three different structural domains are juxtaposed in the study area (Fig. 2):

1. The Mesozoic folded cover of the Jura mountain belt (including the displaced cover underlying the Savoy molasse basin), and the western Chartreuse fold-and-thrust system.
2. The Bresse graben and Bas Dauphiné basin.
3. The reduced autochthonous Mesozoic cover of the Ile Crémieu, overlying a basement high.

Stratigraphic and Paleogeographic Data

The southern Jura and adjacent areas show several stratigraphical units representing the geodynamic evolution of the region from Late Paleozoic to present (Fig. 3).

The Paleozoic basement mainly outcrops to the west of the study area in the French Massif Central, but locally appears as a small high at Chamagnieu in the west of the Ile Crémieu. However, it has been reached in several exploration wells and consists of various deformed metamorphic rocks and intrusive granites associated with the Variscan orogeny (see references and location maps in Debrand-Passard an Courbouleix 1984).

On top of these metamorphic and intrusive units, Late Paleozoic post-orogenic detrital deposits including coal

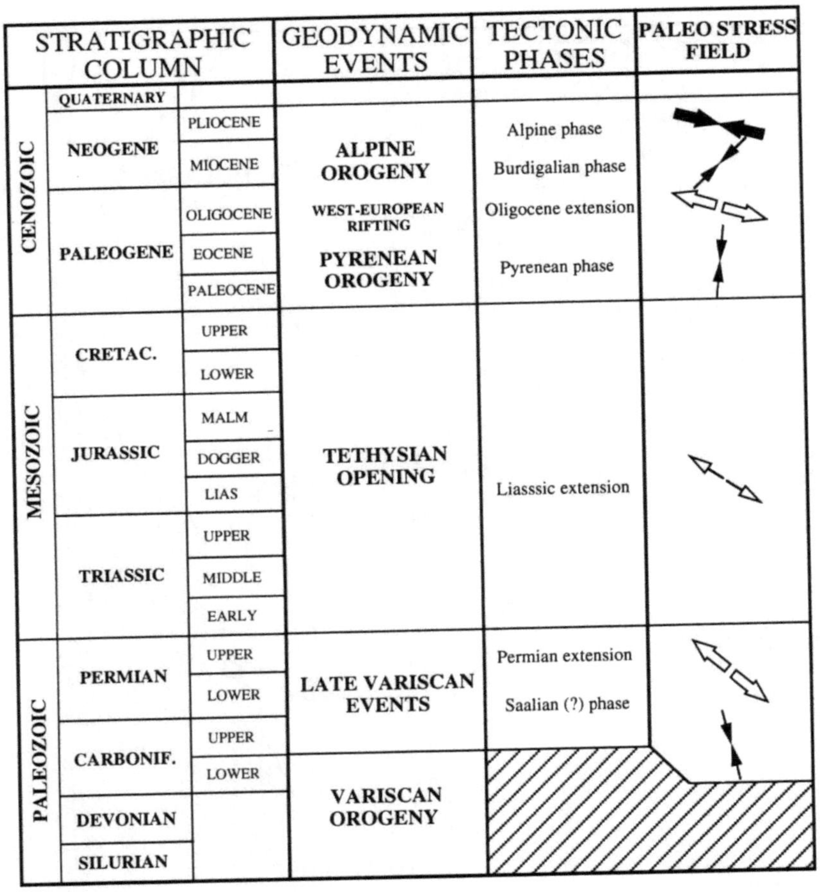

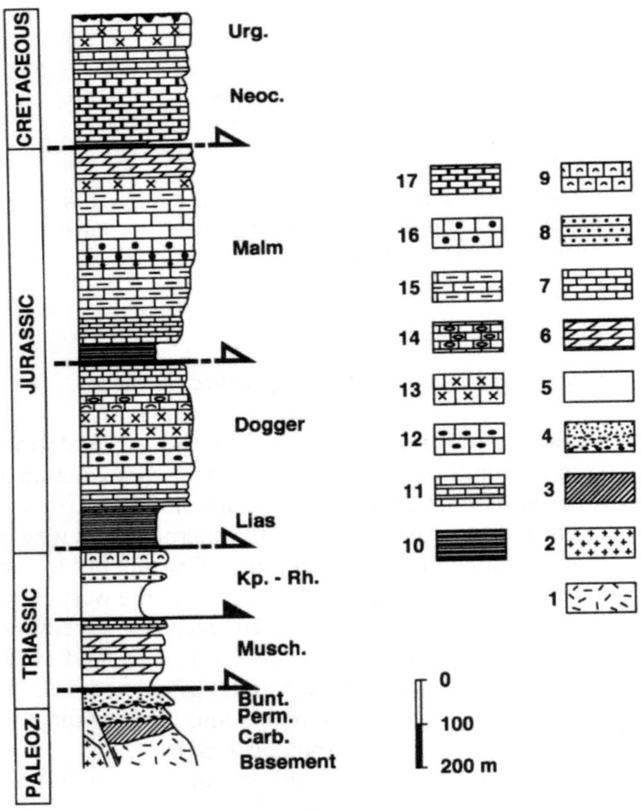

Fig. 3. Structural evolution of the French Alpine foreland (After Cornec 1983, Bergerat 1985) *Large arrows* indicate the significant phases recorded in the southern Jura thrust belt and adjacent areas, *open arrows* extension, *black arrows* compression

Fig. 4. Synthetic lithostratigraphic column of the southern Jura (Enay 1982, Philippe 1991) *Arrows* indicate the potential décollement levels (*black arrow* shows the location of the regional sole thrust) *1* Metamorphic basement, *2* granitic intrusive, *3* coal measures, *4* conglomerates, pebbles, and sandstones, *5* evaporites and massive salt layers, *6* dolomites, *7* limestones, *8* sandstones, *9* shelly limestones, *10* marls, *11* alternate marls and limestones, *12* oncolitic limestones, *13* reef limestones, *14* cherty limestones, *15* argillaceous limestones, *16* nodulous limestones, *17* bioclastic and spathic limestones, *Bunt* Buntsandstein, *Car* Carboniferous, *Kp* -Rh Keuper-Rhetian, *Musch* Muschelkalk, *Neoc* Neocomian, *Perm* Permian, *Urg* Urgonian

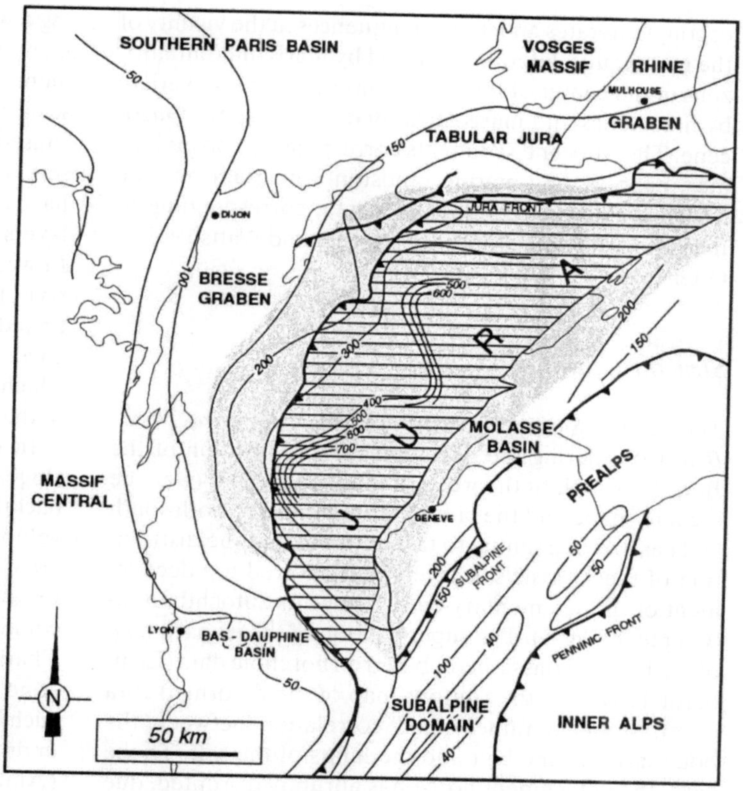

Fig. 5. Relationship between the development of the Jura fold-and-thrust belt and the extension of the Upper Triassic evaporites. The *shaded area* indicates the spatial distribution of the massive salt layers. The *numerical values* indicate the approximate thickness (in m) of the Upper Triassic beds (Lienhardt 1984)

strata and bituminous shales of Permian and Carboniferous age are preserved in the core of complex pull-apart basins linked to deep inherited strike-slip faults (Arthaud and Matte 1977). The Bas-Dauphiné basin, to the east of Lyon, represents the largest of these Late Paleozoic basins, located at depth beneath the Bresse graben and corresponding to the prolongation, towards the northeast, of the outcropping St. Etienne coal basin. It extends at least as far as the Leman Lake under the Jura Mesozoic cover and the Savoy molasse basin, as evidenced by exploration wells (Blyes, Torcieu, Chaleyriat, and Humilly wells, see location of wells in Figs. 8 and 17) and gravity data (Truffert et al. 1990). Permain strata have been encountered in the area by only two wells (Blyes and Charmont wells) and are of sandstone facies with very good reservoir properties. Nevertheless, seismic profiles exhibit strong reflectors at depth which can be attributed to Permain basin fill below the folded cover of Jura belt (see below).

During the Mesozoic, the Jura and Bresse area corresponded to a marginal platform domain with respect to the Tethysian oceanic basin. The Mesozoic cover overlies the whole area and constitutes a relatively isopachous thin series about 1600 m thick, of Early Triassic to Aptian age. The Triassic beds are of Germanic type and represent the gradual changeover from continental to marine environment. The Keuper salt layers are the main detachment horizon above which the Mesozoic pile was displaced. The Jurassic sequence consists of platform carbonate facies rocks, including weak marl levels acting as secondary decollement levels (Fig. 4). The Neocomian limestones have mainly been preserved in the core of large synclines in the inner part of the Jura mountain belt. This is due to extensive erosion prevailing in the vicinity of the Bresse graben during Oligocene times, that could result from a bulge effect, in response to the coeval emplacement of nappes (tectonic overloading) occurring in the Penninic and Subalpine domains (Guellec et al. 1990 b). The Mesozoic series ends with the Barremian-Aptian reef limestones (Urgonian facies), the Upper Cretaceous and Lower Eocene corresponding to a period of nondeposition and erosion.

Tertiary sediments of the Molasse basin, traditionally subdivided into four lithostratigraphical units, are directly linked to the polyphase development of the Alpine foreland. These syn-orogenic deposits recorded and fossilized each tectonic movement resulting from the progressive east to west migration of the deformation front, from Late Eocene to Upper Miocene (Mugnier et Ménard 1986; Guellec et al. 1990 a, b; Deville et al., this Vol.).

The Tertiary fill of the highly asymmetric Bresse graben is bounded towards the east by a major normal fault (Faille bordière), while the thickness of the Cenozoic deposits decreases towards the west (Lefavrais-Raymond 1962; Bergerat et al. 1990). The Bresse fill consists mainly of a thick Upper Eocene-Oligocene succession made up of conglomerates and detrital sequences in the vicinity of the main faults, laterally replaced by marls and salt horizons in the center of the basin This sequence is overlain by limestones and marls attributed to the Upper Oligocene. The Miocene sediments correspond to conglomerates, pebbles, and marine sandstones overlain by Late Miocene-Pliocene continental strata, corresponding to the youngest Alpine foredeep (Karner and Watts 1983, in Guellec et al. 1990b).

Structural Study

The Role of Salt in the Development of the Jura Thrust Belt. Considering the development and position of the Jura thrust belt in the western Alpine foreland (i.e., the arcuate shape and the lateral terminations to both south and east), it is essential to take into account the distribution of the Triassic salt layers that allowed the decollement of the sedimentary cover above its autochthonous basement (Fig. 5). The superposition of the isopach map of the Upper Triassic beds, based on borehole data (Lienhardt 1984) and the contour map of the deformed Jura cover, shows a rather good correlation between the boundaries of the belt and the limits of the salt. To the west, the detachment horizon is abruptly disrupted, due to the development of the highly asymmetric Bresse graben by means of major normal faults that caused a large vertical offset of the Triassic beds, of more than 2000 m in the deep eastern part of the graben (Lefavrais-Raymond 1962; Bergerat et al. 1990). In the south, at the Ile Crémieu, the autochthonous Jurassic cover directly overlies a metamorphic basement and Triassic evaporitic layers are apparently lacking. Moreover, to the west of the Chartreuse fold-and-thrust system, the Paladru borehole, located in the Tertiary Bas-Dauphiné basin, has also demonstrated the lack of Triassic salt layers. Therefore the development of the Jura fold belt seems directly related to the occurrence of Upper Triassic salt layers.

Results of the Field Study The work presented here is mainly based on field observations and well data. Fourteen serial cross sections on a scale of 1/25 000 were carried out and constructed by using the kink method (Suppe 1983), an restored one by one in order to constraint the geometry of thrusts and folds (Figs. 6, 7). Lateral consistency between cross sections has been complied by considering the thin Bathonian unit as a reference horizon. The restoration of the cross sections in the Jura belt (Fig. 7, sections 1 and 2) and the western Chartreuse massif (Fig. 7, section 14) gives a total shortening of about 17 km, along a WNW-ESE direction.

The cross sections clearly reveal the difference in the tectonic style between internal and external parts of the Jura belt, although the accurate location of the limit separating these two zones is sometimes hypothetical. The internal zone is characterized by large-scale structures mainly consisting of ramp anticlines (generally of fault-propagation fold type, see Suppe 1983; Mitra 1990) showing a westwards vergence, and involving the whole sedimentary cover located above the Keuper basal detachment layer. In contrast, the external zone displays small-scale imbricate thrust sheets, with local complications caused by disharmonic phenomena. Moreover, field observations and borehole data in the frontal part of the belt have demonstrated the subordinate role played by marly layers within the displaced cover (Upper Liassic and Lower Oxfordian marls). The difference in deformation style between the two structural zones is likely to be linked to the smaller thickness of the preserved Mesozoic cover in the western and southwestern parts of the belt, which was eroded during Oligocene, i.e., before the onset of the Alpine compressional phase.

In the western frontal parts of the Jura thrust belt (Fig 7, cross sections 1, 2, 3) forethrusts and conjugated backthrusts are often observed, which permitted the development of typical "pop-up" structures. In the study area, the most spectacular backthrust corresponds to the Oyonnax syncline backthrust (Fig. 7, cross section 1), encountered at depth by the Chaleyriat, Chatillon, and Charmont wells. In the southwestern transverse external zone (Bugeysan Faisceau), it is important to notice that such backthrusts are missing, and forethrusts have a lower dip than those located in the frontal NNE striking zone (Ambérieu Faisceau). These characteristics are caused by the southwestwards increase in basal friction due to the abrupt disappearance of the Triassic salt layers in the same direction.

The main result of the field study is that the transfer zone corresponding to the Bugeysan Faisceaux (Figs. 2, 7) does not reveal any strike-slip component, but is characterized by a succession of thin thrust sheets striking N 140 ° (i.e., parallel to the transverse zone), and displaying an unambiguous southwestwards tectonic transport. Furhermore, the tectonic contacts separating these thrust sheets can be followed laterally over the oblique zone from NW to SE. No significant transpressive deformation has been pointed out along reverse fault planes, as demonstrated by microtectonic analysis (see below).

The top of the autochthonous basement dips slightly towards both the Savoy molasse basin, with an average dip of 4–5 ° and the northeast (dip of 1–2 °). The gentle northeastwards dipping of the basement is probably accommodated by deep-seated NW-SE-directed Oligocene normal faults which induced later development of sinistral tear-faults in the displaced cover during Alpine compression. However, without any seismic control, the accurate relationship between lateral and frontal ramps affecting the cover and inherited deep normal faults cannot be evidenced as was done in the eastern termination of the Jura thrust belt (Laubscher 1986).

Microtectonic Study Within a fold-and-thrust belt, the paleostress pattern can be in first approximation deduced from geological macrostructures such as fold axes or movements along large faults. However, using of methods based on measurements of microstructures in sedimentary rocks provides a relatively accurate indicator of

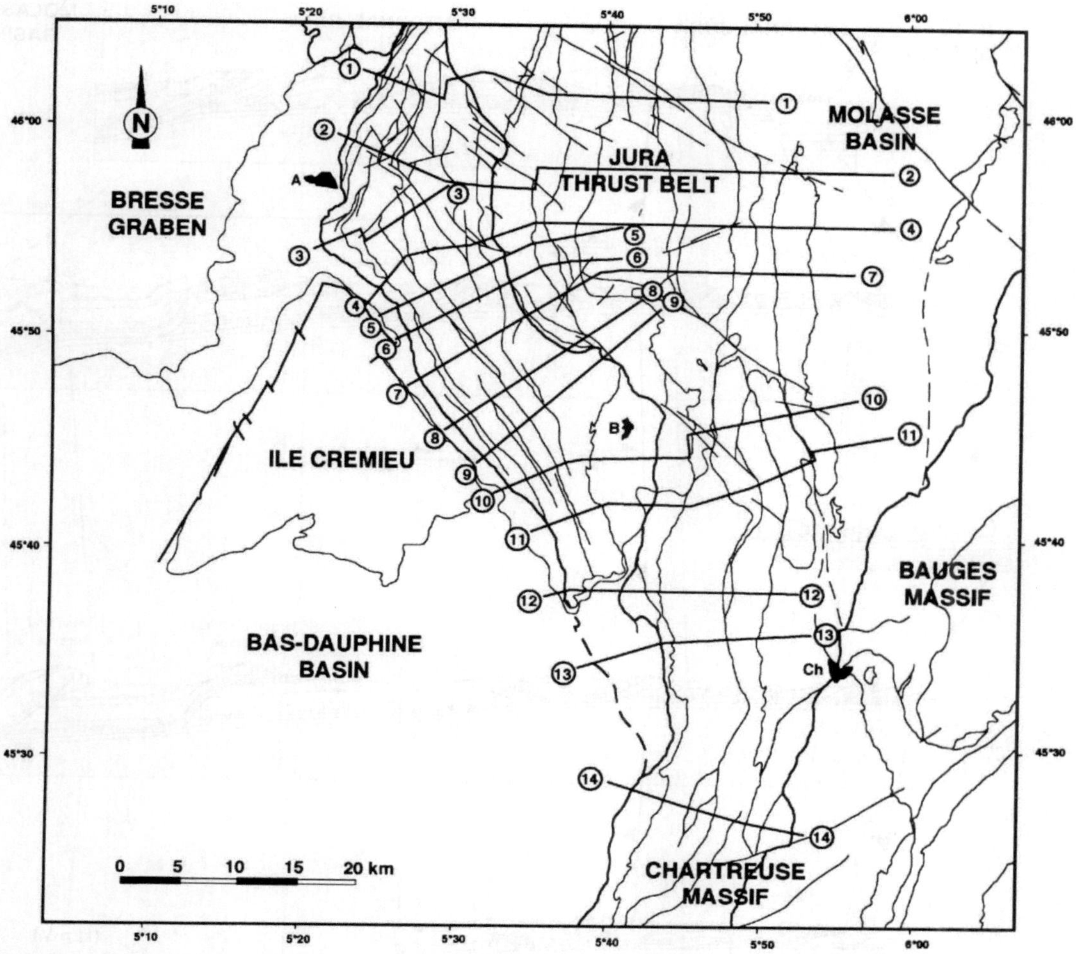

Fig. 6. Location of the cross sections

the paleostress and the direction of movement of tectonic units. These microstructures are subhorizontal stylolitic peaks, which represent the direction of the local maximum principal tectonic stress axis (σ1), tensional joints, and small-scale faults giving an unambiguous indication of the sense of displacement (slickensides on fault planes or bed-to-bed surfaces).

Compiling microtectonic data in the whole area, based on several hundreds of measurements, has revealed a very particular paleostress pattern showing a reorientation of the maximum stress axis in the vicinity of the transfer zone. Slickensides on reverse fault planes and bed-to-bed surfaces give a clear indication of the dip-slip component of thrust faults in both external frontal and lateral parts (Figs. 8, 9). This is with the exception of a particular zone located in the vicinity of the tectonic contact between the internal and external parts of the Jura belt, in the north of the Belley syncline, where both dip-slip reverse and horizontal sinistral strike-slip components have been observed (Fig. 9 "intermediate zone" and Fig. 10b). This divergent transport pattern in the allochthonous cover is consistent with a local N 50–60 ° shortening in the transfer zone (testified by stylolite orientation as well as conjugated strike-slip faults), which is strongly oblique to the shortening in the frontal part of the southern Jura thrust belt (N 105 °) and in the Chartreuse massif (N 110 °).

This disturbance of the stress field in the southern edge of the Jura belt seems to be directly linked to the southwestwards lateral increase in basal friction in response to the abrupt disappearance of the detachment level, which induced the reorientation of the maximum stress axis in map view.

Nevertheless, it is also necessary to take into account the difference in thickness of the Mesozoic cover between the internal and external parts of the belt. As mentioned above, the internal zone displays a complete displaced sedimentary sequence, from Upper Triassic to Middle Cretaceous (Urgonian limestones), whereas in the external part of the belt Upper Jurassic and Cretaceous sediments were eroded during Upper Eocene and Oligocene times, before the onset of the Mio-Pliocene compressional event. Thus, the Internal Jura behaved as a thicker and relatively more rigid block (with respect to the western and southwestern external parts of the belt), and could have acted as an indenter which caused the rotation of stress directions on its lateral boundaries. Moreover, the spatial relationship between the frontal part and the southwestern external zone and the accommodation of divergent transport directions are geometrically and

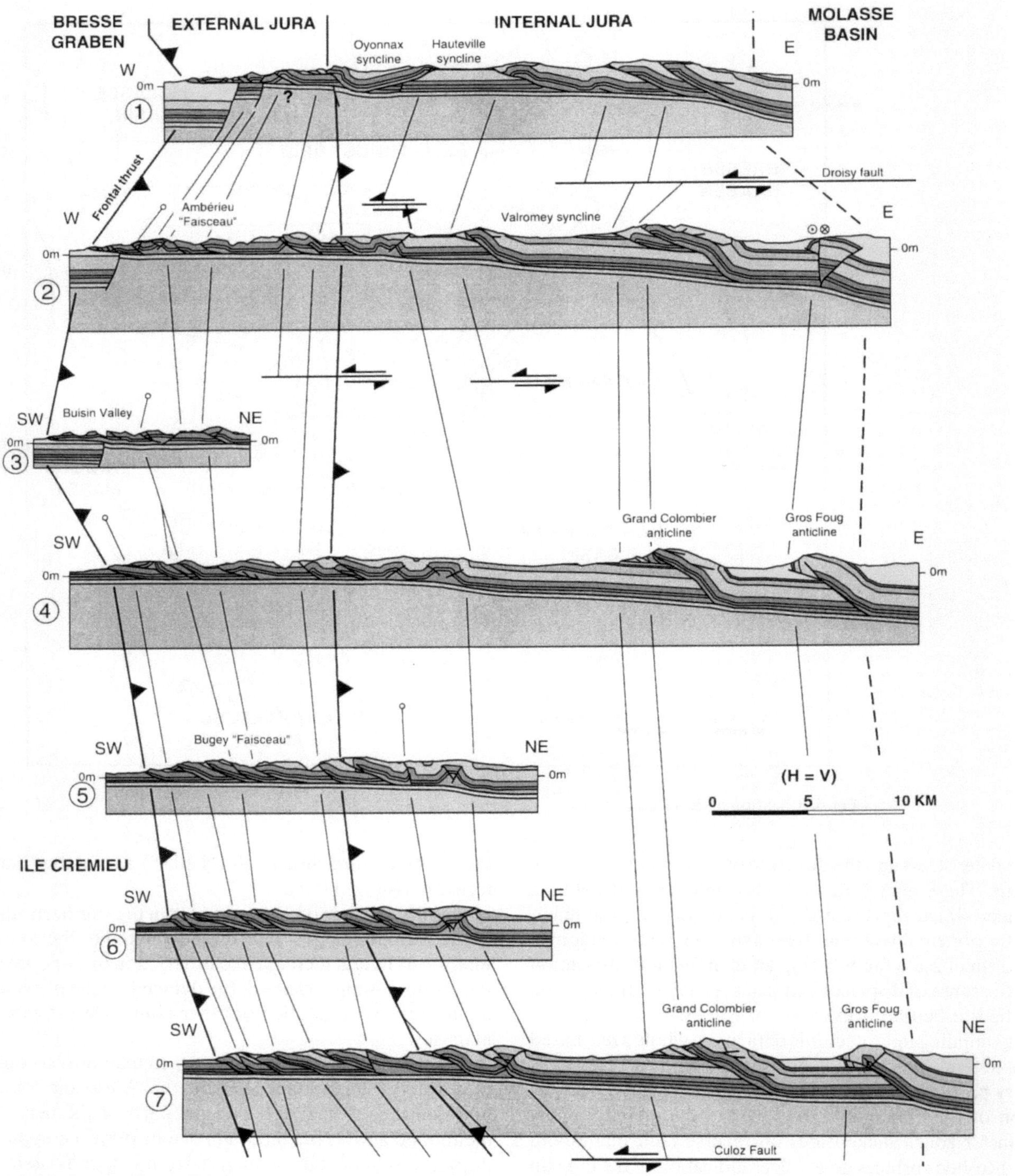

Fig. 7. Serial balanced cross sections of the southern Jura belt and the western Chartreuse massif *Black lines* display the correlation betwen tectonic contacts (See text for further information)

kinematically possible by differential motion on either side of the oblique tectonic contact separating the internal and external zones, which has revealed a significant sinistral strike-slip component This contact is superimposed on an inherited basement fault-zone striking N 130 °, the Culoz-Pont d'Ain lineament (Enay 1982). It exhibited a normal component during Oligocene times and thus preserved the top of the sedimentary cover in the north. This major deep-seated fault, that coincides with the southern limit of the Triassic salt extension (Fig. 5), is part of a system of NW-SE-directed sinistral transverse faults (Cruseille fault, Vuache fault, Droisy fault, Culoz fault, Fig. 2) whose complex and polyphase evolution during Cenozoic times has been demonstrated (Blondel et al. 1988; Wildi et al. 1991; Deville et al., this Vol.) During the Alpine phase, this previous normal fault has induced a differential behavior of the two compartments located on either side and acted as a sinistral

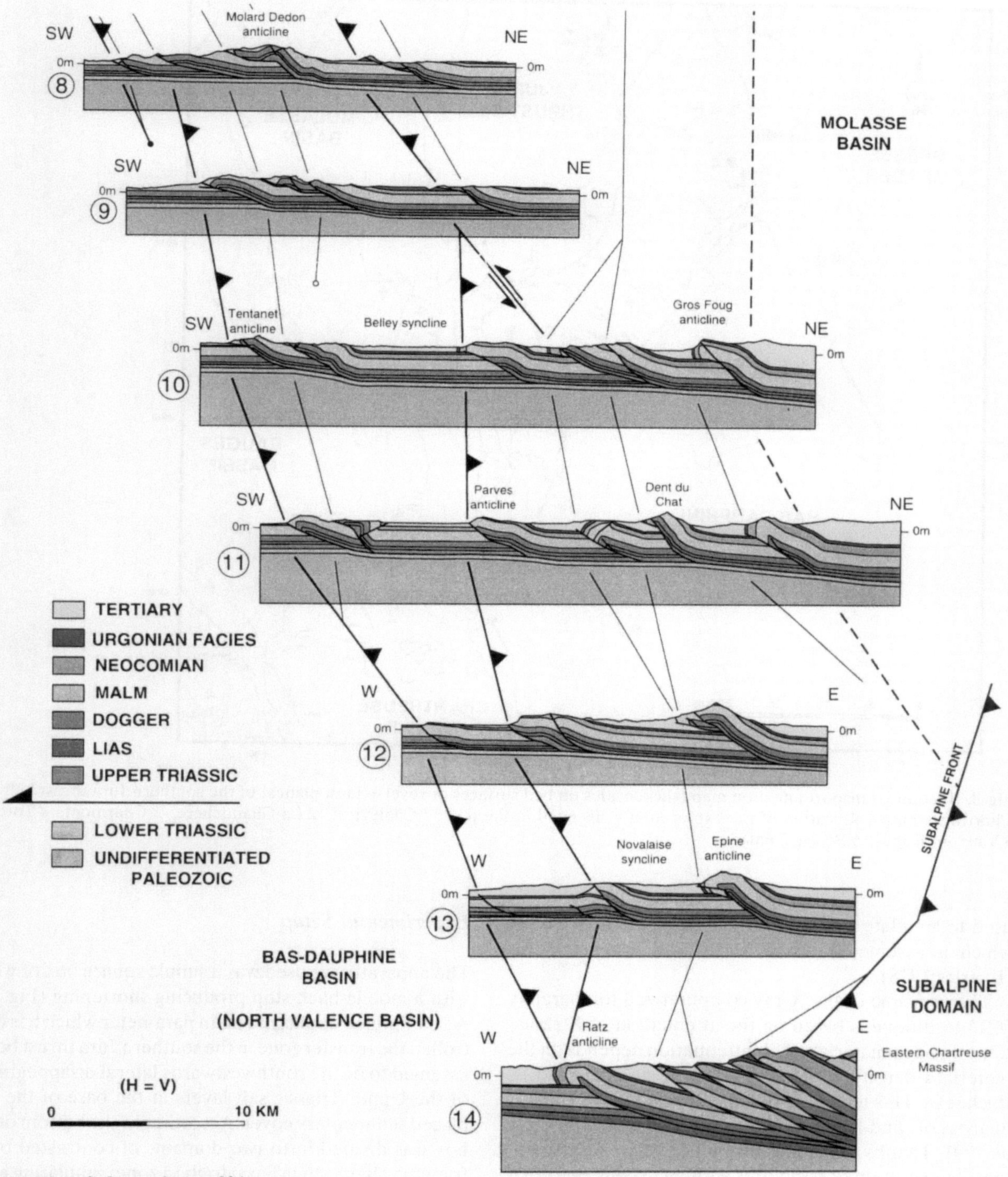

Fig. 7 *(continued)*. Legend see p. 334

transfer fault within the overlying folded cover (the Culoz fault). This transfer fault separates the internal and external zones of the belt where transport directions indicate displacements towards WNW and SW respectively (Fig. 9). This structural setting invokes a strain-partitioning pattern and is in good agreement with the fault-bifurcation model (Lacassin 1989) that can explain the compatibility between coeval and divergent tectonic motions (Fig. 10).

Analogue Modeling Experiment

Analogue modeling experiments have been performed in a normal gravity field in order to understand the mechanisms of deformation and to determine the genetic relationship between imposed boundary conditions and resulting structures.

Granular materials such as sand, glass powder, and glass microbeads with a grain size of about 100 μm were

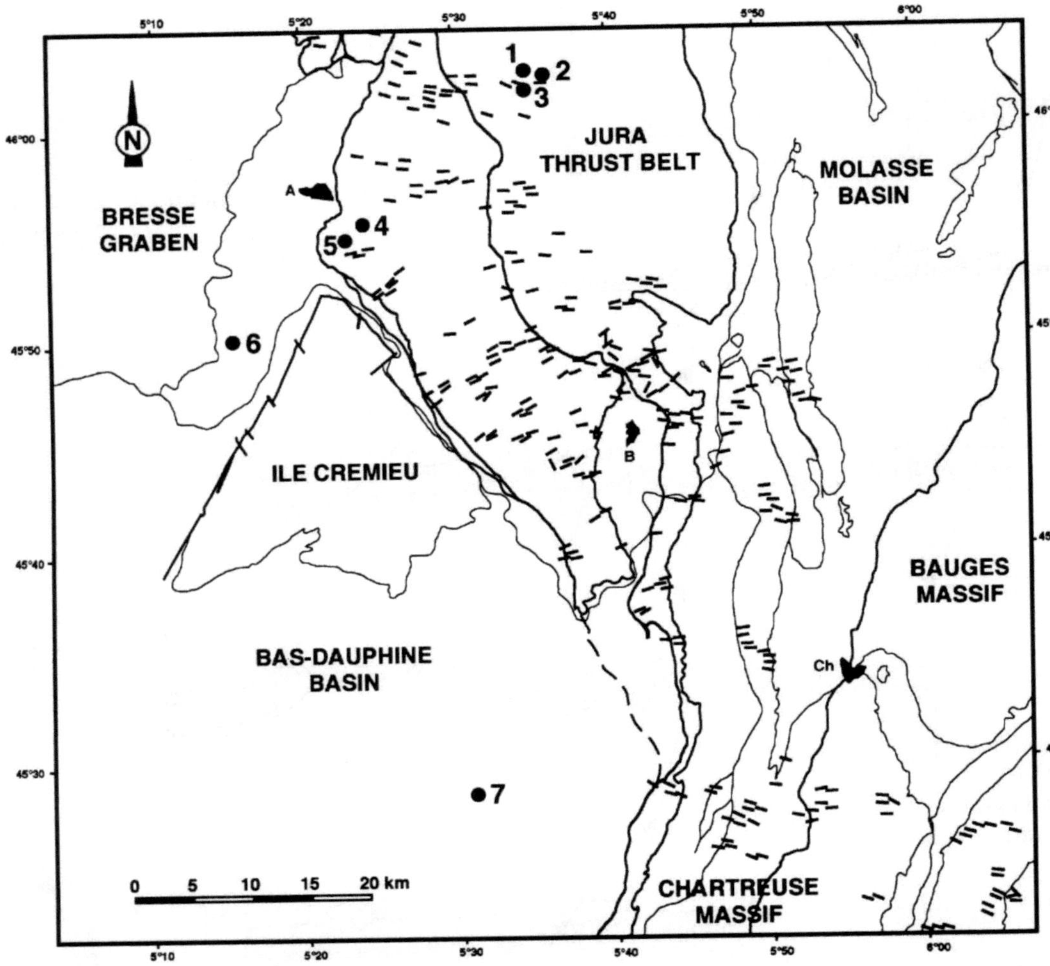

Fig. 8. Tectonic transport lineation map (slickensides on bed surfaces or reverse fault planes) of the southern Jura thrust belt and Chartreuse massif Location of the exploration wells cited in the text *1* Chaleyriat, *2* La Chandelière, *3* Charmont, *4* Torcieu, *5* Vaux-en-Bugey, *6* Blyes, *7* Paladru

used to simulate brittle deformation of the upper crust which is assumed to have Mohr-Coulomb properties (Byerlee 1978).

The principle of the X-ray computerized tomography (CT) technique is based on the attenuation of X radiations through materials. This attenuation depends on the material's density, its effective atomic number, and its thickness. The technical details and results of the CT analysis of sand-box models are presented in Colletta et al. 1991. Thanks to the CT technique, we are able to follow the evolution of structures with increasing deformation of the model, i. e., the development and propagation of each fault. Furthermore, it is possible to obtain, for the final stage of deformation, serial cross-sections of the model spaced out by 3 mm. It is therefore feasible to reconstruct horizontal and longitudinal vertical cross sections by computer processing, to finally obtain a 3-D analysis of structures without destroying the model.

Experimental Setup

The apparatus we used was a simple square box of wood with a mobile back-stop producing shortening (Fig. 11). As mentioned above, the main parameter which has controlled the transfer zone in the southern Jura thrust belt is assumed to be the southwestwards lateral disappearance of the Upper Triassic salt layers at the base of the displaced sedimentary cover. Accordingly, the bottom of the box was divided in to two domains of contrasted basal friction values: a high basal friction zone, simulating a domain without a decollement level representing the external southern Jura belt and the Ile Crémieu, and a lower basal friction zone, simulating a domain provided with a detachment horizon as in the Jura belt. In the first zone, basal friction was increased by sandpaper, whereas in the second, friction was reduced by a thin layer of glass microbeads (internal friction angle of about 20 °). The model itself was made of three layers: two layers of quartz sand and one layer of glass powder, each of them 6–7 mm thick. These materials have the same mechanical properties (equal internal friction angle and shear-failure

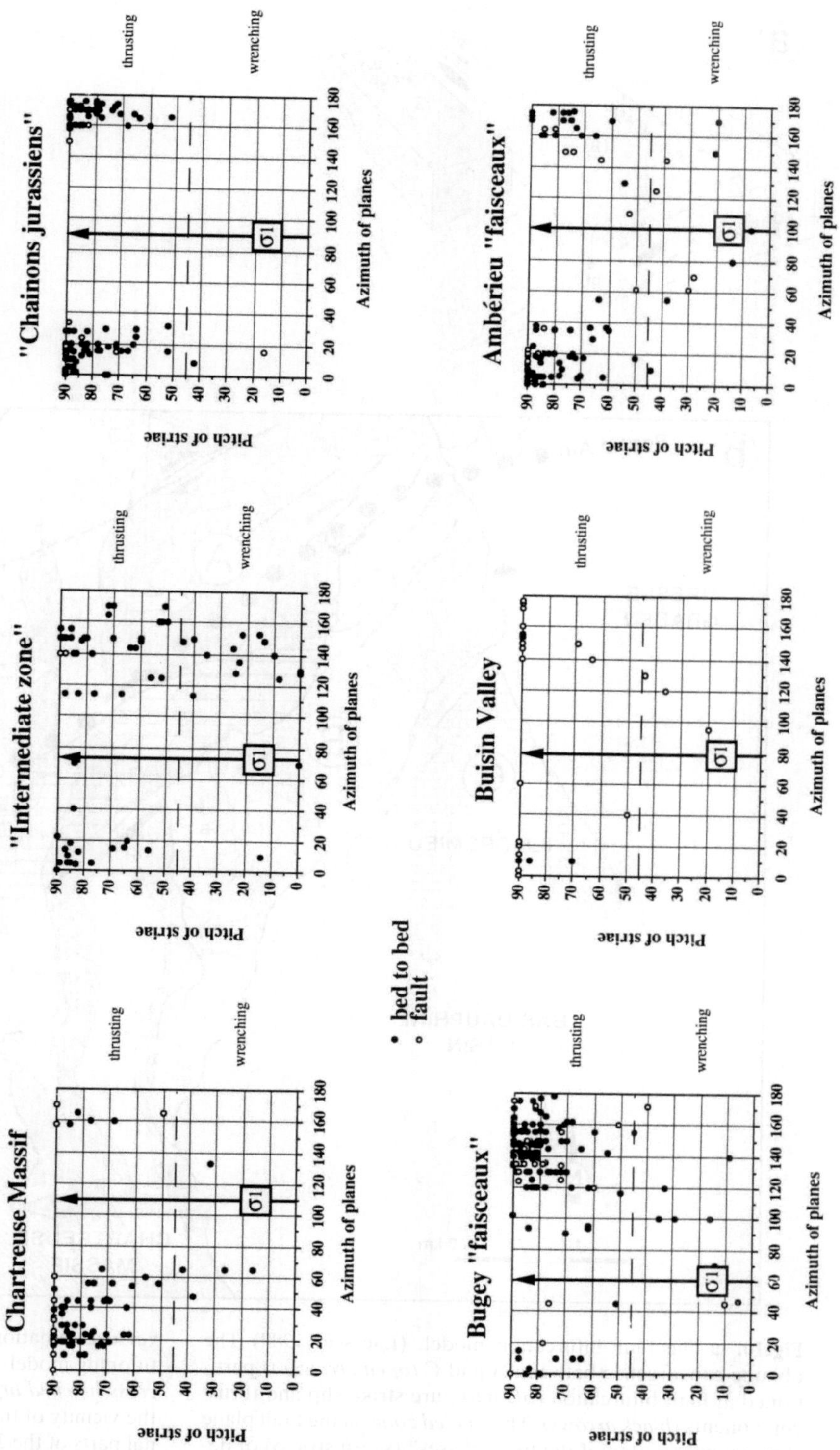

Fig. 9. Pitch of striae vs. strike of reverse fault planes or bed surfaces (pitch of 45 ° separating field where either thrusting or wrenching dominates) *Black circles* correspond to bed-to-bed slickensides, open circles represent reverse fault slickensides. *Vertical arrows* indicate azimuth of the maximum stress axis (σ1) in each domain The Chaînons jurassiens encompass the N-S-directed Epine, Mont du Chat, Parves, and Gros Foug anticlines (see Fig. 2). The *"intermediate zone"* corresponds to the oblique NW-SE structural unit separating the internal and external zones of the southern Jura belt in the north of the Belley syncline, where both sinistral strike-slip and thrust components have been observed (see Fig 10b cross-hatched area). The Buisin Valley corresponds to a small area located in the south of Ambérieu-en-Bugey, in the vicinity of the Vaux-en-Bugey wells (see Fig 8, well 5 and Fig 7, cross section 3)

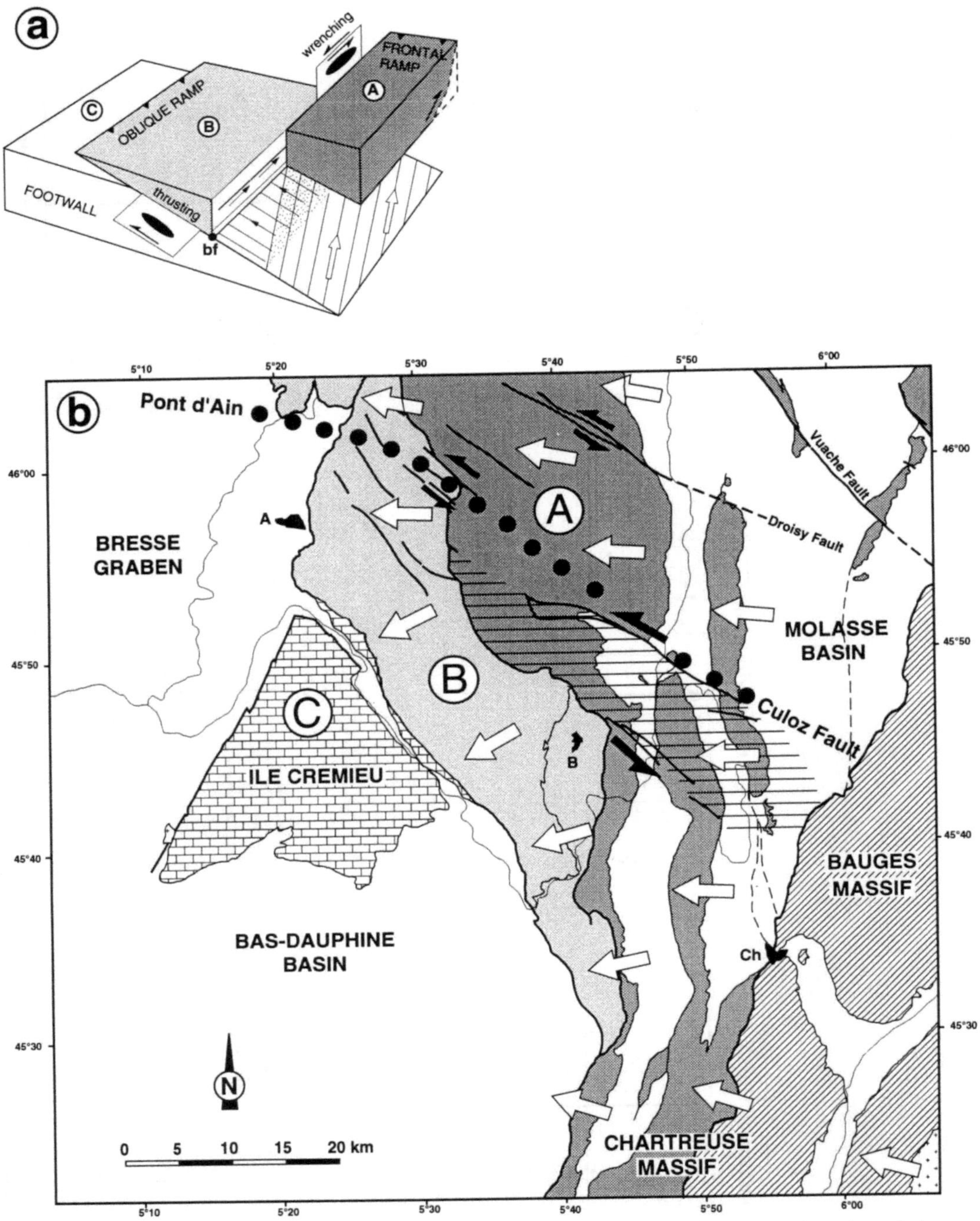

Fig. 10. **a** The fault-bifurcation model (Lacassin 1989) The oblique convergence between *A* and *C* *(open arrows)* is partitioned at fault bifurcation *(bf)* into pure strike-slip and thrust components *(black arrows)* The *shaded zone* on the fault plane shows superposition of the two "phases" (sensu stricto) of deformation **b** Structural interpretation of the southern transverse termination of Jura thrust belt according to the strain partitioning model *White arrows* indicate transport diretion The *cross-hatched area* corresponds to the "intermediate zone" in the vicinity of the tectonic contact between internal and external parts of the Jura thrust belt, where both sinistral strike-slip and thrust components have been observed

angle of about 30 °) but distinct attenuation values which make it possible to visualize the different layers of such a sand and glass powder cake (Colletta et al. 1991)

The modeling was carried out by incremental manual displacement of the mobile back-stop, the behavior of granular materials being independent of the shortening velocity (Ballard et al. 1987). During the experiment, eight successive stages of deformation with increasing shortening were analyzed using the CT technique. At each stage, two cross sections and surface views were recorded (Figs. 12, 13). After the last stage of shortening, a series of 47 contiguous cross sections (every 3 mm) parallel to the transport direction were carried out in the central part of the model in order to reconstruct horizontal and longitudinal vertical cross sections in the vicinity of the transfer zone (Fig 14).

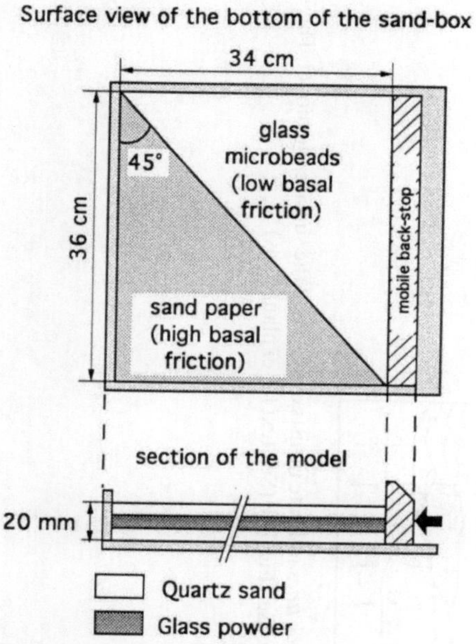

Fig. 11. Experimental setup

Results

The surface views and evolutive cross sections in each domain demonstrate the influence of the basal friction value on thrust propagation in the overlying sand-cake Thus, the deformed model is characterized by two structural domains directly induced by the basal conditions.

Before stage 4, the model develops almost in the same way in both compartments, because in the vicinity of the mobile wall, friction on the back-stop disturbs the stress trajectories (border effects). After stage 4, an oblique ramp cuts the upper surface of the model (Fig. 12, view 5), striking parallel to the boundary separating the two domains, and propagates laterally as frontal thrust (Fig. 12, views 6, 7). In the low basal friction zone, the frontal thrust merges far from the mobile back-stop (Fig. 13 a, cross section 8), with a relatively steep dip, of about 25 °. Moreover, the thrust belt in this domain is wide and low (taper of 8.5 °). In contrast, in the high basal friction zone, the frontal thrusts (i. e., parallel to the mobile wall) are closely spaced and merge in a narrow zone closer to the mobile back-stop (Fig. 13 b, cross section 8), therefore the thrust shows a higher taper (21.7 °). These thrusts display lower dip values, e. g., the most external frontal thrust (Fig. 13 b, thrust 4) dips about 12 °.

Serial horizontal sections make it possible to visualize the internal structures in the vicinity of the transfer zone (Fig. 14 a).

Interpretation

To understand the development and the evolution of noncylindrical zones, it is essential to know the thrust propagation laws in homogeneous media. Thrust propagation is controlled by two main parameters: (1) the thickness of the model and (2) the basal friction value (Fig. 15). The thickness of the displaced sand model determines the spacing of thrusts (Mulugeta 1988; Colletta et al. 1990), whereas the basal friction imposes the dip of the thrusts (Mandl and Shippman 1981; Davis et al. 1983; Calassou and Malavieille 1990), and therefore the vergence of the thrust belt (Davis and Engelder 1985). Thus, a noncylindrical zone must be considered as induced by a lateral change of at least one of these parameters. As a consequence, structures located in a transfer zone accommodate the lateral variations of spacing and/or dipping of thrusts between two domains of distinct characteristics (Philippe and Specht, 1992).

The results of the experimental study has validated these a priori hypotheses: the spacing of thrusts observed in the low friction domain (Fig. 13 a, see thrust 3 on cross section 5) is consistent with the theoretical estimate of thrust fault emergence. Moreover, conjugated backthrust occurs and afterwards is passively transported over the footwall ramp. Therefore, the taper of the thrust belt is low (Figs. 13 a, 16 a). In the other domain, in response to the high value of the basal friction (Fig. 15, *bf*), the maximum stress σ1 trajectories are steep. Therefore thrusts dip gently and backthrusts are lacking because of their higher theoretical dip value. As a consequence, the deformation of the model ist accommodated by a large displacement along thrusts, so the taper of the thrust belt is high (Figs. 13 b, 16 b). The analogy with the natural case of the southern Jura thrust belt and the western Chartreuse massif is quite satisfactory: the Jura belt is characterized by a very low and wide wedge, conjugated thrusts and back-thrusts (pop-up structures) (see Fig. 7, cross sections 1 or 2). On the contrary, the western Chartreuse massif, where efficient basal decollement level is missing, diplays a succession of imbricate thrusts showing a single westwards vergence, which leads to the development of a thick wedge with a high taper (Fig. 7, cross section 14).

Petroleum Geology

Petroleum exploration in the Jura belt started in the early years in this century, e. g., in the southern part of the belt, following the coal-mining industry. Restricted gas fields were discovered and produced in the past but have now been abandoned. The tiny Vaux-en-Bugey gas field (located at the southwestern edge of the Jura belt) produced 13 million m³ from 1922 to 1959. The trap probably consisted of allochthonous Upper Triassic dolomites, but the exact type of the structure proper is not known in detail.

Recently, significant oil shows have been encountered in the Chaleyriat and La Chandelière boreholes, and this has boosted petroleum exploration in the Jura and Bresse area.

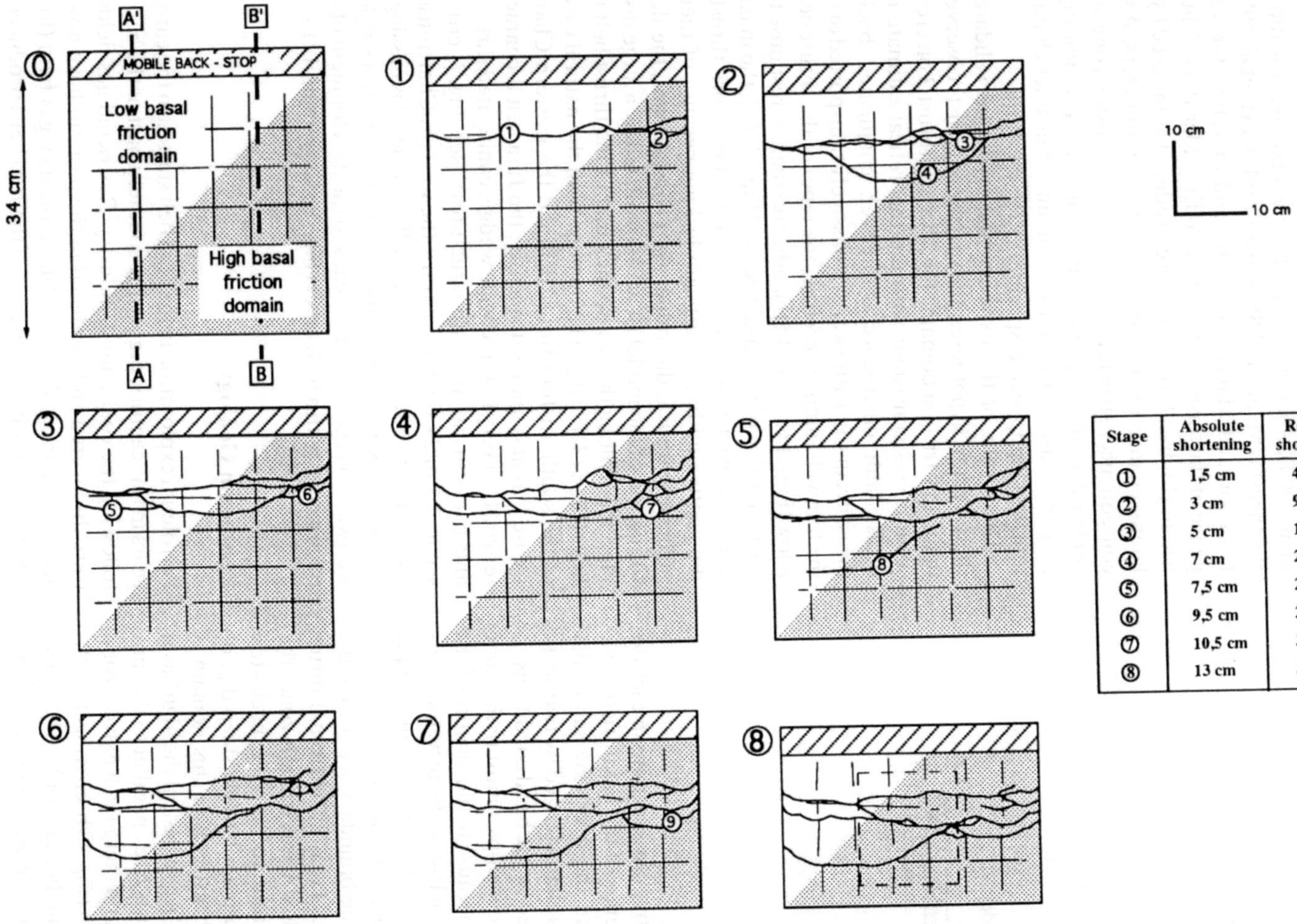

Stage	Absolute shortening	Relative shortening
①	1,5 cm	4,4 %
②	3 cm	9 %
③	5 cm	14,7 %
④	7 cm	20,5 %
⑤	7,5 cm	22 %
⑥	9,5 cm	28 %
⑦	10,5 cm	31 %
⑧	13 cm	38,2 %

Fig. 12. Surface view showing the evolution of the model with increasing shortening *Black lines* indicate thrusts observed at the surface the model The *dottes square* in view *8* delimits the area analyzed by the 3-D X-ray tomography technique (see Fig 14) (See text for explanations)

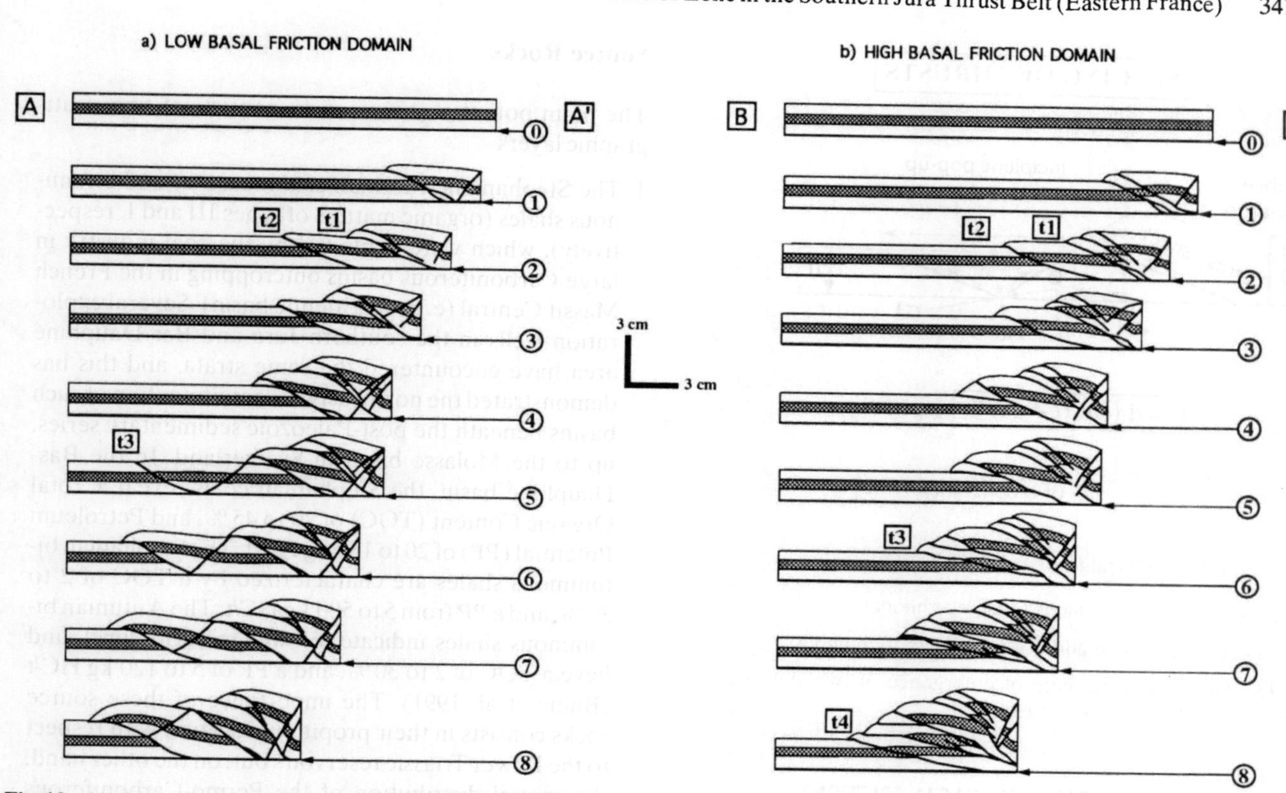

Fig. 13 a, b. Cross sections of the model in each domain showing the evolution of deformation with increasing shortening (see location in Fig 12, view 0) **a** Low basal friction domain **b** High basal friction domain

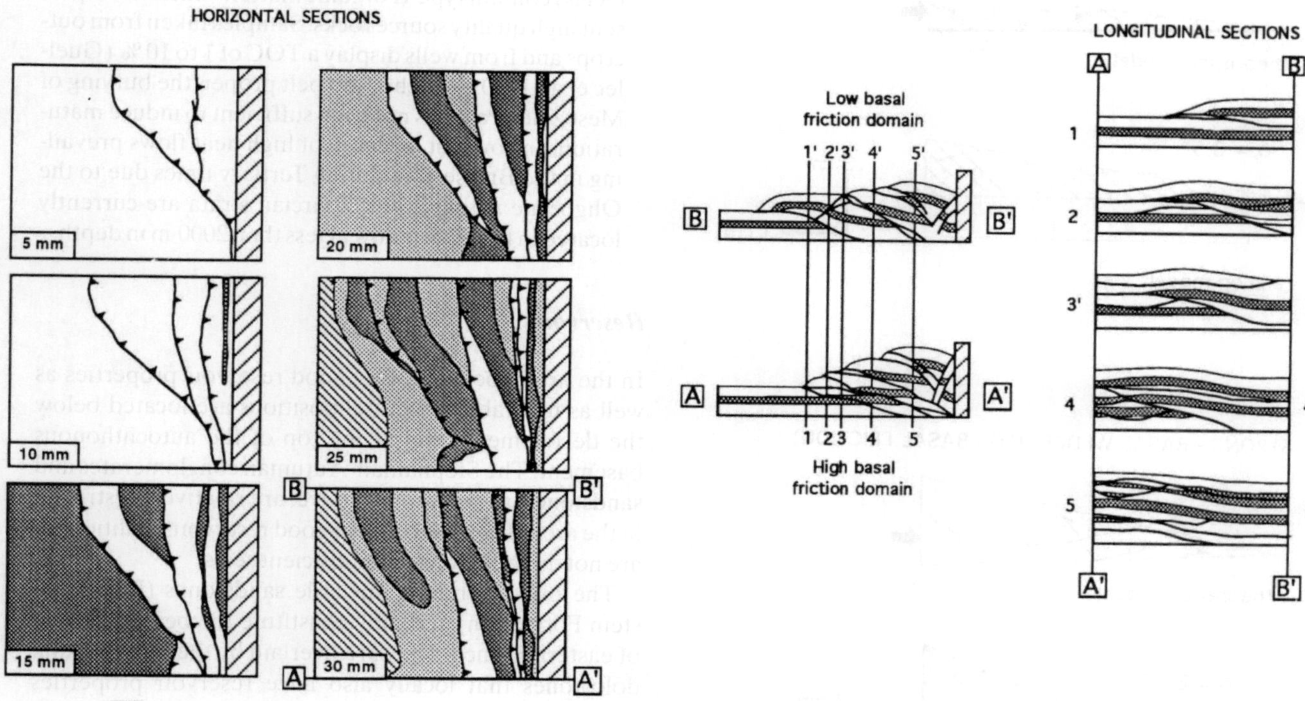

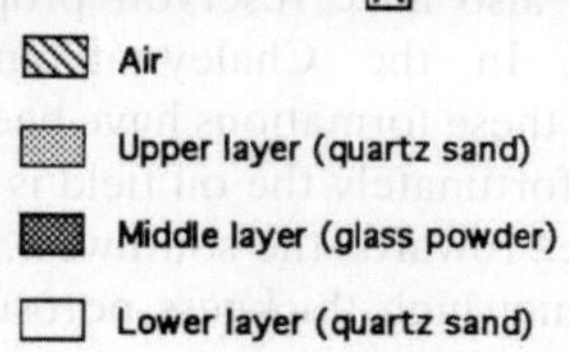

Fig. 14. Horizontal (distance from bottom of the model in cm) and longitudinal cross sections of the transfer zone in the central part of model (see location in Fig 12, view 8)

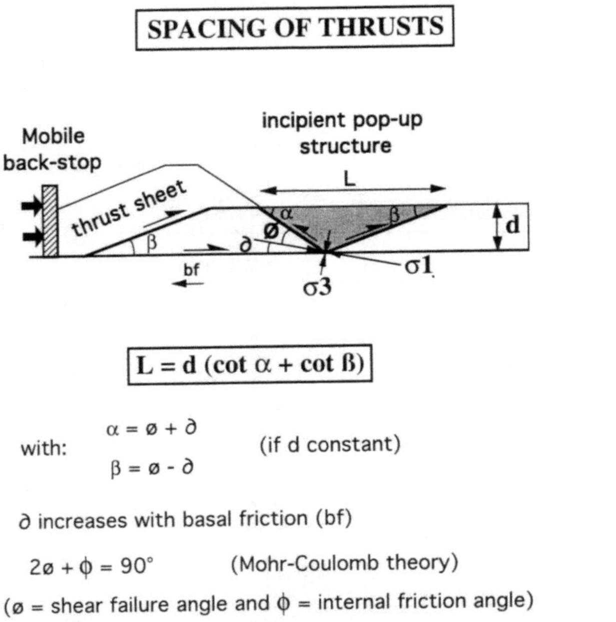

Fig. 15. Theoretical estimate of emergence of two successive thrust. (After Colletta et al. 1990)

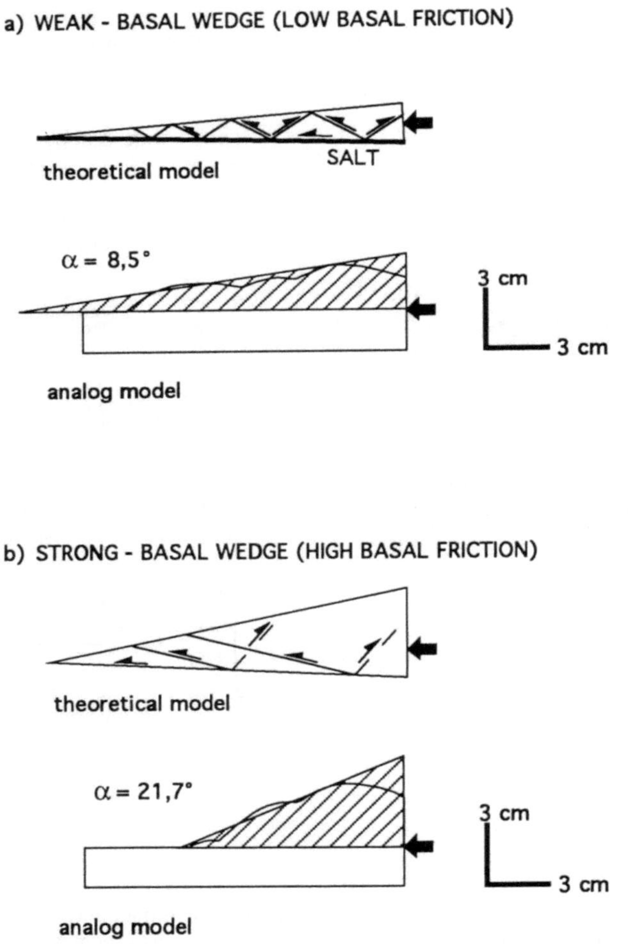

Fig. 16 a, b. Comparison between theoretical (Davis and Engelder 1985) and experimental taper values obtained in the two domains of the model **a** Low basal friction domain **b** High basal friction domain

Source Rocks

The main potential source rocks consists of two stratigraphic layers

1 The Stephanian-Autunian coal measures and bituminous shales (organic matters of types III and I, respectively), which were exploited by the coal industry in large Carboniferous basins outcropping in the French Massif Central (e.g., St. Etienne basin) Several exploration wells in the southern Jura and Bas-Dauphiné area have encountered the same strata, and this has demonstrated the northeastwards continuation of such basins beneath the post-Paleozoic sedimentary series, up to the Molasse basin in Switzerland In the Bas-Dauphiné basin, the Stephanian coals reveal a Total Organic Content (TOC) of 20 to 45%, and Petroleum Potential (PP) of 20 to 180 kg HC/t. The Stephanian bituminous shales are characterized by a TOC of 2 to 30%, and a PP from 5 to 500 kg HC/t. The Autunian bituminous shales indicate a lower maturity level, and have a TOC of 2 to 30%, and a PP of 5 to 120 kg HC/t (Blanc et al. 1991) The importance of these source rocks consists in their propitious location with respect to the Lower Triassic reservoirs but, on the other hand, the spatial distribution of the Permo-Carboniferous basins is relatively restricted.
2 The Toarcian marine black marls (the Schistes cartons facies) contain type II organic matter, which can represent high quality source rocks. Samples taken from outcrops and from wells display a TOC of 1 to 10% (Guellec et al. 1990 a). In the Jura belt proper, the burying of Mesozoic strata has not been sufficient to induce maturation. In contrast, because of high heat flows prevailing in the Bresse area during Tertiary times due to the Oligocene rifting event, Toarcian strata are currently located in the oil-window at less than 2000 m in depth.

Reservoirs

In the area, rocks showing good reservoir properties as well as favorable structural positions are located below the decollement level at the top of the autochthonous basement. The Stephanian-Autunian conglomerates and sandstones, as mentioned above, are relatively restricted in the area. They have shown good reservoir qualities but are not directly overlain by efficient seals.

The Lower Triassic fluviatile sandstones (Buntsandstein Formation) certainly constitute the best reservoirs of eastern France. They are overlain by the Muschelkalk dolostones that locally also have reservoir properties (secondary objective). In the Chaleyriat and La Chandelière boreholes, these formations have been successfully tested, but unfortunately the oil field is not of commercial importance. Towards the southwest, in the direction of the Ile Crémieu high, thickness, porosity, and permeability values of the Buntsandstein sandstones rapidly decrease, as demonstrated by the Chatillon well. Further east towards the Subalpine Domain, they are progressively replaced by carbonates and have additionally lost their reservoir quality. The seal of Buntsandstein reservoirs are of varying quality, and the Upper Muschelkalk Formation (Lettenkhole limestones) where other reservoirs may be found (directly covered by an efficient seal corresponding to the Keuper salt layers) often display low porosity or permeability values.

Structural Traps

The potential hydrocarbon traps which may be expected in such foreland zones, that underwent polyphase and complex development, can be divided in two groups.

1. Extensive structural traps formed during the Late Eocene-Oligocene rifting phase, which produced basement highs and tilted blocks that have been preserved below the Upper Triassic decollement level, especially in the vicinity of the eastern border of the Bresse graben, beneath the Jura allochthonous Mesozoic cover.
2. Structural traps resulting from the Late Miocene-Pliocene Alpine compressional phase. In such thin-skinned fold-and-thrust belts, it is admitted that layers located below the decollement level are not involved in the thrusting. Nonetheless, small sheets of tops of basement highs could have been locally incorporated in the thrusting and displacement of the overlying detached cover. This is the case of the Valempoulières tiny gas field in the central Jura, from which 85 millions m^3 was produced from 1961 to 1983. The reservoir corresponds to the Muschelkalk dolomites, in allochthonous position within the Keuper salt accumulation.

Another type of expected structural traps, also resulting from the Alpine phase could be illustrated by the reinterpretation of part of the ECORS deep seismic profile in the Internal Jura (Figs. 17, 18). The ECORS profile, crossing the Jura thrust belt and the Bresse graben, shot in 1987 (Damotte et al. 1990), has been interpreted in detail by the ECORS team, in addition to surface and other subsurface data (Bergerat et al. 1990; Truffert et al. 1990; Guellec et al. 1990a, b). However, the recent Charmont exploration well, located in the Internal Jura near the profile line, has clarified some ambiguous points, in particular the meaning of the strong reflectors occurring beneath the folded Mesozoic cover in the Internal Jura (Fig. 18). The line drawing of the Upper part of the profile, according to surface data, exhibits a typical thin-skinned deformation style involving a cover thrust above a decollement level. What is important to notice here is the relationship between the displaced cover and its basement, as well as the relatively high position of the Haute-Chaîne with respect to the Molasse basin and the External Jura. Beneath the Oyonnax syncline backthrust, west-dipping reflectors occur between 0.8 and 1 stwt (Fig. 18). Without any complementary data, they have been believed to correspond to Lower Mesozoic markers. This involves a tectonic doubling of the cover (Guellec et al. 1990a, b). However, the recent Charmont well

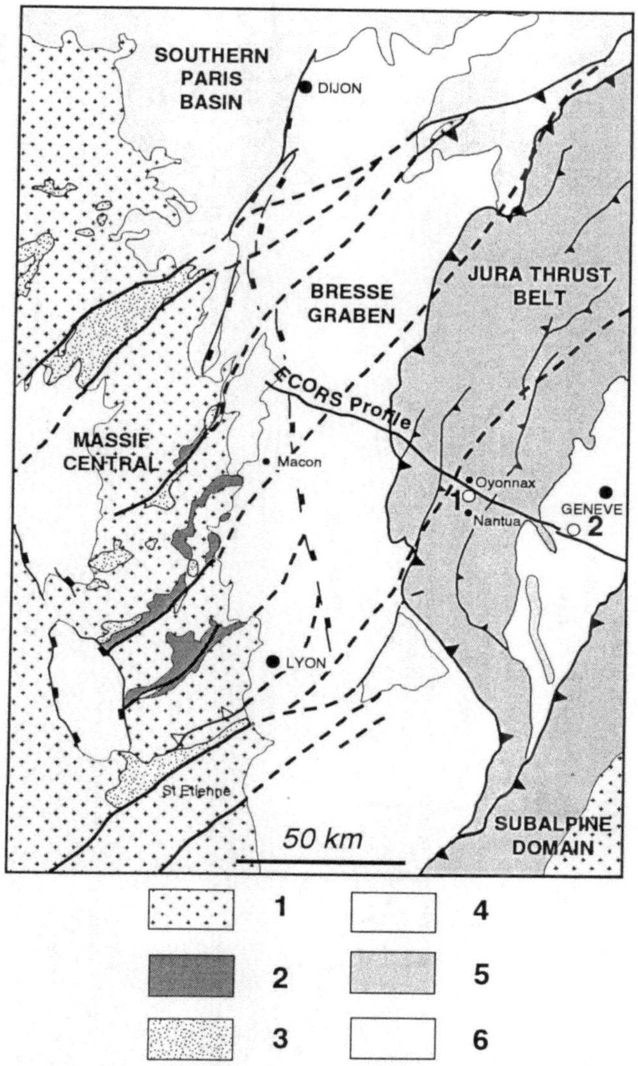

Fig. 17. Location of the ECORS profile and extension of late Variscan lineaments beneath the Bresse graben and Jura thrust belt *(dashed lines)*, according to gravity data (After Truffert et al 1990) *Open circles* indicates the location of exploration wells cited in the text (*1* Charmont, *2* Humilly) *1* Variscan basement, *2* Devono-Dinantian, *3* Permo-Carboniferous, *4* Autochthonous Mesozoic cover, *5* Folded Mesozoic cover, *6* Tertiary

has evidenced that these reflectors correspond to Permian strata and thus demonstrate the presence beneath this part of the belt of Late Paleozoic basins, in an eastwards continuation of the exposed Permo-Carboniferous basins of the French Massif Central (Fig. 17). These horizons, showing both a westwards inclination and an erosional truncation towards the east below uncomformable Lower Triassic strata, suggest an asymmetric Late Paleozoic graben fill, possibly bounded towards the west by a major fault. The basement culmination below the Internal Jura could thus be explained by a late basement shortening postdating the displacement of the overlying cover, by means of reactivation of deep-seated previous normal faults. This tectonic inversion (see Letouzey 1990) can also explain the bending of Late Paleozoic and Lower Triassic strata as shown by this seismic profile.

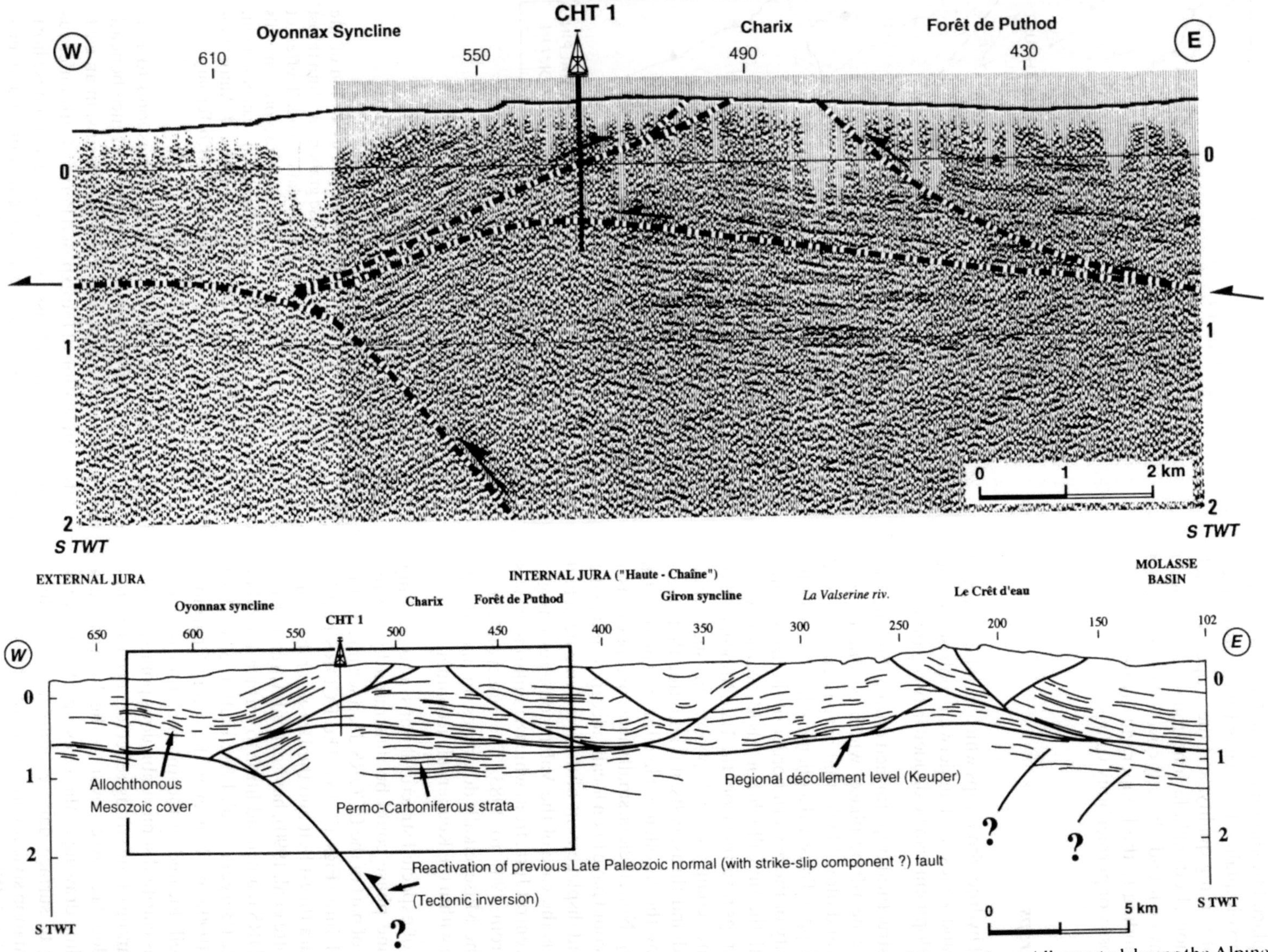

Fig. 18. Line-drawing of the ECORS profile in the Internal Jura Interpreation of the deep strong reflectors as late Paleozoic basin fill inverted during the Alpine compressional phase due to the reactivation of a (NE-SW-directed?) late Variscan fault

Conclusion

The Southern termination of the Jura thrust belt provides a remarkable example of a complex transfer zone located in the most external part of the Alpine foreland. The study of this area was performed with two complementary approaches: (1) a regional field study, and (2) a simple analogue modeling experiment.

The field study, based on regional serial cross sections and microtectonic analysis has made it possible to describe the geometry of structures and to define the tectonic regime associated with this NW-SE-directed transverse zone. The external part of the Jura belt exhibits a divergent transport pattern, i.e., towards the WNW in the NNE-directed frontal part of the belt (Ambérieu Faisceaux), and towards the SW in the NW-SE directed oblique zone (Bugeysan Faisceaux), consistent with the paleostress data. The Internal Jura proper is characterized by a WNW displacement that is in good agreement with the regional shortening (N 100–110 °). The divergent motion pattern is accommodated by partitioning into pure dip-slip and sinistral strike-slip components (in the southwestern external zone and along the NW-SE-directed tectonic contact separating the External and Internal Jura, respectively). The location and development of this oblique zone are mainly caused by the abrupt southwestwards lateral disappearance of the Upper Triassic evaporites and salt layers. Nonetheless, this strain-partitioning pattern is assisted by the difference in thickness of the allochthonous cover between the external and internal parts of the belt (indentation), as well as the presence of NW-SE-striking deep-seated faults inducing sinistral tear-faults in the overlying cover.

Analogue modeling experiments, simulating an abrupt lateral disappearance of basal decollement level, were performed in order to confirm our hypothesis about the boundary conditions which could have induced the development of this transverse zone, and analyzed using the CT technique. The modeling has actually demonstrated the influence of the basal friction on the development of the thrust belt, i.e., oblique thrusts striking parallel to the limit separating domains of constrasted basal friction values were created. Thus this work has also shown that the knowledge of the paleogeographic setup (thicknesses, distribution of weak layers in or at the base of the detached cover, etc.) is necessary to understand the lateral variations of the deformation style in fold-and-thrust belts.

Acknowledgments The author is indebted to Alain Mascle, Bernard Colletta, and Pascal Balé (Institut Français du Pétrole) for supporting the study and improving of the interpretation of field data and analogue modeling results. The chapter has also benefited from critical and helpful suggestions given by Marc Tardy and Walter Wildi. Laurent Vallet (Elf-Aquitaine Production) is also thanked for the correction of the equation in Fig. 15.

References

Arthaud F, Matte P (1977) Late Paleozoic strike-slip faulting in southern Europe and northern Africa: result of a right-lateral shear zone between the Appalachians and the Urals. Geol Soc Am Bull 88: 1305–1320

Ballard JF, Brun JP, Van Den Driessche J, Allemand P (1987) Propagation des chevauchements au-dessus des zones de décollement: modèles expérimentaux. C R Acad Sci Paris 305, II: 1249–1253

Bally AW, Gordy PL, Stewart GA (1966) Structure, seismic data and orogenic evolution of southern Canadian Rockies. Bull Can Pet Geol 14: 337–381

Bergerat F (1985) Déformations cassantes et champs de contrainte tertiaires dans la plate-forme européenne. Thesis, Univ P et M Curie, Paris VI, 315 pp (unpublished)

Bergerat F, Mugnie JL, Guellec S, Truffert C, Cazes M, Damotte B, Roure F (1990) Extensional tectonics and subsidence of the Bresse basin: an interpretation from ECORS data. In: Roure F, Heitzmann P, Polino R (eds) Deep structure of the Alps. Mém Soc Géol Fr Paris 156, Mem Soc Géol Suisse Zurich 1, Vol Spec Soc Geol Ital Roma 1: 145–156

Bitterli P (1972) Erdolgeologische Forschungen im Jura. Bull Ver Schweiz Petrol Geol Ing 39, 95: 13–28

Blanc G, Doligez B, Lajat D, Mascle A (1991) Evaluation du potentiel pétrolier des formations paléozoiques de la Bresse et de sa bordure jurassienne, France. Bull Soc Géol Fr 162, 2: 409–422

Blondel T, Charollais J, Sambeth U, Pavoni N (1988) La faille du Vuache (Jura méridional): un exemple de faille à caractère polyphasé. Bull Soc Vaud Sci Nat 79, 2: 65–91

Boyer SE, Elliott D (1982) Thrust systems. AAPG Bull 66: 1196–1230

Buxtorf A (1916) Prognosen und Befunde beim Hauensteinbasis und Grenchenbergtunnel und die Bedeutung der letzteren fur die Geologie des Juragebirges. Verh Naturforsch Ges Basel 27: 185–254

Byerlee J (1978) Friction of rocks. Pure Appl Geophys 116: 615–626

Calassou S, Malavieille J (1990) Etude des zones de transfert dans les prismes d'accrétion océanique et les avant-pays des chaînes: modélisation analogique. Rapp final Univ Montpellier, IFP 38048

Chauve P, Enay R, Fluck P, Sittler C (1980) Vosges, fossé Rhénan, Bresse, Jura. 26è CGI. Ann Sci Univ Besançon (4) 1: 81–114

Chauve P, Martin J, Petitjean E, Sequeiros F (1988) Le chevauchement du Jura sur la Bresse. Données nouvelles et réinterprétation des sondages. Bull Soc Géol Fr (8) IV, 5: 861–870

Colletta B, Ballard JF, Balé P, Bénard F, Letouzey J, Pinedo R (1990) Thrust propagation and non-cylindrical structures in small-scale models: kinematics and 3D analysis by X-ray tomography. Thrust tectonics 1990, London, April 4–6, Abstracts 64

Colletta B, Letouzey J, Pinedo R, Ballard JF, Balé P (1991) Computerized X-ray tomography analysis of sand-box models: examples of thin-skinned thrust systems. Geology 19, 11: 1063–1067

Cornec JH (1983) Synthèse géologique du Jura et de la Bresse. Direction des Hydrocarbures, Service de Conservation des Gisements d'Hydrocarbures

Dahlstrom CDA (1970) Structural geology in the eastern margin of the Canadian Rocky Mountains. Bull Can Pet Geol 18: 332–406

Damotte B, Nicholich R, Cazes M, Guellec S (1990) Mise en oeuvre, traitement et présentation du profil plaine du Pô-Massif central. In: Roure F, Heitzmann P, Polino R (eds)

Deep structure of the Alps Mém Soc Géol Fr Paris 156, Mém Soc Géol Suisse Zurich 1, Vol Spec Soc Geol Ital Roma 1 145

Davis D, Suppe J, Dahlen FA (1983) Mechanics of fold-and-thrust belts and accretionary wedges J Geophys Res 88 1153–1172

Davis DM, Engelder T (1985) The role of salt in fold-and-thrust belts Tectonophysics 119 67–88

Debrand-Passard S, Courbouleix S (1984) Synthèse géologique du Sud-Est de la France Mém BRGM Fr 126, vol 2

Enay R (1982) Notice explicative de la feuille Saint-Rambert-en-Bugey à 1/50000 BRGM, Carte géologique de la France à 1/50000, 3230

Guellec S, Mugnier JL, Tardy M, Roure F (1990a) Neogene evolution of the western Alpine foreland in the light of ECORS data and balanced cross section In Roure F, Heitzmann P, Polino R (eds) Deep structure of the Alps Mém Soc Géol Fr Paris 156, Mém Soc Géol Suisse Zurich 1, Vol Spec Soc Geol Ital Roma 1 65–76

Guellec S, Lajat D, Mascle A, Roure F, Tardy M (1990b) Deep seismic profiling and petroleum potential in the western Alps constraints with ECORS data, balanced cross sections and hydrocarbon modelling In Pinet B, Bois C (eds) The potential of deep seismic profiling for hydrocarbon exploration Technip, Paris, pp 425–437

Karner GD, Watts AB (1983) Gravity anomalies and flexure of the lithosphere J Geophys Res 88, B12 449–477

Lacassin R (1989) Plate-scale kinematics and compatibility of crustal shear-zones in the Alps In Coward MP, Dietrich D, Park G (eds) Alpine Tectonics, Geol Soc Spec Publ 45 339–352

Laubscher HP (1977) Fold development in the Jura Tectonophysics 37 337–362

Laubscher HP (1986) The eastern Jura relations between thin-skinned and basement tectonics, local and regional Geol Rundsch 75, 3 535–553

Lefavrais-Raymond A (1962) Contribution à l'étude géologique de la Bresse d'après les sondages profonds Mém BRGM 16

Letouzey J (1990) Fault reactivation, inversion and fold-and-thrust belt In Letouzey J (ed) Petroleum and tectonics in mobile belts Technip, Paris, pp 101–127

Lienhardt MJ (1984) Trias – puissance et façiès de la partie supérieure, planche T2 In Debrand-Passard S et al (eds) Synthèse géologique du Sud-Est de la France Mém BRGM 126

Mandl G Shippman GK (1981) Mechanical model of thrust sheet gliding and imbrication In McClay KR, Price NJ (eds) Thrust and nappe tectonics Geol Soc Spec Publ 9 79–97

Maury RC, Varet J (1980) Le volcanisme tertiaire et quaternaire en France In Géologie de la France – 26è CGI, coll C7 – Mém BRGM, Orléans 107 137–159

Mitra S (1990) Fault-propagation folds geometry, kinematic evolution and hydrocarbon traps AAPG Bull 74, 6 921–945

Mitra S, Namson J (1989) Equal-area balancing Am J Sci 289 536–599

Mugnier JL, Ménard G (1986) Le dévelopement du bassin mollasique suisse et l'évolution des Alpes externes un modèle cinématique Bull Cent Rech Expl Prod Elf Aquitaine Pau 10, 1 167–180

Mulugeta G (1988) Modelling the geometry of Coulomb thrust wedges J Struct Geol 10 847–859

Philippe Y (1991) Etude structurale du Jura méridional (région d'Ambérieu-en-Bugey) Rapp IFP-ENSPM 38748

Philippe Y, Specht M (1992) Lateral ramps and transfer zones in fold-and-thrust belts geometry, development and comparison with analogue modelling experiments EAPG 4th Conf, June 1–5, Abstr with Progr 48–49

Suppe J (1983) Geometry and kinematics of fault-bend folding Am J Sci 283 684–721

Thomas WA (1990) Controls of locations of transverse zones in thrust belts Eclogae Geol Helv 83/3 727–744

Truffert C, Burg JP, Cazes M, Bayer R, Damatto B, Rey D (1990) Structures crustales sous le Jura et al Bresse contraintes sismiques et gravimétriques le long des profils ECORS Bresse-Jura et Alpes II In Roure F, Heitzmann P, Polino R (eds) Deep structure of the Alps Mém Soc Géol Fr Paris 156, Mém Soc Géol Suisse Zurich 1, Vol Spec Soc Geol Ital Roma 1 157–164

Wildi W, Blondel T, Charollais J, Jaquet JM, Wernli R (1991) Tectonique en rampe latérale à la terminaison occidentale de la Haute-Chaîne du Jura Eclogae Geol Helv 84/1 265–277

Winnock E (1961) Résultats géologiques du forage Risoux 1 Bull Ver Schweiz Petrol Geol Ing 28, 74 17–26

Geological maps of the study area published by the Bureau de Recherches Géologiques et Minières (BRGM) Ambérieu-en-Bugey, Belley, Bourg-en-Bresse, Chambéry, La Tour-du-Pin, Montagnieu, Montmélian, Saint-Rambert-en-Bugey, Seyssel, Voiron (1/50000), Nantua, Saint Claude (1/80000)

Triassic Transgressive-Regressive Cycles in the Bresse-Jura and Adjacent Basins, Eastern France

G. Dromart, P. Monier, A. Curial, R. Moretto, and F. Guillocheau

Abstract. The Triassic units in the Bresse-Jura area have been extensively drilled for salt and hydrocarbon exploration. Subsurface correlations have recently been refined using current concepts and methods of genetic stratigraphy. The Triassic series in the Bresse-Jura area discloses a net superimposition of two-order Transgressive-Regressive cyclicities of a second-order type. The major Triassic depositional cycle developed from Scythian to Late Carnian times; it started with fluvial sandstones. The transgressive regime climaxed during the Ladinian with regionally extended deposition of marine carbonates. Sulfate-dolomitic depositional facies of the uppermost Muschelkalk, together with overlying Keuper salt-bearing lithologies, marked gentle regressive conditions. The regression was then forced, as evidenced by a clear basinward shift of siliciclastic deposition. Initiation in Late Carnian times of the next major depositional cycle was expressed by dolostones that rapidly spread over the former continental deposits. Transgressive conditions of this upper major cycle culminated much later during Liassic times. Four minor second-order cycles have been identified throughout the Triassic: Scythian, Scythian-Anisian, Anisian-Carnian, and Carnian-Norian. Restored depositional geometries with superimposed depositional environments in the Bresse-Jura area plus interregional comparison with adjacent basins of eastern France have permitted the respective influence of tectonics and eustacy on second-order T-R cycles to be discriminated.

Background and Objectives

Mesozoic strata of the Bresse area unconformably underlie a thick pile of Cenozoic deposits, i.e., West-European Paleogene Rift System, and outcrop along the eastern border of the Massif Central Paleozoic basement. Eastwards, the Bresse Cenozoic Basin itself is thrust by the Mesozoic Jura Mtns (Fig. 1). Thrust-related decollement planes have been found within Triassic salt-bearing lithologies of the lowermost Keuper. Data relative to the Triassic lithologies in the Bresse-Jura come mainly from subsurface investigations. The Triassic units have been extensively drilled for salt and hydrocarbon exploration. Two salt formations of the Muschelkalk and of the Keuper are preserved in the central part of the basin. Oil and gas have been encountered within the Uppermost Muschelkalk carbonates, e.g., Revigny and Briod wells, and within Buntsandstein sandstones, e.g., Cordonnet and La Chandelière wells. Beyond its economic potential, the Bresse-Jura Triassic Basin is favorable to investigate low-order depositional cyclicity because of its extension and lithological composition. Finally, the Bresse-Jura Triassic Basin is critical both in space and time to study the major geodynamic reorganization that marked in western Europe the passage between the Variscan orogeny and the Tethyan opening (see Ziegler 1988).

Initial stratigraphic compilations have been published by Ricour (1962), Courel (1973), and M. J. Lienhardt in Debrand-Passard et al. (1984) Subsurface correlations presented in this chapter have recently been refined (DIASTRATA 1991, 1992), using well-logs and well reports for over 100 wells (Fig. 2). The correlations have been based on current concepts and methods of genetic stratigraphy. Time-equivalent surfaces are lines connecting reversal points of depositional cycles, i.e., transgressive-regressive, since it can be assumed that no block-uplifting occurred during Triassic times. These surfaces together with marker-beds have been tracked, and checked against biostratigraphic data available in well reports and provided by Adloff et al. (1984). Prior to any tentative restoration of depositional geometries, a considerable effort has been made to decipher the intra-Triassic stratigraphic melange associated with the Jura-thrust development.

Data relative to the Paleozoic basement of the Bresse-Jura Triassic Basin are presented below since (1) the basement is partly made up of Stephanian coal-bearing strata thought to be source rocks of the hydrocarbon occurrences, (2) the basement shows a fairly complex organization that may reflect a very specific geodynamic history. The depositional cycles recorded through the Triassic column are then described with specific attention paid to the sandstone and salt lithologies. The nature and architecture of the distinct lithological complexes are interpreted in terms of base-level transit cycles. The origin of these stratigraphic cycles is discussed, and implications proposed. Finally, comparisons with the Triassic stratigraphic record to the North, i.e., western Paris Basin and German Basin, and in adjacent southern Tethyan basins, i.e., northwestern Subalpine Basin and Briançonnais Terrane, are tentatively given.

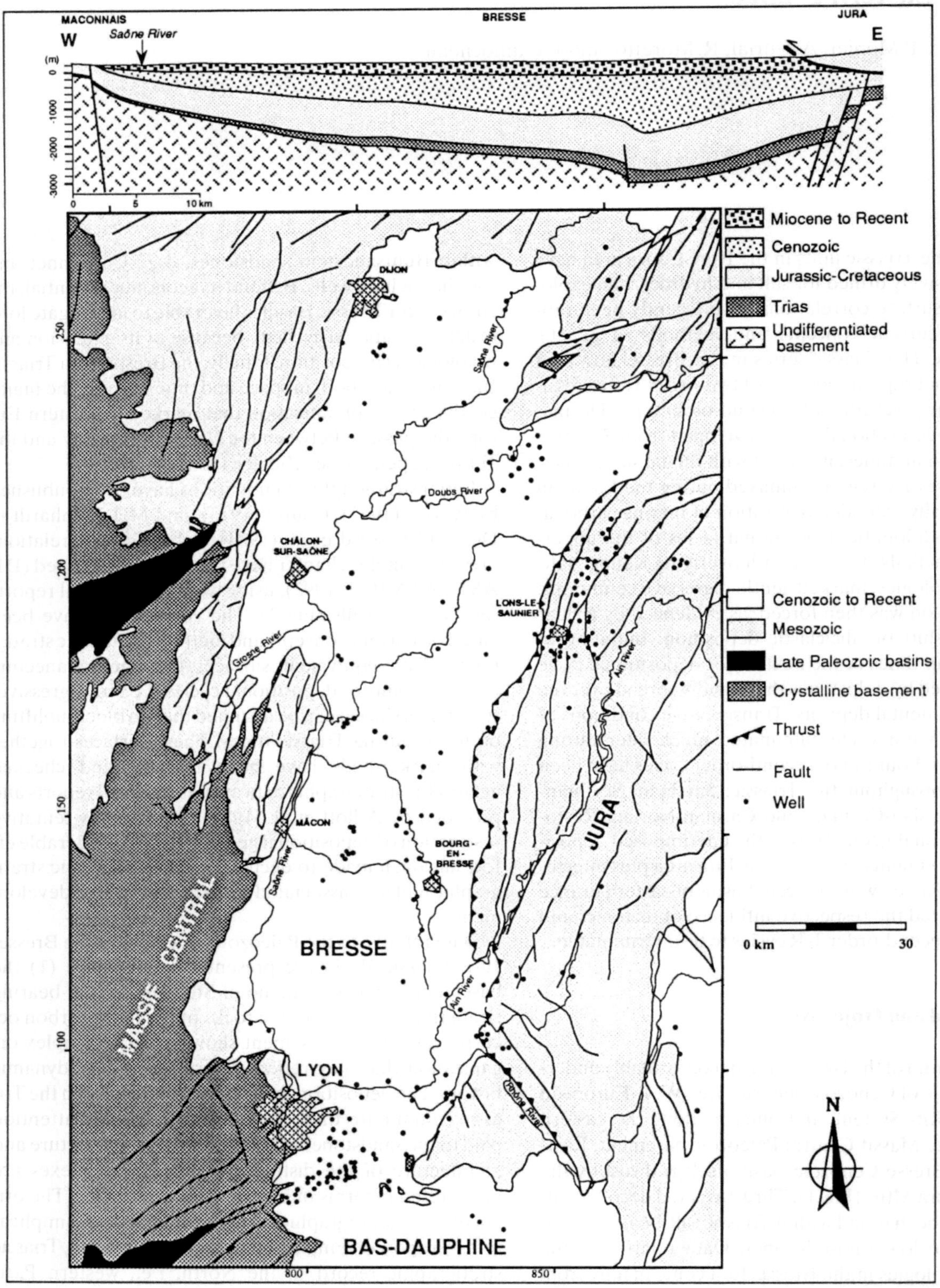

Fig. 1. Simplified geological map of the Bresse-Jura area (eastern France) and W-E cross section of the Bresse Basin

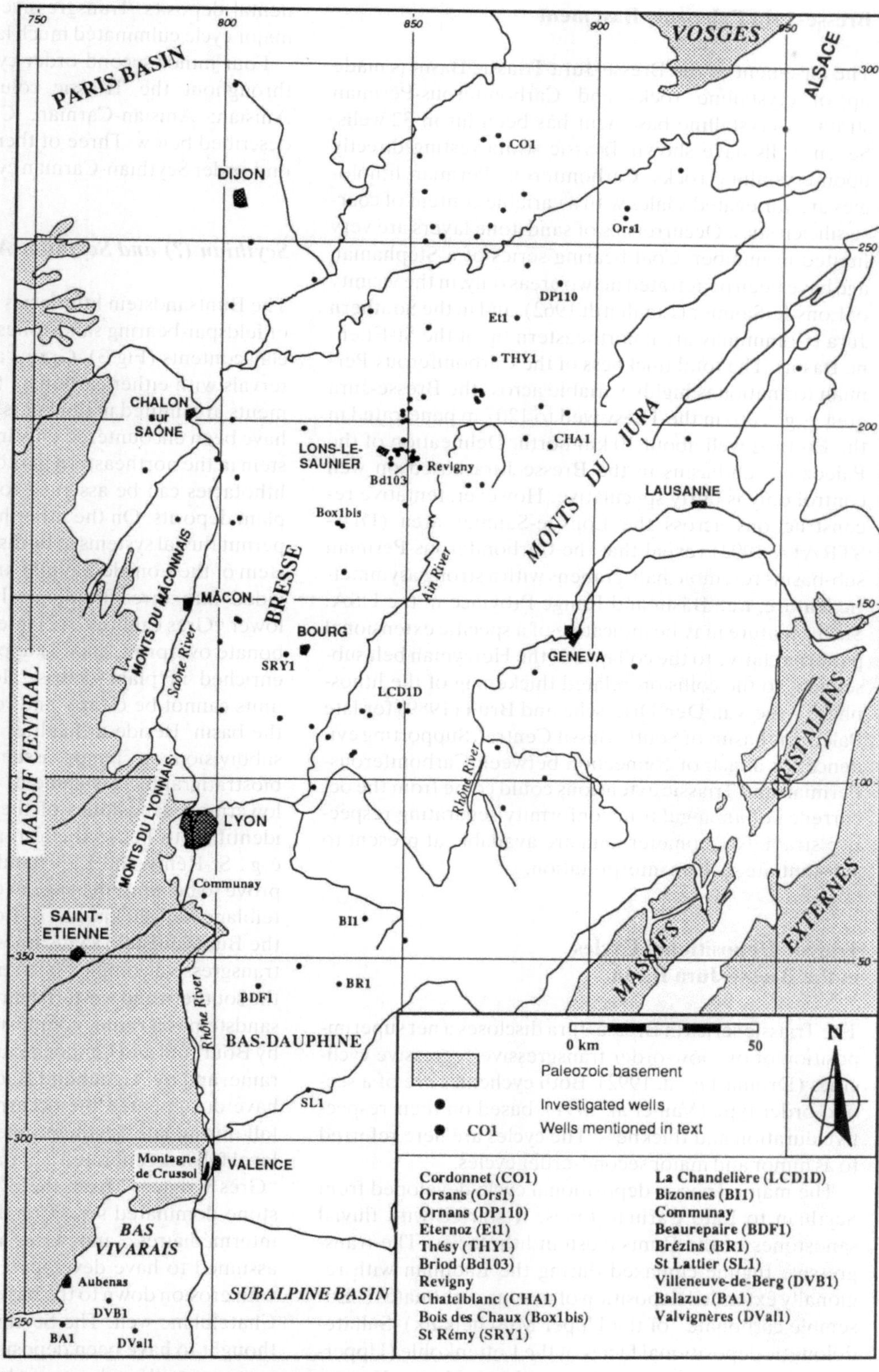

Fig. 2. Outcrop and well distribution index map of the study area

Bresse-Jura Paleozoic Basement

The basement of the Bresse-Jura Triassic Basin is made up of crystalline rocks and Carboniferous-Permian strata. A crystalline basement has been hit in 32 wells Seven wells have shown Triassic strata resting directly upon crystalline rocks Carboniferous-Permian lithologies are variegated shales with a variable content of coarse siliciclastics. Occurrences of sandstone layers are very limited in number Coal-bearing series of a Stephanian age have been penetrated in two areas only, in the vicinity of Lons-le-Saunier (Lienhardt 1962), and in the Southern Jura (Communay area: northeastern tip of the St-Etienne Basin) The total thickness of the Carboniferous-Permian formation is highly variable across the Bresse-Jura area, e. g., zero in the Thesy well to 1207 m penetrated in the Eternoz well about 10 km north. Delineation of the Paleozoic sub-basins in the Bresse-Jura based on well control only is fairly speculative. However, tentative reconstructions across the Lons-le-Saunier area (DIASTRATA 1992) reveal that the Carboniferous-Permian sub-basins resemble half-grabens with a strong asymmetric feature, i. e., Basin and Range Province in the USA. Such a feature may be indicative of a specific extensional pattern relative to the collapse of the Hercynian belt subsequent to the collision-related thickening of the lithosphere – see Van Den Driessche and Brun (1989) for late Paleozoic basins of South Massif Central. Supporting evidence for a lack of connection between Carboniferous-Permian and Triassic extensions could come from the occurrence of an angular unconformity separating respective strata. No dipmeter data are available at present to substantiate such an interpretation.

Triassic Depositional Cycles in the Bresse-Jura Basin

The Triassic series in Bresse-Jura discloses a net superimposition of two low-order transgressive-regressive cyclicities (Dromart et al. 1992). Both cyclicities are of a second-order type (Vail et al. 1991), based on their respective duration and thickness. The cycles are here referred to as minor and major second-order cycles.

The major Triassic depositional cycle developed from Scythian to Late Carnian times. It started with fluvial sandstones of the Buntsandstein lithofacies. The transgressive regime climaxed during the Ladinian with regionally extended deposition of marine carbonates ("Ensemble carbonaté" of the Upper Muschelkalk) Sulfate-dolomitic depositional facies of the Lettenkohle (Uppermost Muschelkalk) together with overlying Keuper "Série salifère" (salt series) marked gentle regressive conditions The regression was then forced, as evidenced by a clear basinward shift of siliciclastic deposition ("Complexe argilo-gréseux" including the "Grès à roseaux" lithologies). Initiation in late Carnian times of the next major depositional cycle was expressed by the "Dolomie moellon" that rapidly spread over the former continental deposits Transgressive conditions of this upper major cycle culminated much later during Liassic times.

Four minor second-order cycles have been identified throughout the Triassic column (Scythian, Scythian-Anisian; Anisian-Carnian, Carnian-Norian), and are described below Three of them make up the major second-order Scythian-Carnian cycle described above.

Scythian (?) and Scythian-Anisian Cycles

The Buntsandstein lithofacies are dominantly composed of feldspar-bearing sandstones having a very low detrital clay contents (Fig. 3). Grains are very poorly sorted. Intervals with either carbonate (dolomite) or siliceous cements are limited in thickness. Conglomerate lithofacies have been encountered only in the lowermost Buntsandstein in the northeastern part of the basin. Buntsandstein lithofacies can be assigned to medium to outer alluvial plain deposits. On the other hand, available data do not permit fluvial systems to be distinguished. The Buntsandstein of the Lons-le-Saunier area has tentatively been divided into three lithological units (Ricour 1962): (1) lower "Grès calibrés", (2) intermediate series with a carbonate overprint, and (3) upper section with sandstones enriched in plant debris. However, these lithological units cannot be clearly and repeatedly identified across the basin Beside difficulties relative to the lithological subdivision, a major issue emerges regarding the biostratigraphic assignment. Based on palynology, the lowermost sandstones of the Buntsandstein have been identified to be Scythian in age in south-located wells, e.g , St-Rémy and La Chandelière wells, whereas they prove to be Anisian in age in one east-located well (Chatelblanc well, 40 km east of Lons-le-Saunier). Therefore, the Buntsandstein sandstones cannot be seen as a single transgressive complex becoming younger as it moves to the southern and western margins of the basin The Buntsandstein is a rather composite complex. Recent studies by Bourquin and Guillocheau (pers. comm. 1992) in Lorraine, and by Aigner and Bachmann (1992) in Germany have documented the occurrence of a major base-level fall during late Scythian times In the Vosges, the base-level fall has generated a hiatal surface that separates the "Grès vosgiens" from the overlying transgressive sandstone-dominated set "Conglomérat principal, Couches intermédiaires", and "Grès à Voltzia". Such a surface is assumed to have developed in the Bresse-Jura as well, with erosion down to the basement in certain places, e g., Chatelblanc well. The Scythian-Anisian sandstones are thought to have been deposited just above the surface in response to the subsequent base-level rise of the next cycle, with no lowstand-complex development It is suggested that the Buntsandstein in the Bresse-Jura is composed of two distinct amalgamated sandstone complexes, the lower Scythian and the upper Scythian-Anisian. The surface separating the complexes has not yet been delineated Finally, it should be stressed that such an interpretation implies the development of a specific Scythian

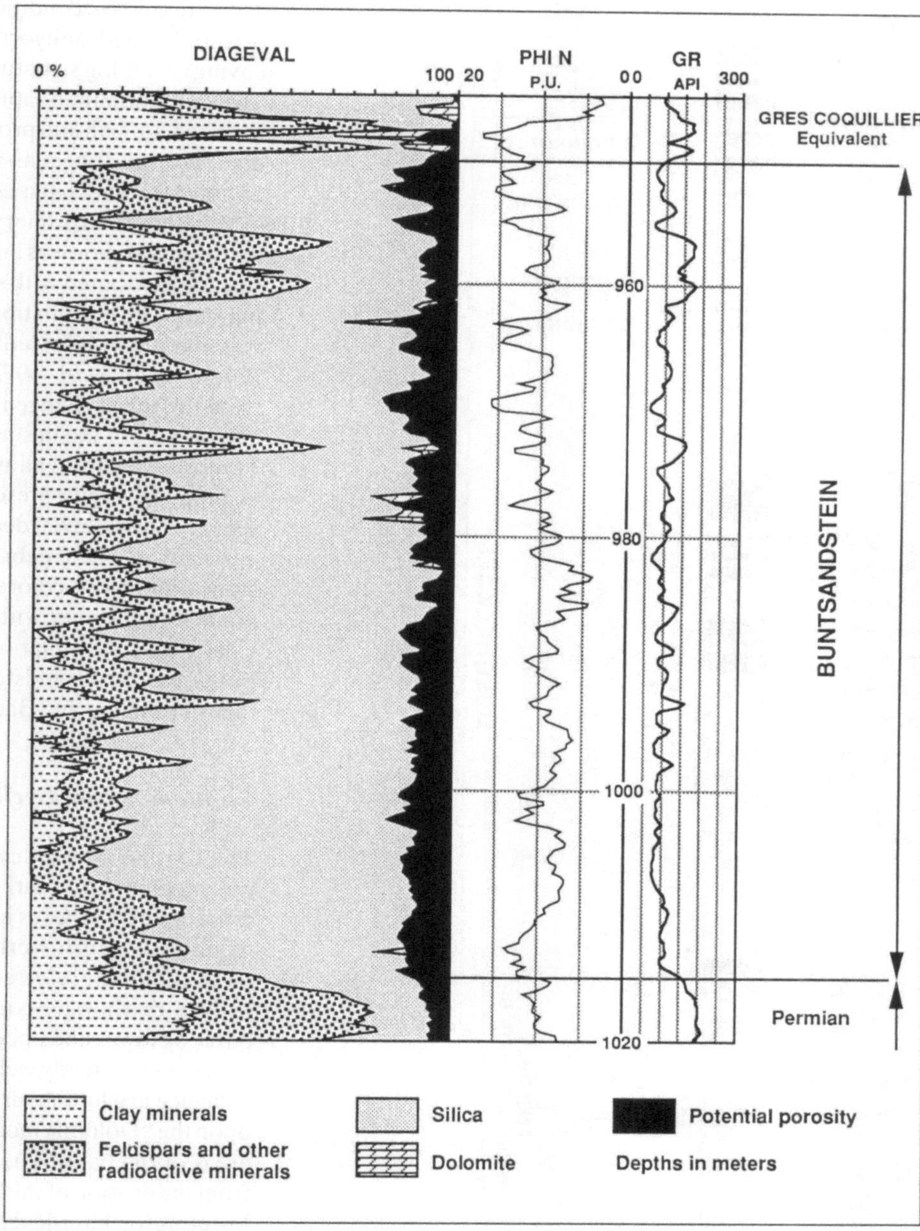

Fig. 3. Well-log-based, computed lithological evaluation of the Buntsandstein lithofacies in Orsans well. Lithologies provided by DiagEval (mark of DIASTRATA) are in relative volumes

base-level transit cycle. The Scythian-Anisian cycle (Fig. 4) was probably initiated with fluvial aggradation of incised-valley fill deposits. The transgression accelerated with the deposition of marine shell-bearing sandstones here referred to as "Grès coquillers". Poor correlation of genetic sequences within the underlying Buntsandstein deposits has not permitted a potential ravinement surface to be documented. The subsequent maximum marine flooding culminated with shaly deposits of the "Zone argileuse". Above, progressive enrichment in carbonate up to the "Dolomie à Myophoria" lithologies reflects gentle regressive conditions. The regression was then marked by salt deposits restricted to the northeastern Bresse-Jura (Fig. 5). Such a localization is fully consistent with the areal extent of the former marine deposits. The cycle terminated with shale-dominated coastal plain deposits interrupted by a thick anhydrite-bearing marker bed (Repère Guillaume) that can be traced much further north over the Lorraine area (Geisler-Cussey 1985).

Anisian-Carnian Cycle

The cycle started with sulfate-shale depositional facies of coastal plain environments. The subsequent marine influence is marked by the "Ensemble carbonaté" lithofacies of a Ladinian age (Fig. 4). The "Ensemble carbonaté" in the Bresse-Jura is lithostratigraphically equivalent to the Upper Muschelkalk in the Alsace-Lorraine even though

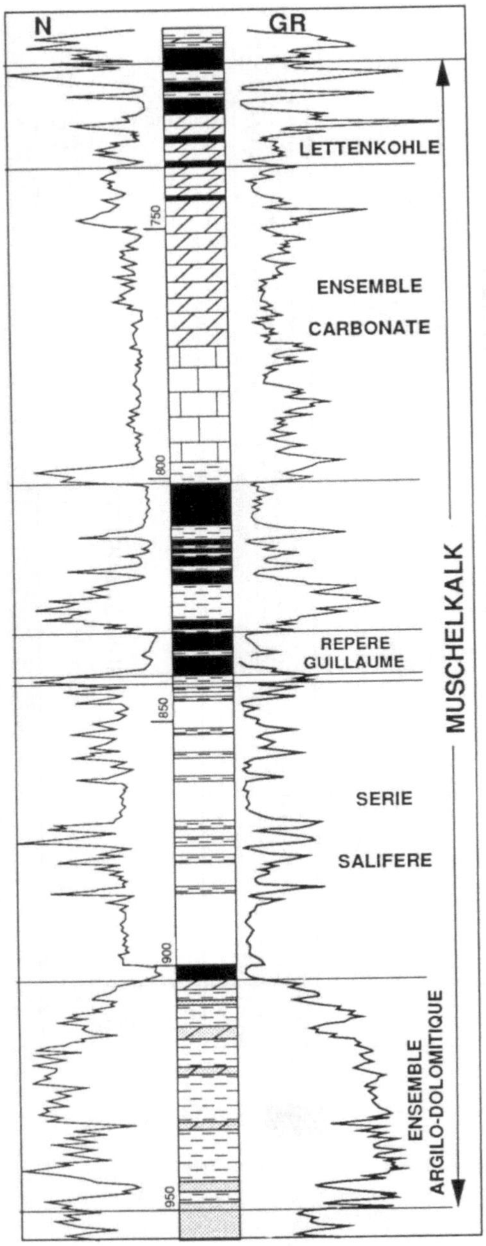

Fig. 4. Well-log signature of the Middle Triassic (Muschelkalk) lithostratigraphic units in the Orsans well (northern Jura) Depths in m

the standard subdivision, i.e., "Calcaires blancs, Calcaires à entroques, Calcaires à Ceratites", and "Calcaires à Térébratules", has not been found in the Bresse-Jura. The "Ensemble carbonaté" lithologies here have a strong dolomitic overprint Crinoidal and siliciclatic carbonates have been encountered. Evidence of the Ladinian marine inundation is available along the western margin of the Bresse-Jura Basin in the Monts du Lyonnais and Monts du Macônnais areas as siliciclastic carbonates supporting storm-related depositional fabrics and marine foraminifera ("Calcaires de la Font-Poivre" and "Calcaires gréseux de Milly", respectively) The Uppermost "Ensemble carbonaté" together with Lettenkohle express initial gentle regressive conditions. The Lettenkohle is made up of thin-bedded anhydrite and dolomite with bed-sets having a well-log signature steady all over the study area Sedimentation then rapidly evolved to evaporitic during the late Ladinian and proceeded through most of the Carnian. The resulting massive Keuper Série salifère is up to 250 m in thickness. It is actually composed of interbeds of shale-sulfate and primary halite, and is free of carbonates. Salt accumulation was limited to the central part of the basin, and associated with shale-sulfate deposition on basin margins. The termination of the cycle in the Late Carnian was marked by a forced regression (see the definition of Posamentier et al. 1992). Supporting evidence comes from the sharp change in the depositional pattern in the central part of the basin· from distal coastal plain shales-evaporites to distal alluvial plain shales-sandstones, i.e., "Complexe argilo-gréseux" including the "Grès à roseaux". These fluvial deposits consist of shaly sandstones enriched in plant debris, clear and fine-grained sandstones, and black-colored shales supporting beds of fine-grained sandstones with little glauconite, e.g., Bois-des-Chaux well. A similar basinward shift of continental deposits, i.e., Schilfsandstein, has been documented in Germany by Aigner and Bachmann (1992)

Carnian-Norian Cycle

The Carnian-Norian cycle commenced with the "Dolomie moellon" that directly and rapidly spread over siliciclastics of the "Grès à roseaux" (Fig. 6) The "Dolomie moellon" is in turn overlain by a package of shale, anhydrite, and dolomite referred to as "Marnes bariolées". An anhydrite-rich horizon is intercalated within this interval, separating lower and upper "Marnes bariolées". The horizon has a very steady well-log signature and is termed "Alternance argilo-anhydritique". This horizon lies directly upon the "Dolomie moellon" in the southern and western areas of the basin. The hiatus is interpreted as resulting from the erosion of the lower "Marnes bariolées" due to a lowering of base-level. Above the horizon, the upper "Marnes bariolées" reflect renewed transgressive conditions that clearly developed throughout the Rhaetian.

Nature and Architecture of Lithologies Versus Base-Level Transit Cycles (Fig. 7)

This section links the nature, overall geometry, and stratal pattern of the distinct lithological complexes encountered to the transit cycles of the base-level (see the definition of Barrell 1917).

Transgressive Amalgamated Sandstone Complexes

The basal sandstones of the Buntsandstein drape the Paleozoic basement all over the Bresse-Jura area, including southern and western margins This complex is thought

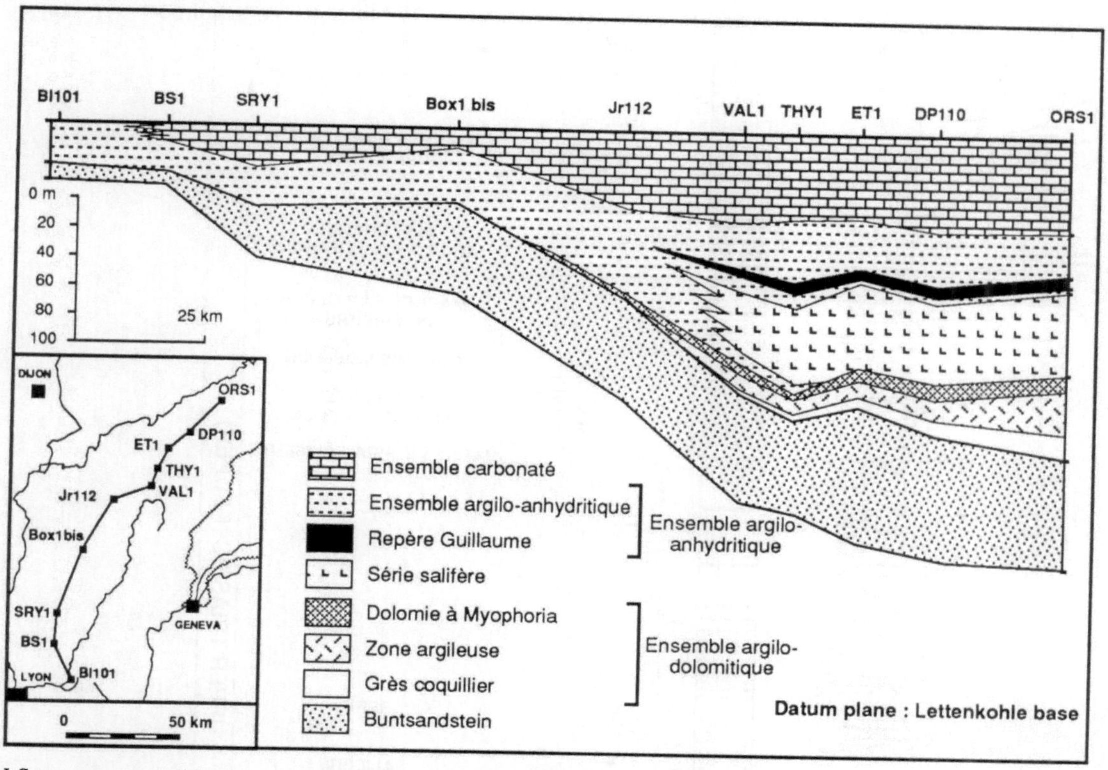

Fig. 5. Simplified N-S cross-section of the Lower and Middle Triassic in the Bresse-Jura Basin

to be composite in nature, and made up of fluvial sandstones successively stacked through the base-level rises of the Scythian and Scythian-Anisian cycles. Regrettably, delineation of these complexes, together with restoration of the stratal termination pattern, have not been possible because of a dearth in biostratigraphic and sedimentologic information.

Flooding Shale-Carbonate Complexes

The Uppermost Scythian-Anisian (Zone argileuse), Ladinian (Ensemble carbonaté), and upper Carnian (Dolomie moellon) are elongated wedges that onlap southern and western basin margins (Figs. 5, 7), and reflect a marine transgressive to maximum flooding regime. The lack of landward thickening of these transgressive complexes suggests that the related transgressions were not caused by any sea-level rises. Rather, the stratal pattern was produced by a change in the flexural pattern of the basin margin that possibly reflects stiffening and cooling of the lithosphere.

Early Highstand Carbonate Fills

The Anisian Dolomie à Myophoria and Ladinian Lettenkohle are specifically sheet-shaped, with an onlap termination upon the margin. Such geometry and stratal pattern express a very slow and uniform creation of accommodation that has prevented any progradation from developing.

Tectonically Induced Late Highstand Salt Complexes

The salt complexes of the Muschelkalk and Keuper succeeded the maximum marine flooding of the Anisian and Ladinian respectively, and are regressive in nature. Both salt complexes are areally restricted to the depocenters, and form high-angle wedges (Figs. 5, 7). The Keuper salt shows a larger areal extent than the Muschelkalk salt. The salt lithologies, i.e., shale-halite interlayers, pass to shale-sulfate deposits on the basins margins. Fine well-log correlations disclose the passage to be a real facies interfingering (at least for the Keuper salt). This interfingering reflects qualitative and volumetric lateral partitioning of depositional products that occurred for each genetic sequence, i.e., basic base-level transit cycle. The areal distribution and shape of the salt complexes are related to differential subsidence that controlled the spatial interfingering position. Lateral shifts of the passage zone have originated apparent onlapping or offlapping stratigraphic patterns for the salt deposits. Both salt complexes directly overlie sheet-shaped carbonate deposits. Accordingly, it can be stated that slow and uniform creation of accommodation abruptly evolved to a rapid and differential regime.

A Lowstand Sandstone Complex

The Upper Carnian fluvial siliciclastics (Complexe argilo-gréseux) have not been encountered on basement

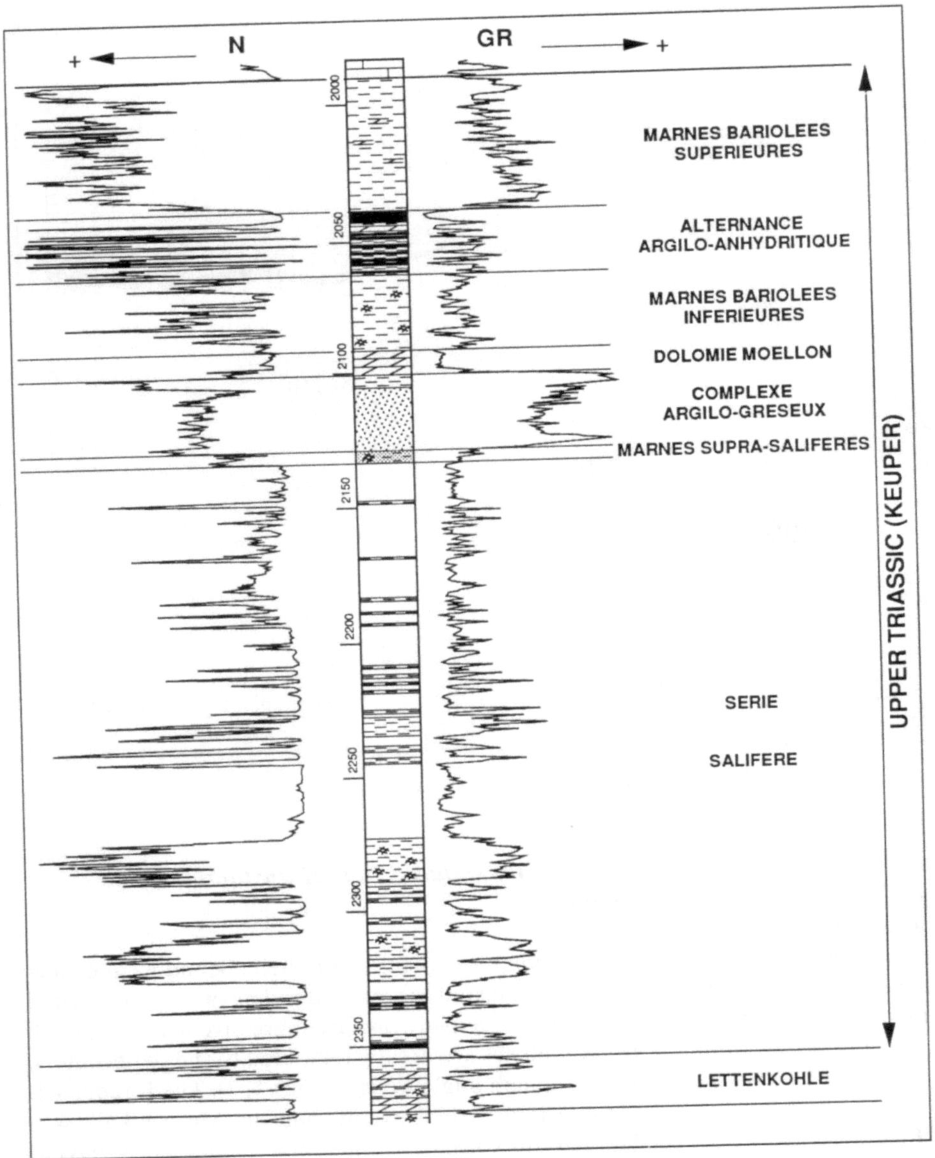

Fig. 6. Stratigraphic column of the Upper Trias in the Shell Box-1bis well (Bresse area) Depths in m

highs, e. g., southern Eperon Lyonnais, and are restricted to the basin center. Restored isopachs show several depocenters and no incised-valley fill configuration (Fig. 8). In addition, it should be noticed that the thicker the complex, the coarser the siliciclastic material. Finally, it should be stressed that there is a lack of evidence for stream incision at the stratigraphic level separating the complex from the underlying "Marnes suprasalifères". These distinct lines of evidence suggest that (1) the land-sourced deposits filled up a leftover space, (2) the space was created because of continuous subsidence together with lack of sediments due to a delay in the basinward shift of fluvial deposition, (3) the "Complexe argilo-gréseux" can be assigned to a lowstand complex, and (4) the net base-level fall recorded in the basin, e. g., deposition of the "Marnes suprasalifères" and nondeposition/by-pass for the lower surface of the complex, was presumably connected to a real abrupt lowering of sea level. However, the surface cannot be labeled a type I unconformity such as defined in the EXXON model (Haq et al. 1987). The surface could be termed a type II unconformity but very little to no fluvial deposits are predicted to develop in the overlying complex referred to as a shelf margin wedge (Vail et al. 1987). In conclusion, it is apparent that the EXXON model is not appropriate for the Bresse-Jura Triassic basin.

Origin of Cycles and Implications

The stratigraphic record of the Triassic in the Bresse-Jura basin can be seen as superimposition of two low-order depositional cyclicities, resulting from the summation of two base-level curves, and producing distortion of the stratigraphic record (see Cross 1988; Guillocheau 1991 a). Consistently, it is a matter of fact that (1) the

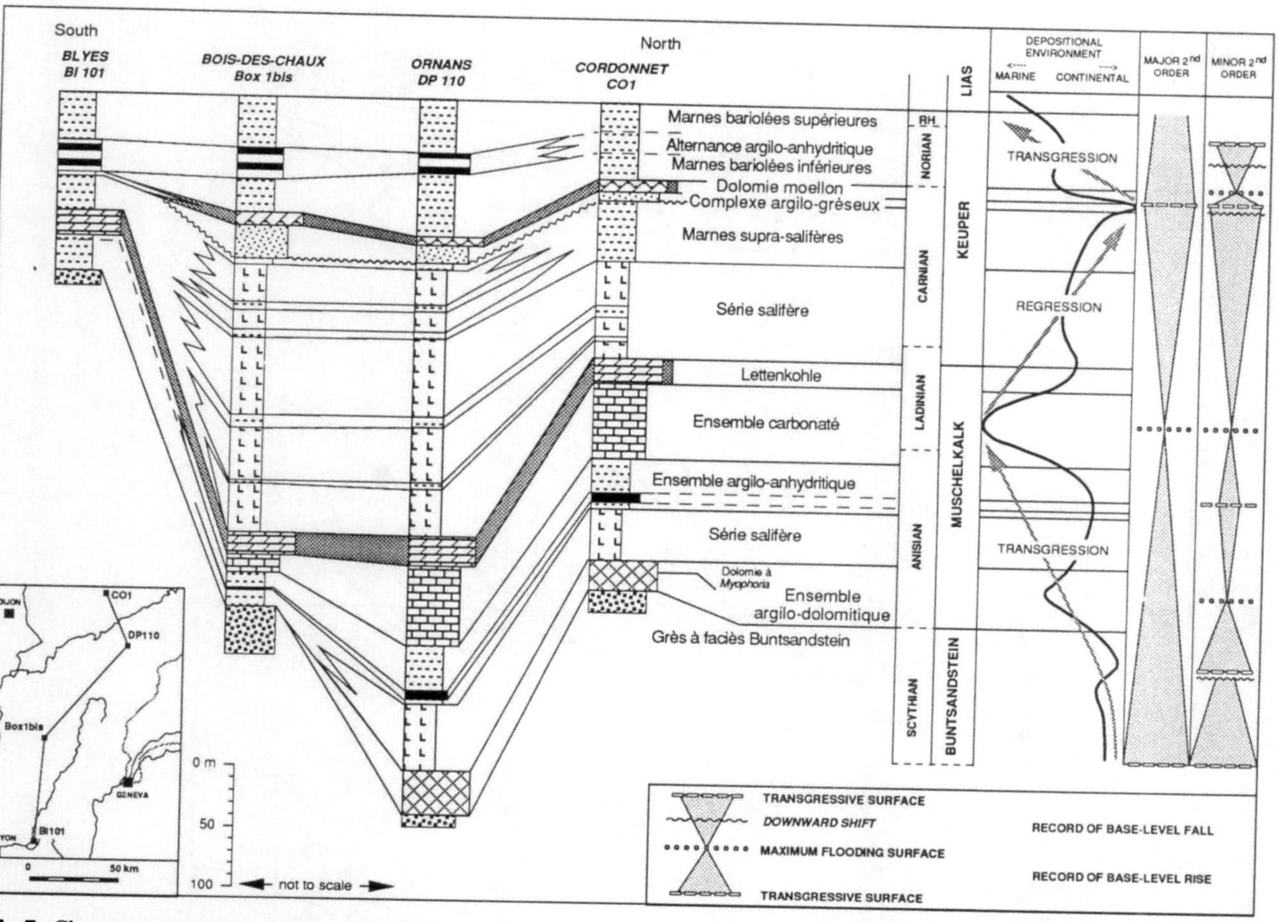

Fig. 7. Chronostratigraphic correlation and depositional cycles through the Triassic in the Bresse-Jura Basin (After Dromart et al 1992)

downward shifts are limited to the regression of the minor cycles included within the relative lowstand of the major cycle (2) the transgressive/regressive volumetric ratio of the minor cycles decreases as it progresses up through the major cycle (Fig. 7). Besides this fairly over-simple interpretation, we are really faced with two basic connected issues (1) what controls the initiation, development, and termination of the minor second-order cycles? (2) does the major second-order cycle have an entity of its own?

It has been suggested above that the arrangement in space of the transgressive complexes, including onlap feature, can be related to cooling/stiffening of the lithosphere. On passive continental margins, such a process usually is subsequent to a syn-rift phase which shows much more localized and much higher rates of subsidence (see Angevine et al. 1990). Such is the case at least for the Muschelkalk salt complex and the Keuper "Série salifère/Complexe argilo-gréseux" package which underlie the transgressive-maximum flooding of the "Ensemble carbonaté" and "Dolomie moellon", respectively. This interpretative assignment in terms of syn-rift and post-rift complexes implies that (1) the minor second-order cycles are geodynamic in nature and illustrate short-term McKenzie cycles, (2) depositional cyclicities (T-R cycles) and genetic geodynamic cyclicities are out of phase here, the higher subsidence rate coinciding with the regressive period of depositional cycles, (3) the Carnian downward shift at least is an excellent candidate for a genuine eustatic fall since it does not fit the geodynamic logic (being within the late syn-rift phase, and (4) the major Scythian-Carnian second-order cycle has no geodynamic significance.

Comparative Triassic Stratigraphic Record of Western Europe

Northwestern Europe

The Triassic sections of the western Paris Basin, Germany, and Bresse-Jura share many lithostratigraphic similarities. Both the diagram of depositional facies zonation for the western Paris Basin (Guillocheau 1991 b; Bourquin and Guillocheau in progress) and the coastal onlap curve for the German basin by Aigner and Bachmann (1992) show a long-term depositional cycle spanning Scythian to Carnian times with a maximum marine influence during the Ladinian. Even though closer comparisons, i. e., minor second-order cycles, are not possible because of a loose biochronologic framework, these three sub-basins can readily be related to a single large-scale basin.

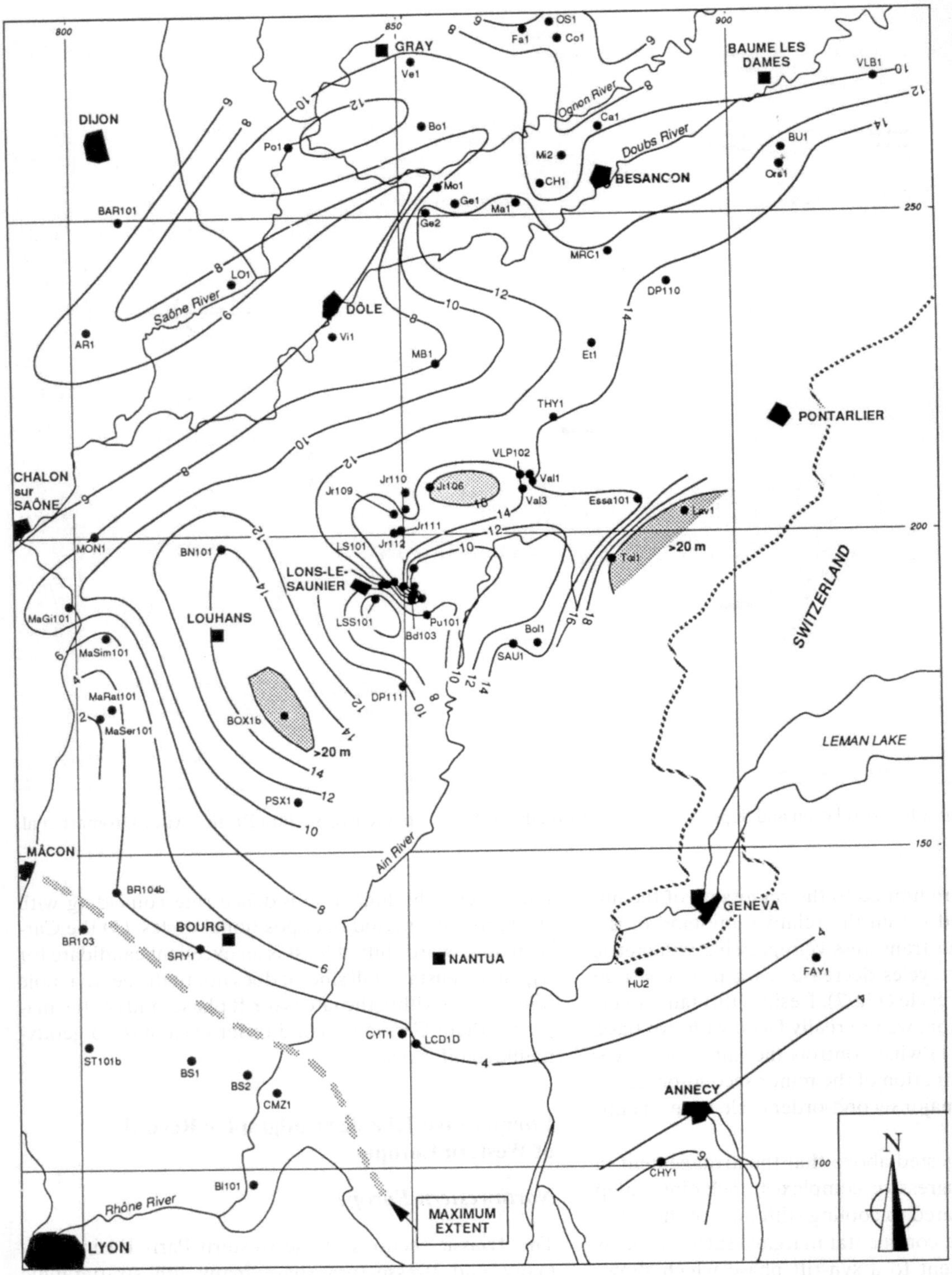

Fig. 8. Isopach map of the "Complexe argilo-gréseux" based on well control (Carnian, Bresse-Jura area). Measured depths in m

Southwestern Europe

Northwestern Subalpine Basin (Bas-Dauphiné and Bas-Vivarais Areas) The Triassic series is still poorly known in the Bas-Dauphiné subsurface because of limited effort in hydrocarbon exploration (Fig. 2). On the other hand, many more data are available on the western margin of the Subalpine Basin, i.e., the Bas-Vivarais, thanks to both major exposures and a large number of shallow exploratory wells for lead, zinc, and uranium.

Synthetic compilations, including age assignments based on palynological data, have been provided by Courel in Debrand-Passard et al. (1984) and Giot et al. (1987). Recently, the Triassic series has been cored in the Balazuc well (Elmi et al. 1991) through a scientific drilling operation conducted by the *Géologie Profonde de la France*

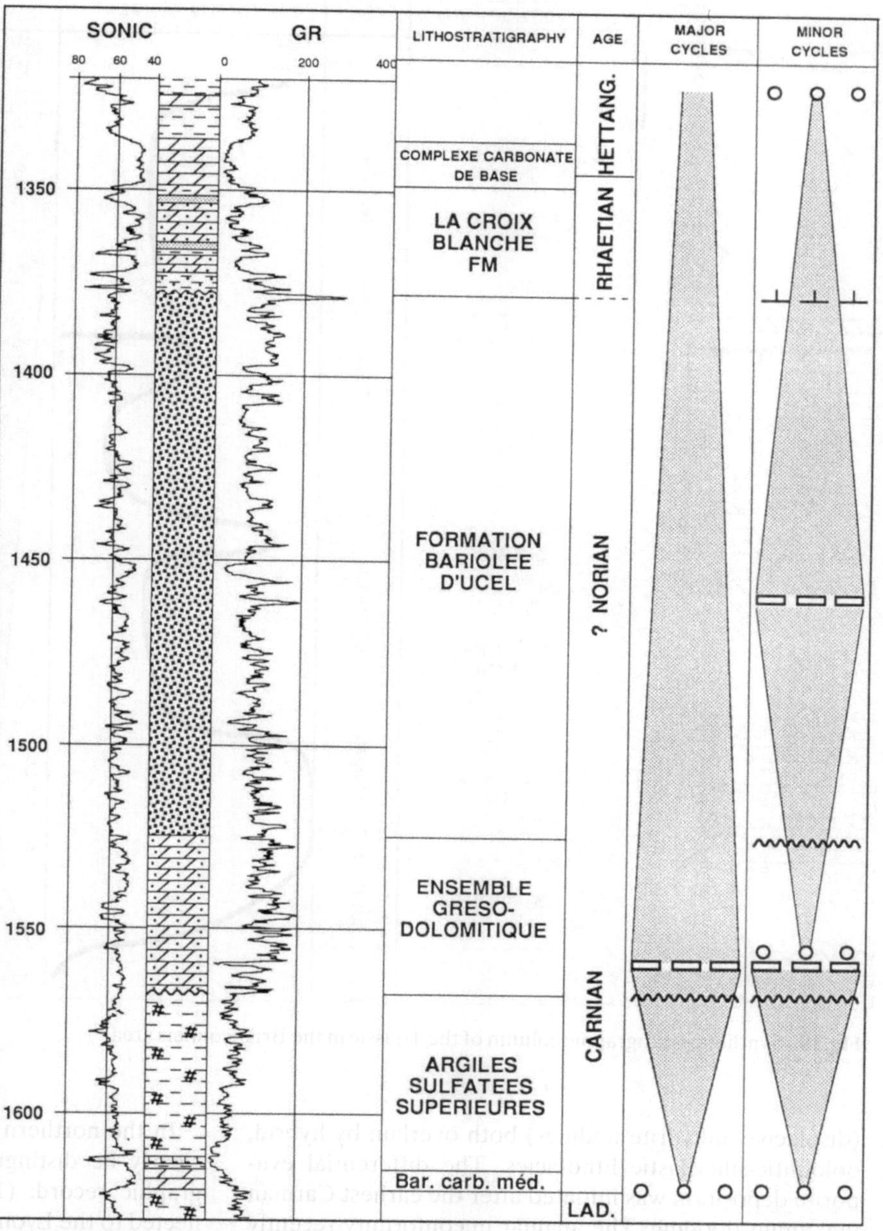

Fig. 9. Composite stratigraphic column of the Upper Triassic in the GPF-Balazuc well (Bas-Vivarais area). Depths in m. Depositional cycles are inferred from sedimentological core inspection only

program. The Triassic of the northwestern Subalpine Basin resembles that of the Bresse-Jura even if there are a number of clear differences. The basal sandstones are of a Buntsandstein lithofacies. They mostly lie upon a crystalline basement. They can also be found above either Stephanian coal-bearing strata, e.g., Balazuc, Valvignères wells in the Bas-Vivarais area, and the Beaurepaire well in northern Bas-Dauphiné, or Permian variegated argillaceous sandstones with an angular unconformity, e.g., Largentière area in the Bas-Vivarais. These basal sandstones are younger in age than the Bresse-Jura counterparts: Late Ladinian in the Bizonnes well and Late Anisian-Early Ladinian on the Bas-Vivarais border (around the city of Aubenas). A marine inundation occurred during the Late Ladinian in the Bas-Dauphiné, i.e., Lower "Formation carbonatée de St-Lattier", St-Lattier-1 well. This marine flooding was discrete as well in the Bas-Vivarais ("Dolomie gréseuse à Lamellibranches", Finelle 1981). The maximum flooding was delayed in the Bas-Vivarais to Earliest Carnian times (northern "Barre carbonatée médiane" and "southern Barre dolomitique du Mas-Dieu"). Both these Ladinian and Carnian carbonate horizons can be found in the intermediate domain that is southern Bas-Dauphiné: St Lattier-1 and Brézins-1 wells, and Montagne de Crussol outcrop.

The Upper Triassic in the Bas-Dauphiné is limited in thickness (15–40 m) and can be attached to the Lyonnais High. In contrast, the Carnian-Norian in the Bas-Vivarais is much thicker, e.g., 250 m in the Balazuc well, Fig. 9, and composed of basinal salt to marginal shale-sulfate

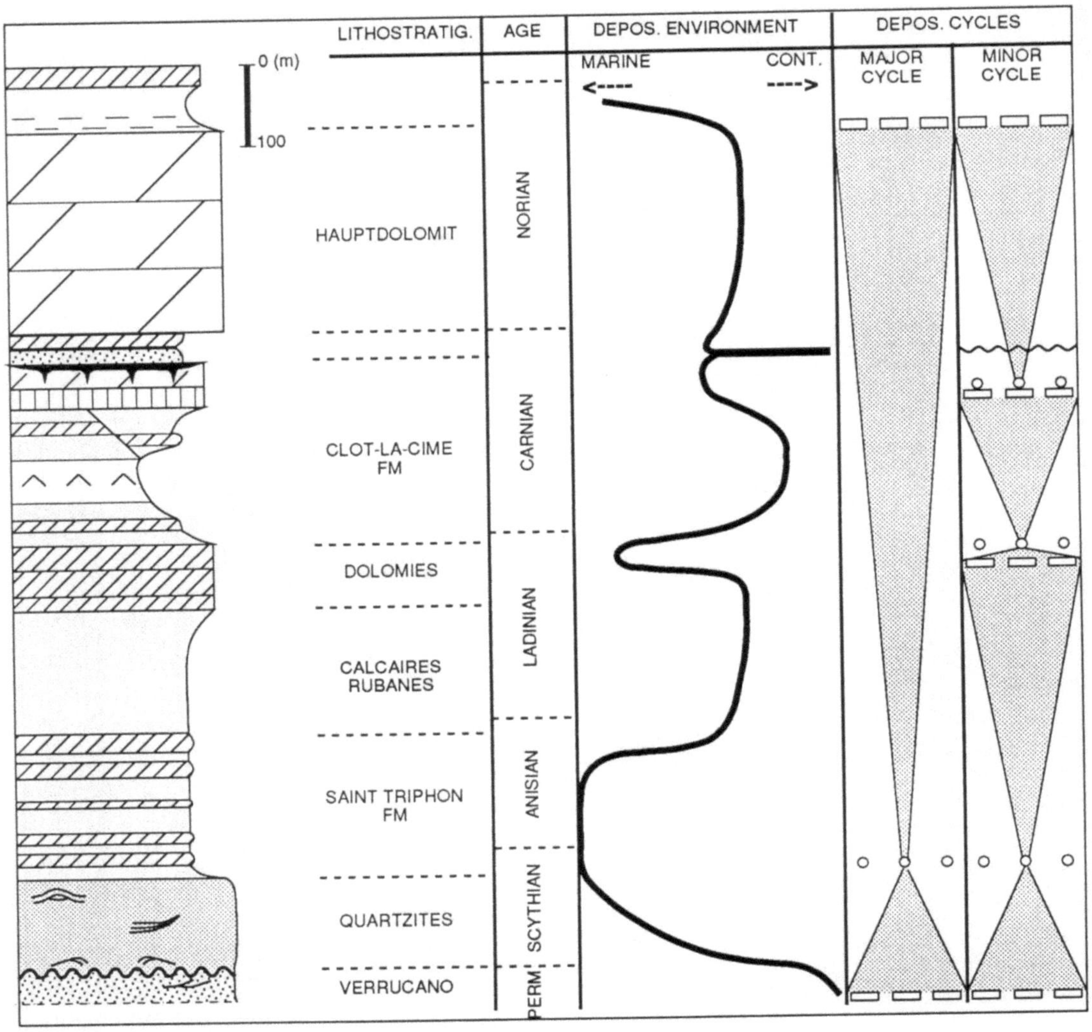

Fig. 10. Synthetic stratigraphic column of the Triassic in the Briançonnais area

(displacive anhydrite nodules) both overlain by hybrid, dolomitic-siliciclastic lithofacies. The differential evaporite deposition was initiated after the earliest Carnian maximum flooding. The angular unconformity recently documented on both sides of the "Barre carbonatée médiane" (Balazuc well; Bergerat and Martin 1993) testifies to a structurally controlled initiation of salt deposition. A major downward shift can be documented above the shale-sulfate deposits as a fluvial lithofacies that contains feldspathic sandstones and lignite. This lithofacies was deposited in the depressed parts of the Subalpine Basin margin (Balazuc, Villeneuve-de-Berg, Valvignères wells). It is thus similar to the "Complexe argilo-gréseux" of the Bresse-Jura basin. This lowstand complex is covered by carbonate-siliciclastic deposits of a moderate marine flooding. This flooding unit in turn is overlain by a fairly thick (170 m in the Balazuc-1 well) siliciclastic alluvial fan referred to as "Formation bariolée d'Ucel" that developed first as an aggradational unit then as a retrogradational unit. The top of this complex is truncated by a regional ravinement surface (first flooding surface) (Dromart et al. 1991) presumably of a Rhaetian age.

In the northern Subalpine Basin, three domains can clearly be distinguished, based on the Triassic stratigraphic record: (1) the northern Bas-Dauphiné connected to the Lyonnais High and thus to the Bresse-Jura basin; (2) the southern Bas-Dauphiné as a hinge zone having a N°45-E direction; (3) the Bas-Vivarais as a specific basin.

The comparative Triassic stratigraphic record between the northern European basin and southern European (Subalpine) basin discloses that (1) marine floodings cannot merely be related to sea-level rises since they are not similar across two adjacent basins either in time, i.e., Anisian and Carnian floodings, or in relative amplitude, i.e., Ladinian versus Carnian flooding; (2) initiation of rifting marked by initiation of salt deposition in depocenter in each basin is clearly offset through time (Late Ladinian against Carnian); (3) interregional basinward shift of fluvial deposition during the Carnian is not due to any slackening in subsidence. All these points are supporting evidence for the interpretation of the origin of the low-order stratigraphic cyclicity proposed above.

Briançonnais Terrane (Alps). Descriptions and interpretations presented below in the text and Fig. 10 are based on direct observations for the Permian-Early Trias plus data available in Mégard-Galli and Faure (1988) for the Middle-Late Triassic.

In the Guil-Valley, Permian lithofacies regionally referred to as Verrucano are indicative of a braided alluvial-fan depositional environment. The Upper Verrucano shows a progressive decrease in the fluvial equilibrium profile with final development of a meandering river system. Higher up, a discrete erosion ravinement surface can be found at the base of the massive and resistant unit composed of light-colored sandstones and referred to as Werfenian quartzites. From the base of this unit upwards, sedimentary structures are diagnostic of successive depositon in tidal (mud drapes) and shoreface (swaley and hummocky cross-stratifications) environments. In summary, the Verrucano-quartzites set discloses fairly gradual change from proximal continental to clear marine conditions, and thus composes a large transgressive complex. The transgression climaxed with marine shales of the St Triphon formation, Scythian-Anisian in age. Upwards through the sequence, the Ladinian "Calcaires rubanés", "Dolomies noires", and "Dolomies blanches" reflect regressive conditions before renewed marine inundation of the "Dolomies grises" Fm. The Lower Carnian is a package of shale-dolomite interbeds, overlain by a gypsum-bearing horizon, plus slumped-brecciated shales and carbonates supporting clear evidences of syndepositional normal faulting (Carnian rifting). The normal faults are sealed by a carbonate-dominant unit made up of massive limestones overlain by stromatolitic dolostones. Traces of aerial exposure, i.e., deep vertical desiccation cracks, can be found at the top and presumably reflect a major sea-level drop. This surface is covered by Carnian hybrid tidal deposits that consist of interbedded fine-grained sandstones, dusky shales with plant debris and oyster shells, and carbonates. Above are stromatolite-bearing lithofacies that grade upwards into the carbonate peritidal deposits of the Norian Hauptdolomit. This massive carbonate package shows vertical stacking of genetic sequences and forms a thick aggradational complex of a relative sea-level low stillstand.

Relative to the Triassic lithostratigraphy, it is apparent that the Briançonnais shares more similarities with the southern Alps in northern Italy than it does with the northwestern European basin: the Verrucano lithofacies is close to the Lombardian Collio Fm; the Werfenian quartzites and the Buntsandstein lithofacies are totally different; the Lower Carnian sulfate-dolomitic lithologies are distinct from the Keuper salt and associated marginal products; the complex Uppermost Carnian/Norian dusky shales Hauptdolomit closely resembles the complex Raibl-Dolomia principale of the Dolomites.

The Triassic sequence in Briançonnais includes three transgressive-regressive cycles as against four in the northwestern European basin and Subalpine basin. Specifically, no marine inundation occurred here around the Carnian-Norian limit, and the maximum marine flooding is much older here, i.e., Scythian-Anisian. The Scythian-Anisian flooding is not recorded either in the Subalpine basin or in the nearby Massif cristallins externes (see Mégard-Galli et al. in Debrand-Passard et al. 1984), both located west of the Penninic thrust. On the contrary, the Scythian-Anisian flooding finds a possible echo in southern Alps (Couches du Campil). Conversely, the Early Carnian flooding phase documented in the western Subalpine basin is not known to exist in the Briançonnais. Finally, a Carnian major sea-level fall recognized in the western European basins is recorded in the Briançonnais as well.

Additional differences between the Briançonnais and Subalpine domains emerge regarding a pure geodynamic matter: continuity versus discontinuity for the Permian-Trias passage; nonsynchronous termination of the Late Triassic syn-rift phase referring to the time-line that is the Carnian sea-level drop surface; and nonsynchronous post-rift climaxes previously suggested to be associated with maximum marine floodings.

In conclusion, it should be stressed that the Triassic sequence in the Briançonnais has a strong southern Alps stratigraphic print in terms of both lithology and depositional cyclicity. The sequence further discloses a geodynamic history with little to no connection to the northwestern European and Subalpine basins. The stratigraphic evidence supports the interpretation of Stampfly and Marthaler (1990) and Stampfly (1992) that the Briançonnais domain is a specific terrane separated from the Subalpine domain by the present Penninic thrust.

Conclusions

A number of points of specific interest have emerged throughout the course of this stratigraphic reappraisal of the Bresse-Jura and adjacent Triassic basins in Eastern France.

1. Both the formations known to be hydrocarbon reservoirs in the Bresse-Jura area have a regional extension: the Buntsandstein is presumably composite, and made up of two stacked transgressive complexes with much variable sedimentological and petrophysical properties; the Lettenkohle is an early highstand sheet-shaped fill that is very constant all over the study area.
2. The salt complexes of the Bresse-Jura area and western Subalpine basins are late highstand high-angle wedges, tectonically induced and controlled. Their original geometry has not been modified by any subsurface dissolution.
3. Restored depositional geometries with superimposed depositional environments in the Bresse-Jura area plus interregional comparison in Eastern France suggest that (a) Ladinian and Carnian inundations were presumably not linked to eustatic rises and thus associated minor second-order T-R cycles (3–15 Ma) are regional, tectonically controlled cycles, (b) depositional cycles and geodynamic cycles can be out of phase, (c) the major Triassic second-order cycle is not the strati-

graphic signature of any geodynamic cycle and the Carnian interregional major lowering of base-level was connected to a real eustatic fall, and (d) Permian and Anisian-Ladinian basins had clear distinct geodynamic patterns in western Europe. Additional data and further investigation are necessary to determine whether Scythian and Scythian/Anisian transgressions were mere effects of eustacy or were already associated with some structural resetting.

Acknowledgments Thanks are directed to Gaz de France Company (DGF-DETN) for permission to publish some data relative to the Buntsandstein of the Bresse-Jura area. Gratitude is due to the Agence Nationale pour la Gestion des Déchets Radioactifs (CEA-ANDRA-DESI) that ordered a survey of the Muschelkalk and Keuper series of the Bresse-Jura area, whose results have already been partially published.

References

Adloff MC, Appia-Lamache C, Doubinger J et al (1984) Zonations palynostratigraphiques dans les séries triasiques traversées par les sondages dans le Jura et le Bas-Dauphiné Géol Fr 1-2. 3–21

Aigner T, Bachman GH (1992) Sequence stratigraphy framework of the German Triassic. Sediment Geol 80 1/2 115–135

Angevine CL, Heller PL, Paola C (1990) Quantitative sedimentary basin modeling AAPG Continuing Education Course Note Ser 32 1–133

Barrell J (1917) Rhythms and the measurements of geological time Bull Geol Soc Am 28 745–904

Bergerat F, Martin P (1993) Mise en évidence d'une tectonique distensive synsédimentaire et caractérisation du champ de contraintes au Trias inférieur-moyen sur la bordure Vivaro-Cévenole du Bassin du Sud-Est de la France. C R Acad Sci Paris 3/6 (II) 1279–1286

Courel L (1973) Modalités de la transgression mésozoique Trias et Rhétien des bordures Nord et Est du Massif Central français Mém Soc Géol Fr 118 1–152

Cross T (1988) Controls on coal distribution in transgressive-regressive cycles, Upper Cretaceous, Western interior, USA. In Wilgus B, Hastings B, Ross C, Posamentier H, Van Wagoner J, Kendal CGSTC (eds) Sea-level changes an integrated approach Soc Econ Paleontol Mineral Spec Publ 42 371–380

Debrand-Passard S, Courbouleix S, Lienhardt MJ (1984) Synthèse géologique du Sud-Est de la France Mém BRGM 125 1–615

DIASTRATA (1991) Synthèse géologique des formations évaporitiques du Trias de l'Ouest du bassin Jurassien Rap ANDRA 624 RP DIA 91-001 1–85

DIASTRATA (1992) Le Trias de la Région Bresse-Jura Rap GDF/92 04-AB 1–40/1–30

Dromart G, Guillocheau F, Curial A et al. (1991) Les séquences du système transgressif Rhéthien-Hettangien basal de la bordure ardéchoise, Forage GPF-Balazuc-1 Spec Publ Assoc Sédiment Fr 15 115–116

Dromart G, Monier P, Curial A et al (1992) Cycles sédimentaires dans le Trias de la Bresse et du Jura C R Acad Sci Paris 315 (II) 609–616

Elmi S, Dromart G, Giot D et al (1991) La succession stratigraphique (Trias-Jurassique) du sondage Balazuc-1 (Ardèche) un exemple d'épaississement remarquable le long d'un escarpement tectoniquement contrôlé C R Acad Sci Paris 313 (II) 427–433

Finelle JC (1981) Contribution à l'étude du Trias de la bordure cévenole Thesis Univ of Dijon 1–240 (unpublished)

Geisler-Cussey D (1985) Données sur le Muschelkalk moyen évaporitique de Lorraine Doc Greco 52 (5) 5–23

Giot D et al (1987) Subsidence et diagenèse de la marge ardéchoise du bassin du Sud-Est Doc BRGM 123 1–43

Guillocheau F (1991 a) Modalités d'empilement des séquences génétiques dans un bassin de plate-forme (Dévonien armoricain) nature et distorsion des différents ordres de séquences de dépôts emboîtées Bull Cent Rech Expl Prod Elf-Aquitaine Boussens 15 (2) 383–410

Guillocheau F (1991 b) Mise en évidence de grands cycles transgression-régression d'origine tectonique dans les sédiments mésozoiques du Bassin de Paris C R Acad Sci Paris 312 (II) 1587–1593

Haq BU, Hardenbold J, Vail PR (1987) Chronology of fluctuating sea levels since the Triassic Science 269 483–489

Lienhardt G (1962) Géologie du bassin houiller stéphanien du Jura et ses morts-terrains Mém BRGM 9 1–449

Mégard-Galli J, Faure JL (1988) Tectonique distensive et sédimentation au Ladinien supérieur-Carnien dans la zone briançonnaise Bull Soc Géol Fr (8) (IV) 5 705–715

Posamentier HW, Allen GP, James DP, Tesson M (1992) Forced regressions in a sequence stratigraphic framework. concepts, examples, and exploration significance AAPG Bull 76 (11) 1687–1709

Ricour J (1962) Contribution à une révision du Tias français Mém Expl Carte Géol Dét Fr Paris 1–461

Stampfly GM (1992) Le Briançonnais, terrain exotique dans les Alpes Géol Alp Sér Spec Rés Col 1 90–91

Stampfly GM, Marthaler M (1990) Divergent and convergent margins in the Northwestern Alps – confrontation to actualistic models Geodinam Acta 4 (3) 159–184

Vail PR, Colin JP, DuChene R et al (1987) La stratigraphie séquentielle et son application aux corrélations chronostratigraphiques dans le Jurassique du Bassin de Paris Bull Soc Géol Fr III (7). 1301–1322

Vail PR, Audemard F, Bowman SA et al (1991) The stratigraphic signatures of tectonics, eustacy, and sedimentology – an overwiew In: Einsele G, Ricken W, Seilacher A (eds) Cycles and events in stratigraphy. Springer, Berlin Heidelberg New York, pp 617–659

Van Den Driessche J, Brun JP (1989) Un modèle cinématique de l'extension paléozoique supérieur dans le sud du Massif Central. C R Acad Sci Paris 309 (II) 1607–1613

Ziegler PA (1988) Evolution of the Arctic-North Atlantic and the western Tethys AAPG Mem 43 1–88

Petroleum Exploration in the French Rhine Graben

A. Richard

The French Rhine Graben (or Alsacian Trough) is part of the West European Rift System stretching from Germany to the Mediterranean Sea. The Rhine Graben, a Tertiary structure, is framed by two Hercynian massifs, i. e., the Schwarzwald and Vosges uplifts.

Petroleum exploration started very early with the exploitation of the Pechelbronn field in the 17th century, since which time approximately 4500 wells have been drilled. The exploration phase was most active during two periods: 1950 to 1960 and 1980 to 1990. It led to the discovery of ten fields of which only three have significant reserves; Pechelbronn (3.5 million MT recoverable oil reserves), Scheibenhardt (0.4), and Schelmenberg (0.1). More than 90 % of the oil is produced from the Tertiary reservoirs. Typical traps are faulted monoclines or faulted anticlines rejuvenated or created during the Alpine orogeny.

Two main petroleum systems have proved productive: the Jurassic ante-rift system and the Oligocene syn-rift system. A third system, consisting of the Triassic objectives, is speculative. The exploration of the Dogger and Triassic objectives has been abandoned: except for minor oil shows, all recent wells drilled to the Trias were found to be dry.

The Oligocene objectives are the Rupelian sandstones sealed by shales of the same age. Present all along the gra-

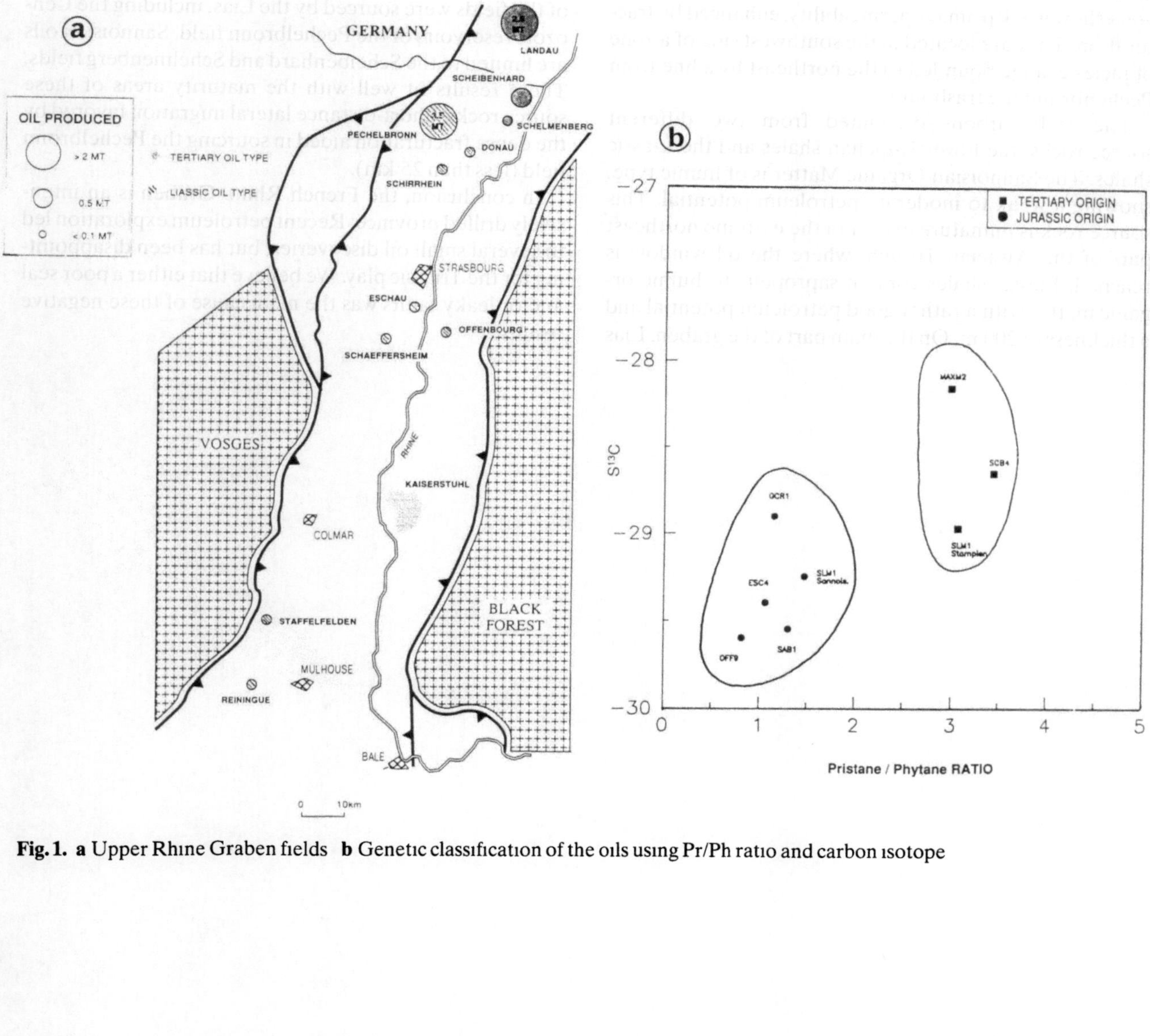

Fig. 1. a Upper Rhine Graben fields **b** Genetic classification of the oils using Pr/Ph ratio and carbon isotope

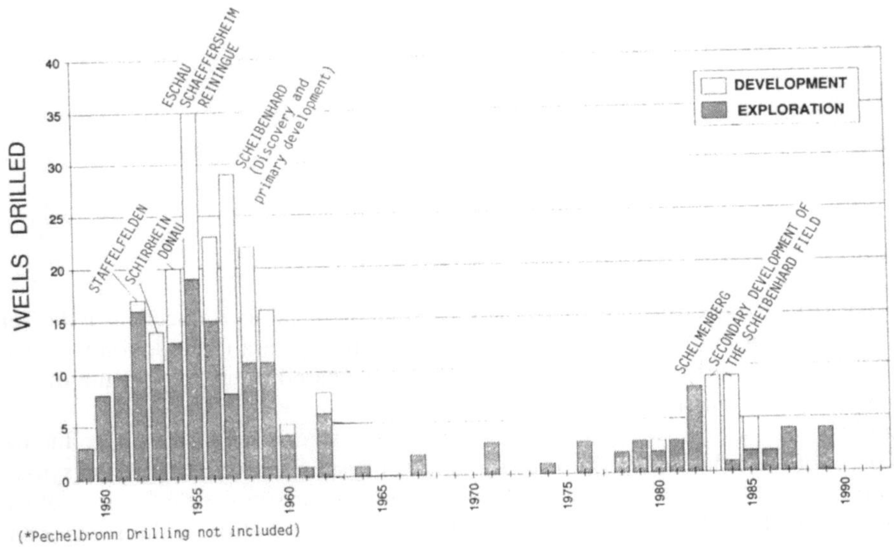

Fig. 2. The Upper Rhine Graben drilling history

ben, these lenticular bodies are of limited lateral extension (less than 1 km) and thickness (less than 5 m). The Jurassic play consists of the oolitic limestones of the Dogger sealed by the Callovo-Oxfordian shales. These reservoirs show weak primary permeability, enhanced by fracturation. They are located in the southwest side of a zone of facies change bounded to the northeast by a line from Pechenbronn to Strasbourg.

The hydrocarbons originated from two different source rocks: the lower Rupelian shales and the Liassic shales. The Sannoisian Organic Matter is of humic type, showing a weak to moderate petroleum potential. This source rock is immature except in the extreme northeast part of the Alsacian Trough, where the oil window is reached. Liassic shales contain sapropelic to humic organic matter with a rather good petroleum potential and a thickness of 200 m. On the main part of the graben, Lias reached the oil window maturity except on the Pechelbronn high, where it is immature.

Modern geochemical analyses have pointed out two different types of oils: Sannoisian oils and Lias oils. Most of the fields were sourced by the Lias, including the Cenozoic reservoirs of the Pechelbronn field. Sannoisian oils are limited to the Scheibenhard and Schelmenberg fields. These results fit well with the maturity areas of these source rocks. Short-distance lateral migration favored by the dense fracturation aided in sourcing the Pechelbronn field (less than 25 km).

In conclusion, the French Rhine Graben is an intensively drilled province. Recent petroleum exploration led to several small oil discoveries, but has been disappointing for the Triassic play. We believe that either a poor seal and/or leaky faults was the main cause of these negative results.

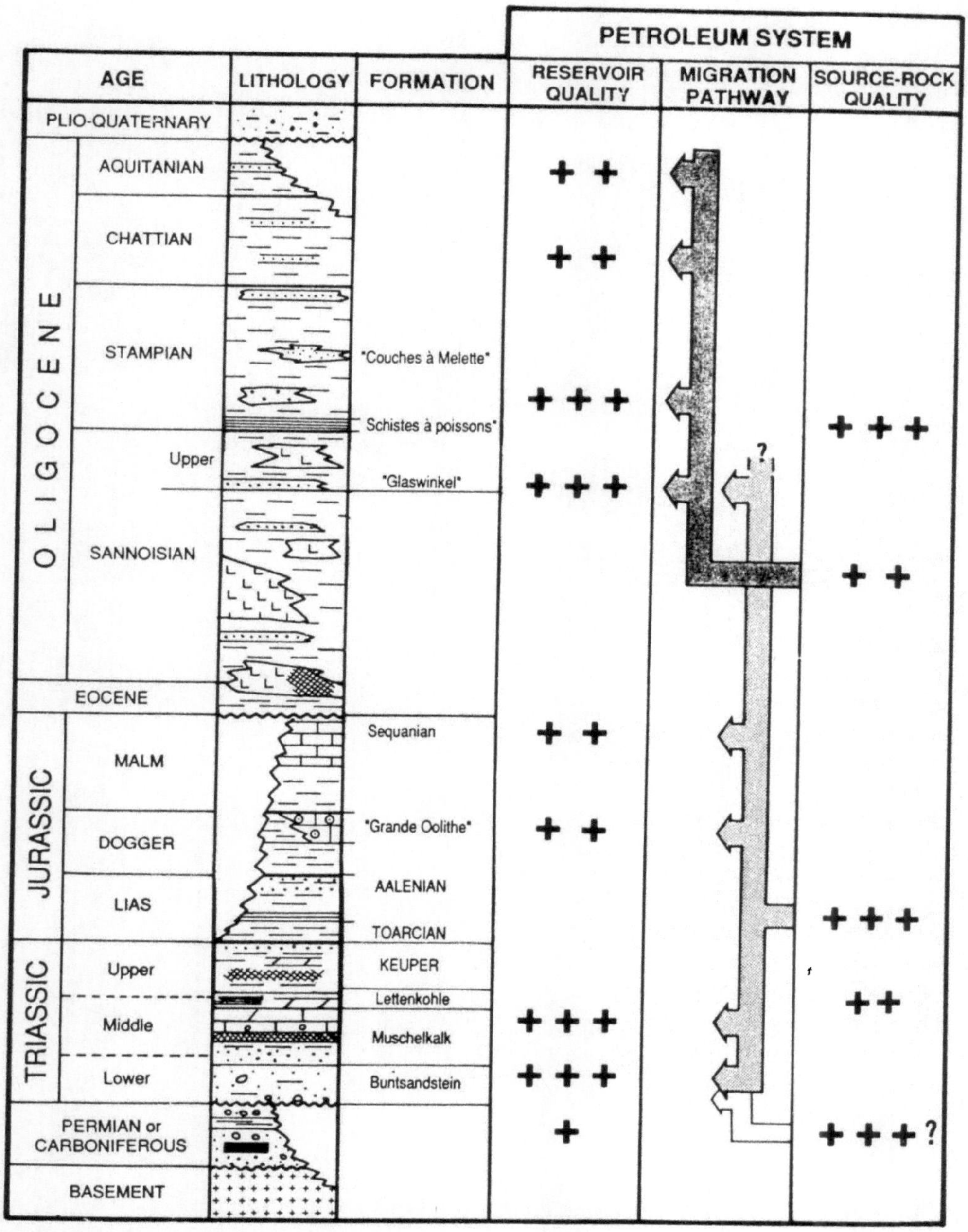

Fig. 3. Type section of the Rhine Graben

Mechanism of Fluid Paleocirculations in Fractured Rocks Determined by Vein Alteration Studies

A. Meunier

Abstract. Vein alteration studies may be very informative for the understanding of fluid circulation processes in fractured rocks. Answers to some fundamental questions are illustrated here using samples taken in the scientific drill-holes or in rocks outcropping in France. The way by which the fluid penetrates inside the fracture and its wall-rocks can be determined by petrographical and mineralogical observations; three vein types can be distinguished: injection, infiltration, and drainage. The chronological sequence in multistage flowing periods can be established by the analysis of the crosscutting relationships and fractal analysis of the fracture network.

The determination of P-T-μX conditions prevailing during the alteration process is made possible by classical thermodynamic calculations describing the fluid-rock interactions. For example, the sequence of alteration mineral in the Echassières albitelepidolite granite was correctly modeled by the temperature decrease effects on the chemical properties of the altering fluids. In an injection vein system, chemical transfers were shown to be controlled by a sequence of interdependent mineral reactions in which the phase growing in the fracture is the end-member. Alteration progresses until this phase can grow. The absolute Si loss is used to calculate the quantity of water which has flown inside the fracture. The presence of a sharp thermal gradient between the vein and the unaltered rock indicates that the flowing regime was pulsated.

The altered wall-rocks are generally characterized by a higher porosity than the original rock. In some cases, the porous spaces are poorly connected and the fluids are more or less trapped inside them: a sequence of clay minerals crystallizes during the cooling period. If the porous spaces are strongly connected, the permeability of the altered zones is several orders of magnitude higher than that of the original rock. Then, new fluid circulations can be established. The altered zones behave as reservoirs.

Introduction

Mineral veins and their altered wall-rocks have long been considered as attractive features indicating the presence of ore. Because of subsequent economic reasons, they were largely studied in mineralized zones which are, for most of them, located in eruptive or metamorphic areas. Little is known of vein systems crosscutting sedimentary formations. The aim of this paper is to review the recent advances in vein alteration studies hoping that, in spite of the fact that they concern nonsedimentary environments, they will be of interest for petrologists involved in petroleum exploration.

Vein alteration mechanisms were early related to the flowing of fluids inside open fractures. The precipitation of secondary minerals in the open space and the mineral reactions observed in the wall-rocks largely depend on the physicochemical properties of the fluids. Consequently, as alteration and precipitation are short-term processes, veins were regarded as instantaneous photographs of the prevailing physicochemical conditions during the flowing period. Most of the published works tried to establish the chronological relations between the different types of veins observed in a hydrothermal system in order to determine the changes in physicochemical conditions during time (Beane 1983; Titley 1983).

In spite of the fact that veins have been so largely studied for half a century, they still remain interesting objects for new research. Indeed, the emerging of image processing techniques allows one to quantify alteration features like dissolution voids or secondary crystals, which were only depicted as qualitative observations until now. Veins are no longer regarded as fossilizing static physicochemical conditions, but rather as recording the dynamics of alteration processes and the flowing regime. Among the questions we ask when observing veins, some of them could be of great interest for the study of fluid circulations in sedimentary formations:

1. Where did the fluid come from?
2. What was the succession of flowing events in a given place?
3. What were the PTμx conditions of the flowing fluids?
4. What are the quantities of altered minerals in a vein system?
5. What is the mechanism of alteration?
6. What was the quantity of fluid which has flown inside the fracture?
7. What was the flowing regime of the fluids in the open fracture?
8. What was the "living time" of the dissolution-precipitation process?
9. What happens after the vein is sealed?

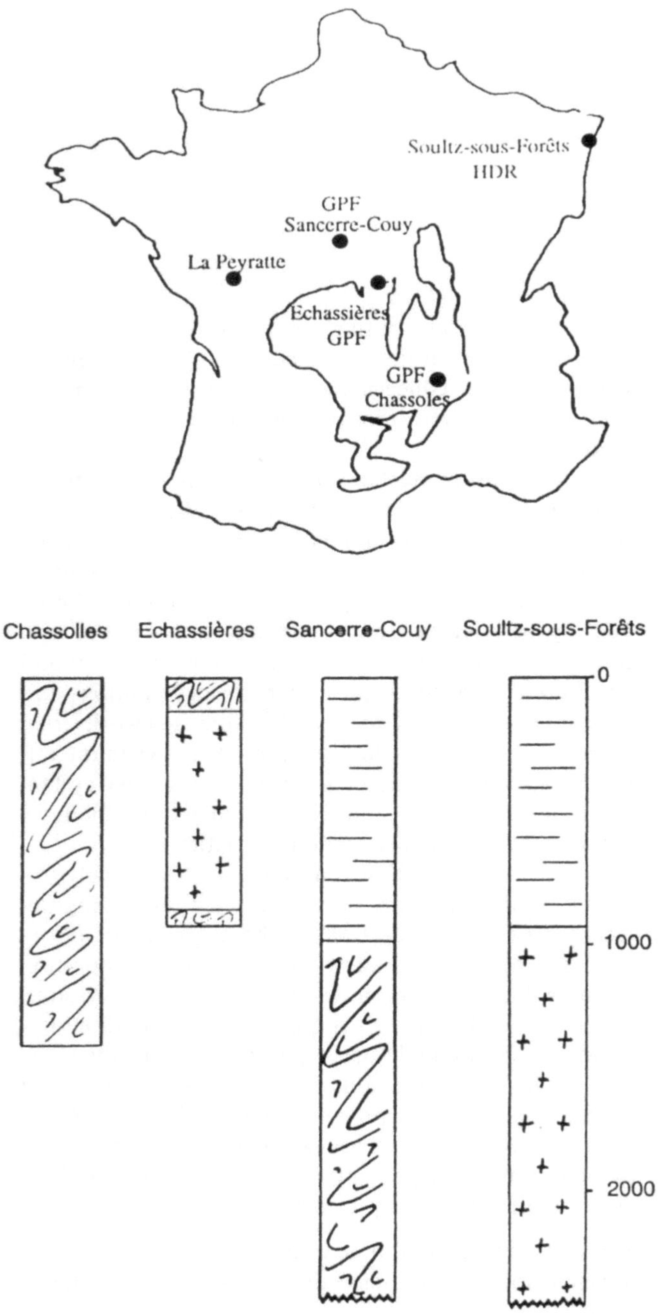

Fig. 1. Location of the studied samples *GPF* French Scientific Deep Drilling program, *HDR* European Hot Dry Rock project Simplified logs are indicated for drill-holes Simplified stratigraphical logs are given for each drill-hole, *curved lines* metamorphic basement, *crosses* granitic rocks, *horizontal lines* Mesozoic sedimentary rocks Depth in m

The aim of this chapter is to present a methodology for vein studies through different samples taken from crystalline basement in order to show how to answer these questions. The samples come from the GPF drill-holes (Géologie Profonde de le France), the Hot Dry Rock European site at Soultz-sous-Forêts (Alsace) and a granite outcropping in the southern part of the Armorican Massif (Deux-Sèvres, Fig 1). It is proposed, through the description of these samples, to enhance a general methodology for vein studies which could be applied in sedimentary petrology.

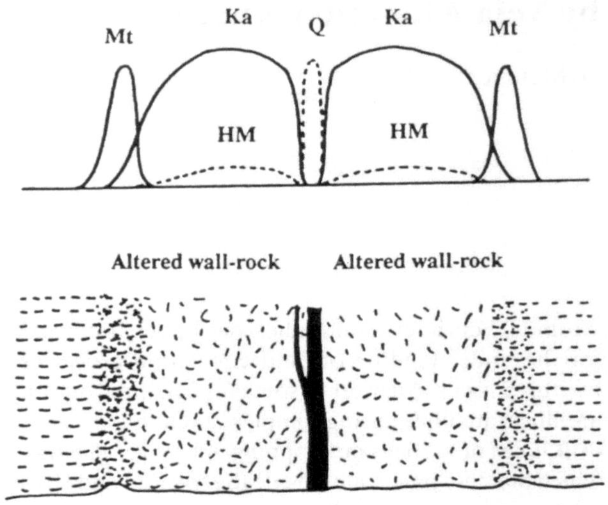

Fig. 2. Cross section of alteration zones at Caribou mine, Idaho Tunnel (After Bonorino 1959) *Q* Quartz deposit sealing the fracture, *Ka* kaolinite, *HM* hydromica, *Mt* montmorillonite *Curves in the upper part of the figure* show approximate proportions of secondary minerals Hydromica and montmorillonite are probably illite/smectite-ordered and random mixed-layered minerals respectively

From Pioneer to Recent Studies: A Brief Review

Because of their fundamental interest for ore prospecting, numerous studies have been devoted to the description of vein alteration systems. Among them, two prophets established the bases of the modern petrological analysis of these systems: Lovering (1950) and Bonorino (1959). An accurate observation of zoning in wall-rock (Fig. 2) led the authors to propose an alteration mechanism based on chemical diffusion processes They considered that temperature may have affected only the position of the alteration fronts or the width of the alteration halo

The very active prospecting of porphyry copper deposits during the past decades led many researchers to study the veins around intrusive bodies. Statistical analysis of vein deposits showed that the geometry of the fracture network in which fluids have circulated at a given time changes during the active hydrothermal period (Titley 1983). Veins were classified according to the secondary parageneses in order to establish a time-dependent sequence based on crosscutting relationships.

The origin of zoning in wall-rocks was reconsidered by Page and Wenk (1979), who stated that alteration is controlled by crystallization kinetics. These authors observed a sequence of phyllosilicates from the vein to the unaltered rock: phengite of 2 M_1 polytype, illite of 1 M or 1 Md polytype and illite/smectite mixed-layered minerals. Assuming isothermal conditions in the wall-rock, illite/smectite mixed layered minerals are considered as metastable precursors of illite and phengite.

Thermal gradients around veins were evidenced at different scales: about 200 °C for 100 m in the Amethyst vein (Horton 1985); 60 °C for 6 m in uranium vein deposits (Al Shaara 1986); 75 °C for 30 mm in porphyry copper vein (Beaufort and Meunier 1983 b); 75 °C for 11 mm in a phengite vein (Turpault et al. 1992 b). From these observations, it is clear that the isothermal assumption cannot be accepted as a general rule for all the vein systems.

Where Did the Fluids Come From?

The Three Vein Types

A vein is composed of two features: a mineral deposit sealing an open fracture and the surrounding altered wall-rocks. Simple petrographical observations show that three vein types can be distinguished according to secondary mineral composition and relative thicknesses of fracture deposit and altered wall-rocks. This is due to the fact that fluids can originate from different source regions

From a theoretical point of view, a fluid will flow in a fracture and penetrate inside a rock if a pressure gradient (ΔP) is established between the source region and the open spaces (fracture and rock porosity). Consequently, in open systems, two pressure gradients are established: (1) $\Delta P_1 = P_{ff} - P_{fr}$ (between the fracture and the unaltered rock); (2) $\Delta P_2 = P_{fs} - P_{ff}$ (between the source region and the fracture). P_{fs}, P_{fr} and P_{ff} represent the fluid pressure in the source region, the porous space in the unaltered rock and the fracture respectively ($P_{fr} < \sigma_3$, Fyfe et al. 1978). Pressure gradients ΔP_1 and ΔP_2 are not fixed with time; they vary from a maximum value to zero and eventually to negative values in a pulsated regime (Grimaud et al. 1990).

Then, the three vein types identified in altered rocks are the following:

1. Injection vein type: the source region is located in high pressure-temperature zones; if $P_{fs} > P_{ff} > P_{fr}$ the fluid is injected in the fracture and the rock.
2. Infiltration vein type: in static open systems, the source region is located in upper formations; $P_{ff} = P_{hs}$ (P_{hs}: hydrostatic pressure) and $P_{ff} = P_{fr}$; fluids circulate by gravity In closed systems, the pressure regime is not hydrostatic but hydrodynamic; a pressure gradient is established between the source regions and the open fracture ($\Delta P = P_{hd} - P_{fr}$; P_{hd} = hydrodynamic pressure). In this case, the veins would be similar to injection ones.
3. Drainage vein type: the source region is located in the rock itself; if $P_{ff} < P_{fr}$ the fluid is drained from the rock towards the fracture until $\Delta P_2 = o$.

Fluid Injection Vein Type

Fluid injection in rocks through fractures is classically illustrated by hot springs in geothermal fields. The hydrothermal fluid, coming from deep high P-T formations, rises up to the surface and permeates through the surrounding rocks. Generally, strong alterations are developed, producing zoned wall-rocks (Bonorino 1959; Fig. 2). The fluid-rock interactions induce the dissolution of the primary minerals and the crystallization of different secondary ones according to the distance to the fracture (Fig. 3 a).

Most often, the reaction is volume conservative and can be written as follows:

Primary mineral → secondary mineral(s) + dissolution voids.

The fluid which impregnates the altered wall-rock cannot move as in the open space of the fracture. The transfer of dissolved components from rock to fracture is controlled by chemical diffusion. Turpault et al. (1992 a) have shown that the transfer is maintained by the nucleation and the growth of secondary minerals on the fracture walls which are the end member of a sequence of interdependent mineral reactions. This process stops when the growth of secondary phases in the fracture stops itself because of free space lacking even if all the other physicochemical conditions remain unchanged (P, T, presence of fluid, etc.). The thickness of the altered zones in the wall-rocks is time-dependent (Fig. 3 b).

Fluid Infiltration Vein Types

Vein alteration studies in the gneiss-amphibolite basement of the Sancerre-Couy GPF drill-hole shows that fluids had percolated from the upper Mesozoic sedimentary formations down to the fractured basement. Stable isotope analysis of natural solutions sampled in the drill-hole indicate an unusual exchange ratio with the surrounding rocks (Boulègue et al. 1990). This means that the residence time of fluids in open fractures was very long and that they were not renewed. For these reasons, their interactions with the surrounding rocks are limited to a thin layer on the wall-rocks and the secondary minerals precipitating in the voids were very well crystallized: large size, euhedral crystals (Beaufort and Meunier 1993). In some places, they did not completely seal the open space and geodas can be observed (Fig. 4).

Except at the sediment-basement interface, where an important fluid circulation was evidenced by the abnormally low maturity of the illite/smectite mixed-layer minerals, the quantities of solutions involved in the infiltration process are weak. The formation of secondary minerals is controlled by the composition of the surrounding rock which is dissolved in a thin layer on the fracture walls: saponite, corrensite, and chlorite were observed in fractures crosscutting pyroxene-rich amphibolites, amphibolites and gneisses respectively.

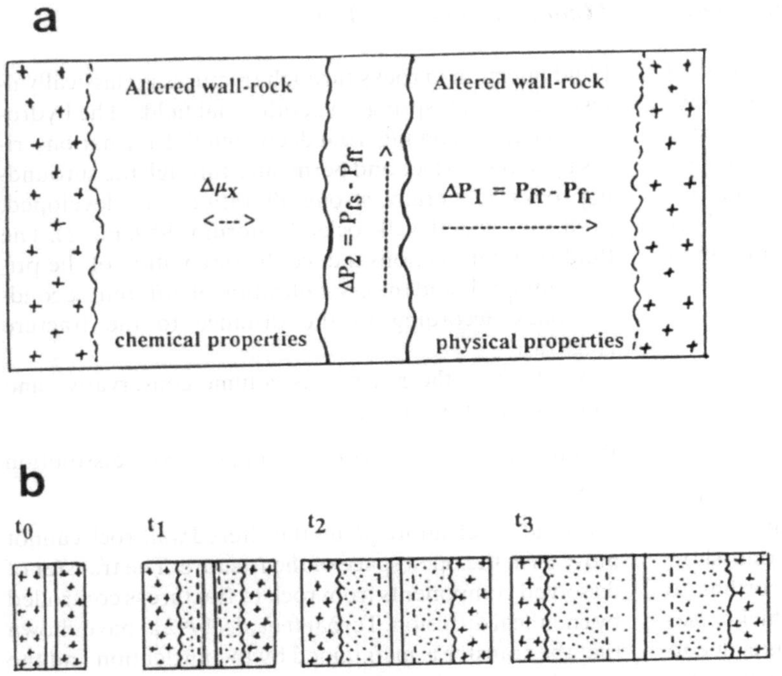

Fig. 3 a, b. Fluid injection vein type The pressure and chemical gradients (ΔP_1, ΔP_2, and $\Delta\mu_x$ respectively) are represented in the fracture-wall-rock system (**a**) The width of altered zones increases with time, as is shown in (**b**)

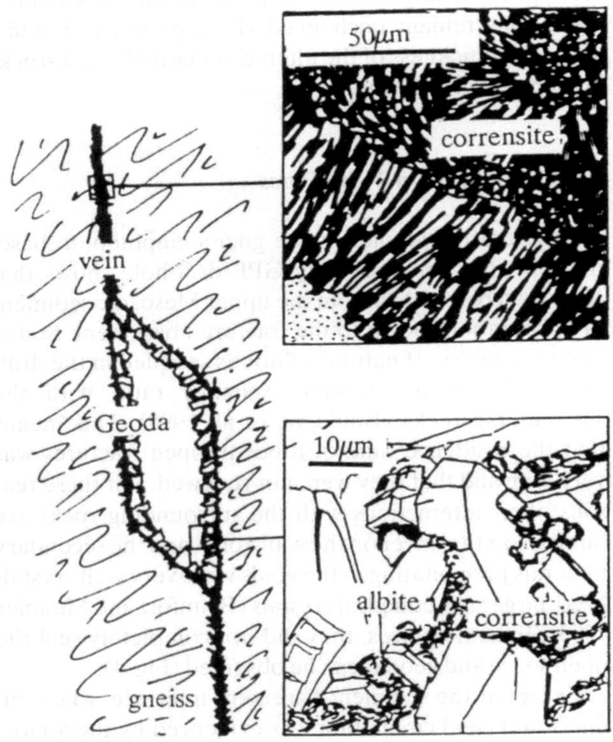

Fig. 4. Infiltration vein type Fluids infiltrate by gravity in a fractured basement from upper geological formations (Sancerre-Couy scientific deep drill-hole) As fluids are not renewed, the wall-rocks of veins and geodas are weakly altered Automorph crystals of albite and corrensite are observed in geodas

Fluid Drainage Vein Type

Drainage veins originate in nonconnected tension fractures which locally create low pressure domains. The fluids trapped in the rocks or liberated by a crystallizing magma (fluid pressure approaching rock pressure) are drained towards these domains. The secondary phases precipitating inside the voids are identical to that existing in the surrounding rocks. Monophased or polyphased assemblages are observed according to the chemical composition of the rocks, which is the dominant parameter in such alteration processes.

The Echassières albite-lepidolite granitic body is crosscut by a series of hydrothermal monophased veins which form the following time-dependent sequence: muscovite, pyrophyllite, donbassite, tosudite, and kaolinite. This sequence was shown to result from reactions between primary minerals and a fluid expelled from the crystallizing magma during the cooling period of the granitic body. Isotopic data and mineralogical analysis showed that there is no contamination from the intruded micaschists (Merceron et al. 1992b). Regarding muscovite veins as an example of a drainage vein type, no crystallochemical differences were observed between crystals inside the fracture and the altered surrounding rock repectively (Merceron 1988). The quantities of muscovite were shown to decrease from the vein to the unaltered rock (Fig. 5).

Drainage veins are common in the pervasive alteration areas like the propylitic zone around magmatic intrusions. In this case, the secondary assemblages which seal the veins are identical to that of the surrounding altered

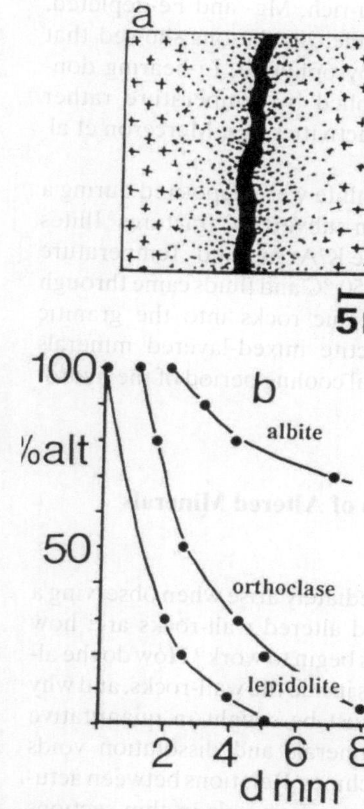

Fig. 5 a, b. Drainage vein type The muscovite deposits in veins which crosscut an albite-lepidolite granite (GPF scientific drill-hole, Echassières) are related to the concentration of post-magmatic fluids in fractures during the cooling period of the granitic body (Merceron 1988 modified) **a** Muscovite vein example at -99 6 m **b** Quantities of primary minerals replaced by muscovite (% alt) versus distance to the vein (estimated by point-counting)

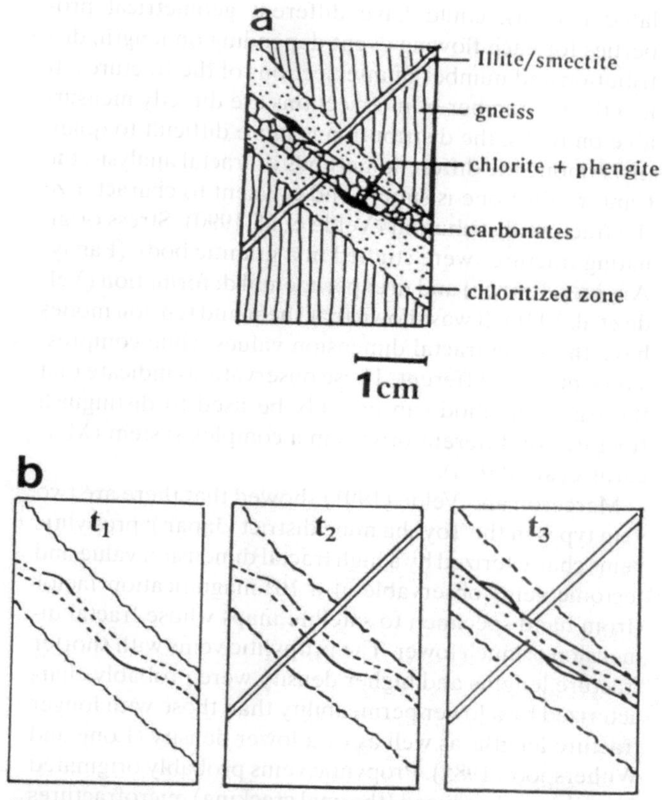

Fig. 6 a, b. Chronology of flowing events **a** Vein crosscutting relationships observed at -351 1 m (GPF scientific deep drill-hole, Cézallier, Turpault et al 1987) **b** Three fracturing-sealing events succeeding in time (t_1, t_2, and t_3) are identified from the observed crosscutting relationships

rock (chlorite + epidote + quartz for example). These assemblages are polyphased because of the chemical composition of the volcanic and plutonic rocks (Beaufort et al. 1992).

What Was the Succession of Flowing Events?

Determination of the Relative Chronology of Vein Alterations

Most often, macroscopic veins show polygenetic mineral deposits indicating that successive sealing-opening events happened. This phenomenon was first described in metamorphic environments and is known as a crack-seal process (Ramsay 1980). In some favorable cases, the older events are crosscut by more recent ones (Beaufort and Meunier 1983 a; Titley 1983). For example, a complex sequence was described in the fractured gneissic basement drilled in the Chassole GPF site (Turpault et al. 1987). In some places, three events are observable at the scale of a thin section (Fig. 6).

Precautions must be taken in the interpretation of these apparently so simple crosscutting relations. The danger lies in the fact that their systematical research is based on a selection of demonstrative microsites. This does not give information on the way in which fluid flows inside the connected fracture network. Indeed, Ledésert et al. (1993 a) showed that fluid circulations are preferentially established in portions of the fracture network presenting a low interconnection coefficient (number of intersections per longer unit). Consequently, the secondary mineral deposits sealing these fractures are different from those precipitating in the highly connected network in spite of the fact that both are contemporaneous. For example, Turpault et al. (1992 a) showed that phengite precipitated in low interconnected large fractures while adularia + quartz precipitated in the highly interconnected network.

Geometrical Changes in Fracture Networks: Fractal Dimension

As stated above, successive flowing events are frequently observed in fractured geological formations. This implies that new fractures are open while pre-existing ones are re-opened by tectonic or thermal constraints. The percolated network could have different geometrical properties for each flowing event depending on length, distribution and number of intersections of the fractures. If length and number of intersections are directly measurable on rocks, the distribution is more difficult to quantify. Among the different methods of fractal analysis, the Cantor's dust one is simple and efficient to characterize the fracture distribution (Velde et al. 1990). Stress originating fractures were studied in a granitic body (Fanay-Augères, France) and in experimental deformation (Velde et al. 1991). It was shown that shear and tension modes have the same fractal dimension values while compressional mode is different. These observations indicate that the fractal method can possibly be used to distinguish fractures of different origins in a complex system (Merceron et al., 1992 a).

Merceron and Velde (1991) showed that there are two vein types in the Toyoha mine district (Japan): propylitic veins characterized by a high fractal dimension value and tectonic veins observable at a 10^6 magnification factor (from hand specimen to satellite map) whose fractal dimension is much lower. The propylitic veins with shorter fracture lengths and higher density were probably characterized by a lower permeability than those with longer fracture lengths as well as by a lower density (Long and Witherspoon 1985). Propylitic veins probably originated from thermal-induced (thermal cracking) microfractures similar to those described in granite by Wang et al. (1989). Locally, intermediate fractal values reflect mixtures of the two groups with different components of each fracture style.

What Were the Fluid P-T-μ_X Conditions?

The determination of P-T-μx conditions in fluid-dominating systems is based on direct and indirect information: fluid inclusion analysis and thermodynamical properties of secondary mineral paragenesis, respectively. Many studies were published on fossil and active geothermal systems; the Echassières GPF drill-hole will be taken as a representative example (Merceron et al. 1992 b). Two successive hydrothermal alteration stages were identified in the albite-lepidolite granitic body. During the early one (273–268 Ma), the following time-dependent series of aluminous phyllosilicate: muscovite, pyrophyllite, donbassite, tosudite, and kaolinite which are observed as vein deposits (< 10 mm wide) and alteration products of primary minerals in wall-rocks. The vein system was sealed by monomineralic assemblages during the cooling period of the granitic body (400 to 150 °C) during which low salinity fluids (1–10 wt. % NaCl equivalent) were expelled out (Aissa 1987). According to $^{18}O/^{16}O$ and D/H stable isotope composition of secondary minerals, these fluids were shown to have a meteoric origin and to have reacted with the magma at high temperature (Fouillac and Rossi 1991).

Considering the particular chemical composition of the secondary mineral phases, it was deduced that the fluids were Li-, K-, and Na-rich, Mg- and Fe-depleted, and pH low. Thermodynamic calculations showed that the sequence: muscovite, pyrophyllite, Li-bearing donbassite, kaolinite is controlled by temperature rather than chemical component activities (see Merceron et al. 1992 b for details).

Fluorite and Fe-Mg rich illite were deposited during a later hydrothermal stage in subvertical fractures. Illites were dated at 151 Ma by the K/Ar method. Temperature conditions did not exceed 250 °C and fluids came through the surrounding metamorphic rocks into the granitic body. At least, illite/smectite mixed-layered minerals precipitated during the final cooling period of the hydrothermal system.

What Are the Quantities of Altered Minerals in a Vein System?

The questions which immediately arise when observing a vein surrounded by zoned altered wall-rocks are: how does the alteration process begin to work? How do the alteration reactions progress inside the wall-rocks, and why does it stop? Answers must be sought in quantitative analyses of secondary minerals and dissolution voids using image processing technics. Relations between actual and apparent dimensions of crystals in thin sections were discussed by Jones and Galwey (1966) and Riss (1988). The vein system crosscutting the La Peyratte granite (Deux-Sèvres, France) has been investigated in detail, it will be described here as a case study for injection vein type.

Secondary Minerals: Species and Quantities

The fracture network of the La Peyratte granite is evidenced by vein deposits whose mineralogical composition varies with the width of the open space: phengite or adularia + quartz or quartz + feldspar overgrowths were observed in veins of 300 to 200, 50, and 20 µm wide, respectively. Whatever the vein width, only oligoclase and biotite are altered in the surrounding rock; microcline remains totally unaltered and quartz is slightly dissolved on the fracture edges. Alteration freatures are zonally distributed in the wall-rocks. Zone 1 (nearest the vein) is intensely altered: it is 4 mm and 1 mm wide for large (300 µm wide) and narrow (< 50 µm wide) veins respectively. Biotite is completely transformed into chlorite + white mica + leucoxene. Oligoclase crystals show numerous dissolution voids, they are partly recrystallized into albite and white mica. White mica is pure muscovite in narrow veins and phengite in larger ones. Both chlorite and albite determine sharp fronts inside the altered rock.

Zone 2 (dissolution voids in oligoclase crystals) extends from the chlorite front to the unaltered rock. It is 12 mm and 10 mm wide for large and narrow veins respectively. Illite/smectite mixed-layered minerals have grown on the walls of some of these voids. Considering increasing distance to the vein, two observations have been made: (1) the smectite percentage decreases from 70% to 40%; (2) the volume of altered oligoclase (replaced by white mica or dissolved) exponentially decreases (Fig. 7). Considering a population of veins, it was shown that the width of the chloritized biotite zone is directly proportional to the width of the zone where oligoclase is replaced by white mica (Fig. 8).

Dissolution Voids: Size and Number

The size and the number of dissolution voids per fixed unit surface (7200 μm^2) in plagioclase crystals of the two zones vary with the distance to the vein (Turpault 1989): the number of voids exponentially decreases (Fig. 9a) while the dimensions (length and width) increase first in the albite zone and then decrease (Fig. 9b).

The formation of dissolution voids is similar to crystal nucleation processes observed in metasomatic environments where the driving force is the chemical diffusion activated by a thermal gradient (Joesten and Fischer 1988). Nucleation is a quick process which depends on the temperature difference established between the heat source and the surroundings. This means that voids "nucleate" very early during the alteration period, just after the opening of the fracture.

The void dimensions do not vary according to a simple exponential function because of the competition between two processes: the growth of nucleated voids and the albitization of oligoclase. Indeed, the dissolution of oligoclase leads to the formation of secondary albite which is the stable state of plagioclase in the alteration conditions. The albite front progresses from the vein towards the unaltered rock as alteration works (Turpault et al. 1992a). Consequently, at a given place in oligoclase crystals, dissolution voids can grow until albite is formed at that place; then, the dissolution stops and the voids are "fossilized" at the reached state. This means that the dissolution voids are "fossilized" very early near the vein and latter in far places. The apparent increasing size of voids versus distance to the vein is due to the migration of the albitization front in the wall-rock (positive slope, Fig. 10). Beyond the albitization front, void dimensions decrease according to an exponential law as do the growth process of crystals in metasomatic environments (negative slope).

What Is the Mechanism of Alteration in a Vein System?

Mass Transfers in a Vein System

The La Peyratte granite vein system will be taken as a case study. Turpault et al. (1992a) showed that alteration proceeded by a sequence of interdependent mineral reactions. The earlier one, which is also the one observed farther inside the rock, is the dissolution of oligoclase.

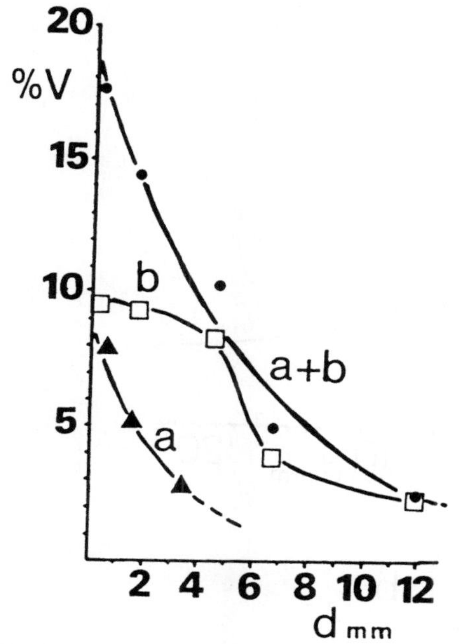

Fig. 7. Injection veins in the La Peyratte granite Two alteration features are observable in oligoclase crystals replacement by phengite flakes and formation of dissolution voids The volume of altered plagioclase *(curve a + b)* is given by adding the volume of phengite flakes *(curve a)* to the volume of dissolution voids *(curve b)* Volume measurements are derived from image analysis (After Turpault et al 1992a)

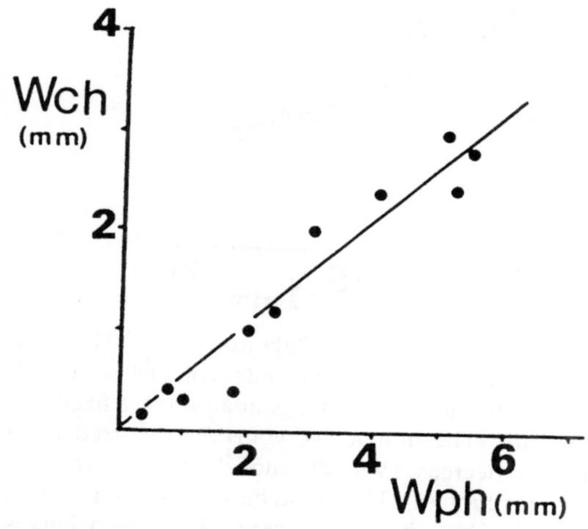

Fig. 8. Injection veins in the La Peyratte granite Altered wall-rocks are zoned, biotite is replaced by chlorite and oligoclase by phengite The observation of a population of veins (20 to 300 µm wide) shows that widths of chlorite and phengite zones (*Wch* and *Wph*, respectively) are related by a linear function This means that the two mineral reactions are interdependent (After Turpault et al 1992a)

The dissolved components are then consumed by different mineral reactions inside the wall-rock. At least, the mineral phases which precipitate inside the fracture, consume the chemical components leached from the altered rock. These reactions can be written in a simplified manner as follows:

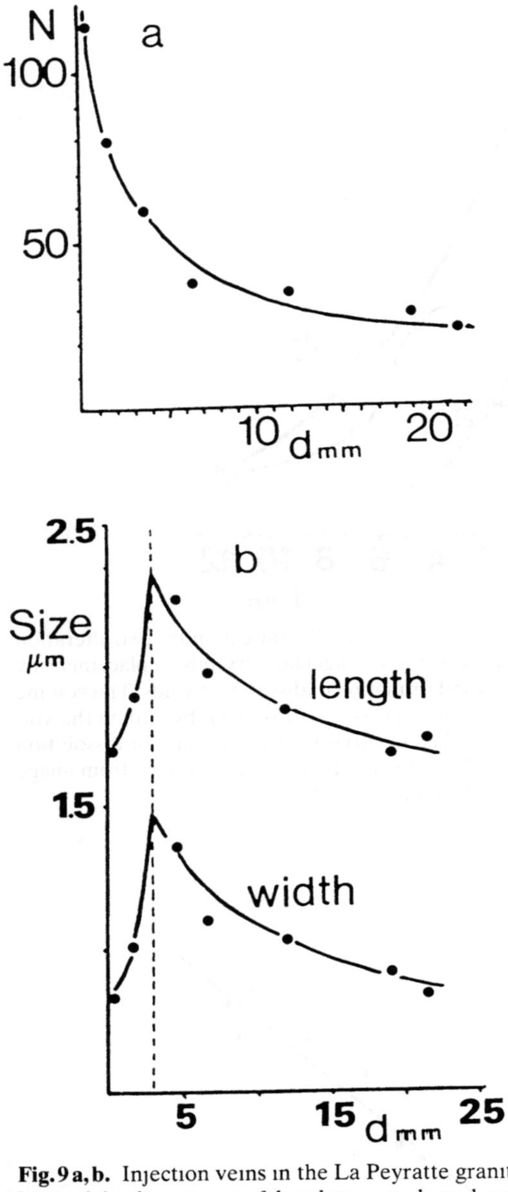

Fig. 9 a, b. Injection veins in the La Peyratte granite. The number and the dimensions of dissolution voids in oligoclase crystals vary with the distance to the vein (image analyses on a fixed unit surface 7200 µm²). The number of voids *N* (reported for a 1000 µm² surface) decreases with distance *d (mm)* according to an exponential function (**a**). This could be similar to nucleation processes controlled by a thermal gradient. The shape of curves in **b** indicates that the growth of voids is controlled by two antagonist processes: dissolution and albitization of oligoclase (Turpault 1989)

oligoclase → Na^{+} + Si^{4+} + Ca^{2+} + Al^{3+} (1)
oligoclase + Na^{+} + Si^{4+} → albite + Ca^{2+} + Al^{3+} (2)
biotite + Al^{3+} → chlorite + K^{+} (3)
Si^{4+} + Al^{3+} + K^{+} → mica or adularia (4)

The quantities of chemical components involved in the interdependent reaction sequence were calculated in a reference volume of rock according to two different ways: (1) the method of Gresens (1967) based on composition-volume relationships; (2) the integration of chemical effects of each individual mineral reaction (comsumption and production of dissolved components) based on quantitative mineralogical data. The two methods give similar results (see Turpault et al. 1992a for calculation details). However, as alteration processes are most often controlled by microsystems (Korzhinskii 1959; Meunier and Velde 1979), the second method is particularly efficient to quantify the chemical transfers between the different microsystems. In the case of volume conservative alteration processes, individual mineral reactions can be written as follows (in volume percentages):
n% Mineral I → x% Mineral II + y% unaltered Mineral I + z% dissolution voids with x% + y% + z% = n%; Mineral I: primary mineral: Mineral II: secondary minerals ($x\% = x_1\% + x_2\% ... + x_n\%$). Volume percentages are determined by surface measurements using image processing on thin sections.

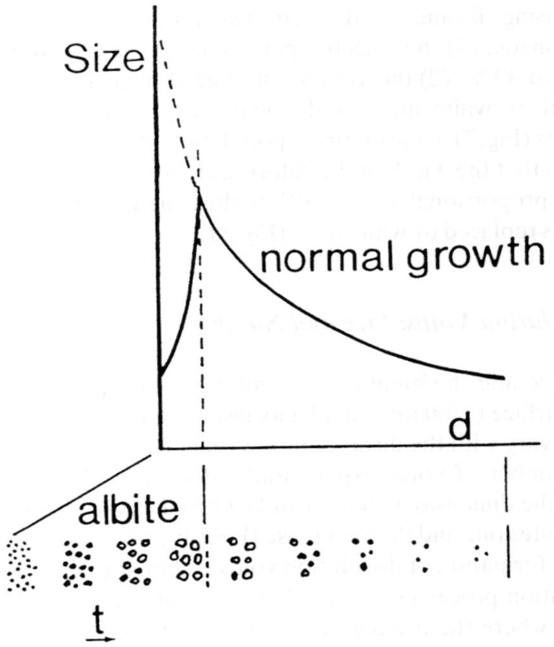

Fig. 10. Schematic explanation of the curves presented in Fig. 9b. The "nucleation" of dissolution voids must occur very early just after the penetration of the hydrothermal fluid in the wall-rocks and the stabilization of the thermal gradient. Then the voids begin to grow until they are "fossilized" by the albitization of oligoclase *(albite)*. As voids far from the vein can grow during a longer time, their dimensions increase in the albite zone until they reach the theoretical exponential growth curve *(normal growth)*. Consequently, the size increase records the duration of the albization process *(t)*

Surprisingly, the mass-balance calculations according to the two methods show that the bulk chemical composition of the rock does not change very much in spite of the strongly altered aspect of the wall-rocks. Only water and small amounts of sulfur were definitely added to the system. This means that all the other elements necessary for the crystallization of secondary phases could have originated in the dissolution of primary minerals (biotite and oligoclase). Some elements, mainly Si and Ca, are exported out of the system. If an oversimplification is authorized, alteration effects seem to be reduced to the

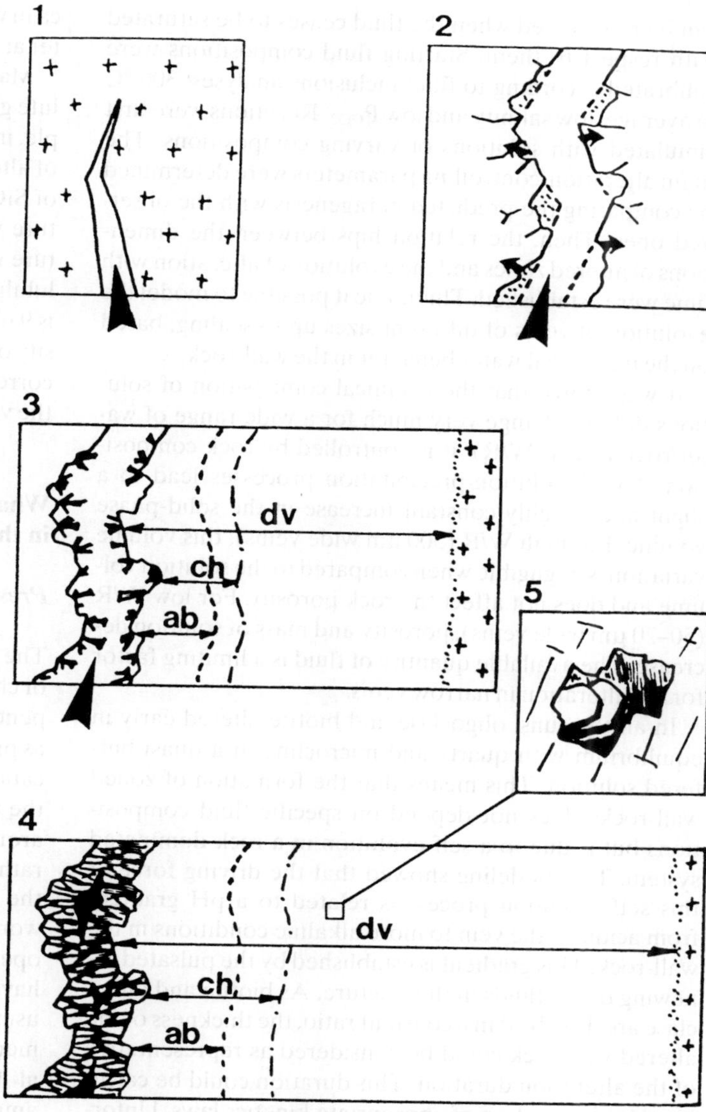

Fig. 11.1–5. Schematic representation of the five stage alteration process in an injection vein type from La Peyratte granite (Turpault et al 1992a) **1** Opening of a fracture and introduction of hot fluid *(large arrow)* in the open space **2** The dissolution of the fracture edges is not homogeneous because microcline remains unaffected. The fluid begins to penetrate inside the wall-rock *(small arrows)*, oligoclase and biotite are partly dissolved **3** A mineral zonation appears in the wall-rocks: *ab* albite zone; *ch* chlorite zone, *dv* dissolution voids in oligoclase crystals The dissolved elements are, for the most part, consumed by the crystallization of secondary minerals albite, phengite (or adularia in the < 50 µm wide veins), and chlorite The alteration phenomenon (dissolution-recrystallization) proceeds then by a sequence of interdependent mineral reactions This is the "living" stage of the hydrothermal alteration during which the temperature and chemical gradients are stabilized **4** The alteration process stops because the secondary phases which precipitate inside the open fracture have no more free space to grow The remaining voids in the vein are sealed by minor nonsilicate minerals fluorite and pyrite (*black deposit* in the center of the vein) Temperature and chemical gradients disappear. **5** The fluids trapped in the secondary porosity of the altered wall-rock (dissolution voids in oligoclase crystals) become saturated with clay minerals when temperature decreases during the collapse of the hydrothermal system Illite/smectite mixed-layered minerals precipitate in these voids

transfer of voids from the open fracture to the dissolution features inside the wall-rocks.

Alteration Mechanism

A modeling of the dissolution-precipitation processes in the La Peyratte vein system was attempted using the EQ3/6 software package in order to approach the composition of fluids and to determine the controlling chemical parameters (see Berger et al. 1992 for details). In spite of the fact that alterations due to hydrothermal fluid circulation are open systems, it was possible to model the La Peyratte system by a run set assuming that the system is closed. Indeed, as will be detailed further, the hydrodynamic regime in the fractures was pulsated. The injection time was very short compared to the residence one. In each run, phases remain in equilibrium with the fluid and can be redissolved when the fluid ceases to be saturated with respect to them. Starting fluid compositions were calibrated according to fluid inclusions analyses: 300 °C in average, low salinity and low P_{CO_2}. Reactions were first simulated with solutions of varying compositions. The main alteration-controlling parameters were determined by comparing the predicted paragenesis with the observed ones. Then, the relationships between the dimensions of altered zones and the evolution of alteration with time were established. This made it possible to model the evolution of veins of different sizes up to sealing, based on the interstitial water behavior in the wall-rock.

It was shown that the chemical composition of solutions did not change very much for a wide range of water/rock ratios (W/R): it is controlled by rock composition. The dissolution-precipitation processes lead to a slight and roughly constant increase in the solid-phase volume. For high W/R (300 µm wide veins), this volume variation is negligible when compared to the solution volume and does not affect the rock porosity. For low W/R (50–20 µm wide veins), porosity and mass of solution decrease. The available quantity of fluid is a limiting factor for the alteration in narrow veins.

In all the runs, oligoclase and biotite altered early in equilibrium with quartz and microcline in a quasi-buffered solution. This means that the formation of zoned wall-rocks does not depend on specific fluid compositions but is due to a self-evolution in a rock dominated system. The modeling showed that the driving force of this self-evolution process is related to a pH gradient from acidic in the vein to more alkaline conditions in the wall-rock. This gradient is established by the pulsated renewing of the fluids in the fracture. As biotite and oligoclase are dissolved in a constant ratio, the thickness of the altered wall-rock could be considered as representative of the alteration duration. This duration could be calculated by application of appropriate kinetics laws. Unfortunately, until now, the kinetics parameters of oligoclase dissolution are not known for near equilibrium dissolution at high temperature conditions. Summarizing the data obtained on the La Peyratte vein system, four successive stages can be distinguished in the alteration process from the opening to the sealing of the initial fracture (Fig. 11).

Quantities of Fluid Involved in Alteration Processes

Silica solubility is known to be temperature-dependent (Fournier and Potter 1982; Michard 1985). Consequently, dissolved silica is classically used as a geothermometer. In the case where the temperature is known, the loss of silica can be used to calculate the quantities (Qw) of solutions which have flowed in the system:

$$Q_w = N_{SiO_2}/N_T x \rho_{wT},$$

where N_{SiO_2} is the number of moles of dissolved silica lost by the system; N_T is the number of moles of dissolved silica in water at temperature T, and ρ_{wT} is the density of water at temperature T.

Mass balance calculation results are expressed as absolute gain or loss for each chemical component. For example, in the La Peyratte granite a 40 dm^3 reference volume of altered wall-rock around a phyllic vein has lost 12 mol of SiO_2 (Turpault et al. 1992a). The corresponding fracture volume is 0.375 dm^3. An average 300 °C temperature was estimated from fluid inclusion analyses. The solubility of silica calculated with the EQ3/6 code for 300 °C is 9.61 10^{-3} mol per kg of water (269 ppm of Si). The density of water is 0.7 at 300 °C, so Qw = 1785 l. This quantity corresponds to 4762 times the volume of solution inside the vein in the reference volume of rock.

What Was the Fluid Flowing Regime in the Fracture?

Presence of Temperature Gradients

The heat diffusion rate is several orders greater than that of chemical diffusion. Thus, mineral reactions, which depend on chemical transfers, were classically considered as proceeding in isothermal conditions whatever their location in the wall-rock (Lovering 1950). Consequently, the formation of zones of different mineral composition around the veins results from chemical gradient effects rather than temperature ones. In fact, this is only true in the case where the heat source is constant, or, in other words, in the case where hot fluid flowing regime in the open fracture is strictly continuous. Some recent studies have shown that thermal gradients could be detected using temperature mineral markers (Beaufort and Meunier 1983b; Horton 1985) or fluid inclusions (Turpault et al. 1992b). Taking the La Peyratte vein system as an example, the thermal gradient evidenced by fluid inclusion study was of about 75 °C between the fracture (300 µm wide) and the altered wall-rock at 11 mm of the fracture (Fig. 12). Such a temperature gradient is established in few seconds (Grimaud et al. 1990); consequently, it is not possible to explain the stabilization of this thermal gradient during long enough periods for mineral reactions to proceed assuming a continuous flowing regime of hot fluids in the open fracture.

If a variable flowing regime is established inside the open fracture, the physicochemical conditions imposed on fluid-rock interactions must also fluctuate. Such fluctuations were suspected in other occurrences because of their effects on the isotopic composition of rocks (Barton et al. 1977) and because of the formation of zoned minerals in vein deposits (Marignac 1988). Using fluid inclusion studies and accurate chemical analyses of zoned epidotes, Marignac (1988) showed that pH, fO_2 and temperature conditions oscillate. Each oscillation comprises two events: injection of hot fluid whose chemical properties are controlled by the source region (active during a short time), and equilibration with the host rock during a longer residence time.

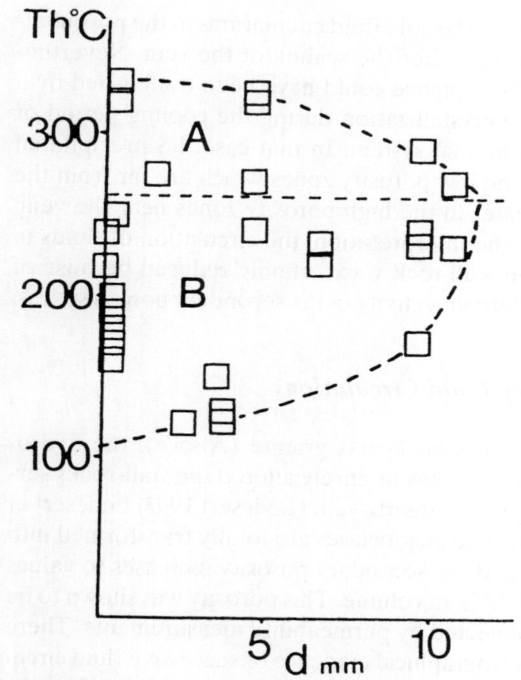

Fig. 12. Injection vein type Homogenization temperatures *(Th)* of secondary fluid inclusions trapped in quartz crystals were measured in the altered wall-rock of a 300 μm wide phengite vein (La Peyratte granite, after Turpault et al 1992b) The injection period is recorded by the 340–265 °C temperature range *(A)*, a thermal gradient is stabilized during this period The 265–100 °C temperature range corresponds to fluid inclusions trapped in quartz during the post-sealing cooling period *(B)*

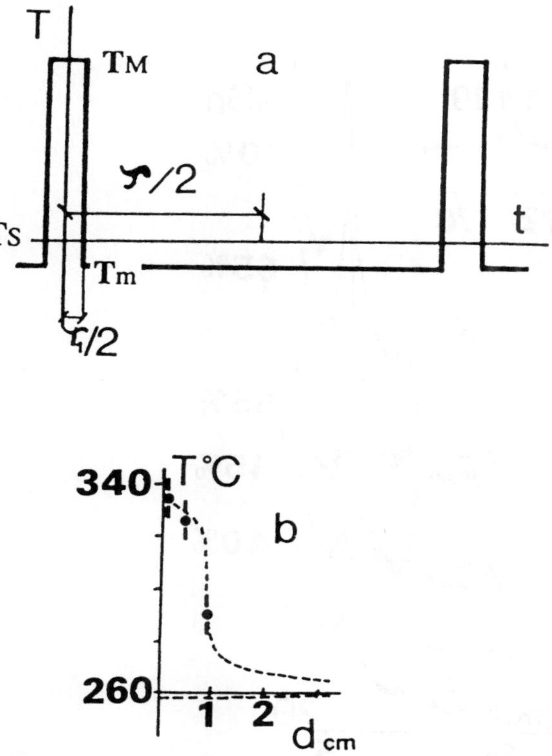

Fig. 13 a, b. Injection vein type Schematic representation of a periodic pulsated regime (After Grimaud et al 1990) **a** Temperature oscillates between T_m and T_m, which are the maximum and minimum values respectively, τ period of the oscillation (injection + residence time). τ_l time during which the temperature is maximum **b** Modeling of the thermal gradient stabilized in the wall-rock (After Turpault et al 1992b)

Modeling a Pulsated Regime

A thermal gradient can be stabilized for period exceeding a few seconds by a variable heat source (Carslaw and Jeager 1986). In order to simplify the modeling, Grimaud et al. (1990) assumed that the variations were periodical as they are in geyser systems. Consequently, these authors described a periodic pulsated hydrodynamic flow at steady state by two equations:

$$T_s = \alpha T_M + (1 - \alpha) T_m, \quad (1)$$

where α is the coefficient which relate τ_1 to τ: $\tau_1 = \alpha\tau$ ($0 < \alpha < 1$); T_s is the mean value around which the temperature oscillates; T_M and T_m are the maximum (340 °C) and the minimum (265 °C) temperature respectively; τ is the period of oscillation, and τ_1 is the time during which the temperature is maximum (Fig. 13a).

$$\delta^2 T/\delta^2 z^2 = 1/a . \delta T/\delta t, \quad (2)$$

where $a = k^*/(\rho c)^*$ is the thermal diffusivity of the porous medium; k^* and $(\rho c)^*$ are the thermal conductivity and the heat capacity of the porous medium respectively. $T = T(z, t)$ is the periodic function of time (t) and of the distance of the vein (z). By using Fourier's method, the solution to this last equation is:

$$T(z, t) = T_s + \sum_{n=1}^{\alpha} T_n(z, t) \quad (3)$$

with

$$T_n(z, t) = C_n \exp(-\sqrt{nw/2a}.z) \text{ x } \cos(nwt - \sqrt{nw/2a}.z),$$

$$Cn = 1/\tau \int_{-\tau/2}^{+\tau/2} (T(0, \xi) - T_s) \exp(-jnw\xi)\, d\xi$$

and

$$w = 2\pi/\tau.$$

The best fit of the measured thermal gradient is obtained with the following values: τ = 1286 s (about 20 min) and α = 0.015 (Fig. 13b). These values are comparable to those of geyser eruptions (Elder 1981). The minimum temperatures obtained are between 255 and 260 °C. This is in good agreement with the observation of phengite or adularia crystal precipitating in the fracture instead of illite/smectite mixed-layered phases. In fact, the oscillatory temperature conditions are known to favor the nucleation process (Baronnet 1980).

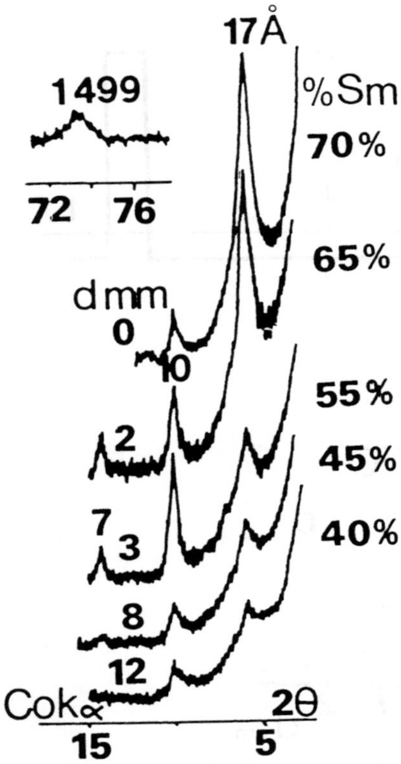

Fig. 14. Injection vein type in the La Peyratte granite Composition change of illite/smectite random mixed-layered minerals with distance to the vein (Turpault et al 1992b)

What Happens After the Vein Sealing?

Precipitation of Secondary Minerals in Dissolution Voids

Clay minerals are frequently observed in hydrothermal veins. Obviously, they cannot form at the same time as the high-temperature minerals except in very short thermal events where both nucleate. Most often, they precipitate during later periods when the temperature conditions have decreased. In the case of La Peyratte granite veins, high-temperature alteration processes stop when the fracture is completely sealed by phengite crystals, pyrite, and fluorite. As the hot fluid is no longer injected, the temperature decreases until it reaches the equilibrium with the surrounding rock. Then, the temperature of the whole system continues to decrease as the hydrothermal activity collapses. This final cooling stage is recorded in secondary fluid inclusions trapped in quartz crystals inside the altered wall-rock (Fig. 12).

During the cooling period, the fluids trapped inside the dissolution voids come progressively to saturation with clay minerals which are the stable silicate phases in low-temperature conditions In the La Peyratte granite veins, a sequence of illite/smectite mixed layers (I/S) was observed in the wall-rocks; I/S are smectite-rich towards the vein (Fig. 14). Assuming that I/S of different compositions crystallized at the same time, Parneix (1992) concluded to the presence of an inverse temperature gradient established by cold fluid circulations in the porous altered wall-rock after the sealing of the vein Nevertheless, the I/S sequence could have been established by a progressive crystallization during the cooling period of the hydrothermal system. In that case, I/S precipitated earlier in the low porosity zones which are far from the vein and later in the high porosity zones near the vein Whatever the interpretation, the circulation of fluids in the altered wall-rock was certainly reduced because of the low interconnectivity of the secondary porous spaces.

Secondary Fluid Circulations

In the Soultz-sous-Forêts granite (Alsace), the hydrothermal activity has intensely altered the wall-rocks surrounding a large quartz vein (Ledésert 1993, Ledésert et al 1993b) The plagioclases are totally transformed into tosudite, and the secondary porosity increases to values as high as 20% in volume This porosity was shown to be highly connected by permeability measurements There is clear petrographical evidence of successive fluid circulations in the altered granite. The first is detected by the presence of hydrocarbon products filling the secondary porous spaces. These organic compounds originated in rocks and reservoirs formations covering the granitic basement (Pechelbronn oil field). Their presence implies that connections were established between the sedimentary reservoir and the porous altered granite located 1000 m below.

A second fluid circulation event is evidenced by illite precipitation in microfractures crosscutting quartz crystals and argillized plagioclases. The lath-shaped illite grew on tosudite flakes and the organics are pushed back around the microcracks (Fig. 15).

At least, saline fluids were sampled in the altered zone during the drilling. Their chemical and isotopic properties show that they originate in the upper sedimentary formation of the Rhine graben margins. This third fluid circulation event indicates that interconnections between the upper reservoir in the sedimentary formations and deep porous altered zones are still working.

Conclusion

It is clear that a complex information is recorded in vein alteration systems. In the present state of knowledge, only a few questions have found an answer (or a partial answer), but many remain unsolved For example, it will be of great interest now to determine the duration of hydrothermal processes. Among the unsolved questions, the following demand our attention:

- could we use the differential dissolution or growth of primary minerals on the fracture surfaces as indicators of initial fluid properties?
- how do the sealing minerals nucleate and grow in the open fracture?
- what is the direction of the flow?

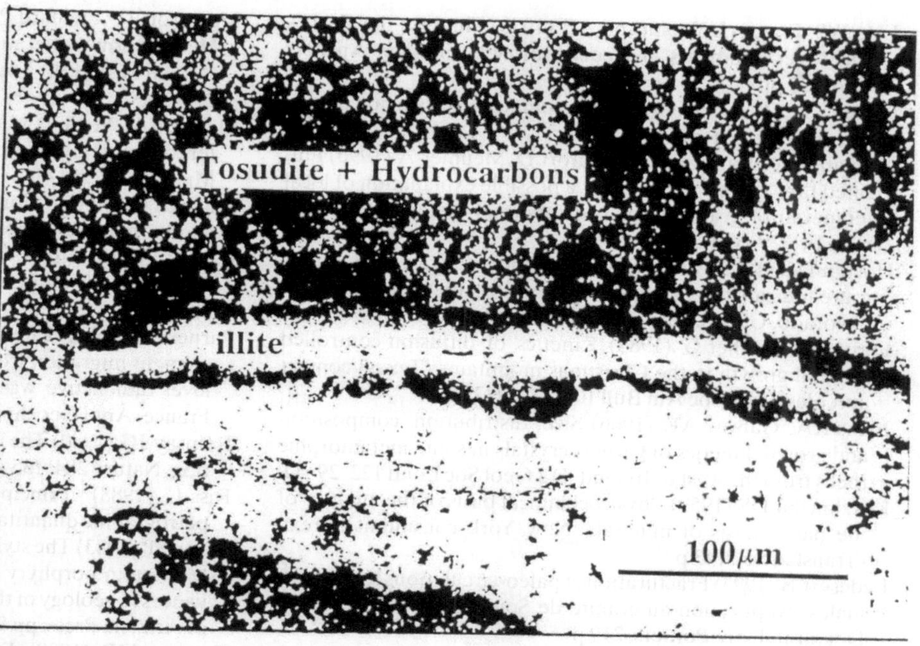

Fig. 15. Post-sealing fluid infiltrations in the porous altered wall-rock of a quartz vein at −2160 m (Soultz-sous-Forêts granite, HDR program, Ledésert et al 1993b) The first infiltration stage is evidenced by hydrocarbon deposits which are pushed back by a latter fluid circulation recorded by illite veinlets

– what are the electrical effects of fluid circulation in a fracture?

The veins described here do not represent the total of the possible cases. For example, in the injection vein type, the fluid can be colder than the rock. In diagenetic series, carbonate or sulfate veins are commonly observed. These minerals have an inverse solubility: they are less soluble in high-temperature conditions than in low ones. It will be of interest to establish an exhaustive study of vein-alteration systems because such information leads to understanding the dynamics of fluid transfers in the rocks.

Acknowledgments. This chapter summarizes the works of several students of mine (T. Merceron, M. P. Turpault, B. Ledésert). I am greatly indebted to them for their communicative enthusiasm and their nonconformist ideas. Financial support came from different scientific programs: PIRSEM/CNRS; Géologie Profonde de la France; Analogues naturels (CEA).

References

Aissa M (1987) Reconstitution d'un système de paléocirculations hydrothermales et ses conséquences métallogéniques le district d'Echassières (Massif Central Français) Thèse INPL, Nancy, 250 pp

Al Shaara M (1986) Etude géochimique et métallogénique des minéralisations (U-Ba) du nord du massif des Palanges (Aveyron, France). Thèse Université Paris 6, 174 pp

Baronnet A (1980) Polytypism in micas a survey with emphasis on the crystal growth aspect In Kaldis E (ed) Current topics in material science, 5 North Holland, Amsterdam, pp 447–548

Barton PB, Bethke PM, Roedder E (1977) Environment of ore deposition in the Creede Mining District, San Juan Mountains, Colorado Part III Progress toward interpretation of the chemistry of ore-forming fluid from the OH vein Econ Geol 72 1–24

Beane RE (1983) Hydrothermal alteration in silicate rocks In Titley SR (ed), Advances in geology of the porphyry copper deposits University of Arizona Press, pp 117–137

Beaufort D, Meunier A (1983a) Petrographic characterization of an argillic alteration containing illite, K-rectorite, K-beidellite, kaolinite and carbonates in a cupromolybdenic porphyry at Sibert (Rhône, France) Bull Min 106 535–551

Beaufort D, Meunier A (1983b) A petrographic study of phyllite alteration superimposed on potassic alteration the Sibert Porphyry deposit (Rhône, France) Econ Geol 78 1514–1527

Beaufort D and Meunier A (1993) Saponite, corrensite and Chlorite/Saponite Mixed-Layer in the Sancerre-Covy Deep Drill-Hole (France) Clay Min (in press)

Beaufort D, Patrier P, Meunier A, Ottaviani MM (1992) Chemical variations in assemblages including epidote and/or chlorite in the fossil hydrothermal system of Saint Martin (Lesser Antilles) J Volc Geotherm Res, 51 95–114

Berger G, Turpault MP, Meunier A (1992) Dissolution-precipitation processes induced by hot water in a fractured granite Part 2 Modelling of water-rock interaction Eur J Mineral 4 1477–1488

Bonorino FG (1959) Hydrothermal alteration in the front range mineral Belt, Colorado Bull Geol Soc Am 70 53–90

Boulègue J, Benedetti M, Gauthier B, Bosch B (1990) Les fluides dans le socle du sondage GPF Sancerre-Couy Bull Soc Geol Fr 8 789–795

Carslaw H, Jaeger J (1986) Conduction of heat in solids Clarendon, Oxford, Elder J (1981) Geothermal systems Academic Press, London, 508 pp

Fouillac AM, Rossi P (1991) Near solidus ^{18}O depletion in a Ta-Nb bearing albite granite the Beauvoir granite, France Econ Geol 86 1704–1720

Fournier RO, Potter RW (1982) A revised and expanded silica (quartz) geothermeter Geotherm Resour Counc Bull 12 3–9

Fyfe WS, Price NJ, Thompson AB (1978) Fluids in the Earth's crust Developments in Geochemistry 1, Elsevier, Amsterdam, 383 pp
Gresens RL (1967) Composition-volume relationships of meteasomatism Chem Geol 2 47–65
Grimaud PO, Touchard G, Beaufort D, Meunier A (1990) Pulsated flow regime in fractures a possible explanation of local temperature gradients in hydrothermal systems Geothermics 19 329–339
Horton DG (1985) Mixed layer illite/smectite as a paleotemperature indicator in the amethyst vein system, Creede district, Colorado, USA Contrib Mineral Petrol 91 171–179
Joesten R, Fischer G (1988) Kinetics of diffusion-controlled mineral growth in the Christmas mountains (Texas) contact aureoles Geol Soc Am Bull 100 714–732
Jones KA, Galwey AK (1966) Size, distribution, composition and growth kinetics of garnets crystals in some metamorphic rocks from the west of Ireland Q J Geol Soc Lond 122 29–44
Korzhinskii DS (1959) Physicochemical basis of the analysis of the paragenesis of minerals New York consultant bureau (Translation), 143 p
Ledésert B (1993) Fracturation et paléo-circulations hydrothermales Application au granite de Soultz-sous-Forêts Thèse Doctorat, Univ Poitiers, 219 p
Ledésert B, Dubois J, Velde B, Meunier A, Genter A, Badri B (1993a) Geometrical and fractal analysis of a three-dimensional hydrothermal vein network in a fractured granite J Volc Geotherm Res 56, 267–280
Ledésert B, Dubois J, Genter A, Meunier A (1993b) Fractal analysis of Fractures applied to Soultz-sous-Forêts hot dry rock geothermal program J Volc Geotherm, 57, 1 17
Long JCS, Witherspoon PA (1985) The relationship of the degree of interconnection of to permeability in fracture networks J Geophys Res 90 (B 4) 3087–3098
Lovering TS (1950) The geochemistry of argillic and related types of rock alteration Colo School Mines Q 45 (1 B) 231–260
Marignac C (1988) Composition des minéraux et évolution des phases fluides le cas des filons polymétalliques d'Ain Barbar, (Algérie) Bull Mineral 111 183–206
Merceron T (1988) Les altérations hydrothermales de la coupole granitique d'Echassières et de son environement (sondage GPF ECHA 1) Thèse Univ Poitiers, 167 pp
Merceron T, Velde B (1991) Application of Cantor's method for fractal analysis of fractures in the Toyoha mine, Hokkaido, Japan J Geophys Res 96 16641–16650
Merceron T, Velde B, Nakashima S, Badri A, Touchard G (1992a) Fractal analysis of fractures and microstructures in rocks 3rd Int Symp Advanced Nuclear Energy Research – Global Environment and Nuclear Energy, pp 1–9
Merceron T, Vieillard P, Fouillac AM, Meunier A (1992b) Hydrothermal alterations in the Echassières granitic cupola (Massif Central, France) Contrib Mineral Petrol 112 279–292
Meunier A, Velde B (1979) Weathering mineral facies in altered granites the importance of local small-scale equilibria Min Mag 43 261–268
Michard G (1985) Equilibres entre minéraux et solutions géothermales Bull Min 108 29–44
Page R, Wenk HR (1979) Phyllosilicate alteration of plagioclase studied by transmission electron microscopy Geology 7 393–397
Parneix JC (1992) Effects of hydrothermal alteration on radioelement migration from a hypothetical disposal site for high level radioactive waste example from the Auriat granite, France Appl Geochim 1 253–258
Ramsay JG (1980) The crack-seal mechanism of rock deformation Nature 284 135–139
Riss J (1988) Principes de stéréologie des formes en pétrographie quantitative Thèse Univ Orléans, 2 vol, 420 pp
Titley SP (1983) The style and progress of mineralizarion and alteration in porphyry copper systems In Titley SP (ed) Advances in geology of the porphyry copper deposits University of Arizona Press, pp 93–116
Turpault MP (1989) Etude des mécanismes des altérations hydrothermales dans les granites fracturés Thèse Doctorat, Univ Poitiers, 117 pp
Turpault MP, Fouillac AM, Cathelineau M, Lespinasse M, Meunier A, Feybesse JL, Leroy J (1987) Circulations hydrothermals récentes et anciennes dans le Cézallier les données des phases solides et fluides et des isotopes stables Géol Profonde Fr 2, 4 81–89
Turpault MP, Berger G, Meunier A (1992a) Dissolution-precipitation processes induced by hot water in a fractured granite Part 1 wall-rock alteration and vein deposition processes Eur J Mineral 4 1457–1475
Turpault MP, Meunier A, Guilhaumou N, Touchard G (1992b) Analysis of hot fluid infiltration in fractured granite by fluid inclusion study Appl Geochim Suppl 1 269–276
Velde B, Dubois J, Touchard G, Badri A (1990) Fractal analysis of fractures in rocks the Cantor's dust method Tectonophysics 179 345–352
Velde B, Dubois J, Moore D, Touchard G (1991) Fractal patterns of fractures in granites Earth Planet Sci Lett 104 25–35
Wang HF, Bonner BP, Carlson SR, Kowallis BJ, Heard HC (1989) Thermal stress cracking in granite J Geophys Res 94 1745–1758

Deformation of the Nord-Pas-de-Calais Carboniferous Coalfield (France) in the Variscan Frontal Tectonic Pattern

B. Le Gall

Abstract. Geometrical and kinematic evidence supplied by both reinterpreted mining and geophysical data emphasizes the significance of thrust tectonics in the overall structure of the Nord-Pas-de-Calais Carboniferous coalfield (NPC). On a balanced section passing through the Marchiennes antiform area, parallel to the regional shortening, the deformed NPC coalfield appears to be a northerly-thinning tectonic wedge, detached above a 25 km long gently S-dipping decollement that lies close to the Caledonian basement roof. To the S, it is overstepped by the Dinant Devonian unit along a major polyphase thrust contact referred to here as the Condroz-Midi thrust. The overall internal deformation of the NPC thrust wedge is related to a piggyback thrust sequence, strongly influenced by both stratigraphy and pre-existing basinal morphotectonic features, and involving, in its innermost part, a complex pile of dominantly Dinantian limestone thrust sheets. These latter are assumed to be underlain at depth by a basal hinterland-dipping duplex which may have been facilitated by easy slip along anhydritic layers. In fact, it gives rise farther N, beyond the evaporitic pinchout line, to a gentle N-verging imbricate fan causing decreasing displacements as far N as a blind tip line located close to the Evin frontal merging thrust.

One of the most suitable aspects of our kinematic model is to relate the overall contrasted NPC surface structures to distinct mechanical processes acting during a single, but long-lived, thrust emplacement event. Hence, progressive rotation of hangingwall strata in frontal fault-propagation folds satisfactorily accounts for the downward-facing structures lying in the Epinoy and Barrois thrust sheets, whereas the poorly understood low-angle fault set transecting the Coal Measures is interpreted here in terms of extensional gravity structures developed on top of deeper Dinantian-cored thrust culminations migrating to the N.

Lastly, the enigmatic Roucourt conglomeratic formation is likely to correspond to syntectonic deposits, supplied by active erosion of foreland-propagating thrust sheets, and which later unconformably overlap the entire thrust stack sequences.

Quantitative estimates have also been attempted along the balanced Epinoy-Marchiennes transect, and they lead to a bulk horizontal shortening rate of 50%, e.g., about 40 km, mainly achieved by thrusting, as along the Epinoy or the Barrois thrusts that induce northerly translations of 5 and 4 km, respectively. In our opinion, the concept of the Midi fault as a continuous and one-stage thrust contact is an oversimplication, which is better substituted for by a Condroz-Midi thrust concept that implies a multistage emplacement with an early large-scale overthrust later disrupted by a minor thrust (the so-called Midi fault of previous authors).

On a regional scale, the entire NPC fold-thrust pattern seems to us to be indicative of a large-scale compressional impingement process including a mobile fold-thrust cover that migrates to the N and comes to mold and collide against a buried S-plunging asymmetric high (the Marchiennes antiform), acting at depth as a sort of steady frontal taper.

Introduction

The northern branch of the NW European Variscides extends over more than 2000 km from Ireland eastwards to N Germany, and it is fringed by Upper Carboniferous foredeep deformed coal basins (SW Wales, Boulonnais, Nord-Pas-de-Calais, Hainaut, Rhur, Fig. 1) that have registered the ultimate effects of a collision-related crustal shortening during the late Carboniferous. With respect to the Variscan tectonic framework, the Nord-Pas-De-Calais coalfield (NPC in the text) lies along a large-scale curved belt, in a frontal position, flanked laterally by two oblique segments that extend NW- and NE-wards up to the Boulonnais and the Rhur coal districts, respectively.

Although completely concealed beneath thin Cretaceous deposits of the Paris basin, the shallow tectonic framework of the NPC coalfield is fairly well known in detail thanks to a considerable amount of mining data, gathered during some 100 years of intense coal mining, and that have long notably recognized large-scale thrust tectonics (Cornet and Briart 1863; Barrois 1905; Deflines 1913; Pruvost and Bertrand 1932). Surprisingly, very few synthetic structural approaches have, up to now, been devoted to the NPC coalfield, and the first complete tectonic map of the basin (Fig. 2 A) was compiled in 1963 by Bouroz et al. on the basis of mining data. More recently, basic advances in understanding the deformation history of the NPC coalfield were achieved by Houchen (1988), who built a dominantly thrusting model by applying modern tectonic principles stated elsewhere by Boyer and Elliott (1982) or Elliot (1983), for example. Although broadly in accordance with our own conceptions,

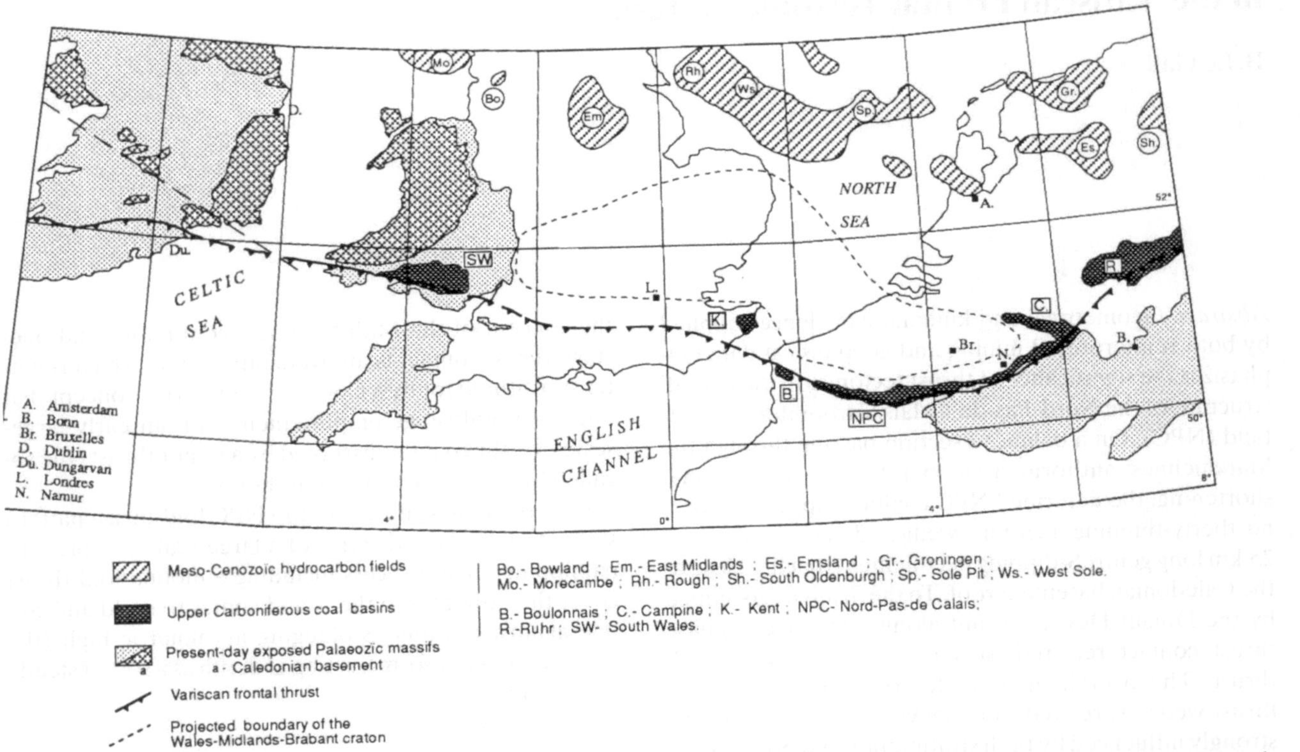

Fig. 1. Location of both Meso-Cenozoic hydrocarbon fields and Upper Carboniferous coal basins of NW Europe with respect to the Variscan deformation front

Houchen's thrusting model differs significantly in some important aspects, so that it justifies proposing and presenting an alternative interpretation here

Geologic Context – Problems

As shown on the tectonic map and associated cross-sections in Fig. 2, the NPC Carboniferous coalfield represents a 15×85 km elongate and arcuate asymmetric syncline, largely overturned to the N and involving, for most of it, a thick sequence of Dinantian carbonates and their overlying Namuro-Westphalian Coal Measures. The less deformed strata of the gently S-dipping northern normal limb give way gradually northwards to undeformed rocks of the Brabant foreland, while to the S, as a result of increasing strain, steep inverted thrust series occur and are, in turn, overthrust by the Dinant Devonian nappe along the Midi Fault. The immediate hangingwall of the Midi thrust locally involves a thin Caledonian basement thrust sheet, the Condroz slice, whose structural meaning is still controversial, giving interpretations that vary from that of an autochthonous basement-cored anticline (Raoult and Meilliez 1985) to that of a blind decollement-related structure (Houchen 1988).

As also shown in Fig. 2, the geometry of the NPC syncline is highly disrupted by numerous N-verging thrusting accident (Cran de Retour, Barrois, Chalard, Pruvost, Vicoigne, etc., thrusts), whose curved cartographic trace mimics the present-day limits of the coalfield, outlining to the E a broad S-directed indentation, the Marchiennes antiform (Becq-Giraudon 1983), and swinging into a N 120 ° oblique direction farther W through the adjoining Marles-Lens syncline. In addition, the youngest folded Coal Measures sequences that occur along the southern, and more deformed, part of the coalfield are intensely offset by a low-angle listric-shaped fault system facing to the N.

Likewise, on the cross sections in Fig. 2B, the entire NPC fold-thrust pattern is cut by steep S-dipping normal faults that are sealed upwards by Mesozoic deposits, and that are considered here as possible Late Variscan extensional structures. Consequently, these late structures have been cancelled out on all the interpretative cross sections, in order to restore the final stack geometry of the Variscan fold-thrust pattern.

Many of the above-mentionned Variscan tectonic features, such as those concerning, for example, (1) the morphology and the deep structure of the entire thrust system, (2) the significance of the low-angle extensional faults, (3) the southerly extent of the NPC coalfield beneath the Dinant Devonian unit, or (4) the total amount of regional horizontal shortening, still have to be discussed and furthermore integrated into a unified regional kinematic model The alternative tectonic model proposed in the present chapter is mainly based on an exhaustive review of all the available mining and geophysical data, that have been reinterpreted for most of them, and then used to define geometrical and kinematic frameworks from which a dynamic evolution of the whole Variscan frontal belt can be proposed. After a detailed analysis of the subsurface geologic pattern, our

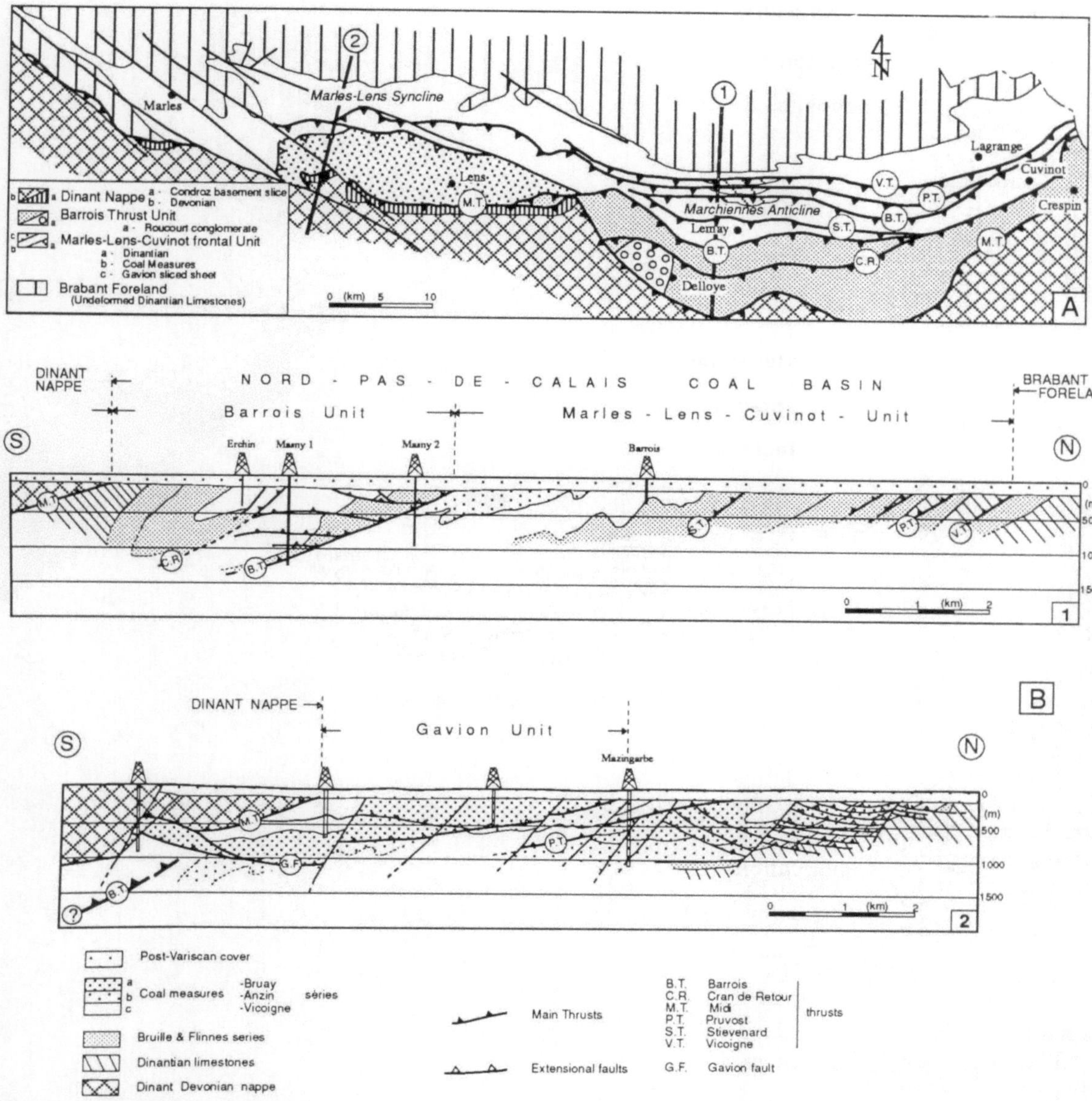

Fig. 2 A, B. Structure of the Nord-Pas-de-Calais Carboniferous coalfield **A** Tectonic sketch map of the Paleozoic basement (–300 m level) elaborated from mining data (After Bouroz et al 1963) **B** Seriated NS cross sections showing the shallow structure of the NPC coalfield through the Marchiennes antiform *(1)* and the adjoining Marles-Lens syncline *(2)* (After Bouroz et al 1963) On the Lens section, the low-angle fault system of the Gavion sliced sheet is not drawn

model will be illustrated by a deep large-scale N-S cross section, passing through the Marchiennes antiform, and constrained at depth by industrial seismic lines as well as by deep boreholes.

Lithostratigraphy

From a paleogeographic point of view, the Devonian and, to a lesser extent, the Carboniferous series of the NPC coalfield, are classically assumed to have been deposited on a shallow outer platform, extending farther E up to the Namur area, and bounded to the N by the emerged Brabant landmass (Fig. 1) (Fourmarier 1954). To the S, it passes, along the Condroz normal-faulted zone, into the Dinant subsiding basin where, in contrast, thicker dominantly clastic deposits were laid down.

Due to their poor economical promise, the Devonian and the Lower Carboniferous rocks that underlie the NPC Coal Measures have been little investigated by mining work, so that their stratigraphy is only indirectly known from some deep boreholes (Epinoy), as well as from lateral correlations with the type-sequences exposed in the Namur synclinorium (Fig. 1). A simplified stratigraphic column of the NPC Upper Paleozoic succession is shown in Fig. 3 and is briefly commented on below (for further details, see Waterlot et al. 1973 or Paproth et al. 1983).

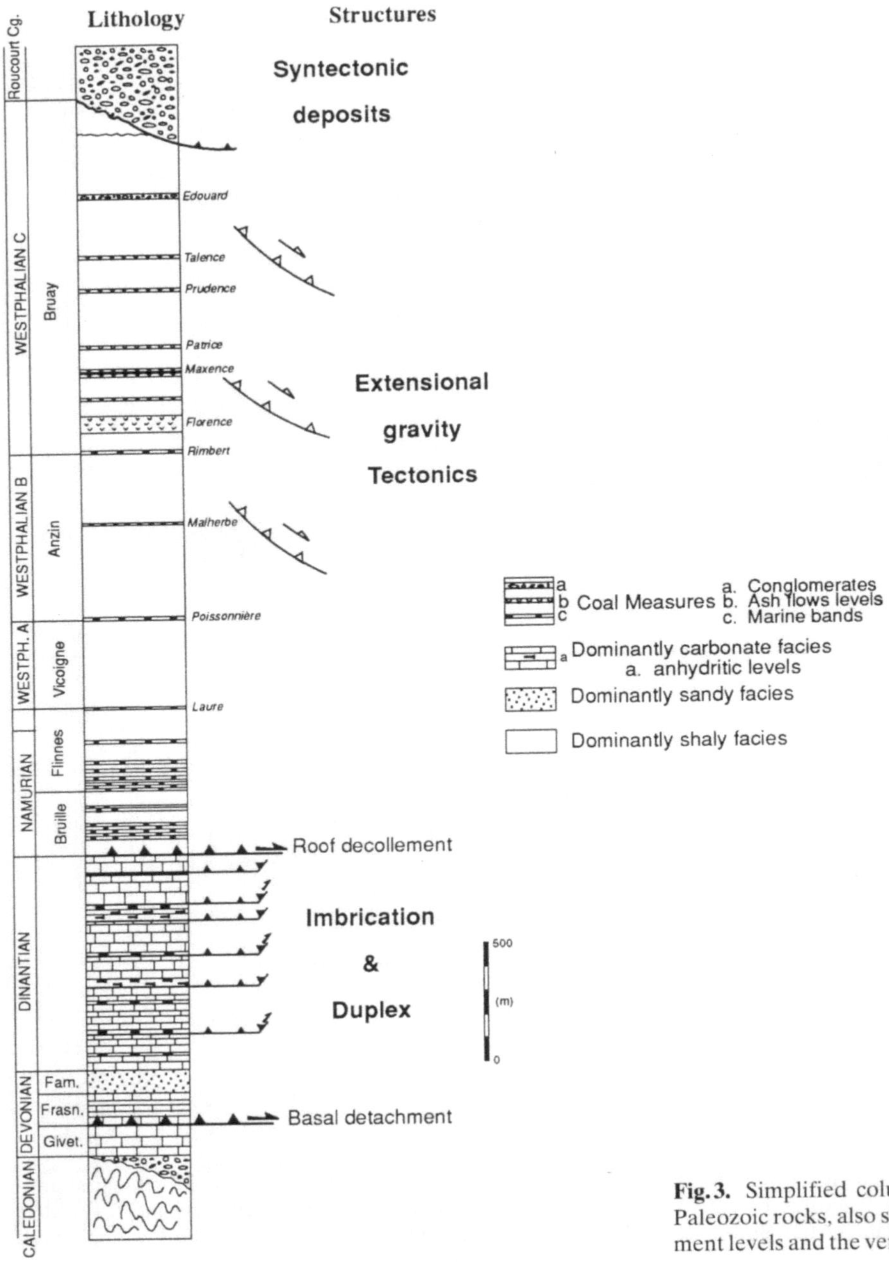

Fig. 3. Simplified columnar stratigraphic section of the NPC Paleozoic rocks, also showing the position of the major decollement levels and the vertical variations in tectonic style

Devonian

The Devonian is only represented by a thin (about 2–300 m) dominantly clastic series, Middle-Upper Devonian in age, underlain by Couvinian conglomeratic deposits that directly overlap the Brabant basement rocks. Some evaporitic (Givetian) and calcareous (Givetian, Frasnian) sequences are also present, and are usually related to regional transgression-regression cycles.

Dinantian

The Devonian series grade conformably upwards into thick marine Tournaisian-Visean limestones (mainly reef-building and oolithic facies) that include interbedded breccia deposits (Pirlet 1972) and, above all, thick anhydrite levels. These latter show dramatic lateral thickness changes, thinning, for example, from about 1000 m in the Epinoy borehole (Bouquillon 1984) to 600 m in the St-Ghislain borehole (Delmer 1977), and finally dying out farther E in the Jeumont area (see location in Fig. 10). Their approximate western boundary is generally assimilated with the Marchiennes high, which is therefore regarded as a major intrabasinal active tectonic feature (Becq-Giraudon 1983).

Coal Measures

The Coal Measures series start with fine-grained detrital marine sediments, Namurian in age (Bruille and Flinnes series), which transit progressively upwards into a thick molasse-type coal-bearing complex, the Coal Measures sensu stricto, subdivided into three series (Vicoigne, Anzin, and Bruay), and ranging in age from Westphalian A to C (Bouroz 1969). The Coal Measures (s. s.) consist of monotonous grit-shale alternations that enclose about 80 productive coal seams, as well as subordinate interbedded marine bands and tonstein horizons with large areal extent, both of which constitute excellent regional stratigraphic markers.

Rapid lateral (E–W) and longitudinal (N–S) thickness variations of the Coal Measures are well documented throughout the basin, and they seem to reflect a relative instability of the basin floor during sedimentation. As already stated for the Dinantian episode, the Marchiennes area is likely to have acted again at this stage as a "positive" zone, capped by condensed coal-bearing deposits that thicken gradually westwards over the Marles-Lens subsiding domain, where about 3000 m of sediments were accumulated (Bouroz 1969).

The stratigraphy of the NPC Carboniferous succession also includes poorly known conglomeratic series, the Roucourt Formation (Barrois et al. 1930), which is restricted to the western oblique termination of the Barrois overturned syncline (Fig. 2 A), where it unconformably overlies steep inverted Lower Carboniferous sequences (Fig. 5 A). Ascribed to the Westphalian C (in Becq-Giraudon 1983), the Roucourt conglomerates can be tentatively regarded, in agreement with recent tectonic work devoted to depositional conditions in Alpine frontal thrust belts (Puigdefabregas and Souquet 1986; Specht 1990), as syntectonic deposits genetically related to the NPC thrust stack emplacement, and they therefore probably represent the last detrital infilling stage of the basin (Fig. 3).

Shallow Tectonic Pattern

On the basis of cartographic evidence, further complemented by revised mining data, the geometry, and kinematics of the overall NPC fold/thrust system can be defined from selected demonstrative examples, believed to be representative of the regional tectonic style, as a whole.

Fold-and-Thrust Structures

Fold/thrust relationships can be deciphered, for example, from the Lagrange colliery cross section that extends through the Marles-Cuvinot frontal Unit, E of the Marchiennes antiform (Figs. 2 A, 5 A). There, the gently S-inclined Lower Coal Measures of the Vicoigne series are involved in parasitic asymmetric N-facing folds, that are closely associated with S-dipping thrusts, such as the Nord Fault (structure X, Fig. 4 A). The incompetent coal-bearing sequences do not show any apparent thickening through the hinge zones, so that the folded structures may be assigned to the concentric fold model of Ramsay (1974).

According to Bouroz et al. (1963), the Nord Fault is disrupted on either side of an anticlinal fold hinge, while keeping an almost constant southerly dip; but such a conception leads to an unrealistic geometrical fold-thrust pattern that can be more satisfactorily described, in agreement with recent thrust tectonics principles (Suppe and Medwedeff 1984), in terms of fault-propagation anticlines related here to N-directed imbricate thrust ramps (Fig. 4 B). In addition, some other minor folds may have developed above blind ramp sticking points. Due to the lack of any significant pervasive deformation, stratal shortening caused by folding, and estimated at about 25% on some continuous coal seams, could hence correspond to the total displacement having occurred along the imbricate thrusts.

This kinematic model can be confidently extrapolated, with some slight differences, on a wider scale to the Barrois syncline unit as a whole (Figs. 2, 5). In fact, by considering the surface geometry and the initial thickness of its folded hangingwall sequences, as well as the depth of the Barrois thrust as deduced from some mining boreholes (Masny, see Fig. 5 B for location) (Bouroz 1950), the overturned Barrois syncline is thought here to be soled at depth by a 3–4-km-long shallow detachment that lies close the bottom of the Flinnes series along the northern normal limb, and that transects, to the S, the steep sequences of the inverted flank (mainly Dinantian and Upper Devonian strata).

The Barrois detachment should therefore represent a sort of shearing plane along which an early syncline would later be displaced northwards over at least 3–4 km. The origin and initial geometry of the primary syncline can be deduced from deep borehole and seismic data, and are hence discussed in the final kinematic model.

Our quantitative estimate, although half smaller than the value predicted in Houchen's model (1988), leads us to regard the Barrois thrust as the most important NPC intrabasinal structure, in accordance with complementary stratigraphic data, provided by the Delloye and Lemay collieries (see location in Fig. 2 A), that reveal significant facies changes between its hangingwall and footwall Coal Measures sequences, respectively (Bouroz et al., 1963).

In the Crespin area, to the E (Fig. 2 A), the shallow Barrois thrust is, in turn, involved in elongate and partially eroded synform-antiforms, which actually occur as tectonic klippes and windows, and that hence could be ascribed to subsequent blind ramping at depth during the regional N-directed thrust propagation.

On a regional scale, some significant lateral changes in tectonic styles take place along most individual thrusts, and in the Marchiennes antiformal area, they are, in addition, associated with a large-scale virgation of the structures (Fig. 5 A).

One of the best examples is observed along the Barrois thrust merging line which appears as a composite tectonic feature comprising a middle E-W-trending segment that

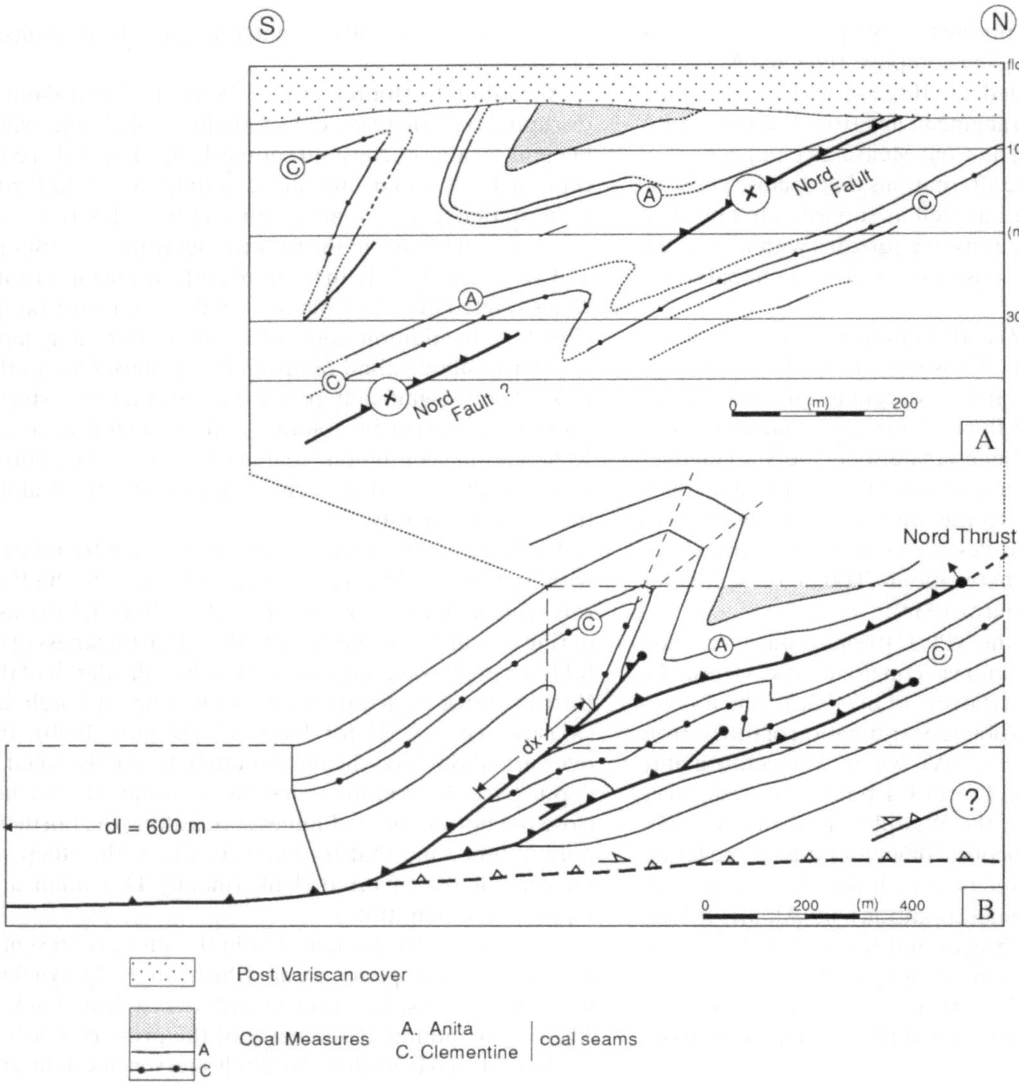

Fig. 4 A, B. Geometry of the frontal fold-thrust system along the La Grange colliery cross section (see location in Fig 2 A) **A** Mining data showing N-directed folds and thrusts in the Vicoigne basal series According to Bouroz et al (1963), the Nord thrust (labeled *X*) could be disrupted through an anticline fold hinge **B** Alternative model considering the folded structures as fault-propagation anticlines associated with N-directed imbricate thrust ramps

extends at the rear of the S-plunging Marchiennes antiform, and forms a narrow lens-shaped imbricate fan composed of seven thin and closely spaced thrust slices of the lower Flinnes and Bruille series (Fig. 5 B). Laterally, the frontal imbricate thrust stack pinches out into a single surface thrust fault, the Barrois thrust sensu stricto, which concomitantly shows a gradual change of its strike along the flanks of the Marchiennes antiform. Along its eastern segment, the Barrois thrust propagates obliquely with respect to its folded hangingwall and footwall sequences, whilst remaining at the same topographic level (Fig. 5 A).

Farther N, the outermost structures developed in the footwall of the Barrois thrust display a quite similar organization with imbricate thrusts, in a frontal position (Bouroz, Pruvost and Vicoigne thrusts, respectively, from S to N), that decapitate the S-inclined Marchiennes axial zone, and then extend laterally across the Coal Measures sequences before dying out, or branching into a single thrust plane, within the Anzin series (Fig. 5 A) Along the eastern flank, this anastomosed fault zone corresponds to a thick shear zone, known as "la zone failleuse" (Bouroz 1950).

Now, if considering the stratigraphy of each successive frontal ramp hangingwall, we can observe that the amount of lower stratigraphical levels progressively increases forelandwards from the Bouroz thrust up to the Vicoigne frontal ramp. Such an atypical thrust pattern is likely to result from flat-lying thrusts moving northwards at a constant depth through southerly-dipping Carboniferous strata (Marchiennes antiform)

According to mining data, the northward displacement along each ramp does not exceed 1–2 km, and that precludes assimilating the Pruvost thrust with a major regional structure, as previously stated (Bouroz and Stievenard 1958)

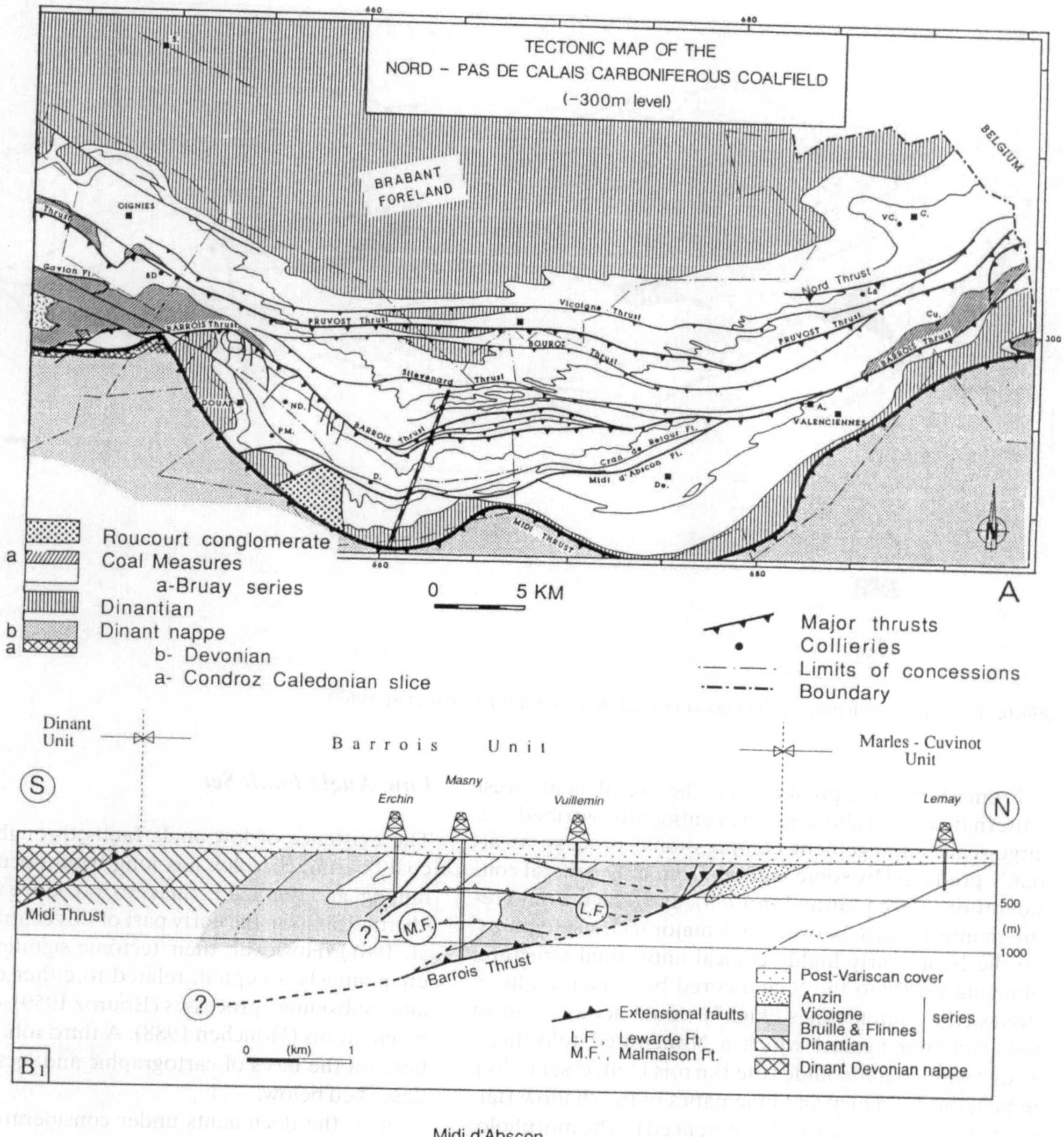

Fig. 5 A, B. Structure of the Barrois thrust unit **A** Tectonic map of the eastern part of the NPC coalfield (After Bouroz et al 1963) The trace of section **B** is drawn **B$_1$** Subsurface mining constraints along a N-S section crossing through the central part of the Barrois syncline (see location in **A**) **B$_2$** Interpretative cross section illustrating the geometry of the structures at an early stage of the shortening The structure of the Marles-Lens unit, in the footwall of the Barrois thrust, is not described *Numbers* refer to the sequential development of the structures

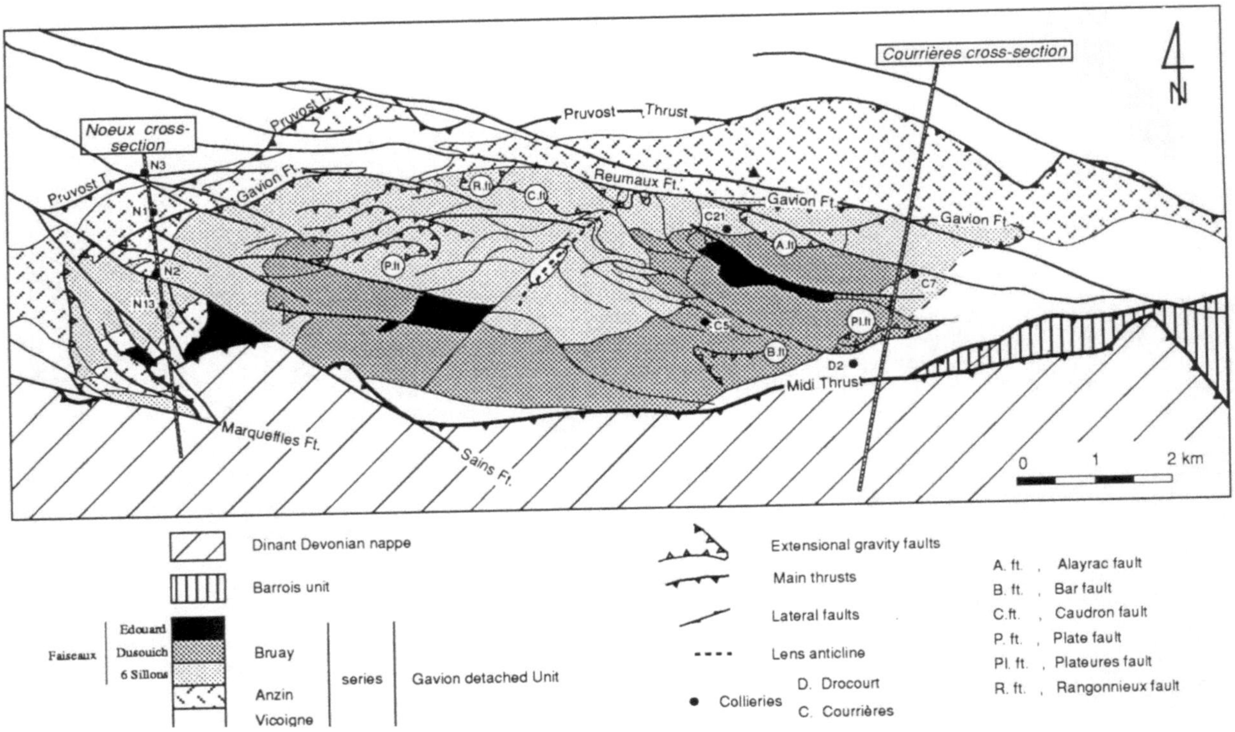

Fig. 6. Tectonic sketch map of the Gavion sliced sheet (After Bouroz et al 1963)

From a kinematic point of view, the overall bent thrust pattern described above can be confidently ascribed to a large-scale impingement compressional process, as already predicted by some authors from only general considerations (Becq-Giraudon 1983), and which would result in interference between two major tectonic features To the N, an early highly conical antiformal structure, plunging gently to the S, and cored by Dinantian limestones (the Marchiennes "high"), could act as a sort of rigid indenter against which a N-migrated fold-thrust wedge would later collide (the Barrois Unit, essentially). In fact, the geometry and kinematics of the shallow Barrois thrust sheet are strongly influenced by the morphology of the Marchiennes "high", as expressed by (1) the cartographic deflection of the Barrois merging line, and (2) the development, at the rear of the taper edge, of frontal imbricate ramping in response to a buttress effect. Farther N, the apex of the Marchiennes S-inclined indenter is itself caught up in more external ramp thrusts which cut down-section stratigraphically northwards in the Lower Carboniferous whilst remaining at an almost constant structural level. One consequence of this frontal ramping is to have amplified the culminations of the Dinantian limestone roof, and then to have induced, in the overlying Coal Measures thrust sheets (Barrois Unit, mainly), some specific structures, discussed below, and referred to as low-angle faults.

Low-Angle Fault Set

The existence of low-angle faults throughout the NPC coal-bearing complex has long been documented from mining, and that led E. Suess to introduce the concept of "listric" faults in the early part of this century (in Bally et al. 1981). However, their tectonic significance was not unanimously accepted, related to either compressional and "subsiding" processes (Bouroz 1950), or back-thrust mechanisms (Houchen 1988). A third solution is favored here on the basis of cartographic and tectonic evidence described below.

In all the documents under consideration, the entire NPC low-angle fault set exhibits some common features, which are partially visible in cross sections 2B and 5B, and which consist as follows:

- they preferentially occur in Upper Coal Measure-scored synclines;
- they usually dip to the N or NW, with a well-pronounced listric shape, truncating sharply to the S or SE the generally gently inclined strata, before flattening gradually N- or NW-wards;
- their hangingwall sequences are systematically downthrown to the N or NW by rigid rotation, without any associated internal deformation, as small-scale tilted fault blocks;
- the resulting N or NW-directed extensional displacement along each individual fault never exceeds 150–200 m.

Using kinematic indicators, indirectly deduced from tectonic maps and specific mining cross sections, leads us to

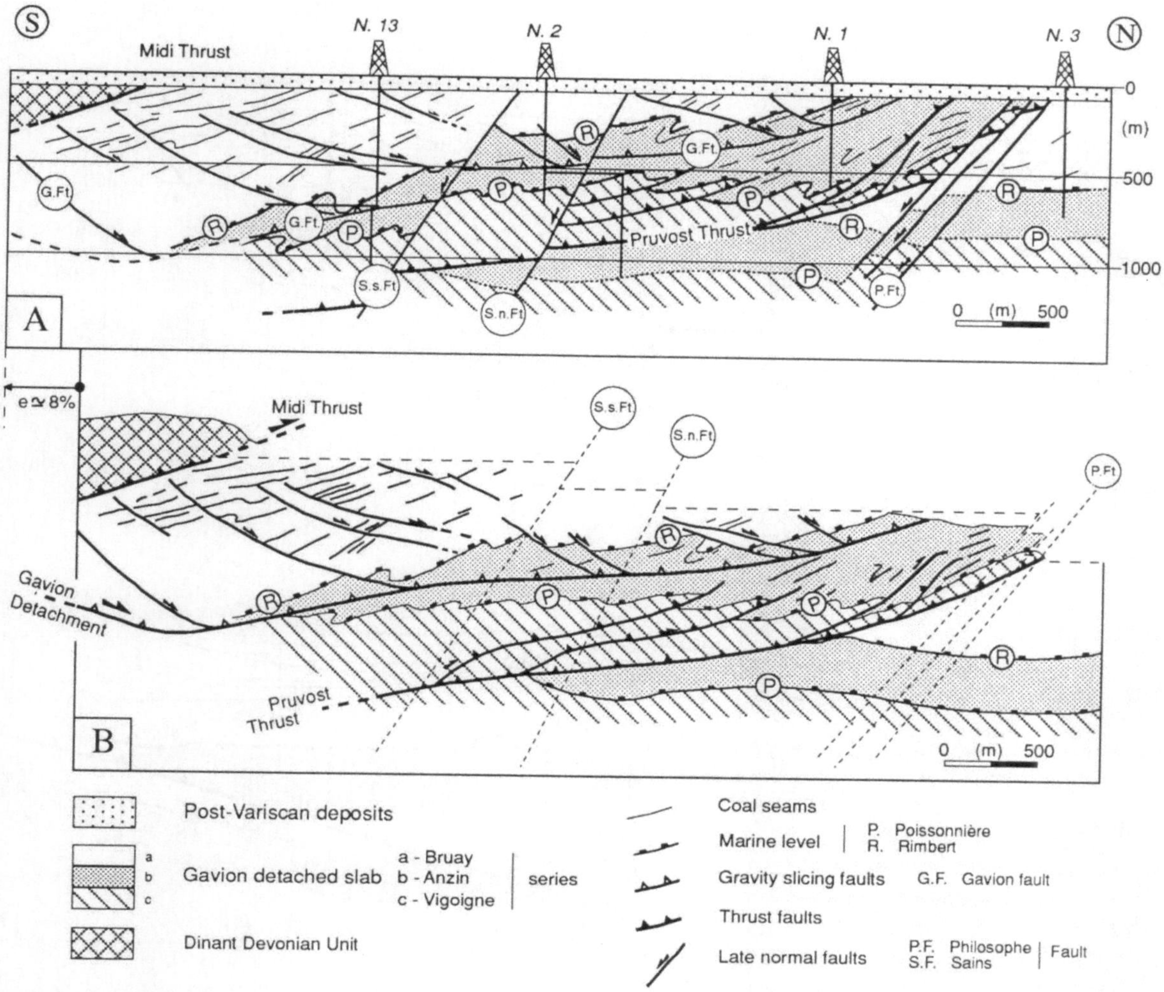

Fig. 7 A, B. Geometry of the Gavion-type low-angle extensional faults along the Noeux cross section (see location in Fig 6) **A** Present-day geometry of the Gavion fold-thrust pattern deduced from mining data (After Bouroz and Stievenard 1958) **B** Restored section obtained after cancelling the offset of (late Variscan?) steep-normal faults Low-angle extensional faults are confined to the hangingwall of the Gavion detachment The total amount of extension is estimated to be about 15 %

distinguish two classes of extensional faults – down dip-slip and oblique structures – that are furthermore separately distributed throughout the NPC coalfield.

Down Dip-Slip Extensional Faults This first class of faults represents the most common extensional features of the NPC coalfield. Whereas some of them occur in the SW termination of the Barrois unit (see the cross section in Fig. 5 B 1), they gain considerably in size farther W, in the inner part of the Marles-Lens syncline, where they mainly develop in a lens-shaped sheet of high Coal Measures sequences (Bruay series) surrounded by a major accident, the Gavion Fault (Fig. 6). As shown in the tectonic map in Fig. 6, the Gavion fault never carries older sequences over younger ones, and that constant footwall/hangingwall relationships definitely contradicts the back-thrust hypothesis of Houchen (1988).

The internal structure of the thus-defined Gavion faulted sheet can be seen along the Noeux cross section (Fig. 7) where coal-bearing sequences, with a general moderate southerly dip, are crosscut by an intense synthetic listric-shaped fault set, dipping to the N. These low-angle normal structures probably sole at depth and branch along the Gavion Fault, which also shows a quite similar asymmetric morphology, and can therefore be considered as a master listric decollement. The direction of displacement along each individual low-angle fault is not well constrained due to the lack of kinematic criteria, but from indirect cartographic and tectonic evidence, it is inferred to occur in a fairly constant N down-dip-slip direction, perpendicular to the regional fold-thrust pattern, and hence parallel to the principal shortening direction of the thrust belt.

The total offset of the hangingwall rocks never exceeds 150–200 m, and it progressively increases downwards towards the Gavion basal detachment. Along the Noeux cross section, the overall bulk extension measured for continuous coal seams is about 15 %, and that requires a total northward translation of about 1 km for the sliced Gavion sheet.

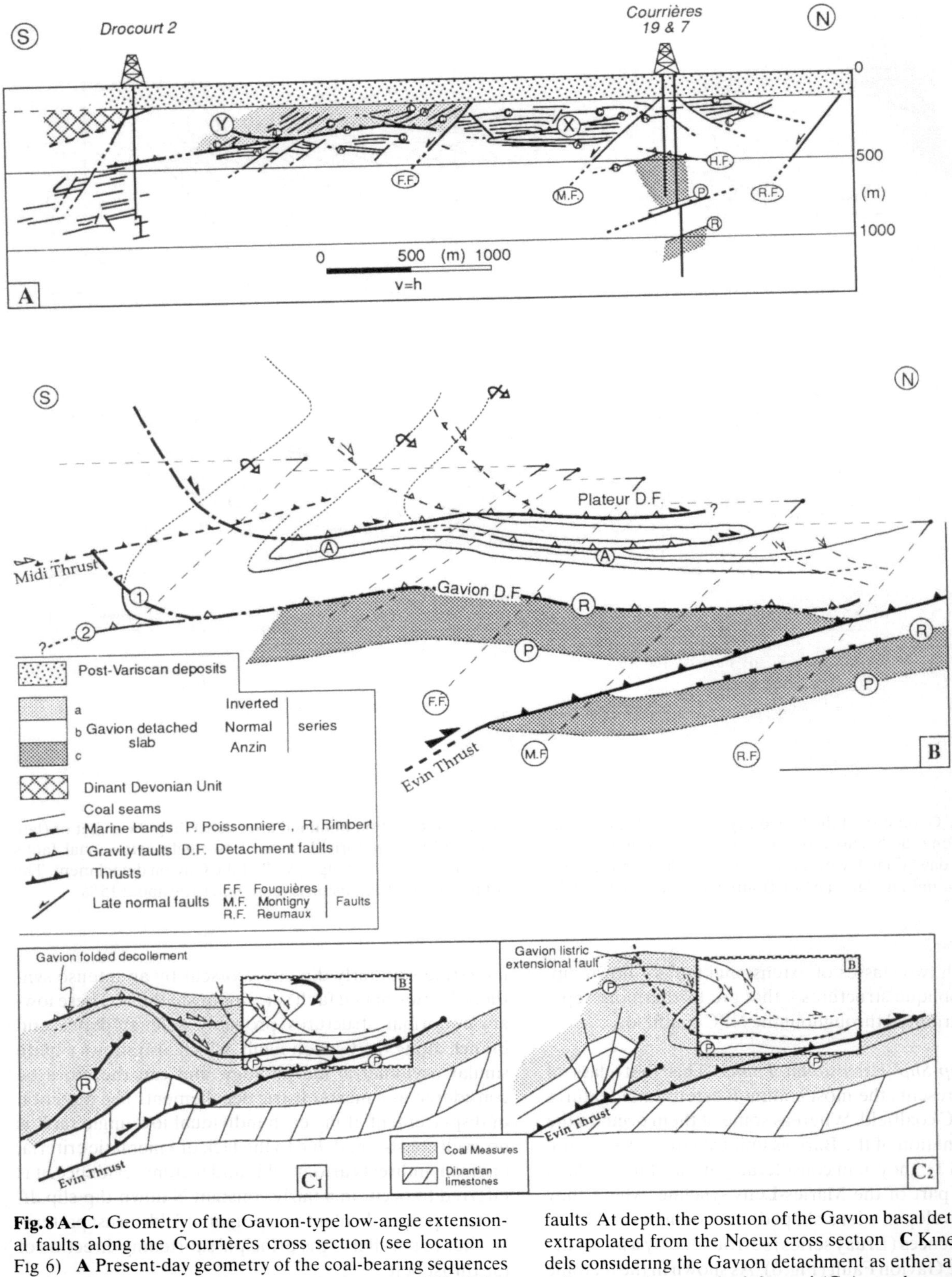

Fig. 8 A–C. Geometry of the Gavion-type low-angle extensional faults along the Courrières cross section (see location in Fig 6) **A** Present-day geometry of the coal-bearing sequences from mining data (After Bouroz and Stievenard 1958) **B** Restored geometry of the large-scale fold-thrust structures obtained after cancelling the offset of (Late Variscan?) normal faults At depth, the position of the Gavion basal detachment is extrapolated from the Noeux cross section **C** Kinematic models considering the Gavion detachment as either an early decollement subsequently deformed (**C** $_1$) or a large-scale listric normal fault (**C** $_2$) In both cases, the Coal Measures high-level structures develop above deeper ramp-related culminations

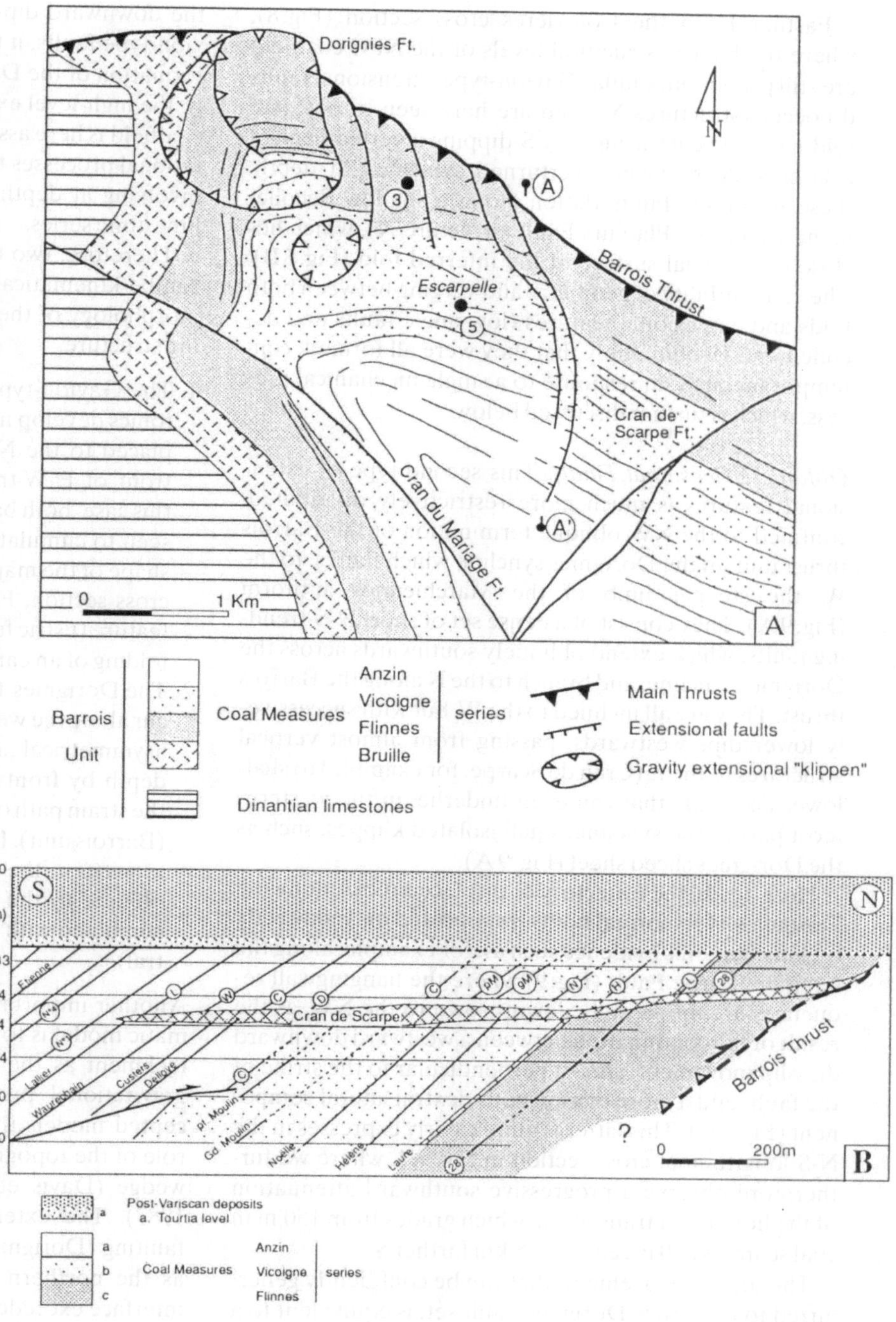

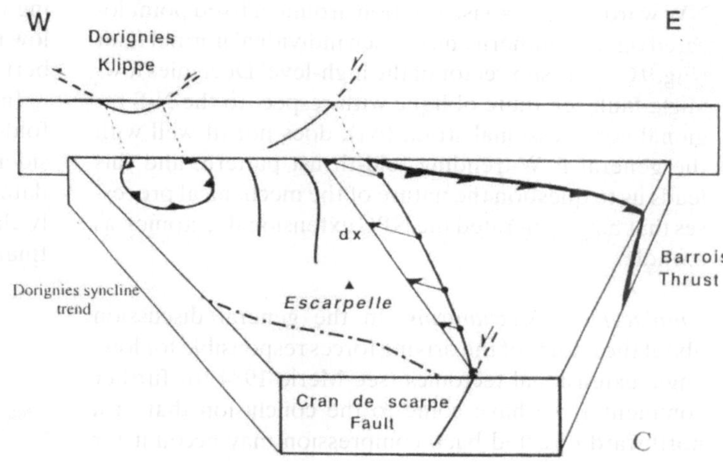

Fig. 9 A–C. Geometry of the Dorignies-type low-angle extensional faults **A** Tectonic sketch map of the western oblique termination of the Barrois thrust unit (After Bouroz et al. 1963) **B** N-S section along the Cran de Scarpe Fault showing a progressive southerly decrease of the lateral offset which passes from about 100 m in coal seam 28 to 0.1 km farther S. in seam 1 **C** Three-dimensional sketch diagram showing the projected trajectory *(dx)* of the displacement along the Cran de Scarpe Fault

Farther E, in the Courrières cross section (Fig. 8), where the highest structural levels of the NPC coalfield are still preserved, similar Gavion-type extensional faults do occur (structures Y), and are here seen to postdate folding and to cut through the S-dipping inverted flank of a large-scale N-facing overturned syncline. At depth, these low-angle faults flatten along a shallow decollement, called the Plateurs Fault (structure X), which lies close to the axial surface of the inferred fold (Fig. 8 B) The compatibility of geometry and verging between both folds and extensional features (low-angle faults and decollements) would imply that they were all formed, contemporaneously, in response to a single mechanical process, which is further discussed below.

Oblique Extensional Faults. This second type of extensional features is much more restrictively distributed, confined to the NW oblique termination of the Barrois thrust unit, in the Dorignies syncline which flanks, to the W, the steeper limb of the Marchiennes antiform (Fig. 9 A). They consist of a dense set of short N-S-trending faults, which extend obliquely southwards across the Dorignies syncline, and branch to the N along the Barrois thrust. They are all inclined to the W, but with successively lower dips westwards, passing from almost vertical structures to the E (Cran de Scarpe, for example) to shallower accidents that can even underlie, in the westernmost part of the syncline, small isolated klippes, such as the Dorignies sliced sheet (Fig 9 A).

Here again, the trajectory of displaced rocks along the Dorignies-type normal faults is supported only indirectly by some tectonic evidence such as, for example, along the Cran de Scarpe Fault (Fig. 9) where the hangingwall sequences are moderately downthrown to the NW, as the result of the combination between a westward downward dip-slip movement, almost perpendicular to the strike of the fault, and a more conspicuous dextral lateral component (Fig. 9 C). This latter is quite clearly expresses in the N-S longitudinal cross section in Fig. 9 B, where we furthermore observe a progressive southward attenuation of the horizontal translation, which grades from 150 m in coal seam 28 to 0 in seam 1, 1.5 km farther S.

The total displacement, that can be confidently generalized to the entire Dorignies fault set, is equivalent to a NW ward anticlockwise rotation around a fixed point located on the southern end of each individual normal fault (Fig. 9 C). The slip vector of the high-level Dorignies low-angle fault set, quite oblique with respect to the N-S regional compressional strain (σ_1), does not fit well with the general E-W-trending fold-thrust pattern, and this leads us to question the nature of the mechanical processes that have generated the NPC extensional tectonics, as a whole.

Emplacement Mechanisms. In the general discussion about the nature of the driving forces responsible for low-angle extensional tectonics (see Merle 1984 for further comments), we have come to the conclusion that, if a northward-directed back compression may account for the downward dip-slip sliding occurring along the Gavion-type faults, it totally fails to explain the oblique deformation of the Dorignies thrust syncline. Consequently, the high-level extensional tectonics of the entire NPC coalfield is here assumed to result from large-scale gravitational processes that took place on top of culminations involving at depth the thick and competent Dinantian limestone series.

Therefore, two types of gravity-related deformations can be kinematically distinguished, as a function of the morphology of the underlying Dinantian-cored disturbing structure.

1. The Gavion-type downward dip-slip extensional tectonics develop in a high-level detached megaslide, displaced to the N above a master decollement, at the front of E-W-trending thrust-related uplifts, and, in this case, both back compression and gravity forces are seen to cumulate their effects. The present-day listric shape of the major Gavion detachment (see the Noeux cross section, Fig. 7) may, either represent a primary feature (listric fault model, $8C_2$), or have resulted in refolding of an early decollement (model $8C_1$)
2. The Dorignies-type oblique extensional structures occur along the western and steeper flank of a preexisting asymmetrical antiform, later slightly duplicated at depth by frontal ramping, and which highly disturbs the strain path of the overlying N-migrated thrust sheet (Barrois unit). In this more complex interfering tectonic pattern, the gravity component is therefore quite oblique to the regional strain, and it is furthermore combined with a significant lateral displacement (dextral).

Another important consequence arising from our kinematic model is to regard the dip of the Gavion basal detachment as the leading factor controlling the overall gravitational process, contrary to conventionally accepted models that, by contrast, rather emphasize the role of the topographic slope presented by the detached wedge (Davis et al. 1983; Lallemand and Le Pichon 1987). The extensional gravitary sliding (Gavion) or faulting (Dorignies) processes may have started as soon as the northern dip of the Dinantian-Coal Measures interface exceeded a certain critical angle which, according to general theoretical modeling, could be relatively low if compensated in turn by high fluid pressures (Hubbert and Rubey 1959).

In order to depict the deep geometry of the surface fold-thrust structures described above, we can now consider industrial seismic documents, as well as drilling data, both of which supply first-hand constraints for finally elaborating the balanced cross section discussed in the final section.

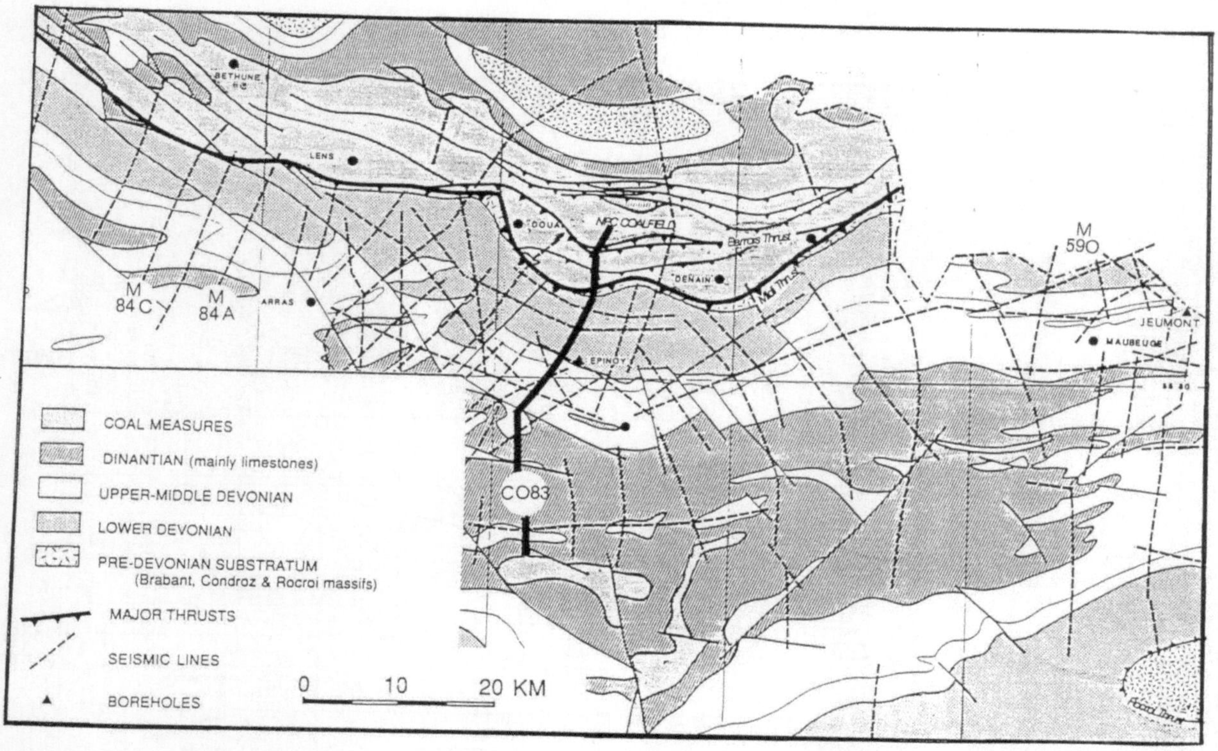

Fig. 10. Tectonic sketch map of the NPC Paleozoic basement roof, with the location of the main deep boreholes and industrial seismic lines

The Deep Structure of the NPC Coalfield

Seismic Data

Over the last three decades, a large amount of shallow seismic has been performed by oil companies (CFP, SNEA, AGIP, BP, TOTAL) throughout the Boulogne-Maubeuge concession (Fig. 10), as part of an extensive oil and gas exploration program mainly focused on the concealed NPC coalfield.

Here, we are concerned only with the northern end of the CO 83 seismic line, which extends over about 30 km through the southern part of the coalfield, S of the Marchiennes antiformal axis (Fig. 10). This section has so far never been fully interpreted in detail, and the time-migrated section in Fig. 11 A, recorded at 5 s two-way-travel time (TWTT), provides an accurate two-dimensional picture of the NPC Paleozoic substratum. Beneath the thin Paris basin deposits, the structure of the Paleozoic rocks is dominated by two strong and continuous reflection bands, which extend over all the length of the line with an apparent shallow southerly dip, and which separate three domains with distinctive seismic signatures.

- The upper reflector dips regularly towards the S at about 15° from shot point 1250 down to the southern end of the line where it lies close to 2.2 S (TWTT), e. g., about 5–6 km on a conventional depth-converted section. From subsurface mining data, furthermore corroborated with the Epinoy deep drilling, it is confidently regarded as the trace of the thrust contact between the NPC domain and the overlying Dinant Devonian thrust unit. Though this large-scale thrust-like structure strictly correlates with the Midi Fault at the (sub)surface, it is here referred to, along its deeper portion, as the Condroz-Midi thrust with regard to more general considerations discussed in the last section. Over the Condroz-Midi thrust, the almost featureless character of the Dinant nappe could be the seismic expression of its dominantly folded structure (domain I in Fig. 11 A).
- The deeper reflector crosses all the profile between 2–3 S (TWTT), and it rises slightly to the N via two minor ramps, called R_1 and R_2 on the interpretative section in Fig. 11 B.

Due to the lack of any direct geologic control, its tectonic meaning is more problematic, but from lateral correlations with similar seismic data recorded farther E, in the Jeumont area (Fig. 10), and calibrated by deep drilling, it could coincide with the base of a Frasnian limestone series, and could therefore be confidently regarded as the basal detachment drawn in Fig. 3. In the general stratigraphy of the NPC-Namur Devonian outer platform (Waterlot et al. 1973), this horizon could lie very close to the top of the underlying Caledonian basement (less than 100 m), so that these two surfaces can reasonably be confused on the interpretative structural section (Fig. 11 B). In the lowest part of the profile, the thus-inferred Brabant Caledonian basement, labelled III in Fig. 11 B, appears as a transparent seismic zone totally devoid of any apparent dipping thrust-like features, and that leads us to consider the present-day southerly dip of the basement

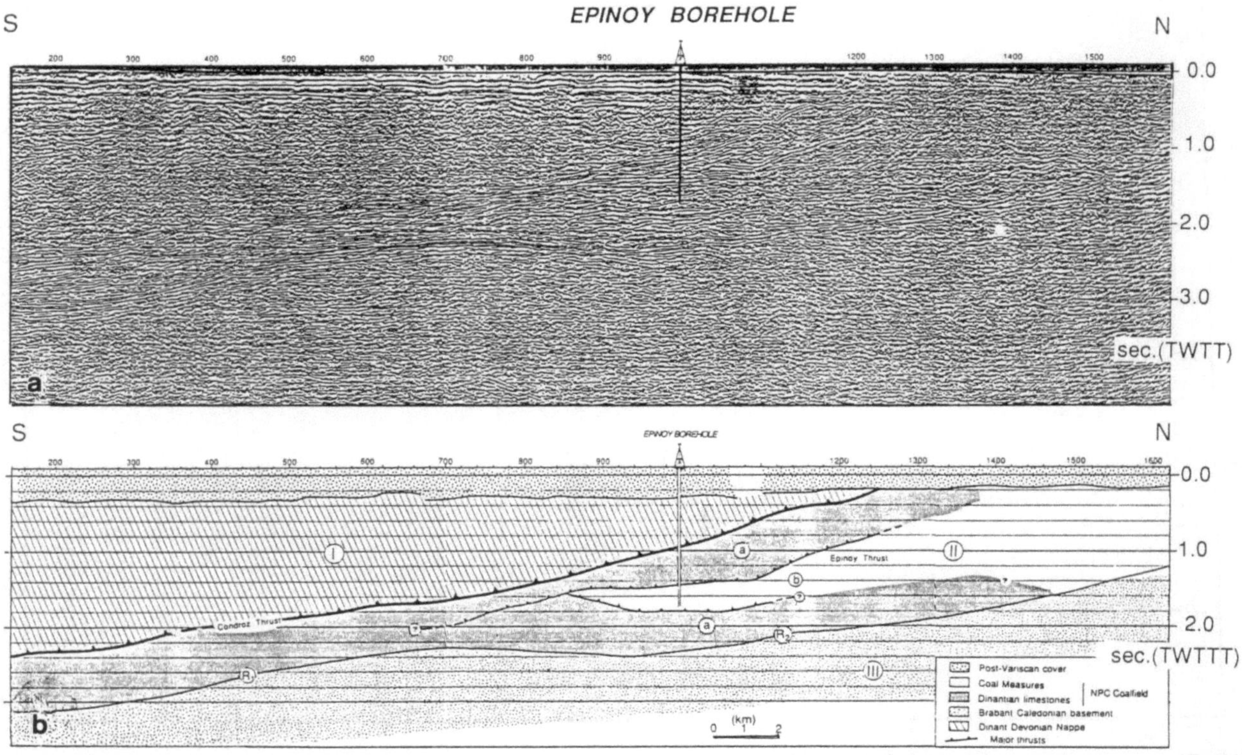

Fig. 11. The CO 83 industrial seismic line (Boulogne-Maubeuge concession) Northern end of the stack section revealing the deep structure of the NPC coalfield beneath the Mesozoic cover Vertical scale (in seconds) is two-way-time (TWT) Interpretative cross section, partially calibrated by the Epinoy borehole We observe the rapid southerly closure of the NPC coalfield which is overstepped by imbricate limestones (Epinoy thrust), and the Dinant Nappe (Midi-Condroz thrust) The *labeled structures* are discussed in the text

roof, either as a primary feature that mimics the topography of the Devonian basin floor along the southern edge of the Brabant emerged landmass, or, more probably, as resulting from some inflexion caused by tectonic loading

- Finally, the NPC coalfield and its Devonian substratum appear on the CO 83 seismic line as a wedge-shaped zone that thicks rapidly southwards beneath the Dinant thrust unit, and whose relatively high reflectivity chiefly results from both flat-lying and moderately S-dipping events, which often display sharp angular truncations similar to that observed in classical ramp-flat thrust systems (Bally et al. 1966; Dahlstrom 1970; for example).

Deciphering the overall internal tectonic organization of the thus-defined NPC thrust wedge then requires the calibration of its constitutive reflective and adjoining transparent zones, and such an attempt can be performed with the help of the Epinoy deep borehole data

Epinoy Borehole Data

The Epinoy deep borehole was drilled through the Dinant nappe, about 10 km S of the Midi fault projected merging line (Fig. 10). The simplified columnar log in Fig. 12 A shows that, beneath the bottom of the Dinant Devonian thrust unit, crossed at a depth of 2100 m, the well penetrates 2830 m NPC-type series comprising from the top downwards (1) Middle-Upper Devonian, (2) Visean limestones with numerous anhydritic interbedded layers, and lastly (3) Namurian Coal Measures ending at 4962 m depth (Bouquillon 1984).

From a seismic point of view, we first observe that the competent Dinantian limestones are expressed on the profile C0 83 by strong well-defined reflection events, whereas the more incompetent Coal Measures are, by contrast, imaged by more transparent zones.

Secondly, extending this rough seismo-stratigraphic calibration to the whole NPC zone leads us to identify a large-scale recumbent fold-like structure, cored with Coal Measures sequences, that can be explained in different ways depending on the interpretations of the Epinoy well.

Stratigraphically speaking, whereas the NPC series in the Epinoy borehole undoubtedly form an overall inverted succession, in turn, far less is known about their internal tectonic arrangement, so that asserting, like Raoult (1988), that they consist of continuous downward facing sequences lying on the southern inverted limb of a large-scale recumbent syncline (Fig. 12 B_1) is considered here as an overinterpretation.

In our opinion, the validity of the recumbent fold concept is furthermore highly weakened and contradicted by some evidence supplied for the most by the Epinoy well:

- a critical review of the above-mentioned micropaleontological data reveals that only eight horizons in the

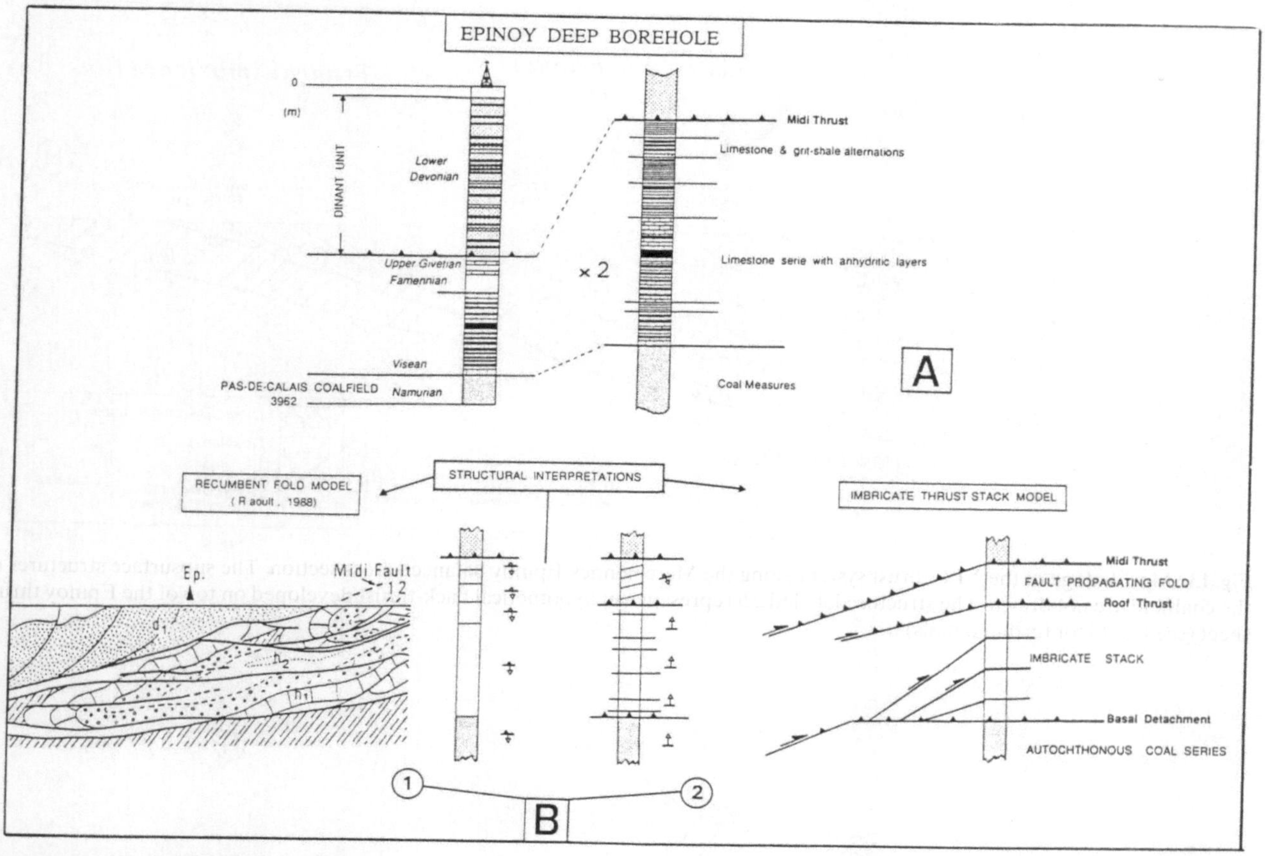

Fig. 12 A, B. The Epinoy deep borehole **A** Simplified columnar section of the entire Devono-Carboniferous succession (After Bouquillon 1984) The right-hand log shows the more detailed lithology of the Midi thrust footwall sequences **B** Structural interpretative models implying, either a large-scale recumbent fold *(1)* (Raoult 1988), or an imbricate thrust stack sequence *(2)*, (this chapter)

2830 m NPC drilled series have yielded unquestionable ages, and these limitations preclude a firm statement for the stratigraphic continuity of the whole succession;

- due to the "destructive" nature of the drilling process, no sedimentological evidence has been preserved, which could only prove the younging direction of the NPC sequences;
- lastly, the resulting interpretative structural cross-section of Raoult (Fig. 12 B_1) does not comply with the conventional rules for section balancing (in Houchen, 1988).

In contrast, we propose an alternative structural interpretation which, in good agreement with the subsurface tectonic pattern, regards the entire Epinoy drilled succession as part of a large-scale imbricate thrust stack involving, for the most part, upward facing sequences. Therefore the NPC "inverted" series could include three superposed thrust sheets comprising, from bottom upwards, (1) Namurian parautochthonous Coal Measures, (2) a thick imbricate fan of Dinantian limestone slices, and (3) a high-level Devonian thrust slab which, according to the above-mentioned micropaleontological data, could constitute the only plausible downward facing series in the Epinoy borehole (Fig. 12 B_2). Our thrust stack model is also corroborated, to a lesser extent, by some regional tectonic considerations which suggest that the NPC thrust system development seems to have been directly controlled by the lithostratigraphy, and in fact, we must note that the intra-Dinantian duplications identified to the W, in the Epinoy "evaporitic" province, have never been evidenced farther E on seismic lines in the Jeumont area (Fig. 10) (Clément 1963) where, concomitantly, the Dinantian carbonates are known to be totally devoid of any anhydritic intercalations (Robineau 1983).

Now, combining both these available geophysical constraints with subsurface geology enables us to elaborate a complete and deep structural cross section of the NPC deformed coalfield that illustrates the final geometry of the deformation along the Marchiennes antiform.

Interpretative Structural Cross Section

Although little surface evidence exists for structural complexity at depth, both borehole and seismic data lead us to consider the overall deformation of the NPC coalfield as resulting from a relatively imbricate thrust-dominated tectonics that we propose to illustrate by a balanced cross section elaborated by applying the general rules emitted by Elliott and Johnson (1980) (Figs. 13, 14).

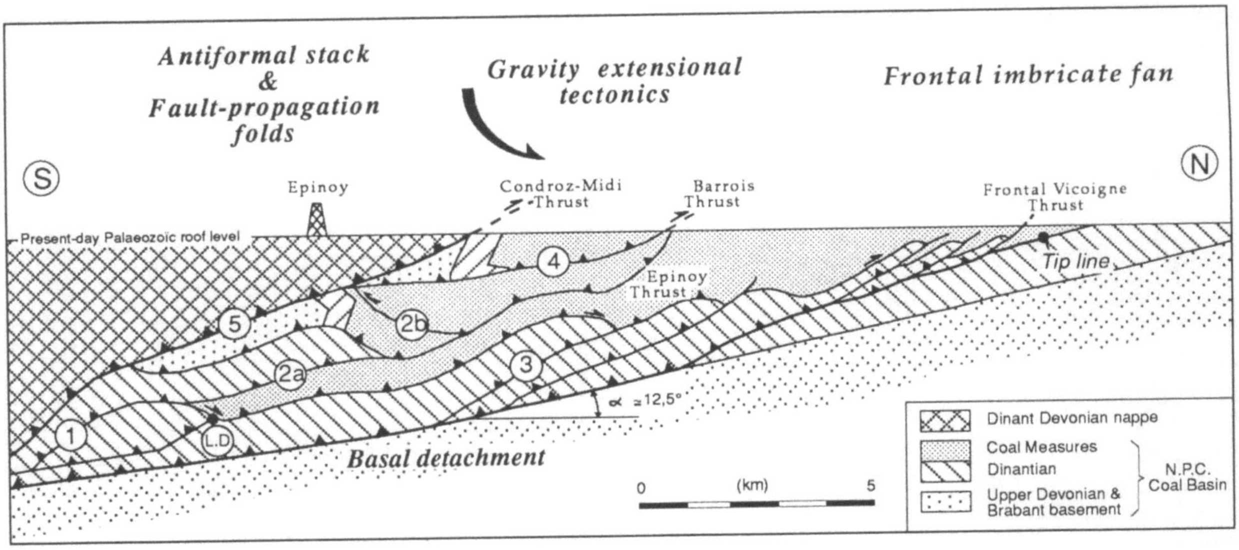

Fig. 13. Morphology of the NPC thrust system along the Marchiennes-Epinoy balanced cross section The subsurface structures of the coalfield are not drawn The structure labeled *2b* represents an hypothetical back-thrust developed on top of the Epinoy thrust sheet (see Fig 14 for further details)

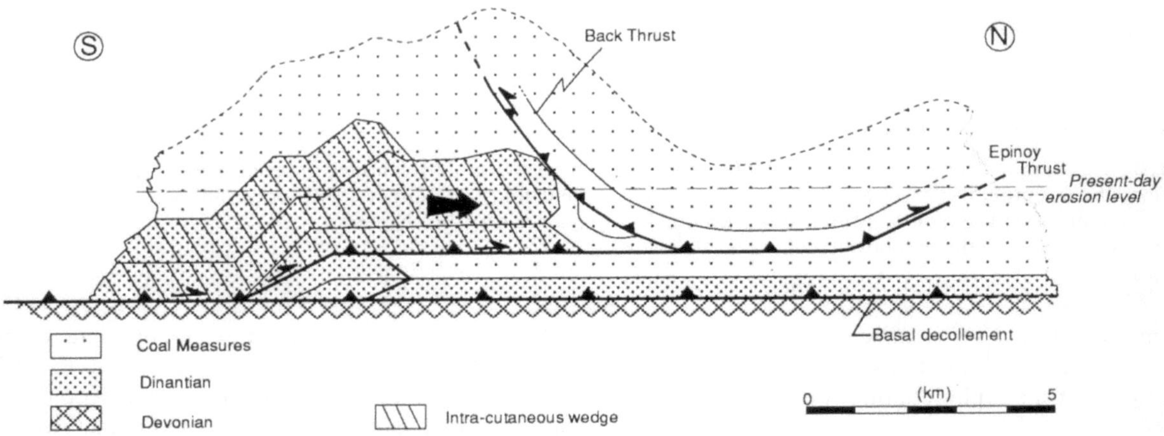

Fig. 14. Sketch section illustrating some hypothetical backthrusting within Coal Measures, on top of a Dinantian-cored intra-cutaneous wedge (intermediate stage of shortening)

The strict involvement of Devonian and Carboniferous strata in the NPC thrust system, without any apparent Pre-Devonian rock slicing, indicates that thrusting took place above a master decoupling level lying, at least along its southern course, close to the Caledonian basement roof surface. To the N, significant ramping of the sole thrust through the Dinantian succession is required in order to join the Dinantian-Coal Measures interface, and then die out just N of the Vicoigne frontal thrust merging line.

The restored stratigraphic thickness of the thrust sequences has been compiled from published data, either supplied by mining word (Coal Measures) and drilling, or extrapolated from the Namur key section (Fig. 1). In a first approximation, a simplified layer-cake template has been applied for the whole succession (Fig. 15 A), despite some significant longitudinal (NS) facies changes that are definitely evidenced as, for example, within the Dinantian evaporitic facies.

Due to the lack of any pervasive deformation within the entire NPC thrust wedge, the proposed section has been only length balanced (isopach layer method) and, as shown in Fig. 13, it displays a well-marked increasing complexity towards the S, while a strong stratigraphic control on deformation style is also likely to have operated.

Before describing these main structural features, it should be stressed that our balanced section is not the only solution and that it has been selected among many others because it implies a minimum bulk shortening rate. Shortening estimates were deduced, for the most, from hangingwall/footwall cutoffs imaged by the strongly reflective Dinantian limestones along specific thrust-like structures of the CO 83 profile. From a mechanical point

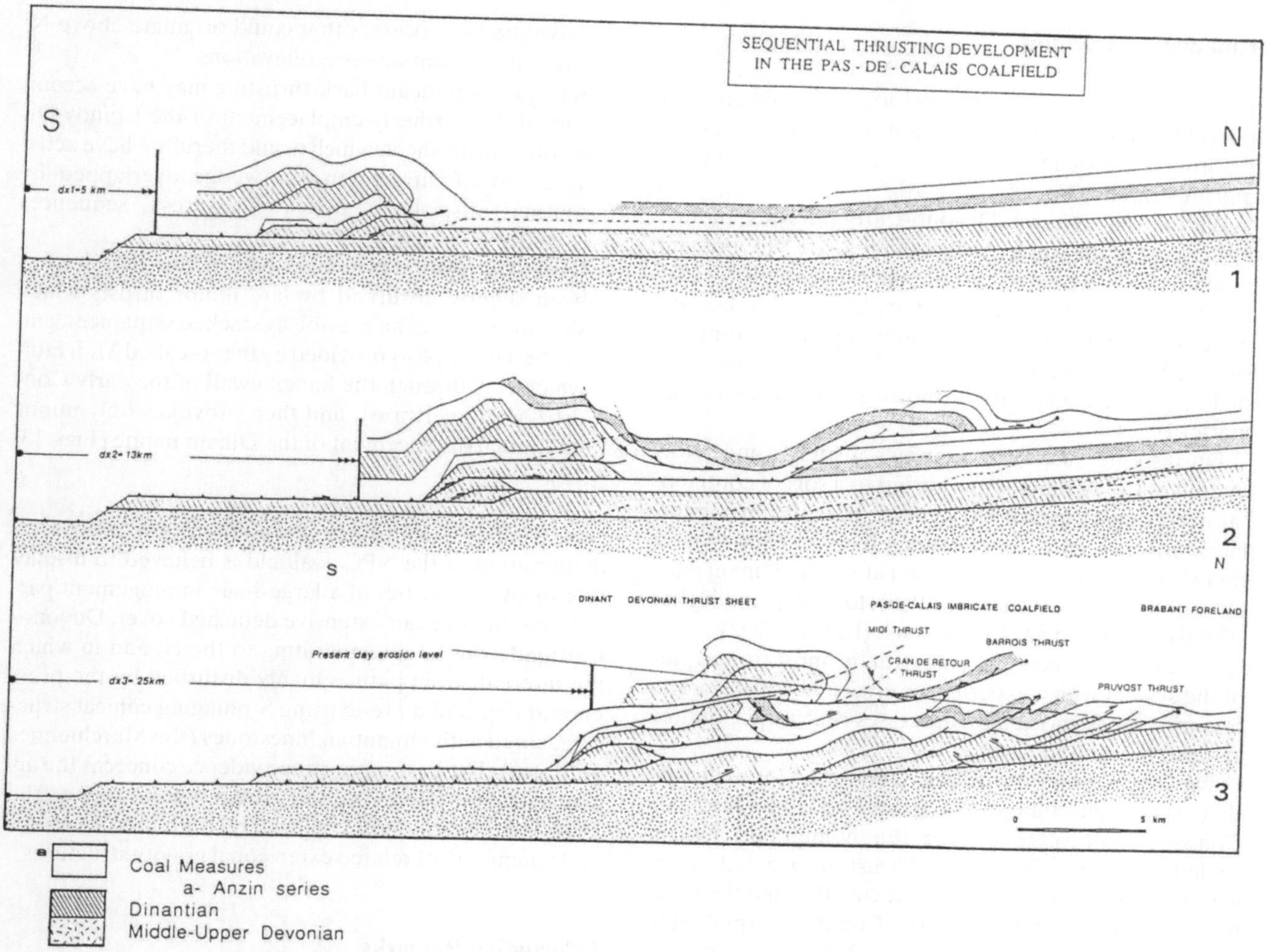

Fig. 15. Three schematic cross sections illustrating the sequential development of the NPC thrust-related structures along the Marchiennes-Epinoy transect, from an early stage *(1)* to the final stack geometry *(3)*

of view, the overall NPC stratigraphy can be roughly subdivided into two superposed sequences with highly contrasting competences, and consisting as follows:

- In mechanically strong Dinantian limestones, shortening was mainly achieved by ramping and bedded-parallel slip, that have probably been facilitated by the occurrence of anhydritic layers, and that lead to a large-scale heterogeneous imbricate stack. This latter comprises, in the S, a sort of hinterland-dipping duplex whose leading edge, labelled L. D. in Fig. 13, should occur close to the supposed northerly depositional pinch-out of the evaporitic-rich facies. Beyond that point, the duplex gives rise to a series of S-dipping splay fault thrusts that cut upsection through the gently S-inclined Dinantian series. To the S, the thick imbricate stack also contains higher thrust sheets that involve mostly Dinantian carbonates, and that were emplaced in the N along the duplex roof thrust before propagating through the Coal Measures via a minor ramp that passes farther N into a 4-km-long decollement. The northern tip line of the so-called Epinoy thrust is no longer visible because it was subsequently overlapped by the Barrois thrust sheet, but it would have been located at least 6–7 km N of the Dinantian cut-off line.
- In the dominantly shale Coal Measures complex, the deformational style changes slightly and shortening here is accommodated, in addition to thrusting, by buckling of the hangingwall-displaced sequences. Some of the resulting folds, regarded as fault-propagation structures with a uniform northerly sense of verging, are sharply truncated, such as the Barrois thrust syncline, by shallow shear thrust planes, and both the structures testify to the multistage emplacement of thrusting.

However, the most prominent structural change is expressed by high level extensional gravity tectonics that affect Upper Coal Measures strata, progressively rotated to steeper dips above deep thrust-related culminations.

In any of the above-mentioned tectonic processes the possible activation of coal seams as easy slip planes is not steadily evidenced by the observed mining data.

Kinematic Evolution

Balancing a section implies restoring concomitantly a plausible undeformed basinal template, via successive deformational stages which are briefly discussed below and illustrated in Fig 15.

Within the general Devono-Carboniferous basinal framework of the whole Ardennes area, the NPC domain is generally accepted as forming an outer shallow E-W-trending platform (whose initial width is still to be estimated) giving way in the S to a more open subsiding basin, the Dinant basin, along a major S-downthrown normal active fault scarp, the Condroz faulted zone (Fourmarier 1954)

During Late Carboniferous times, inversion and deformation of the Ardennes margin led to a sort of continental accretionary prism thickening dramatically hinterlandwards, to the S, and evolving from a shallow outer thin-skinned thrust belt, detached along the Dinant master decollement (Gazes et al. 1985), to a deep and higher strained basement antiformal stack (Le Gall 1992).

Here we are concerned only with the most frontal part of the system which experiences inversion and deformation as soon as shortening processes reach the Condroz extensional faulted zone

The first assumption in our kinematic model is to assert that, as a result of a buttress effect exerted by the steep Condroz fault scarp against the N-migrated Dinant folded nappe, the Dinant sole thrust propagated northwards as a new steeper ramp that cuts through the basement lying in the footwall of the Condroz normal fault, before branching along the shallow NPC basement roof surface. A lens-shaped basement slice was hence accreted onto the hangingwall of the frontal thrust, referred to here as the Midi-Condroz thrust, and later transported to the N with its thick overlying Devonian folded cover along the NPC shallow detachment.

The footwall short-cut evolution applied here to the Condroz normal fault, and commonly evidenced elsewhere by either theorical modeling or field Alpine examplec (Gillcrist et al. 1987), would satisfactorily account for the thin Condroz basement slice that actually occupies the immediate hangingwall of the Condroz-Midi thrust. On a restored section, the Condroz basement footwall ramp would lie about 25 km S of the present-day eroded leading edge of the Dinant unit (Fig. 15, stage 3). Later, as shortening proceeded northwards through the NPC coalfield, deformation was mainly achieved by a general forward-propagating thrust sequence, actively controlled by stratigraphy, and displaying many common features with other orogenic thrust belts. However, in addition to conventional structures, such as the duplex, the imbricate fans or the fault-propagation folds described above, some more specific structures have also been inferred or identified in the NPC thrust stack

- Firstly, the large-scale extensional (sliding or faulting) and contractional (folding) structures involving high level Coal Measures are here thought to result from gravitational processes that could originate above N-migrating Dinantian-cored elevations
- Secondly, significant back-thrusting may have accompanied the northerly emplacement of the Epinoy Dinantian thrust sheet, which would therefore have acted as a sort of intra-cutaneous wedge overlapped by southerly-displaced Coal Measures sequences (Fig 14).
- Lastly, the general piggyback NPC thrust complex has been slightly disturbed by late minor thrusts which sharply truncate the previously stacked sequences, and the best example is provided by the so-called Midi fault which cuts through the hangingwall of the early Condroz-Midi overthrust, and then provokes only minor additional displacement of the Dinant nappe (Figs 13, 15).

Now, from a regional kinematic point of view, the overall deformation of the NPC coalfield is believed to display typical characteristics of a large-scale impingement pattern that involves an extensive detached cover, Devono-Carboniferous in age, migrating to the N, and in which the internal strain path is highly disturbed by the presence at depth of a pre-existing S-plunging conical structure, cored with Dinantian limestones (the Marchiennes antiform) The best supporting evidence concerns the arcuate shape of the entire fold-thrust belt, as well as the large lateral variations displayed by both thrust profiles and kinematics of related extensional gravity structures

Concluding Remarks

A critical review of subsurface (mining work) and geophysical data (drilling, seismic) supplies convincing evidence for us to apply a unified thrust-dominated kinematic model to the entire NPC deformed Carboniferous coalfield.

Deformation of the Devono-Carboniferous detached sequences is related here to a general N-propagating piggyback thrust sequence whose overall morphology is assumed to have been strongly controlled by early basinal features, such as the stratigraphy of the infilling deposits or large-scale irregularities in the morphotectonic pattern. On our interpretative balanced section that crosses through the Marchiennes antiform, the NPC fold-thrust complex appears as a lens-shaped S-thickening wedge, oversteped to the S by the Dinant Devonian unit, and underlain, at a depth of about 6 km, by a 25-km-long gently S-inclined master detachment lying close to the Caledonian basement roof.

In common with many other thrust belts, its internal structure displays a well-marked increasing complexity hinterlandwards, e g., towards the S, and in fact, the rocks occurring in the immediate footwall of the Condroz-Midi thrust are assumed to be involved, at depth, in a much more complex thrust stack than expected from subsurface geology. The inferred thrust stack develops for the most part within the thick evaporite-rich Dinantian limestones series, and it includes a basal hinterland-dipping duplex, in turn largely overlapped by a thin sheet of carbonates (Epinoy thrust slice) that moves 5 km farther N over the Coal Measures. To the N, the thrust pile gives rise to minor imbricate fans, whereas the underlying sole thrust is transferred to higher Dinantian stratigraphic levels via minor ramps, as far N as a blind tip line supposed to occur just N of the Vicoigne frontal thrust.

Many parasitic structures are likely to have accompanied thrust emplacement, and they may account for some specific surface structures of the NPC coalfield.

Thus, large-scale fault-propagation folds may have accommodated shortening within the hangingwall thrust sequences, and their progressive northerly rotation leads to downward facing fold structures such as, for example, those observed in the Epinoy or Barrois units.

The resulting Dinantian-cored culminations are considered to be responsible for setting of large-scale high level gravity tectonics within the overlying Coal Measures complex. These gravitational deformations are exacerbated in the lens-shaped Gavion faulted slice that we regard as a sort of large-scale klippe, transported to the N, e.g., parallel to the principal shortening direction, above a master listric decollement, and where early N-facing recumbent folds are sharply truncated and slightly displaced with a down dip-slip component along an intense set of low-angle listric extensional faults, dipping to the N.

Additional gravity faulting and sliding have been also evidenced in the western part of the Barrois thrust unit, but in this case, they show an oblique extensional movement, with a significant dextral component, which could result from the interference between an early asymmetrical "high" at depth, the Marchiennes antiform, and the N-migrating thrust sheets.

As a working hypothesis, we propose interpreting the poorly understood Roucourt conglomerate as syn-thrust deposits that unconformably overlie the imbricate sequences before being gently folded during a final shortening stage.

The overall bulk shortening required on the Epinoy-Marchiennes balanced section is 40 km, e.g., about 50%, and due to the lack of any pervasive deformation, it is assumed to have been achieved, for the most, by thrusting and ramp-related folding. From this point of view, the shallow Barrois thrust, as well as the inferred Epinoy blind thrust, are regarded here as the most important structures of the entire NPC thrust wedge, implying northerly translations of about 5 and 4 km, respectively.

Lastly, our tectonic model considers the Dinant-NPC thrust contact, classically referred to as the Midi Fault, and in turn termed here the Condroz-Midi thrust, as a relatively complex structure, emplaced during a multistage process. It could therefore represent an early large-scale overthrust, first carrying the Dinant unit over the NPC domain as the result of the Condroz fault scarp rejuvanation, before being passively translated about 25 km northwards and then cut by a late minor thrust merging along the present-day exposed Midi fault. This latter induces only a minor additional displacement (about 4 km) of the Dinant folded thrust unit.

Within the general framework of the N France Variscan belt, the overall structure of the NPC coalfield is assumed to be indicative of a large-scale impingement process involving, to the N, a deep S-plunging taper, cored with thick Dinantian limestones, the Marchiennes high, which would later act, during the late Carboniferous compression, as a buttress against N-migrating high-level thrust sheets. However, further refined kinematic studies would have to be performed in order to firmly support such a model, and the construction of additional cross sections, notably through the adjoining Marles-Lens syncline, would also provide complementary information on the possible control of basement dip or salt thickness, for example, on the NPC deformation style, as a whole.

Acknowledgments The present chapter synthesizes the main tectonic results of a recent study, funded by the E.A.P. oil company (Division Exploration-Production France), and devoted to the structural and paleogeothermal evolution of the Carboniferous coalfields of N France and SE Britain. E. Deville and Y. Philippe are thanked for improvements in the initial manuscript. M.N. Deniel and J.L. Travers are kindly acknowledged for typing and drawing the figures, respectively. The chapter is published by permission of E.A.P.

References

Bally AW, Gordy PH, Stewart GA (1966) Structure, seismic data and orogenic evolution of the Southern Canada Rocky Mountains Bull Can Petrol Geol 14 337–381

Bally A, Bernoulli D, Davis G, Montadert L (1981) Listric normal faults Oceanol Acta, 26th Geol Int Congr Paris 87–101

Barrois C (1905) Observations sur le bassin houiller du Nord de la France C R 6ème Cong Int Mines, Liège, 501 pp

Barrois C, Bertrand P, Pruvost P, Corsin P (1930) Le conglomérat houiller de Roucourt C R 6ème Congr Int Mines, Liège, 147 pp

Becq-Giraudon JF (1983) Synthèse structurale et paléogéographique du bassin houiller du Nord et du Pas-de-Calais Mém Bur Rech Géol Min 123 72

Bouquillon A (1984) Stratigraphie, paléoenvironnement et diagenèse dans le Primaire sédimentaire des forages profonds du Nord de la France Diplôme d'Etudes Approfondies, Université de Lille

Bouroz A (1950) Sur quelques aspects du mécanisme de la déformation tectonique dans le Bassin houiller du Nord de la France Ann Soc Géol Nord 70 1–23

Bouroz A (1969) Le Carbonifère du Nord de la France Ann Soc Géol Nord 89 47–65

Bouroz A, Stievenard M (1958) La structure du gisement des charbons gras du Pas-de-Calais et la notion de Faille Reumaux Ann Soc Géol Nord 77 146–172

Bouroz A, Charlard J, Dalinval A, Stievenard M (1963) Carte des zones stratigraphiques à la cote –300 Inst Geog Nat Paris

Boyer SE, Elliott D (1982) Thrust systems Bull Am Assoc Petrol Geol 66 1196–1230

Cazes M, Torreilles G, Bois C, Damotte B, Galdeano A, Hirn A, Mascle A, Matte P, Pham VN, Raoult JF (1985)

Structure de la croûte hercynienne du Nord de la France premiers résultats du profil ECORS Bull Soc Géol Fr 8 925–941
Clement J (1963) Résultats préliminaires de campagnes géophysiques de reconnaissance dans les permis de recherches "Arras et Avesnes" de l'Association Shell Française – PCRB-SAFREB Objectifs du forage profond Jeumont-Marpent Ann Soc Géol Nord 83 237–241
Cornet F, Briart A (1963) Procès-verbaux des séances mensuelles de la Soc Anc Elèves de l'Ecole Spec de Comm, d'Ind et des Mines de Heinaut 11 9–13
Dahlstrom CDA (1970) Structural geology in the eastern margin of the Canadian Rocky Mountains Bull Can Pet Géol 18, 332–406
Davis D, Suppe J, Dahlen FA (1983) Mechanics of fold-and-thrust belts and accretionary wedges J Geophys Res 88 1153–1172
Deflines (1913) Note sur la constitution de la partie méridionale du Bassin Houiller du Nord dans la région de Valenciennes Ann Mines 14 479
Delmer A (1977) Le bassin du Hainaut et le sondage de St-Ghislain Serv Géol Belg 143
Elliott D (1983) The construction of balanced cross-sections J Struct Géol 5 106
Elliott D, Johnson MRW (1980) Structural evolution in the northern part of the Moine thrust belt, NW Scotland Trans R Soc Edinb 71 69–96
Fourmarier P (1913) Les phénomènes de charriage dans le bassin de Sambre et Meuse et le prolongement du terrain houiller sous la Faille du Midi dans le Hainaut Bull Soc Géol Belg 40 192–235
Fourmarier P (1954) Prodrome d'une description géologique de la Belgique Bull Soc Géol Belg 53 1–123
Gillcrist R, Coward MP, Mugnier JL (1987) Structural inversion and its controls examples from the Alpine Foreland and the French Alps Geodin Acta 1 5–34
Houchen MA (1988) Structural modelling of the external Variscides of France and Belgium Ph D thesis, University of Cork, Ireland (unpublished)
Hubbert MK, Rubey WW (1959) Role of fluid pressure in mechanics of overthrust faulting Geol Soc Am Bull 70 115–166
Lallemand S, Le Pichon X (1987) Coulomb wedge model applied to the subduction of seamounts in the Japan Trench Geology 15 1065–1069
Le Gall B (1992) The deep structure of the Ardennes Variscan thrust belt from structural and ECORS seismic data J Struct Geol 14, 5 531–546
Merle O (1984) Déplacement et déformation des nappes superficielles Rev Geol Dyn Géog Phys 25, 1 3–17
Paproth, Conil R, Bless M, Boonen P, Carpentier N, Coen M (1983) Bio- and lithostratigraphic subdivision of the Dinantian in Belgium, a review Ann Soc Géol Belg 106 185–239
Pirlet H (1972) La "Grande Brèche" viséenne est un olistostrome Son rôle dans la constitution du géosynclinal varisque en Belgique Ann Soc Géol Belg 95 53–134
Pruvost G, Bertrand P (1932) Quelques résultats des récentes explorations géologiques du bassin houiller du Nord de la France RIM 282 365
Puigdefabregas C, Souquet P (1986) Tecto-sedimentary cycles and depositional sequences of the Mesozoic and Tertiary from the Pyrenees Tectonophysics 129, 1–4 173–203
Ramsay JG (1974) Development of chevron folds Geol Soc Am Bull 85 1741–1754
Raoult JF (1988) Le Font Varisque du Nord de la France Interprétation des principales coupes d'après les profils sismiques, la géologie de surface et les sondages In Etude de la croûte terrestre par sismique profonde Profil Nord de la France Technip, Paris, pp 171–196
Raoult JF, Meilliez F (1985) Commentaires sur une coupe structurale de l'Ardenne selon le méridien de Dinant Ann Soc Géol Nord 105 97–109
Robineau JM (1983) Contributions à l'étude sédimentologique des carbonates dinantiens du bord Nord du bassin de Namur, Hainaut, Belgique Thèse 3ème Cycle, Université de Paris
Specht M (1990) Tectonique de chevauchement le long du profil ECORS-PYRENEES un modèle d'évolution de prisme d'accrétion continentale Thèse nouveau régime, Université de Brest, 353 pp
Suppe J, Medvedeff DA (1984) Fault-propagation folding Abs Prog Geol Soc Am 16 670
Waterlot G, Beugnies A, Bintz J (1973) Guides géologiques régionaux Ardenne, Luxembourg Masson, Paris

Polyphase Coalification in French Intermontane Coal Basins of the Late Carboniferous

L. Courel, X. B. Liu, and H. Wang

Abstract. In intermontane coal basins the organic matter maturation rate varies greatly with geodynamic conditions. The Stephanian basin of the Cevennes is a small hyperthermic basin which underwent early major faulting. The Westphalian-Stephanian basin of the Lorraine is larger, with minor early tectonic activity.

In the Cevennes basin, coalification was controlled during the filling stage by basement topographic irregularities and heat-flow distribution in fault zones. The Rouvergue paleohorst acted as a hot zone. In the Lorraine basin, heterogeneity in coal-rank distribution is directly linked to variations in geothermal conductivity, which depends on lithology and mainly on the dip of the strata.

In intermontane coal basins, geothermal gradients vary significantly in space and time. Hyperthermic conditions are characteristic of the filling stage and are chiefly linked to heat-flow circulation controlled by lineaments. In the Lorraine basin, present geothermal gradients are distinctly higher along anticlinal axes than along syncline axes. This characteristic is mainly concerned with coalification of younger seams which underwent major post-folding maturation.

The geothermal history of intermontane coal-basins is polyphase. During successive filling, folding and covering stages, maturation conditions varied considerably, depending on where the coal seams were situated in the basin and also according to their stratigraphic position

Introduction

In recent years, numerous papers have dealt with organic matter maturation conditions and how they relate to basin dynamics. The main controls of basin geothermal history that have been discussed include: rifting (Robert 1985, 1989; Bois et al., this Vol.); magmatic intrusions (Teichmuller et al. 1984; Snyman and Barclay 1989; Schmitz and Wenzlow 1990); and polyphase tectonics (Bustin and England 1989; Bustin and Moffat 1989; Hacquebard and Cameron 1989; Kalkreuth et al. 1989; Middleton and Hunt 1989; Takahashi and Aihara 1989; Courel and Liu 1991; Langenberg and Kalkreuth 1991; Sachsenhofer 1991). Geothermal computed models have been used in order to evaluate the oil potential of basins (Doligez et al. 1987; Blanc et al. 1991; Gaulier and Burrus, this Vol.). Interest in them must not lead to the local geodynamic controls of organic matter maturation in coal basins being overlooked. Heterogeneous distribution in space and time of temperature and heating duration are the most significant processes.

The aim of this chapter is to discuss the rank distribution of two intermontane coal basins of the French Variscides which are different in terms of geodynamic conditions. The Cevennes basin is characterized by marked early and late tectonic activity and a high coalification gradient (Wang 1991). In the Lorraine basin, in contrast, faulting and folding are moderate, as is the coalification rate (Liu 1990).

Rank parameters were collected from these coal basins: C% in the Lorraine basin and Volatile Matter% in the Cévennes basin (Liu 1990; Wang 1991). Data on rank distribution provide important information on the relationships between the paleogeothermal history of the basins and their sedimentary dynamics (Courel and Liu 1991). Commercial measurements of parameters are sometimes questionable; C% in the Lorraine basin and especially Volatile Matter% (measured on a dry ash free basis) do not sufficiently take into account variations in maceral composition (Van Krevelen 1961; Teichmuller 1981). However, within supervised limits, such parameters are useful, because they provide a quantity of data under similar experimental conditions in one basin.

The coalification patterns inferred from the two examples may, moreover, be proposed as models for explaining past conditions of organic matter maturation in the basins. Finally, these results must be taken into account when evaluating the oil and gas potential of coal-bearing strata (Clayton et al. 1991; Fowler et al. 1991; Pickel and Gotz 1991).

Geological Setting

The Stephanian coal basin of the Cevennes, in the SE part of the Massif central (Fig. 1), is a small hyperthermic strike-slip coal basin which underwent early major faulting (Arthaud et al. 1977; Frère 1984; Delénin et al. 1988; Delénin 1990; Wang 1991). The coal-bearing strata are about 3000 m thick and are strongly faulted and locally folded in a basin that is up to 15 km wide and lies between two major bounding faults.

The Westphalian-Stephanian coal basin of the Lorraine (Fig. 1), which is close to the French-German bor-

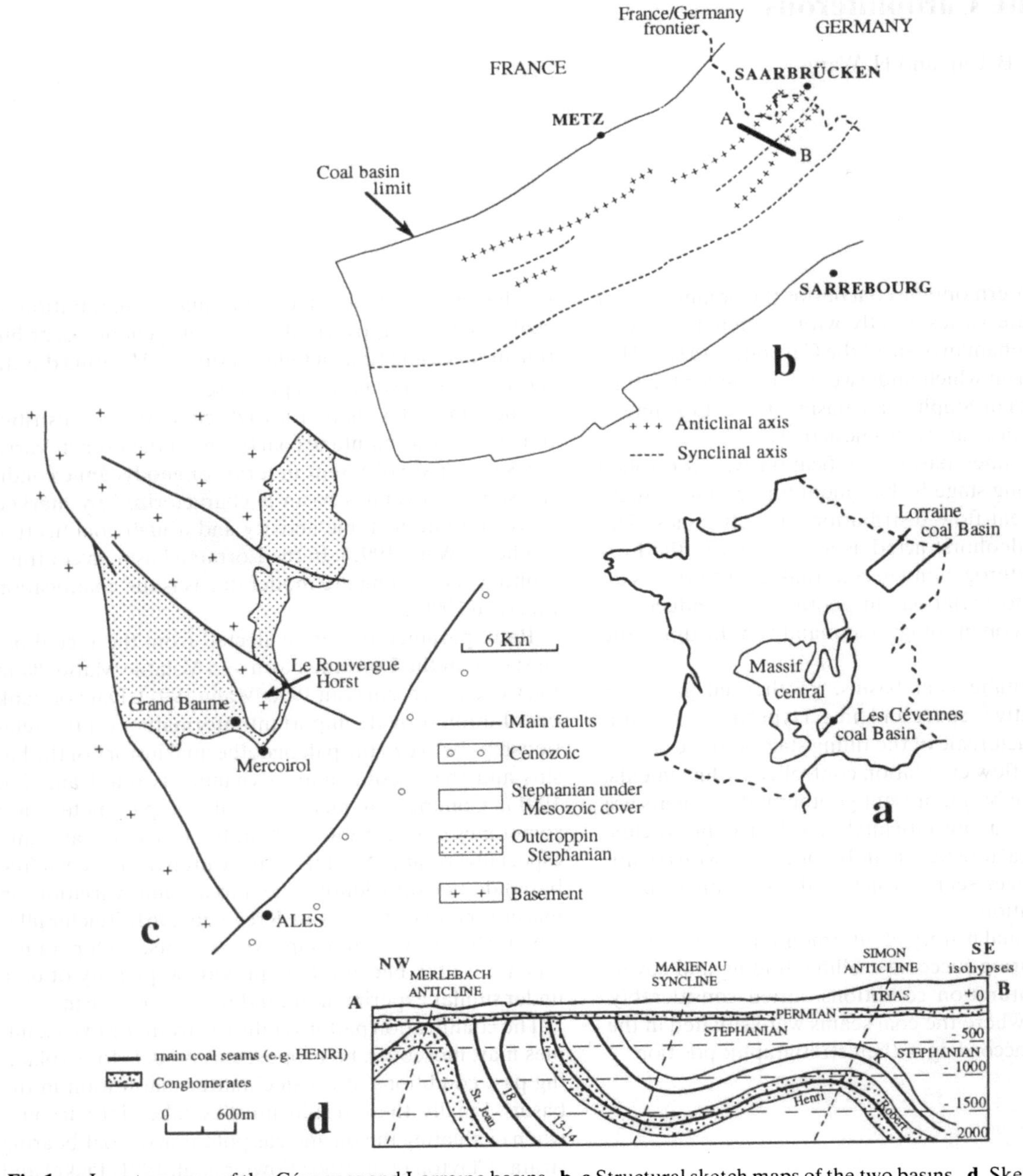

Fig. 1. a Location map of the Cévennes and Lorraine basins **b, c** Structural sketch maps of the two basins **d** Sketch section across the Lorraine basin (*A–B* on **b**) showing main coal seams

der, is larger (140 km long and 80 km wide) with early minor structures and large late Stephanian anticlines and synclines in the NE part of the basin (Donsimoni 1981; Liu 1990; Courel and Liu 1991).

Paleogeothermal Heterogeneities and Their Main Controls

Four main geodynamic controls of paleogeothermal heterogeneities in basins will be discussed from the Cevennes and Lorraine coal basins. a convective thermal regime linked with major faults; a paleohorst acting as hot zone; variations in the thermal conductivity in relation to different lithologies; and the dip of strata.

Fault Control of the Rank Distribution

This is inferred from the "Mercoirol" sketch profile in the central part of the Cévennes basin (Fig 2). Volatile matter isovalue lines show deflexion of curves along the fault planes suggesting higher coalification linked to tectonically controlled water circulation patterns. This example is one of numerous others in coal basins of the French Massif Central where major Variscan faults were active over a long period, inducing local hyperthermic anomalies (Berquer 1986; Robert et al. 1988; Robert 1989).

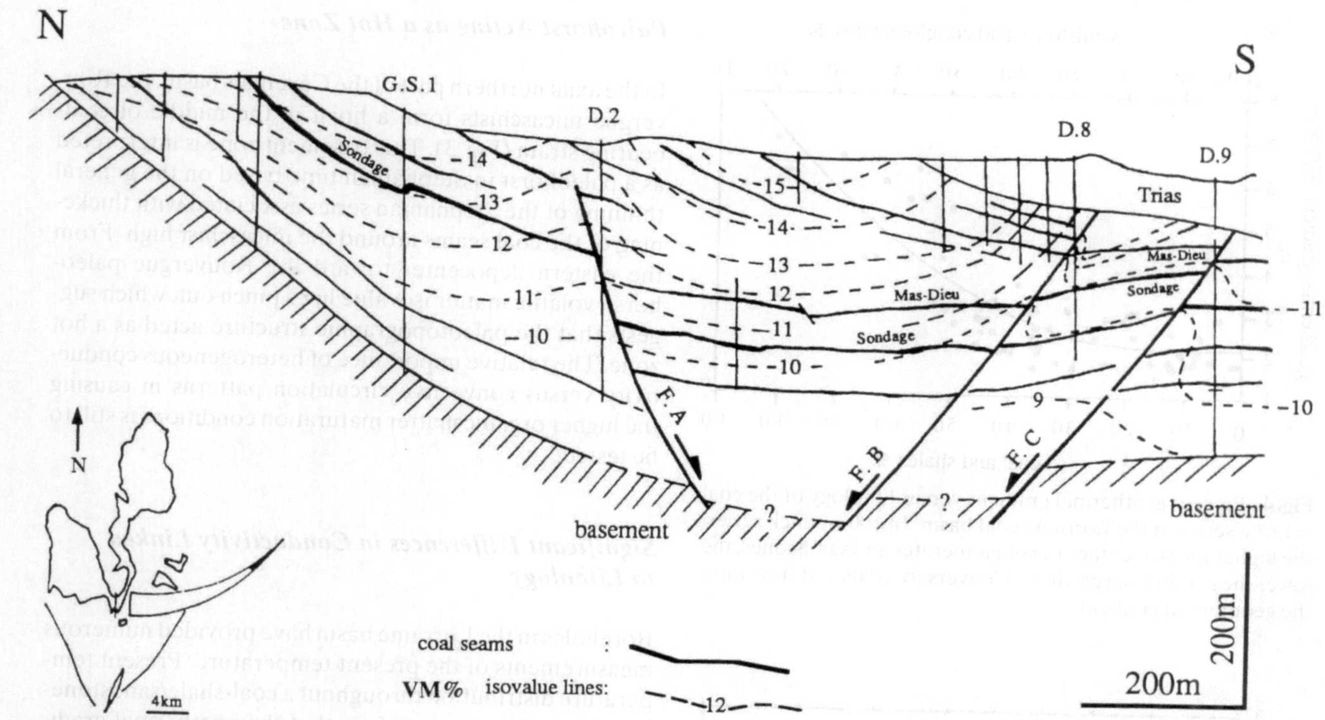

Fig. 2. Fault control of heat flow circulation in the Cévennes coal basin area of the Mercoirol opencast mine *FA, FB, FC* faults, *G S 1, D 2, D 8, D 9* boreholes

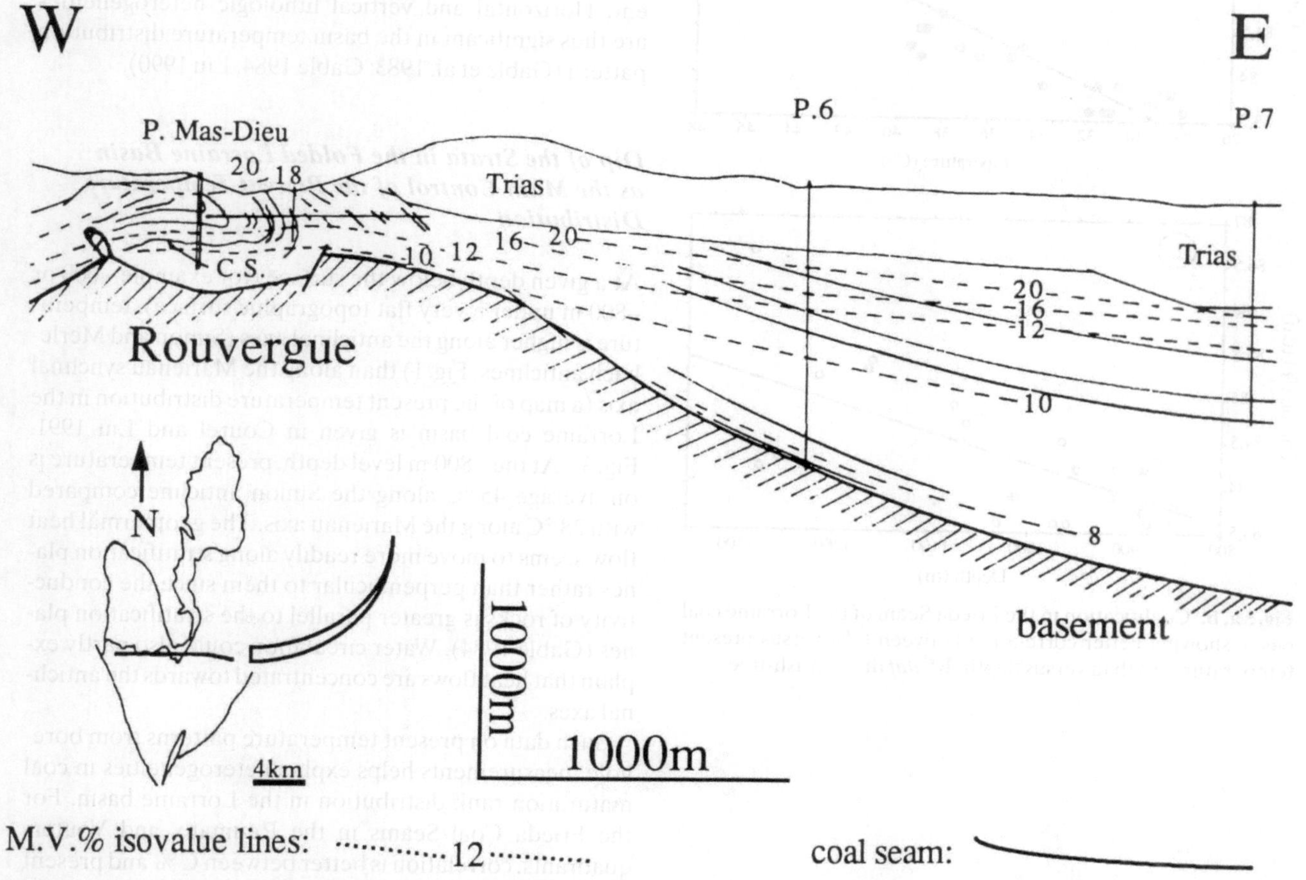

Fig. 3. Le Rouvergue paleohorst acting as hot zone in the Cévennes coal basin. *P Mas-Dieu, P6, P7* boreholes

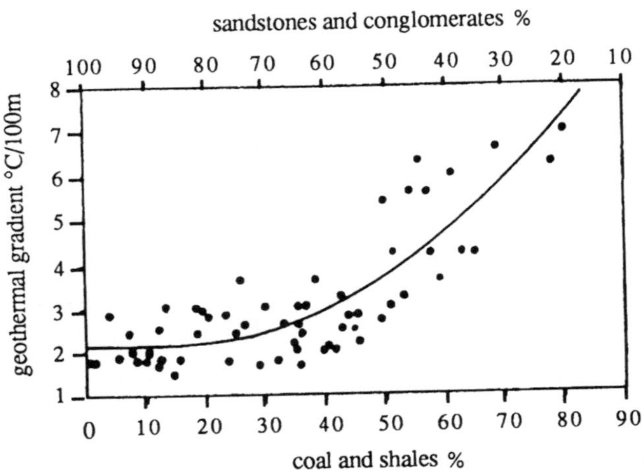

Fig. 4. Present geothermal gradient versus lithology of the coal bearing series in the Lorraine coal basin In 100-m thick series, the higher the percentage of conglomerates and sandstones, the lower the geothermal gradient Conversely, coals and clays raise the geothermal gradient

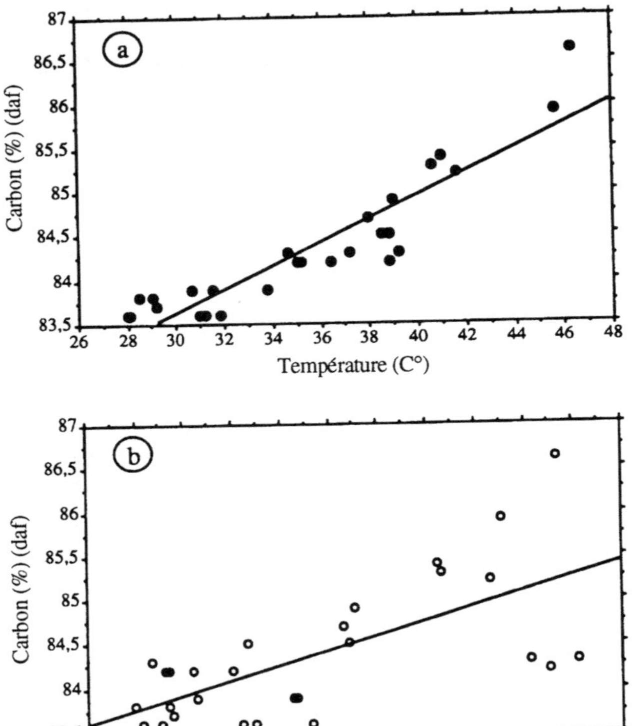

Fig. 5 a, b. Coalification in the Frieda Seam of the Lorraine coal basin, showing better correlation between C % versus present temperature (**a**) than versus depth (**b**), *daf* dry and ash-free

Paleohorst Acting as a Hot Zone

In the axial northern part of the Cévennes basin, the Rouvergue micaschists form a horst in the middle of coal-bearing strata (Fig 3). This basement zone is interpreted as a paleohorst in Stephanian time based on the general thinning of the Stephanian series associated with thickening of the coal seams around the micaschist high From the eastern depocenter toward the Rouvergue paleohorst, volatile matter isovalue lines pinch out, which suggests that the paleotopographic structure acted as a hot zone. The relative importance of heterogeneous conductivity versus convective circulation patterns in causing the higher organic matter maturation conditions is still to be tested.

Significant Differences in Conductivity Linked to Lithology

Boreholes in the Lorraine basin have provided numerous measurements of the present temperature. Present temperature distribution throughout a coal-shale/sandstone-conglomerate series confirm that the geothermal gradient is in inverse ratio to the thermal conductivity. For a 100-m-thick series, the greater the amount of conglomerates and sandstones, the lower the present geothermal gradient will be (Fig. 4). On the other hand, greater amounts of coal and clay increase the geothermal gradient. Horizontal and vertical lithologic heterogeneities are thus significant in the basin temperature distribution pattern (Gable et al. 1983; Gable 1984, Liu 1990).

Dip of the Strata in the Folded Lorraine Basin as the Main Control of the Present Temperature Distribution

At a given depth below the surface (for example −600 or −800 m under a very flat topographic surface), temperature is higher along the anticlinal axes (Simon and Merlebach anticlines, Fig. 1) than along the Marienau synclinal axis (a map of the present temperature distribution in the Lorraine coal basin is given in Courel and Liu 1991, Fig. 3) At the −800 m level depth, present temperature is on average 45 °C along the Simon anticline compared with 28 °C along the Marienau axis. The geothermal heat flow seems to move more readily along stratification planes rather than perpendicular to them since the conductivity of rocks is greater parallel to the stratification planes (Gable 1984). Water circulation could also partly explain that heat flows are concentrated towards the anticlinal axes.

Such data on present temperature patterns from borehole measurements helps explan heterogeneities in coal maturation rank distribution in the Lorraine basin. For the Frieda Coal Seams in the Reumaux and Vouters quadrants, correlation is better between C % and present temperature than between C % and depth (Fig. 5). In this

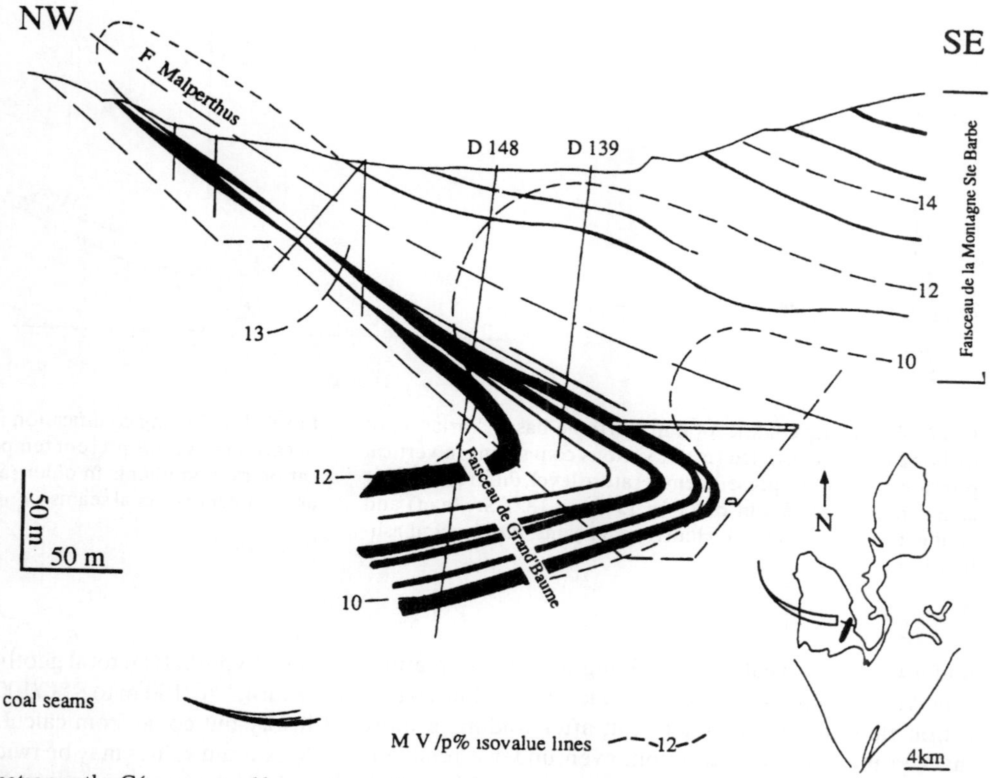

Fig. 6. Early prefolding coalification in the Cévennes coal basin Grand'Baume opencast mine in the Rouvergue zone *F Malperthus* main fault, *D 148, D 139* boreholes

case the paleotemperature distribution pattern was strongly controlled by folding, as is present-day temperature distribution by similar present tectonic conditions. Coalification has occurred in this way since the late Stephanian, mainly under Mesozoic cover. The present temperatur pattern in the Lorraine coal basin acts then as a model for understanding past maturation controls in folded basins.

Timing of the Paleogeothermal Distribution

Where coal seams are deformed, comparison between geometry of rank isosurfaces and coal seam limits enables the identification of pre- and postfolding coalification.

In the Cévennes coal basin evidence of early coalification is inferred in the strongly folded Rouvergue area. In the Grand'Baume opencast mines (Fig. 6) a complex folded structure, running along the whole length of the fold axis, is superimposed on a former early faulted structure. Early faulting is inferred from sharp differences in seam thickness and sandstone characteristics and overall thickness on both sides of the supposed fault. Volatile matter isovalue lines are deformed along the synclinal axis and were deformed at the same time as coal seam folding. Thus in this site, there is evidence of early intensive coalification before folding. Such conditions may commonly be observed in small intermontane Stephanian coal basins (Robert 1985, 1989; Robert et al. 1988).

In the Lorraine coal basins, according to Alpern's (1967) rule, and taking in account the observations of Teichmuller and Teichmuller (1966), the degree of coalification of vertically dipping series serve to distinguish pre- and postfolding coalification. Prefolding coalification may be recognized where, at the same depth or/and the same present temperature level, vertically dipping older coal seams have undergone more intense maturation than younger ones. Postfolding coalification is easily inferred in a vertically dipping coal seam from vertical rank variations versus depth or/and present temperature isolines. In the Lorraine coal basin, the correlation is sometimes better in younger coal seams (Frieda coal seam, Fig. 5), between rank (C %) and present temperature than between rank and present depth. Such data suggest, for younger coal seams, a greater importance of maturation under post-folding conditions.

It is helpful to distinguish pre- and post-folding coalification but more realistic to analyze the superimposition of the maturation stages in one coal seam. In Lorraine coal basins, the coalification pattern differs in older and younger coal seams depending on whether they underwent major pre- or postfolding maturation. A major postfolding stage is inferred from a marked increase in the coalification rate (C %) versus present temperature in younger coal seams (e.g., Henry and K seams, Fig. 7). Conversely, in older coal seams (e.g., Saint-Jean seam), a more weakly increasing coalification rate implies a major prefolding stage. Older coal seams attained a high prefolding coalification rank at an early stage and thus

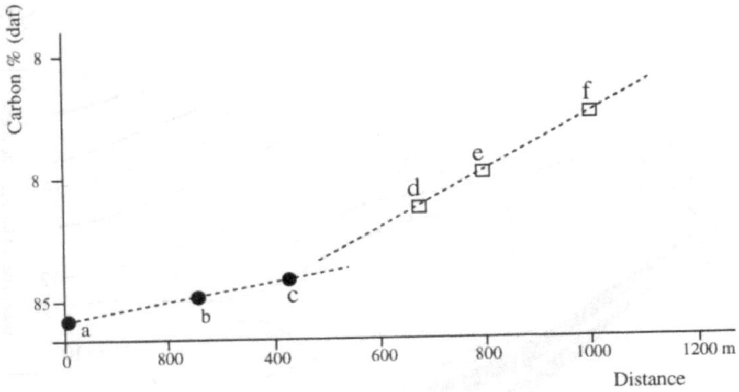

Fig. 7. Prefolding coalification in Lorraine basin Variation of carbon % versus distance (m) of various coal seams in vertical position at the 40 °C present temperature level, different coalification rank gradients in older (*d* 13–14, *e* 18, *f* St Jean) and younger (*a* K, *b* Robert, *c* Henri) coal seams, *daf* dry and ash-free

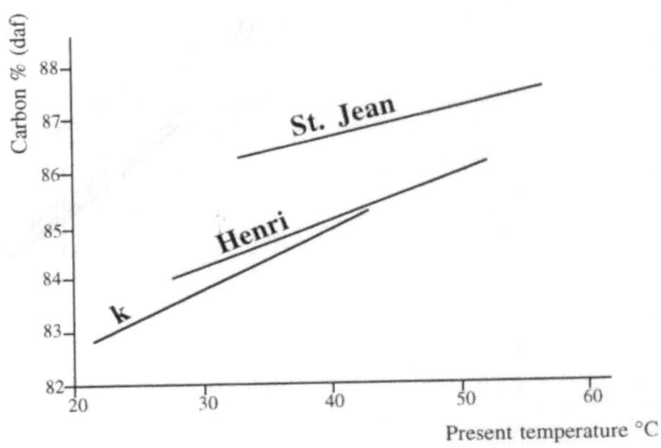

Fig. 8. Postfolding coalification in the Lorraine basin Increase of carbon % versus present temperature in °C, different coalification rank gradients in older *(St Jean)* intermediate *(Henri)*, and younger *(K)* coal seams

underwent only weaker postfolding increase in maturation. Younger coal seams reached a low prefolding coalification rank and later underwent, after folding, a more marked increase in maturation, even under a relatively thin cover. A major prefolding stage is inferred from variations in rank (C %) versus distance of vertical dipping coal seams at the 40 °C present temperature level in the Lorraine basin (Fig. 8). Older coal seams show a strong increase with depth in maturation from one ot the next according to their position in the lower part of the coal-bearing series before folding. Younger coal seams show a weaker increase in maturation from one to the next depending on their position in the upper part of the infilling series.

Polyphase Coalification Linked to Geodynamic Conditions

The Stephanian coal basin of the Cevennes is also a good example of polyphase coalification linked to basin geodynamics. Four sections (Fig. 9) represent the tectonics and associated polyphase coalification in the basin (Wang 1991). During the Stephanian period (1st stage, Fig. 9) the usual pattern of increasing maturation versus depth may be slightly corrected by pinching out of the Volatile Matter % isolines towards the Rouvergue paleohorst which acted as a hot zone. In relation to folding of the Stephanian series before the Mesozoic (2nd stage, Fig. 9) Volatile Matter % isolines became steeper. During the Mesozoic, after an erosion stage, new coalification occurred, particularly affecting younger seams (3rd stage, Fig. 9) Volatile Matter % isolines lie parallel to the base of the Mesozoic. During the Tertiary period, slight deformation affected Volatile Matter % isolines (4th stage, Fig. 9)

Using a Karweil diagram, from C % or Volatile Matter % values and coalification duration it is tentatively possible to infer paleotemperatures and thus geothermal paleogradient values. In the Cévennes basin (Fig. 10) such hypothetical total geothermal paleogradients range from 4.4 °C/100 m to 8 °C/100 m (decimal numbers are illusory but come from calculation of arithmetic means) Maximum values may be twice the minimum ones, which indicates a wide dispersion. Moreover, paleogradient values are very high in the Cévennes basin as well in numerous intermontane Stephanian coal basins of the French Massif Central (Mascle 1989; Robert 1989; Courel and Liu 1991). In the Cévennes basin, postfolding coalification gradient values range from 3 ° 5 to 7 ° 2/100 m. High values and a wide dispersion are still found in spite of the assumed equal thickness of the Mesozoic-Tertiary cover. Such data sugest a heterogeneity of the paleogradients and local hyperthermies. Geodynamics obviously controlled paleogeothermal heterogeneities (see Sect. 3). A comparison between the values of total and postfolding geothermal paleogradients (Fig. 10) suggests that various geodynamic controls affected either pre- or postfolding coalification stages. Inferred heterogeneities in the Mesozoic coalification stage were probably related to the occurrence of numerous local metalliferous concentrations in the Cévennes Mesozoic cover.

In the Lorraine basin, strong variations in the postfolding coalification rate depend on whether Mesozoic maturation affected older or younger seams (Sect. 4 and Fig. 8). The slope of the coalification rate curves is clearly different in both cases. A slow gradual increase in the covering thickness would hardly explain such variation. A short hyperthermic Mesozoic event would acount for the spatio-temporal coalification heterogeneities reported. This acted particularly in the cover and the younger coal seams which did not undergo strong maturation during the prefolding stage. Hyperthermic diagenetic events have recently been analyzed specifically in Mesozoic cover from numerous French basins (Pagel 1992). Such models would fit in with the inferred polyphase coalification process linked to geodynamic conditions in the Lorraine basin.

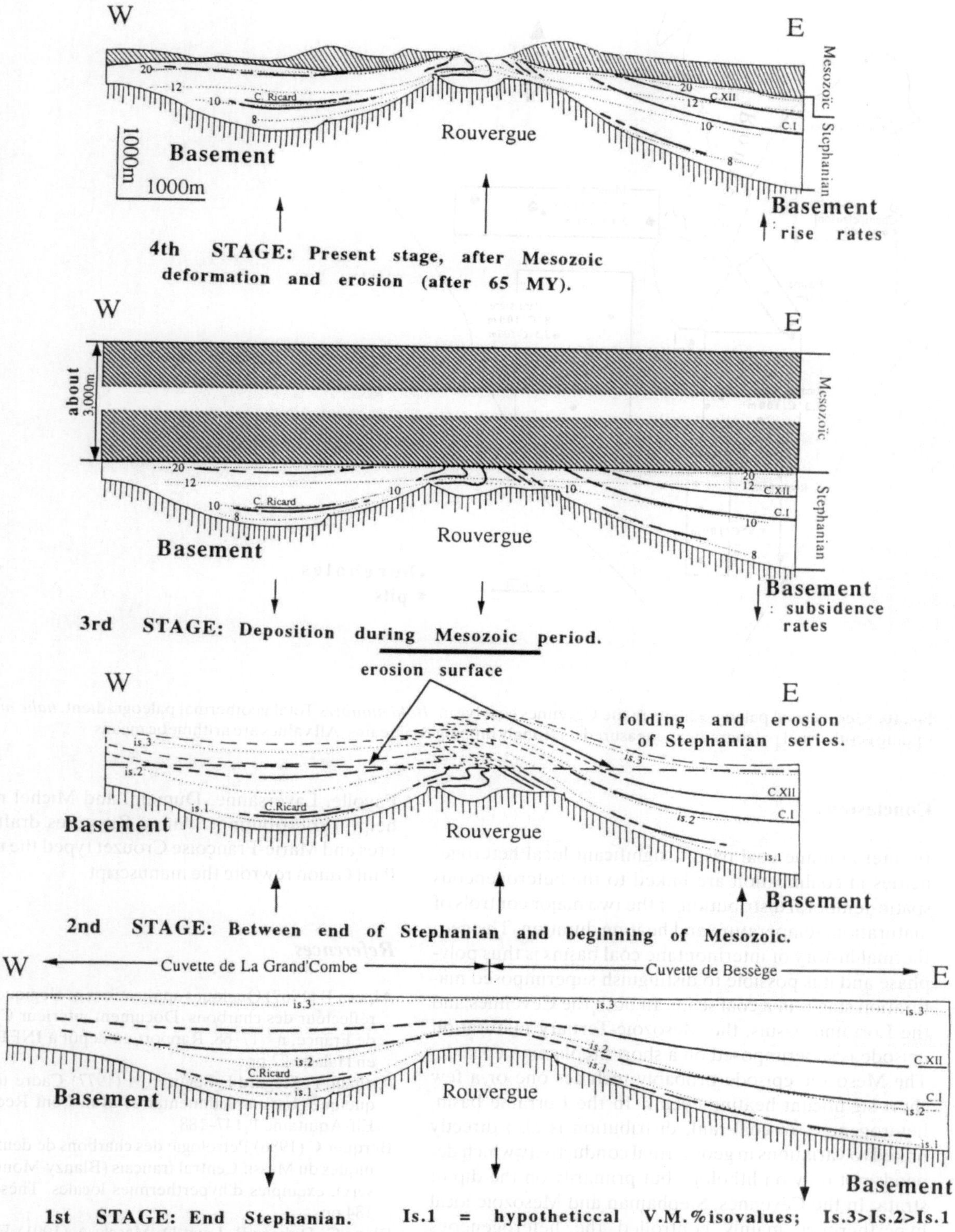

Fig. 9. Polyphase coalification stages in the Cévennes coal basin See text for full explanation

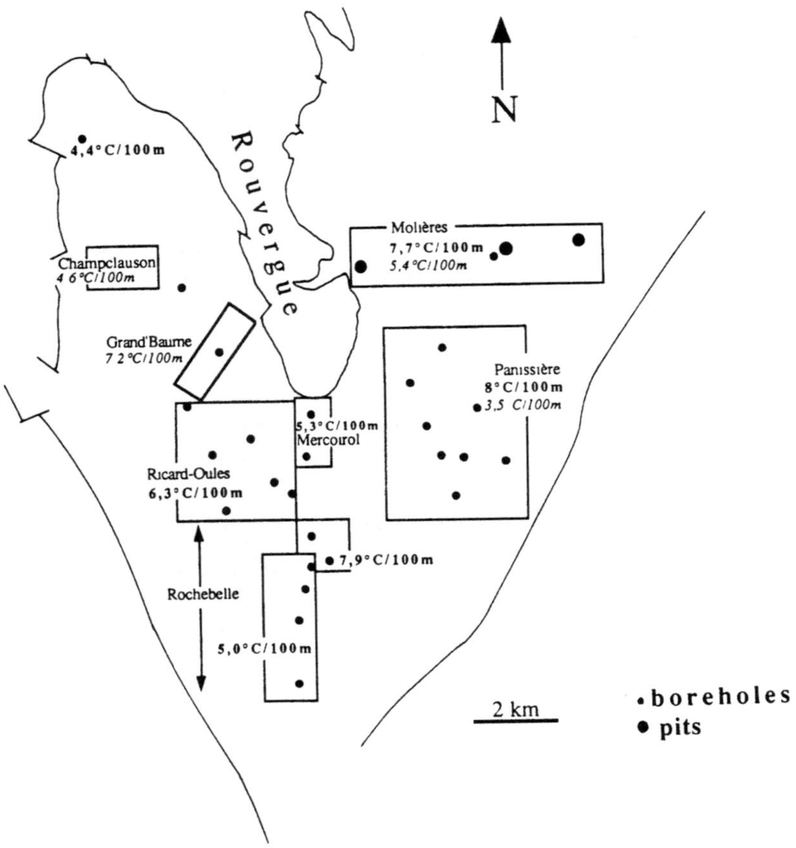

Fig. 10. Geothermal paleogradients in the Cévennes coal basin *Bold numbers* Total geothermal paleogradient, *italic numbers* post-folding geothermal paleogradient, measured on stongly dipping coal seams All values are arithmetic means

Conclusions

In intermontane coal basins, significant local heterogeneities in coalification are linked to the heterogeneous spatio-temporal distribution of the two major controls of maturation: temperature and heating duration. The geothermal history of intermontane coal basins is thus polyphase and it is possible to distinguish superimposed maturation stages in a coal seam. In both the Cévennes and the Lorraine basins, the Mesozoic-Tertiary maturation episode is superimposed on a short Stephanian episode. The Mesozoic episode probably includes one or a few short significant heating events. In the Lorraine basin, heterogeneity in coal-rank distribution is also directly linked to variations in geothermal conductivity which depends not only on lithology but primarily on the dip of strata. In the Cévennes, Stephanian and Mesozoic local hyperthermies mainly controlled the heterogeneous coalification pattern. These data have important implications in reservoir geology and also in the evaluation of the oil/gas potential of coal-bearing series.

Acknowledgments This chapter would not have been possible without the data supplied by coal mining corporations: Charbonnages de France, Houillères de Bassin du Centre Midi, Houillères de Lorraine, Unité d'exploitation du Gard. Messrs Schroeter, Blanchet, Fayolle, Laversanne, Dumain, and Michel made other helpful contributions. Annie Bussières drafted the figures and Marie-Françoise Crouzet typed the manuscript. Paul Guion rewrote the manuscript.

References

Alpern B (1967) Quelques applications géologiques du pouvoir réflecteur des charbons Document intérieur Charbonnages de France, n ° 17–68, Rapport en dépôt à INERIS, Verneuil en Halatte 22 pp

Arthaud F, Mégard F, Séguret M (1977) Cadre tectonique de quelques bassins sédimentaires Bull Cent Rech Expl Prod Elf-Aquitaine 1 147–188

Berquer C (1986) Pétrologie des charbons de deux bassins limniques du Massif Central français (Blanzy-Montceau et Messeix), exemples d'hyperthermies locales Thèse Univ Lille, 184 pp

Blanc G, Doligez B, Lajat D, Mascle A (1991) Evaluation du potentiel pétrolier des formations paléozoiques de la Bresse et de la bordure jurassienne Bull Soc Géol Fr 162 409–422

Bustin RM, England JD (1989) Timing of organic maturation (coalification) relative to thrust faulting in the southeastern Canadian Cordillera Int J Coal Geol 13 327–339

Bustin RM, Moffat I (1989) Semiathracite, anthracite and metaanthracite in the central Canadian Cordillera their geology, characteristics and coalification history Int J Coal Geol 13 303–326

Clayton JL, Rice DD, Michael GE (1991) Oil generating coals of the San Juan basin, New Mexico and Colorado, USA Org Geochem 17, 6 735–742

Courel C, Liu XB (1991) Variations in the geothermal history in coal basins, relationship with basin dynamics. Bull Soc Géol Fr 162, 2 363–370

Delénin B (1990) Dynamique du remplissage du bassin stéphanien des Cévennes Structuration et mise en cause des charriages Thèse Univ Dijon, 217 pp

Délénin P, Clermonté J, Courel L, Dumain M, Laversanne J (1988) Remise en cause des charriages dans le bassin houiller stéphanien des Cévennes (Gard, France) C R Acad Sci Paris 307, 11 1237–1243

Doligez B, Ungerer P, Chenet PY, Burrus J, Bessereau G (1987) Numeral modeling of sedimentation, heat transfer, hydrocarbon formation and fluid migration in the Viking graben, North Sea. In Brooks J, Glennie KW (eds) Petroleum geology of north west Europe Graham & Trotman, London, pp 1039–1048

Donsimoni M (1981) Le bassin houiller lorrain Synthèse géologique Mém Bur Rech Géol Min 17 68

Fowler MG, Gentzis T, Goodarzi F, Foscolos AE (1991) The petroleum potential of some Tertiary lignites from Northern Greece as determined using pyrolysis and organic petrological techniques Org Geochem 6 805–826

Frère I (1984) Le bassin stéphanien des Cévennes (Gard) Dynamique du remplissage – place du charbon – Cinérites Thèse Univ Dijon, 172 pp

Gable R (1984) Température, gradient et flux de chaleur terrestre, mesures et interprétation Thèse Univ Paris VI, 183 pp

Gable R, Disson D, Donsimoni M (1983) Etude thermique du bassin houiller lorrain Approche théorique et expérimentale Convention BRGM (HBL, 15 pp Book in deposit in Bureau Recherches Géologiques et Minières, Orléans)

Hacquebard PA, Cameron AP (1989) Distribution and coalification patterns in Canadian bituminous and anthracite coals Int J Coal Geol 13 207–260

Kalkreuth W, Langenberg W, McMechan M (1989) Regional coalification pattern of Lower Cretaceous coal-bearing strata, Rocky Mountain Foothills and foreland, Canada – implications for future exploration Int J Coal Geol 13 261–302

Langenberg W, Kalkreuth W (1991) Tectonic controls on regional coalification and vitrinite reflectance anisotropy of Lower Cretaceous coals in the Alberta Foothills, Canada Bull Soc Géol Fr 162, 2 375–383

Liu XB (1990) La qualité des charbons dans le bassin houiller de Lorraine Thèse Univ Dijon, 258 pp

Mascle A (1989) Géologie pétrolière des bassins permiens français Comparaison avec les bassins permiens du Nord de l'Europe Chron Rech Min 499 69–96

Middleton MF, Hunt JW (1989) Influence of tectonics on Permian coal-rank patterns in Australia Int J Coal Geol 13 391–411

Pagel M (1992) Paleofluids and thermal history along the Ardèche margin through the Balazuc – 1 drill hole (GPF program) Int Symp Continental Scientific Drilling Programs, Paris, 1992, Abstr

Pickel W, Gotz GKE (1991) Investigations on the petroleum generation potential of bituminous coals from the Saar region Org Geochem 6 695–704

Robert P (1985) Histoire géothermique et diagenèse organique Bull Cent Rech Expl Prod Elf-Aquitaine 8 345

Robert P (1989) The thermal setting of Carboniferous basins in relation to the Variscan orogeny in central and western Europe In Lyons PC, Alpern B (eds) Coal classification, coalification, mineralogy, trace elements, chemistry, and oil and gas potential Int J Coal Geol 13 171–206

Robert P, Berquer C, Courel L, Kubler B, Robert P (1988) Anomalie thermique précoce dans le bassin houiller stéphanien de Messeix-Singles Massif Central français Sci Géol Bull 41, 3–4 333–349

Sachsenhofer RF (1991) Geological setting and coalification of intermontane basins of the eastern Alps (Tertiary, Austria) Bull Soc Géol Fr 162, 2 371–374

Schmitz U, Wenzlow B (1990) Maturity anomalies of the western Lower Saxony Basin in their regional geological context Z Geol Palaontol I, 8 1091–1103

Snyman CP, Barclay J (1989) The coalification of South African coal Int J Coal Geol 13 375–390

Takahashi R, Aihara A (1989) Characteristic nature of sedimentation and coalification in the Tertiary system of the Japanese Islands Int J Coal Geol 13· 437–453

Teichmuller M (1981) Origin of the petrographic constituents of coal In Stach E (ed) Coal Petrology Gebr Borntrager, Berlin, 535 pp

Teichmuller M, Teichmuller R (1966) Die Inkohlung im saarlothringischen Karbon, verglichen mit der im Ruhr-Karbon Z Dtsch Geol Ges 117 243–279

Teichmuller M, Teichmuller R, Bartenstein H (1984) Inkohlung und Erdgas – eine neue Inkohlungskarte der Karbon-Oberflache in Nordwestdeutschland Fortschr Geol Rheinl Westfalen 32 11–34

Van Krevelen DW (1961) Coal typology – chemistry – physics – constitution Elsevier, Amsterdam

Wang H (1991) Dynamique sedimentaire structuration et houillification dans le bassin houiller stéphanien des Cévennes Thèse Univ Dijon, 266 pp

French Overseas Departments and Territories

The French Guyana Margin and the Demerara Plateau: Geological History and Petroleum Plays

S. Gouyet, P. Unternehr, and A. Mascle

Abstract. The French Guyana margin is located between 4 and 6° N on the western side of the Equatorial Atlantic Ocean. The geologic history of the margin can be divided into two main stages. In Jurassic and early Cretaceous times, the northwestern part of the margin (Demerara Rise) is believed to have represented the southern extremity of the Central Atlantic Rift and Ocean. Sedimentation at that time on the Demerara rise occurred mainly in an inner shelf environment, with significant continental influxes. As early as middle Albian times, shallow marine to open marine environments of deposition prevailed in response to the opening of the E-W-trending Equatorial Ocean and margins along a dextral shear zone between South America and Africa.

Exploration on the Demerara Rise was active in the 1960s and 1970s, with only two wells drilled in French waters (Sinna Mary 1 and FG2-1). The first well reached the basement at a depth of 2104 m below early Cretaceous sandstones and younger sediments. The second one bottomed out in basaltic lava flows of Barremian-early Aptian age. A still untested play may be present in the Jurassic (?) section clearly depicted on the seismic below the Early Cretaceous interval. The southeastern side of the French Guyana margin is still very little explored, with no well ever having been drilled. The margin is topped off by a continental shelf about 120 km wide, but the slope is narrow with water depths rapidly deepening to 3000 m. A few seismic lines shot 15 years ago show, below the outer shelf and slope, a 3 s-thick sedimentary wedge of presumed early Cretaceous to Present age. The presence or absence of older Mesozoic troughs below the shelf cannot be inferred owing to the poor quality of seismic data at depth.

Introduction

The French Guyana margin is located on the western side of the Central Atlantic Ocean, between 4° and 10° north latitude and 50° to 56 west longitude, i.e., between the margin of northern Brazil to the east, and the margins of Surinam and Guyana to the west. This margin can be divided into five main physiographic areas, corresponding to different structural units (Fig. 1): The Southern Monocline (1) and the Eastern Margin (2) correspond to the present continental shelf lying under less than 200 m of water. While the Eastern Margin is bounded to the north by a very steep northwest-southeast-trending continental slope leading to the Demerara abyssal plain (3), the Southern Monocline passes gradually into the broad Demerara Plateau (4), a major feature of the western half of French Guyana, with an average water depth of 1200 m The plateau is bounded by steep slopes and escarpments along its eastern and northern edges, while to the west it deepens gradually (to 3500 m) towards the North Coronie zone (5) located in Surinam waters and therefore out of the scope of this study (Figs. 1 and 2).

A first phase of petroleum exploration along the French Guyana-Surinam margin started in 1957 and was active until 1978. The first wells were spudded on onshore Surinam in 1966 and found excellent reservoirs in the Eocene-Oligocene carbonate platform. They were infilled, however, with fresh water (Morange et al. 1992). Exploration then moved offshore, and more than 20000 km of seismic were shot and eight wells (all of them tight) were drilled. Reservoirs were found in early Cretaceous sandstones and source rocks in Albian strata. Sinna Mary 1 (1975) and FG2-1 (1978) were the only two holes located in French waters. The first one encountered the early Cretaceous water bearing sandy reservoirs. The second one bottomed in early Cretaceous basalts, where reef carbonates were expected (Morange et al. 1992).

Since petroleum company proprietary data have never been published, very little was known until now about the geology and remaining petroleum potential of the French Guyana margin. Some information from neighboring countries, however, was published in the 1980s, such as on the Guyana offshore area (Lawrence and Coster 1985), the onshore Guyana area and northern Brazil (Takutu graben, Eiras and Kinoshita 1988), and the Northern Brazil offshore area (Campos 1983; Gouyet 1988).

The geodynamic framework of the French Guyana margin is relevant to the opening of both the Central and South Atlantic Oceans, in a former and later stage, respectively. Numerous fits between South America and Africa have been proposed in past years (on one hand: Bullard et al. 1965; Sibuet and Mascle 1978; Rabinowitz and Labrecque 1979; on the other hand: Curie 1985; Unternehr et al. 1988). The first authors proposed kinematic reconstructions which imply a very large gap (several hundred kilometers) between the Demerara Plateau and its conjugate African margin. The second authors proposed a different reconstruction based on structural studies of the margins, where these two complementary

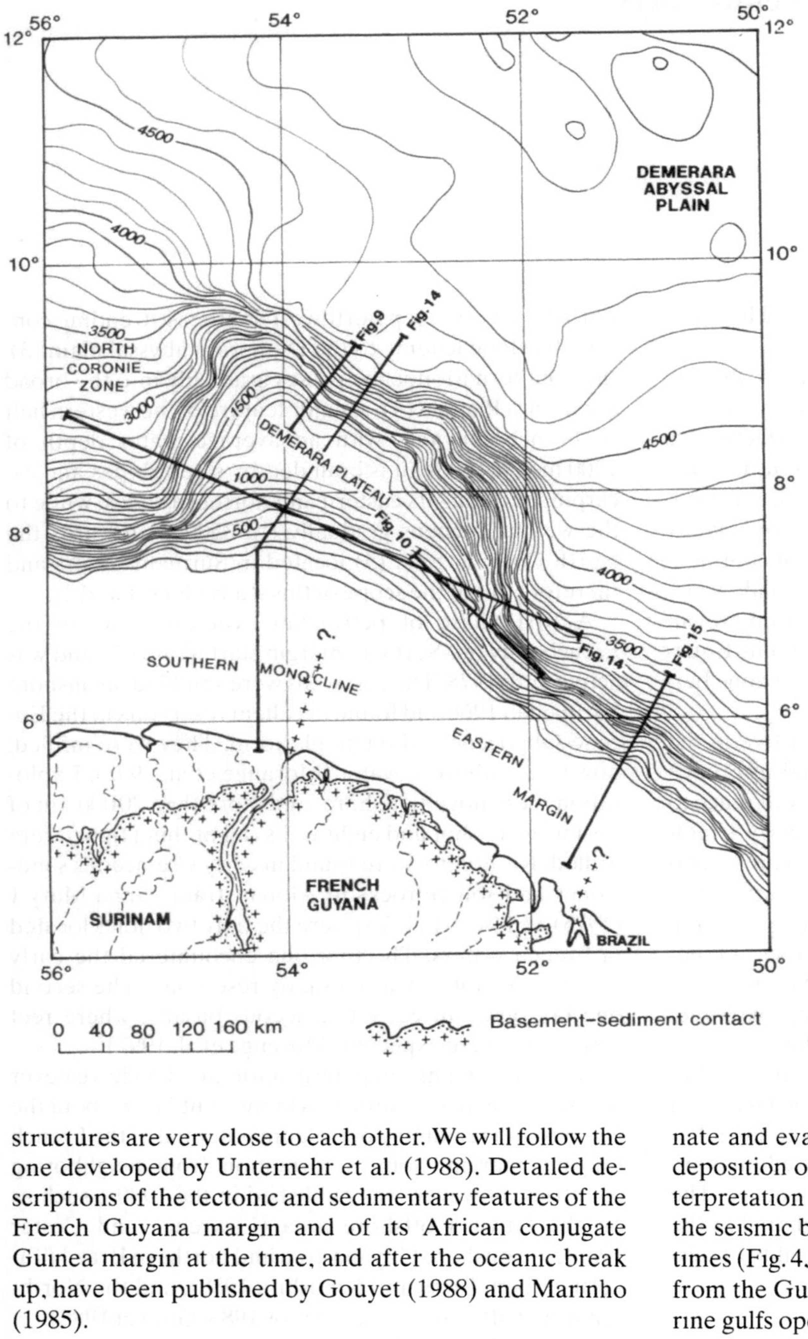

Fig. 1. Main physiographic features of the French Guyana and adjacent Surinam margins with the location of Sinna Mary 1 and FG 2-1 wells Water depths in m

structures are very close to each other. We will follow the one developed by Unternehr et al. (1988). Detailed descriptions of the tectonic and sedimentary features of the French Guyana margin and of its African conjugate Guinea margin at the time, and after the oceanic break up, have been published by Gouyet (1988) and Marinho (1985).

Geodynamic Framework

The geologic history of the margin can be divided into two main stages.

Lias to Aptian Times

The Demerara Plateau together with the Guinea plateau on the conjugate African margin were the southernmost margins of the Central Atlantic Ocean (Fig. 3, p. 414). The mid-Atlantic rift extended progressively between North America and Africa during Jurassic times. Carbonate and evaporitic conditions prevailed, leading to the deposition of some intervals of salt, according to our interpretation of the deepest undrilled sequences seen on the seismic below the Demerara Plateau. In Neocomian times (Fig. 4, p. 415), a local high separated the Demerara from the Guinea plateau, thus isolating two shallow marine gulfs open to the north, in which sedimentation consisted mainly of inner shelf carbonates. Farther east and south, continental clastics from the African-American supercontinent were spread into the newly created depressions. The southernmost graben related to this initial phase of rifting would be the Takutu graben in both Guyana and northern Brazil, where more than 2000 m of Cretaceous strata have been deposited above a thick pile of Jurassic sediments (Fm. Manari, Pirara, and Takutu) and lava flows (Fm. Apoteri; Eiras and Kinoshita 1988). Neocomian times also see the start of the opening of the South Atlantic Ocean in its southern part (Argentina and South Africa margins) with the rift progressively propagating to the north (Herbin et al. 1986).

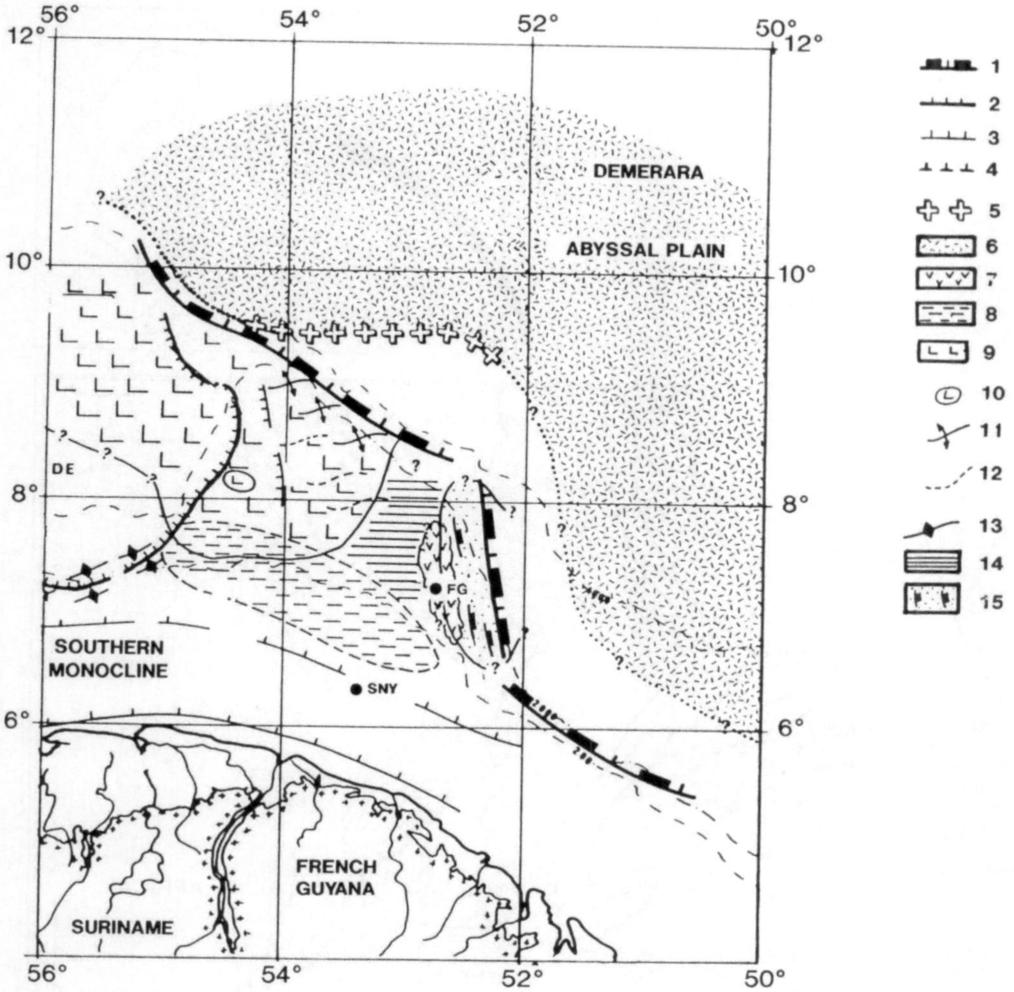

Fig. 2. Simplified structural map of the Demerara Plateau *1* Submarine scarp with faults, *2* submarine scarp, *3* fault, *4* basement downwarping, *5* volcanic ridge, *6* oceanic basement, *7* basalts (125 Ma); *8* sedimentary wedge, *9* inferred distribution of Jurassic salt, *10* diapirs (?), *11* en-echelon folds in pre-Albian strata, *12* gentle folds in pre-Albian strata, *13* wide-low amplitude folds in pre-Albian strata, *14* area of regularly dipping pre-Albian strata, *15* area with tilted blocks in pre-Albian strata

Post-Aptian Times

The breakup between the Demerara and the Guinea Plateau resulted in the final opening of the South Atlantic Ocean. This event was initiated in early Albian times (Fig. 5, p. 415). Oceanic crust appeared to the east of the Demerara Plateau, probably with a large component of dextral shearing since the newly created western Sierra Leone margin was moving eastward with respect to its conjugate South American margin. Shallow marine sediments of mid-Albian age were deposited over the margin in a still restricted environment and lie unconformably above the older series. Deposition of open marine shales followed during the separation of the Demerara and Guinea plateau, which was completed by the end of Cenomanian times (Fig. 6, p. 416).

The Southern Monocline and the Demerara Plateau

Structural Framework

Nine horizons have been identified from seismic data (Fig. 7, p. 416). The seven youngest are tied to wells, but the age of the two oldest is still speculative.

Sediments below the Southern Monocline show regular dips to the north (Fig. 8, p. 417). Two gentle east-west-trending basement flexures have controlled their distribution and facies. The northern one is located at about 7 °N, and corresponds to the southern limit of deposition of pre-Albian strata. The southern flexure is located onshore, close to the present coastline, and has been transgressed to the south by the Paleocene sea only. It thus represents the southern limit of Late Cretaceous deposits.

The structure of the Demerara plateau appears to be more complex. Seismic profiles in the north show several isopach folds trending southwest-northeast as well as positive flower structures (Fig. 9, p. 417). They affect the

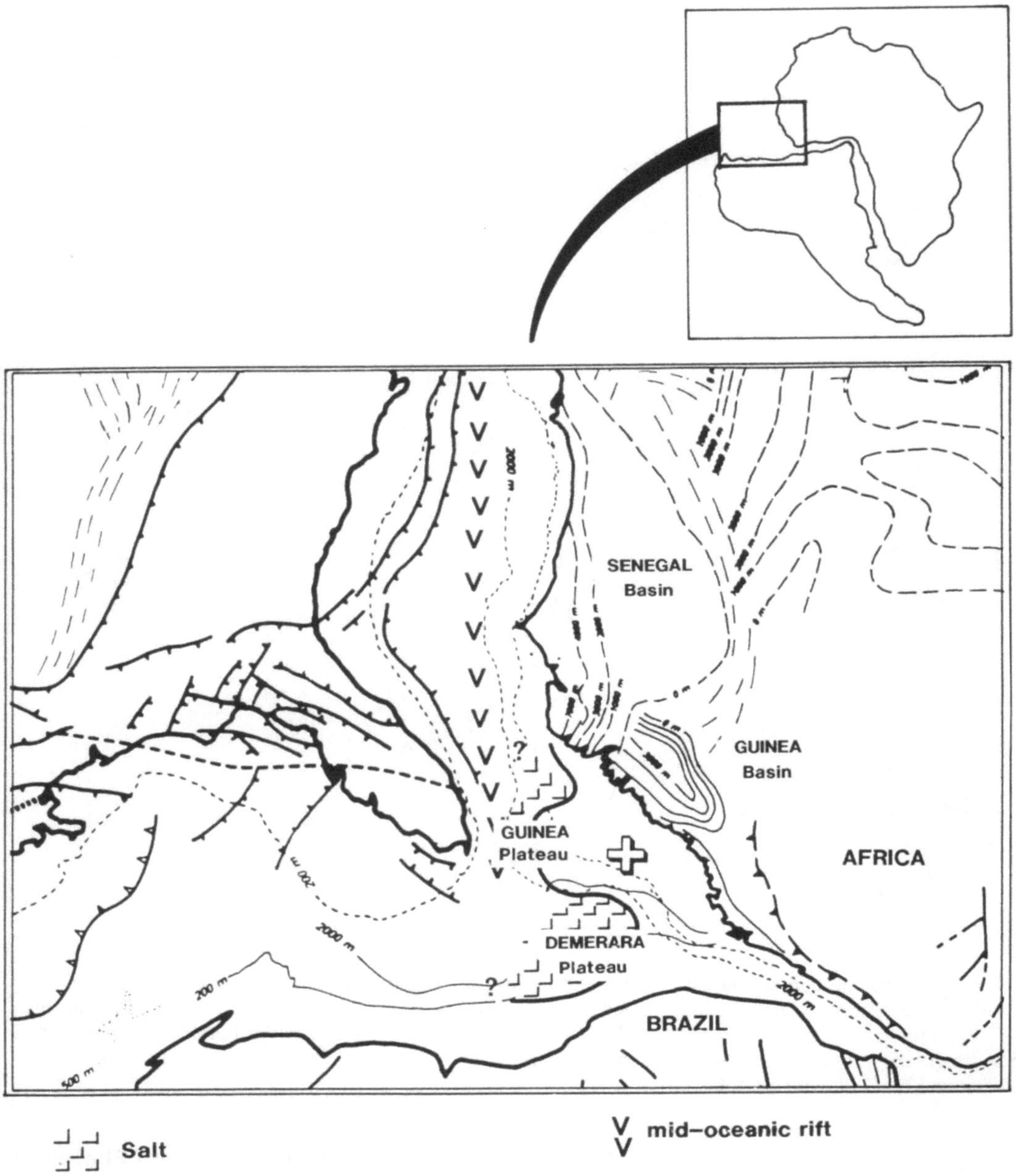

Fig. 3. Reconstruction of the Central Atlantic Ocean in Liassic times

thick Jurassic to early Cretaceous series. A striking angular unconformity (horizon H4) underlines by numerous toplaps below, reveals the erosion of these strata before the deposition of Albian series. In the southwest, the folds give way to gently undulating structures and then to tilted blocks bounded by northwest-southeast-trending normal faults. In the eastern part of the plateau, the seismic shows that the pre-Albian series were eroded by the Albian unconformity, but they have only undergone block tilting (Fig. 10, p. 418). The related normal faults also trend northwest-southeast. The striking structural features of pre-Albian strata below the Demerara Plateau can thus be summarized as follows (Fig 11, p. 418).

- northwest-southeast-trending normal faults,
- northeast-southwest-trending fold axis,
- a grid of extensional faults trending orthogonally to the folds.

The coexistence of extensional and compressional structures of apparently similar age can be interpreted as resulting from the specific ways in which the French Guyana and Sierra Leone margins have rifted apart. Similar structures have actually been described along the South

Fig. 4 ***(above).*** Reconstruction of the Central Atlantic Ocean ▶ in Neocomian times

Fig. 5 ***(below).*** Reconstruction of the Central Atlantic Ocean in Albian times *1* Relative motion of South America with respect to Africa, *2* relative motions along the shear zones, *3* main shear zones, *4* en echelon folds

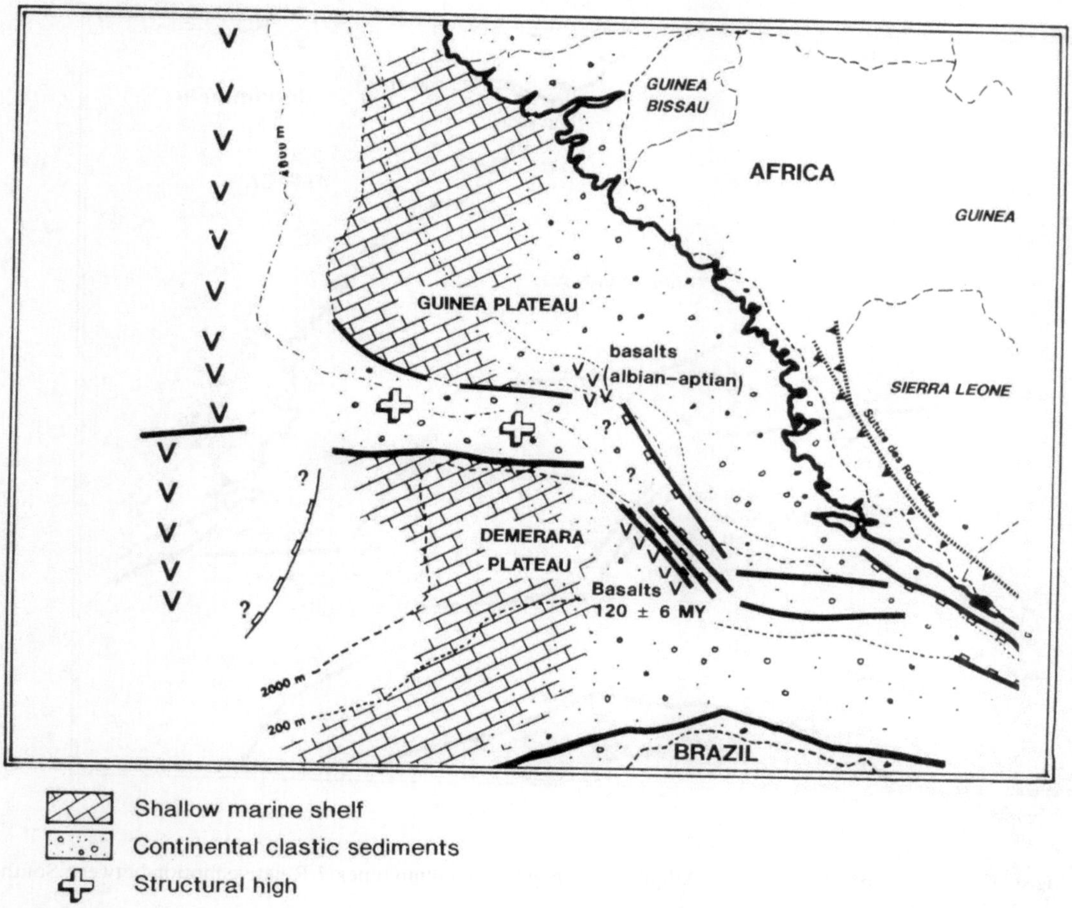
GUINEA BISSAU
AFRICA
GUINEA
GUINEA PLATEAU
basalts (albian–aptian)
SIERRA LEONE
Suture des Rockelides
DEMERARA PLATEAU
Basalts 120 ± 6 MY
4000 m
2000 m
200 m
BRAZIL
Shallow marine shelf
Continental clastic sediments
Structural high

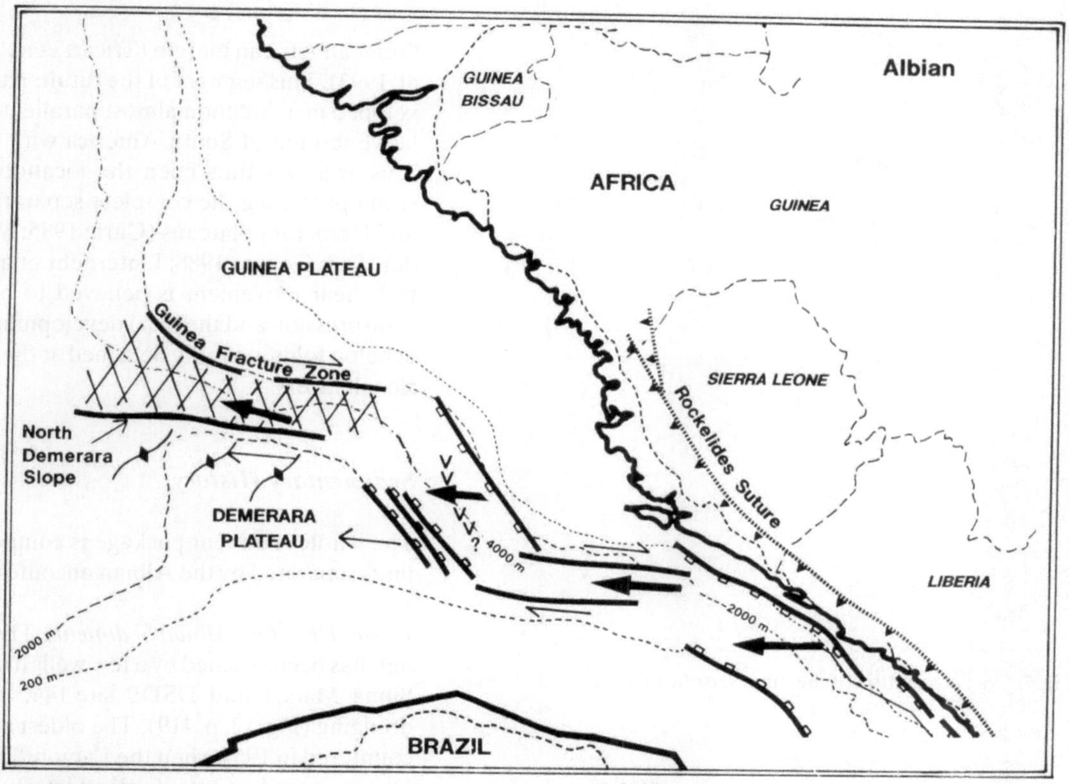
Albian
GUINEA BISSAU
AFRICA
GUINEA
GUINEA PLATEAU
Guinea Fracture Zone
SIERRA LEONE
Rockelides Suture
North Demerara Slope
DEMERARA PLATEAU
4000 m
LIBERIA
2000 m
200 m
BRAZIL
1
2
3
4

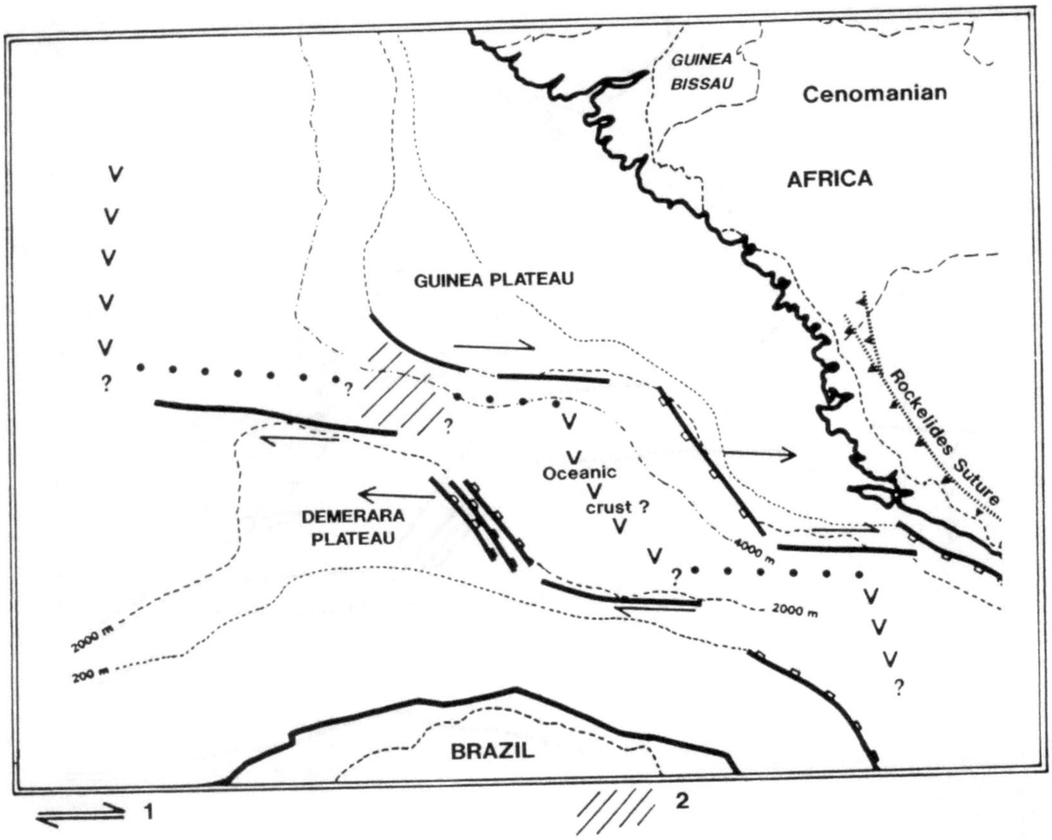

Fig. 6. Reconstruction of the Central Atlantic Ocean in Cenomanian times *1* Relative motion between South America and Africa. *2* remnant active shear zones

HORIZON	STRATIGRAPHY	AGE
H1	BASE LATE MIOCENE	10.3 MA
H2	BASE EOCENE	51 MA
H3	BASE CENOMANIAN	95 MA
H4	BASE ALBIAN	107 MA
HB	TOP HAUTERIVIAN BASALTS	120 ± 6 MA
H5	INTRA-HAUTERIVIAN	116 MA
H6	INTRA-VALANGINIAN	125 MA
H7	BASE TOARCIAN ?	195 MA ?
BASEMENT	PRECAMBRIAN	

Fig. 7. Age attribution of the main seismic markers

Guinean African margin (Tricart et al. 1991; Moullade et al. 1993). This segment of the future Atlantic margins developed in a direction almost parallel to the vector of relative motion of South America with respect to Africa. This area has thus been the location of dextral shear strain preceding the complete separation of the Guinea and Demerara plateaus (Curie 1985; Marinho 1985; Pindel 1985; Gouyet 1988; Unternehr et al. 1988). This dextral shear movement is believed to be at the origin of transpression and the local development of a trend of en-echelon folds, possibly detached at the base above Jurassic salt layers.

Sedimentary History

The whole sediment package is composed of two major units separated by the Albian unconformity.

Lower Unit Pre-Albian Sediments The upper part of this unit has been reached by a few wells in Surinam waters, at Sinna Mary 1 and DSDP site 144, and at the Kane 9 dredging (Fig. 12, p. 419). The oldest sediments were encountered in 1964 when the Lamont Doherty Geological Observatory Institute dredged late Jurassic shallow marine sandstones (?) from 4400 m of water depth at the foot of the northern Demerara escarpment (Kane 9 cruise). DSDP site 144 was located on the northern edge

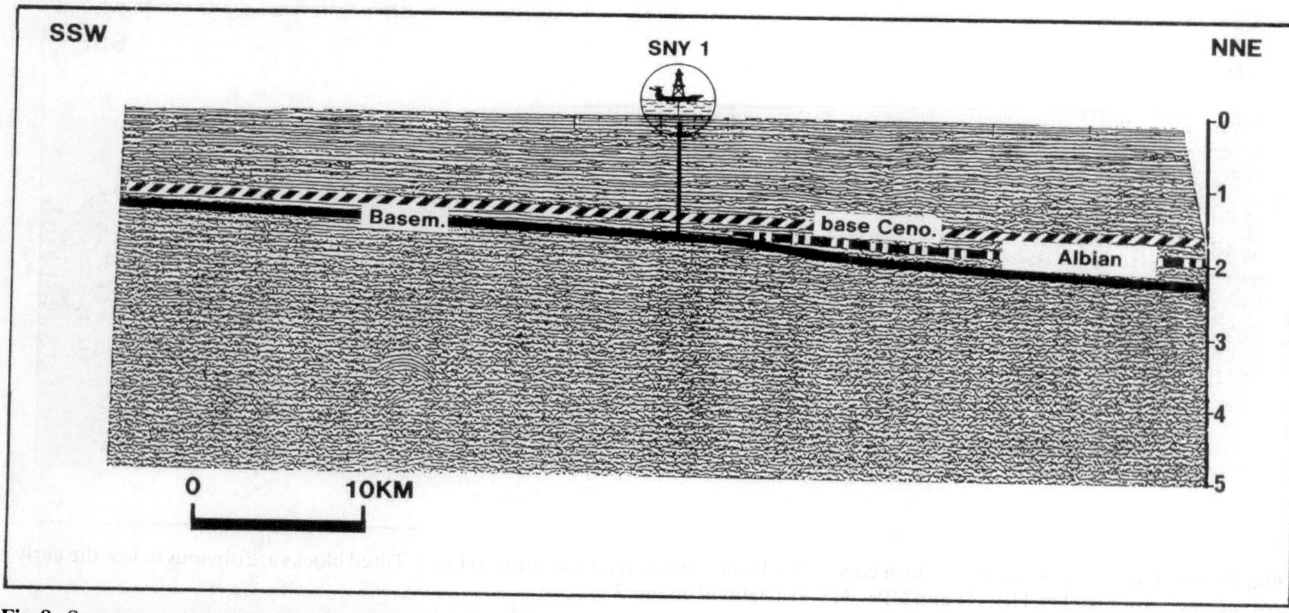

Fig. 8. Seismic section at the location of Sinna-Mary 1 well

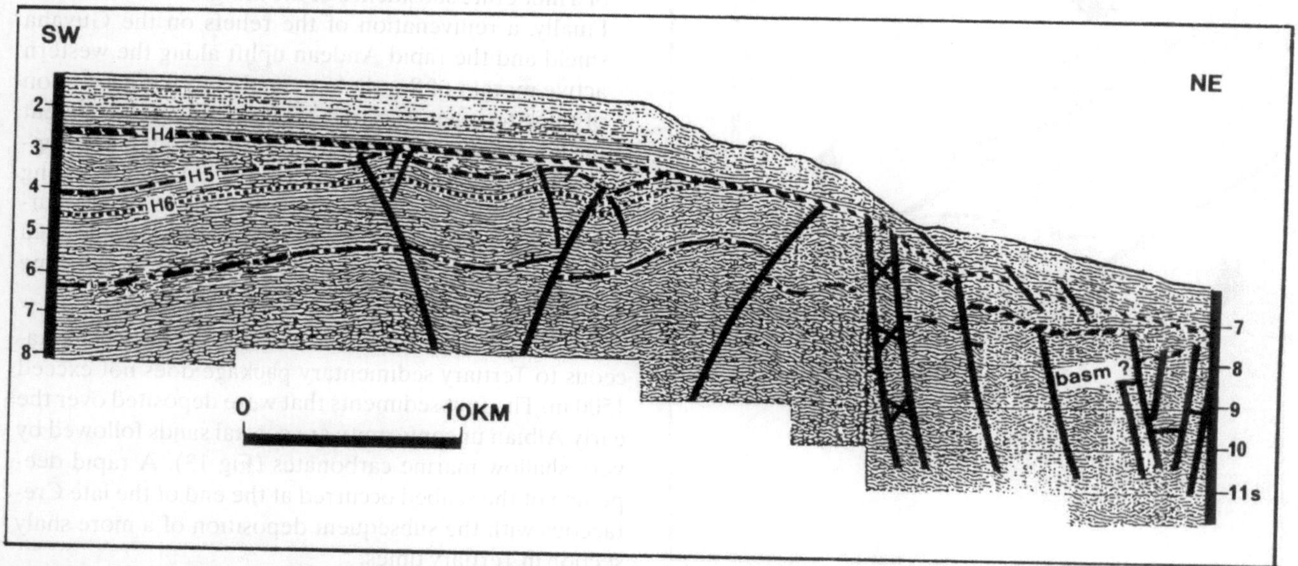

Fig. 9. Seismic section across the northern edge of the Demerara Plateau Location in Fig 1 *H4* is the early Albian unconformity below which en-echelon folds developed in response to the dextral shearing between the Demerara rise and Guinea Plateau just before the final opening of the Southern Atlantic Ocean H4, H5, and H6 are defined in Fig 7

of the Demerara plateau and penetrated a nearly complete Oligocene to Aptian succession. The Aptian sediments consist of open marine marls (Hayes et al. 1972). A thick section of carbonates of Berriasian to Hauterivan age was drilled on top of the plateau. They were followed by Barremian to Aptian sediments becoming gradually more shaly upward, as a possible result of a slight deepening of the sea floor at that time.

Sediments become progressively more clastic to the east and southeast. The Sinna Mary 1 well penetrated early Cretaceous red continental and fluvial sandstones, while well FG2-1 bottomed in basalts of Hauterivian age (120–126 Ma). These lava flows are probably related to the first rifting events between French Guyana and Sierra Leone. The paleogeographic map shown in Fig. 12 is the first attempt to gather and synthesize all available data relevant to the early Cretaceous (pre-Albian) time interval.

The western part of the French Guyana margin appears as a relatively shallow marine platform corresponding to the southernmost extension of the Central Atlantic Ocean, with depositional environments becoming gradually more continental to the east and southeast.

Upper Unit: Albian to Present Time. On the Southern Monocline, where the upper unit reaches thicknesses of

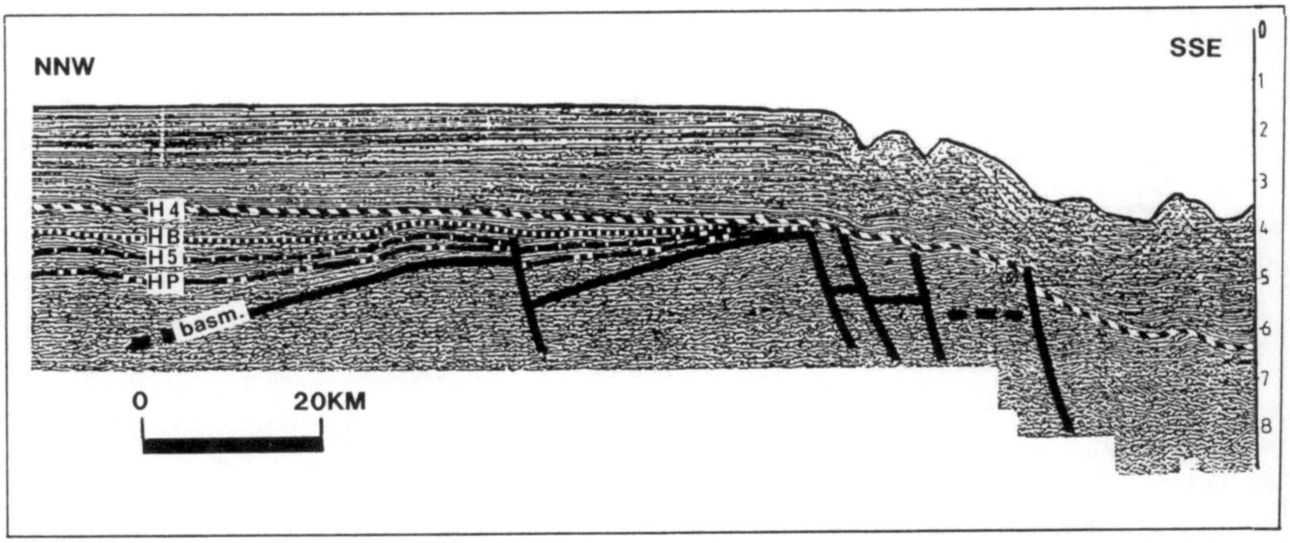

Fig. 10. Seismic section across the eastern edge of the Demerara Plateau Location in Fig 1 Tilted blocks are obvious below the early Albian unconformity H4, HB, H5 and HP (=H7) are defined on Fig 7

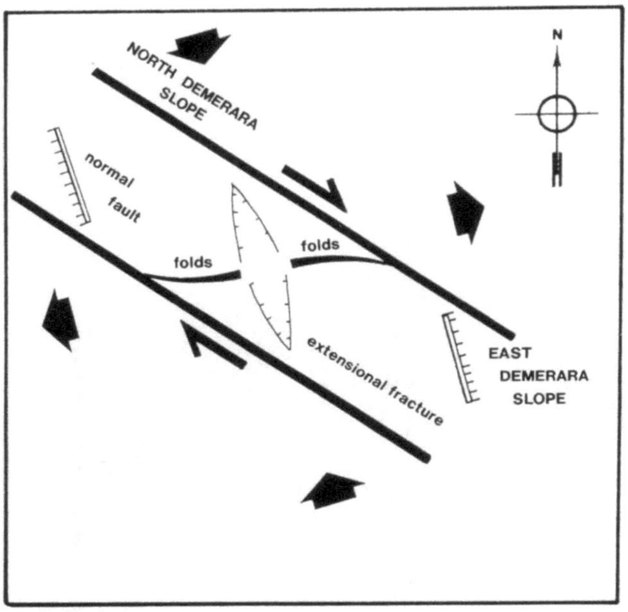

Fig. 11. Schematic interpretation of the structures related to dextral shearing along the northern edge of the Demerara Plateau when the northern tip of the South Atlantic Ocean began to open

4000 to 6000 m, three intervals can be defined. They are, from bottom to top·

- A transgressive, shaly and terrigenous shallow marine interval at the base, that includes the Albian and most of the late Cretaceous interval Subsidence of the margin at that time is believed to reflect the thermal cooling of the margin following the final continental break-up between South America and Africa
- A regional unconformity corresponds to the absence of early Paleocene strata An extended high-energy carbonate platform developed above from Paleocene to late Early Miocene times, corresponding to a period of a moderate subsidence of the margin.
- Finally, a rejuvenation of the reliefs on the Guyana shield and the rapid Andean uplift along the western active margin of South America led to the deposition of thick terrigenous series transported by the present onshore drainage system. Part of these sediments, supplied by the Amazon River, were caught by strong coastal currents going to the northwest. These currents, together with the relative fluctuations of the sea level, probably controlled the deposition of Neogene to Recent sediments.

On the Demerara Plateau, the thickness of the late Cretaceous to Tertiary sedimentary package does not exceed 1500 m. The first sediments that were deposited over the early Albian unconformity are coastal sands followed by very shallow marine carbonates (Fig 13). A rapid deepening of the seabed occurred at the end of the late Cretaceous with the subsequent deposition of a more shaly section in Tertiary times.

Two synthetic sections showing the superimposition of the Albian to Present sedimentary cover over the more restricted, but thicker pre-Albian trough, are shown in Fig 14.

Eastern Margin

The eastern margin has been much less explored than the western margin. Only a few regional seismic profiles were shot in the 1970s. The processing done at that time was of poor quality As a result, it is quite difficult below the continental shelf to separate the geological markers from multiples. The line shown in Fig. 15 is located close to the Brazilian border. The stratigraphic attribution of the seismic sequences is still speculative, as no well has been drilled in this area. We have extrapolated the results

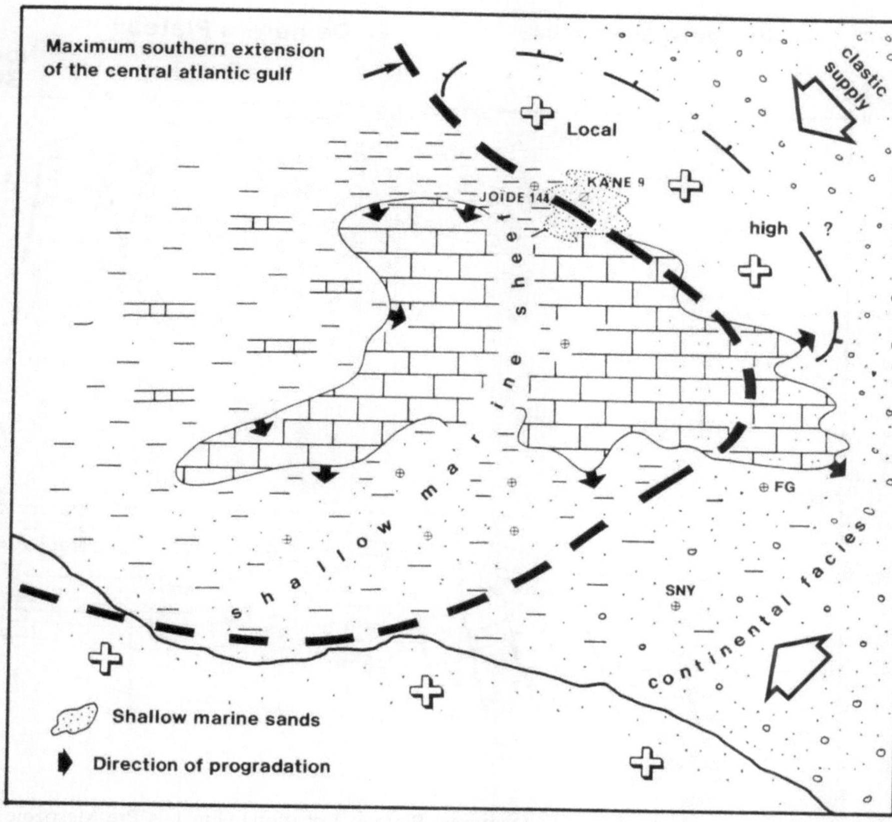

Fig. 12. Paleoenvironment of sedimentation on the margin in early Cretaceous times

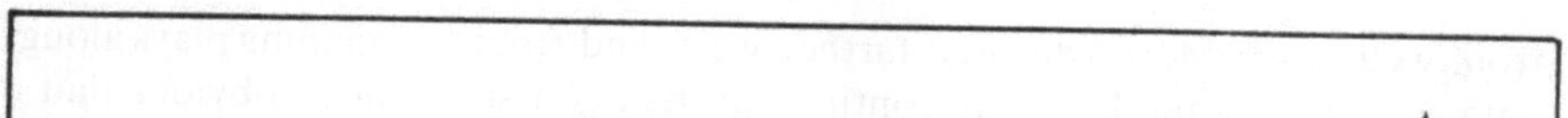

Fig. 13. Paleoenvironment of sedimentation on the Demerara Plateau in late Cretaceous times

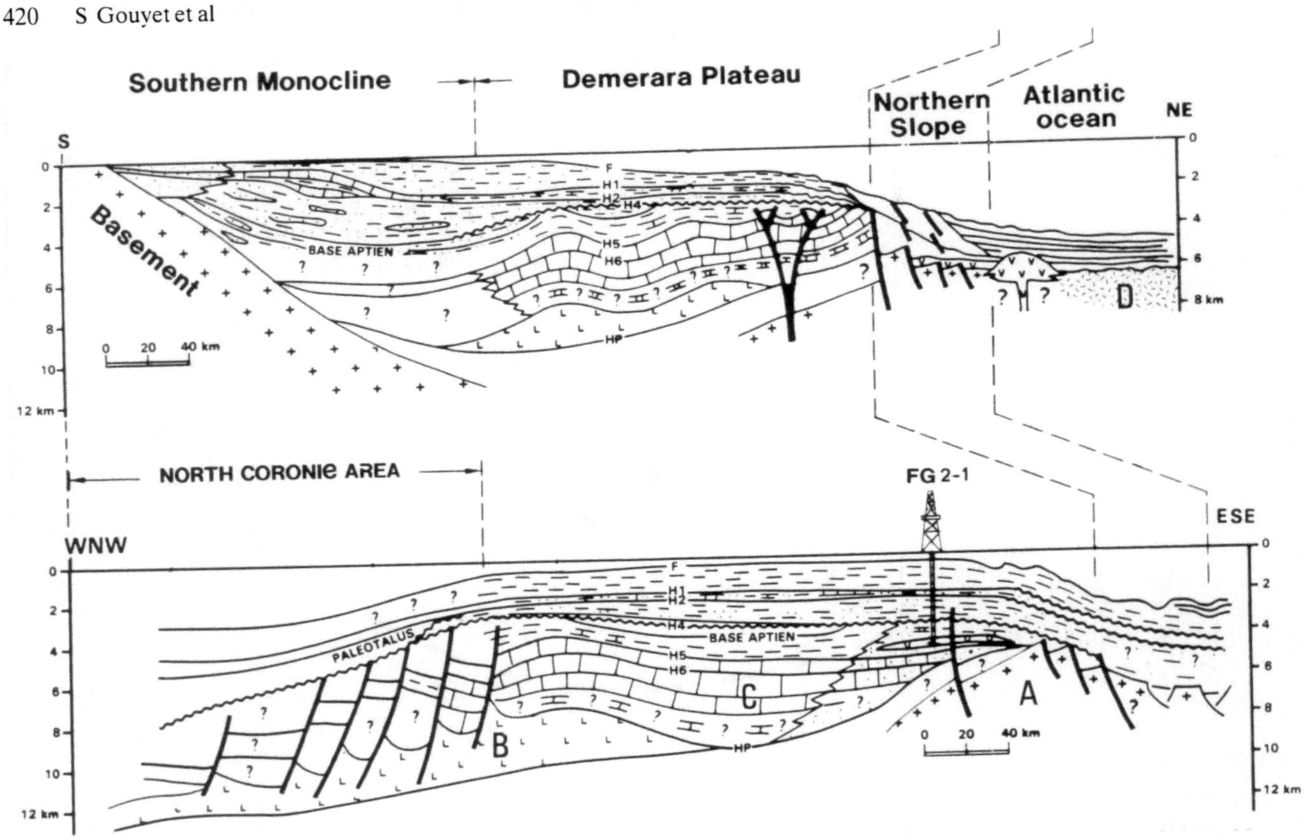

Fig. 14. Synthetic sections across the Demerara Plateau Location in Fig 1 *A* Pre-Mesozoic continental basement, *B* inferred Jurassic salt, *C* predominantly carbonate series of Jurassic (?) Neocomian age (inferred *below H6*), *D* Late Cretaceous oceanic crust

from well Sinna Mary 1 located farther west, and from data collected on the Brazilian continental shelf (Cassipore Basin and Amapa Platform) on the other side of the border. The upper sequence, of Middle Miocene to Present age, is very thin below the shelf, but has formed a thick prograding wedge below the continental slope. It overlies a Paleocene to Middle Miocene sequence, which is believed to represent the carbonate platform drilled both at Sinna Mary 1 and on the Brazilian shelf. However, at this location the platform may have been initiated slightly later, in early Eocene times. This sequence shows a constant thickness of about 0.4 s below the shelf. The underlying sequence gradually thickens toward the shelf edge. A major unconformity of early Cenomanian age has been described in the Cassipore Basin, with Cenomanian marine strata above and continental Albian beds below (Gouyet 1988). The unconformity seen on our line at the bottom of the third sequence could be of similar age, and we thus tentatively propose a late Cretaceous age for the whole sequence The poorly defined reflectors seen below would be of early Cretaceous and/or older age, i.e., pre-rift and/or syn-rift units.

Petroleum Plays

From the summarized data above (but more details can be found in Gouyet 1988), can we define any untested remaining plays along the French Guyana margin? It is first of all obvious that the first phases of exploration in the past 30 years have been disappointing, as no significant shows have been observed from the two wells drilled in French waters. The results have not been any more exciting in Surinam waters, but this country, however, has a small onshore production from the Tambaredjo oil field. The production of about 4500 bopd of low sulfur and naphthenic heavy oil (15.5° API) is achieved from almost 100 shallow wells. The producing horizons are fluvial-deltaic to coastal marine Paleocene sands (Anonymous 1988). On the Brazilian side of the offshore French Guyana border, a single undeveloped gas field, Pirapema, has been discovered, at a distance of about 230 km from the coastline, and under 135 m of water. The gas could be of biogenic origin (Campos 1983; Gouyet 1988). We can also mention the onshore Takutu graben, which straddles the Guyana-Brazil border, and where a subcommercial accumulation of oil was encountered in 1982 from Jurassic fractured basalts in the Karanambo well (Guyanna; Eiras and Kinoshita 1988). Mature source rocks have furthermore been described in other wells in the early Cretaceous section. This graben is supposed to have some genetic affinities with the pre-Albian trough cancelled below the late Cretaceous-Tertiary cover of the Demerara plateau.

We thus do not have any guideline to define remnant plays in French Guyana We shall therefore mention only

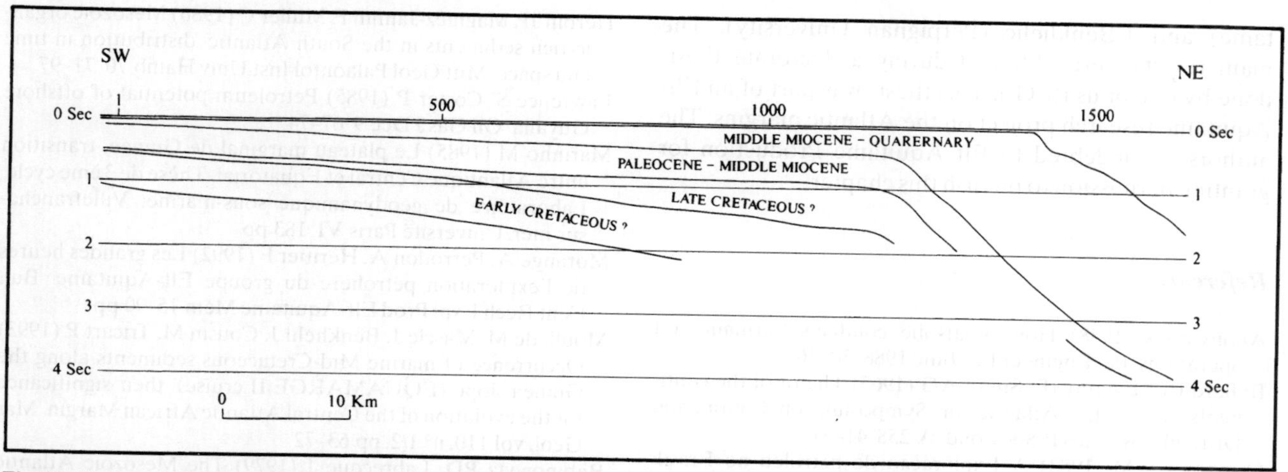

Fig. 15. Time section across the southeastern segment of the French Guyana Margin

the place where no drilling has ever been done, with the possibility that still undisclosed traps and petroleum systems could be present there. Three of them should be mentioned:

On the Demerara Rise, the deepest wells never reached the lowest sequences seen on the seismic, of presumed Triassic-earliest Cretaceous age. A major uncertainty will be the relative importance of volcanic versus sedimentary series that could be found there. Potential plays are early Cretaceous and Jurassic carbonates with source rocks of similar or younger (Albian) age.

On the eastern margin no well has ever been spudded in, and the present available seismic is of poor quality below the Albian-early Cenomanian unconformity. Potential plays could be early Cretaceous sandstones with source rocks provided by Cenomanian-Turonian black shales, if we refer to the strata of this age found along the South Atlantic margins (Herbin et al. 1986). Deeper Jurassic troughs could be present below, although never observed on the available seismic.

Along the eastern margin a thick prograding wedge also developed after late Cretaceous times and is at the origin of the structure of the present continental slope. Stratigraphic traps with clastic reservoirs could be present in upper or lower slope fans of late Cretaceous or Paleogene ages with potential hydrocarbons still being provided by the Cenomanian-Turonian black shales.

Conclusions

The French Guyana Margin is located on a unique segment of the Atlantic margins where two major episodes of rifting and basin subsidence are superimposed. In Jurassic and Early Cretaceous times this area, together with the Guinean Plateau in Africa, was the southern tip of the Central Atlantic Ocean. The deep basin cancelled below Cenomanian and younger sediments on the Demerara Rise (Fig. 14) is currently interpreted as a major rift graben, probably similar in many aspect to the Takutu Graben to the west (NE Brazil-Guyana). They both resulted from a restricted amount of stretching of the Gondwana continental crust, which never led to a complete separation of the rifted margins and to oceanic crust accretion.

In late Albian and younger times, the final opening of the South Atlantic Ocean at its northern tip was related to dextral shearing along what was to become the Equatorial transform faults. As a result, no major rifted basin related to this second phase of crustal stretching has ever been recognized so far, although the poor quality of the available seismic on the eastern part of the French Guyana margin does not allow a definitive statement to be made for this area. The subsidence of the margin, when the oceanic crust was accreted and the seaway between the Southern and Central Atlantic Ocean finally opened, led to the deposition of the late Cretaceous-Tertiary cover over the previous rift graben on the Demerara Rise, and to the formation of a new passive continental margin in the east.

The French Guyana margin can be considered as underexplored, as only two wells, Sinna Mary 1 and FG2-1, have ever been drilled, in 1975 and 1978, respectively. From the above summarized geologic history, two distinct areas should be considered for future petroleum evaluation and exploration. To the west, on the Demerara Rise, deep Jurassic and early Cretaceous potential plays remain untested. A careful reprocessing of some of the line shots in the 1970s could possibly lead to better mapping and resolution of pre-Cenomanian structures present in this area (Fig. 10). To the east, the late Cretaceous-Tertiary passive continental margin is unexplored. Potential plays on the shelf may include shallow marine Cretaceous clastics, or Paleogene carbonates, but also deeper marine silicoclastic or bioclastic slope wedges of late Cretaceous-Tertiary age along the outer shelf (Fig. 15). Traps could predominantly be stratigraphic only. New regional seismic profiles would be necessary to obtain the decisive data necessary for a correct evaluation of this area.

Acknowledgments. The final version of this chapter has benefited from the comments of F. Schein (Elf Aquitaine) and J Benkhelic (Perpignan University). The main results were obtained during a doctorate thesis done by one of us (S G). This thesis was part of an Elf-Aquitaine research project on the Atlantic margins The authors are indebted to Elf Aquitaine Production for granting permission to publish this chapter

References

Anonymous (1988) How Staatsolie conducts Surinam's oil operations Pet Engineer Int. June 1988 30–36
Bullard EC, Everett JE, Smith AG (1965) The fit of the continents around the Atlantic In Symposium on Continental Drift Philos Trans R Soc Lond. A 258 41–51
Campos CWM (1983) A Exploraçao de petroleo no Brasil situaçao actual e perspectivas 2 Congr Brasileiro de Petroleo Anais, Rio de Janeiro, 1982, pp 201–239
Curie D (1985) Ouverture de l'Atlantique Sud et discontinuités intraplaques Thèse de 3ème cycle, Université de Bretagne Occidentale, Centre de Brest, DERO-GM, IFREMER, 192 pp
Eiras JF, Kinoshita EM (1988) Evidencias de movimentos transcorrentes na bacia do Tacutu Bol Geocie Pet 2, 2/4 193–208
Gouyet S (1988) Evolution tectono-sédimentaire des marges Guyanaises et Nord-Brésilienne au cours de l'ouverture de l'Atlantique Sud Thèse de Doctorat, Université de Pau et des Pays de l'Adour, 374 pp
Hayes DE, Pimm AC, Beckman JB, Benson WE, Berger WH, Roth PH, Supko PR, von Rad U (1972) Initial reports of the Deep Sea Drilling Project, vol 14, US Government Printing office, Washington DC, 975 p
Herbin JP, Magniez-Jannin F, Muller C (1986) Mesozoic organic-rich sediments in the South Atlantic distribution in time and space Mitt Geol Palaontol Inst Univ Hamb 70 71–97
Lawrence S Coster P (1985) Petroleum potential of offshore Guyana Oil Gas J Dec 9 67–74
Marinho M (1985) Le plateau marginal de Guinée, transition entre Atlantique Central et Equatorial Thèse de 3ème cycle Laboratoire de géodynamique sous-marine, Villefranche-sur-mer, Université Paris VI, 183 pp
Morange A, Perrodon A, Heritier F (1992) Les grandes heures de l'exploration pétrolière du groupe Elf-Aquitaine Bull Cent Rech Expl Prod Elf-Aquitaine Mém 15 90 pp
Moullade M, Mascle J, Benkhelil J, Cousin M, Tricart P (1993) Occurrence of marine Mid-Cretaceous sediments along the Guinea slope (EQUAMARGE II cruise) their significance for the evolution of the Central Atlantic African Margin Mar Geol, vol 110, n° 1/2, pp 63–72
Rabinowytz PD, Labrecque J (1979) The Mesozoic Atlantic Ocean and evolution of its continental margin J Geophys Res 84 5973–6002
Sibuet JC, Mascle J (1978) Plate kinematic implications of Atlantic Equatorial fracture zones trends J Geophys Res 83 3401–3421
Tricart P, Mascle J, Basile C, Benkhelil J, Ciais G, Villeneuve M (1991) La tectonique d'inversion médio-crétacée de la marge sud-guinéenne (campagne EQUAMARGE II) Bull Soc Géol Fr 162, 1 91–100
Unternehr P, Curie D, Olivet JL, Goslin J, Beuzart P (1988) South Atlantic fits and intraplate boundaries in Africa and South America Tectonophysics 155 169–179

Structure and Tectonic History of the Kerguelen Plateau

R. Schlich, M. P. Angoulvant, B. Fritsch, M. Munschy, and M. Schaming

The Kerguelen Plateau in the Southern Indian Ocean is one of the largest submerged plateau in the world. The origin and subsequent geologic and tectonic history of the imposing feature have long been a matter of debate and speculation. These outstanding problems were the primary topics addressed by several French and Australian multichannel seismic reflection surveys carried out on the plateau between 1981 and 1991 and by deep sea drilling performed in 1988 by the Ocean Drilling Program (Barron, Larsen et al. 1989; Schlich, Wise et al. 1989).

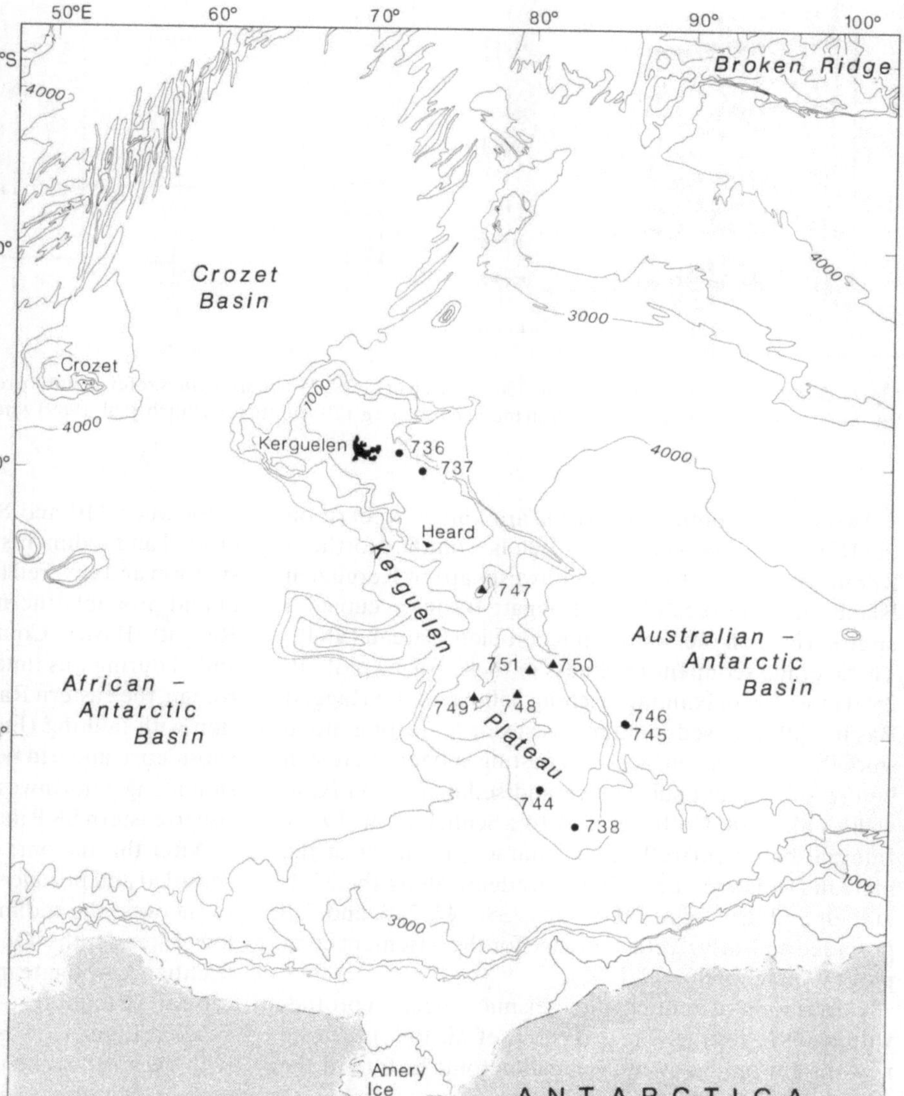

Fig. 1. Bathymetric chart of the Kerguelen Plateau (After Schlich et al 1987) The contour interval is 500 m Location of Leg 119 *(dots)* and Leg 120 *(triangles)* ODP drill sites

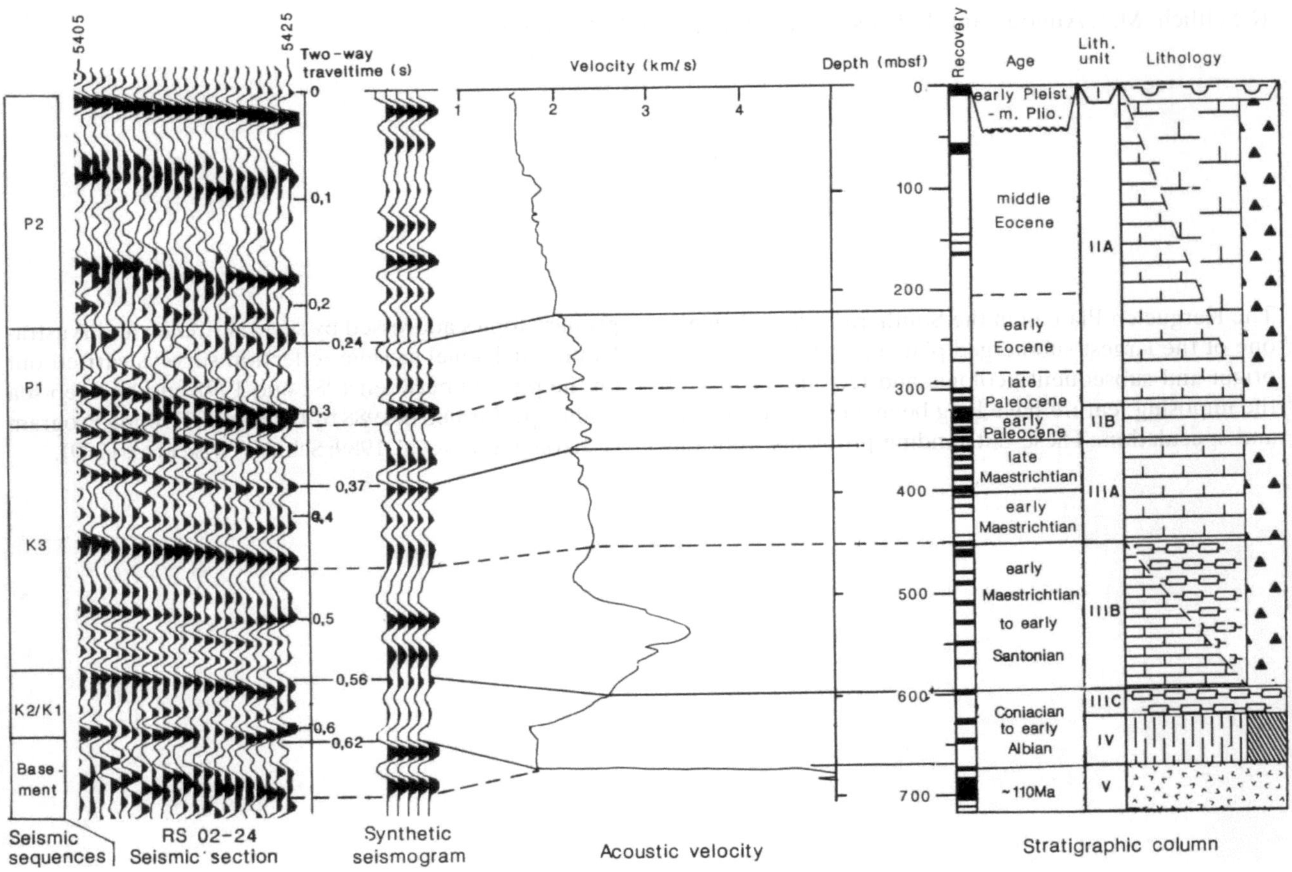

Fig. 2. Depth-time correlation for Site 750 (Fritsch et al 1992) The *solid lines* refer to the correlations proposed by Fritsch et al 1992, the *dashed lines* refer to the correlation made by the Leg 120 scientists (Schlich et al 1989) when they are different

Two major sedimentary basins are known to exist on the Kerguelen Plateau (Fig. 1); one lies on the Northern Kerguelen Plateau (NKP), southwestward of Kerguelen Island, and the second, the Raggatt Basin, occupied a large surface on the Southern Kerguelen Plateau (SKP). The average sediment thickness in these basins is about 2500 m with a maximum reaching 3500 m in the Raggatt Basin. All the sedimentary basins are resting upon smooth basement surfaces, suggesting subaerial erosion before subsequent subsidence and sedimentation (Colwell et al. 1988; Coffin et al. 1990; Schlich et al. 1988). Internal basement reflections characterize most of the plateau (Rotstein et al. 1990). Dredging along the 77°E graben and drilling at ODP sites 738, 747, 749, and 750 provided an Early Cretaceous age for the basement complex (Whitechurch et al. 1992).

Correlation of multichannel seismic reflection profiles with drilling results (Fig. 2; Fritsch et al. 1992) provide new insight on the evolution and tectonic history of the Kerguelen Plateau (Fig. 3).

Before or during Albian time (110 Ma), thick basalt flows formed the Kerguelen Plateau during successive volcanic episodes. As a result of the volcanic activity, basement synforms and basement ridges shaped the plateau At this time much of the plateau was above sea level (Fig. 3 A).

Between 110 and 88 Ma, the basement was strongly eroded and sediments filled the Raggatt Basin basement synform and covered the eastern basement ridge. Gradual and probably thermal subsidence affected the entire Raggatt Basin. Continental-type sedimentation prevailed during this time span. At the beginning of the Turonian, the eastern Raggatt Basin was uplifted in association with faulting (Fig. 3 B). The uplift was followed by subsidence and can be interpreted as the result of extension along a northwest-southeast direction that will give rise to eastern SKP margin.

After this tectonic event, the eastern part of the SKP subsided and pelagic sediments were deposited, whereas to the west the sedimentation regime remained at shallow water depths. During the Campanian and the Maestrichtian, carbonate mounds were developed in the western part of the Raggatt Basin, suggesting a very slow subsidence. Later, the basin was uplift (66 Ma) in association with the north-south tectonic event that formed the 77°E graben (Munschy et al. 1992 a). The western Raggatt Basin represent one of the rift flanks associated with the 77°E graben (Fig. 3 C).

During the Paleocene, the western part of the Raggatt Basin subsided rapidly to 1000 m water depths. From the late Paleocene (58 Ma) to about 40 Ma, a slow thermal subsidence affected the entire SKP, which reached a

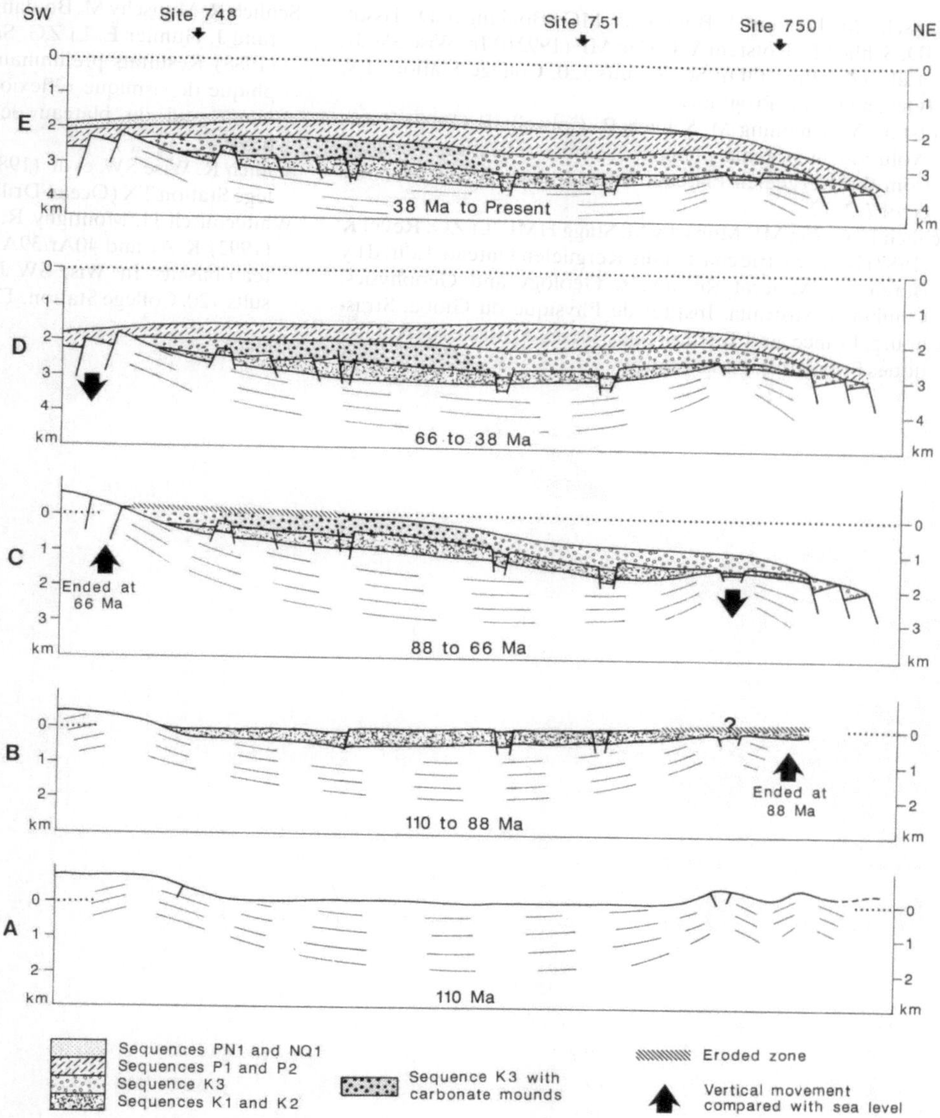

Fig. 3 A–E. Schematic representation of the evolution of the Raggatt Basin (Fritsch et al 1992)

depth of 1000–2000 m. At 54-52 Ma, the sedimentation rate changed abruptly over most of the Raggatt Basin (Fig. 3 D).

At 38 Ma, an erosional event was observed on the entire plateau. This event is related to the separation by seafloor spreading of the Kerguelen Plateau-Labuan Basin and Broken Ridge-Diamantina Zone dated at 45-42 Ma (Munschy et al. 1992b). No sediment was deposited in the eastern part of the basin between middle Eocene and middle Pliocene times. In the central and western parts of the basin, Oligocene and Miocene sediments are present, and two erosional events were observed in the Miocene between 17 and 12 Ma and between 8 and 5 Ma (Fig. 3 E).

References

Barron J, Larsen B, et al (1989) Proc ODP, Init Rep 119 College Station, TX (Ocean Drilling Program)

Coffin MF, Munschy M, Colwell JB, Schlich R, Davies HL, Li ZG (1990) Seismic stratigraphy of the Raggatt Basin, Southern Kerguelen Plateau tectonics and paleoceanographic implications. Geol Soc Am Bull 102 563–579

Colwell JB, Coffin MF, Pigram CJ, Davies HL, Stagg HMJ, Hill PJ (1988) Seismic stratigraphy and evolution of the Raggatt Basin, Southern Kerguelen Plateau. Mar Pet Geol 5 75–81

Fritsch B, Schlich R, Munschy M, Fezga F, Coffin MF (1992) Structure and evolution of the Southern Kerguelen Plateau deduced from seismic stratigraphic studies and ODP drilling at site 748 and 750 In: Wise SW Jr et al (eds) Proc ODP Sci Results 120, College Station, TX (Ocean Drilling Program)

Munschy M, Fritsch B, Schlich R, Fezga F, Rotstein Y, Coffin MF (1992 a) Structure and evolution of the Central Kerguelen Plateau deduced from seismic stratigraphic studies and ODP drilling at site 747. In. Wise SW Jr et al (eds) Proc ODP Sci Results 120, College Station, TX (Ocean Drilling Program)

Munschy M, Dyment J, Boulanger MO, Boulanger D, Tissot JD, Schlich R, Rotstein Y, Coffin MF (1992b) In Wise SW Jr et al (eds) Proc ODP Sci Results 120, College Station, TX (Ocean Drilling Program)
Rotstein Y, Schaming M, Schlich R, Colwell JB (1990) Basin evolution in oceanic plateaus seismic reflection evidence from the Kerguelen Plateau, South Indian Ocean Mar Pet Geol 7 2–12
Schlich R, Coffin MF, Munschy M, Stagg HML, Li ZG, Revill K (1987) Bathymetric chart of the Kerguelen Plateau Edited by Bureau of Mineral Resources, Geology and Geophysics, Camberra Australia, Institut de Physique du Globe, Strasbourg, France, and Territoire des Terres Australes et Antarctiques Françaises, Paris, France
Schlich R, Munschy M, Boulanger D, Cantin B, Coffin MF, Durand J, Humler E, Li ZG, Savary J, Schaming M, Tissot JD (1988) Résultats préliminaires de la campagne océanographique de sismique réflexion multitrace MD47 dans le domaine sud du plateau de Kerguelen CRAS Paris 306 635–642
Schlich R, Wise SW, et al (1989) Proc ODP, Init Rep 120, College Station TX (Ocean Drilling Program)
Whitechurch H, Montigny R, Sevigny J, Storey M, Salters V (1992) K-Ar and 40Ar/39Ar ages of central Kerguelen Plateau basalts In Wise SW Jr et al (eds) Proc ODP Sci Results 120, College Station, TX (Ocean Drilling Program)

The Sedimentary Basins of the New Caledonia Region

Y. Lafoy

New Caledonia and the regions around it are located in a zone halfway between the oceanic Pacific domain in the east and the continental Australian domain in the west Bathymetry (Missègue et al. 1991) shows that the region is made up of a succession of elongated submarine basins and ridges. Two main trends occur within the New Caledonia region:

- a north-south-trending zone which, subparallel to the Australian continental slope, includes the alignments of the Lord Howe Guyots, Lord Howe Rise, and the central part of the New Caledonia Basin;
- a north-west to south-east-trending zone which comprises Fairway Basin and Ridge, the northern part of New Caledonia Basin, and Loyalty Basin.

A synthetic chronological scheme can be proposed for the region (Collot et al. 1982; Maillet et al. 1982):

- during the Eocene, a northeast-dipping subduction involved the New Caledonia-Loyalty region;
- the upper Eocene (40-38 Ma) is marked by the blocking of the subduction, then by obduction of a sheet of oceanic lithosphere in the form of the peridotite nappes overthrust the New Caledonia basement;
- the end of the compressive phenomena was followed, after a tectonic respite, by a tensional phase beginning in the middle Miocene (15 Ma).

In a belt nearly 1000 km wide, three great sedimentary basins occur, the Fairway Basin and the New Caledonia Basin, which lie west of New Caledonia, and the Loyalty Basin to the east (Fig. 1).

The Fairway Basin (1500–2000 m average depth), located between Lord Howe Rise and Fairway Ridge, trends northwest-southeast and shows a sedimentary fill 2000 m thick on average. Its basement is delimited in some places by a strong reflector revealing considerable diffraction In other places the substratum is poorly defined, masked by diffracting reflectors (volcanic series?) or by thick series.

The northern part of the New Caledonia Basin is located between Fairway Ridge and Norfolk Ridge. Within New Caledonia Basin (3500 m deep), the basement, highlighted by a strong reflector, occurs under a sedimentary blanket 2000 to 3000 m thick. The Chesterfield area, north of New Caledonia Basin, is characterized by a thick sedimentary series (2500 m), structural traps (tilted blocks) and high heat flow values (Guignard and Ravenne 1982).

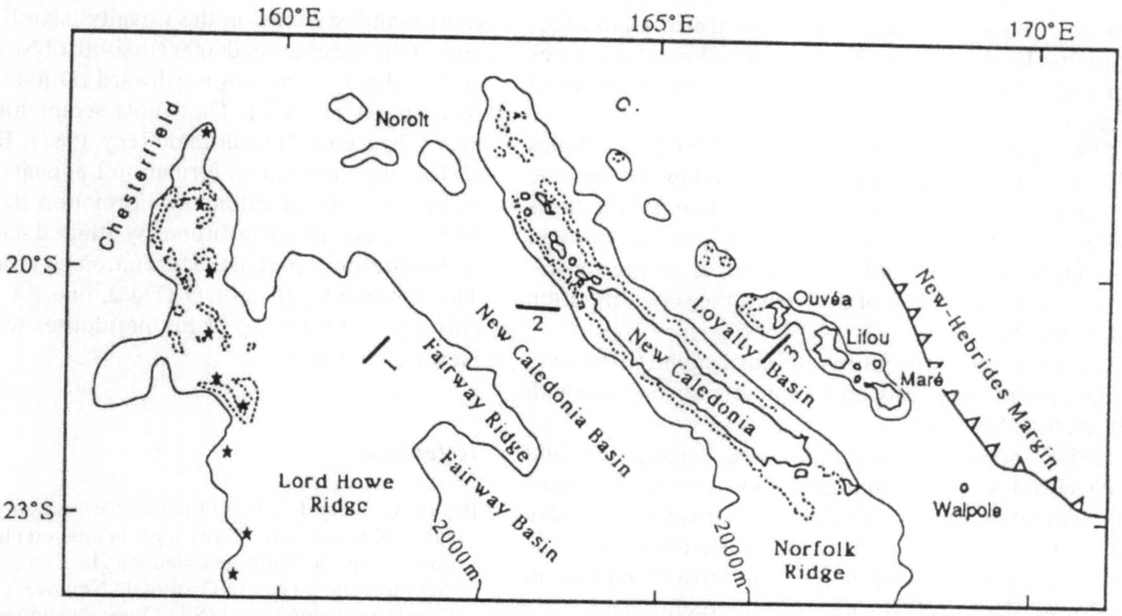

Fig. 1. Main geological units around New Caledonia and location of multichannel seismic profiles *1*, *2*, and *3*

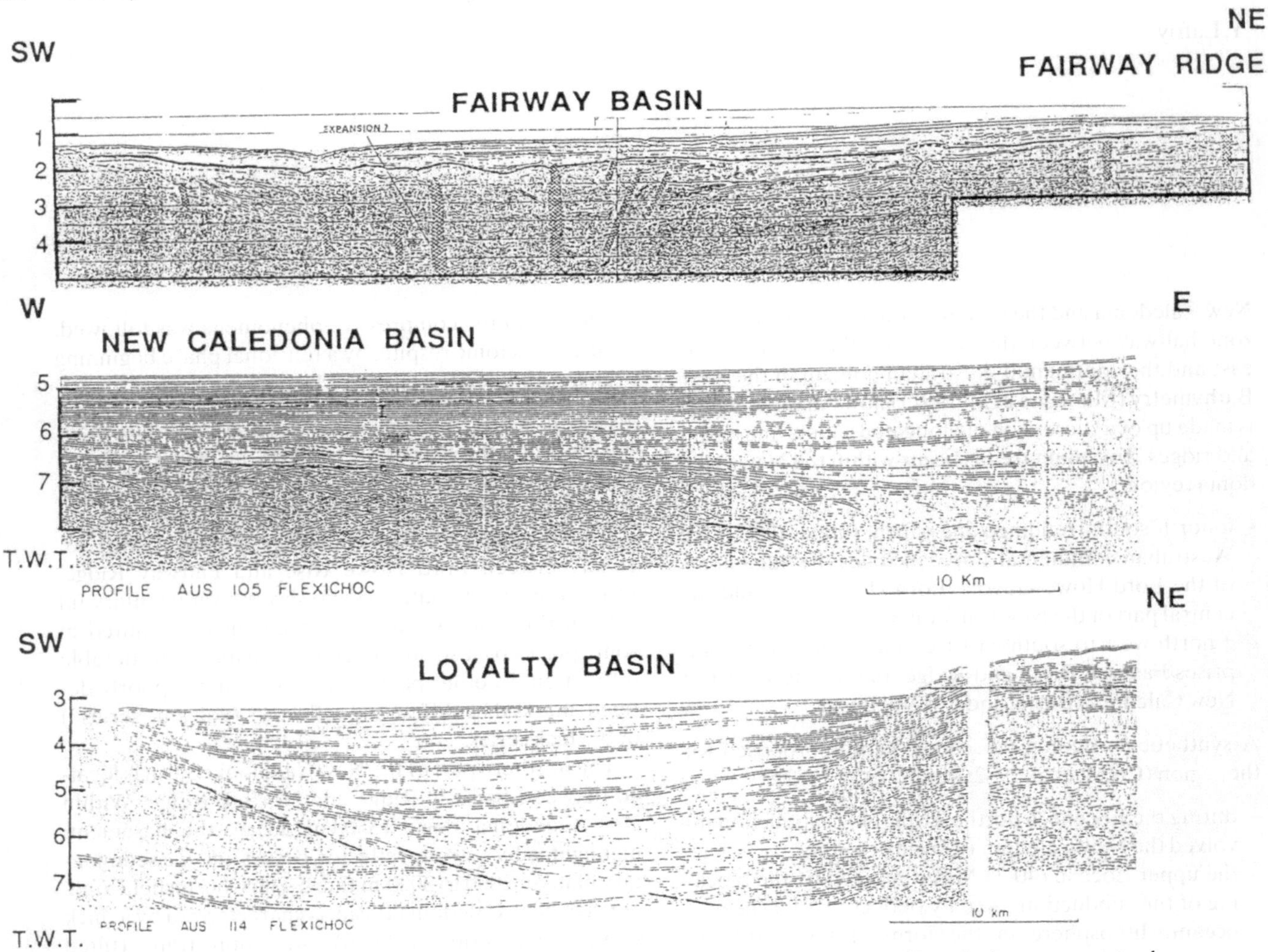

Fig. 2. Multichannel seismic profiles *1*, *2*, and *3* carried out within the Fairway, New Caledonia, and Loyalty basins, respectively

Both Fairway and New Caledonia Basins are characterized by their substratum which reveal no organized reflections. In the basins, three main sedimentary series separated by two unconformities can be recognized (Fig. 2).

The upper discontinuity is dated by DSDP Boreholes of Leg 21 (Burns and Andrews 1973) and by the tectonic phases known on land (Lillie and Brothers 1970); this is the Eocene-Oligocene unconformity, which is obvious on the ridges and noticeable on the edges of the basins. However, in the middle of the basins, the series are often conformable.

The lower unconformity would correspond to the level of the upper Cretaceous, and is sometimes confused with the acoustic basement.

DSDP boreholes, seismic facies (Austradec and WNC80 lines), and formations known on land suggest that sediments are mainly pelagic in Fairway Basin and in the central part of New Caledonia Basin (Ravenne et al. 1977), and essentially of detrital and terrigenous type in the New Caledonia Basin near the emergent island.

Loyalty Basin (2000 m deep), which lies between New Caledonia and Loyalty Ridge (Fig. 1), is characterized by a thick sedimentary series. This fill is provided simultaneously from the erosion of New Caledonia and the volcanic manifestations on the Loyalty Islands. The thickness of the series exceeds 6000 m south of New Caledonia and gradually decreases northward (Dubois et al. 1974; Ravenne et al. 1977). The whole sedimentary series is post-Oligocene (Bitoun and Récy 1982). Between the surface and horizon C, formation I appears constituted by unconsolidated sediments. Formation II, between reflectors C and D, is constituted by lithified sediments, and within the lower part of the formation, by stratified rocks. The basement, reflector D (Fig. 2, line AUS114) could correspond to the top of the peridotites which overlap New Caledonia.

References

Bitoun G, Récy J (1982) Origine et évolution du bassin des Loyauté et de ses bordures après la mise en place de la série ophiolitique de Nouvelle Calédonie In Equipe de Géologie-Géophysique du centre Orstom de Nouméa Contribution à l'étude géodynamique du Sud-Ouest Pacifique Trav Doc Orstom 147 505–539

Burns RE, Andrews JE (1973) Regional aspects of deep sea drilling in the Southwest Pacific In Burns RE, Andrews JE et al (eds) Initial Reports of the Deep Sea Drilling Project, vol 21, Washington, 897–906

Collot JY, Missègue F, Malahoff A (1982) Anomalies gravimétriques et structure de la croûte dans la région de la Nouvelle Calédonie enracinement des péridotites In Equipe de Géologie-Géophysique du centre Orstom de Nouméa Contribution à l'étude géodynamique du Sud-Ouest Pacifique Trav Doc Orstom 147 549–564

Dubois J, Ravenne C, Aubertin A, Louis J, Guillaume R, Launay J, Montadert L (1974) Continental margins near New Caledonia In Burk CA, Drake CL (eds) The geology of continental margins. Springer, Berlin Heidelberg New York, pp 521–535

Guignard JD, Ravenne C (1982) Interprétation géologique et géophysique de la campagne Ouest-Nouvelle Calédonie (WNC80), Comité d'Etudes Pétrolieres Marines, DCEG 1849, 16 pp

Lillie AR, Brothers RN (1970) The geology of New Caledonia NZ J Geol Geophys 13 145–183

Maillet P, Monzier M, Selo M, Storzer D (1982) La zone d'Entrecasteaux (Sud-Ouest Pacifique) nouvelle approche pétrologique et géochronologique In Equipe de Géologie-Géophysique du centre Orstom de Nouméa. Contribution à l'étude géodynamique du Sud-Ouest Pacifique Trav Doc Orstom 147. 441–458

Missègue F et al (1991) Carte bathymétrique de la Zone Economique de la Nouvelle Calédonie, 1/1 095 708 á l'équateur, équidistance 200 m, set complet de 3 feuilles Orstom, Nouméa

Ravenne C, de Broin CE, Dupont J, Lapouille A, Launay J (1977) New Caledonia Basin-Fairway Ridge structural and sedimentary study In Int Symp Geodynamics in the South West Pacific Nouméa 1976 Technip, Paris, pp 63–78

Sedimentary Basins and Petroleum Plays Around the French Antilles

Ph. Bouysse and A. Mascle

Abstract. The French Antilles are part of the Lesser Antilles island arc, an active intraoceanic volcanic arc resulting from the westward subduction of the Atlantic crust at a very slow rate from early Eocene times. Reflections profiles have shown the presence of relatively thick sedimentary basins around the volcanic islands. Shallow platforms and moderate water depths predominate around St Martin and Guadeloupe, whereas greater depths are rapidly reached around Martinique. These basins are virtually unexplored, with the exception of the Saba Bank, where two wells drilled in 1977 and 1982 encountered only small oil and gas shows. Reservoir and immature source rocks have, however, been reported in Eocene and Oligocene carbonates and volcanoclastics sandstones. Similar plays could be attractive around St Martin and Guadeloupe, as heat flows higher than on the Saba Bank have recently been measured. They could in recent times have significantly enhanced the maturation of any Paleogene source rocks.

General Setting of the Lesser Antilles Island Arc

The Lesser Antilles Island Arc (LAIA) is of volcanic origin and located at the eastern edge of the Caribbean plate beneath which the Atlantic oceanic crust (North and/or South American plates) is being presently subducted (Bouysse et al. 1990; Figs. 1 and 2) The arc forms a quite regular curve, about 850 km long, from Grenada Island in the south, to the tiny islet of Sombrero in the north, linking the Venezuelan borderland against which the arc is colliding (Pindell 1991), with the eastern tip of the Greater Antilles (Puerto Rico and the Virgin Islands). The Greater and Lesser Antilles are separated by the Anegada Passage, an extensional complex of Neogene age including small pull-apart basins (Gill 1990; Jany et al. 1990; Speed and Larue 1991).

In the east of the LAIA the contact between the Caribbean and American plates is underlined by a strong negative anomaly, about 170 km away from the present volcanic line. This axis corresponds to the north with the Puerto Rico Trench (more than 6000 m deep) which is becoming progressively shallower southward as the

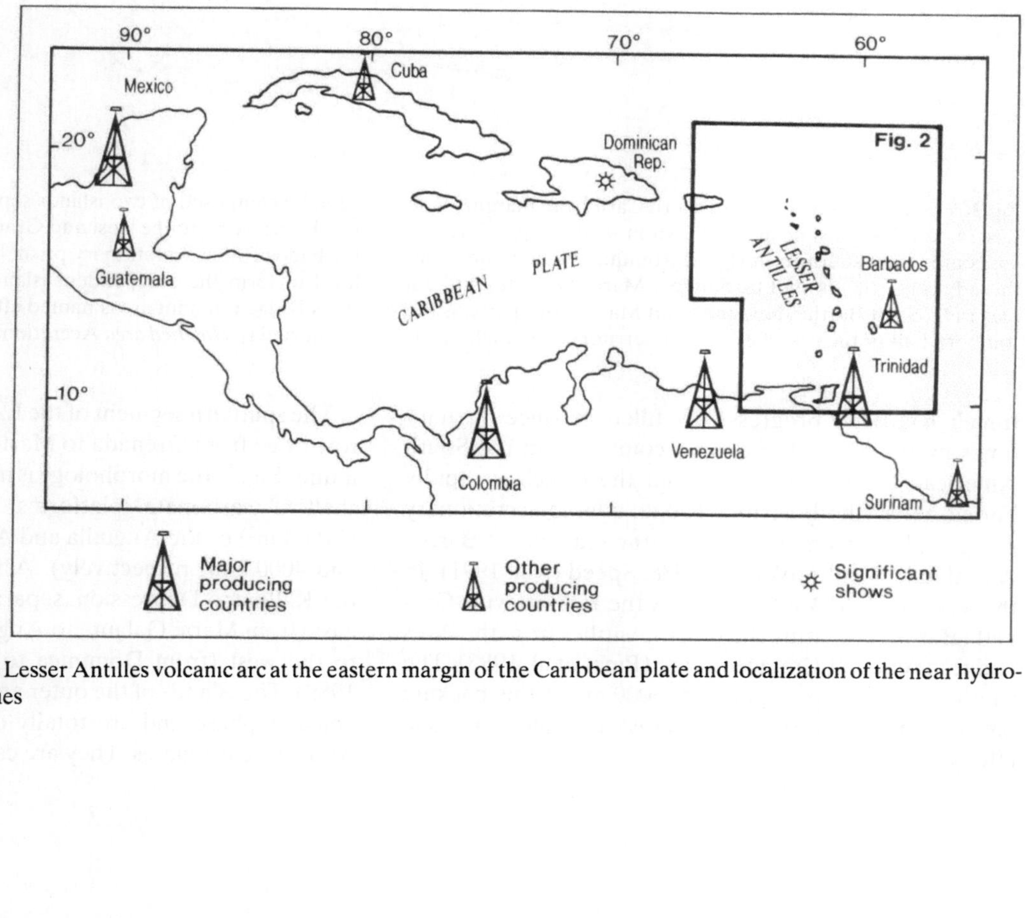

Fig. 1. Localization of the Lesser Antilles volcanic arc at the eastern margin of the Caribbean plate and localization of the near hydrocarbon-producing countries

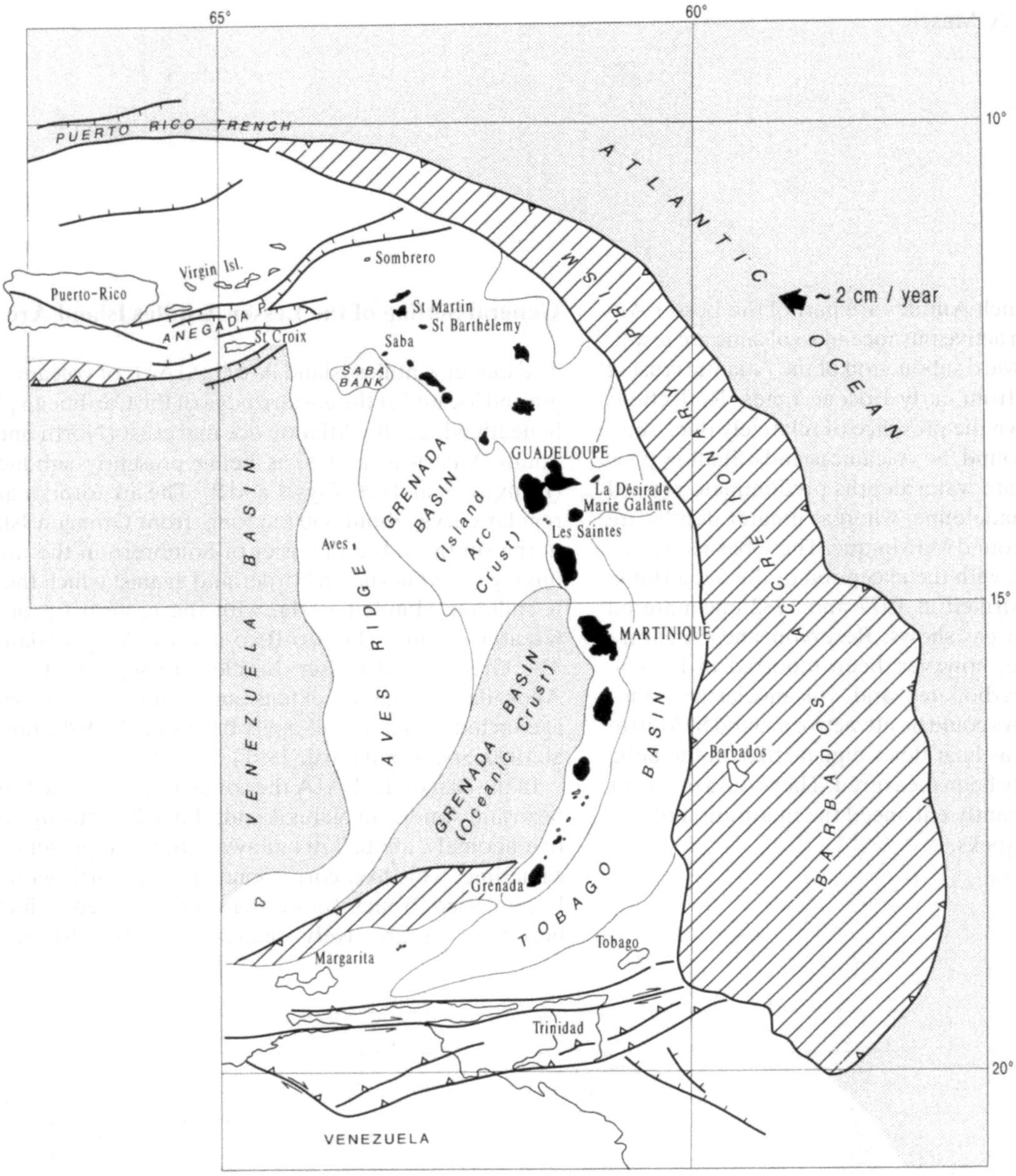

Fig. 2. General features of the eastern Caribbean margins The Lesser Antilles volcanic arc is shown in *black* The French islands are, from south to north Martinique, Guadeloupe, and the adjacent islands of Les Saintes, Marie Galante, and La Désirade, Saint Barthélémy, and Saint Martin (pro parte as the southern half of the island is a Dutch territory) Guadeloupe is actually composed of two islands separated by a narrow channel Basse-Terre to the west and Grande-Terre to the east The Barbados Ridge accretionary prism locally emerges above sea level to form the independent island of Barbados, when the Aves Ridge remnant arc is named after the tiny island of Aves (Venezuela) *Hatched area* Accretionary wedges

trench has been progressively filled in since Tertiary times by the large clastic input coming from the South American Continent. As a result the trench is rapidly buried below the Barbados Ridge, a large accretionary complex, locally emerging above the sea level at Barbados Island (Westbrook et al. 1988; Speed et al. 1991) In the west, the LAIA is bounded by the 150-km-wide Grenada Basin, separating the Lesser Antilles from the Aves Ridge, a remnant Cretaceous arc (Pinet et al. 1985). The southern and deepest part (ca. 3000 m) of this backarc basin is made of oceanized crust of Paleocene age (Bouysse 1988).

The southern segment of the LAIA forms a rather narrow ridge from Grenada to Martinique. North of Martinique (Fig. 3) the morphology is more complex with large shallow carbonate platforms such as Saba Bank (2100 km^2) or the Anguilla and Antigua platforms (4500 and 4000 km^2, respectively). A narrow intra-arc basin, the Kallinago Depression, separates an outer arc to the east (from Marie Galante to Anguilla) from an inner arc to the west (from Dominica to Saba; Andreieff et al 1989). The islands of the outer arc belong to a Paleogene volcanic phase and are totally or partially overlain by Neogene carbonates. They are called the Limestone Caribbees. The islands of the inner arc are Neogene and Recent volcanoes. The two arcs are superimposed from Martinique southward so that it is possible to define an older arc from Grenada to Anguilla and a Recent arc from Grenada to Saba, with this latter also being known as the Volcanic Caribbees. On the easternmost island, La Désirade, the basement of both Tertiary arcs is exposed, with primitive Mesozoic island arc meta-basalts and intercalated cherts (Bouysse et al. 1983), the cherts having been recently ascribed to the upper Tithonian (Montgomery et al. 1992).

Evolution of the Lesser Antilles Since the Late Mesozoic

Three main stages of arc activity, interrupted by two periods of quiescence each 10 Ma long, characterize the evolution of the LAIA.

Stage 1: from the latest Jurassic (ca. 140 Ma) to the latest Cretaceous (ca. 65 Ma), a proto-arc was in activity. It corresponds to the so-called Mesozoic Caribbean arc, which included the present Aves Ridge and the basement of the Lesser Antilles. Evidence of Upper Cretaceous volcanics and volcanoclastics is provided by outcrops dredged on the slope of the Aves Ridge and on the northern and northeastern slopes of the LAIA. They consist of granodiorites, metabasalts, meta-andesites, brecciae, conglomerates, tuffites, and tuffaceous carbonates. Micropaleontological data indicate that both shallow platform and slope deposits are contemporaneous with the volcanic activity (Bouysse et al. 1985b). The oldest onshore rocks of the whole proto-arc outcrop on La Desirade and St Croix islands (Bouysse et al. 1983; Speed and Joyce 1990; Montgomery et al. 1992). The thickness of these Mesozoic formations can reach several thousand meters.

Interstage 1/2: there was a general cessation of the arc's volcanism of about 10 Ma during the Paleocene. This could have been in relation with the opening of the Grenada Basin and the correlative westward shift of Aves Ridge. The only contemporaneous strata presently known are 20 m-thick dark marly limestones and volcanoclastics of late Paleocene age, outcropping in Anguilla beneath Eocene volcanoclastics (Andreieff et al. 1984).

Stage 2 the volcanic activity resumed during the earliest Eocene all along the older arc, and continued until the mid-Oligocene. The extinct Aves Ridge subsided, and from middle Eocene onwards shallow water carbonates were deposited on top of topographic highs (Bouysse et al. 1985b). On the LAIA, Eocene and lower Oligocene formations are found on the islands of Grenada, the Grenadines, Martinique, Antigua, St Barthélemy, St Martin, and Anguilla.

The largest section of Eocene rocks outcrops on St Martin (Andreieff et al. 1988). The so-called Pointe Blanche formation forms a monocline dipping 30° SSE and about 3000 m-thick. The formation ranges in age from the earliest Eocene to the early late Eocene, representing a time lapse of 20 Ma. The deposits are represented by volcanoclastics (hyaloclastites, tuffites, etc.), andesitic lava flows, and interbedded tuffaceous limestones, limestones, and marls. The volcanism is entirely submarine, and sediments are shallow platform deposits at the bottom, grading upwards to essentially pelagic facies. The volcanism stopped at the middle to late Eocene limit, but all Eocene series were subsequently intruded by late early Oligocene plutons. It should be noted that a 2100 m thick section, of similar facies and age, has recently been drilled in the north coast basin of Puerto Rico, thus suggesting that the older arc extended far more to the west than the Recent arc (Frost and Schellekens 1991; Larue 1991; Smith et al. 1991).

On St Barthélemy island, the Eocene series are much more reduced in thickness and were deposited during an interval of only 3 Ma (46-43 Ma, middle Eocene). The volcanism is also submarine (hyaloclastites, brecciae, lava flows), and is interbedded with six neritic limestones horizons, with the younger of them being about 180 m thick (Westercamp and Andreieff 1983). All of them were intruded by porphyritic andesites and microdiorites in early Oligocene times.

In Antigua, the volcanic basement with interbedded limestone lenses of early Oligocene age is overlain first by a 500 m-thick volcanoclastic unit with interbedded carbonates, and secondly by a 500 m-thick shallow-water limestone and marly limestone sequence grading upward to pelagic facies. This latter sequence is of late early Oligocene to mid (?) late Oligocene age (Mascle and Westercamp 1983; Multer et al. 1986).

In Martinique, the oldest rocks belonging to Stage 2 outcrop along the eastern and southern coasts and form the basal complex of the island made of massive lava flows and hyaloclastic tuffs overlain by upper Oligocene neritic limestones (Westercamp et al. 1989).

Interstage 2/3: a generalized interruption of volcanism occurred during the late Oligocene and the earliest Miocene. This event has been interpreted as resulting from the interaction of a buoyant anomalous Atlantic ridge with the Lesser Antilles subduction zone (Bouysse et al. 1990).

Stage 3: the volcanic activity resumed in the early Miocene (Burdigalian) all along the Recent arc, after a slight decrease of the angle of dip of the Benioff slab in the northern half of the LAIA. Mainly volcanics were produced along the Recent arc, with only a limited amount of limestone lenses. On the now extinct outer arc, the previous volcanic reliefs underwent a process of erosion and subsidence. Grande Terre of Guadeloupe and Marie Galante were thus covered by 200-m-thick (at least) subreefal limestones ranging in age from the early Pliocene to the early Pleistocene, when about 300-m-thick reefal limestones of late early Miocene age (Terre Basse forma-

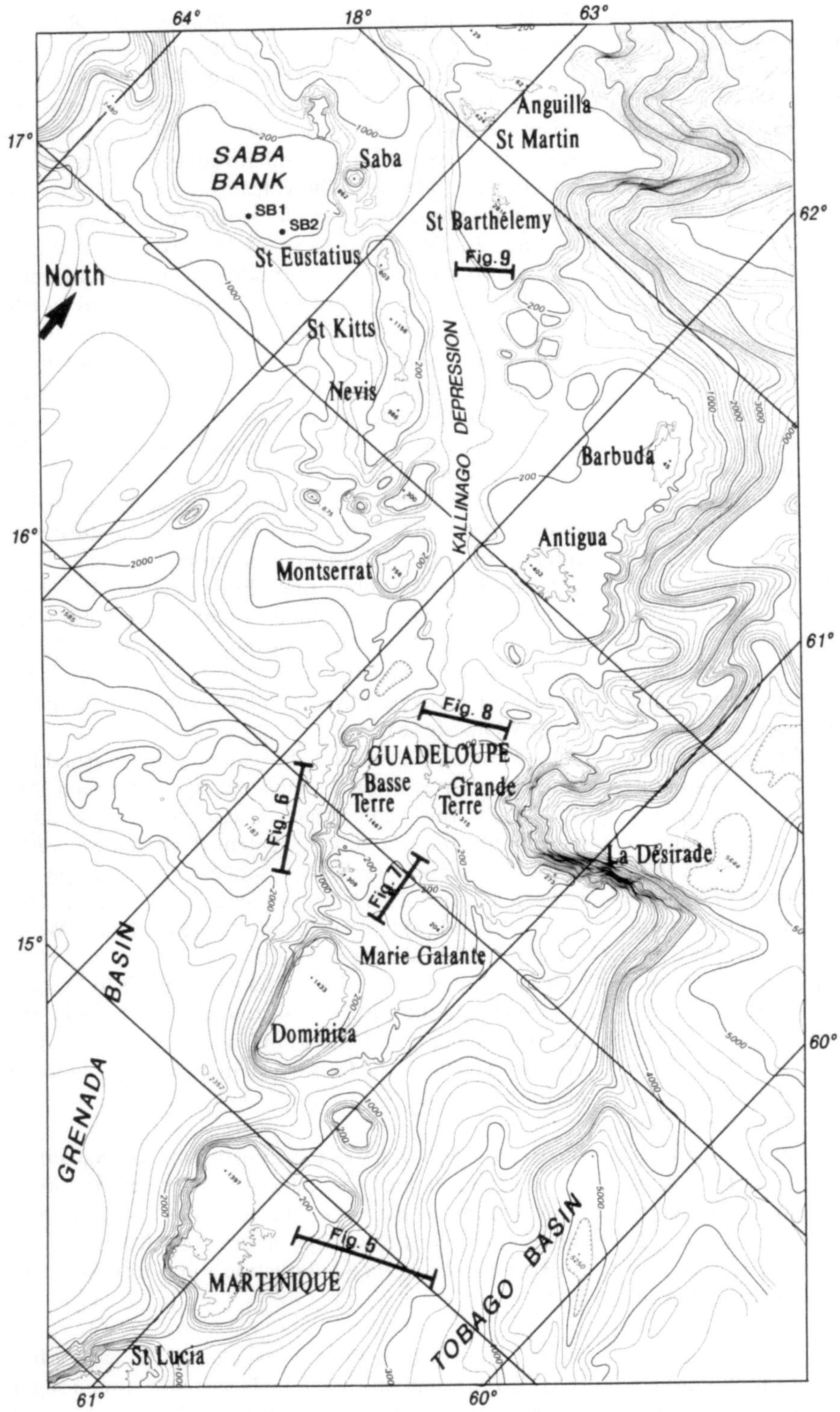

Fig. 3. Bathymetric map around the French islands with the location of the seismic sections and of the two Saba Bank wells *SB1* and *SB2* Water depths in m (Bouysse et al 1985b)

tion) and 80-m-thick upper slope limestone of late middle to latest Miocene (Cupecoy formation) are present on St Martin).

Offshore Sedimentary Basins

Relatively thick sedimentary basins are present in the offshore area of the French Lesser Antilles (Fig. 4). They were discovered and mapped about 10–15 years ago during an initial phase of exploration undertaken by the In-

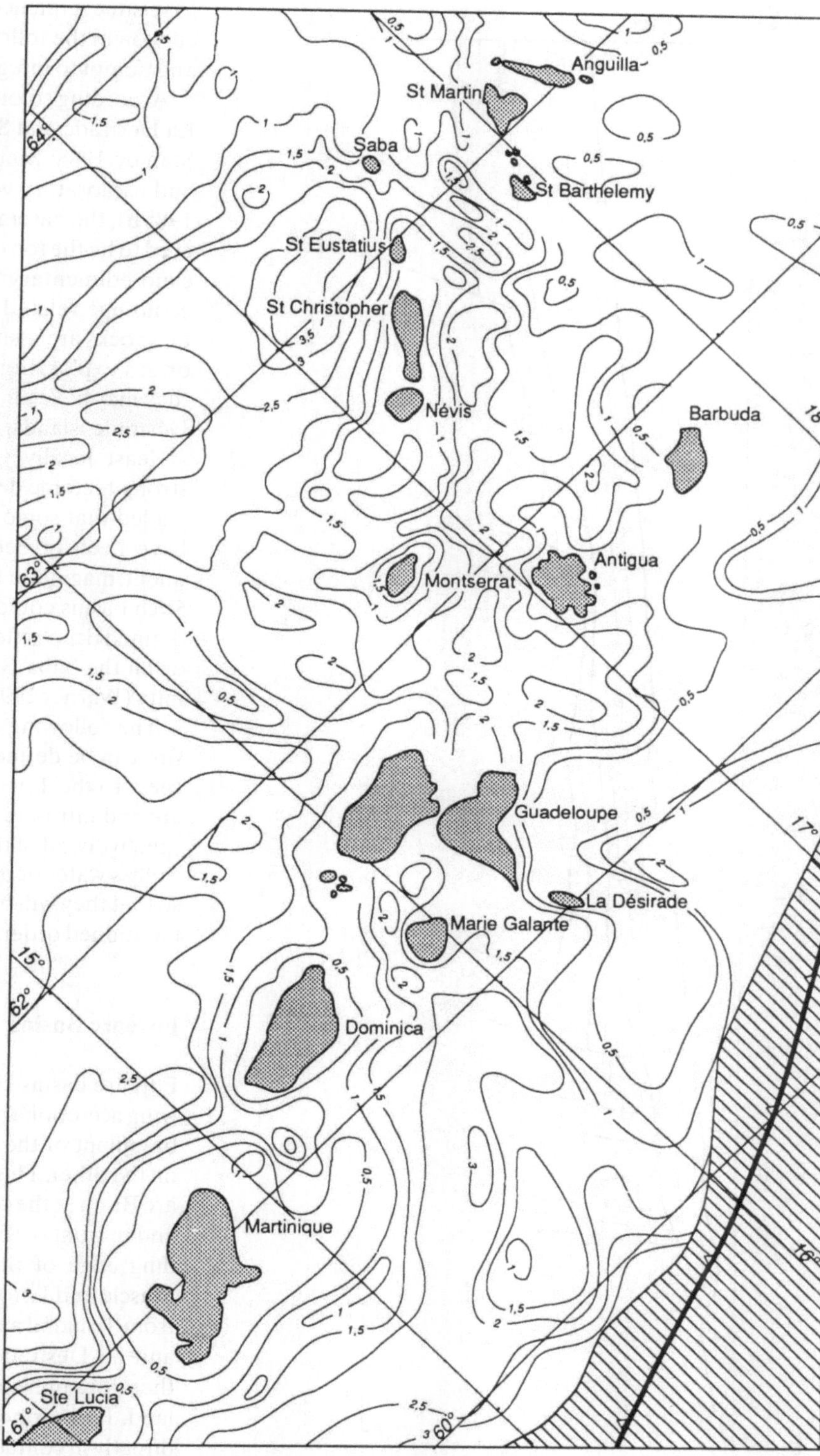

Fig. 4. Thickness in seconds (DWTT) of sedimentary basins of presumed Tertiary age around the French islands (Pinet et al 1985)

stitut Français du Pétrole, Elf Aquitaine and Total with intensive seismic surveying (Pinet et al. 1985; Mascle and Letouzey 1990). Very few well controls were available at the beginning with the exception of long-distance correlations from the Northern Trinidad shelf through the Tobago Forearc Basin (Robertson and Burke 1989). Two wells were drilled on the Saba Bank in 1977 and 1982 that brought important data for the interpretation of seismic data in this area (Despretz et al. 1985). Seismic correlation to the French islands offshore appears, however, quite difficult because of lateral facies changes, faulting, and volcanic events. Stratigraphic and lithologic attribution

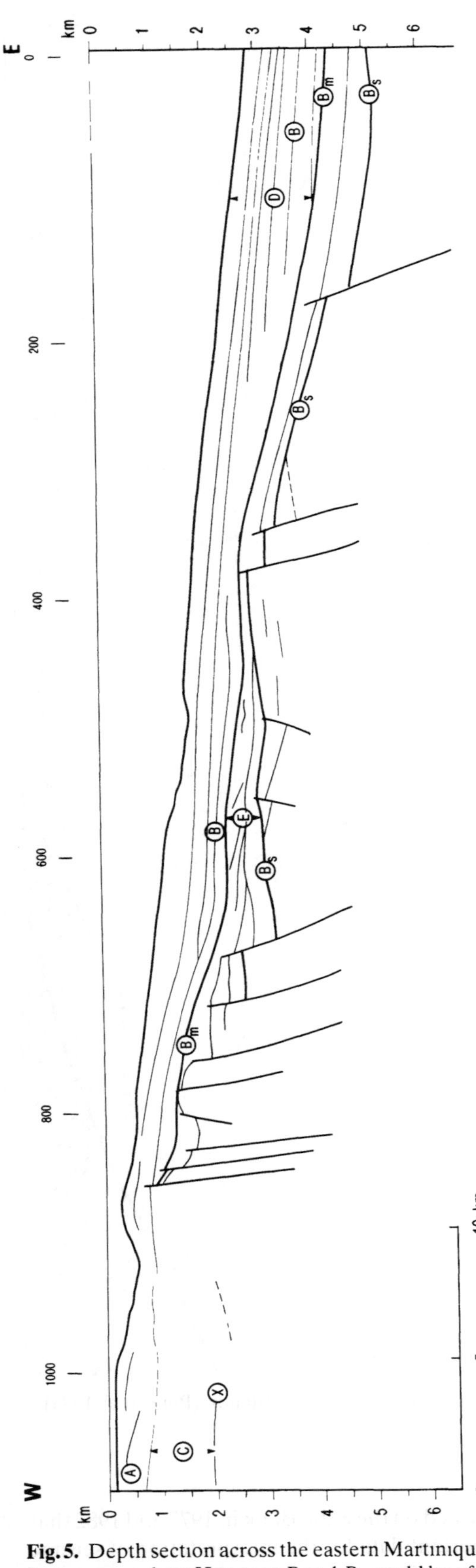

Fig. 5. Depth section across the eastern Martinique island slope from seismic data Horizons *B* and B_m could be of late Miocene and early Miocene age, respectively, and horizon B_s could be the top of an acoustic basement of Paleogene age Sequences *A*, *D*, *E*, and *C* could be of Plio-Pleistocene, late early Miocene-Pleistocene, late Oligocene-early Miocene, and Oligo-Miocene age, respectively

to seismic sequences thus remain speculative and our discussion in the following sections should be considered as an attempt to integrate all available data

According to outcrops on neighboring islands such as La Désirade and St Croix (Westercamp 1980; Lidz 1988; Stanley 1988; Maury et al. 1990), submarine dredgings, and exploratory wells on the Saba Bank (Bouysse et al 1985b), the basement of these basins is generally considered to be the top of a thick accumulation of volcanic, volcanosedimentary and sedimentary strata of late Cretaceous age related to Stage 1 of activity. These sedimentary rocks are assumed to have very little potential for oil or gas exploration because of the slight metamorphism they may have undergone as observed on St Croix and La Désirade islands, and because they have been subjected, at least locally (St Croix; Speed and Joyce 1990), to strong tectonic deformations. It cannot, however, be excluded that some sedimentary basins of Cretaceous age have been preserved from any synchronous or subsequent magmatic activity and still have some potential. Such basins could be at the origin of the relatively well-defined deep reflectors present locally on the Aves Ridge or on the Saba Bank below the base Tertiary unconformity (Warner 1990; Larue and Warner 1991).

The following categories of offshore sedimentary basins can be defined if we consider their location with respect to the Tertiary eruptive centers: the forearc, backarc and intra-arc basins in the east, west, and center, respectively, all of them of Tertiary age. Petroleum plays, as well as water depths, are quite different for each of them, so that they will be described separately and in the above-mentioned order.

Forearc Basins

Forearc basins commonly develop at the back of a growing accretionary prism on one side, and on the subsiding basement of the leading edge of the overthrusting plate on the other. This is typically the case of the Tobago Forearc Basin at the back of the Barbados accretionary prism, and in front of the Lesser Antilles volcanic arc at the leading edge of the Caribbean plate (Speed et al. 1989; Mascle and Letouzey 1990). This basin is well developed from Trinidad and Tobago in the south, to Marie Galante and La Desirade in the north. A large forearc basin is therefore present in French waters east of Martinique island. Farther north, the Barbados accretionary prism is directly in contact with the Caribbean basement outcropping along steep submarine slopes.

East of Martinique, water-depths rapidly deepen from a narrow shelf to more than 3000 m, with the thickest sedimentary section being found in the deepest water at the western edge of the accretionary prism (Pinet et al 1985). A section shown in Fig. 5 is located immediately east of the Caravelle peninsula where the oldest sediments of the island outcrop (shallow-water limestones of late Oligocene age) on top of an undated volcanic complex generated at Stage 2 (Westercamp et al. 1989). This complex is, in turn, overlain by lava flows and tuffs interbedded with volcanoclastic sandstones and limestones of Miocene age. The shelf east of Martinique is therefore interpreted as made of a few-hundred-meter-thick sedimentary cover of Plio-Pleistocene age, resting on a volcanic and sedimentary complex of Miocene and older age. The sedimentary cover rapidly thickens beneath the slope, and a major angular unconformity (Bm) may be defined, with a gently onlapping cover above and more complexly arranged sedimentary sequences below. The Bm unconformity is not dated, but if we take into account the long distance correlation from the northern Trinidad shelf through the Tobago Basin, an early Miocene age can be tentatively proposed. Consequently, the sequences below would be of Oligo-early Miocene age and be time-equivalent to the lower volcanic and sedimentary complex outcropping on the eastern flank of Martinique. Their seismic configuration below 1000 to 2000 m water-depth suggests that they are carbonate platforms deposits overlain by a clastic prograding shelf. If this interpretation is correct, it would imply about 2500 m of subsidence of the leading edge of the Caribbean plate since middle Miocene times. This would conform to dredging results close to La Désirade island, farther north, where shallow-water Paleocene-Eocene and Oligocene limestones have been encountered at depths of 2150 and 1000 m, respectively (Garrabé et al. 1988; dredge stations ST 18 and 43 D).

Large and still undeveloped gas fields have been discovered in the south of the Tobago Basin in Venezuelan waters (the Patao and Mejillones fields are the largest of them with reserves of 170 and 140 billions m^3 respectively) and Trinidadian waters (Chaconia and Point Settia fields with reserves said to be about 4 tcf). The reservoirs are Middle Miocene to Pliocene sands and sandstones, and the gas could be of biogenic origin. Along the eastern edge of the basin, a small amount of oil and gas (1200 bopd and 3000 mcfd) is currently being produced on Barbados island from Eocene deep-water sandstones in complex structural traps. Source rocks are generally believed to be interbedded marine clays with a large amount of terrestrial organic matter (Speed et al. 1991).

Such proven plays cannot reasonably be proposed for the Tobago Basin western margin off Martinique because of the different geologic setting. However, Neogene turbiditic sands are probably at the origin of the well-bedded reflectors seen on the seismic in the Tobago basin, and the organic-rich Eocene marls may possibly extend far to the north of the Tobago basin. Prospective plays would then include drowned carbonate platforms, as proposed in our interpretation in Fig. 10, as well as the pinchout of Neogene turbiditic sands, as seen on the same section above the Bm horizon. Source rocks could hopefully be present in the deepest part of the Tobago Basin. The great water depths, however, will unfortunately delay any risky drilling in this area to the distant future.

Backarc Basins

There is no significant intra-arc basin at the latitude of Martinique (except to the north between Martinique and Dominica islands) and in the west the island faces the wide and deep-water (up to 3000 m) Grenada backarc basin. The insular shelf on this side of the island is nearly nonexistent, and the insular slope is steeper than to the east as well. Since Pliocene and Quaternary times, volcanic centers are all located along this western side of the island (including the Mont Pelée historic eruptive center), the shelf and slopes are believed to be made of primary lava and pyroclastic flows and/or reworked volcanoclastic material with poor petroleum potential. The potential of any Paleogene strata possibly preserved beneath these recent volcanic and volcanoclastic series remains highly hypothetical. A sedimentary section of significant thickness is present in the Grenada Basin. The water depths, however, approach 3000 m and the almost complete ignorance we have about the distribution of potential reservoirs, covers, and source rocks will probably prevent any exploration development in the near future.

The backarc basin at the latitude of Guadeloupe, Saint Barthelemy, and Saint Martin is located on top of the remnant Mesozoic volcanic arc, since the early Tertiary oceanic backarc opening is not believed to have extended further north than Martinique. As a result, water depths are shallower and generally not deeper than 2000 m (Bouysse et al. 1985 a, b). The Saba Bank, in the northern area, has very shallow water depths thus enabling conventional exploration to have been conducted in the 1970s and early 1980s. Two holes have been drilled (SB 1 and 2, in 1977 and 1982 respectively). Although both of them were dry, they encountered small oil and gas shows in Paleogene calcareous and volcanoclastic strata (Despretz et al. 1985; Pinet et al. 1985; Warner 1990). Of more significant interest, potential reservoirs and source rocks have been said to be present in Eocene carbonates and Oligocene volcanoclastic intervals. The low geothermal gradient in SB 1 (2.2 C/100 m) could explain the nonmaturity of the source rocks encountered. The two wells bottomed in andesites of early Paleocene age (Stage 1), but deeper reflectors on the seismic have been interpreted as originating from a deeper Cretaceous basin (Larue and Warner 1991). Gas shows would have originated from this basal sedimentary interval since potential source rocks of a similar late Cretaceous age have actually been described from strata outcropping on the neighboring island of Puerto Rico (Hayes et al. 1986).

The backarc area immediately west of Guadeloupe includes a very narrow shelf and steep slope down to a plateau between 1000 and 2000 m deep, including the Vieux Fort and Pointe Noire ridges. Two late Pliocene submarine volcanoes (Directeur, ca. 3.5 Ma, and Vieux Fort, ca. 2 Ma; Bouysse et al. 1985 c) have been recognized on the upper slope, and the acoustic response of the sea bottom on the shelf is indicative of volcanic materials (Got et al. 1985). Plio-Quaternary volcanic eruptive centers are located on the adjacent mountainous western half of

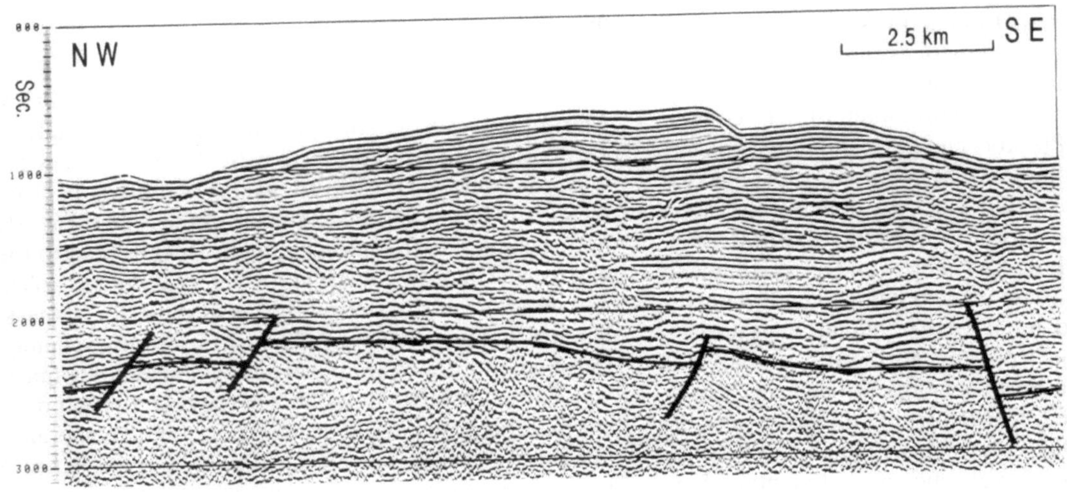

Fig. 6. Seismic section to the west of Guadeloupe The inferred basement *(black line)* could be the top of Cretaceous volcanics, and the sediments above show internal and external configurations strongly suggesting alternating depositional and erosional processes of clastic sedimentation on the eastern edge and slope of the backarc basin

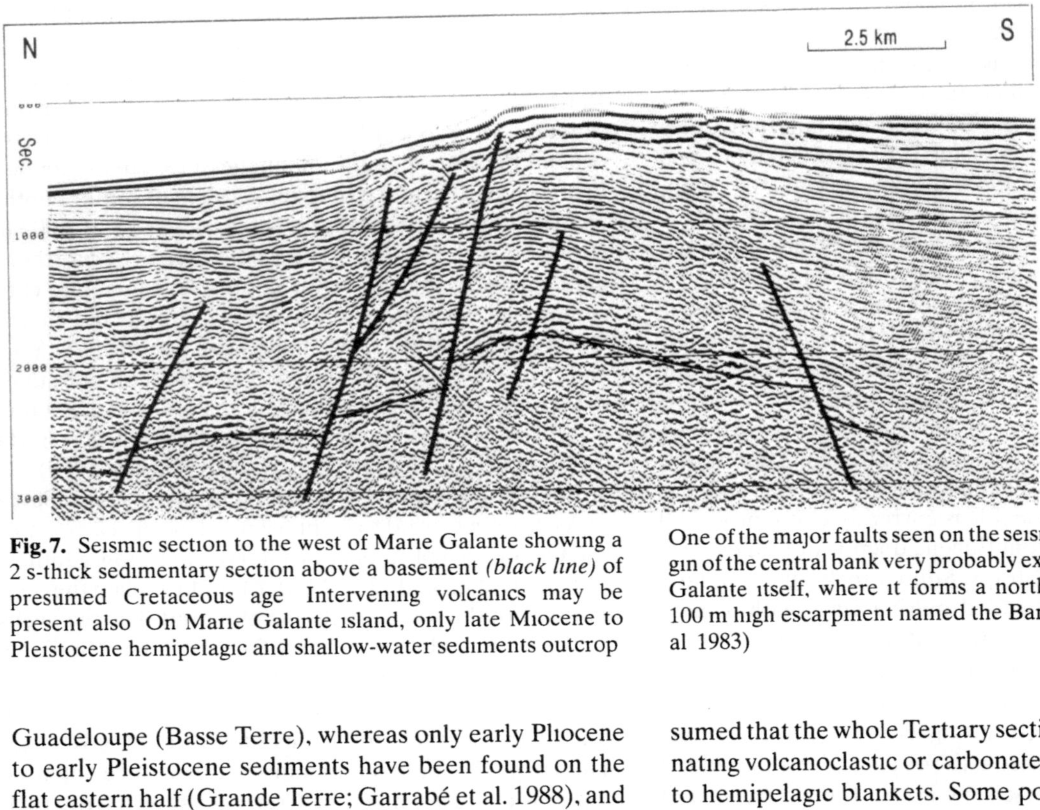

Fig. 7. Seismic section to the west of Marie Galante showing a 2 s-thick sedimentary section above a basement *(black line)* of presumed Cretaceous age Intervening volcanics may be present also On Marie Galante island, only late Miocene to Pleistocene hemipelagic and shallow-water sediments outcrop One of the major faults seen on the seismic on the northern margin of the central bank very probably extends eastward on Marie Galante itself, where it forms a northward facing more than 100 m high escarpment named the Barre de l'Ile (Andreieff et al 1983)

Guadeloupe (Basse Terre), whereas only early Pliocene to early Pleistocene sediments have been found on the flat eastern half (Grande Terre; Garrabé et al. 1988), and late Miocene to early Pleistocene sediments in Marie-Galante (Andreieff et al. 1989; Bouysse et al. 1993). West of Basse Terre, the sedimentary section above the acoustic (Mesozoic?) basement is generally between 1 and 2 s thick and show little tectonic deformations (Pinet et al. 1985). The sea bottom, however, is severely cut by submarine canyons originating from the island. Clastic influxes from the island are also recorded on the whole sedimentary column as seen in a representative section (Fig. 6). The external and internal configurations of seismic sequences strongly suggest sediment supply through the plane of the section. It can thus be reasonably assumed that the whole Tertiary section is made up of alternating volcanoclastic or carbonate turbidites and pelagic to hemipelagic blankets. Some potential reservoirs and covers may thus be present in the section, but the question of the presence or not of mature potential source rocks remains open.

Intra-Arc Basins

This area probably holds the best potential for oil and gas exploration in the Lesser Antilles Arc. Water depths are quite diversified, with shallow to very shallow banks (such as the Anguilla or Antigua Banks), and large areas of moderate (less than 1000 m) water-depths (Kallinago

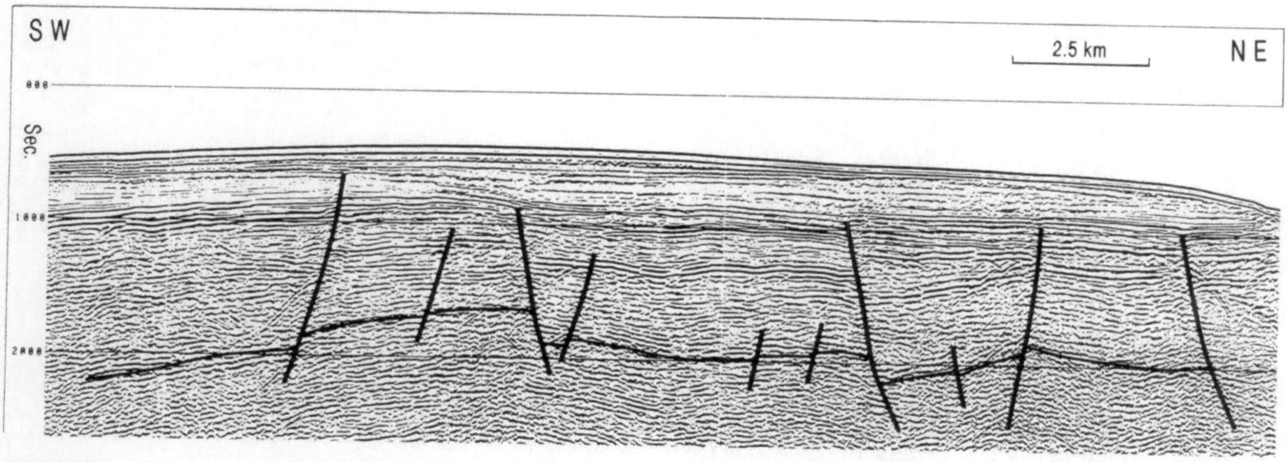

Fig. 8. Seismic section north of the island of Guadeloupe The deepest reflectors recognized with some confidence *(black line)* can be either the top of the Mesozoic basement or, alternatively, a deep volcanic or sedimentary level of Paleogene age

Trough; Bouysse and Guennec 1983). Seismic data show the presence of large sedimentary basins with thicknesses often in excess of 1 and sometimes 2 s, even below the present shallow banks (Figs. 7, 8, and 9). Their sedimentary content is, however, poorly constrained since no well has ever been drilled in this area. The islands to the east (from Anguilla to Grande Terre of Guadeloupe) are made up of late Paleocene to late Oligocene carbonates and volcanoclastic sediments, lava flows, tuffs, and intrusive rocks (Mascle and Westercamp 1983; Westercamp and Andreieff 1983; Andreieff et al. 1984; Garrabé et al. 1988; Dagain et al. 1989). They are locally covered by a thin Neogene-Pleistocene cover. The islands to the west (Inner arc) have been built by Neogene volcanoes (and bear a few still active eruptive centers) that erupted through what was in Paleogene times the backarc basin of the eastern volcanic arc (outer arc). As a matter of fact, blocks of limestones of probable Eocene age have been reported from Pliocene pyroclastic flows on Nevis island (Hutton 1968). Therefore the sedimentary packages that these islands are exposing are probably not representative of the 1500- to 3000-m-thick Tertiary infilling of the offshore basins. We believe that these sediments are probably much more similar to those drilled on the Saba Bank, or to the ones outcropping in the Northern Coast and Southern Coast basins of Puerto Rico, i.e., away from any major volcanic flows or intrusives. The wells drilled in these two areas were dry (Larue et al. 1990; Warner 1990), but all of them were located in areas of present low heat flow, thus preventing any recent significant maturation of organic matter et depths less than 3000 m. This is confirmed by 1-D Matoil maturity modeling using a constant heat flow of 1.25 HFU (52.3 mW/m^2). The results (Fig. 10) show that the three selected potential source-rocks (two of them of early-middle Eocene age and the younger of late Oligocene age) have still not reached the oil window. If the same Tertiary series were located north of the island of Guadeloupe, the maturity history would have been quite different, since the present heatflow is much higher, i.e., about 105 mW/m^2 according to recent geothermal data (Polyak et al. 1992). This is confirmed by the Matoil modeling, since the potential Eocene source rocks enter the oil window as early as early Miocene times for the deepest ones. From these results we therefore believe that there is some untested oil potential in the intra-arc basins around the French (and neighboring) northern Lesser Antilles islands. Structural traps seem to be present in shallow to moderately deep waters, and only the quality of potential reservoirs and covers cannot be predicted with the available data.

Conclusion

Volcanic arcs do not have the reputation of being major oil and gas provinces, although significant production has locally been achieved in such a geological setting, such as in Indonesia (Cameron et al. 1980), Japan (Suzuki 1983) or Georgia (Vernik 1990) for instance. Seismic exploration of the French offshore areas, however, has revealed the presence of a few-thousand-meter-thick sedimentary basins of presumed Tertiary age (Fig. 11). Present relatively high heat flows related to the proximity of magmatic chambers in the crust may have favorably enhanced the maturity of potential source rocks since our modeling indicates that the present oil window may have been reached at depth of 1700 m in the intra-arc basins. The distribution and quality of reservoirs, source rocks, and covers is more difficult to assess. The presence of carbonate platforms and margins of Paleogene age can reasonably be assumed in the intra-arc basins, with covers made of intervening pelagic or hemipelagic strata. Volcanoclastic sandstones are probably widespread throughout the arc and in the whole Tertiary section. High resolution seismic could be a very efficient tool for mapping these two types of potential reservoirs because of their specific seismic configuration. The distribution of source rocks remains the most hypothetical parameter to be taken into account, but the presence of organic-rich intervals in Pa-

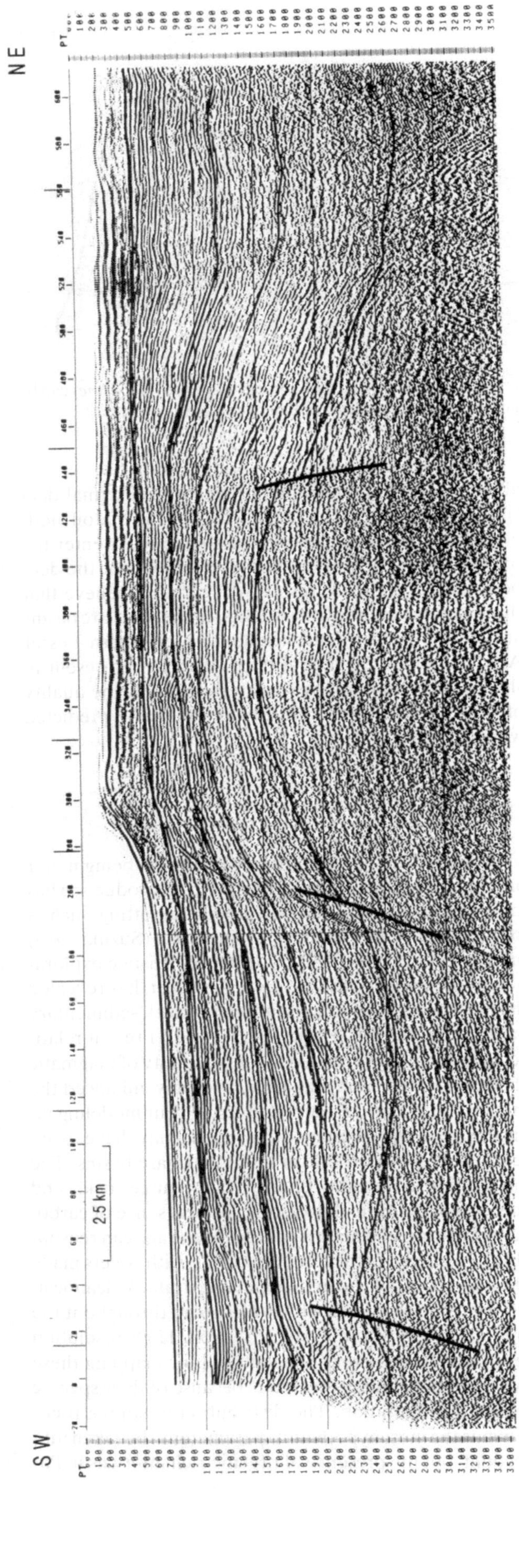

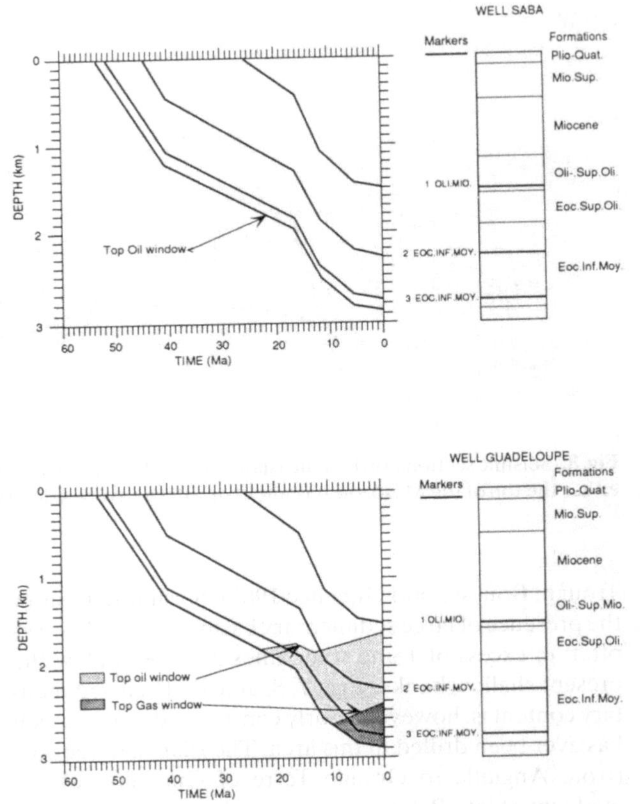

Fig. 10. Numerical modeling of subsidence and maturity of potential source-rocks at SB 1 well (Saba Bank) and to the north or south of Guadeloupe island with similar strata (fictitious well)

leogene marine strata on Saba Bank, Puerto Rico (Larue et al. 1991), and Barbados islands (Speed et al. 1991) should be considered as an encouraging indication that similar rocks are present around or in the immediate vicinity of the French islands. Present petroleum licenses in the Lesser Antilles arc are restricted to the Antigua-Barbuda Bank.

From a more general point of view, this survey has shown that volcanic island arcs may develop through and contemporaneously with thick sedimentary basins, a point already emphasized by Larue et al. (1991). This implies a regional subsidence of the island arc platform and of the backarc area as well, possibly because of the adding of dense mantellic magmas to the crust, and their subsequent cooling. An additional reason could be, for the southern segment of the arc at least, the load of the thick infilling of the Tobago forearc basin and of the backthrust western edge of the Barbados Ridge accretionary prism on the eastern edge of the Caribbean plate basement. Such regional subsidence, however, is not uniform, since

◀ **Fig. 9.** Seismic section south of St Barthelemy The reflectors between the two deepest reflectors could be the lateral equivalent of middle Eocene volcanics and carbonates outcropping on the island where the youngest formation is the 180 m-thick Plateau du Lurin limestone (Westercamp and Andreieff 1983)

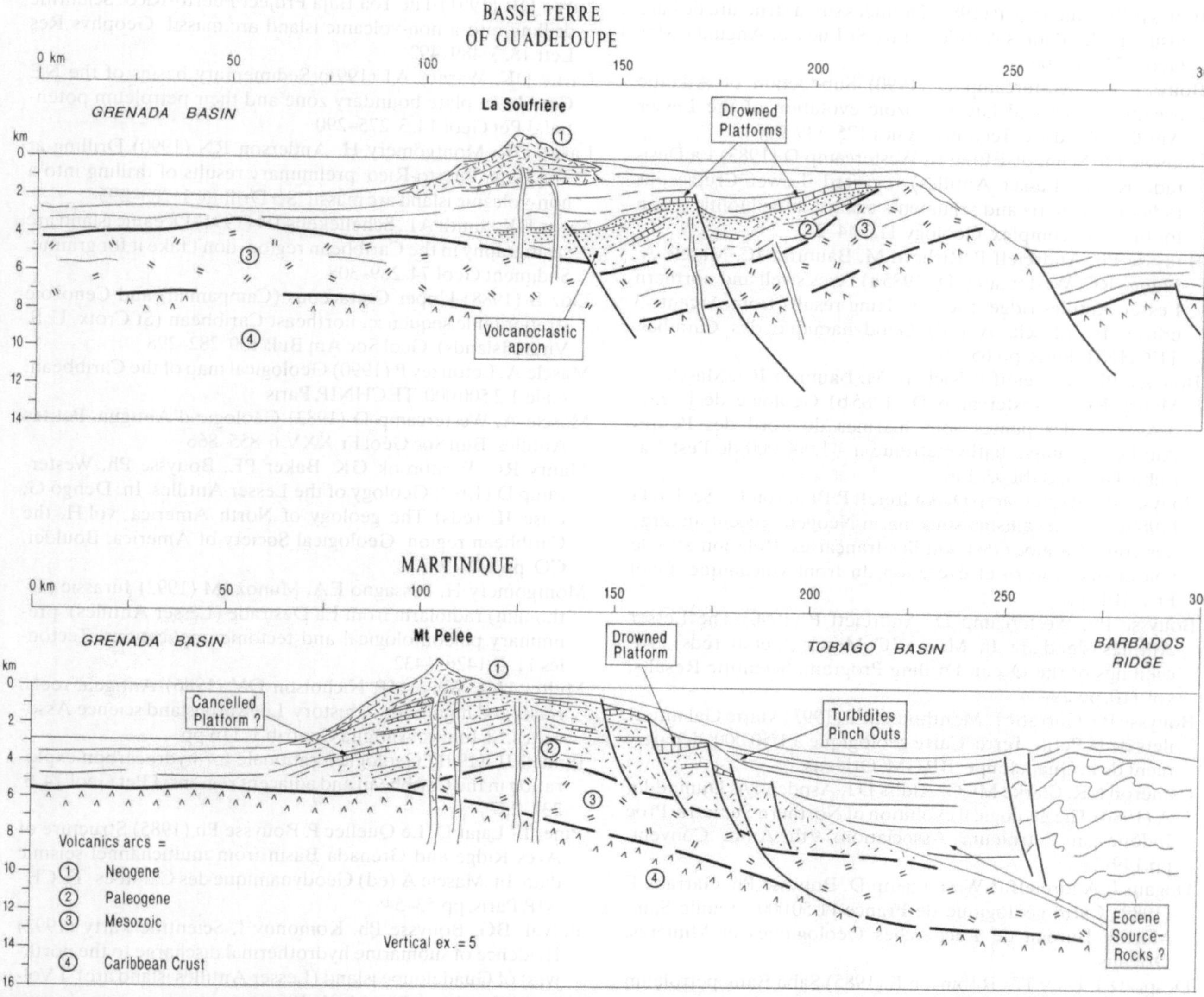

Fig. 11. Synthetic sections across the Lesser Antilles arc at the latitude of Guadeloupe and Martinique with potential petroleum plays

some parts of the island arc plateau have obviously been uplifted, specially along the northern segment, as observed by the present reliefs of the Limestone Caribbees (e.g., La Désirade). A complex arrangement of stresses due to both the local deep and superficial magmatic activities, and to the interference of stresses originating at the front of the margin (compression), to the south (transpression), and to the north (transtension), should probably be taken into account and quantitatively modeled in the future if some progress is to be made in the understanding of the geodynamic processes controlling island arc evolutions.

Acknowledgments. The final version of this chapter benefited from the comments of D. Larue (EXXON) and B. Van den Berghe (AGIP-France). The seismic surveys around the French Lesser Antilles were shot by the CEPM (Elf-Aquitaine, Total and IFP) from 1973 to 1983. Most of these lines can be purchased to CGG. The petroleum evaluation data presented here are preliminary results from a regional report on the French West Indies prepared by Institut Français du Pétrole.

References

Andreieff P, Bouysse Ph, Westercamp D (1983) Révision géologique de l'île de Marie Galante (Petites Antilles). Bull Soc Géol Fr 7, 25, 6: 805–810

Andreieff P, Bonneton JR, Vila JM, Westercamp D (1984) Découverte de Paléocène supérieur à Anguilla, à l'extrémité nord de l'arc des Petites Antilles (abs.). 10è Réunion Annuelle des Sciences de la Terre Société Géologique de France, Paris, p 15

Andreieff P, Westercamp D, Garrabé F, Bonneton JR, Dagain J (1988) Stratigraphie de St Martin, Petites Antilles septentrionales. Géol Fr 2/3: 401–411

Andreieff P, Bouysse Ph, Westercamp D (1989) Géologie de l'arc insulaire des Petites Antilles et évolution géodynamique de l'Est-Caraïbe Doc BRGM 171: 250

Bouysse Ph (1988) Opening of the Grenada backarc basin and evolution of the Caribbean plate during the Mesozoic and early Paleogene. Tectonophysics 149: 121–143

Bouysse Ph, Guennoc P (1983) Données sur la structure de l'arc insulaire des Petites Antilles, entre St Lucie et Anguilla Mar Geol 53 131–166
Bouysse Ph, Westercamp D (1990) Subduction of Atlantic oceanic ridges and late Cenozoic evolution of the Lesser Antilles island arc Tectonophysics 175 349–380
Bouysse Ph, Schmidt-Effing R, Westercamp D (1983) La Désirade island (Lesser Antilles) revisited Lower Cretaceous radiolarian cherts and arguments against an ophiolitic origin for the basal complex Geology 11 244–247
Bouysse Ph, Andreieff P, Richard M, Baubron JC, Mascle A, Maury RC, Westercamp D (1985 a) Aves swell and northern Lesser Antilles ridge rock-dredging results from Arcante 3 cruise In Mascle A (ed) Géodynamique des Caraïbes TECHNIP, Paris, pp 65–76
Bouysse Ph, Andreieff P, Richard M, Baubron JCl, Mascle A, Maury RCh, Westercamp D (1985 b) Géologie de la ride d'Aves et des pentes sous marines du nord des Petites Antilles Esquisse bathymétrique à 1 1,000,000 de l'est Caraïbe Doc BRGM 93 146
Bouysse Ph, Westercamp D, Andreieff P, Baubron JC, Scolari G (1985 c) Le volcanisme sous-marin Néogène récent au large des côtes Caraïbes des Antilles françaises Relation avec le volcanisme à terre et évolution du front volcanique Géol Fr 1 101–114
Bouysse Ph, Westercamp D, Andreieff P (1990) The Lesser Antilles island arc In Moore JC, Mascle A et al (eds) Proceedings of the Ocean Drilling Program, Scientific Results, vol 110, pp 29–44
Bouysse Ph, Garrabé F, Montboussin T (1993) Marie Galante et ilets de la Petite Terre, Carte géologique à 1/50 000, departement de la Guadeloupe BRGM, Orléans
Cameron NR, Clarke MCG, Aldiss DT, Aspden JA, Djunuddin A (1980) The geological evolution of Northern Sumatra Proc Indonesian Petroleum Association, 9th Annu Convent, pp 149–187
Dagain J, Andreieff P, Westercamp D, Bouysse Ph, Garrabé F (1989) Carte géologique de France (1 50000), feuille Saint Martin Bureau de Recherches Géologiques et Minières, Orléans
Despretz J, Daly TE, Robinson E (1985) Saba Bank petroleum geology, NE Caribbean Oil Gas J Nov 4. 112–118
Frost CD, Schellekens JH (1991) Rb-Sr and SM-Nd isotopic characterization of Eocene volcanic and volcanoclastic rocks from Puerto-Rico Geophys Res Lett 18, 3 545–548
Garrabé F, Andreieff P, Bouysse Ph, Rodet J (1988) Carte géologique à 1/50000, département de la Guadeloupe, Grande Terre Bureau de Recherches Géologiques et Minières, Orléans, France
Gill IP (1990) The evolution of Tertiary St Croix Ph D , Louisiana State University and Agricultural and Mechanical Col , 1989, 289 pp
Got H, Aloisi JC, Inoubli H, Perret M, Mirabile L (1985) Etude structuro-sédimentaire des marges sud et ouest de la Guadeloupe. In Mascle A (ed) Symposium sur la géodynamique des Caraïbes TECHNIP, Paris, pp 161–172
Hayes J, Larue D, Joyce J, Schellekens J (1986) Puerto-Rico reconnaissance study of the maturation and source-rock potential of an oceanic arc involved in a collision J Mar Pet Geol 2 126–138
Hutton C (1968) The mineralogy and petrology of Nevis, Leeward Islands, British West Indies A progress report Trans 4th Caribbean Geological Conf, pp 383–388
Jany I, Scanlon KM, Mauffret A (1990) Geological interpretation of combined Seabeam, Gloria, and seismic data from Anegada Passage (Virgin Islands, North Caribbean) Mar Geophys Res 12 173–196
Larue DK (1991) The Toa Baja Project Puerto-Rico Scientific drilling into a non-volcanic island arc massif Geophys Res Lett 18, 3 489–492
Larue DK, Warner AJ (1991) Sedimentary basins of the NE Caribbean plate boundary zone and their petroleum potential J Pet Geol 14, 3 275–290
Larue DK, Montgomery H, Anderson RN (1990) Drilling at Toa Baja, Puerto-Rico preliminary results of drilling into a non-volcanic island arc massif Sci Drilling 1 283–295
Larue DK, Smith AL, Schellekens JH (1991) Oceanic island arc stratigraphy in the Caribbean region don't take it for granite Sediment Geol 74 289–308
Lidz B (1988) Upper Cretaceous (Campanian) and Cenozoic stratigraphic sequence, northeast Caribbean (St Croix, U S Virgin Islands) Geol Soc Am Bull 100 282–298
Mascle A, Letouzey P (1990) Geological map of the Caribbean, scale 1 2 500 000 TECHNIP, Paris
Mascle A, Westercamp D (1983) Géologie d'Antigua, Petites Antilles Bull Soc Géol Fr XXV, 6 855–866
Maury RC, Westbrook GK, Baker PE, Bouysse Ph, Westercamp D (1990) Geology of the Lesser Antilles In Dengo G, Case JE (eds) The geology of North America, vol H, the Caribbean region Geological Society of America, Boulder, CO, pp 141–166
Montgomery H, Pessagno EA, Munoz IM (1992) Jurassic (Tithonian) radiolaria from La Desirade (Lesser Antilles) preliminary paleontological and tectonic implications Tectonics 11, 6 1426–1432
Multer HG, Weiss MP, Nicholson DV (1986) Antigua, reefs, rocks and highroads of history Leeward island science Associates, St John's, Antigua Contrib 1, 116 pp
Pindell JL (1991) Geological rationale for hydrocarbon exploration in the Caribbean and adjacent regions J Pet Geol 14, 3 237–257
Pinet B, Lajat D, Le Quellec P, Bouysse Ph (1985) Structure of Aves Ridge and Grenada Basin from multichannel seismic data In Mascle A (ed) Géodynamique des Caraïbes TECHNIP, Paris, pp 53–54
Polyak BG, Bouysse Ph, Komonov I, Scientific Party (1992) Evidence of submarine hydrothermal discharge to the northwest of Guadeloupe island (Lesser Antilles island arc) J Volcanol Geotherm Res 54 81–105
Robertson P, Burke K (1989) Evolution of Southern Caribbean plate boundary, vicinity of Trinidad and Tobago AAPG Bull 73 490–509
Smith AL, Severin K, Larue DK (1991) Stratigraphy, geochemistry and mineralogy of Eocene rocks from the Toa Baja drillhole Geophys Res Lett 18, 3 521–524
Speed RC, Joyce J (1990) Cretaceous tectonic complex, St Croix In Larue DK, Draper G (eds) Transactions of the 12th Caribbean Geological Conf, St Croix, U S Virgin Islands Miami Geological Society, pp 301–313
Speed RC, Larue DK (1991) Extension and transtension in the plate boundary zone of the northeastern Caribbean Geophys Res Lett 18, 3 573–576
Speed RC, Torrini R, Smith PL (1989) Tectonic evolution of the Tobago Trough forearc basin J Geophys Res 94 2913–2936
Speed RC, Barker LH, Payne P (1991) Geologic and hydrocarbon evolution of Barbados J Pet Geol 14, 3 323–342
Stanley DJ (1988) Deep-sea current flow in the late Cretaceous Caribbean measurements on St Croix, U S Virgin islands Mar Geol 79 127–133
Suzuki U (1983) The volcanic mound In Bally AW (ed) Seismic expression of structural styles AAPG Stud Geol 15 1 3 15–1 3 18
Vernik L (1990) A new type of reservoir rock in volcanoclastic sequences AAPG Bull 74, 6 830–836

Warner AJ (1990) The Cretaceous age sediments of the Saba Bank and their petroleum potential. In: Larue DK, Draper G (eds) Trans 12th Caribbean Conference, Miami Géol Soc, South Miami, pp 341–354

Westbrook GK, Ladd JW, Buhl P, Bangs N, Tiley GJ (1988) Cross-section of an accretionary wedge Barbados Ridge complex Geology 16. 631–635

Westercamp D (1980) Carte géologique de La Désirade à 1·25000 et notice explicative Bureau de Recherches Géologiques et Minières, France

Westercamp D, Andreieff P (1983) Carte géologique à 1/20000, Saint Barthélemy et ses ilets Bureau de Recherches Géologiques et Minières, Orléans

Westercamp D, Pelletier B, Thibaut PM, Traineau H (1989) Carte géologique de la France (1:50000), feuille Martinique. Bureau de Recherches Géologiques et Minières Orléans Notice explicative par Westercamp D, Andreieff P, Bouysse P, Cottez S, Battistini R (1989), 246 pp

Subject Index

Springer-Verlag and the Environment

We at Springer-Verlag firmly believe that an international science publisher has a special obligation to the environment, and our corporate policies consistently reflect this conviction

We also expect our business partners – paper mills, printers, packaging manufactures, etc – to commit themselves to using environmentally friendly materials and production processes

The paper in this book is made from low- or no-chlorine pulp and is acid free, in conformance with international standards for paper permanency

MIX
Papier aus verantwortungsvollen Quellen
Paper from responsible sources
FSC® C105338

If you have any concerns about our products,
you can contact us on
ProductSafety@springernature.com

In case Publisher is established outside the EU,
the EU authorized representative is:
Springer Nature Customer Service Center GmbH
Europaplatz 3, 69115 Heidelberg, Germany

Printed by Libri Plureos GmbH
in Hamburg, Germany